Waste-to-Energy

Abd El-Fatah Abomohra · Qingyuan Wang ·
Jin Huang
Editors

Waste-to-Energy

Recent Developments and Future Perspectives towards Circular Economy

Springer

Editors
Abd El-Fatah Abomohra
School of Architecture and Civil Engineering
Chengdu University
Chengdu, Sichuan, China

Qingyuan Wang
School of Architecture and Civil Engineering
Chengdu University
Chengdu, Sichuan, China

Jin Huang
School of Architecture and Civil Engineering
Chengdu University
Chengdu, Sichuan, China

ISBN 978-3-030-91569-8 ISBN 978-3-030-91570-4 (eBook)
https://doi.org/10.1007/978-3-030-91570-4

This Springer imprint is published by the registered company Springer Nature Switzerland AG
The registered company address is: Gewerbestrasse 11, 6330 Cham, Switzerland

For all the children around the world, to whom we are trying to ensure a clean and healthy future

Preface

The continuous increase in global population and concurrent fast industrialization led to massive generation of municipal wastes, with raised challenges of safe disposal. The proper management of municipal wastes through recycling is an essential approach for global sustainable development. So far, many countries established regulatory guidelines for different waste management routes and pollution control measures. However, most of the applied routes are waste dumping, composting or direct discharge in water bodies without adequate pretreatment, which seriously threaten the environment and human health. Despite the applicability of incineration for waste disposal, the low thermal value and high organic load of the waste streams are the main obstacles. However, proper waste segregation and separation provide an efficient option for waste conversion into energy. Thus, waste-to-energy is nowadays a widely used statement for efficient waste management, which is getting much attention. Therefore, recent advances in waste-to-energy and the necessity to achieve a circular economy are a timely topic that deserves in-depth discussion.

There are many established methods for waste conversion into different forms of energy and valuable by-products. After proper segregation for separation of undesired wastes such as metal and concrete, all other wastes can be thermochemically converted into crude bio-oil, biochar and/or syngas. This pathway is favourable for some hazardous wastes, such as those from hospitals, as an alternative to incineration, which provides simultaneous sterilization while producing energy. Biological conversion using anaerobic digestion or fermentation can be applied for organic wastes to produce biogas, bio-ethanol, biohydrogen, and/or biobutanol. After conversion, the produced slurry can be converted by thermochemical methods to crude bio-oil or used as a soil fertilizer. Moreover, lipid-rich wastes such as fat, oil, and grease (FOG) can be used efficiently for biodiesel production due to the high theoretical energy yield. After biodiesel production, the glycerol by-product can be recycled for many other industrial purposes. Despite the recent development in biofuel production, the rapid progress in this research field necessitates integration of different conversion routes in order to introduce all the latest novel technologies of continuous R&D. Such integration requires concise summary of the recent cutting-edge research and industrial applications in this field.

This book aims to cover the recent updates in the waste-to-energy field, starting from waste generation to large-scale application and industrialization. The book summarizes and evaluates the recent R&D results which provide remarkable competences in terms of design, performance, efficiency and implementation of different systems used for energy recovery from wastes, revamping the existing technologies and feedback along with techno-economic analysis for case studies in different countries. The cutting-edge research topics of this book were achieved through contributions from professionals and experts engaged in research, education and industry of the corresponding topics. This book can be considered as a primary reservoir for a reader with any scientific background exploring waste management, biomass conversion or bioenergy from any angle, including undergraduate students, teachers, researchers and consulting professionals in renewable energy, biotechnology, environmental engineering and biomass conversion.

Chengdu, China

Abd El-Fatah Abomohra
Qingyuan Wang
Jin Huang

Contents

About the Editors

Prof. Abd El-Fatah Abomohra is a Professor of Environmental Engineering at Chengdu University, Chengdu, China since 2019. He received his Ph.D. in 2012 through a cooperation program between Tanta University (Egypt) and Hamburg University (Germany) funded by Deutscher Akademischer Austauschdienst (DAAD). He has work experience of 20 years at different universities as research assistant, lecturer, associate professor, and professor. He authored seven books and of more than 100 contributions published in SCI index journals. He was one of the World's Top 2% of Scientists List published by Stanford University in 2020 and 2021. His research group is primarily working on bioenergy production from different biomass feedstocks and wastes.

Prof. Qingyuan Wang is a full Professor at Sichuan University since 2001, and currently he is the president of Chengdu University. He received his Ph.D. in 1998 from Ecole Centrale Paris, France, followed by postdoctoral experience from 1999 to 2003 in Faculty of Engineering of Purdue University, and JSPS fellow at Kagoshima University, Japan. He served as a visiting Professor at University of Paris-Sud, TU-Clausthal, University of New Hampshire and Honorary Professor of University of Liverpool and RMIT. He is also a fellow of International Association of Advanced Materials (IAAM). Prof. Wang has published over 200 papers in SCI index journals, and is one of the Most Cited Chinese Researchers (Elsevier 2014, 2015, 2016, 2017, 2018, 2019, 2020). He was one of the World's Top 2%

of Scientists List published by Stanford University in 2020 and 2021. His research team is focusing on Green Materials, Material Science and Solid Mechanics, Very High Cycle Fatigue (VHCF), Mechanical Behavior of Structural Materials and Structures, Composite Repairs of damaged Structures.

Prof. Jin Huang is a Professor of Environmental Engineering at Chengdu University, Chengdu, China since 2010. She received her Ph.D. in 2008 from Sichuan University, Chengdu, China. She has a work experience of 19 years at Chengdu University as Associate Professor, and Professor. In addition, she has a work experience in industry sector for 7 years as engineer to enhance the municipal waste recycling and wastewater treatment towards green net-zero utilization. She authored 13 papers related to municipal waste management published in international journals.

Chapter 1
An Overview of Municipal Wastes

Eslam G. Al-Sakkari, Mahmoud M. Habashy, Mai O. Abdelmigeed, and Mohammed G. Mohammed

Abstract Recently, waste management became a tremendous global concern due to the high rate of waste materials production. This rate leads to wastes accumulation in the environment without proper management or valorization. Consequently, more problems appeared to the surface, such as global warming and other dangerous phenomena on the whole ecosystem. For instance, in 2018, the total municipal solid waste production was 292.4 million tons, only 38.2 wt.% of which were managed mainly through mechanical recycling and composting. The remaining amounts were incinerated for energy recovery (11.8 wt.%) and landfilled (50 wt.%, which accounts for 146 million tons). The massive amounts of landfilled wastes consume large land areas; therefore, more waste valorization techniques should be applied to achieve a zero waste point in the circular economy system. This chapter defines and classifies solid wastes according to their sources and compositions. It also discusses the types of recyclable, hazardous, and hard-to-control wastes besides briefly discussing the efforts to manage them. Additionally, waste production statistics of different countries are presented to give a real figure about waste amounts and emphasize the urgent need for their management. Finally, the circular economy practices and business models that consider the socioeconomic impact on the whole economy are discussed.

Keywords Municipal wastes · Hazardous wastes · Recyclable wastes · Waste management · Circular economy

E. G. Al-Sakkari (✉) · M. O. Abdelmigeed · M. G. Mohammed
Faculty of Engineering, Chemical Engineering Department, Cairo University, Giza 12613, Egypt
e-mail: eslam.ibrahim@polymtl.ca

M. G. Mohammed
e-mail: mgmohammed@eng.cu.edu.eg

M. M. Habashy
Department of Environmental Engineering and Water Technology, UNESCO-IHE Institute for Water Education, Delft, The Netherlands
e-mail: mab007@uni-ihe.org

A. E.-F. Abomohra et al. (eds.), *Waste-to-Energy*,
https://doi.org/10.1007/978-3-030-91570-4_1

1.1 Introduction

The significant increase in wastes products in the last two decades has become a global challenge and concern that requires urgent and serious actions. For instance, in 2018, around 38 percent by weight of global solid waste was managed mainly through mechanical recycling and composting. Smaller amounts of the waste were incinerated for energy recovery and landfilled. Therefore, a thorough investigation of the amounts and characteristics of wastes is needed to have reliable and feasible waste management strategies. Unfortunately, the systematic surveying of the quantities and types of wastes produced, future trends of wastes generation, characteristics, and seasonal variations are poorly understood. However, general trends and common elements are observable (UN-ESCAP 2006).

Wastes differ according to many factors such as their sources, nature, the medium they affect, their ability to accumulate, their ability to transform, and the region they affect. The classification of wastes according to their sources is commonly applied since it helps to understand the nature of the wastes and dictates the most proper waste management methods. Table 1.1 provides an example of wastes classification by showing the different sources of solid wastes and the types of wastes produced from each source (UN-ESCAP 2006). This chapter reviews the different sources of wastes and the types of wastes produced from each source. Then, it gives a brief overview of the standard management methods such as biological methods, including composting. After that, it presents some statistics on waste production and methods of waste management used worldwide. By the end of this chapter, the circular economy practices and business models based on this type of economy are summarized, introducing some real-world case studies.

Table 1.1 Types and sources of solid wastes (UN-ESCAP, 2006)

Waste source	Solid wastes examples
Residential	Food wastes, paper, cardboard, plastics, wood, glass, metals, and household hazardous wastes such as batteries and electric light pulps
Industrial	Packaging wastes including plastics, hazardous wastes, e.g., solid chemicals, and workers' food wastes
Commercial	Paper, cardboard, plastics, wood, food wastes, glass, metals, and hazardous wastes
Institutional	Paper, cardboard, plastics, wood, food wastes, glass, metals, and hazardous wastes
Construction and demolition	Wood, metals, concrete, rocks, dirt, and dust

1.2 Classification and Types of Wastes

In order to obtain reliable and feasible waste management strategies, the amount and characteristics of the wastes should be investigated. The characterization and quantification of wastes are considered as the cornerstone for intervention and management in some developed countries. However, the systematic surveying of the quantities and types of wastes produced, future trends of wastes generation, characteristics, and seasonal variations are poorly understood. Even though there is insufficient consistent or comprehensive information at the country level, general trends and common elements are observable (Bui et al. 2020).

Generally, much higher amounts of waste per capita are produced by developed countries than developing ones (Bundhoo 2018). For example, modest economic activity and small populations have guaranteed that relatively small amounts of wastes are produced like in the south pacific subregion's small islands. However, managing small waste amounts could be a challenge in some situations, such as in Marshall Islands, Tuvalu, and Kiribati where the small land areas reduce the disposal options.

Solid wastes are mainly produced from agricultural and residential households, construction, commercial, institutional, and industrial sources. Table 1.1 provides a breakdown of different solid wastes sources and their types.

The solid wastes sources are divided into four main classes: agricultural waste, municipal solid waste, industrial waste, and hazardous waste (Abdel-Shafy and Mansour 2018); each waste type is discussed in detail separately throughout this chapter.

1.2.1 Agricultural Wastes

Agricultural wastes refer to any type of waste that is generated from different agricultural activities. Examples of agricultural wastes are animal manure, post-harvest waste such as rice husks, rotten or bad fruits, and vegetables, corn stover and husks, and wheat straw (Nagendran 2011). There are two main types of residues (field residues and process residues) produced from agriculture activities. Stems, seed pods, stalks, and leaves represent the field residues that are left in the field after the crop harvesting process. On the other hand, roots, peel, stubble, pulp, shell, stalk, straw, stem, leaves, seeds, bagasse, husks, molasses, etc., represent the process residues that are present even after the crop is processed into valuable alternate products. Agricultural residues can be differentiated based on their availability and characteristics, different from other solid fuels like charcoal, wood, and char briquette (Sadh et al. 2018). Several industries use these process residues as raw materials to produce other products such as fertilizers, soil improvement additives, animal fodder, paper, synthetic wood, and others. Nevertheless, a big portion of the generated field residues is underutilized, which in some cases results in the accumulation of residues in the fields, preventing the farmers from utilizing the land. In such cases, farmers seek cheap, easy, and fast methods to eliminate the residues, such as burning the residues and generating large amounts of smoke and greenhouse gases in the air (Marey et al. 2010). Research has been going on to develop new applications of agricultural waste such as using them as a precursor for activated carbon synthesis (Köseoğlu and Akmil-Başar 2015; Teo et al. 2016; Yahya et al. 2015), cement additive (Sathiparan and De Zoysa 2018), and source for biofuels production (Li et al. 2011; Stephen and Periyasamy 2018).

The rapid population growth leads to extensive expansion in agricultural production that naturally exhibited a noticeable increase in agro-wastes, livestock wastes, and by-products of agro-industrial activities. For example, in the Asian and Pacific region, China produces the largest amounts of agro-waste or crop residues of 842 million tons/year, followed by India, which produces 560 million tons/year. China produces 587 million tons of agro-wastes per year, and more than 80% of these residues are from rice, corn, and wheat (Chen et al. 2019).

1.2.2 Municipal Solid Waste

The term "municipal wastes" refers to household wastes or any other wastes that are similar in nature and composition. Municipal wastes are commonly recognized as "garbage" or "trash". In other words, municipal solid waste (MSW) is produced from different facilities, including households, schools, hospitals, hotels, offices, and shops, where its main components are food waste, metal, rags, paper, glass, and plastic (Environmental Statistics and Accounts 2016). It also includes market cleansing waste, the content of litter containers, street sweepings, grass clippings, leaves, yard waste, and bigger wastes such as mattresses, old furniture, and white goods. The municipal solid waste is either collected by traditional door-to-door collection (mixed household waste) or collected separately for recovery operations (through the door-to-door collection and/or through voluntary deposits) (Yadav and Karmakar 2020). These wastes can be gathered by municipalities or on their behalf. In addition, they can be collected directly by the private sector including business or private non-profit institutions instead of municipalities for energy or material recovery. It is worth noting that, by definition, MSW does not include municipal sewage network and treatment wastes or those produced from municipal construction and demolition activities.

Generation rates for MSW vary from season to season and from city to city and strongly correlate with levels of activity and economic development. The highest production rates of MSW (kg/capita/day) are found in high-income countries (Makarichi et al. 2018). Based on the economic development, the composition of MSW can change drastically across the same region. Differences in the reporting and characterization of waste types also change based on the responsible authorities and their definition of municipal waste, including industrial waste, demolition, and construction waste (Alzamora and Barros 2020). The source of energy used and climate conditions influence the MSW. For instance, cities with cold winter that rely on coal as their heating source tend to produce a larger quantity of ash in the waste this season. Other differences are raised due to the basic infrastructure. For example, the amount of dirt and dust from street sweeping is large in the unpaved or poorly paved streets. Other differences are raised related to the numbers of shrubs and trees, which increase the number of organic wastes. In the high-income cities, bulky and large wastes like furniture abandoned motorcars and packaging are found, unlike in low-income cities (Iyamu et al. 2020).

1.2.3 Industrial Waste

Industrial wastes involve a broad range of materials with different environmental toxicity. Typically, this range would involve packaging materials, paper, waste from food processing, solvents, oils, paints, resins, sludge, metals, glass, stones, ceramics, plastics, leather, rubber, wood, straw, cloth, abrasives, etc. The exact production rates are not precisely known owing to the lack of regularly systematic and updated tracking of industrial wastes. The vast differences in the raw materials, industrial process, final product, and environmental consideration between different industries make it very difficult to define common properties that characterize industrial wastes in general; however, they can be categorized according to their nature. Table 1.2 classifies the most common industrial wastes according to their nature and gives examples of each category.

Food processing industries like fruit industries, confectionery, meat, chips, and juice are one of the industrial sectors that produce large amounts of organic waste every year. The demand for food products increases continuously due to the increase in population; that is why a lot of beverage and food industries have expanded remarkably all over the world to fulfill the food demand. As a result, larger amounts of wastes are generated annually. There is a trend to use fruit industrial wastes as raw materials to produce other valuable products since they are rich in lignin, hemicellulose, cellulose, nitrogen, carbon, ash, and moisture, which can be biochemically digested to produce bio-ethanol, biogas, and other products.

The production of industrial wastes varies even among developing countries, not only between countries at different stages of development. For example, the production ratio of MSW to industrial wastes is one to three in China; however, this ratio is significantly lower in other countries with similar income per capita. On the other hand, in the high-income developed country, the ratio reaches one to eight. This quantity of industrial waste will be expected to be doubled in the next 20 years at the current growth rates. This incremental growth is considered a serious challenge since many countries suffer from the insufficient capacities of the existing industrial waste collection, processing, and disposal systems (Zhang et al. 2019).

Due to the increasing risks of industrial wastes on communities and the environment, governments enact laws to force owners of industrial activities to perform environmental impact assessment studies to get the project approved or renew their license. In this study, all the materials used, products, amounts, and compositions

Table 1.2 Classification of industrial wastes

Category of wastes	Examples
Cinder	Incineration residues, including ashes
Oil, grease, and fats	Waste mineral, animal or plant-based oil, including grease (lubricant), waste animal fats, and oily wastes from soybean oil production
Acid	Waste organic and inorganic acids, including acidic wastes generated from phosphate rock-related industries or waste acids generated from the commercial acids production lines
Alkali	Alkali solids and liquids, such as waste soap solution
Plastics	Waste styrene foam, waste synthesized fibers, and synthesized polymers (solid, liquid), including synthesized rubber
Metal	Waste iron, waste aluminum, and other waste metals generated in polishing and cutting
Slag	Sand, sandblast waste, low-quality coal, and blast furnace wastes
Rubble	Concrete pieces, brick pieces, etc., generated in the construction, reconstruction, or removal of structure
Dust and soot	Dust and soot generated from different chemical and industrial plants
Paper	Paper waste discharged from construction activities, pulp, and paper manufacturing as well printing facilities
Wastes of animal or plant origin	Animal and plant wastes are used as raw materials in food, pharmaceuticals, and spices production. They include fish or animals bones, brewing, and fermentation wastes

of wastes produced must be reported and revised by the responsible authorities to decide by approving or refusing the project (Glasson and Therivel 2019). This will potentially reduce the generated industrial waste amounts and provide a database of generated wastes from current and future industries.

1.2.4 *Hazardous Waste*

The production of hazardous wastes is increasing continuously due to the developments in various sectors, including agricultural activities, industrial plants, and healthcare institutions. This development results in the consumption of large amounts of toxic chemicals. For example, nowadays, there are about 110,000 different toxic chemicals available in the market. However, each year, about one thousand new chemicals appear for being utilized for different purposes. Wastes of chemicals, electric light bulbs, batteries, automotive parts, and leftover medicines are classified as hazardous wastes (Letcher and Slack 2019).

There are several issues related to the availability of data about the global generation of hazardous wastes worldwide as the reliability of hazardous waste production limits it. It is doubtful due to several reasons, including human error during performing the essential assessment, the possible unwillingness of industrial plants to provide process information (including waste arising data), the poor awareness of the dangerous effects of these wastes, absence of a clear, and unified definition of hazardous wastes from a country to another.

Most hazardous wastes are introduced to the environment as by-products of different processes in various sectors, including agricultural and industrial processes. Hospitals, nuclear facilities, and healthcare institutions have their distinct share of producing this dangerous type of wastes. The highest generation rates of hazardous materials come from petrochemicals, chemicals, and petroleum plants. In addition, energy production plants, pulp, and paper facilities, metals manufacturing and machining facilities, and wood treatment plants are heavy producers of this type of waste. Leather production lines are also a famous source of hazardous wastes, e.g., chromium ions in wastewater effluents, with a high production rate (Prakash and Gowtham 2019). The pesticides production and utilization sectors have a significant share in hazardous wastes production as well.

There is a strong relation between hazardous waste type and production rates and the production region where the industrial activities' variety and degree of modernization influence the production (Abdel-Shafy and Mansour 2018). For instance, the major hazardous wastes in the Asian and Pacific Region are waste solvents due to extensive use of solvents, chlorine wastes, pesticides, e.g., organophosphates, and even the wastes bearing considerable amounts of solvents, chlorine, and pesticides.

1.3 Municipal Waste Management Systems

The practiced municipal waste management systems worldwide are many and diverse. Among the adopted solutions are source reduction, recycling, composting, incineration, dumping, and landfilling. Furthermore, some of these systems require large investments that are more suitable for high-income per capita regions and more effective and safer for people and the environment. However, some practices, like waste burning and dumping on the domestic level, have a catastrophic impact on the environment. The environmentally appropriate management solutions are summarized in Fig. 1.1 where they will be discussed in more detail in Chap. 2.

Choosing a particular solution depends on social, economic, environmental, and technical factors that vary from one place to another. Table 1.3 summarizes the advantages and disadvantages of the discussed methods. The most desirable solution should have the potential to fully treat different waste categories with valuable raw material and maximum energy recovery. Composting and open burning of waste methods are not commonly used methods as they both lack energy recovery. However, waste management systems with energy recovery, i.e., anaerobic digestion, pyrolysis, and gasification, combined with combustion of evolved products, are considered

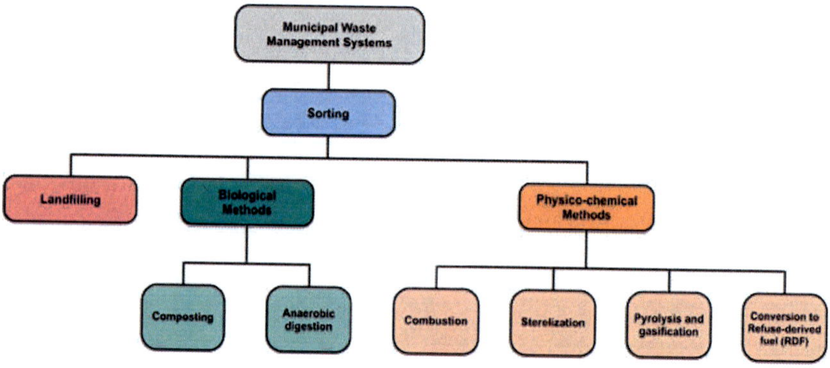

Fig. 1.1 Current technologies for the municipal waste management system

the most favorable solutions. Produced biogas and syngas can cover the energy requirements for these technologies, which makes the investment very attractive economically.

1.4 Statistics of Wastes Production and Management

Population growth, economic development, urbanization, and the type of activities in a certain country or region directly impact the type and amounts of waste produced. They also directly impact the waste management capabilities and hence the waste management methods used in this country or region. There is a growing interest in studying the common factors affecting waste production and management; however, it is very challenging since the data about waste production and management are not available in all the countries. Therefore, most of the published studies are performed by international organizations (such as The World Bank) that have the authority to collect data from governments. The reliability of the gathered data in many cases is questionable for many reasons, such as missing dates and methodologies, incomplete definitions, undefined words, and missing and inconsistent units. The process of data collection, checking, and analysis is very time consuming. Therefore, published data from reliable sources are limited. It is worth mentioning that these studies aim to highlight the trends of waste production and management in countries, cities, and regions that have common economic and sociological status and not for the sake of ranking.

This chapter presents the important data and statistics on quantities and types of waste produced in different regions worldwide and the methods of management in each region. It highlights the common factors in the regions and countries with similar waste production and management data.

Table 1.3 Comparison of the current technologies for the municipal waste management systems

Technology	Advantages	Disadvantages
Landfilling	Ease of operation and management Low fixed and operating costs High disposal capacity	Secondary pollution and a huge footprint Neither energy nor raw material is recovered
Composting	Ease of operation Resource utilization at a high rate	Hard maintenance of operational parameters Longer retention time Large footprint No energy and raw material recovery Generation of unpleasant smells and presence of insects
Anaerobic digestion	Low operational cost Energy and raw material recovery Acceptable condition of hygiene	Elevated costs, i.e., fixed and maintenance Longer retention time Pretreatment and post-treatment of waste Continuous monitor of operational parameters
Combustion	Easy operation Heat recovery Waste volume reduction	High capital and operational costs Low calorific value No raw material recovery Generation of harmful gases
Pyrolysis and gasification	Production of high-energy density fuels High calorific value Energy and raw material recovery	Complex waste separation systems Waste transportation High capital and operational costs
Sterilization	Easy operation Shorter retention time Treatment of biological, medical, and hazardous waste	High capital and maintenance costs No energy recovery
Refuse-derived fuel (RDF)	High calorific value No need for waste sorting Waste collection frequency reduction Energy and raw material recovery	Emissions of toxic substances Further research is needed to calculate the CO_2 footprint

1.4.1 Waste Production

The World Bank performed an extensive study on waste production and management in 367 countries and grouped the world countries according to their geographical locations (Ekström 2014; Kaza et al. 2018). Table 1.4 summarizes the total amounts and the classification of the solid wastes produced in each region in 2016. The

Table 1.4 Total quantities and classification of waste produced in different regions worldwide in 2016 (Kaza et al. 2018)

Waste type/population	East Asia and Pacific	Europe and central Asia	South Asia	North America	Latin America and Caribbean	Sub-Saharan Africa	Middle east and north Africa
Food and green	53%	36%	57%	28%	52%	43%	58%
Metal	3%	3%	3%	9%	3%	5%	3%
Paper and cardboard	15%	19%	10%	28%	13%	10%	13%
Wood	2%	2%	1%	6%	1%	1%	1%
Glass	3%	8%	4%	5%	4%	3%	3%
Plastics	12%	12%	8%	12%	12%	9%	12%
Rubber and leather	1%	1%	2%	9%	0%	0%	2%
other	12%	21%	15%	4%	15%	30%	8%
Population (million person)	2.29	0.91	1.76	0.37	0.64	1.04	0.44
Total waste production in 2016 (million tons)	468	392	334	298	231	174	129

available data show that more than 2 billion ton of solid waste were generated in 2016. It is expected that this number becomes 3.40 billion tons by 2050 due to the growing population and urbanization.

The data show that East Asia and the Pacific region produced the maximum annual production amount of waste at 468 million tons with an average of 0.56 kg per capita per day. Europe and Central Asia come in second place by producing 392 million tons, corresponding to 1.18 kg per capita per day. The waste is mainly composed of organics and solid recyclables such as paper and plastics. The least waste-producing region is the Middle East and North Africa (MENA) region, with 129 million tons which corresponds to 6% of the global waste production (Kaza et al. 2018).

The amount of waste produced in a region is a function of many factors; the most obvious factors are population, degree of urbanization (migration of people from rural to urban regions), and economic development. With the exceptions of Europe and Central Asia and North America, it can be observed from the data in Table 1.4 that large amounts of wastes are produced in highly populated regions. As a general trend, the countries with a high degree of urbanization and high gross national income (GNI) per capita produce large amounts of waste compared to countries with low degree of urbanization and low GNI. This is apparent in Europe and Central Asia and North America, where the average degree of urbanization and the average GNI per capita are higher than other countries, and the waste produced is high relative to the population in these regions. The abovementioned study demonstrated the relation between the wastes produced in 2016 in 237 cities with different degree of urbanization and confirmed the relation between them. The waste generated in highly urbanized cities can be double or triple the waste generated in cities with low degree of urbanization and similar population; this has been confirmed by other studies done on cities in different regions(Yiougo et al. 2013; Chen 2018).

From another perspective, Wilson et al. deduced a relation between the waste generation and the GNI per capita after collecting the data of waste generation in twenty cities in different countries with various income levels (Wilson et al. 2012). The study clearly showed that the waste generation per capita increases with the city's income level (GNI per capita) (Wilson et al. 2012). In particular, high-income countries account for 16% of the world's population; however, they generate 34%, or 683 million tons, of the world's waste. On the other hand, low-income countries account for 9% of the world's population but generate only about 5% of global waste or 93 million tons.

Many studies performed on different cities have confirmed the relationship between the GNI and waste production (Hoornweg and Thomas 1999; Lacoste and Calmin 2010; Gardiner and Hajek 2017; Khajuria et al. 2010). The composition of waste differs according to the income level as it reflects the consumption pattern. For instance, waste from low-income and middle-income countries consists mainly of organic materials such as food and green waste and low recyclable waste percent with an average of 16%. High-income countries produce comparative amounts of organic wastes; however, the large amounts of recyclable wastes such as packaging materials and cardboard reduce the percent of the organic waste to around 32%. The quantity of recyclable wastes increases as the income level rises, with the amount of paper waste most significantly increasing (Kaza et al. 2018).

Waste generation is expected to grow with economic development and population growth. The greatest increase is expected in low-income countries. For instance, it is expected that the waste levels will double in Sub-Saharan Africa and triple in South Asia regions in the next thirty years. The wastes generated in countries with higher income levels are expected to increase at a lower rate. The expectations for future waste generation are worrying and alarming since the increasing amounts of wastes will put more load on the environment, require fast actions to reduce the amounts of wastes, and encourage the different methods of waste management.

1.4.2 Waste Management

The waste management industry (municipal, industrial, and hazardous wastes) is rapidly growing with market size of 2080 billion dollars in 2019 and is expected to reach 2339.8 billion dollars in 2027. The waste management industry includes waste collection, waste treatment, and disposal (Pawar and Sawant 2020).

The budgets allocated and accordingly the extent of waste management are highly dependent on the income level. For instance, Wilson et al. summarized the solid waste management budget per capita, both in total and as a percentage of GNI per capita in twenty cities in 2010 (Wilson et al. 2012). They showed that the annual allocated budget per capita rises from USD 1.4 for the low-income cities to USD 75 for high-income cities. In the first glance, the high waste management budget reflects a large economic load on high-income communities; however, the trend is reversed when the data are expressed as a percentage of income per capita. The average for low-income countries is 0.32%, while it is only 0.17% for high-income countries. The middle-income countries have the highest percentage of an average of 0.65%. The ratios are low in high-income and low-income countries due to the high GNI for high-income countries and the low allocated budgets in the case of low-income countries. This clearly shows that the most challenging cases are in the low- and middle-income countries where the current expenditures on solid waste management are low compared to developed high-income countries, in addition to the extremely limited chances of expansion and increase in waste management as a result of the inability of citizens to economically afford that (Wilson et al. 2012).

Waste collection is considered the first stage in waste management and indicates the degree of efforts done in waste management. It is a service provided by the governments or by private parties and is accomplished by many methods such as the door-to-door collection where truck or waste collection vehicles pick up the wastes from homes and markets. In other cases, waste is disposed of in a central collection point and is picked up from this point to management or disposal sites. Income level has a great impact on waste collection rates. It was found that countries with similar income levels have similar waste collection rates. The extended study performed by Kaza et al. for the World Bank showed that waste collection percentage in high-income cities is close to 100%. While this value decreases at lower-income cities, for instance, it has 82%, 51%, and 39% in upper-middle-income, low-middle-income, and low-income cities, respectively (Kaza et al. 2018). It was also found that waste collection rates in urban areas are higher than in rural areas. The rates in some urban areas are more than twice the rates in rural areas in the same cities. Communities usually get rid of uncollected waste by open dumping and burning, which negatively impacts the environment and human health (Yoada et al. 2014). The uncollected wastes are estimated to be around 33% of the global waste in 2016 (Kaza et al. 2018).

Collected wastes are treated using the methods mentioned earlier in this chapter. Globally, almost 40% of waste is disposed of in landfills, about 19% are recovered through recycling and composting, 11% are treated by modern incineration, and around 37% of the global wastes are disposed of in landfills.

The World Bank survey also summarized the waste management methods used in different regions and the ratios of waste managed by each method. The income level highly influences waste disposal and treatment methods. For instance, in low-income, where landfill is not available, burning, open dumping in waterways, open lands, and roads is the common method of getting rid of wastes and is applied for around 70% of the wastes; however, only 2% of the waste in high-income countries is burned or dumped. The dumped waste in Sub-Saharan Africa and South Asia regions represents more than 66% of the total dumped waste worldwide (Kaza et al. 2018).

Proper waste disposal and treatment such as controlled landfilling and recycling are widely applied in high and upper-middle-income countries. However, the dependence on landfilling is higher in upper-middle-income countries, representing the highest percentage with an average of 54%. High-income countries are less dependent on landfilling (39% of waste), due to the expansion in economically profitable methods such as recycling and compositing (25% of wastes) and incineration (22% of wastes).

Paper and paperboard constitute around 67% of the total recycled wastes in 2018, metals comprised about 13%, while glass, plastic, and wood made up around 5%. The most recycled wastes are corrugated boxes, mixed non-durable paper products, newspapers/mechanical papers, lead–acid batteries, major appliances, wood packaging, glass containers, and tires (United States Environmental Protection Agency (US-EPA) 2020). The efforts exerted in the field of waste management are still not sufficient to counter the current and the expected increase in waste production mentioned in the previous section due to many challenges such as the steady increase in waste generation due to the increase in population and consumption, limited financial resources and great financial demand, limited access to technical knowledge, limited technical expertise and awareness of best practices, lack of planning and evaluation, limited or lack of government coordination, difficult working conditions, and limited available land compared to large areas needed for some methods such as landfilling (United States Environmental Protection Agency (US-EPA) 2020; Guerrero et al. 2013; McAllister 2015; Taherzadeh and Rajendran 2014).

1.5 Impact of Applying Circular Economy Principles

1.5.1 Relation Between Circular Economy and Waste to Energy WTE

In the circular economy, unlike the linear economy, wastes are being processed by different methods such as recycling, reusing, and converting to energy to reduce the quantities and negative environmental impacts of wastes (Neczaj and Grosser 2018). Figure 1.2 briefly shows the difference between these two types of linear and circular economies. It is worth noting that the advances in the circular economy will be presented and discussed in more detail in Chap. 3.

As it can be depicted from Fig. 1.2, all the materials or products are disposed of after finishing their lifetime in a linear economy. However, part or even all of the waste in the circular economy can be recycled to produce raw materials or lower-quality products. The application of circular economy practices has positive environmental, economic, and social impacts. Regarding the environmental impacts, it reduces the negative environmental impacts by eliminating part of the emissions and water footprints resulting from extraction and upgrading of raw materials (Maina et al. 2017; Rada et al. 2018). Waste to energy (WTE) is considered one of the most

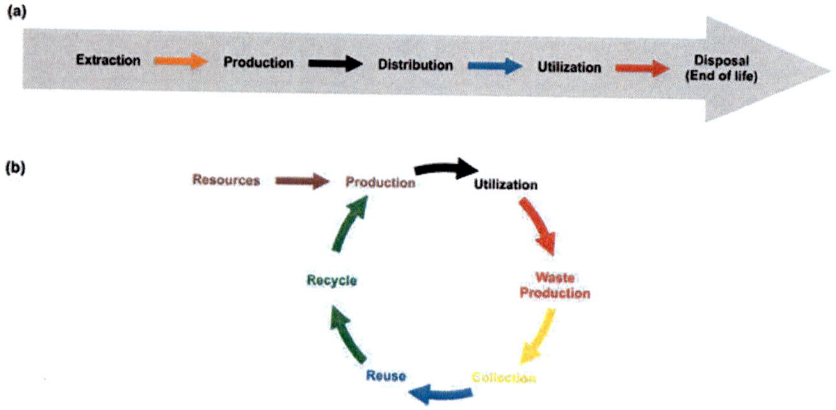

Fig. 1.2 Linear (**a**) versus circular (**b**) economies

important figures of circular economy practices (Asian Development Bank 2020). Additionally, the landfilled or disposed of materials will be diminished. From the economic perspective, a circular economy has a positive impact by decreasing the costs of extraction and disposal at the end of life of different products. Socially, a circular economy leads to job vacancies for unemployed people in poor areas or in the areas assigned to perform the waste conversion (Aguilar-Hernandez et al. 2021). Besides, applying these concepts has other obvious social effects such as changing the lifestyle, demography, and public health ("Socio-economic Impact Assessment of Waste-to-Energy" 2020).

Despite the diverse advantages of a circular economy, some challenges must be addressed first. One example is social acceptance and the attitude of communities toward circular economy practices, including WTE. Social acceptance is a dynamic term that changes with time and is also a function of the society and culture and believes of people (Baxter et al. 2016). Awareness and engagement campaigns are necessary to change the communities' ideas and accept circular economy practices. This can be performed through different channels, including social media platforms, to reach different communities and people with various backgrounds.

1.5.2 Different Circular Economy Business Models

Before presenting the detailed impacts of the circular economy and WTE on the whole economy and society, it is important to briefly review different circular economy business models. Generally, there are five business models for the circular economy to be applied in real-life practices ("Circular Business Models" 2021; Gerholdt 2015). A summary of these models is presented below.

1.5.2.1 Circular Supplies Model

This model involves the introduction of new renewable materials to the manufacturing matrix instead of not sustainable resources. This model is preferred when the other traditional or conventional raw materials are scarce, limited, or not continuously available. For instance, Royal DSM company accounts for the lignocellulosic agricultural wastes as a renewable source for bio-ethanol production ("Putting our Energy into Renewables" 2021). This sustainable biofuel can reduce emissions to reach a carbon neutral state besides creating job vacancies as an important positive social impact. Moreover, it should have an economic impact by adding revenues and return from processing no-cost waste materials.

1.5.2.2 Resource Recovery Model

In this model, any waste or side product at any stage of manufacturing, utilization, and end of life can be reused as a resource or energy production. Walt Disney World Resort applies this model by collecting food waste, converting it into biogas, and fertilizing in a designated plant under the supervision of Harvest Power company (Coughlin 2012; "Disney Citizenship 2014 Performance Summary" 2014). This model has similar positive economic and social impacts to those of the previous model.

1.5.2.3 Product Life Extension Model

This model aims to reuse waste products or improve the quality of the fresh ones to live more instead of being wasted. This is not a figure of WTE; however, it targets the reduction of the energy required for extraction and other downstream processes. The landfilled materials will also be reduced, reducing the negative impacts on the environment and the surrounding communities. Caterpillar company is well known for applying this model during the last 40 years as a part of its sustainability plan through Reman Project ("2020 Annual Report" 2020).

1.5.2.4 Other Models

"Sharing platforms" and "product as a service" models are two famous bridging circular economy models where the process expenses can be minimized. This is achieved by sharing the production facilities/equipment or using certain products for short times without owning them, i.e., leasing the product. These two models do not involve WTE practices; yet, energy, resources, and production costs will be decreased upon applying them. Hence, the environmental and socioeconomic state will be improved.

1.5.3 Examples of Positive and Negative Impacts

As an example of job creation, in 2013, it was reported that the number of workers in US WTE plants was 7000 working in 86 facilities ("Waste to Energy—Social Impacts" 2013). In Maine state in USA, the WTE plants offered 597 full and part-time jobs inside and outside the plants ("The Science Behind Your Landfill" 2013). The annual mass of processed wastes is 830 million tons that were used to generate 475,000 MWh of electricity. The total labor income of these projects was estimated to be approximately 34 million USD. Fortunately, as evidence of the increased interest in WTE plants, about 420 facilities in Europe compared to 87 units in the USA in 2010. The numbers are rapidly growing (Reigosa 2010). The increase in the number of facilities and depending on the WTE and recycling practices will positively impact increasing the employment rate, which has a direct positive influence on the socioeconomic state of different areas and their population ("Waste-to-Energy Research and Technology Council" 2013). Other positive social impacts include the improvement of public infrastructure and schools. Besides, the communities will benefit from the improvement of environmental conditions due to the reduction of waste disposal and emissions ("Waste-to-Energy Ash Reuse" 2013). It should be noticed that the maximum possible percentage of waste conversion to energy is related to the type of the collected wastes, which varies according to the season and region of production.

Although WTE and circular economy, in general, have promising impacts socially and economically, they also possess some drawbacks or, in other words, challenges that should be taken into consideration and solved (Michael 2013). In contrast to what was reported previously, some studies stated that some units are not feasible, and the returns do not cover the expenses of installation and processing (Tan et al. 2015). Besides, in other reports, these facilities were reported to have a counter environmental impact. For instance, the design of some combustion and incineration plants produces large amounts of flue gases and emissions that are hazardous and not environmentally friendly such as dioxins ("Negative Impacts of Incineration-based Waste-to-Energy Technology" 2021; Vehlow 2013). If this problem is not solved, the social interest or acceptance of building incineration-based WTE facilities will decrease dramatically.

1.6 Conclusions

This chapter is an introductory one to a vital and hot topic that currently captures global attention: waste management and conversion to valuable figures such as energy. It presented an overview of the classification of different wastes produced from various sources. In addition, the currently utilized waste management techniques were discussed briefly and their advantages and drawbacks. It also showed the essential statistics of the annual global production of wastes. In this regard, the relations between the production and composition of the wastes and the income levels of different regions were well defined. Due to the high production rates, it is highly recommended to pay more attention to future research studies on waste management and conversion. In these studies, researchers should prioritize inventing new feasible and more efficient methods to overcome the drawbacks of the existing ones. By the end of the chapter, circular economy principles and their relation with waste management techniques, particularly waste to energy ones, were briefly stated. Additionally, the famous circular economy models were mentioned, including examples of their application in the real world. The conversion from a linear economy to a circular one should be a high priority for the researchers, stakeholders, and decision makers due to its promising environmental, social, and economic merits.

References

2020 Annual Report [WWW Document] (2020) Caterpillar. URL: https://sustainabilityguide.eu/methods/circular-business-models/

Abdel-Shafy HI, Mansour MSM (2018) Solid waste issue: sources, composition, disposal, recycling, and valorization. Egypt J Pet 27(4):1275–1290. https://doi.org/10.1016/j.ejpe.2018.07.003

Aguilar-Hernandez GA, Dias Rodrigues JF, Tukker A (2021) Macroeconomic, social and environmental impacts of a circular economy up to 2050: a meta-analysis of prospective studies. J Cleaner Prod **278**, 123421. https://doi.org/10.1016/j.jclepro.2020.123421

Alzamora BR, de Barros RTV (2020) Review of municipal waste management charging methods in different countries. Waste Manage 115:47–55. https://doi.org/10.1016/j.wasman.2020.07.020

Asian Development Bank (2020) Waste to energy in the age of the circular economy: best practice handbook. Asian Development Bank, Philippines

Baxter J, Ho Y, Rollins Y, Maclaren V (2016) Attitudes toward waste to energy facilities and impacts on diversion in Ontario, Canada. Waste Manage 50:75–85. https://doi.org/10.1016/j.wasman.2016.02.017

Bui TD, Tsai FM, Tseng M-L, Ali MH (2020) Identifying sustainable solid waste management barriers in practice using the fuzzy Delphi method. Resour Conserv Recycl 154:104625. https://doi.org/10.1016/j.resconrec.2019.104625

Bundhoo ZMA (2018) Solid waste management in least developed countries: current status and challenges faced. J Mater Cycles Waste Manag 20(3):1867–1877. https://doi.org/10.1007/s10163-018-0728-3

Chen J, Gong Y, Wang S, Guan B, Balkovic J, Kraxner F (2019) To burn or retain crop residues on croplands? An integrated analysis of crop residue management in China. Sci Total Environ 662:141–150. https://doi.org/10.1016/j.scitotenv.2019.01.150

Chen Y-C (2018) Effects of urbanization on municipal solid waste composition. Waste Manage 79:828–836. https://doi.org/10.1016/j.wasman.2018.04.017

Circular Business Models [WWW Document] (2021) Sustainability guides. URL https://sustainabilityguide.eu/sustainability/circular-economy/

Coughlin C (2012) Sowing seeds, reaping energy with Harvest Power [WWW Document]. Greenbiz. URL https://www.greenbiz.com/article/sowing-seeds-reaping-energy-harvest-power

Disney Citizenship 2014 Performance Summary [WWW Document] (2014) Walt Disney. URL: https://thewaltdisneycompany.com/app/uploads/FY14-Performance-Summary.pdf#waste_

Ekström K (2014) Waste management and sustainable consumption, 1st edn. Taylor & Francis group, London

Environmental statistics and accounts (2016) Guidance on municipal waste data collection. European Commission, EU

Gardiner R, Hajek P (2017) Impact of GDP, capital and employment on waste generation—the case of France, Germany and UK Regions. In: Proceedings of the 8th international conference on e-business, management and economics (ICEME 2017). Association for Computing Machinery, New York, NY, USA, pp 94–97. https://doi.org/10.1145/3157754.3157761

Gerholdt J (2015) The 5 business models that put the circular economy to work [WWW Document]. GreenBiz. URL https://www.greenbiz.com/article/5-business-models-put-circular-economy-work

Glasson J, Therivel R (2019) Introduction to environmental impact assessment, 5th edn. Routledge, London. https://doi.org/10.4324/9780429470738

Guerrero LA, Maas G, Hogland W (2013) Solid waste management challenges for cities in developing countries. Waste Manage 33(1):220–232. https://doi.org/10.1016/j.wasman.2012.09.008

Hoornweg D, Thomas L (1999) What a waste : solid waste management in Asia

Iyamu HO, Anda M, Ho G (2020) A review of municipal solid waste management in the BRIC and high-income countries: a thematic framework for low-income countries. Habitat Int 95:102097. https://doi.org/10.1016/j.habitatint.2019.102097

Kaza S, Yao L, Bhada-Tata P, Van Woerden F (2018) What a waste : a global review of solid waste management to 2050. World Bank Publications, Washington

Khajuria A, Yamamoto Y, Morioka T (2010) Estimation of municipal solid waste generation and landfill area in Asian developing countries. J Environ Biol 31(5):649–654

Köseoğlu E, Akmil-Başar C (2015) Preparation, structural evaluation and adsorptive properties of activated carbon from agricultural waste biomass. Adv Powder Technol 26(3):811–818 (Special issue of the 7th World Congress on Particle Technology). https://doi.org/10.1016/j.apt.2015.02.006

Lacoste E, Calmin P (2010) 2009 World Waste Survey. From Waste to Resource. Economica, Paris

Letcher TM, Slack R (2019) Chemicals in waste: Household hazardous waste, in: Waste. Elsevier, pp 337–352. https://doi.org/10.1016/B978-0-12-815060-3.00017-7

Li Q, Zheng L, Cai H, Garza E, Yu Z, Zhou S (2011) From organic waste to biodiesel: Black soldier fly, Hermetia illucens, makes it feasible. Fuel 90(4):1545–1548. https://doi.org/10.1016/j.fuel.2010.11.016

Maina S, Kachrimanidou V, Koutinas A (2017) A roadmap towards a circular and sustainable bioeconomy through waste valorization. Curr Opin Green Sustain Chem 2017(8):18–23. https://doi.org/10.1016/j.cogsc.2017.07.007

Makarichi L, Jutidamrongphan W, Techato K (2018) The evolution of waste-to-energy incineration: a review. Renew Sustain Energy Rev 91:812–821. https://doi.org/10.1016/j.rser.2018.04.088

Marey HS, Gille JC, El-Askary HM, Shalaby EA, El-Raey ME (2010) Study of the formation of the "black cloud" and its dynamics over Cairo, Egypt, using MODIS and MISR sensors. J Geophys Res Atmos 115(D21). https://doi.org/10.1029/2010JD014384

McAllister J (2015) Factors influencing solid-waste management in the developing world. Utah State University, Utah

Michael T (2013) Environmental and social impacts of waste to energy (WTE) conversion plants. In: Klinghoffer NB, Castaldi MJ (eds) Waste to energy conversion technology. Woodhead Publishing Series in Energy. Woodhead Publishing, pp 15–28. https://doi.org/10.1533/9780857096364.1.15

Nagendran R (2011) Agricultural waste and pollution. In: Letcher TM, Vallero DA (eds) Waste. Academic Press, Boston, pp 341–355. https://doi.org/10.1016/B978-0-12-381475-3.10024-5

Neczaj E, Grosser A (2018) Circular economy in wastewater treatment plant–challenges and barriers. Proceedings 2(11):614. https://doi.org/10.3390/proceedings2110614

Negative Impacts of Incineration-based Waste-to-Energy Technology [WWW Document] (2021) AENews. URL http://www.alternative-energy-news.info/negative-impacts-waste-to-energy/

Pawar D, Sawant R (2020) Waste management market report (Market report No. A00327). Allied Market Research

Prakash R, Gowtham M (2019) Hazardous waste and its treatment process. In: Energy from toxic organic waste for heat and power generation. Elsevier, Amsterdam, pp 119–138. https://doi.org/10.1016/B978-0-08-102528-4.00009-2

Putting our energy into renewables [WWW Document] (2021) DSM bio-based products & services. URL: https://www.dsm.com/corporate/about/businesses/dsm-bio-based-products-services.html

Rada EC, Ragazzi M, Torretta V, Castagna G, Adami L, Cioca LI (2018) Circular economy and waste to energy. AIP Conf Proc 1968(1):030050. https://doi.org/10.1063/1.5039237

Reigosa P (2010) Energy from waste across Europe: current statistics and trends [WWW Document]. Tetra Tech. URL: https://www.tetratechcoffey.com/Uploads/Documents/Energy-from-Waste-across-Europe-Current-Statistics-and-Trends-paper_20110513135927.pdf

Sadh PK, Duhan S, Duhan JS (2018) Agro-industrial wastes and their utilization using solid state fermentation: a review. Bioresour Bioprocess 5(1):1. https://doi.org/10.1186/s40643-017-0187-z

Sathiparan N, De Zoysa HTSM (2018) The effects of using agricultural waste as partial substitute for sand in cement blocks. J Build Eng 19:216–227. https://doi.org/10.1016/j.jobe.2018.04.023

Socio-economic Impact Assessment of Waste-to-Energy [WWW Document] (2020) Word Press. URL: https://prigemconcepts.wordpress.com/2020/04/08/socio-economic-impact-assessment-of-waste-to-energy/

Stephen JL, Periyasamy B (2018) Innovative developments in biofuels production from organic waste materials: a review. Fuel 214:623–633. https://doi.org/10.1016/j.fuel.2017.11.042

Taherzadeh MJ, Rajendran K (2014) Factors affecting development of waste management: experiences from different cultures. In: Waste management and sustainable consumption. Routledge, London, pp 79–99. https://doi.org/10.4324/9781315757261-12

Tan ST, Ho WS, Hashim H, Lee CT, Taib MR, Ho CS (2015) Energy, economic and environmental (3E) analysis of waste-to-energy (WTE) strategies for municipal solid waste (MSW) management in Malaysia. Energy Convers Manage 102:111–120. https://doi.org/10.1016/j.enconman.2015.02.010

Teo EYL, Muniandy L, Ng E-P, Adam F, Mohamed AR, Jose R, Chong KF (2016) High surface area activated carbon from rice husk as a high performance supercapacitor electrode. Electrochim Acta 192:110–119. https://doi.org/10.1016/j.electacta.2016.01.140

The Science Behind Your Landfill [WWW Document] (2013) Ecomaine. URL https://www.ecomaine.org/wp-content/uploads/2020/06/15-Science-Behind-Landfill-2.pdf

UN-ESCAP (2006) Types of wastes, in: State of the Environment in Asia and the Pacific 2005 Synthesis: economic growth and sustainability. UN-ESCAP, New York, USA, p 256.

United States Environmental Protection Agency (US-EPA) (2020) Best practices for solid waste management: a guide for decision-makers in developing countries (Policy report No. EPA 530-R-20-002). United States Environmental Protection Agency

Vehlow J (2013) Transformation of waste combustion facilities from major polluters to pollution sinks. In: Klinghoffer NB, Castaldi MJ (ed) Waste to Energy Conversion Technology, Woodhead Publishing Series in Energy. Woodhead Publishing, pp 179–203. https://doi.org/10.1533/9780857096364.3.179

Waste to energy—Social Impacts [WWW Document] (2013) Word Press. URL: https://wastetoenergy13.wordpress.com/solution-waste-to-energy/social-impacts/

Waste-to-Energy Ash Reuse [WWW Document] (2013) Energy Recovery Council. URL http://www.energyrecoverycouncil.org/ash-reuse-a2974&gt

Waste-to-Energy Research and Technology Council [WWW Document] (2013) Columbia University. URL: http://www.seas.columbia.edu/earth/wtert/&gt

Wilson DC, Rodic L, Scheinberg A, Velis CA, Alabaster G (2012) Comparative analysis of solid waste management in 20 cities. Waste Manag Res 30(3):237–254. https://doi.org/10.1177/0734242X12437569

Yadav V, Karmakar S (2020) Sustainable collection and transportation of municipal solid waste in urban centers. Sustain Cities Soc 53:101937. https://doi.org/10.1016/j.scs.2019.101937

Yahya MA, Al-Qodah Z, Ngah CWZ (2015) Agricultural bio-waste materials as potential sustainable precursors used for activated carbon production: a review. Renew Sustain Energy Rev 46:218–235. https://doi.org/10.1016/j.rser.2015.02.051

Yiougo LSA, Oyedotun TDT, Some CYC, Da ECD (2013) Urban cities and waste generation in developing countries: a GIS evaluation of two cities in Burkina Faso. J Urban Environ Eng 7(2):280–285

Yoada RM, Chirawurah D, Adongo PB (2014) Domestic waste disposal practice and perceptions of private sector waste management in urban Accra. BMC Public Health 14(1):697. https://doi.org/10.1186/1471-2458-14-697

Zhang A, Venkatesh VG, Liu Y, Wan M, Qu T, Huisingh D (2019) Barriers to smart waste management for a circular economy in China. J Cleaner Prod 240:118198. https://doi.org/10.1016/j.jclepro.2019.118198

Chapter 2
Different Waste Management Methods, Applications, and Limitations

Bahram Barati, Fatemeh Fazeli Zafar, and Shuang Wang

Abstract Approximately, half a ton of waste is produced per person annually; thus, waste management is essential to avoid environmental issues. The waste management system encompasses the entire set of activities related to treating, handling, recycling, or disposing of waste materials. This chapter aims to represent the history of waste management, as well as the current commonly used methods for waste management, together with their pros and cons. The commonly applied methods, including sanitary landfill, composting, safe disposal of biomedical wastes, recycling of industrial CO_2 emissions, incineration of hazardous wastes, sludge recycling in the cement industry, direct combustion of sludge, wastewater treatment, and construction waste recycling, are discussed. In this chapter, the problems associated with landfills, such as lack of efficient systems for gas utilization from the landfill and lack of proper leachate management and recycling CO_2 from industrial flue gas and wastewater treatment methods, are more highlighted. Also, the need for law enforcement to control the negative environmental impacts properly is highlighted.

Keywords Conventional waste management · Landfill · Recycling · Incineration · Combustion

2.1 Municipal Solid Wastes

The load of municipal solid waste (MSW) produced worldwide from urban regions is overgrowing due to the rapid urbanization and growing human population (America 2010). MSW includes a substantial fraction of paper, food waste, plastic, textile, and wood. The current global MSW production is around 2.01 BT annually, and the magnitude is expected to increase to around 3.40 BT by 2050 (Zhang et al. 2021). The top 10 MSW producing countries, together with the share of the main waste category, are presented in Fig. 2.1. China, India, and the USA are the leading producers of MSW by generating 15.55, 11.95, and 11.65% of the total MSW worldwide.

B. Barati · F. F. Zafar · S. Wang (✉)
School of Energy and Power Engineering, Jiangsu University, Jiangsu 212013, China
e-mail: alexjuven@ujs.edu.cn

A. E.-F. Abomohra et al. (eds.), *Waste-to-Energy*,
https://doi.org/10.1007/978-3-030-91570-4_2

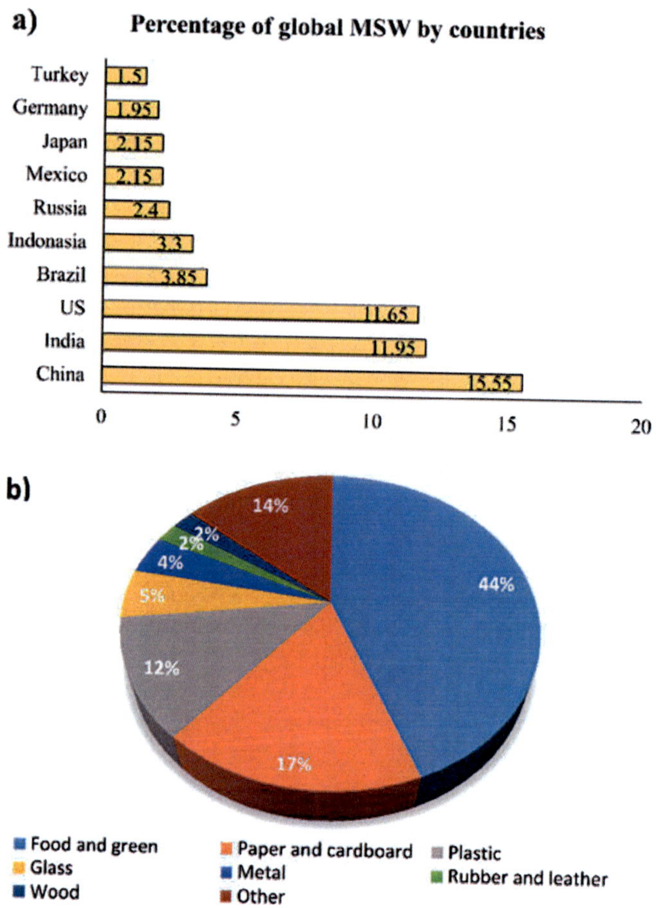

Fig. 2.1 Overview of MSW production worldwide. **a** Top 10 MSW producing countries; **b** percentage of the main MSW materials. *Source* statistica.com

Various methods have been employed to manage the produced MSW, including landfills, incineration and combustion, reduce, reuse, and recycle. To decide which method to apply depends on the characteristics of waste, available budget and technology availability. The conventional routes of MSW management are summarized in Fig. 2.2.

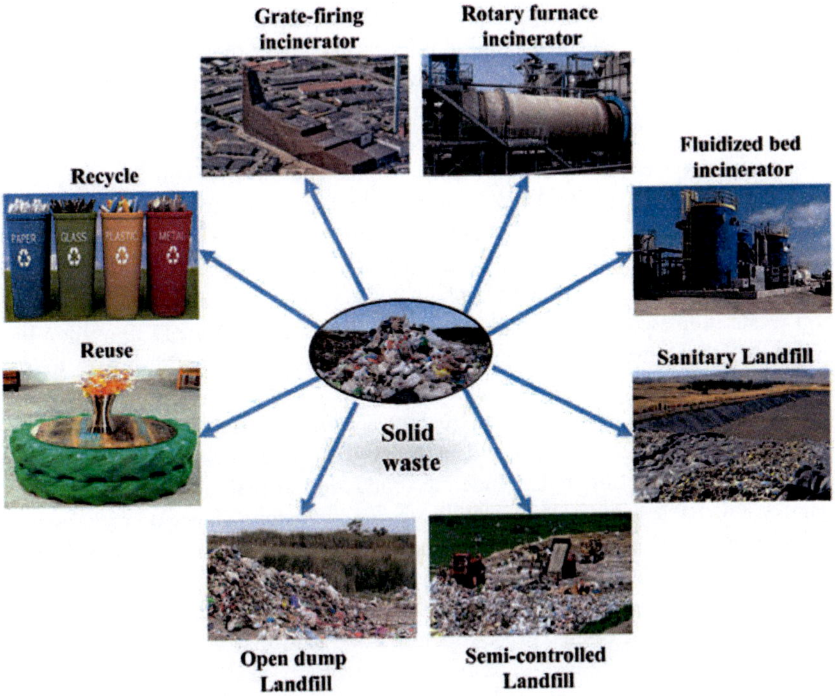

Fig. 2.2 Conventional methods of solid waste management

2.2 Reduce, Reuse, and Recycle

Plastic and construction waste are the main class of waste that can be managed by recycling, reducing, and reusing strategies. The plastics manufacturers have developed noticeably due to numerous polymer production methods from petrochemical sources. They comprise considerable advantages such as durability, low weight, and lower cost in relation to many other types of materials. Nowadays, plastics are nearly entirely made from petrochemicals. Roughly 50% of plastics are consumed for one-time uses, such as agricultural films and packaging, and around 20–25% are used for durable infrastructure, including cable coatings, pipes and structural materials, and the leftovers for long-lasting consumer uses, such as in vehicles, furniture, electronic goods, and others. Packaging is the leading waste plastic; however, it is evident that other sources, including waste electrical and electronic equipment, are becoming important bases of future waste plastics (van Heek et al. 2017).

Debris and materials such as steel, concrete, wood, and mixed materials produced in the process of demolition, excavation, road construction, and the construction or reconstruction of buildings are called "construction waste" (Shen and Tam 2002). Because the composition and types of construction waste are very diverse, the exact amount of this waste produced during the construction process cannot be predicted.

However, this waste typically includes materials such as concrete, brick, soil and stone, gypsum, wood and pipe parts, and electrical components (Gavilan and Bernold 1994). Based on the available data, it can be estimated that in 2014, about 333 million tons of construction waste was produced in Europe except for soil (Menegaki and Damigos 2018). In this estimate, countries such as Germany with 85 million tons of construction waste, France with 65 million tons, and the UK with 58 million tons were the three countries with the highest production of construction wastes. Waste produced in these countries also included materials such as ferric and non-ferric metals, wood, plastic and glass, asbestos-containing minerals, and polycarbonate biphenyl wastes (Menegaki and Damigos 2018). As the first producer of construction waste in the world rankings in 2014, China produced about 1130 million tons of construction waste. At the same time, in our country, only 534 million tons of construction wastes have been produced through activities such as road construction, bridge construction, dam construction and construction-related activities, of which 28.9 million tons are related to construction activities. 505.1 million tons are related to demolition and include the production of materials such as concrete, steel, wood, brick, asphalt parts, metal parts, and gypsum (Menegaki and Damigos 2018). Among the five levels of waste management, the 3Rs strategy, which relies on reducing waste generation through the three processes of reuse, reduction, and recycling, is recognized as the most important principles in construction wastes management (Huang et al. 2018). It can also be said that the implementation of these three processes (reduction, reuse, and recycling) has been useful in research objectives in the field of construction wastes heritage. Therefore, in this chapter, we mainly focus on construction materials that can be reused for constructions.

2.2.1 *Reduce*

Reduce is one of the strategies used for waste management. It is ideal for reducing the destructive and adverse effects of waste on the environment; therefore, it is a priority (Huang et al. 2018). It is important to reduce waste generation, identify and learn how to reuse materials, and recycle non-reusable materials. By this strategy, several important benefits can be achieved, such as reduced emissions of CO_2, less purchase of materials that lead to lower costs, reduced cost of transporting waste to the disposal site due to reduced volume of waste generated and revenue generation from the collection. Collecting some materials is a straightforward and economical solution to protect the environment (Ding et al. 2016). Lack of stakeholder awareness of this management strategy and the unwillingness of housing industry activists to cooperate with this strategy are important obstacles to implementing the waste reduction process. Therefore, in order to encourage industry authorities to pool resources to achieve this goal and obtain its benefits, a waste reduction strategy should be included in their management strategy (Esa et al. 2017).

2.2.2 *Reuse*

Reuse is another important strategy in waste management. It can be applied to waste from different sectors associated with wood, plastic, metals, and construction materials. Since most construction materials waste can be reused after demolition, reducing and reusing waste effectively protects the environment and reduces costs. Reducing will help in reducing greenhouse gas emissions, protect the environment, and prevent adverse global climate change (Marzouk and Azab 2014). Usually, different building materials can be used in various ways, such as selling or storing, for reuse in subsequent construction projects after the completion of construction projects, demolition, or reconstruction of buildings. It is important to note that some of these materials, such as adhesives, latex, or solvents, are toxic and classified as hazardous waste, so they will usually not be reusable (Marzouk and Azab 2014). Another point to consider when reusing waste from a demolition or reconstruction project is the age of the destroyed building (Akinade et al. 2015). This is because, in the past, some buildings were usually made of materials such as asbestos, which are no longer allowed to be used today due to their toxic and carcinogenic effects. Another vital point to consider in the reuse of construction waste is the employment of workers who are skilled in demolishing the building and collecting the resulting waste. Consideration of incentives and planning to encourage construction industry managers to participate in the reuse of construction waste and create a market for the purchase and sale of these materials can also be an effective step (Marzouk and Azab 2014).

2.2.3 *Recycle*

Recycling means the destruction of materials and their transformation into new usable materials. However, this process faces some obstacles such as inefficient management, the existence of unsuitable technologies for recycling, and the lack of a suitable market for the supply of recycled materials (Marzouk and Azab 2014). The recycling process of construction materials can be done on the same site or in another location, depending on the capacity and facilities of the construction projects. Various construction materials such as concrete, various metals, asphalt, wood parts, materials used to construct roofs, and plaster can be recycled at construction sites. The major benefits of recycling construction materials are the reduction of CO_2, which, if not recycled, will be released in large quantities during the disposal of waste and the production and manufacture of new materials (Huang et al. 2018). Also, emissions of other pollutant greenhouse gases can be decreased due to the reduced need to produce new raw materials and building materials. There is no need for new sites for construction waste disposal, which reduces the harmful effects of the environment and reduces costs and energy savings (Pham and Hargreaves 2003). In addition to the benefits mentioned above, recycling has been very effective in various areas such as job creation and opportunity creation in economic activities.

As opposed to most convenience goods and packaging, long-lasting plastics are thrown away after one-time use, while they tend to comprise three or further years of lifespan. The plastics used in household appliances, computers, automobiles, fabrics, and carpets belong to this classification. However, the recovery of these plastics is a very complex task, as they are combined with some other plastics or non-plastic materials. Their isolation, purification, and recovery involve a number of phases, and commonly, the sizes of such compounds accessible for recovery are inadequate. Methods to recycle the plastics wastes include the production of building blocks and monomers in great purity, which enable the remanufacturing of the new or original plastics. Glycolysis, pyrolysis, and ammonolysis are novel recycling methods that could offer existing mechanical recycling techniques (Subramanian 2000). Plastic recycling is completed in four steps: primary (mechanical reprocessing to a martial with comparable characteristics), which is also referred to as closed-loop recycling; secondary (mechanical reprocessing to materials needing lower characteristics), known as downgrading, tertiary (chemical elements recovery) and quaternary (energy recovery). Tertiary recycling is described as either feedstock or chemical recycling and is employed when the polymer is depredated to its chemical elements. (Hopewell et al. 2009). Biodegradable plastics can be composted, an example of tertiary recycling, and are also defined as biological or organic recycling. On the other hand, building materials are widely used in the construction of soundproof walls, cityscapes, roads, sports fields, and foundations, and there is a large market of supply and demand for this group of recycled products (Fatemi and Imaninasab 2016). However, in order to develop the recycling industry and achieve the desired results in this area, the participation and cooperation of governments are also necessary (Kabirifar et al. 2020).

2.3 Landfill

Landfilling is the controlled disposal of non-biodegradable and biodegradable wastes during a selected terrestrial cemetery, which is found far away from a municipality's residential regions. Landfilling has been a standard and most profitable waste disposal method practiced in several countries. The amount of active and sealed landfills inside the European Union is flanked by 150,000 and 500,000, which is the warehouse for vast volumes of urban solid waste produced in the EU (Jones et al. 2013). Primarily in Europe, 150,000 landfills encompass 30–50 × 10^9 m^2 of MSW (Wagner and Raymond 2015). More than 33 MT of MSW is incinerated in the USA, and around 136 MT of MSW is landfilled per annum (USEPA 2019). Nevertheless, in the USA, landfilling of MSW reduced from 89% in 1980 to 53% in 2014 because of the developments in incineration, composting, energy recovery, and recycling technologies.

2.3.1 Landfill Leachate and Gases

Municipal waste leachate or landfill leachate is a derivative from landfills, anaerobic digesters, solid waste treatment plants, or compost piles (Neczaj et al. 2005). It comprises concerning environmental alarms for treatment because of the high concentration of organic matters such as dissolved solids, toxic chemicals, and carboxylic acids, heavy metals, inorganic salts, minerals, ammonia, and xenobiotic organic molecules (Wiszniowski et al. 2006). Also, the non-biodegradable or refractory compounds like humic substances, such as humic acids and fulvic, encompass the organic segment of landfill leachate (Kumar et al. 2011). These environmental contaminants exist at a high rate within the leachate, mainly from care products, pharmaceuticals, households, and industrial chemicals. Landfill leachate displays chronic and acute toxicity, and it is hazardous as it can transmit into groundwater, causing biomagnification (Mishra et al. 2019). The diffusion of the leachate to the soil may be a typical challenge in landfills globally. When the semi-controlled and open dump landfills are located in low-lying coastal zones, the leachate might seep and pollute the water. Also, heavy thawing and downpours of permafrost in polar regions can mix leachate with surface water or leakage into groundwater. This leachate penetration into the soil and groundwater is more evident in third-world and developing countries, where the landfills are deprived of the leachate collection systems and/or baseliners and treatment plants (Han et al. 2014). Landfill leachate is mostly stopped from discharging and percolating into the groundwater by using a proper leachate collection setup at the landfills' bottom bed. After collection, it requires specific treatment prior to being considered safe for discarding into nature. A couple of such leachate remediation paths consists of anaerobic biological treatment (anaerobic lagoons and anaerobic bioreactors), aerobic biological treatment (activated sludge and aerated lagoons), physicochemical treatment (chemical precipitation, air stripping, oxidation, pH, and reduction), coagulation (via lime, ferric chloride, and alum), also as adsorption (natural process resins and activated charcoal adsorption) (Raghab et al. 2013). The treatment of leachate of MSW mostly according to biological oxygen demand (BOD), chemical oxygen demand (COD), absorbable organic halides, and ammonium. COD is the ratio of oxygen used throughout reactions in a solution while BOD is the ratio of oxygen used by microbes during the aerobic degradation of organic matter at a certain temperature (Han et al. 2014). It was evaluated that the environmental effects of a landfill without any proper leachate collection system in China (Han et al. 2014). It was shown that the groundwater in 30 m depth adjacent to the landfill was not proper for drinking owing to the concentration of the exceedingly high pollutants derived from leachate. There is no doubt that landfill leachate contains high levels of pollutants, which may pose significant health risks to humans as well as aquatic and terrestrial animals in the neighborhood of landfill sites. The metalloids, heavy metals, and even minerals beyond their limits can cause diseases like arsenicosis, Minamata, baby syndrome, pneumoconiosis, pulmonary fibrosis, Wilson disease, encephalopathy, hepatic cirrhosis, alopecia, cancer, and argyria.

Landfills also generate gases due to chemical, thermal, and biological reactions. Because of the temperature rise in landfills, the development of gases can take place from volatile compounds existing in MSW, like naphthalene and alcohol. Also, several chemical reactions among diverse wastes may occur after being mixed through disposal, leading to gas releases. Moreover, microbial decomposition, hydrolysis, and fermentation can take place. The activity of methanogenic bacteria is comparatively greater in the landfill bed due to oxygen supporting anaerobic decomposition. Landfill gas forms through a spread of anaerobic oxidation, fermentation, acidogenesis, methanogenesis, and acetogenesis. Some bacteria like hydrogenotrophic methanogens, proton-reducing acetogens, hydrolytic fermentative bacteria, and acetoclastic methanogens bacteria participate throughout anaerobic digestion of wastes and produce gases (Demirel and Scherer 2008). The preliminary hydrolysis of organic substances in MSW generates amino acids, sugars, and fatty acids via anaerobic oxidizers and fermentative microbes. These organic monomers are subsequently converted to acetate, propionate, butyrate, and lactate by acidogenic bacteria. Acid fermentation is increased by high organic matter and moisture ratio during the first phases of a landfill's lifespan, which generates volatile fatty acids (VFAs) like acetic, butyric, lactic acids, and propionic (Kumar et al. 2011). As the landfill ages, methanogenic and acetotrophic phases take place. The acetate-oxidizing bacteria can transform acetate to H_2 and CO_2 via oxidative homoacetogenesis or alter H_2 and CO_2 to acetate via a reverse reaction, such as reductive homoacetogenesis (Demirel and Scherer 2008). In addition, acetate-oxidizing bacteria, via acetogenesis, transform acetate to CO_2 and CH_4, which are the most constituents of landfill gas.

Moreover, the rival species of acetotrophic methanogens similarly transform CO_2 and H_2 to CH_4. Thus, making landfills a key contributor to global warming. It should be noted that CH_4 is 25 times more potent to global warming compared to CO_2 and includes a lifetime of 12 years within the atmosphere (Nanda et al. 2016). Besides, in the USA landfills are known as the third-largest supplier of CH_4 afterward fossil fuels and livestock farming (USEPA 2019). Furthermore, the high CH_4 concentration poses potential risks of unintended fires and blasts (Narayana 2009). The non-methane organic components in landfill gases comprise hazardous air contaminants, odorous compounds, and volatile organic carbons, which could include around 39% of entire gas releases from the landfills (Davoli et al. 2010). Furthermore, there is a fear about the bad odors formed by landfill gases, which is concerned rather as a problem than a health risk. However, landfill operators and residents nearby are assumed to be susceptible to possible negative health impacts from long-term contact with gases (Davoli et al. 2010; Narayana 2009).

2.3.2 *Landfill Classification*

Landfills are categorized into three types: sanitary landfills, semi-controlled, and open dump (Narayana 2009). An open dump might be a part of the land, where the

MSW wastes are dumped and exposed to the air. They are frequently established in developing countries, where the MSW are discarded into low-lying air exposed areas. As they are managed poorly, they become a separate section for scavengers, e.g., vultures, eagles, falcons, crows, mosquitoes, flies, rodents, pests, worms, and pathogenic microbes. Nevertheless, these operational problems are absent in anaerobic digestion due to oxygen-deficient environments. Residents in the neighborhood criticize about continuous bad odor from the open dump landfills. Therefore, in several developed countries, governments prevent open dumping. Although 79% of MSW is landfilled in Canada, the Ministry of Environment and global climate change imposed severe regulations and restrictions on illegal dumping and waste disposal. Operated landfills and semi-controlled landfills are placed in selected dumpsites where MSW is sorted, shredded, and compressed prior to disposal.

The disposed thrashes crumpled and leveled with crawlers or bulldozers and enclosed with a coating of topsoil every day to avoid the reproduction of animals, scavenging birds, pests, and microbes. Although the semi-controlled landfills are comparatively less smelly because of the topsoil shield, they are not planned to cope with the leachate discharge and gas emission (Narayana 2009). Conversely, sanitary landfills are progressive types of semi-controlled landfills. In addition to solid waste sorting, segregation, size reduction, densification, and covering by topsoil, they consist of facilities to collect landfill gas and liquid leachate and are also located remote from the residential zones. These kinds of landfills are primarily used in developed nations with amenities for treatment and leachate interception.

2.3.3 Modern Landfills

The engineered modern landfill is known as a bioreactor landfill that changes the standard of landfilling from storing to treatment. In comparison with typical landfills, bioreactor landfills comprise several benefits, including improved quality of leachate, (ii) storing and partial on-site treatment of leachate, higher yields rate of landfill gases, effective gas rescue for on-site burning, early waste stabilization, improved breakdown of biodegradable materials in MSW resulting in the quicker settlement, (vii) cost-effective (viii) lower wastes toxicity due to anaerobic and aerobic digestions, lesser environmental effects due to lower of greenhouse gas release and groundwater pollution and higher possibility for waste-to-energy transformation (Kumar et al. 2011; Reinhart et al. 2002). Also, bioreactor landfills seem to be comparatively efficient for the elimination of hazardous organic pollutants. This can be attained by conditions optimization for biodegradation, stripping volatiles by elevated gas yield, and pollutant immobilization through humification (Reinhart et al. 2002). Bioreactor landfills contain leachate recirculater in a number of configurations, temperature controllers, microbial growth nutrients suppliers, pH buffer and are equipped to recover gases for storing and on-site burning. Leachate recirculation enables the transformation of biodegradable constituents of MSW into intermediary gases and products, such as CH_4 and CO_2, via improved hydrolysis, fermentation, acetogenesis, acidogenesis, methanogenesis, and anaerobic oxidation (Kumar et al. 2011).

Bioreactor landfills are categorized into aerobic, semi-aerobic, and anaerobic landfills. In anaerobic bioreactors, the anaerobic condition is provided for the activity of anaerobic and facultative bacteria, mainly acidogenic, methanogenic, and acetogenic bacteria. Anaerobic bacteria alter biodegradable wastes to VFAs and ultimately to landfill gases, such as CO_2 and CH_4. Carbon, phosphorus, and nitrogen present naturally within the disposed of wastes can considerably affect the growth of microbes throughout the anaerobic biodegradation. Perforated wells or transport systems usually recirculate leachate to enhance the moisture ratio. Leachate produced from bioreactor landfills also has an elevated level of ammonia and VFAs, demanding additional treatments at extra costs.

Aerobic bioreactor landfills increase the speed of waste breakdown via making oxygen-enriched environments for aerobic microbes. In these landfills, aerobic microbes obtain energy through the oxidation of organic materials to generate water and CO_2. Aerobic digestion is more rapid than anaerobic degradation since aerobic microbes reproduce at a higher speed owing to aerobic respiration, which is more competent in energy production compared to anaerobic respiration. In aerobic bioreactor landfills, ventilation is done via air injection to the soil layers and waste mass. The air injection into landfills hinders the development of anaerobic microbes and pulls down methanogenesis and consequently CH_4 formation (Ritzkowski and Stegmann 2012). The most benefits of the aerobic bioreactor are lesser waste stabilization times, quicker biodegradation, and humidity elimination through the air stream. Also, the leachate from aerobic bioreactor has decreased the requirement for biological oxygen, chemical oxygen, organic carbon, and lesser ammonia production.

Nonetheless, the gas produced from such landfills as aerobic degradation produces mainly CO_2 and only a very low amount of CH_4, which is not sufficient for burning. Due to low CH_4 and ammonia production, gas and leachate are comparatively less odorous, which might be a substantial social gain. On the other hand, sufficient and uniform aeration adds to the overall process cost. The arrangement of perforated injection wells for leachate and air separates the semi-aerobic and aerobic bioreactor landfills. Semi-aerobic bioreactor landfills offer moderately oxygen-deficient environments to favor both anaerobic and aerobic microbes. In aerobic bioreactors, the air is injected into the aerial section of the landfill. Also in several structures, air can naturally flow through the leachate-saving pipes (Yang et al. 2012). In such systems, the additional cost of air addition is less (Huang et al. 2008).

On the other hand, methane in the gas and consequently the air addition will be possibly hazardous. One of the main drawbacks is swelling and separation of plastic surface liners as the closed system is making an anaerobic condition. This challenge also can be tackled by constantly collecting the created gases (Reinhart et al. 2002). Carry on the leachate infiltration/collection piping is an additional technical problem. During dry periods, the leachate formed via the bioreactor landfill may not be adequate to add moisture to the wastes, and thus demanding extra water resources.

2.4 Incineration and Pyrolysis

Primary, simple deployment of wastes into the landfills was a typical technique of disposal. Then incineration was presented to decrease the waste size and provide more space in the landfills. In the 1970s, fears about the ecological impacts of incineration enhanced the cleaning systems, and lastly, a few years back, energy recovery (Waste-to-Energy, WtE) and other valued constituents of the waste came into attention. Three forms of combustion platforms for the conversion of MSW are available, including grate-firing (GF), rotary furnace, and fluidized bed (FB). The developments related to energy recovery and the environment improved the typical GF platform, comprising the transfer of heat and cleaning segments. Later, FB boilers, presented for coal burning and biomass, also were used for MSW. Recently, FB is commercially available for operating with MSW (Makarichi et al. 2018). FB is used in China by around 30%, Canada, and the USA by about 20%; however, the share in Europe is around 5% (Bösenhofer et al. 2015). FB has the standing of being a new and environmentally favorable combustion platform, while GF is considered to be incompetent and old-fashioned (Makarichi et al. 2018). In combustion furnaces, due to the heat of the fuel, volatile molecules are blister massively and create noticeable flame (Hunsinger et al. 2002). Air functions for the finishing combustion by mixing the gases, enhancing the mixing among oxygen and gases. The emitted gas proceeds to the secondary combustion sector, causing the second mechanism (Waldner et al. 2013). In a new GF platform, many actions modify and control the efficiency of the grate. For example, they might comprise a laser scanner to notify the control system regarding the sort of waste arriving and the pressure variance among the fuel substrate to ensure its height and compaction (Strobel et al. 2018).

The incineration of MSW with possible energy recovery and managing MSW incineration (MSWI) ashes are drawing worldwide attention. Numerous countries have followed the advantages of MSWI ash by applying strategic management regulations and programs. For instance, some European nations use MSWI bottom ash (BA) as viable construction constituents to extend its environmental and economic benefits (Bösenhofer et al. 2015). The USA generates more MSW than any other countries, while its recycling rate is meager (Woodward 2004). As the quantity of waste production continues to upsurge, incineration technologies have been exploited to manage MSW, reducing the volume and load of waste by around 90% and 60%, respectively (Sakai and Hiraoka 2000). MSWI plays a crucial role in dealing with this escalating size of wastes and recovering energy, supplementing traditional supplies (Cucchiella et al. 2017). The MSWI is commonly separated into three key parts: energy recovery, incineration, and air pollution control (Singh et al. 2011). MSWI can demonstrate a significant role in balancing fuel usage and improving renewable energy ratio while contributing to waste treatment (Idzorek 1991). So far, about 1179 MSWI plants globally exist with a capacity of more than 700,000 MT/d (Lu et al. 2017). Most of the plants are situated within the USA, the EU, and East Asia (Lu et al. 2017; Scarlat et al. 2015).

Since the middle of the nineteenth century, it had become obvious that the increased waste sizes can be addressed through incinerating waste and utilizing various elements existing in the waste (Makarichi et al. 2018). The main purpose of initial incinerators was only to manage waste with no target to use the produced heat. There was a rising concern that landfills would rapidly refill, and the lands to build extra would quickly end due to the fast expansion of cities and populations (Lu et al. 2017). Recovery of energy from the heat of incinerators began at the end of the nineteenth century in Europe. While in the USA, it began in the early twentieth century due to rising oil values. Initial WtE plants were simple water-wall and modular incinerators, which did not have a flue gas treatment module. Also, methane recovery over co-digestion of sewage sludge and wastes and was later introduced. The MSWI plants had mechanical difficulties, which made typical plant pack-up and undue rundown hours (Makarichi et al. 2018). The initial purpose was to utilize the ferrous and non-ferrous elements, particularly glass, paper fiber, iron, and aluminum. Heat recovery for electric power was initiated around the middle twentieth century, with the main plant being constructed in Paris, France (Mourad 2016).

New incorporate various levels of MSW preprocessing, which did not exist in primary incinerators. For many years, preprocessing methods have been established to exclude unsafe and bulky materials as non-combustibles, making MSW more combustible and improving emission control (Sun et al. 2016). Screening by using trommel screens, magnetic separators, and air classifiers to scale back the MSW heterogeneity prior to incineration (Fitzgerald 2013). Fluidized bed incinerators are able to manage mainly waste that has gone through size decrease and shredding aside from primary separation (Pisupati and Tchapda 2015). Moving grate incinerators have shown to be greater to either fluidized bed or rotary kiln incinerators mainly due to their capability to deal huge size of MSW deprived of prior shredding or sorting but requires the elimination of massive materials and explosive or unsafe materials, which damages the MSWI apparatus (Wissing et al. 2017). Although the fluidized bed and rotary kiln incinerators have equally been utilized since the mid-twentieth century, only moving grates are mainly established for large-scale performance (Fitzgerald 2013). A comparison made in 2012 showed that within Germany, the EU, and the USA, the amount of MSWI plants using moving grate is 94%, 88%, and 76%, individually, whereas the rest are rotary kiln or fluidized bed incinerators (Lu et al. 2017). The main shortcoming of moving grates is the high investment requirement (Pisupati and Tchapda 2015). Although fluidized bed incinerators require lower capital investment and operational costs, their strict necessities regarding feedstock homogeneity and their high susceptibility to variations in the calorific value of waste making their process challenging (Fitzgerald 2013). For many years, air emission control systems that manage the flue gas MSWI faced variations in particulate filling and dissimilarities in gas flow rates as the feed materials are not homogeneous. Consequently, new APC systems include control for combustion and post-combustion to limit trace and conventional contaminants and increase the flue gas standard (Shi et al. 2018).

A modern platform of MSWI is presented in Fig. 2.3. In this design, the fly ash (FA) and bottom ash (BA) are collected, and the gas emissions are controlled using

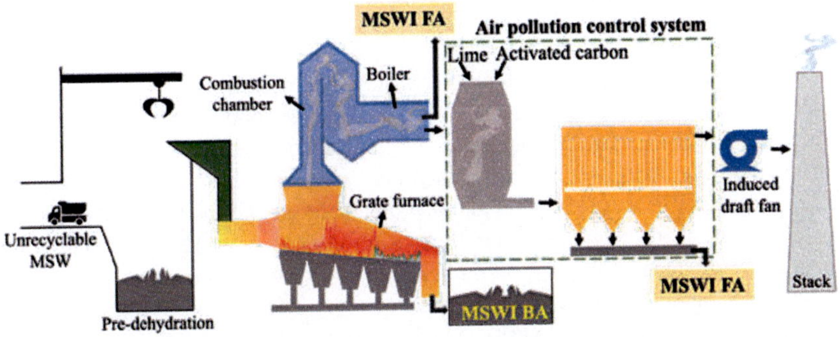

Fig. 2.3 Schematic diagram of the MSWI system and the formation of MSWI FA and MSWI BA. *Source* Zhang et al. (2021) with permission number 5158581326703

activated carbon and lime. MSWI BA is the residual material with a higher weight, which resulted from incineration, while MSWI FA, due to lower weight, can move into the air. The MSWI FA production typically accounts for around 3–15% of the amount of MSW subjected to furnaces differences. MSWI FA comprises exceedingly irregular structure and form, containing a considerable quantity of particles rods and irregular, and is combined with a minor quantity of filamentous crystals (Mao et al. 2020).

Globally, the tendency in the generation of MSW proposes that waste load that needs MSWI will still increase as this is mostly an effective method of decreasing residual waste that requires landfilling, with the extra advantage of making renewable energy. As land for constructing additional landfills becomes limited in most cities, more need for MSWI is assumed. Thus, the quantity of MSWI plants is possibly going to rise in the coming years to meet the rising request for waste transformation to energy. The majority of MSWI platforms are MB systems that blister the MSW subsequent to the separation of metals for recycling. Refuse derived fuel (RDF) plants enable the preprocessing of MSW to discard non-combustibles and to shred the MSW to equal size fuel pellets. Also, FBC systems incinerate RDF in a fluidized bed at high temperatures for non-combustible granule-like sand within a furnace. According to the literature issued over the last 20 years, the key concern linked to MSWI is the pollution produced by furan dioxin ($C_4H_4O_2$), (C_4H_4O), and heavy metals attained from MSW (Tasneem 2014). By means of APCs, toxic flue gases can be treated as they usually are equipped with dry/semidry and wet scrubber setups (Tasneem 2014). The application of the ash in road subgrade/subbase and backfill materials offers good performance (Oppelt 1987).

2.4.1 Hazardous Wastes

Hazardous wastes are other important wastes that need special treatments. The example of these waste are solvents, acids, alkalinize, photochemical, pesticides, fluorescent tubes, mercury-containing wastes, detergents containing dangerous substances, cytotoxic and cytostatic medicine, mixed batteries and accumulators containing batteries, etc. (Slack et al. 2009). The aforementioned wastes are treated by various methods such as incineration, landfilling, and respective ultimate treatment activities depending on the composition of the wastes. For example, waste with the calorific value ≥ 2500 kcal/kg is disposed of by incineration. While waste with a low calorific value (≤ 2500 kcal/kg) and comprising heavy metals is disposed into secured landfills (Singh and Budarayavalasa 2021), the incineration of hazardous waste is usually performed in the rotary kiln, and the wastes are incinerated at temperatures between 1000 and 1200 °C. The inclinator ashes are treated to recycle its iron, and the residual is utilized in construction materials or treated in landfills (Block et al. 2015). Solidification and stabilization of hazardous waste have been known as frequently practiced for several decades. This technology has been demonstrated to be an efficient tool for the immobilization of the pollutants inside a solid matrix and constructing value-added resources as construction materials (Singh and Budarayavalasa 2021). Also, the residue according to the ratio of calcium oxide, iron oxide, aluminum oxide, and silica is used to prepare SiO_2–Fe_2O_3–Al_2O_3–CaO glass–ceramic samples that immobilized Zn, Cr, and Cu effectively over the melting–sintering procedure (Chen et al. 2021).

Medical waste (typically made from healthcare centers, clinics and hospitals) biomedical waste (made during medical diagnosis, treatment, or immunization of disease, and biological research activities on animals). Pandemic situations, particularly the current one, the COVID-19 pandemic, upsurge extensive consumption of plastics and make environmental problems (Sharma et al. 2020). Medical waste that is released from healthcare centers or hospitals carries more or less infected materials and is consequently known as hazardous waste (HW). These HW or MW needs special care and consideration for the safe disposal and might lead to reinfection. Biomedical waste is typically made from medical facilities and research institutes, particularly places where the vaccine or drug trial stages are commenced to detect and treat the diseases (WHO 2018). Commonly, standard disinfection methods are applied to treat medical waste such as incineration, chemical, physical, and approaches. Incineration can be the promising or efficient method that can be applied, and it would be an adequate investment to disinfect pharmaceutical and pathological wastes. Chemical disinfection together with other methods like steam disinfection or microwave, in which the macromolecular composites are broken down to simpler molecules in form of carbonaceous materials, synthetic gases, and liquid or solid fuels. Incineration is predominantly adopted for large-scale platforms with a huge volume of wastes while steam disinfection and chemical treatments are mostly adopted for minor investments and places such as healthcare centers or hospitals for storing or temporary treatment of wastes, which can lessen the pathogens or infections transmission (Dharmaraj et al. 2021). Alternatively, pyrolysis can be performed. Pyrolysis is a thermochemical procedure wherein the organic compounds (liquid and solid waste) are broken down at high temperatures (400–700 °C) under anaerobic conditions. The liquid or solid derivatives have high commercial value and can be utilized as a substitute for the existing fossil fuels (Dharmaraj et al. 2021; Wang et al. 2020b).

Radioactive wastes consist of waste that containing radioactive elements known as radionuclides. It can produce ionizing radiations that damage biological life upon exposure and therefore are hazardous. Radionuclides exist naturally or are generated over the process of nuclear energy production. Radionuclides are typically big unstable atoms that incline to attain stability via discharging energy as radioactivity. Several methods comprising chemical precipitation, distillation, acid digestion, evaporation, wet oxidation, and dumping have been used to treat these wastes (Darda et al. 2021). The first method was geological disposal introduced more than five decades ago by geoscientists collaborating with nuclear scientists and has taken many years of intensive scientific, societal, and political effort to reach the present point (Chapman and Hooper 2012). Also, radioactive waste can be treated through incineration yet it needs a gas-filtering setup to regulate radioactive releases (Darda et al. 2021) and is doable for 98% of radioactive wastes including very low level waste and low level waste (Darda et al. 2021). A comprehensive review article by Darda et al. (2021) has described all categories of radioactive waste and proper management strategies.

2.4.2 Plastic Waste

Plastic materials break down slowly over their lifespan as a result of sunlight, water, air, temperature stresses from hot and cold conditions, biological and chemical degradation from diverse contact materials, etc. (Idumah and Nwuzor 2019). These changes in the characteristics of the plastics make them unsuitable for recycling to high-end uses and even unsuitable for mechanical recycling entirely. Therefore, substitute treatments need to be applied for the great volume of plastics that are not recycled mechanically. Incineration of plastic waste with energy recovery, as a crucial aspect of the circular economy, is utilized in both developing and developed countries. Nevertheless, incineration produces ending bottom ash that is discharged into the environment. Previously, it was commonly believed that incineration is the terminator for eradicating plastic waste by ultimately transforming polymers into mineral fractions and CO_2, while the unburned substances from the bottom ash encompass synthetic fibers of plastics, which indicates that plastics and microplastics could be conveyed into the ecosystem through its dumping or reuse. Also, incinerating plastics can lead to numerous environmental issues, such as fly ash, dioxins formation, generation of SO_x and NO_x, and other toxins (Yang et al. 2021).

In recent years, by developing thermochemical conversion methods such as pyrolysis, hydrothermal liquefaction, and gasification, various wastes are transformed into high-value products such as biochar, bio-oil, catalyst, and syngas (Cao et al. 2020; Hu et al. 2020, 2021; Wang et al. 2019, 2018). Pyrolysis has presented substantial benefits over the others methods, since it makes lesser gaseous pollutants, because of O_2 absence, which is contrary to incineration (Fivga and Dimitriou 2018). Based on the pyrolysis conditions, a blend of molecules in the form of wax or liquid as key products is produced. Therefore, pyrolysis of plastic valorizes such plastic waste into valuable fuel monomers/oil. However, there is a limitation in fed compatibility to pyrolysis procedure, compared to mechanical recycling it is able to tolerate significantly higher intensities of impurities in the feed that is economically attractive due to lesser steps of pretreatment steps. It is a rather straightforward procedure, and it has the competence to treat numerous sorts of plastic waste covering from packaging waste to more complex ones, including plastics and rubber, hospital waste, and end-of-life vehicle which are polluted with hazardous and toxic substances. Pyrolysis of plastics makes a hydrocarbon-rich gas. This gas based on conditions and feed can have a heating value ranging from 25 to 45 MJ/kg that making it a good option for energy recovery. Consequently, in the viable scale, the made energy is circulated back into the procedure, making the whole procedure self-sustainable (Qureshi et al. 2020).

Availability of feedstock is one of the issues of wastes thermochemical conversion. Also, a relatively consistent quality feedstock is essential to have economical pyrolysis of plastic waste. Plastics obtained from diverse waste sources are very heterogeneous. Therefore, pretreatment might be essential, as plastic items have different sizes and shapes which is essential to make it uniform size via crushing in advance. This is an extra step that acquires an additional expense to the entire process. Also, the selection of reactor plays a vital role in determining the pyrolysis products ratio (Papari et al. 2021; Qureshi et al. 2020). Pyrolysis of plastics, particularly polyolefins, derives wax as the key product. To recover the waxes conveniently, a recovery setup must be considered to handle waxes. As often the inefficient condensing setups accumulate wax on the inner sides, which cause challenging recovery. Potentially the waste can comprise different toxic constituents, such as biohazardous materials, halogens, and other harmful additives. Thus, it is essential to carefully analyze the feedstock prior to pyrolysis. Oil derived from the pyrolysis of plastics is not a consistent product; hence, standard testing techniques are practically non-existent. The aging and stability of pyrolysis liquids are crucial characteristics that define the quality of the liquid. Pyrolysis liquid is unstable thermodynamically and inclines to repolymerize. As a result, post-treatments such as dewaxing and blending of liquids are essential to maintain its quality over a long period (Fivga and Dimitriou 2018; Qureshi et al. 2020). Table 2.1 summarizes the advantages and disadvantages of various solid waste management methods.

Table 2.1 Comparison of different solid waste management methods

Method	Technology	Advantages	Disadvantages
Landfill	Leachate control system	• Less energy input • Simple • Not influence by waste composition changes • Large processing capacity	• Land requirement • Prolonged decomposition • Low degree of waste volume reduction • Finding a suitable site is challenging • Covers a large area • Produces secondary pollution, which is difficult to control
Incineration	High temperature burning	• Matured technology • Energy recovery • Suitable for the disposal of waste comprising more combustibles (plastic, food waste packaging, paper, and cloth, etc.) • High waste volume reduction	• Generation of toxic gases • Air pollution • Ash production • Requires purification technology for flue gas
Thermochemical conversion	High temperature input	• Production of high-value products • Transformation of waste into new materials	• High energy requirement • Further upgrading of end products required
Reusing		• Cost saving • Reduced disposal need • Saving raw material and energy • Manufacturing of fess materials	• More durable items need to be produced • The container needs to be washed or cleaned • Reduces jobs
Recycling	Changing the physical shape	• Saving resources • Product can be utilized in making new products • Reduces litter • Less pollutant production	• Require high tech system • Comparatively more time consuming • Low economic benefit

2.5 Wastewater Treatment

Water contamination by chemicals has developed a widespread concern and an urgency for both people and community authorities, and importantly for the whole industrial world (Sonune and Ghate 2004). The sources of pollution are various:

industrial wastes, sewage, mining activities, chemical fertilizers, pesticides, energy use, urban development, radioactive material, and others. Any activities, encompassing agricultural, domestic and industrial, generate effluent comprising undesirable contaminants that can be toxic. Thus, an ongoing effort must be made to protect water resources.

Throughout the last 30 years, several chemicals, physicals, and biological approaches are reported, such as flotation, oxidation, precipitation, evaporation, solvent extraction, membrane filtration, carbon adsorption, natural biodegradation process, phytoremediation, and electrochemistry (Samer 2015). However, there is no sole technique capable of sufficient treatment, primarily owing to the complicated nature of industrial effluents. In reality, a combination of several techniques is usually required to attain the desired water quality. A purification procedure generally involves five consecutive steps (1) pretreatment or preliminary treatment (mechanical and physical); (2) primary treatment (chemical and physicochemical); (3) purification or secondary treatment (biological and chemical); (4) final treatment or tertiary (chemical and physical); and (5) sludge treatment (incineration or recycling, supervised tipping,) (Anjaneyulu et al. 2005). Table 2.2 lists the advantages and disadvantages of different treatment methods.

2.5.1 Physical Methods

Physical methods are mostly used owing to their flexibility, simplicity, pollutant recyclability, high efficiency, and minimum chemical consumption. Researchers have recently employed adsorption due to its lesser cost of operation, reliability, and high efficiency (Foroutan et al. 2019; Liu et al. 2020). During adsorption, removable species or solutes are conveyed from a liquid part onto the superficial part of a solid material. The adsorbed materials via physiochemical interactions are attached to the solid material surface (Manchisi et al. 2020). The adsorption method is introduced by the United States Environmental Protection Agency (USEPA) as the most effective wastewater treatment method compared to others (Anil et al. 2020). It is a well-developed method for eliminating dyes and colors from wastewater owing to its cost-effectiveness and simplicity in comparison with the other methods. The most frequently used adsorbents for dye and pollutants elimination are silica, bio-adsorbents, alumina, metal oxides, activated carbon, titania, and clay (Wadhawan et al. 2020). Activated carbon (AC) is known as the most effective adsorbent for eliminating many forms of contaminants from wastewater (Azam et al. 2020). AC is low-cost and non-toxic material with great efficiency due to the massive surface area and porous structure (Kamaraj et al. 2020). Silica is another efficient adsorbent because of its identical pore size, big surface area, and potential catalysis uses (Slatni et al. 2020). Recently, nano-porous composites made from TiO_2, ZnO, and Fe_3O_4 have been the primary choice because of their suitable pore structure, chemical and physical stability, non-toxicity, exclusively enormous specific surface capacity, and demonstrating excellent catalytic efficiency (Jaseela et al. 2019). Moreover, these

Table 2.2 Advantages and disadvantages of different wastewater treatment methods

Methods	Advantages	Disadvantages
Chemical precipitation	• Simple equipment • Combined physicochemical method • Economically viable • Adaptable to elevated pollutant loads • Very effective for eliminating fluoride and metals • Non-selective to metal • Significant decline in the COD	• Consumption of chemical (e.g., oxidants, lime, and H_2S) • Unsuccessful in the elimination of low concentrated metal ions • Physicochemical observing of the discharge (pH) • Produces high sludge load • Needs an oxidation phase if the metals are relatively complexed
Coagulation/flocculation	• Simplicity • Efficient and swift for insoluble pollutants • Combined physicochemical method • Capable of bacterial inactivation • Reasonable capital cost • A widespread range of chemicals are commercially available • Strongly efficient for colloidal particles • Suitable characteristics for dewatering and sludge settling • Significant reduction in COD and BOD	• Physicochemical monitoring of the effluent (pH) • Necessitates adjunction of chemicals such as coagulants and flocculants • Low elimination rate of arsenic • Higher sludge size
Flotation	• Efficient for the elimination of small size particles and able to eliminate low-density particles, which would need extended settling periods • Diverse kinds of collectors (ionic or nonionic) • Combined physicochemical method • Suitable for initial clarification • Low retention time • Selective to metal • Applied as an effective treatment in the paper and pulp industry	• High capital cost • High energy expenses • Considerable operation and maintenance costs • Chemicals are necessary

(continued)

Table 2.2 (continued)

Methods	Advantages	Disadvantages
Chemical oxidation	• Integrated physicochemical process • Simple, efficient and rapid process • *On-site* formation of ozone (no storage-related dangers) • Effective degradation of the contaminants and color reduction and odor • Effective for removal of sulfide and cyanide • Suitable for hypochlorite treatment as it initiates and speeds up azo bond cleavage • Rises product biodegradability • High throughput • No sludge creation • Water recycle possibility • Sterilization	• Chemicals are required • Formation, carrying and supervision of other oxidants • Pretreatment is essential • Type of oxidant substantially influences the efficiency • Short half-life of ozone • Unknown intermediates may be produced • No diminution of COD values or limited effect • No impact on salinity • Release of aromatic amines and volatile compounds • Generates sludge
Adsorption	• Simple equipment • Adaptable to various treatment designs • Global removal but perhaps selective reliant on adsorbent • Extensive collection of commercial products • Wide diversity of target pollutants • Exceptional quality of the outcome • Highly effective method with fast reaction	• Costly materials and investment • Non-destructive and non-selective • Performance is subject to the form of material • Requisite several kinds of adsorbents • Rapid clogging and saturation of the reactors • Regeneration is costly and results in material loss • Economically non-viable for industries such as textile, paper, and pulp
Membrane filtration	• Lesser space necessity • Simple, efficient and rapid, even at elevated concentrations • Broad choice of commercial membrane existing from numerous manufacturers • No chemicals needed • Less solid waste formation • Removes all forms of mineral, salts and dyes • Efficient removal of suspended solids microorganisms, and particles (UF, MF, NF, and RO), nonvolatile and volatile organics (NF, RO) • Selective to metal	• Extraordinary costs for medium and small industries • High energy demand • High operation and maintenance expenses • Removal of the concentrate • Early clogging of the membrane • Low throughput • Restricted flow rates • Specific processes

(continued)

Table 2.2 (continued)

Methods	Advantages	Disadvantages
Biological methods	• Economically attractive • Acknowledged by the public • Large quantity of species employed • Bacteria and fungi produce a broad variety of extracellular enzymes with strong biodegradability capability • Efficiently removes NH_4^+, iron, NH_3, and biodegradable organic matter • High removal of BOD and suspended solids	• Essential to make an optimal environment • Needs maintenance and management of the microbes • Ineffective on non-degradable complexes or when noxious compounds are existing • Slow and uncontrolled process • Poor biodegradability of specific molecules • Poor decolonization • Sludge creation and foaming • Knowledge of the enzymatic processes is essential

composites are reusable and recyclable, suggesting them as cost-effective and highly attractive materials (Singh et al. 2018).

Additionally, the separation or elimination of organic matter and dyes via membrane filtration is also an efficient and economically viable wastewater treatment method. Nevertheless, membrane development with adequate thermal stability and enhanced efficiency is a complex task. The surface charge and hydrophilicity of the membrane define its rejection, antifouling, and permeability efficiency (Yang et al. 2020). The nanofiltration (NF) platform has been shown as an effective separation and purification technique in treating wastewater. Nonetheless, the effectiveness of the NF membrane might severely be limited by fouling and concentration polymerization. Fouling reduces the permeability of the membrane since it blocks membrane pores by creating a coating of organic compounds. In addition, elevated salinity negatively affects the flux of the membrane due to high osmotic pressure. Therefore, the NF membranes that pose hydrophilic surfaces are preferred under massive throughputs for the treatment of textile wastewater (Zhu et al. 2020). Also, ultrafiltration (UF) is another interesting method in dye and salt fractionation through high salt diffusion with upgraded permeation flux; it comprises high efficiency for separation because it offers improved salt permeation sustains high throughputs owing to reduced osmotic pressure. To make highly effective UF membranes, various polymeric materials such as cellulose acetate (CA), polysulfone (PSF), polyacrylonitrile (PAN), and poly(vinylidene fluoride) (PVDF), polyethersulfone (PES) have been exploited widely (Yang et al. 2020).

Coagulation–flocculation is another approach that is adequate for dye removal while it comprises low decolorization performance and efficiency for vat and reactive dye. It produces a significant amount of sludge, which is one of the biggest drawbacks of this method (Liang et al. 2014). Coagulation is a method in which a dye or organic matter solution is destabilized to produce agglomerates and flocs.

Flocculation is a process of destabilizing the floated particles and connecting aggregated flocs to more oversized agglomerates that precipitate due to gravity (Lee et al. 2012). The coagulation–flocculation method is done to offset charges through bridging or trapping the suspended materials to produce gelatinous agglomerates; these particles are usually big enough to be removed by a filter. Coagulation–flocculation is commonly used in textile manufacturing. In these techniques, coagulants like ferric chloride ($FeCl_3{\cdot}7H_2O$), lime ($Ca(OH)_2$), ferric sulfate ($Fe_2(SO_4)_3{\cdot}7H_2O$), and aluminum sulfate ($Al_2(SO_4)_3{\cdot}18H_2O$) attach to the contaminants to remove them by sorption, bridging, and electrostatic force. The bridging and sorption assist coagulate-flocculating pollutants from wastewater owing to the protonated polymer (high molecular weight) and amine groups. Coagulation–flocculation decreases the suspended matter, colloidal particles, dissolved substances, non-settable particles, and coloring agents from the wastewater effluents (Lee et al. 2012).

2.5.2 *Chemical Methods*

A number of chemical oxidation procedures are described for a series of catalysis uses. Advanced oxidation processes (AOPs) designate all procedures applied in wastewater treatment that comprise a regular basis in producing oxidizing species, like hydroxyl radicals (•OH) (Shafiq et al. 2020). Oxidation can include electrochemical oxidation, UV-assisted Fenton's oxidation, and photo-electrochemical oxidation, and ozonation. Mostly, pH and catalysts play a critical role in these processes (Ahmad et al. 2015). The electrochemical advanced oxidation processes (EAOPs) have been developed as a useful method for wastewater treatment. In general, pulse electrolysis is complemented with a stream of square-wave pulse current, in which the pulse cycle (T) contains a turn-on time (Ton) and a turn-off time (Toff). Current is produced as continuous amplitude and intermitted at the Ton and the Toff, individually (Wang et al. 2020a). Increasingly among AOPs, •OH-based electrochemical methods are becoming the main route (Geng et al. 2018). In the photo-electrocatalysis method, a chemical reaction takes place at the semiconductor anode to raise light efficiency according to the electric current. A photoanode semiconductor is illuminated with light to remove pollutants such as methyl orange (MO). •OH radicals, which are made through water oxidation, facilitate MO oxidation. •OH radicals are reactive sufficiently to yield aromatic ring-opening, small molecules, and azo bond interruption (Mais et al. 2020). AOPs can aid in attaining partial or whole breakdown of dyes or organics in textile wastewater. The oxidation procedures create free radicals such as •OH radicals comprising more substantial oxidation power. Fenton's oxidation is deliberated as the advanced and practical water treatment process (Sözen et al. 2020) as it uses ferrous sulfate ($FeSO_4$) and hydrogen peroxide (H_2O_2) to produce robust oxidizing agents, such as •OH radicals. Also, ozone participates in numerous chemical reactions with inorganic and organic compounds due to its strong oxidizing capacity. •OH radicals are formed from the breakdown of ozone. Ozone can produce hydroxyl radicals in water; thus, it is deliberated a developed oxidation method (Cao et al. 2021).

2.5.3 *Biological Methods*

Microorganisms break down organic compounds in biological treatment through the anaerobic or aerobic cycle. They utilize organics as a source of energy through degradation. Biofilms are usually produced during biological treatment. Numerous dyes that are used in textile manufacturing are toxic to aerobic microbes and generate sludge. Hence, the biological method that utilizes the aerobic approach appears inadequate to digest textile dyes, particularly azo dyes (Ibrahim et al. 2009). Furthermore, it also needs extra space and a further hydraulic retention period. Aeration is compulsory to produce unknown oxidation complexes that may introduce color to the effluent. While anaerobic microorganisms are sluggish, and so need an extended time to adjust. In anaerobic conditions, the treatment of textile dyes produces additional noxious aromatic amines throughout azo dye degradation by azoreductase. Aromatic amine formed in an anaerobic condition mineralizes to the environment-friendly complex when open to the air; therefore, an anaerobic platform combined with the aerobic platform can be applied to attain effective treatment (Wainaina et al. 2020). The efficiency of biological procedures for degradation relies on the performance of selected microbes. Thus, numerous enzymes and microorganisms comprising algae, fungi, and bacteria have been isolated and tried to break down dyes (Holkar et al. 2016).

2.6 CO_2 Emission

Due to increasing energy consumption worldwide, greenhouse gas emissions, especially CO_2, are constantly increasing. As a result, the National Oceanic and Atmospheric Administration (NOAA) reported that the average CO_2 level increased to 410 ppm and its emission rate elevated to 33.1 GT. The largest polluter in terms of CO_2 emissions is the fossil fuel power stations, which in 2018 they accounted for 30% of the world's total CO_2 emissions (Zhang 2018). In other words, the gas released from fuel power plants after desulfurization operation by removing SO_2 and nitrous oxide contains a volume equal to 10–20% of CO_2 and N_2 gas (Herzog et al. 1997). While this produced CO_2 can be absorbed using various methods such as hydrate-based methods, physical and chemical adsorption methods, membrane, and cryogenics (Lal 2008), the methods of CO_2 sequestration are summarized in Fig. 2.4.

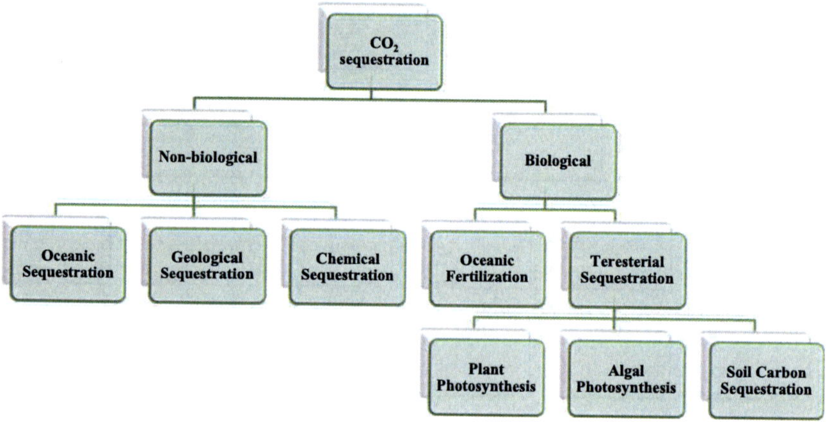

Fig. 2.4 Various approaches of CO_2 sequestration

2.6.1 *Chemical Absorption*

One of the efficient and economical methods that have high efficiency in terms of CO_2 absorption, and storage is the chemical adsorption method (Mondal et al. 2012; Zhang et al. 2018b). The amine scrubbing method is one of the main methods performed by reacting CO_2 with an amine solution after the flue gas enters an adsorbent, and it can be considered an acceptable and effective method for CO_2 adsorption. In this method, after sending a solvent containing large amounts of CO_2 gas to a repellent, the solvents are heated to release CO_2, and the solvents with fewer CO_2 are returned to the adsorbent to absorb CO_2. Researchers have been trying to optimize this method in terms of performance and cost reduction using four approaches. These four approaches are (1) improving the flowsheet method, (2) integrating energy with other plant facilities, (3) using different solvents, and (4) modifying and optimizing the CO_2 uptake process.

Alternatively, physical absorption can be done, in which solvents such as water and alcohol (methanol and ethanol) are used (Zhang et al. 2014). Compared to chemical solvents, physical solvents have various advantages: cheapness and easy accessibility, non-corrosive, low energy consumption, and features such as heating, pressure reduction, and noble gas stripping. The process of physical absorption of CO_2 also depends on Henry's law, which works well due to its high solubility of CO_2 at low temperatures and high altitudes (Ban et al. 2014). Additionally, nanofluids such as Al_2O_3 and SiO_2 can be employed, which contain a liquid base with suspended particles of the size of 1–100 nm. These nanofluids comprise several advantages: better permeability and thermal conductivity, viscosity, and convective heat transfer coefficient than the bottom solution (Zhang et al. 2018a). Nanofluids also form an extensive area between the nanoparticles and the liquid. Therefore, the presence of small amounts of nanoparticles in the solution can increase the thermal efficiency and heat transfer, and mass. These lampblack nanoparticles can also be used to increase the absorption and reproduction of CO_2 in monoethanolamine (MEA) solutions (Goetz et al. 2016). Another type of chemical solvent is alkaline solvents such as NaOH and KOH, which are highly alkaline and good adsorbents for CO_2. Other solutions such as carbonate (K_2CO_3) and washing soda (Na_2CO_3) have long been used as CO_2 adsorbents. In addition, these solutions reduce reproductive energy consumption in the disposal phase because they produce less heat than primary and secondary amines (Penders-van Elk et al. 2013).

One of the most important chemical solvents is the group of amines, which have a high capacity to absorb CO_2. The other group is ionic liquids (ILs), which have high CO_2 solubility and high thermal stability due to low vapor pressure, which makes them easier to use and have many fans due to the reduction of harmful and toxic compounds for the environment. Another group of solvents is anions, which have a higher solubility for CO_2 than ionic liquids. Also, alkanoic amino acid salts, which include a-, b-, g-, and d-amino acids and alkaline solutions, have advantages over amino solvents such as rapid chemical reaction, high stability against oxygen, vapor pressure (Zhang et al. 2018b). Recently, researchers have found that mixing different solvents optimizes solvent performance and reduces system energy consumption. This solvent mixture contains physical, chemical solvents and a combination of two types of solvents. For example, when primary/secondary amine solvents are combined with tertiary/hindered amine solvents, it results in a rapid reaction with low energy consumption for reproduction and increased adsorption capacity (Borhani and Wang 2019). Because each type of solvent has different strengths and weaknesses, researchers have constantly been investigating new techniques to enhance these solvents' efficiency. Typically, an optimal solvent should have properties such as high heat capacity, high thermal stability, and high reaction capacity with CO_2. Also, in preparing solvents, their impact on the environment should be considered (Zhang et al. 2020).

2.6.2 *Oceanic Sequestration of CO_2*

The amount of inorganic carbon in the oceans around the world is 50 times higher than in the atmosphere. Thus, the oceans can act as reservoirs for storing atmospheric carbon (Raven and Falkowski 1999). There are different types of carbon sequestration and storage from the oceans, including abiotic and biotic, which means the direct injection of CO_2 into the ocean and the fertility of the ocean (Farrelly et al. 2013). Temperature, pH, and salt and water concentrations affect the CO_2 distribution between the atmosphere and the ocean. The solubility of CO_2 in the ocean decreases at high concentrations of salt and heat (Raven and Falkowski 1999).

CO_2 leakage from the ocean must be minimized; therefore, CO_2 must be injected into the deepest possible point in the ocean. Direct injection of CO_2 into ocean water is known as one of the non-biological methods of CO_2 sequestration in the ocean (Lal 2008). The strategy of direct injection of CO_2 can have adverse effects on marine organisms because by injecting CO_2 into the ocean water, the pH decreases and, in the long run, causes acidification and poisoning of water for microorganisms, which can affect the life of deep-sea fish (Auerbach et al. 1997). However, this method is effective in reducing CO_2 leakage. A depth of 3000 m can be suitable for CO_2 injection to prevent its leakage into the air. In recent studies, CO_2 separation through direct carbonation of minerals has been investigated. In addition to the injection depth, storage time is also essential and affects the adsorption efficiency.

Another method of CO_2 sequestration is the geological method, which provides the chance of CO_2 storage by using different geological formations in the depths of the Earth. This method has two vital advantages: (1) reducing greenhouse gas emissions and (2) helping the process of oil recovery. These two advantages have also been used in various oil industries for a long time (Klara et al. 2003). CO_2 is absorbed and stored in diverse geological formations such as oil and gas reservoirs, deep saltwater aquifers, and cracks in coal (Klara et al. 2003). This type of carbon capture and storage process is effective for a long time. Since CO_2 injection replaces oil and gas, thus it can be economically beneficial for oil recovery. However, despite intensive research, the safety and stability of CO_2 storage and absorption methods are still debated. As reducing CO_2 is more critical relatively, geological formations are still a suitable method. Although CO_2 sequestration is an effective and suitable method in geological formations, coal cracks are formed at great depths and are not extractable; thus, this method is non-economic. However, CO_2 sequestration using deep-sea coal cracks has advantages such as high CO_2 sequestration efficiency and potential adsorption methane transport, meaning that two CO_2 molecules replace the methane molecule in coal. Consequently, the methane in the coal sequestration is recovered, and CO_2 is absorbed (Gunter et al. 1997). One of these geological formations is the saline aquifers, which are made of permeable rocks saturated with water and salt. When CO_2 is injected into these aquifers, CO_2 dissolves in saltwater and reacts with the dissolved mineral or adjacent rocks. However, the usage of this method is limited due to the likelihood of CO_2 leakage and the high capital cost.

2.6.3 *Biological Method of CO_2 Sequestration*

One of the ways CO_2 sequestration is made by living organisms such as microorganisms and plants, in which biological processes cause the absorption and storage of carbon, is biological sequestration. The process itself involves ocean sequestration through fertility and terrestrial sequestration and is mainly the result of natural processes. Terrestrial sequestration also includes soil carbon sequestration and phytosequestration.

Terrestrial carbon sequestration means the storage of carbon in the soil, biomass above and below ground, and vegetation (Post et al. 2009). These processes include soil carbon sequestration and phytosequestration, respectively (Post et al. 2009). Because this process has a high capacity for long-term storage of CO_2, it is widely used. Living and non-living organic matter, in which the process of photosynthesis mainly takes place, acts as the primary sink for carbon storage (Lal 2008). Various methods such as carbon uptake and storage by soil, vegetation, forests, and wetlands for terrestrial sequestration have been introduced (Lal 2008). Also, it has been found that plants have a high capacity to absorb and store carbon in the soil. Thus, the atmospheric CO_2 and the increase in soil carbon can affect the balance of total carbon in the environment. Plant photosynthesis converts part of atmospheric CO_2 into organic matter.

Meanwhile, another part of this atmospheric CO_2 is transferred to the soil by plant roots and stored in the soil in organic and inorganic forms (Jansson et al. 2010). Both phytosequestration and soil carbon sequestration are, to some extent, related. The quality of the soil carbon sequestration process depends on various conditions such as soil moisture, type of mineral, amount of rainfall, climate, temperature, and amount of clay in the soil (Metting et al. 2008). One of the most effective ways to strengthen soil in agricultural fields is to create a physical disturbance in the soil, which enhances soil respiration as well as the decomposition of organic matter in it (Ibrahim Khalil et al. 2020). This approach, known as no-till agriculture, is widely used by farmers today to absorb and store carbon in agricultural soils and produce crops without disturbing the soil. This no-till farming method also makes the soil more fertile, reduces erosion, increases the soil's water-holding capacity, and produces sustainable organic matter in the soil. These results are obtained by comparing two agricultural lands without using the no-till method through 13C labeling. Also, the no-till cultivation method greatly contributes to the carbon sequestration process, increases organic reserves, and increases storage capacity (Lal 2008). Another optimal approach that is associated with high growth yields in annual agricultural products is the season extension approach, in which perennial agricultural species can have an excellent response to cultivation with the least amount of soil turbulence (Luo et al. 2010).

A plant species that can be applied as a renewable energy resource for the future is called bio-energy. Upon the comprehensive characterization, these are often utilized for increasing energy and biomass potential, and therefore, improved production of bio-energy (Lemus and Lal 2005). It was stated that at the current time, oil yields, sugarcane, and a few of the cereals like wheat and maize are being operated as bio-energy yields (Jansson et al. 2010). Bio-energy plants have the capacity to exchange current fossil fuel energy resources and are capable of decreasing CO_2 concentration because of their high biomass accumulation potential (Huth et al. 2012), and therefore the contribution of carbon sequestration. Bio-energy products also sequester carbon through soil and roots. It was reported that the effectiveness and efficiency of bio-energy yields mostly relied on the kind of species by considering high output, insemination, and finally harvest management (Lemus and Lal 2005).

Atmospheric CO_2 capture through photosynthesis, viz. phytosequestration can be a more useful means of carbon sequestration. It was stated that 99.9% of carbon biota is being subsidized by microbial biomass and vegetation (Jansson et al. 2010). Phytosequestration can work as the greatest essential carbon reservoir for centuries, provided the plantation residues are undisturbed and appropriately preserved. Forests reduction is the key cause of the rise in atmospheric CO_2; therefore, eliminating the source could reverse the impact. Photosynthesis is a natural route in which plants alter inorganic carbon into organic carbon (CO_2 into sugar) using solar energy. In this method, atmospheric carbon (CO_2) is assigned to various plant sections such as roots, stems, seeds, and many other organs. Therefore, CO_2 fixation procedures increase the sequestration of universal atmospheric CO_2. It has been proposed that by using engineering approaches, the carbon uptake and storage in the atmosphere can be enhanced, which is usually by improving photosynthetic efficiency (Barati et al. 2021; Jansson et al. 2010). The different photosynthetic organisms such as cyanobacteria, C4 plants, C3 plants, microalgae, crassulacean acid metabolism (CAM) plants, and several others utilize diverse types of CO_2 fixation pathways. Apart from C3 plants, carbon concentrating mechanisms (CCMs) is the main contributor to photosynthetic carbon fixation (Nogia et al. 2016).

2.6.4 Mineralization of CO_2 as Inorganic Carbonates

CO_2 mineralization is sometimes done through accelerated carbonation. Gaseous CO_2 is regularly mineralized as a thermodynamically stable precipitate throughout this process, thus being seldom discharged subsequent to the mineralization (Lackner 2002; Verma et al. 2015). CO_2 mineralization through accelerated carbonation is often considered into three core procedures: (1) direct carbonation, which is linked to the creation of green cement/concretes like supplementary cementitious complexes, (2) indirect carbonation, which is linked with the creation of high-value substances like precipitated calcium carbonates, and (3) carbonation curing for cement mortar and/or concrete block to reinforce their strength and sturdiness.

Natural carbonate and/or silicate ores are appropriate substances for accelerated carbonation because of their great level of magnesium and/or calcium oxides, like diopside/amphibolite (Koivisto et al. 2016), and serpentine (Veetil et al. 2015). Accelerated carbonation consuming natural ores might offer great capture volume and a long storage duration for anthropogenic CO_2 (Bobicki et al. 2012; Lackner 2003). Carbonate minerals are actively preferred to make from the reaction of silicates and CO_2 like serpentine, olivine, and anorthite (Lackner 2002). It is also recognized that there are sufficient natural ores on Earth to sequester entire CO_2 discharges from fossil-based resources (Lackner 2003). Nevertheless, because of the requirement for large-scale mining of natural ores (Kelly et al. 2011), discovering and pre-treating materials are expensive. Thus, alkaline solid trashes from coal-fired power plants or industries are flattering more considered as the desired material for enhanced carbonation due to their comparatively low-cost ores. While an integrated method to combining CO_2 valorization with alkaline leftovers treatment could be obtained simultaneously, a summary of different CO_2 sequestration strategies is presented in Table 2.3.

Table 2.3 Comparison of different CO_2 sequestration methods

Method of CO_2 sequestration	Pro	Cons
Chemical	• Long-term storage • Minimum leakage of stored CO_2 • Thermodynamically favorable	• Less efficient at the industrial level • Reactions are very slow and extremely high temperature is required
Oceanic	• Large sink capacity for carbon storage • Minimum leakage of stored CO_2	• Carbon injection • Adverse effect on marine biota
Geological	• Great depth of occurrence favors stable carbon sequestration • Helps in oil and methane recovery process	• High cost is associated • Storage volume is not predictable • Chances of leakage
Biological	• Enhancement in phytoplanktonic CO_2 fixation is an added advantage • Occurs naturally and can be improved easily with proper management of the ecosystem • Large sink capacity • Natural process • Cost-effective	• Affects ocean ecology • Efficiency is being limited by various environmental factors • Deforestation • Photosynthetic efficiency and plantation needs to be enhanced at the global level

2.7 Challenges and Future Perspectives

With increased population and industrialization huge amounts of waste including solid waste and wastewater as well as CO_2 emissions are released into the environment. In the first step, minimizing strategies to reduce wastes are in the frontline. Maximizing reusability can be a promising approach likewise to reduce the generation of waste materials. While still some products such as plastics, woods, metals, and glass are produced which can be recycled. In recent years, technologies to improve the efficiency of recycling have been increasing considerably. However, the separation of recyclable material from garbage is an enormous challenge and requires appropriate infrastructure and cooperation of the public together with proper legislation. While still there are a huge amount that ends up in landfills or incinerators as it is not viable to recycle them. In recent years by advances in energy recovery from incarnation, landfills process are being replaced with modern incinerations. Although incineration produced ashes, this ash can be further utilized in constructions. Also, thermochemical conversion methods such as pyrolysis have been shown promising results, as transforming waste into new high-value products, although the produced materials are not consistent and still need further upgrading process.

Many treatment approaches mainly the physicochemical treatment techniques have displayed the capacity to decrease both organic pollutants and color. However, these techniques apply excessive amounts of chemicals, which cause sludge formation as secondary pollutants and high operational costs. Activated carbons are very efficient and cost-effective, particularly biochars as are obtained from biomass. In the treatment of wastewater, biological methods such as aerobic and anaerobic digestion and microalgae-assisted treatment have been a recent trend lately. Through anaerobic digestion, biogas and high-value products can be produced which adds economic value to the treatment process and reduces the cost, and more importantly, chemicals are not used. Biological treatments have not been applied largely; however, this approach has presented great capacities for the removal of nutrients and organic pollutants. The difficulty could be the slowness at which these treatment approaches operate. While microalgae have also been demonstrated to be an ecological-friendly and cost-effective wastewater treatment technique and can be promising organisms in this approach.

Increased CO_2 levels are influencing our planet, and it is not likely to overcome this issue by using any of the sole methods due to several downsides linked to each strategy. Therefore, it can be comprehended that for efficient CO_2 sequestration further than one approach is required to be applied to attain sustainability and combat global warming. Geological and oceanic sequestration require specific locations, and therefore, they are not feasible everywhere. Mineralization also is energy-intensive; it is suggested that the natural mineralization pathway would need to be optimized to become more energy efficient in order for this process to become feasible. Although biological approach, particularly the utilization of microalgae demonstrated promising result, still this technology is in its infancy and requires more research to achieve a sustainable and economically viable platform.

2.8 Conclusions

Recent rapid developments and increase in population have resulted in massive waste that damaged our environment, which these pollutants are found as solid waste, gas emissions, and wastewater. In recent years, various methods have been applied to manage these wastes, and progressively a new technique is introduced. Landfills have been used for a long time; however, they are replacing with incinerators as the volume of waste can be reduced, and energy can also be recovered. Also, novel methods are introduced to reduce waste generation and enhance their usability and recyclability. Sequestrating CO_2 also has been the main topic of research for the past decade, and its importance is becoming more evident gradually. Moreover, wastewater management is broadly applied in many countries as almost all countries face freshwater scarcity. There is no single solution to manage the waste and emission, and therefore depending on countries geographical region, budget and the type and magnitude of waste, different strategies have been applied.

References

Ahmad A, Mohd-Setapar SH, Chuong CS, Khatoon A, Wani WA, Kumar R, Rafatullah M (2015) Recent advances in new generation dye removal technologies: novel search for approaches to reprocess wastewater. RSC Adv 5:30801–30818. https://doi.org/10.1039/c4ra16959j

Akinade OO, Oyedele LO, Bilal M, Ajayi SO, Owolabi HA, Alaka HA, Bello SA (2015) Waste minimisation through deconstruction: a BIM based Deconstructability Assessment Score (BIM-DAS). Resour Conserv Recycl 105:167–176. https://doi.org/10.1016/j.resconrec.2015.10.018

America N (2010) ISWA key issue paper on waste prevention, waste minimization and resource management. Africa (Lond) 2007

Anil I, Gunday ST, Bozkurt A, Alagha O (2020) Design of crosslinked hydrogels comprising poly(vinylphosphonic acid) and bis[2-(methacryloyloxy)ethyl] phosphate as an efficient adsorbent for wastewater dye removal. Nanomaterials 10. https://doi.org/10.3390/nano10010131

Anjaneyulu Y, Sreedhara Chary N, Samuel Suman Raj D (2005) Decolourization of industrial effluents—available methods and emerging technologies—a review. Rev Environ Sci Biotechnol 4:245–273. https://doi.org/10.1007/s11157-005-1246-z

Auerbach DI, Caulfield JA, Adams EE, Herzog HJ (1997) Impacts of ocean CO_2 disposal on marine life: I. A toxicological assessment integrating constant-concentration laboratory assay data with variable-concentration field exposure. Environ Model Assess 2:333–343. https://doi.org/10.1023/A:1019029931755

Azam K, Raza R, Shezad N, Shabir M, Yang W, Ahmad N, Shafiq I, Akhter P, Razzaq A, Hussain M (2020) Development of recoverable magnetic mesoporous carbon adsorbent for removal of methyl blue and methyl orange from wastewater. J Environ Chem Eng 8:104220. https://doi.org/10.1016/j.jece.2020.104220

Ban ZH, Keong LK, Shariff AM (2014) Physical absorption of CO_2 capture: a review. Adv Mater Res 917:134–143. https://doi.org/10.4028/www.scientific.net/AMR.917.134

Barati B, Zeng K, Baeyens J, Wang S, Addy M, Gan S, El-Fatah Abomohra A (2021) Recent progress in genetically modified microalgae for enhanced carbon dioxide sequestration. Biomass Bioenerg 145:105927. https://doi.org/10.1016/j.biombioe.2020.105927

Block C, Van Caneghem J, Van Brecht A, Wauters G, Vandecasteele C (2015) Incineration of hazardous waste: a sustainable process? Waste Biomass Valorization 6:137–145. https://doi.org/10.1007/s12649-014-9334-3

Bobicki ER, Liu Q, Xu Z, Zeng H (2012) Carbon capture and storage using alkaline industrial wastes. Prog Energy Combust Sci 38:302–320. https://doi.org/10.1016/j.pecs.2011.11.002

Borhani TN, Wang M (2019) Role of solvents in CO_2 capture processes: the review of selection and design methods. Renew Sustain Energy Rev 114:109299. https://doi.org/10.1016/j.rser.2019.109299

Bösenhofer M, Purgar A, Winter F (2015) The role of fluidized bed technology for waste to energy, its current status and potential—an Austrian perspective. In: 22nd FBC Finland

Cao B, Sun Y, Yuan J, Wang S, Gong X, Barati B, Zheng A, Jiang D, Hu Y, Yuan C, He Z (2020) Co-pyrolysis characteristics of polysaccharides-cellulose and the co-pyrolyzed compound distributions over two kinds of zeolite catalysts. Int J Energy Res 44:6350–6362. https://doi.org/10.1002/er.5360

Cao Q, Sang L, Tu J, Xiao Y, Liu N, Wu L, Zhang J (2021) Rapid degradation of refractory organic pollutants by continuous ozonation in a micro-packed bed reactor. Chemosphere 270:128621. https://doi.org/10.1016/j.chemosphere.2020.128621

Chapman N, Hooper A (2012) The disposal of radioactive wastes underground. Proc Geol Assoc 123:46–63. https://doi.org/10.1016/j.pgeola.2011.10.001

Chen H, Lin H, Zhang P, Yu L, Chen L, Huang X, Jiao B, Li D (2021) Immobilisation of heavy metals in hazardous waste incineration residue using SiO_2–Al_2O3–Fe_2O_3–CaO glass-ceramic. Ceram Int 47:8468–8477. https://doi.org/10.1016/j.ceramint.2020.11.213

Cucchiella F, D'Adamo I, Gastaldi M (2017) Sustainable waste management: waste to energy plant as an alternative to landfill. Energy Convers Manag 131:18–31. https://doi.org/10.1016/j.enconman.2016.11.012

Darda SA, Gabbar HA, Damideh V, Aboughaly M, Hassen I (2021) A comprehensive review on radioactive waste cycle from generation to disposal. J Radioanal Nucl Chem 329:15–31. https://doi.org/10.1007/s10967-021-07764-2

Davoli E, Fattore E, Paiano V, Colombo A, Palmiotto M, Rossi AN, Il Grande M, Fanelli R (2010) Waste management health risk assessment: a case study of a solid waste landfill in South Italy. Waste Manag 30:1608–1613. https://doi.org/10.1016/j.wasman.2009.10.013

Demirel B, Scherer P (2008) The roles of acetotrophic and hydrogenotrophic methanogens during anaerobic conversion of biomass to methane: a review. Rev Environ Sci Biotechnol 7:173–190. https://doi.org/10.1007/s11157-008-9131-1

Dharmaraj S, Ashokkumar V, Pandiyan R, Halimatul Munawaroh HS, Chew KW, Chen WH, Ngamcharussrivichai C (2021) Pyrolysis: an effective technique for degradation of COVID-19 medical wastes. Chemosphere 275:130092. https://doi.org/10.1016/j.chemosphere.2021.130092

Ding Z, Yi G, Tam VWY, Huang T (2016) A system dynamics-based environmental performance simulation of construction waste reduction management in China. Waste Manag 51:130–141. https://doi.org/10.1016/j.wasman.2016.03.001

Esa MR, Halog A, Rigamonti L (2017) Developing strategies for managing construction and demolition wastes in Malaysia based on the concept of circular economy. J Mater Cycles Waste Manag 19:1144–1154. https://doi.org/10.1007/s10163-016-0516-x

Farrelly DJ, Everard CD, Fagan CC, McDonnell KP (2013) Carbon sequestration and the role of biological carbon mitigation: a review. Renew Sustain Energy Rev 21:712–727. https://doi.org/10.1016/j.rser.2012.12.038

Fatemi S, Imaninasab R (2016) Performance evaluation of recycled asphalt mixtures by construction and demolition waste materials. Constr Build Mater 120:450–456. https://doi.org/10.1016/j.conbuildmat.2016.05.117

Fitzgerald GC (2013) Pre-processing and treatment of municipal solid waste (MSW) prior to incineration. Waste Energy Convers Technol. https://doi.org/10.1533/9780857096364.2.55

Fivga A, Dimitriou I (2018) Pyrolysis of plastic waste for production of heavy fuel substitute: a techno-economic assessment. Energy 149:865–874. https://doi.org/10.1016/j.energy.2018.02.094

Foroutan R, Mohammadi R, Farjadfard S, Esmaeili H, Saberi M, Sahebi S, Dobaradaran S, Ramavandi B (2019) Characteristics and performance of Cd, Ni, and Pb bio-adsorption using *Callinectes sapidus* biomass: real wastewater treatment. Environ Sci Pollut Res 26:6336–6347. https://doi.org/10.1007/s11356-018-04108-8

Gavilan RM, Bernold LE (1994) Source evaluation of solid waste in building construction. J Constr Eng Manag 120:536–552. https://doi.org/10.1061/(asce)0733-9364(1994)120:3(536)

Geng H, Du P, Zhang Z, Yao L, Cao K, Li S, Sheng P (2018) Architecting Bi_2S_3/graphene quantum dots/TiO_2 photoelectrodes for aqueous Cr(VI)/methyl orange removal. Mater Lett 214:146–149. https://doi.org/10.1016/j.matlet.2017.11.126

Goetz SA, Nguyen DT, Esser-Kahn AP (2016) Surface modification of carbon black nanoparticles enhances photothermal separation and release of CO_2. Carbon NY 105:126–135. https://doi.org/10.1016/j.carbon.2016.03.053

Gunter WD, Gentzis T, Rottenfusser BA, Richardson RJH (1997) Deep coalbed methane in Alberta, Canada: a fuel resource with the potential of zero greenhouse gas emissions. Energy Convers Manag 38:217–222. https://doi.org/10.1016/s0196-8904(96)00272-5

Han D, Tong X, Currell MJ, Cao G, Jin M, Tong C (2014) Evaluation of the impact of an uncontrolled landfill on surrounding groundwater quality, Zhoukou China. J Geochem Explor 136:24–39. https://doi.org/10.1016/j.gexplo.2013.09.008

Herzog H, Drake E, Adams E (1997) CO_2 capture, reuse, and storage technologies. Initiatives 1–70

Holkar CR, Jadhav AJ, Pinjari DV, Mahamuni NM, Pandit AB (2016) A critical review on textile wastewater treatments: possible approaches. J Environ Manag 182:351–366. https://doi.org/10.1016/j.jenvman.2016.07.090

Hopewell J, Dvorak R, Kosior E (2009) Plastics recycling: challenges and opportunities. Philos Trans R Soc B Biol Sci 364:2115–2126. https://doi.org/10.1098/rstb.2008.0311

Hu S, Barati B, Odey EA, Wang S, Hu X, Abomohra AEF, Lakshmikandan M, Yerkebulan M, Esakkimuthu S, Shang H (2020) Experimental study and economic feasibility analysis on the production of bio-oil by catalytic cracking of three kinds of microalgae. J Anal Appl Pyrolysis 149:104835. https://doi.org/10.1016/j.jaap.2020.104835

Hu Y, Li J, Wang S, Xu L, Barati B, Cao B, Wang H, Xie K, Wang Q (2021) Catalytic fast hydropyrolysis of seaweed biomass with different zeolite catalysts to produce high-grade bio-oil. Process Saf Environ Prot 146:69–76. https://doi.org/10.1016/j.psep.2020.08.019

Huang B, Wang X, Kua H, Geng Y, Bleischwitz R, Ren J (2018) Construction and demolition waste management in China through the 3R principle. Resour Conserv Recycl 129:36–44. https://doi.org/10.1016/j.resconrec.2017.09.029

Huang Q, Yang Y, Pang X, Wang Q (2008) Evolution on qualities of leachate and landfill gas in the semi-aerobic landfill. J Environ Sci 20:499–504. https://doi.org/10.1016/S1001-0742(08)62086-0

Hunsinger H, Jay K, Vehlow J (2002) Formation and destruction of PCDD/F inside a grate furnace. Chemosphere 46:1263–1272. https://doi.org/10.1016/S0045-6535(01)00256-9

Huth T, Porder S, Chaves J, Whiteside JH (2012) Soil carbon and nutrient changes associated with deforestation for pasture in Southern Costa Rica. Biotropica 44:661–667. https://doi.org/10.1111/j.1744-7429.2012.00863.x

Ibrahim Khalil M, Francaviglia R, Henry B, Klumpp K, Koncz P, Llorente M, Emoke Madari B, Muñoz-Rojas M, Nerger R (2020) Strategic management of grazing grassland systems to maintain and increase organic carbon in soils. CO_2 Sequestration 1–20. https://doi.org/10.5772/intechopen.84341

Ibrahim Z, Amin MFM, Yahya A, Aris A, Umor NA, Muda K, Sofian NS (2009) Characterisation of microbial flocs formed from raw textile wastewater in aerobic biofilm reactor (ABR). Water Sci Technol 60:683–688. https://doi.org/10.2166/wst.2009.440

Idumah CI, Nwuzor IC (2019) Novel trends in plastic waste management. SN Appl Sci 1:1–14. https://doi.org/10.1007/s42452-019-1468-2

Idzorek J (1991) Municipal solid waste incineration. Munic Waste Combust 111

Jansson C, Wullschleger SD, Kalluri UC, Tuskan GA (2010) Phytosequestration: carbon biosequestration by plants and the prospects of genetic engineering. Bioscience 60:685–696. https://doi.org/10.1525/bio.2010.60.9.6

Jaseela PK, Garvasis J, Joseph A (2019) Selective adsorption of methylene blue (MB) dye from aqueous mixture of MB and methyl orange (MO) using mesoporous titania (TiO_2)–poly vinyl alcohol (PVA) nanocomposite. J Mol Liq 286:110908. https://doi.org/10.1016/j.molliq.2019.110908

Jones PT, Geysen D, Tielemans Y, Van Passel S, Pontikes Y, Blanpain B, Quaghebeur M, Hoekstra N (2013) Enhanced Landfill Mining in view of multiple resource recovery: a critical review. J Clean Prod 55:45–55. https://doi.org/10.1016/j.jclepro.2012.05.021

Kabirifar K, Mojtahedi M, Wang C, Tam VWY (2020) Construction and demolition waste management contributing factors coupled with reduce, reuse, and recycle strategies for effective waste management: a review. J Clean Prod 263:121265. https://doi.org/10.1016/j.jclepro.2020.121265

Kamaraj M, Srinivasan NR, Assefa G, Adugna AT, Kebede M (2020) Facile development of sunlit ZnO nanoparticles-activated carbon hybrid from pernicious weed as an operative nano-adsorbent for removal of methylene blue and chromium from aqueous solution: extended application in tannery industrial wastewater. Environ Technol Innov 17:100540. https://doi.org/10.1016/j.eti.2019.100540

Kelly KE, Silcox GD, Sarofim AF, Pershing DW (2011) An evaluation of ex situ, industrial-scale, aqueous CO_2 mineralization. Int J Greenh Gas Control 5:1587–1595. https://doi.org/10.1016/j.ijggc.2011.09.005

Klara SM, Srivastava RD, McIlvried HG (2003) Integrated collaborative technology development program for CO_2 sequestration in geologic formations—United States Department of Energy R&D. Energy Convers Manag 44:2699–2712. https://doi.org/10.1016/S0196-8904(03)00042-6

Koivisto E, Erlund R, Fagerholm M, Zevenhoven R (2016) Extraction of magnesium from four Finnish magnesium silicate rocks for CO_2 mineralisation—Part 1: Thermal solid/solid extraction. Hydrometallurgy 166:222–228. https://doi.org/10.1016/j.hydromet.2016.07.005

Kumar S, Chiemchaisri C, Mudhoo A (2011) Bioreactor landfill technology in municipal solid waste treatment: an overview. Crit Rev Biotechnol 31:77–97. https://doi.org/10.3109/07388551.2010.492206

Lackner KS (2003) A guide to CO_2 sequestration. Science (80) 300:1677–1678. https://doi.org/10.1126/science.1079033

Lackner KS (2002) Carbonate chemistry for sequestering fossil carbon. Annu Rev Energy Environ 27:193–232. https://doi.org/10.1146/annurev.energy.27.122001.083433

Lal R (2008) Carbon sequestration. Philos Trans R Soc B Biol Sci 363:815–830. https://doi.org/10.1098/rstb.2007.2185

Lee KE, Morad N, Teng TT, Poh BT (2012) Development, characterization and the application of hybrid materials in coagulation/flocculation of wastewater: a review. Chem Eng J 203:370–386. https://doi.org/10.1016/j.cej.2012.06.109

Lemus R, Lal R (2005) Bioenergy crops and carbon sequestration. CRC Crit Rev Plant Sci 24:1–21. https://doi.org/10.1080/07352680590910393

Liang CZ, Sun SP, Li FY, Ong YK, Chung TS (2014) Treatment of highly concentrated wastewater containing multiple synthetic dyes by a combined process of coagulation/flocculation and nanofiltration. J Memb Sci 469:306–315. https://doi.org/10.1016/j.memsci.2014.06.057

Liu Q, Zhou Y, Lu J, Zhou Y (2020) Novel cyclodextrin-based adsorbents for removing pollutants from wastewater: a critical review. Chemosphere 241:125043. https://doi.org/10.1016/j.chemosphere.2019.125043

Lu JW, Zhang S, Hai J, Lei M (2017) Status and perspectives of municipal solid waste incineration in China: a comparison with developed regions. Waste Manag 69:170–186. https://doi.org/10.1016/j.wasman.2017.04.014

Luo Z, Wang E, Sun OJ (2010) Soil carbon change and its responses to agricultural practices in Australian agro-ecosystems: a review and synthesis. Geoderma 155:211–223. https://doi.org/10.1016/j.geoderma.2009.12.012

Mais L, Vacca A, Mascia M, Usai EM, Tronci S, Palmas S (2020) Experimental study on the optimisation of azo-dyes removal by photo-electrochemical oxidation with TiO_2 nanotubes. Chemosphere 248:125938. https://doi.org/10.1016/j.chemosphere.2020.125938

Makarichi L, Jutidamrongphan W, Techato K-a (2018) The evolution of waste-to-energy incineration: a review. Renew Sustain Energy Rev 91:812–821. https://doi.org/10.1016/j.rser.2018.04.088

Manchisi J, Matinde E, Rowson NA, Simmons MJH, Simate GS, Ndlovu S, Mwewa B (2020) Ironmaking and steelmaking slags as sustainable adsorbents for industrial effluents and wastewater treatment: a critical review of properties, performance, challenges and opportunities. Sustainability 12:1–47. https://doi.org/10.3390/su12052118

Mao Y, Wu H, Wang W, Jia M, Che X (2020) Pretreatment of municipal solid waste incineration fly ash and preparation of solid waste source sulphoaluminate cementitious material. J Hazard Mater 385:121580. https://doi.org/10.1016/j.jhazmat.2019.121580

Marzouk M, Azab S (2014) Environmental and economic impact assessment of construction and demolition waste disposal using system dynamics. Resour Conserv Recycl 82:41–49. https://doi.org/10.1016/j.resconrec.2013.10.015

Menegaki M, Damigos D (2018) A review on current situation and challenges of construction and demolition waste management. Curr Opin Green Sustain Chem 13:8–15. https://doi.org/10.1016/j.cogsc.2018.02.010

Metting FB, Smith JL, Amthor JS, Izaurralde RC (2008) Science needs and new technology for increasing soil carbon sequestration 1. Soil Carbon Sequestration Potential Land use options for enhanced C sequestration at the landscape and regional scales include protection and selective management of native ecos. Clim Change 5:11–34

Mishra S, Tiwary D, Ohri A, Agnihotri AK (2019) Impact of Municipal Solid Waste Landfill leachate on groundwater quality in Varanasi, India. Groundw Sustain Dev 9. https://doi.org/10.1016/j.gsd.2019.100230

Mondal MK, Balsora HK, Varshney P (2012) Progress and trends in CO_2 capture/separation technologies: a review. Energy 46:431–441. https://doi.org/10.1016/j.energy.2012.08.006

Mourad M (2016) Recycling, recovering and preventing "food waste": competing solutions for food systems sustainability in the United States and France. J Clean Prod 126:461–477. https://doi.org/10.1016/j.jclepro.2016.03.084

Nanda S, Reddy SN, Mitra SK, Kozinski JA (2016) The progressive routes for carbon capture and sequestration. Energy Sci Eng 4:99–122. https://doi.org/10.1002/ese3.117

Narayana T (2009) Municipal solid waste management in India: from waste disposal to recovery of resources? Waste Manag 29:1163–1166. https://doi.org/10.1016/j.wasman.2008.06.038

Neczaj E, Okoniewska E, Kacprzak M (2005) Treatment of landfill leachate by sequencing batch reactor. Desalination 185:357–362. https://doi.org/10.1016/j.desal.2005.04.044

Nogia P, Sidhu GK, Mehrotra R, Mehrotra S (2016) Capturing atmospheric carbon: biological and nonbiological methods. Int J Low-Carbon Technol 11:266–274. https://doi.org/10.1093/ijlct/ctt077

Oppelt ET (1987) Incineration of hazardous waste a critical review. J Air Pollut Control Assoc 37:558–586. https://doi.org/10.1080/08940630.1987.10466245

Papari S, Bamdad H, Berruti F (2021) Pyrolytic conversion of plastic waste to value-added products and fuels: a review. Materials (Basel) 14. https://doi.org/10.3390/ma14102586

Penders-van Elk NJMC, Hamborg ES, Huttenhuis PJG, Fradette S, Carley JA, Versteeg GF (2013) Kinetics of absorption of carbon dioxide in aqueous amine and carbonate solutions with carbonic anhydrase. Int J Greenh Gas Control 12:259–268. https://doi.org/10.1016/j.ijggc.2012.10.016

Pham L, Hargreaves R (2003) Sustainability and the building code of Australia

Pisupati SV, Tchapda AH (2015) Thermochemical processing of biomass. Adv Bioprocess Technol. https://doi.org/10.1007/978-3-319-17915-5_15

Post WM, Amonette JE, Birdsey R, Garten CT, Izaurralde RC, Jardine PM, Jastrow J, Lal R, Marland G, McCarl BA, Thomson AM, West TO, Wullschleger SD, Metting FB (2009) Terrestrial biological carbon sequestration: science for enhancement and implementation. Geophys Monogr Ser 183:73–88. https://doi.org/10.1029/2008GM000753

Qureshi MS, Oasmaa A, Pihkola H, Deviatkin I, Tenhunen A, Mannila J, Minkkinen H, Pohjakallio M, Laine-Ylijoki J (2020) Pyrolysis of plastic waste: opportunities and challenges. J Anal Appl Pyrolysis 152:104804. https://doi.org/10.1016/j.jaap.2020.104804

Raghab SM, Abd El Meguid AM, Hegazi HA (2013) Treatment of leachate from municipal solid waste landfill. HBRC J 9:187–192. https://doi.org/10.1016/j.hbrcj.2013.05.007

Raven JA, Falkowski PG (1999) Oceanic sinks for atmospheric CO_2. Plant Cell Environ 22:741–755. https://doi.org/10.1046/j.1365-3040.1999.00419.x

Reinhart DR, McCreanor PT, Townsend T (2002) The bioreactor landfill: Its status and future. Waste Manag Res 20:172–186. https://doi.org/10.1177/0734242X0202000209

Ritzkowski M, Stegmann R (2012) Landfill aeration worldwide: concepts, indications and findings. Waste Manag 32:1411–1419. https://doi.org/10.1016/j.wasman.2012.02.020

Sakai SI, Hiraoka M (2000) Municipal solid waste incinerator residue recycling by thermal processes. Waste Manag 20:249–258. https://doi.org/10.1016/S0956-053X(99)00315-3

Samer M (2015) Biological and chemical wastewater treatment processes. Wastewater Treat Eng 1–50. https://doi.org/10.5772/61250

Scarlat N, Motola V, Dallemand JF, Monforti-Ferrario F, Mofor L (2015) Evaluation of energy potential of municipal solid waste from African urban areas. Renew Sustain Energy Rev 50:1269–1286. https://doi.org/10.1016/j.rser.2015.05.067

Shafiq I, Shafique S, Akhter P, Yang W, Hussain M (2020) Recent developments in alumina supported hydrodesulfurization catalysts for the production of sulfur-free refinery products: a technical review. Catal Rev Sci Eng 00:1–86. https://doi.org/10.1080/01614940.2020.1780824

Sharma HB, Vanapalli KR, Cheela VS, Ranjan VP, Jaglan AK, Dubey B, Goel S, Bhattacharya J (2020) Challenges, opportunities, and innovations for effective solid waste management during and post COVID-19 pandemic. Resour Conserv Recycl 162:105052. https://doi.org/10.1016/j.resconrec.2020.105052

Shen LY, Tam VWY (2002) Implementation of environmental management in the Hong Kong construction industry. Int J Proj Manag 20:535–543. https://doi.org/10.1016/S0263-7863(01)00054-0

Shi B, Hu J, Peng H, Ishizuka S (2018) Effects of internal flue gas recirculation rate on the NO_x emission in a methane/air premixed flame. Combust Flame 188:199–211. https://doi.org/10.1016/j.combustflame.2017.09.043

Singh NB, Nagpal G, Agrawal S, Rachna (2018) Water purification by using adsorbents: a review. Environ Technol Innov 11:187–240. https://doi.org/10.1016/j.eti.2018.05.006

Singh R, Budarayavalasa S (2021) Solidification and stabilization of hazardous wastes using geopolymers as sustainable binders. J Mater Cycles Waste Manag 23:1699–1725. https://doi.org/10.1007/s10163-021-01245-0

Singh RP, Tyagi VV, Allen T, Ibrahim MH, Kothari R (2011) An overview for exploring the possibilities of energy generation from municipal solid waste (MSW) in Indian scenario. Renew Sustain Energy Rev 15:4797–4808. https://doi.org/10.1016/j.rser.2011.07.071

Slack RJ, Gronow JR, Voulvoulis N (2009) The management of household hazardous waste in the United Kingdom. J Environ Manag 90:36–42. https://doi.org/10.1016/j.jenvman.2008.03.007

Slatni I, Elberrichi FZ, Duplay J, Fardjaoui NEH, Guendouzi A, Guendouzi O, Gasmi B, Akbal F, Rekkab I (2020) Mesoporous silica synthesized from natural local kaolin as an effective adsorbent for removing of Acid Red 337 and its application in the treatment of real industrial textile effluent. Environ Sci Pollut Res 27:38422–38433. https://doi.org/10.1007/s11356-020-08615-5

Sonune A, Ghate R (2004) Developments in wastewater treatment methods. Desalination 167:55–63. https://doi.org/10.1016/j.desal.2004.06.113

Sözen S, Olmez-Hanci T, Hooshmand M, Orhon D (2020) Fenton oxidation for effective removal of color and organic matter from denim cotton wastewater without biological treatment. Environ Chem Lett 18:207–213. https://doi.org/10.1007/s10311-019-00923-8

Strobel R, Waldner MH, Gablinger H (2018) Highly efficient combustion with low excess air in a modern energy-from-waste (EfW) plant. Waste Manag 73:301–306. https://doi.org/10.1016/j.wasman.2017.06.049

Subramanian PM (2000) Plastics recycling and waste management in the US. Resour Conserv Recycl 28:253–263. https://doi.org/10.1016/S0921-3449(99)00049-X

Sun R, Ismail TM, Ren X, El-Salam MA (2016) Influence of simulated MSW sizes on the combustion process in a fixed bed: CFD and experimental approaches. Waste Manag 49:272–286. https://doi.org/10.1016/j.wasman.2015.12.019

Tasneem K (2014) Beneficial utilization of municipal solid waste incineration ashes as sustainable road construction materials 2004–2019

U.S. Environmental Protection Agency (USEPA) (2019) Overview of greenhouse gases. https://www.epa.gov/ghgemissions/overview-greenhouse-gases. Accessed 10 Jul 2021

van Heek J, Arning K, Ziefle M (2017) Reduce, reuse, recycle: Acceptance of CO_2-utilization for plastic products. Energy Policy 105:53–66. https://doi.org/10.1016/j.enpol.2017.02.016

Veetil SP, Mercier G, Blais JF, Cecchi E, Kentish S (2015) Magnetic separation of serpentinite mining residue as a precursor to mineral carbonation. Int J Miner Process 140:19–25. https://doi.org/10.1016/j.minpro.2015.04.024

Verma M, Brar SK, Tyagi RD, Surampalli RY (2015) Carbon sequestration via mineral carbonation: overview and assessment. Carbon Capture Storage Phys Chem Biol Methods 281–302. https://doi.org/10.1061/9780784413678

Wadhawan S, Jain A, Nayyar J, Mehta SK (2020) Role of nanomaterials as adsorbents in heavy metal ion removal from waste water: a review. J Water Process Eng 33:101038. https://doi.org/10.1016/j.jwpe.2019.101038

Wagner TP, Raymond T (2015) Landfill mining: case study of a successful metals recovery project. Waste Manag 45:448–457. https://doi.org/10.1016/j.wasman.2015.06.034

Wainaina S, Awasthi MK, Sarsaiya S, Chen H, Singh E, Kumar A, Ravindran B, Awasthi SK, Liu T, Duan Y, Kumar S, Zhang Z, Taherzadeh MJ (2020) Resource recovery and circular economy from organic solid waste using aerobic and anaerobic digestion technologies. Bioresour Technol 301:122778. https://doi.org/10.1016/j.biortech.2020.122778

Waldner MH, Halter R, Sigg A, Brosch B, Gehrmann HJ, Keunecke M (2013) Energy from waste—clean, efficient, renewable: transitions in combustion efficiency and NO_x control. Waste Manag 33:317–326. https://doi.org/10.1016/j.wasman.2012.08.007

Wang S, Jiang D, Cao B, Qian L, Hu Y, Liu L, Yuan C, Abomohra AEF, He Z, Wang Q, Zhang B (2018) Bio-char and bio-oil characteristics produced from the interaction of *Enteromorpha clathrate* volatiles and rice husk bio-char during co-pyrolysis in a sectional pyrolysis furnace: a complementary study. J Anal Appl Pyrolysis 135:219–230. https://doi.org/10.1016/j.jaap.2018.08.030

Wang S, Shang H, Abomohra AEF, Wang Q (2019) One-step conversion of microalgae to alcohols and esters through co-pyrolysis with biodiesel-derived glycerol. Energy Convers Manag 198:111792. https://doi.org/10.1016/j.enconman.2019.111792

Wang J, Yao J, Wang L, Xue Q, Hu Z, Pan B (2020a) Multivariate optimization of the pulse electrochemical oxidation for treating recalcitrant dye wastewater. Sep Purif Technol 230:115851. https://doi.org/10.1016/j.seppur.2019.115851

Wang S, Zhao S, Uzoejinwa BB, Zheng A, Wang Q, Huang J, Abomohra AE-F (2020b) A state-of-the-art review on dual purpose seaweeds utilization for wastewater treatment and crude bio-oil production. Energy Convers Manag 222:113253. https://doi.org/10.1016/j.enconman.2020.113253

World Health Organization (2018) Healthcare waste. https://www.who.int/news-room/fact-sheets/detail/health-care-waste

Wissing F, Wirtz S, Scherer V (2017) Simulating municipal solid waste incineration with a DEM/CFD method—influences of waste properties, grate and furnace design. Fuel 206:638–656. https://doi.org/10.1016/j.fuel.2017.06.037

Wiszniowski J, Robert D, Surmacz-Gorska J, Miksch K, Weber JV (2006) Landfill leachate treatment methods: a review. Environ Chem Lett 4:51–61. https://doi.org/10.1007/s10311-005-0016-z

Woodward R (2004) The organisation for economic cooperation and development. New Polit Econ 9:113–127. https://doi.org/10.1080/1356346042000190411

Yang C, Xu W, Nan Y, Wang Y, Hu Y, Gao C, Chen X (2020) Fabrication and characterization of a high performance polyimide ultrafiltration membrane for dye removal. J Colloid Interface Sci 562:589–597. https://doi.org/10.1016/j.jcis.2019.11.075

Yang Y, Yue B, Yang Y, Huang Q (2012) Influence of semi-aerobic and anaerobic landfill operation with leachate recirculation on stabilization processes. Waste Manag Res 30:255–265. https://doi.org/10.1177/0734242X11413328

Yang Z, Lü F, Zhang H, Wang W, Shao L, Ye J, He P (2021) Is incineration the terminator of plastics and microplastics? J Hazard Mater 401:123429. https://doi.org/10.1016/j.jhazmat.2020.123429

Zhang X (2018) Microalgae removal of CO_2 from flue gas. https://doi.org/10.13140/RG.2.2.26617.77929

Zhang Y, Wang L, Chen L, Ma B, Zhang Y, Ni W, Tsang DCW (2021) Treatment of municipal solid waste incineration fly ash: state-of-the-art technologies and future perspectives. J Hazard Mater 411:125132. https://doi.org/10.1016/j.jhazmat.2021.125132

Zhang Z, Borhani TN, Olabi AG (2020) Status and perspective of CO_2 absorption process. Energy 205:118057. https://doi.org/10.1016/j.energy.2020.118057

Zhang Z, Cai J, Chen F, Li H, Zhang W, Qi W (2018) Progress in enhancement of CO_2 absorption by nanofluids: a mini review of mechanisms and current status. Renew Energy 118:527–535. https://doi.org/10.1016/j.renene.2017.11.031

Zhang Z, Li Y, Zhang W, Wang J, Soltanian MR, Olabi AG (2018) Effectiveness of amino acid salt solutions in capturing CO_2: a review. Renew Sustain Energy Rev 98:179–188. https://doi.org/10.1016/j.rser.2018.09.019

Zhang Z, Wang B, Sun Q, Zheng L (2014) A novel method for the preparation of CO_2 sorption sorbents with high performance. Appl Energy 123:179–184. https://doi.org/10.1016/j.apenergy.2014.02.012

Zhu Z, Liu D, Cai S, Tan Y, Liao J, Fang Y (2020) Dyes removal by composite membrane of sepiolite impregnated polysulfone coated by chemical deposition of tea polyphenols. Chem Eng Res Des 156:289–299. https://doi.org/10.1016/j.cherd.2020.02.001

Chapter 3
Recent Advances in Circular Bioeconomy

Muhammad Adil, Amar Nasir, Arbab Sikandar, and Noor Muhammad Khan

Abstract The progressively growing patterns of global population, urbanization, industrialization, environmental degradation and associated depletion of natural reserves are paving the way towards the conservation and recycling of resources. Contrary to the traditional linear economy, based on procurement-production-usage-disposal, the novel circular economy model corresponds to sustainable flow of materials, products and energy, in terms of growth-manufacturing-utilization-restoration. Likewise, bioeconomy deals with the production and subsequent conversion of renewable biological residues into bioenergy and value-added products. Accordingly, circular bioeconomy constitutes the integration of circular economy and bioeconomy for the sustainable and cascading use of bioresources into reusable/recyclable bio-based products. In addition to biomass, the organic waste containing plentiful amounts of proteins, carbohydrates, lipids and other essential substances can be efficiently converted into useful and eco-friendly products. Therefore, circular bioeconomy represents a promising and effective strategy for resolving the global issues of food scarcity, constant dependence on fossil fuels, waste management, energy deficit, limited employment opportunities and environmental pollution. Nevertheless, limitations in supply chain, technological advancement, legislative measures and consumer compliance necessitate thorough consideration for the implementation of circular bioeconomy approach in various sectors of socioeconomic significance.

M. Adil (✉)
Pharmacology & Toxicology Section, University of Veterinary & Animal Sciences, Lahore, Jhang Campus, Pakistan
e-mail: muhammad.adil@uvas.edu.pk

A. Nasir
Department of Clinical Sciences, University of Veterinary & Animal Sciences, Lahore, Jhang Campus, Pakistan

A. Sikandar
Anatomy & Histology Section, University of Veterinary & Animal Sciences, Lahore, Jhang Campus, Pakistan

N. M. Khan
Physiology & Biochemistry Section, University of Veterinary & Animal Sciences, Lahore, Jhang Campus, Pakistan

A. E.-F. Abomohra et al. (eds.), *Waste-to-Energy*,
https://doi.org/10.1007/978-3-030-91570-4_3

Keywords Circular bioeconomy · Biomass · Biowaste valorization · Biorefining

3.1 Circular Bioeconomy: Concepts, Elements and Significance

3.1.1 Economic Growth in the Milieu of Finite Natural Resources

Sustainable economic growth is an essential entity for the survival and development of human population. As the available resources on our planet are finite, the increasing trend in human population demands equilibrium between economic growth and management of resources. At present, our economy is subjected to instability and unsustainability, and by 2050, the projected world population is expected to reach 9.2 billion, further raising concerns about the increasing demands of energy and resources (Stephens et al. 2010). Among the natural resources, fossil fuels are considered as the most storable and transformable energy reserves which have significantly improved the quality of life for years (Kalmykova et al. 2018). Unfortunately, the fate of such fossil fuels lies under declining because of the rapid economic growth at an annual rate of 1.5–3% (Stephens et al. 2010). Our natural resources are constantly depleting, thereby increasing the environmental pollution and associated risk of global warming, while, reducing the biodiversity (Mohan et al. 2016a). The challenges of climate change, greenhouse gases emissions and constant depletion of natural resources in the form of diminishing fossil fuels have stressed upon the policymakers to devise strategies for environmental protection and sustainable development, which is the key for the survival of mankind. To cope with these challenges, we need to make a shift towards an economy that is least concerned with fossil fuels, in order to maintain an inhabitable planet (Mardani et al. 2019). We are also in need of "decoupling "the economic growth which refers to a sustainable economic growth without any simultaneous consumption of resources and energy, which may contribute effectively towards a sustainable economic growth and preservation of the planet resources. An economy of decoupled nature should neither deleteriously affect the air, water, soil and biodiversity nor exhaust the resources. Decoupling will provide the economy with some extra time before it is finally trapped into a state of diminished resources and environmental pollution. Absolute decoupling is the need of hour which enables the economy to grow in a sustainable manner (Wijkman and Skanberg 2015). Globally, the resource-rich nations circulate the materials to powerhouses required for the manufacturing of different products and then supply these products through complex supply chains towards the global markets for recycling. Although these supply chains are of complicated nature and harbour numerous challenges, however, such challenges can be overcome by formulating friendly global trade policies.

3.1.2 Circular Economy and Bioeconomy for Sustainable Growth

Sustainable economy is the hallmark of economic growth, and during its transition, the concept of circular economy is widely accepted. European Commission (EC) envisages the circular economy as "The economic model where the value of resources, materials and products is maintained for an extended period of time along with least generation of waste products" (European Commission 2016). In circular economy, the raw materials are being recycled and reused, whereas the restorative potential of natural resources is also maintained in parallel. In addition, circular economy ensures the existence of life through its restorative potential by minimizing the usage of less toxic substances and encouraging the trend towards the use of renewable energy resources. The materials being used are recycled in the form of manufacturing of useful products of human significance which helps in a reduced generation of waste products (Philp and Winickoff 2018).

Several other concepts are in place regarding the contributions and pivotal roles of circular economy in the execution of a sustainable economic growth. The Ellen MacArthur Foundation, which is working closely with the EC on circular economy, describes the circular economy as "a model with regenerative and restorative potential in which the constant availability of resources is ensured". According to this model, the nonliving (abiotic) materials are separated from the living (biotic) materials because of the regenerative or restorative nature of the biological resources. This biological cycle keeps on extracting chemicals from the environment and adding nutrients to the biosphere through different processes like fermentation and composting. The materials circling for longer time lead to maximization of consecutive cycles in terms of their number or duration. It has been observed that the materials or products which remain in the cycle for a longer duration minimize the use of new energy, raw materials and labour in order to generate new products. These cycles particularly of closed-loop nature are helpful in the circulation of materials which can be recycled and rendered completely safe and nontoxic for consumption (MacArthur 2013).

The second component for a sustainable economic growth is circular bioeconomy. Circular bioeconomy is presented in many ways by different organizations around the world. According to the Organization for Economic Co-operation and Development (OECD), circular bioeconomy is "a planet where there is a significant contribution of biotechnology in the economic output". It utilizes the knowledge about biological materials which can enter the biological cycles and can be processed and recycled. It also encompasses the knowledge about bioprocesses, renewable biomass, genetics and simultaneous integration of these disciplines into biotechnology for the enhanced production and recycling of useful products (OECD 2009). According to the German Bioeconomy Council, "bioeconomy is the efficient utilization of all the available bioresources in the generation of useful products, establishment of processes and providing services in all the economic sectors in the vicinity of a sustainable economic growth". The EC defines circular bioeconomy as "The generation of value-added products (bioenergy, food and feed) by the utilization of renewable energy resources and efficient waste recycling". Bioeconomy is anticipated to undertake the demands and challenges for energy and food because of its potential to implement new biological approaches in collaboration with all the stakeholders, scientists and masses involved in policy drafting (European Commission 2012).

Bioeconomy is concerned with feed and food sectors, whereas the "bio-based economy" incorporates bio-based materials, products, biofuels production and industries which make partial or complete utilization of the biological feedstock (Carus 2017). The European Economic and Social Committee (EESC) are of the view that circular economy is valued by bioeconomy through constant generation, conversion and efficient consumption of the natural resources. In terms of sustainability, the circular bioeconomy is defined as the manufacturing, consumption, preservation and reprocessing of biological resources for sustainable products and processes across all economic sectors. A sustainable bioeconomy can be developed by carbon-neutrality and circularity, which can directly add benefits to the social, economic and climatic conditions. Bioeconomy plays an important role in climate control by acquiring carbon dioxide through photosynthesis, storage of carbon in the form of biological products and consumption of bio-based feedstock. Bioeconomy also contributes significantly by incorporating the bioenergy chemicals in the current industrial and energy sectors and simultaneously reducing the load of fossils. Significance of bioeconomy is now widely acknowledged, and many countries are closely working on these strategies and concepts to achieve the economic goals (Mohan et al. 2019).

3.1.3 *Circular Bioeconomy: The Integration of Bioeconomy and Circular Economy*

Bioeconomy and circular economy are always considered in parallel in the context of sustainable economic growth. Bioeconomy has the potential to replace fossils, while circular economy cements the efficiency of resources. The amalgamation of bioeconomy and circular economy has thus led to the emergence of "circular bioeconomy" and is defined as "The sustainable conversion of biological residues and wastes into useful biological products which can be recycled/reused or released to the environment in a safer manner through organic and nutrient cycles". Bioeconomy is concerned with biological factors and is circular in nature; however, the utilization and application of biomachinary and bioprocesses may not be circular, and hence, there is a wide scope for the intersection of circular economy and bioeconomy to introduce the concept of circular bioeconomy (Carus and Dammer 2018). There is a close association between circular economy and circular bioeconomy Fig. 3.1, and most of their developmental and economic targets are in common. An example is an economic goal towards a sustainable and resource-efficient planet having low carbon footprint. In this way, the challenges of climate change are tackled by the circular

Fig. 3.1 Circular bioeconomy; the integration of bioeconomy and circular economy. Modified from Carus (2017)

economy and the bioeconomy through reduced fossil fuel consumption. The resource efficiency of processes is strengthened by circular economy, and the utilization of recycled materials is reduced to eliminate the consumption of additional fossil carbon. Under the concept of bioeconomy, biomass-derived carbon from different sectors like forestry, agriculture and marine environment (including wastes and by-products from water) substitutes the fossil carbon.

The circular bioeconomy pathway has been presented in Fig. 3.2. In general, the circular economy and circular bioeconomy strive for achieving the following economic and ecological targets as mentioned elsewhere (Carus and Dammer 2018).

- Better resource efficiency and eco-friendliness
- Lowering greenhouse gases emissions
- Least dependence on fossil-based carbon
- Exploitation of wastes.

Fig. 3.2 Circular bioeconomy pathways indicating the cyclic and sustainable use of biomass in terms of product manufacturing, consumption, reprocessing and re-usage

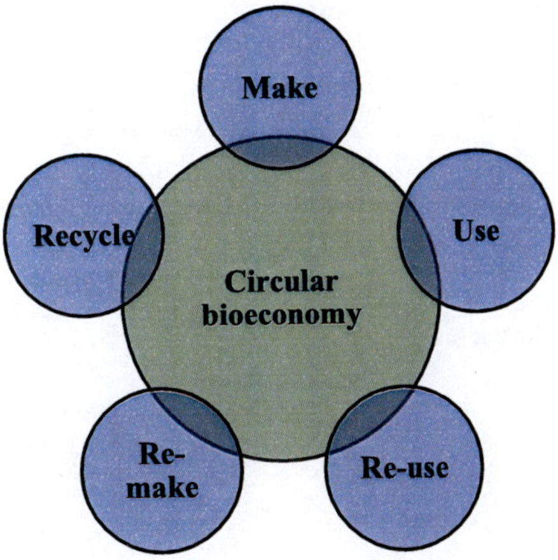

3.1.4 *Elements of Circular Bioeconomy*

The basic elements of circular bioeconomy have been discussed in the following section and illustrated in Fig. 3.3.

3.1.4.1 Biomass/Bio-Based Products

Biomass is considered as an integral component of circular bioeconomy. Production of bio-based products requires energy, irrespective of the stage of the cascade. Accordingly, the preferential utilization of renewable bioenergy for bio-based manufacturing, whereas, avoidance of fossil fuel consumption and carbon sequestration are recommended in compliance with the circular bioeconomy principles (Carus and Dammer 2018).

3.1.4.2 Resource Efficiency

It includes technical efficiency in terms of material output/input, productivity of resources and intensity of emissions. Keeping in view these three interpretations of resource efficiency, there is a need for balanced approach because increasing one type of efficiency might also increase the other (e.g. raising the technical efficiency may also increase the intensity of emission (Stegmann et al. 2020).

3.1.4.3 Cascading Use of Biomass

Cascading can be defined as the sequential consumption of accessible resources for diverse purposes (Olsson et al. 2018). This definition emphasizes on the judicious and step by step use of available resources for the synthesis of valuable products. Fehrenbach et al. (2017) describe the cascading use of biomass as "the synthesis of

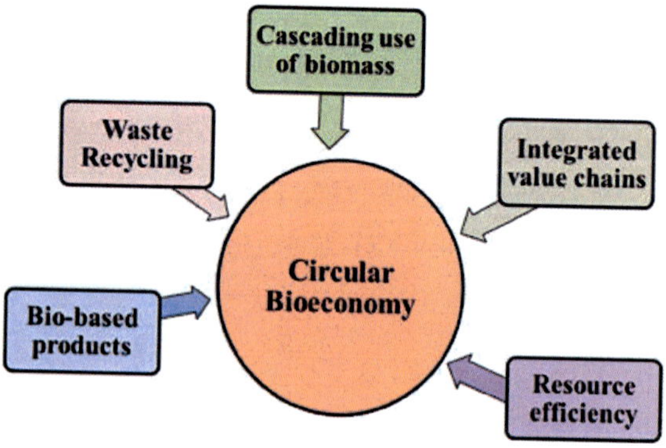

Fig. 3.3 Main elements of circular bioeconomy

bio-based products by the processing of biomass which is used at least once more either for energy or material purposes".

3.1.4.4 Waste Management/Recycling

More than 50% of the food wasted in European Union along the value chain is comprised of household food waste (Stenmarck et al. 2016). The quantity of food, its availability, affordability, accessibility and caloric density represent the determining factors towards an increase in over-consumption and wastage of food in the developed countries (Rozin 2005). Therefore, researchers in the discipline of circular bioeconomy are focusing on the reuse/recycling processes of organic waste. A wide range of bio-based products can be synthesized from renewable feedstock through multidisciplinary studies in science, engineering and management (Amulya et al. 2015).

3.1.4.5 Integrated Production Chains

Integrated product chains constitute an indispensable component of circular bioeconomy by promoting the utility and resource efficiency of biomass (Zabaniotou, 2018). Combined biorefining technologies facilitate the concurrent manufacturing of high-grade and lower-value products from several types of biomass, utilization of waste streams and cost reduction (McCormick and Kautto, 2013).

3.1.5 Significance of Circular Bioeconomy

3.1.5.1 Contribution of Circular Bioeconomy in Sustainable Development

Circular bioeconomy is playing a pivotal role in the accomplishment of several sustainable development goals (SDGs). Out of the total 17 SDGs, 11 SDGs are addressed by circular bioeconomy Fig. 3.4. SDGs linked with the socioeconomic targets are 1, 2, 3 and 8, while the five SDGs (6, 13, 14 and 15) fall under the umbrella of ecological targets. The remaining three SDGs (7, 9 and 12) are concerned with economic and clean industry (Heimann 2019). It is widely accepted that in order to attain these determined global targets, a business-as-usual model is not supposed to work. Policies need to be revived, and production and consumption practices will need to be changed. These global policies and agreements say a go ahead to transform our current economic model into one which includes natural capital in the way we advance societal well-being. One of the important elements in this change is the move towards circular bioeconomy, which encourages the use of renewable, non-fossil raw materials and products in a circular, resource-efficient

Fig. 3.4 Sustainable developmental goals addressed by circular bioeconomy. Clean industry and economic targets, ecological targets and socioeconomic targets are represented by red, green and blue squares, respectively

and sustainable pattern (Hetemaki et al. 2017). Overall, the circular bioeconomy has been correlated with both positive as well as negative impacts on SDG targets. Although targets for cleaner industrial production are strongly reinforced, socioeconomic targets are subjected to varied effects, while, the environmental targets are considerably deteriorated (Heimann 2019).

3.1.5.2 Economic and Ecological Roles of Circular Bioeconomy

Current bioeconomy strategies have been successful in demonstrating the use of renewable biomass to substitute to fossil-based raw materials and products for a more sustainable society. A bioeconomy has powerful contributions to sustainable economy in general and engages many services and industries, such as food, health, housing, clothing and transportation. The developing cities around the world are working on incorporating the principles of bioeconomy in order to reach a standard level of sustainable living, consumption and providing services, from fresh drinking water and food supply to recreation and urban cooling (Hetemaki et al. 2017).

According to the Joint Research Centre (JRC) report published in 2017, the EU28 bioeconomy yielded an annual turnover of about EUR 2.2 trillion in 2014 and has provided employment opportunities to 18.6 million people (JRC 2017). This report highlights the contributions of bioeconomy and its employment potentials in different sectors like agriculture, food, tobacco, beverage and forestry. The forest-based bioeconomy has significant contributions in different sectors like biofuels, bio-based electricity, chemicals, pharmaceuticals, plastics, manufacture of paper and paper-based products, textiles, wood products and furniture. In 2014, a turnover of around EUR 700 billion with an employment potential of 4.3 million was recorded for these sectors by adhering to the principles of bioeconomy (Hetemaki et al. 2017). In addition, bioeconomy also plays a significant role in food sector. Food-based industries valorize their biogenic waste/side streams. It includes side streams of the milk, cheese and alcohol industries to generate organic acids and bio-based plastics and side streams of orange juice and olive industries, which can be used for the extraction of high value organic components as well as biomolecules including proteins and fatty acids from the fish processing industry (Carus and Dammer 2018). Circular bioeconomy is playing its due role in minimizing the greenhouse gases emission and fossil fuels consumption in order to establish an eco-friendly planet for the survival of mankind.

3.2 Diversity of Biomass in Circular Bioeconomy Context

3.2.1 Significance of Biomass in Circular Bioeconomy

The total animal- and plant-derived waste mass in a specific location per unit time is termed as biomass (Rasheed et al. 2021). Usually, its amount and mass decrease with frequent consumption by the end users. The sustainable biomass is nurtured and picked in a way that becomes sustainable to the environment, and no more remains as mere assembly and devastation of waste (Titus et al. 2021). The biomass is extremely substantial in a circular economy for the production of useful items and energy. For efficient waste management, it is needed to set up a valuable circular bioeconomy for the inferences of biomass uses (Sherwood 2020). Circular bioeconomy focuses on the resource-efficient and sustainable valorization (waste conversion into valuable products) of biomass in integrated chains of biorefineries. The bioeconomy transforms the renewable biological products leading to energy production. Furthermore, the resources are also converted to useful and value-added products like feed, food and bioenergy.

3.2.2 *Variety of Biomass in Circular Bioeconomy*

3.2.2.1 Aquatic Plants and Microalgae

Aquatic biomass includes microalgae, seaweeds and aquatic plants. Hydrothermal liquefaction process of aquatic biomass has received ample attention for microalgae because they offer remarkable per area yields of biomass (Hajinajaf et al. 2021). The aquatic biomass is rich in protein and oil with a potential for rapid regrowth. The actual size of microalgae ranges from one micron to two millimetres. Among other aquatic plants, the *Spirulina* and *Chorella* species are famous for potential application in circular bioeconomy. The microalgae yield per unite area is higher than any other agricultural product and can be used as an entity for animal/poultry and fish feed. Furthermore, they are used for high-quality applications such as medications and food supplements and also provide raw material for cosmetics and energy. Alga is a biofuel feedstock and is a resourceful foundation of biomass for the production of biohydrogen which is most reliable carrier of sustainable energy (Nogueira et al. 2020).

Biomass-built biofuels are substitute to fossil fuels. The major hindrance in developing this biofuel is the availability of water and a handful source of carbon. Furthermore, harvesting of the biomass in liquid medium is another challenge. The circular bioeconomy is a smart idea for 3Rs representing reduce, reuse and recycle. The nutrient regaining waste product via microalgae may confer a demand-driven significant role in boosting the existing energy sources and delivery of feed material to the rising population.

One of the main sources of nutrient and carbon is the wastewater from various livestock and agriculture industries, and this asset of the nature can be recovered and employed via circular bioeconomy. Yet, the wastewater utilization appears to be a desired approach for bulk farming and culturing of microalgae in order to reduce the consumption of freshwater, removal of pollutants from wastewater and producing sustainable biomass (Hajinajaf et al. 2021). In the presence of sunlight (photosynthesis), the microalgae consume micronutrients, CO_2, nitrogen, phosphorus, and potassium, while release O_2 and biomass to the environment (Lorentz et al. 2020). It provides additional benefits as a prebiotic which supports the aquaculture in boosting immunity and emending gut physiology. The products of microalgae have favourable utility in the fields of cosmetics, pharmaceuticals, nutraceuticals, feed, health and renewable energy (Vieira et al. 2020).

3.2.2.2 Plant-Derived/Lignocellulosic Biomass

Plant products are being used extensively for the energy production. The woods are reused and cycled for feed, food, nutraceuticals, energy and other synthetic products (Mettu et al. 2020). Wood effectively supports to the economy, and several commodities including papers household gas, coal, medication (such as aspirin), and highly demanded furniture are developed from it (Devi et al. 2020). Some the associated disadvantages are around 50% water content of its stuff making it difficult to carry due to heavy weight and high transportation cost. Its burning can pollute the environment and produces smell and smokes (Cheng et al. 2021).

The term cash crop is used for crops which are basically cultivated to earn revenue like wheat, rice, sugarcane, cotton, corn, peanut and coconut. The biomass of such crops is mainly thrown in the field not being consumed for some productive purpose, rather subjected to fire in the open field. Such practice may pollute the air for humans and other living organisms. While no more trend is witnessed applying technology for biomass conversion of cash crop waste like sugarcane bagasse, wheat straw, rice husk, etc., to other useful residues. Such biomass may be upgraded to gain energy such as electricity, steam, gas, heat, oil, and alcohol (Wang et al. 2021). Furthermore, such products may be converted into paper, textiles, insulator and feed. Most of the soybeans, rapeseed and vegetables waste oil are used for biodiesel production, and mixed waste is exploited to generate electricity, gas, methanol and thermal energy, while wood waste, corn and rice straw and sugar crops are used for ethanol and charcoal production. As far as non-energy products are concerned, the cane residues, wheat straw and cotton are converted into animal feed, beverages and fertilizers. Rice husk is used for the production of synthetic lubricants, silica and water-soluble oil, cane bagasse and wheat straw for panel boards and lumber materials (Crini et al. 2020; Mettu et al. 2020).

3.3 Biorefineries for Sustainable Waste Valorization: The Mainstay of Circular Bioeconomy

3.3.1 Types of Biorefineries

A biorefinery is a facility enabling the biomass conversion processes and apparatus to yield fuels, power and chemicals from biomass (Soetaert 2009). The present-day mechanized world is seriously confronting with the issues of global warming and greenhouse effect due to high carbon emissions especially in extensively industrialized countries. The emergence of the concept and application of "biorefineries" to meet the ever-rising human needs has led to the exploration of this field in the various aspects of life. Biorefineries are divided into several types, and the salient aspects of some important biorefineries have been discussed in the following lines.

3.3.1.1 Agricultural Biorefinery

This type of biorefinery has its application in the various agricultural wastes and biomass produced in the developing countries such as wastes of crops, cereals, vegetables, fruits, tea seeds, tobacco seeds, date seeds, palm trees, linseed, rice straw, wheat straws and leaves of plants (Cardoen et al. 2015). These wastes can be transformed by agricultural biorefinery into products like bioethanol and biohydrogen from cellulose and hemicellulose, respectively (Kaparaju et al. 2009). Likewise, it can also be used to produce activated carbon, biochar, organic fertilizer, natural fibre composites and nanocomposites under agriculture waste biorefinery concept (Rathore et al. 2016).

3.3.1.2 Forestry Biorefinery

Forests occupy almost 32% of the entire area of land producing nearly 89.3% of the total existent biomass and 42.9% of total annual biomass (Ismail and Nizami 2016). Though the forestry biomass is currently being utilized for preparing lumber, paper and certain wood products, it still has the potential to furnish persistent source of feed resources to waste biorefinery. It possesses lignocellulosic materials, such as lignin, cellulose, hemicellulose and various extractives. This biorefinery can transform waste pulp and paper mill residues to several fuels and chemicals with value addition increasing the revenues without damaging forest flora (Pande and Bhaskarwar 2012).

3.3.1.3 Algae-Based Biorefinery

Microalgae have been employed for the production of fuels and feed for animals as the sources of polyunsaturated fatty acids, polysaccharides, polyamines and carotenoid pigments (Rathore et al. 2016). Their high lipid content is an excellent opportunity for biodiesel, bioethanol and biogas synthesis especially useful in areas with abundant water for their growth. Such a biorefinery would be a significant source of alternate energy in the coastal areas of developing countries (Nizami et al. 2016b). The wastewater from another agriculture-based activity in an integrated biorefinery system can also be utilized. Microalgae have also the ability of bio-fixation using CO_2 and thus reduce the atmospheric CO_2 emissions.

3.3.1.4 Industrial or Municipal Waste-Based Biorefinery

This type of biorefinery has a pivotal role in the contemporary circular economy, and as an example, cassava-based industrial waste under the concept of the waste biorefinery has been investigated (Zhang et al. 2016). Cassava is a low-priced source of food carbohydrate energy after rice, sugarcane and maize and is used as a staple food in African countries. It is the third largest mean of carbohydrate available to humans, 60% used for humans, 33% as animal feed and 7% in textile, paper, food and fermentation industries. It has also evolved as a chief source of starch for bioethanol production and other bio-based chemicals owing to being cheap and surplus in availability, and in China, cassava-based alcohol entrepreneurs produce ethanol > 0.4 million metric tons every year. This biorefinery can be very beneficial in transforming the waste food, catered food and expired foods to various fuels and other bio-based products.

3.3.1.5 Animal Waste Biorefinery

The potential aspect of fat extraction in animal waste biorefinery has been evaluated on the municipal solid waste particularly animal waste and cooking oil waste. It was concluded that overall net savings based on landfill diversion, carbon credits, fuel savings, and electricity generation may touch to US $163 million and US $340 million in 2014 and 2050, respectively, only from the fat fractions of municipal solid waste in the city of Makkah, Saudi Arabia (Shahzad et al. 2017).

3.3.2 Significance of Biorefining in Circular Bioeconomy

The usage of biomass as feedstock has been there in the past; but a renewed focus towards the efficient and effective manipulation of inevitable organic wastes is aimed at reducing eco-footprint and ensuring consistent availability of sustainable resources (Mohan et al. 2016b). The human goal of valorizing wastes has led towards the flourishing of bioeconomy via squeezing the wasted mass. The production of ethanol through dark fermentation of biohydrogen is one such relevant example having significant commercial value. In addition, certain by-products obtained during the process may act as precursors for the synthesis of other products of economic value such as polyhydroxyalkanoates (bioplastic), biohythane, lipids and electricity, besides serving as phosphate-soluble fertilizers (Sarma et al. 2015).

3.3.3 Socioeconomic and Environmental Impacts of Sustainable Waste Valorization

Installing waste biorefineries in developing countries is highly anticipated to change the linear economies towards circular economies along with paying to rectify public health and environment. This can be had via installing more advanced and maintainable biorefinery set ups and changing the gears from current fossil fuels utility to recyclable and environmentally friendly options. Applying biorefinery set-up can yield the type of fuel, based on nature and accessibility of resources (Mohan et al. 2016a, b). In short, bringing such set-ups connected to an integrated waste biorefinery approach can help to process varied type of feedstock into diverse products like food, feed, fuel, power and heat, besides the value addition of chemicals (Posada et al. 2013).

These approaches can be transformed into novel models and scopes at various disciplines of agriculture, food industry, chemicals, pharmaceuticals and logistics as well as health care (Amulya et al. 2016). Among many developing nations, it is pertinent and essential in the scenario of environmental and economic loads inflicted by the existing waste management approaches and coping up the rising energy requirements besides providing new entrepreneurs, job markets and enhancing public health standards and the regional ambience (Ismail and Nizami 2016).

The setting up of biorefineries is highly anticipated to mitigate the environmental contaminants using locally available recyclable resources to yield lasting energy and goods.

The chief environmental gain would be the mitigation of greenhouse gases emissions and reducing the environmental change on their part (Mohan et al. 2016b). Moreover, decreased landfilling and minimizing the harmful effects of biowaste on ambience and inhabitant's health, uplifting agricultural industry and recyclable energy synthesis are additional benefits of biorefining. The energy conserved through waste biorefining would otherwise be consumed for the production and carriage of raw materials, and manufacturing of final products (Ouda et al. 2016).

3.4 Implementation of Circular Bioeconomy: Opportunities and Challenges

3.4.1 Opportunities

Contrary to the higher processing cost and rapid depletion of fossil-derived carbon sources, the relatively lower cost of renewable biomass and technological advancement in biorefining indicate optimal conditions for the development of circular bioeconomy (Kircher 2012; Brown and Feuvre 2017). The potential opportunities for transition to circular bioeconomy have been summarized in Table 3.1. Various useful outcomes in terms of improved crop productivity, food security, efficient use of natural resources and provision of job opportunities have been attributed to bio-based products such as biochar, biofertilizers, biofuels, biopesticides and biological agents for aflatoxin neutralization (Feleke et al. 2021). The use of biological resources and resultant bio-based products are considered to mitigate climate change through the considerable reduction of greenhouse gases emission (Leskinen et al. 2018; Spierling et al. 2018). Biofuel represents an excellent candidate for substituting or supplementing the fossil-based fuel to fulfil the progressively growing energy needs of electrical, thermal and transport sectors. Although, higher food prices, food insecurity and poverty have been attributed to biofuel production. Nevertheless, the negative impacts of biofuel generation on food security can be minimized by exploiting feedstock with labour-intensive production and technological advancement for second-generation biofuels. Industrial symbiosis represents a successfully implemented, typical model of circular bioeconomy. Likewise, the recycling and cascading use of glass, plastics,

Table 3.1 Prospects for transition to circular bioeconomy

• Decomposable, environment-friendly and sustainable nature of bio-based products
• Expansion of niche markets for bio-based health supplements, cosmetics and biofuels
• Analogies between petroleum refinery and biorefinery in terms of feedstock and products and potential substitution of several petrochemical products by comparable bio-based products
• Convenient and effective assessment of operational processes, socioeconomic attributes and environmental impacts of integrated biorefineries by means of computer-aided simulations
• Improved efficiency of woody biomass-driven plants through the concomitant combustion of microalgal biomass and coal
• Reduced dependency on biomass production through the efficient utilization of organic waste
• Proven valorization methods for the conversion of biowaste into diverse bioactive compounds with potential applications in food, agriculture, pharmaceutical, industrial and bioenergy sectors
• Conversion of gaseous carbon-rich molecules into low-carbon fuels, polymers and commodity chemicals using gas fermentation technology
• Integrated farming and soilless culture systems for the avoidance of food scarcity biodiversity disruption, soil degradation and deforestation
• Opportunities for public and private investments and employment generation in terms of sustainable value chains and new business models

ceramics and construction materials have already been practiced. Fortunately, the well-established niche markets for many bio-based cosmetics, health supplements and biofuels are progressively rising (Nagarajan et al. 2021). Biorefining constitutes an effective value chain through the manufacturing of different valuable products from a variety of biological sources with diverse applications and market values (Gundekari et al. 2020; Ummalyma et al. 2020). Besides capable of providing the alternatives for nearly all fossil-based products, the biorefineries are also known to minimize the production cost and pollution (Nagarajan et al. 2021). Moreover, the continuous input of low-cost feedstock and regular supply of products into the market can provide opportunities for public and private investments in terms of sustainable value chains and new business models.

Huge amounts of biomass can be consumed either directly or with minimal pre-treatment in various industries associated with the generation of electricity, construction material, food, furniture, home appliances, packaging and toys (Ballesteros et al. 2018; Evans 2019; Churkina et al. 2020). For example, the isolated biomass constituents including peptides, lignin, starch, cellulose and hemicellulose can be utilized in the manufacturing of furniture, paper and stabilizers (Ayyachamy et al. 2013). Likewise, treated ingredients, such as dissolved polymers, microcrystalline cellulose and nano-cellulose have prospective applications in the production of aerogels, plastics, clothing and composites (Ferreira et al. 2019; Bilal et al. 2020). Whereas, platform chemicals can be employed in the cosmetic and pharmaceutical industries.

Several studies have exploited organic waste valorization for the synthesis of established products such as bioethanol, biofertilizers, biosurfactants, bioplastics, butanol and lactic acid (Vea et al. 2018; Ramesh Kumar et al. 2020; Sari et al. 2020). Moreover, organic waste can also yield bioactive compounds including carotenoids, flavonoids, pectin, polyphenols and vitamins for food and pharmaceutical industries (McElroy et al. 2018). Besides, semi-solid and solid-state fermentation techniques have been successfully used for the bioconversion of food waste into various types of enzymes like cellulase, glucoamylase, lipase and protease, and biopesticides (Lee et al. 2020; Sharma et al. 2020). A wide range of bioenergy products including biodiesel, biogas, char, electricity, hydrogen, methane and syngas can be generated from organic waste using appropriate and systemic approaches (Tsang et al. 2018; Algapani et al. 2019). Such products can be either turned into industrially useful chemicals like furfurals, guaiacols and levoglucosenone or further processed as biofuels for the substitution of fossil-derived products (Maina et al. 2017; Pang 2019). Besides decreasing the cost of feedstock, the biowastes and residues have been linked with more efficient diminution of greenhouse gases emission than the primary biomass (Creutzig et al. 2015).

Although, typically focusing on carbon-containing bio-based materials, the circular bioeconomy can potentially utilize the waste feedstock of non-biological origin. For instance, the gaseous carbon-rich molecules can be converted into low-carbon fuels for aviation and road transport, single-celled proteins, specialty chemicals, polymers and commodity chemicals through the application of gas fermentation technology (Brandao et al. 2021).

Contemporary scientific advancements and technological innovations can provide rational strategies for biomass production, improved biorefining processes and generation of superior quality bio-based products. Circular bioeconomy-based models can minimize the dependence on biomass production through effective utilization of municipal and agro-industrial organic waste (de Schouwer et al. 2019; Gajula et al. 2019; Arun et al. 2020). The application of multifunctional technologies and integrated approaches for the conversion of organic waste into marketable bio-based products is gaining considerable attention. Substantial amounts of agro-residual and municipal waste are available for energy production, and more promising bioenergy utilization systems are being developed (Tollefson 2017). Minor changes in the infrastructure can enable the simultaneous combustion of coal and microalgal feedstock for alleviating the lower efficiency of woody biomass-operated plants (Lane et al. 2014; Choi et al. 2019). More efficient catalysts and preservatives are being offered for optimizing the bioconversion reactions and stabilizing the shelf life of bio-based products, respectively (Patel et al. 2020; Gajula and Reddy 2021). Modern information technologies such as artificial intelligence and software tools can facilitate the collection and analysis of data for smooth implementation and integration of circular bioeconomy process. Moreover, a digital platform organization known as circularity broker has been suggested for the convenient and adequate retrieval of discarded material from food waste (Ciulli et al. 2020). Computer-assisted simulations can be employed for investigating the operational, socioeconomic and environmental feasibilities of integrated biorefineries (Julio et al. 2017; Ioannidou et al. 2020). Integrated farming and soilless culture systems are available to avoid the potential issues of biodiversity disruption, deforestation, food scarcity and soil degradation, particularly in regions with limited resource capacity and unfavourable climatic conditions.

Table 3.2 Challenges of circular bioeconomy implementation

• Relatively expensive procurement and pre-treatment of biomass
• Uninterrupted supply of biomass for the operation of biorefineries
• Unsuitability of currently available biomass for biorefining on account of its undesirable composition, inferior growth rate, low energy potential, putrescible nature and seasonal occurrence
• Complicated processing and substandard yield of biowaste
• Shrinkage of forests, grasslands and savannahs, with subsequent increase in greenhouse gas emissions and biodiversity losses
• Conflicting food versus non-food consumption of biomass
• Technological advancement and development schemes to facilitate the cascading use of biomass for manufacturing bio-based products
• Optimization of emissions resulting from biowaste processing
• Organization of market demands and supply chain
• Consumer awareness and legislative support

3.4.2 Challenges

The shift towards a circular bioeconomy is assumed more challenging than the previous transformation of pre-modern bioeconomy into recent fossil-based economy (Kircher 2012). Despite the technically feasible substitution of fossil resources by bio-based feedstock in nearly all industries, the proper adoption of this value chain necessitates thorough consideration of certain constraints. The challenges of circular bioeconomy implementation have been outlined in Table 3.2.

Being the most indispensable resource for circular bioeconomy, the biomass should be either produced locally or imported in sufficient quantities. Uninterrupted supply of biomass is essential for biorefining (Dale 2017). Moreover, the quality and type of biomass also differ with the nature of bio-based production process. The currently available biomass is less suitable for biorefineries owing to its undesirable composition, inferior growth rate and low energy potential. The food versus non-food utilization of biomass demonstrates a major conflict that requires strategic solution in terms of circular bioeconomy (Issa et al. 2019). Another drawback is the putrescible nature of biomass, which necessitates its prompt collection, transportation and consumption to avoid decomposition (Lin et al., 2013). Besides, the accessibility of sufficient feedstock is also affected by market volatility and seasonal occurrence of biomass.

Further expansion of croplands is requisite to fulfil the higher demand of non-food biomass for the development of circular bioeconomy. The shrinkage of forests, grasslands and savannahs, with subsequent increase in greenhouse gases emissions and biodiversity losses, has been projected as the potential consequence of growing competition for croplands, predominantly in tropical regions (Bringezu et al. 2009).

The randomly disseminated, huge amounts of specific organic waste necessitate organized relocation to a central place that further augments the cost of transportation and ultimately depreciates the economic feasibility of biorefining (Lin et al. 2013). Moreover, the complicated processing, high production cost and substandard yield have been ascribed to physico-chemical variability of biowaste streams (Yamakawa et al. 2018). The emissions resulting from collection, segregation and processing of biowaste stream and those associated with the generation of bio-based products should be optimized to minimum possible levels.

The cost-effectiveness and regulation of bio-based production are also influenced by unstable oil prices. Accordingly, the sustainability assessment of renewable energy is recommended in the circular bioeconomy framework (Mohan et al. 2019). Additionally, the recycling and cascading processes can be hampered by the probable accumulation of toxic substances. Therefore, physico-chemical compatibility and ecological sustainability should be emphasized in conjunction with the circularity and cascading use of biological sources (Mohan et al. 2019). Ideally, the bio-based products should be either superior or at least equivalent to petrochemical counterparts in terms of quality, production cost and market value. Nevertheless, eutrophication and high-water content have been implicated in deteriorating the sustainability and environmental safety of bio-based products (Van den Oever et al. 2017).

Besides, the adoption of circular bioeconomy is further confounded by the dependence of global economy on a well-established carbon-based industry (de Assis et al. 2017). Therefore, strong policies, regulatory interventions and legislative measures are vital for upgrading the socioeconomic and environmental impacts of bio-based products. Most circular bioeconomy strategies are confronting difficulties in the production of bio-based materials following the application of latest technologies under strict regulatory restrictions (Purnhagen and Wesseler 2020). Consequently, suitable reforms are urgently required for the approval of modern techniques and new products at international level (Eriksson et al. 2019). Finally, the lack of adequate public awareness and consumer interest have also been identified as the potential grey areas for further improvements (de Besi and McCormick 2015).

3.5 Conclusions and Future Perspectives

Circular bioeconomy is capable of accomplishing the increasing needs for the subsistence of growing population in the global context of finite natural resources. The circular bioeconomy-based efficient and cascading use of biological feedstock, limited consumption of finite resources and development of eco-safe, bio-based production system require suitable alternatives for fulfilling the needs of communities. Hence, the technical analysis and modification of existing fossil-dependent global economy can facilitate its conversion into circular bioeconomy. For that reason, circular bioeconomy is considered as a priority area for extensive research, sustainable development and regulatory actions in various regions of the world. The European Union has emphasized the significance of bio-based industrial development for the mitigation of environmental pollution. Several fossil-derived products are either completely replaced by bio-based counterparts or minimally used in many countries.

However, the existing technological and production systems cannot support the development of a circular bioeconomy for the total replacement of non-renewable and fossil-based sources by biomass. Still, the society shows a general agreement on minimizing the generation of waste to ensure the long-term productivity of natural systems. Comprehensive research is needed for upgrading the efficiency, sustainability and environmental friendliness of biorefinery technologies. The cascade use of biological resources can be difficult to execute particularly in small domestic economies. Although few countries of the world have abundant resources, capital, technology and infrastructure for immediate transition to a circular bioeconomy, other regions are still deficient in basic needs. Therefore, circular bioeconomy models should be formulated to minimize the discrepancies among countries and those between the urban and rural areas. Circular bioeconomy can also contribute significantly towards the sustainable growth in developing countries. The growing crisis of organic waste with potential impacts on health and environment can be efficiently turned into valuable opportunities for recycling and bioprocessing. The circular bioeconomy principles should be implemented in global north and south in accordance with the universal projection of SDGs. Additionally, distinct strategies focusing the public and private sectors are suggested for the progression of intra-territorial circular bioeconomy.

Interdisciplinary, multidisciplinary as well as transdisciplinary approaches are critical for the progression of circular bioeconomy. This can be accomplished through mobilizing the investments towards scientific innovation, technology transfer and social development. The concerns in circular bioeconomy implementation can be tackled by means of improved biomass production and biorefining models. Moreover, the development of improved methods for collection, separation, storage, handling and processing of biological feedstock should be targeted by future research. The seasonal availability of different feedstocks and prolonged downtime period of biorefineries can be addressed by the introduction of emerging technologies with capability to process multiple kinds of feedstock. Appropriate policies and incentives are requisite concerning the types of biomass and cascading pathways associated with highest reduction in greenhouse gases emission. The cumulative impact of different biomass consumptions and potential advantages of various end-of-life strategies should be evaluated in the framework of broader economy through an integrated analysis approach. Global contribution towards the scientific, technological and socioeconomic advancements is proposed for circular bioeconomy development in accordance with SDGs. The explicit participation of all stakeholders including academic institutions, private companies and government authorities as well as their interrelationships also requires comprehensive assessment.

References

Abd Manaf L, Samah MA, Zukki NI (2009) Municipal solid waste management in Malaysia: practices and challenges. Waste Manag 29(11):2902–2906

Algapani DE, Qiao W, Ricci M, Bianchi D, Wandera SM, Adani F, Dong R (2019) Bio-hydrogen and bio-methane production from food waste in a two-stage anaerobic digestion process with digestate recirculation. Renew Energy 130:1108–1115

Al-Khatib IA, Arafat HA, Daoud R, Shwahneh H (2009) Enhanced solid waste management by understanding the effects of gender, income, marital status, and religious convictions on attitudes and practices related to street littering in Nablus-Palestinian territory. Waste Manag 29(1):449–455

Amulya K, Dahiya S, Mohan SV (2016) Building a bio-based economy through waste remediation: innovation towards sustainable future. In: Prasad MNV (ed) Bioremediation and bioeconomy. Elsevier, pp 497–521

Amulya K, Jukuri S, Mohan SV (2015) Sustainable multistage process for enhanced productivity of bioplastics from waste remediation through aerobic dynamic feeding strategy: process integration for up-scaling. Bioresour Technol 188:231–239

Arun KB, Madhavan A, Sindhu R, Binod P, Pandey A, Reshmy R, Sirohi R (2020) Remodeling agro-industrial and food wastes into value-added bioactives and biopolymers. Ind Crops Prod 154:112621

Ayyachamy M, Cliffe FE, Coyne JM, Collier J, Tuohy MG (2013) Lignin: untapped biopolymers in biomass conversion technologies. Biomass Convers Bior 3(3):255–269

Ballesteros LF, Michelin M, Vicente AA, Teixeira JA, Cerqueira MA (2018) Use of lignocellulosic materials in bio-based packaging. In: Navard P (ed) Briefs in molecular science—biobased polymers. Springer, Cham, pp 65–85

Bayer IS, Guzman-Puyol S, Heredia-Guerrero JA, Ceseracciu L, Pignatelli F, Ruffilli R, Cingolani R, Athanassiou A (2014) Direct transformation of edible vegetable waste into bioplastics. Macromolecules 47(15):5135–5143

Bilal M, Rasheed T, Nabeel F, Iqbal HM (2020) Bionanocomposites from biofibers and biopolymers. In: Khan A, Rangappa SM, Siengchin S, Asiri AM (eds) Biofibers and biopolymers for biocomposites. Springer, Cham, pp 135–157

Bocken NM, Olivetti EA, Cullen JM, Potting J, Lifset R (2017) Taking the circularity to the next level: a special issue on the circular economy. J Ind Ecol 21:476–482

Brandao AS, Gonçalves A, Santos JM (2021) Circular bioeconomy strategies: from scientific research to commercially viable products. J Clean Prod 295:126407

Bringezu S, Schutz H, Arnold K, Merten F, Kabasci S, Borelbach P, Michels C, Reinhardt GA, Rettenmaier N (2009) Global implications of biomass and biofuel use in Germany—recent trends and future scenarios for domestic and foreign agricultural land use and resulting GHG emissions. J Clean Prod 17:S57–S68

Brown A, Le Feuvre P (2017) Technology roadmap: delivering sustainable bioenergy. International Energy Agency, Paris, France, p 94

Brunner PH, Rechberger H (2015) Waste to energy—key element for sustainable waste management. Waste Manag 37:3–12

Cardoen D, Joshi P, Diels L, Sarma PM, Pant D (2015) Agriculture biomass in India: Part 1. Estimation and characterization. Resour Conserv Recycl 102:39–48

Carus M (2017) Biobased economy and climate change—important links, pitfalls, and opportunities. Ind Biotechnol 13(2):41–51

Carus M, Dammer L (2018) The circular bioeconomy—concepts, opportunities, and limitations. Ind Biotechnol 14(2):83–91

Chandrasekhar K, Kumar S, Lee BD, Kim SH (2020) Waste based hydrogen production for circular bioeconomy: current status and future directions. Bioresour Technol 302:122920

Cheng F, Small AA, Colosi LM (2021) The levelized cost of negative CO_2 emissions from thermochemical conversion of biomass coupled with carbon capture and storage. Energy Convers Manag 237:114115

Choi HI, Lee JS, Choi JW, Shin YS, Sung YJ, Hong ME, Kwak HS, Kim CY, Sim SJ (2019) Performance and potential appraisal of various microalgae as direct combustion fuel. Bioresour Technol 273:341–349

Chokshi K, Pancha I, Ghosh A, Mishra S (2016) Microalgal biomass generation by phycoremediation of dairy industry wastewater: an integrated approach towards sustainable biofuel production. Bioresour Technol 221:455–460

Churkina G, Organschi A, Reyer CP, Ruff A, Vinke K, Liu Z, Reck BK, Graedel TE, Schellnhuber HJ (2020) Buildings as a global carbon sink. Nat Sustain 3(4):269–276

Ciulli F, Kolk A, Boe-Lillegraven S (2020) Circularity brokers: digital platform organizations and waste recovery in food supply chains. J Bus Ethics 167(2):299–331

Creutzig F, Ravindranath NH, Berndes G, Bolwig S, Bright R, Cherubini F, Chum H, Corbera E, Delucchi M, Faaij A, Fargione J (2015) Bioenergy and climate change mitigation: an assessment. Glob Change Biol Bioenergy 7(5):916–944

Crini G, Lichtfouse E, Chanet G, Morin-Crini N (2020) Applications of hemp in textiles, paper industry, insulation and building materials, horticulture, animal nutrition, food and beverages, nutraceuticals, cosmetics and hygiene, medicine, agrochemistry, energy production and environment: a review. Environ Chem Lett 18:1451–1476

Cuetos MJ, Gomez X, Otero M, Morán A (2008) Anaerobic digestion of solid slaughterhouse waste (SHW) at laboratory scale: influence of co-digestion with the organic fraction of municipal solid waste (OFMSW). Biochem Eng J 40(1):99–106

Dahiya S, Kumar AN, Sravan JS, Chatterjee S, Sarkar O, Mohan SV (2018) Food waste biorefinery: sustainable strategy for circular bioeconomy. Bioresour Technol 248:2–12

Dale BE (2017) Feeding a sustainable chemical industry: do we have the bioproducts cart before the feedstocks horse? Faraday Discuss 202:11–30

de Assis CA, Gonzalez R, Kelley S, Jameel H, Bilek T, Daystar J, Handfield R, Golden J, Prestemon J, Singh D (2017) Risk management consideration in the bioeconomy. Biofuel Bioprod Biorefin 11(3):549–566

de Besi M, McCormick K (2015) Towards a bioeconomy in Europe: national, regional and industrial strategies. Sustainability 7(8):10461–10478

de Schouwer F, Claes L, Vandekerkhove A, Verduyckt J, De Vos DE (2019) Protein-rich biomass waste as a resource for future biorefineries: state of the art, challenges, and opportunities. Chemsuschem 12(7):1272–1303

Devi VC, Mothil S, Raam RS, Senthilkumar K (2020) Thermochemical conversion and valorization of woody lignocellulosic biomass in hydrothermal media. In: Kumar RP, Bharathiraja B, Kataki R, Moholkar VS (eds) Biomass valorization to bioenergy. Springer, Singapore, pp 45–63

Eriksson D, Kershen D, Nepomuceno A, Pogson BJ, Prieto H, Purnhagen K, Smyth S, Wesseler J, Whelan A (2019) A comparison of the EU regulatory approach to directed mutagenesis with that of other jurisdictions, consequences for international trade and potential steps forward. New Phytol 222(4):1673–1684

European Commission (2012) Bioeconomy strategy: innovating for sustainable growth: a bioeconomy for Europe. European Commission, Brussels, Belgium

European Commission (2016) Mapping of EU member states'/regions' research and innovation plans and strategies for smart specialization (RIS3) on bioeconomy

Evans J (2019) Building with biosolids. Biofuels Bioprod Biorefin 13:434

Fehrenbach H, Köppen S, Kauertz B, Detzel A, Wellenreuther F, Brietmayer E, Essel R, Carus M, Kay S, Wern B (2017) Biomass cascades: increasing resource efficiency by cascading use of biomass—from theory to practice. German Environmental Agency, Heidelberg, Germany, p 29

Feleke S, Cole SM, Sekabira H, Djouaka R, Manyong V (2021) Circular bioeconomy research for development in sub-Saharan Africa: innovations, gaps, and actions. Sustainability 13(4):1926

Ferreira FV, Pinheiro IF, de Souza SF, Mei LH, Lona LM (2019) Polymer composites reinforced with natural fibers and nanocellulose in the automotive industry: a short review. J Compos Sci 3(2):51

Gajula S, Antonyraj CA, Odaneth AA, Srinivasan K (2019) A consolidated road map for economically gainful efficient utilization of agro-wastes for eco-friendly products. Biofuel Bioprod Biorefin 13(4):899–911

Gajula S, Reddy CR (2021) More sustainable biomass production and biorefining to boost the bioeconomy. Biofuels Bioprod Biorefin

Gregg JS, Jürgens J, Happel MK, Strøm-Andersen N, Tanner AN, Bolwig S, Klitkou A (2020) Valorization of bio-residuals in the food and forestry sectors in support of a circular bioeconomy: a review. J Clean Prod 11:122093

Gundekari S, Mitra J, Varkolu M (2020) Classification, characterization, and properties of edible and non-edible biomass feedstocks. In: Hussain CM, Sudarsanam P (eds) Advanced functional solid catalysts for biomass valorization. Elsevier, Oxford, pp 89–120

Gupta N, Yadav KK, Kumar V (2015) A review on current status of municipal solid waste management in India. J Environ Sci 37:206–217

Hajinajaf N, Mehrabadi A, Tavakoli O (2021) Practical strategies to improve harvestable biomass energy yield in microalgal culture: a review. Biomass Bioenergy 145:105941

Heimann T (2019) Bioeconomy and SDGs: does the bioeconomy support the achievement of the SDGs? Earth's Future 7(1):43–57

Henry RK, Yongsheng Z, Jun D (2006) Municipal solid waste management challenges in developing countries—Kenyan case study. Waste Manag 26(1):92–100

Hetemaki L, Hanewinkel M, Muys B, Ollikainen M, Palahí M, Trasobares A, Aho E, Ruiz CN, Persson G, Potoćnik J (2017) Leading the way to a European circular bioeconomy strategy, vol 5. European Forest Institute

Hoornweg D, Bhada-Tata P (2012) What a waste: a global review of solid waste management. Urban Development Series Knowledge Papers, World Bank

House TW (2012) National bioeconomy blueprint. Ind Biotechnol 8(3):97–102

Iniguez ME, Conesa JA, Fullana A (2016) Marine debris occurrence and treatment: a review. Renew Sustain Energy Rev 64:394–402

Ioannidou SM, Pateraki C, Ladakis D, Papapostolou H, Tsakona M, Vlysidis A, Kookos IK, Koutinas A (2020) Sustainable production of bio-based chemicals and polymers via integrated biomass refining and bioprocessing in a circular bioeconomy context. Bioresour Technol 307:123093

Ismail IM, Nizami AS (2016) Waste-based biorefineries in developing countries: an imperative need of time. In: Canadian Society for Civil Engineering: 14th international environmental specialty conference, London, Ontario, Canada, pp 1–4

Issa I, Delbrück S, Hamm U (2019) Bioeconomy from experts' perspectives—results of a global expert survey. PLoS ONE 14(5):e0215917

Jambeck JR, Geyer R, Wilcox C, Siegler TR, Perryman M, Andrady A, Narayan R, Law KL (2015) Plastic waste inputs from land into the ocean. Sci 347(6223):768–771

Jayathilakan K, Sultana K, Radhakrishna K, Bawa AS (2012) Utilization of byproducts and waste materials from meat, poultry and fish processing industries: a review. J Food Sci Technol 49(3):278–293

John VM, Zordan SE (2001) Research & development methodology for recycling residues as building materials—a proposal. Waste Manag 21(3):213–219

Joint Research Centre (2017) Jobs and Ministry of Economy and competitiveness turnover in the European Union Bioeconomy

Jordahl JL, Foster L, Schnoor JL, Alvarez PJ (1997) Effect of hybrid poplar trees on microbial populations important to hazardous waste bioremediation. Environ Toxicol Chem: Int J 16(6):1318–1321

Julio R, Albet J, Vialle C, Vaca-Garcia C, Sablayrolles C (2017) Sustainable design of biorefinery processes: existing practices and new methodology. Biofuels Bioprod Biorefin 11(2):373–395

Kalmykova Y, Sadagopa M, Rosado L (2018) Circular economy—from review of theories and practices to development of implementation tools. Resour Conserv Recycl 135:190–201

Kaparaju P, Serrano M, Thomsen AB, Kongjan P, Angelidaki I (2009) Bioethanol, biohydrogen and biogas production from wheat straw in a biorefinery concept. Bioresour Technol 100:2562–2568

Kircher M (2012) The transition to a bio-economy: national perspectives. Biofuels Bioprod Biorefin 6:240–245

Kronenberg J, Winkler R (2009) Wasted waste: an evolutionary perspective on industrial by-products. Ecol Econ 68:3026–3033

Lane DJ, Ashman PJ, Zevenhoven M, Hupa M, van Eyk PJ, de Nys R, Karlström O, Lewis DM (2014) Combustion behavior of algal biomass: carbon release, nitrogen release, and char reactivity. Energy Fuel 28(1):41–51

Lee JK, Patel SK, Sung BH, Kalia VC (2020) Biomolecules from municipal and food industry wastes: an overview. Bioresour Technol 298:122346

Leskinen P, Cardellini G, González-García S, Hurmekoski E, Sathre R, Seppala J, Smyth C, Stern T, Verkerk PJ (2018) Substitution effects of wood-based products in climate change mitigation: from science to policy 7. European Forest Institute, Joensuu

Lin CS, Pfaltzgraff LA, Herrero-Davila L, Mubofu EB, Abderrahim S, Clark JH, Koutinas AA, Kopsahelis N, Stamatelatou K, Dickson F, Thankappan S (2013) Food waste as a valuable resource for the production of chemicals, materials and fuels. Current situation and global perspective. Energy Environ Sci 6(2):426–464

Liu Y, Dong L, Guo D, Xu Y, Liu J (2016) Pollution status and environmental sound management (ESM) trends on typical general industrial solid waste. Procedia Environ Sci 31:615–620

Lorentz JF, Calijuri ML, Assemany PP, Alves WS, Pereira OG (2020) Microalgal biomass as a biofertilizer for pasture cultivation: plant productivity and chemical composition. J Clean Prod 276:124130

MacArthur FE (2013) Towards the circular economy, vol 1: an economic and business rationale for an accelerated transition. Ellen McArthur

Maina S, Kachrimanidou V, Koutinas A (2017) From waste to biobased products: a roadmap towards a circular and sustainable bioeconomy. Curr Opin Green Sustain Chem 8:18–23

Mardani A, Streimikiene D, Cavallaro F, Loganathan N, Khoshnoudi M (2019) Carbon dioxide (CO_2) emissions and economic growth: a systematic review of two decades of research from 1995 to 2017. Sci Total Environ 649:31–49

McCormick K, Kautto N (2013) The bioeconomy in Europe: an overview. Sustainability 5(6):2589–2608

McElroy CR, Attard TM, Farmer TJ, Gaczynski A, Thornthwaite D, Clark JH, Hunt AJ (2018) Valorization of spruce needle waste via supercritical extraction of waxes and facile isolation of nonacosan-10-ol. J Clean Prod 171:557–566

Mettu S, Halder P, Patel S, Kundu S, Shah K, Yao S, Hathi Z, Ong KL, Athukoralalage S, Choudhury NR, Dutta NK (2020) Valorisation of agricultural waste residues. In: Lin CSK, Kaur G, Li C, YangJohn X (eds) Waste valorisation: waste streams in a circular economy, Wiley, pp 51–85

Miandad R, Rehan M, Ouda OKM, Khan MZ, Shahzad K, Ismail IMI, Nizami AS (2017) Waste-to-hydrogen energy in Saudi Arabia: challenges and perspectives. In: Singh A, Rathore D (eds) Biohydrogen production: sustainability of current technology and future perspective. Springer, India, pp 237–252

Milea A (2009) Waste as a social dilemma: Issues of social and environmental justice and the role of residents in municipal solid waste management, Delhi, India. Master's thesis, Lund University, Lund, Sweden

Modak P, Wilson DC, Velis C (2015) Waste management: global status. In: Wilson DC, Rodic L, Modak P, Soos R, Carpintero A, Velis K, Iyer M, Simonett O (eds) Global waste management outlook. UNEP, Athens, Greece, pp 51–79

Mohan SV, Butti SK, Amulya K, Dahiya S, Modestra JA (2016a) Waste biorefinery: a new paradigm for a sustainable bioelectro economy. Trends Biotechnol 34(11):852–855

Mohan SV, Dahiya S, Amulya K, Katakojwala R, Vanitha TK (2019) Can circular bioeconomy be fueled by waste biorefineries—a closer look. Bioresour Technol Rep 7:100277
Mohan SV, Nikhil GN, Chiranjeevi P, Reddy CN, Rohit MV, Kumar AN, Sarkar O (2016b) Waste biorefinery models towards sustainable circular bioeconomy: critical review and future perspectives. Bioresour Technol 215:2–12
Mosler HJ, Tamas A, Tobias R, Rodríguez TC, Miranda OG (2008) Deriving interventions on the basis of factors influencing behavioral intentions for waste recycling, composting, and reuse in Cuba. Environ Behav 40(4):522–544
Nagarajan D, Lee DJ, Chang JS (2021) Circular bioeconomy: an introduction. In: Pandey A, Tyagi RD, Varjani S (eds) Biomass, biofuels, biochemicals, Elsevier, pp 3–23
Nizami AS, Mohanakrishna G, Mishra U, Pant D (2016a) Trends and sustainability criteria for the liquid biofuels. In: Singh RS, Pandey A, Gnansounou E (eds) Biofuels: production and future perspectives. CRC Press, New York, pp 59–95
Nizami AS, Ouda OK, Rehan M, El-Maghraby AM, Gardy J, Hassanpour A, Kumar S, Ismail IM (2016b) The potential of Saudi Arabian natural zeolites in energy recovery technologies. Energy 108:162–171
Nogueira LA, Souza GM, Cortez LA, de Brito Cruz CH (2020) Biofuels for Transport. In: Letcher TM (ed) Future energy: improved, sustainable and clean options for our planet, Elsevier, pp 173–197
OECD (2009) The bioeconomy to 2030: designing a policy agenda. OECD Publishing, Paris
Olsson O, Roos A, Guisson R, Bruce L, Lamers P, Hektor B, Thrän D, Hartley D, Ponitka J, Hildebrandt J (2018) Time to tear down the pyramids? A critique of cascading hierarchies as a policy tool. Wiley Interdiscip Rev Energy Environ 7(2):e279
Ouda OK, Raza SA, Nizami AS, Rehan M, Al-Waked R, Korres NE (2016) Waste to energy potential: a case study of Saudi Arabia. Renew Sustain Energy Rev 61:328–3240
Pande M, Bhaskarwar AN (2012) Biomass conversion to energy. In: Baskar C, Baskar S, Dhillon RS (eds) Biomass conversion: the interface of biotechnology, chemistry and materials science. Springer, Berlin Heidelberg, pp 1–90
Pang S (2019) Advances in thermochemical conversion of woody biomass to energy, fuels and chemicals. Biotechnol Adv 37(4):589–597
Patel AK, Dixit P, Pandey A, Singhania RR (2020) Promising enzymes for biomass processing. In: Singh SP, Pandey A, Singhania RR, Larroche C, Li Z (eds) Biomass, biofuels, biochemicals. Elsevier, Oxford, pp 245–271
Philp J, Winickoff D (2018) Realising the circular bioeconomy, OECD science, technology and industry policy papers, No. 60. OECD Publishing, Paris
Posada JA, Patel AD, Roes L, Blok K, Faaij AP, Patel MK (2013) Potential of bioethanol as a chemical building block for biorefineries: preliminary sustainability assessment of 12 bioethanol-based products. Bioresour Technol 135:490–499
Purnhagen, K.; Wesseler, J (2020) EU regulation of new plant breeding technologies and their possible economic implications for the EU and beyond. Appl Econ Perspect Policy
Ramesh Kumar S, Shaiju P, O'Connor KE (2020) Bio-based and biodegradable polymers—state-of-the-art, challenges and emerging trends. Curr Opin Green Sustain Chem 21:75–81
Rasheed T, Anwar MT, Ahmad N, Sher F, Khan SU, Ahmad A, Khan R, Wazeer I (2021) Valorisation and emerging perspective of biomass based waste-to-energy technologies and their socio-environmental impact: a review. J Environ Manag 287:112257
Rathi S (2006) Alternative approaches for better municipal solid waste management in Mumbai, India. Waste Manag 26(10):1192–1200
Rathore D, Nizami AS, Pant D, Singh A (2016) Key issues in estimating energy and greenhouse gas savings of biofuels: challenges and perspectives. Biofuel Res J 3(2):380–393
Rozin P (2005) The meaning of food in our lives: a cross-cultural perspective on eating and well-being. J Nutr Educ Behav 37:S107–S112

Sadef Y, Nizami AS, Batool SA, Chaudary MN, Ouda OK, Asam ZU, Habib K, Rehan M, Demirbas A (2016) Waste-to-energy and recycling value for developing integrated solid waste management plan in Lahore. Energy Sources Part B 11(7):569–579

Sanneh ES, Hu AH, Chang YM, Sanyang E (2011) Introduction of a recycling system for sustainable municipal solid waste management: a case study on the greater Banjul area of the Gambia. Environ Dev Sustain 13(6):1065–1080

Sari SK, Trikurniadewi N, Ibrahim SN, Khiftiyah AM, Abidin AZ, Nurhariyati T (2020) Bioconversion of agricultural waste hydrolysate from lignocellulolytic mold into biosurfactant by *Achromobacter* sp. BP (1) 5. Biocatal Agric Biotechnol 24:101534

Sarma SJ, Pachapur V, Brar SK, Le Bihan Y, Buelna G (2015) Hydrogen biorefinery: potential utilization of the liquid waste from fermentative hydrogen production. Renew Sustain Energy Rev 50:942–951

Schmitt CC, Moreira R, Neves RC, Richter D, Funke A, Raffelt K, Grunwaldt JD, Dahmen N (2020) From agriculture residue to upgraded product: The thermochemical conversion of sugarcane bagasse for fuel and chemical products. Fuel Process Technol 197:106199

Shahzad K, Nizami AS, Sagir M, Rehan M, Maier S, Khan MZ, Ouda OK, Ismail IM, BaFail AO (2017) Biodiesel production potential from fat fraction of municipal waste in Makkah. PLoS ONE 12(2):e0171297

Sharma P, Kumar Gaur V, Kim SH, Pandey A (2020) Microbial strategies for bio-transforming food waste into resources. Bioresour Technol 299:122580

Sherwood J (2020) The significance of biomass in a circular economy. Bioresour Technol 300:122755

Soetaert W (2009) Defining biorefineries and different concepts. In: Presentation in BioreFuture 2009 conference, Brussels, Belgium

Spierling S, Knüpffer E, Behnsen H, Mudersbach M, Krieg H, Springer S, Albrecht S, Herrmann C, Endres HJ (2018) Bio-based plastics—a review of environmental, social and economic impact assessments. J Clean Prod 185:476–491

Stegmann P, Londo M, Junginger M (2020) The circular bioeconomy: its elements and role in European bioeconomy clusters. Resour Conserv Recycl 6:100029

Stenmarck A, Jensen C, Quested T, Moates G, Buksti M, Cseh B, Ostergren K (2016) Estimates of European food waste levels. IVL Swedish Environmental Research Institute

Stephens E, Ross IL, Mussgnug JH, Wagner LD, Borowitzka MA, Posten C, Kruse O, Hankamer B (2010) Future prospects of microalgal biofuel production systems. Trends Plant Sci 15(10):554–564

Titus BD, Brown K, Helmisaari HS, Vanguelova E, Stupak I, Evans A, Clarke N, Guidi C, Bruckman VJ, Varnagiryte-Kabasinskiene I, Armolaitis K (2021) Sustainable forest biomass: a review of current residue harvesting guidelines. Energy Sustain Soc 11(1):1–32

Tollefson J (2017) The wooden skyscrapers that could help to cool the planet. Nat News 545(7654):280–282

Torretta V, Rada EC, Ragazzi M, Trulli E, Istrate IA, Cioca LI (2015) Treatment and disposal of tyres: two EU approaches. A review. Waste Manag 45:152–160

Tsang YF, Kwon EE, Ok YS, Gao MT, Cao B (2018) Biowaste for energy recovery and environmental remediation. Process Saf Environ 115:1–38

Tue NM, Goto A, Takahashi S, Itai T, Asante KA, Kunisue T, Tanabe S (2016) Release of chlorinated, brominated and mixed halogenated dioxin-related compounds to soils from open burning of e-waste in Agbogbloshie (Accra, Ghana). J Hazard Mater 302:151–157

Ummalyma SB, Sahoo D, Pandey A (2020) Microalgal biorefineries for industrial products. In: Yousuf A (ed) Microalgae cultivation for biofuels production. Academic, London, pp 187–195

Usmani Z, Kumar V, Varjani S, Gupta P, Rani R, Chandra A (2020) Municipal solid waste to clean energy system: a contribution towards sustainable development. In: Varjani S, Pandey S, Gnansounou E, Khanal SK, Raveendran S (eds) Resource recovery from wastes. Elsevier, Amsterdam, Netherlands, pp 217–231

Van den Oever M, Molenveld K, van der Zee M, Bos H (2017) Bio-based and biodegradable plastics: facts and figures: focus on food packaging in the Netherlands. No. 1722. Wageningen Food and Biobased Research, Wageningen, The Netherlands

Vea EB, Romeo D, Thomsen M (2018) Biowaste valorisation in a future circular bioeconomy. Procedia CIRP 69:591–596

Vieira MV, Pastrana LM, Fuciños P (2020) Microalgae encapsulation systems for food, pharmaceutical and cosmetics applications. Mar Drugs 18(12):644

Wang F, Ouyang D, Zhou Z, Page SJ, Liu D, Zhao X (2021) Lignocellulosic biomass as sustainable feedstock and materials for power generation and energy storage. J Energy Chem 57:247–280

Wijkman A, Skanberg K (2015) The circular economy and benefits for society. Club of Rome

Wilson DC, Rodic L, Scheinberg A, Velis CA, Alabaster G (2012) Comparative analysis of solid waste management in 20 cities. Waste Manag Res 30:237–254

Yadav V, Karmakar S (2020) Sustainable collection and transportation of municipal solid waste in urban centers. Sustain Cities Soc 53:101937

Yamakawa CK, Qin F, Mussatto SI (2018) Advances and opportunities in biomass conversion technologies and biorefineries for the development of a bio-based economy. Biomass Bioenergy 119:54–60

Yevtushenko A, Tolstoj-Sienkiewicz J (2006) Distribution of friction heat between a stationary pin and rotating ring. Arch Appl Mech 76:33–47

Zabaniotou A (2018) Redesigning a bioenergy sector in EU in the transition to circular waste-based bioeconomy—a multidisciplinary review. J Clean Prod 177:197–206

Zaman AU, Lehmann S (2013) The zero waste index: a performance measurement tool for waste management systems in a 'zero waste city.' J Clean Prod 1(50):123–132

Zhang M, Xie L, Yin Z, Khanal SK, Zhou Q (2016) Biorefinery approach for cassava-based industrial wastes: current status and opportunities. Bioresour Technol 215:50–62

Chapter 4
Biofuels: An Overview

Abdelrahman S. Zaky, Kamaljit Moirangthem, and Radziah Wahid

Abstract Biofuels have a long history but have recently gained increasing attention and demand as a renewable, environmentally friendly, and sustainable energy source. Different kinds of biofuels can easily replace traditional fossil fuels with positive environmental impact and potential for net-zero or even negative carbon emissions. Hence, biofuel is vital for tackling the current global warming crisis, which has resulted from our overreliance on fossil fuels. The varied types and forms of biofuels include liquid fuel (such as bioethanol, biodiesel, and jet fuel), gaseous fuel (such as biogas, biomethane, syngas, and biohydrogen), and solid fuels (such as charcoal, biochar, briquettes, and pellets). Bioelectricity is also introduced briefly as another source of bioenergy. These biofuels are generated from a range of different biomass feedstocks, which are divided into four generations based on their biochemical composition, typical usage, and cultivation methodologies. This chapter presents an overview of the major aspects of biofuels, including the generations of biomass feedstocks, types of biofuels, and the main conversion technologies applied to generate the biofuels from biomass.

Keywords Biofuel · Biomass · Bioenergy · Bio economy · Carbon emissions

---

A. S. Zaky (✉)
School of Biological Sciences, University of Edinburgh, King's Buildings, Edinburgh E9 3FF, UK
e-mail: A.Zaky@ed.ac.uk

K. Moirangthem
Department of Engineering, Aarhus University, N 8200 Aarhus, Denmark
e-mail: kamaljit.moirangthem@teagasc.ie

Food Biosciences, Teagasc Food Research Centre, Ashtown, Dublin 15 KN3K, Ireland

R. Wahid
Antec Biogas AS, Olaf Helsets vei 5 A, 0694 Oslo, Norway
e-mail: radziah@antecbiogas.com

A. E.-F. Abomohra et al. (eds.), *Waste-to-Energy*,
https://doi.org/10.1007/978-3-030-91570-4_4

4.1 Introduction

The excessive use of fossil fuels (coal, petrol, diesel, and natural gas) since the Industrial Revolution resulted in the massive accumulation of greenhouse gases (GHGs) currently trapped in the atmosphere. This has resulted in the current global warming debate and the search for solutions to avert the catastrophic environmental crisis that is expected by the end of this century. On the other hand, biofuels provide a renewable and eco-friendly source of energy, not only for road and aviation, but also for marine transport (Moirangthem and Baxter 2016). They are generated from biomass in different forms, including liquids such as bioethanol and biodiesel, gases such as biogas and syngas, and solids such as charcoal and biochar. Biofuels are considered the best alternative to fossil fuels, as they can directly substitute fossil fuel with almost no modification to the existing systems. In addition, biofuels are a vital element of the global effort for climate change mitigation as they can be produced with a net negative carbon emission value, acting as a carbon sink (Röder et al. 2019; Field et al. 2020).

Bioenergy and biofuels have a long history, perhaps since the discovery of fire about 2 million years ago, as human has been using biomass as a source of energy for heating and cooking for thousands of years (Gowlett 2016). Biofuels, mainly biodiesel, bioethanol, and biogas, have been used in internal combustion engines for several decades as an alternative to petrol. The development in biofuel production was classified according to the feedstock type into four generations. The first-generation biofuels (G1) utilise crops—mainly edible—that are rich in sugar, starch, or oil content. The second-generation biofuels (G2) utilise lignocellulosic biomass, which can be agricultural residues or waste or dedicated biomass plants. The third-generation biofuels (G3) are produced from microalgae (mainly for biodiesel) and seaweed (mainly for bioethanol). The fourth-generation biofuels (G4) are produced from genetically modified microalgae.

There are several methods for biomass conversion to biofuels which are classified under three main categories: a) pre-treatment and hydrolysis, b) biochemical conversion, and c) thermochemical conversion. The pre-treatment and hydrolysis category includes many methods such as milling, hydrothermal pre-treatment, and enzymatic hydrolysis. The biochemical conversion mainly includes fermentation, anaerobic digestion (AD), and lipid extraction and transesterification. The thermochemical conversion includes hydrothermal liquification (HTL), pyrolysis, gasification, combustion, carbonisation, and torrefaction.

This chapter provides an overview of the biofuel generations based on their biomass feedstocks, types of biofuels currently being produced, and the main conversion technologies applied to generate these biofuels.

4.2 Types of Biofuels

4.2.1 Liquid Biofuels

Liquid biofuels are the most convenient substitute for fossil fuel because they can be easily stored and transported using the existing infrastructure made for the distribution of petrol. There are several liquid biofuels (already in market or in development), but the most important and commercially available are bioethanol and biodiesel.

4.2.1.1 Bioethanol

Bioethanol is the ethanol (C_2H_5OH) that has been produced from variety of biomasses such as sugarcane, maize, and wheat straw. With a high-octane number, bioethanol replaced lead (traditional octane enhancer) in petrol. Due to its renewable and cleaner combustion (releases only CO_2 and H_2O), ethanol has been a gasoline alternative for several decades. The Ford Model T, which was manufactured from 1908 to 1927, was the first commercial vehicle that could run on pure ethanol.

Currently, bioethanol blended with petrol, with different blend percentages, is widely used around the world. E10 fuel blend, which contains 10% ethanol and 90% petrol, is the most common blend being sold in the USA and the UK. E10 can be used in traditional vehicles without affecting their warranties. Since the petrol crisis in early 1970s, Brazil has been using E20 (contains up to 20% ethanol) and E25 (contains up to 25% ethanol). However, E85 blend (contains up to 85% ethanol) is commonly used for flex-fuel vehicles (FFVs) in Europe and the USA (Yusoff et al. 2018). Bioethanol blended with gasoline can also be oxygenated, resulting in a more complete burn and lower waste gas emissions. In Brazil, E100, the pure hydrous ethanol fuel, is being used for the neat gasoline vehicles and FFVs (Yusoff et al. 2018; Rico 2007; Hogarth 2017).

The contribution of bioethanol is expected to increase further over the next decades due to the increasing demand for energy with an ever-growing global population and the limited supply of fossil fuels. Worldwide bioethanol production exceeded 120 billion litres/year in 2016 and is projected to reach nearly 137 billion L by 2026, which will be ~2.4% of the global transport fuel consumption (OECD 2016; Zaky et al. 2018a). According to the International Energy Agency (IEA), bioethanol and advanced biofuels contribution to the total global transportation fuels could reach up to 9.3% by 2030 and up to 27% by 2050 (IEA-ETSAP 2013).

4.2.1.2 Biodiesel

Biodiesel is an alternative diesel engine fuel derived from renewable biomass sources such as plant oils and fats of animal origin (Demirbas 2007; Gerpen 2005; Hoekman et al. 2012; Vasudevan and Briggs 2008). Biodiesel is made up of long-chain fatty acid alkyl esters produced from triglycerides (via transesterification) or from free fatty acids (via esterification) with light alcohol in the presence of a catalyst (generally base, acid, or enzyme). However, under supercritical circumstances, the reaction can proceed in the absence of the catalyst (Aghbashlo et al. 2021). Biodiesel is renewable, biodegradable, less harmful, and safer for storage and handling. It also has high lubricity and can give equivalent energy density to diesel (Siddiquee and Rohani 2011).

Rudolph Diesel, the inventor of the compression ignition (CI) engine, pioneered the use of biodiesel in diesel engines during the 1900 World Exhibition in Paris, using peanut oil as fuel (Ma and Hanna 1999; Sharma and Singh 2009). However, due to the abundance of petroleum diesel at the time, research and development on vegetable oil fuels were not explored significantly (Yusuf et al. 2011). Vegetable oils were utilised in emergencies to replace diesel during World War II (Shahid and Jamal 2011). Currently, there is a rise in interest in producing biodiesel due to fossil oil's surges in prices, limited resources, and negative environmental impact (Ma and Hanna, 1999; Shahid and Jamal 2011; Sharma and Singh 2009).

Pure biodiesel can be used directly in engines or after being blended with diesel in substantial ratios to provide an alternative fuel solution in the internal combustion engines (Singh et al. 2020). Compared to petroleum-based diesel fuel, biodiesel is biodegradable, non-toxic, with better lubricity and lower flammability (Balat and Balat 2010; Ma and Hanna, 1999; Yusuf et al. 2011), while also causing less air pollution (Aghbashlo et al. 2021; Gerpen 2005).

Utilisation of biodiesel as an alternative to diesel could provide energy security, improve rural industries, and decrease the adverse environmental effects of fossil fuels (Mahlia et al. 2020). However, biodiesel has certain drawbacks, including greater viscosity than conventional diesel, which can cause issues with fuel pumping, combustion, and atomisation. Furthermore, the utilisation of biodiesel might result in the formation of derivatives in the fuel injectors at the head of the engine and the pistons. Catalytic distillation, dilution, micro-emulsion, pyrolysis, and transesterification are used to generate biodiesel from plant oil. Transesterification is the most cost-effective of these conversion processes, and the biodiesel produced from it has characteristics similar to diesel (Singh et al. 2020).

Edible feedstocks such as oils from rapeseed, soybean, and palm are used to manufacture the first-generation biodiesels. In contrast, lignocellulosic materials (such as straw, wood, and husks) and non-edible oils (such as from jatropha and mahua) are used in the second-generation biodiesels production (Mahlia et al. 2020). Biodiesel generated from algae biomass is referred to as third-generation biodiesel. In comparison, the fourth-generation biodiesel is mainly generated from genetically modified microalgae (Abdullah et al. 2019). Table 4.1 compares the benefits and drawbacks of the different generations of biodiesel (Singh et al. 2020; Vignesh et al. 2021). When employing vegetable oils, the feedstock price can account for 70–80% of the entire biodiesel manufacturing cost. Hence, the cost of producing biodiesel can be decreased by up to 50–70% when using waste oils or animal fats.

Table 4.1 Advantages and drawbacks of different biodiesel generations

Generations	Advantages	Drawbacks
First generation	1. Low greenhouse gas emissions 2. Simple conversion technology and low cost	1. Impact on food supply 2. Low biomass yield 3. High land footprint
Second generation	1. Food supply is unaffected 2. Feedstocks can be grown on non-arable land 3. Production costs are reduced	1. Expensive pre-treatment technology 2. Crop yields for some feedstocks are low
Third generation	1. Algal growth is rapid 2. Does not compete with food crops for resources 3. Wastewater and seawater can both be utilised	1. Algal cultivation requires high energy consumption 2. Low lipid level in open pond systems 3. Costly algae oil extraction technique
Fourth generation	1. Higher biomass and biodiesel yield 2. Greater CO_2 absorption capacity 3. Faster growth rate	1. Bioreactors are costly 2. High initial expenditure at the early research stage 3 Health and safety concerns if reached the surrounding environment

4.2.2 Gaseous Biofuels

4.2.2.1 Biogas

Biogas is a renewable energy carrier produced by the anaerobic digestion (AD) of various organic feedstocks such as municipal organic wastes, agricultural residues, and energy crops. Methane (CH_4) is the main component of biogas, contributing 50–70% depending on the substrate, followed by carbon dioxide (CO_2), which accounts for 30–50% (Angelidaki et al. 2018; Wahid et al. 2019). Both CH_4 and CO_2 constitute most biogas, although it also contains water vapour, trace amounts of nitrogen, hydrogen sulphide, and other gases (Angelidaki et al. 2018; Weiland 2010). The biogas yield and content of the substrates vary greatly depending on their origin, organic compound content, and substrate composition (Weiland 2010).

Biogas has been recognised to exist since the seventeenth century, and the development of biogas systems and plants began as early as the mid-nineteenth century. One of the first examples of biogas systems is the septic tank, which has been used for wastewater treatment since the late 1800s (Jørgensen 2009). In the 1890s, Donald Cameron constructed a septic tank that could collect biogas and be used to power streetlights (Jørgensen 2009). In 1859, the first AD facility was constructed in a leper colony in Bombay, India (Kougias and Angelidaki 2018). Commercial usage of biogas in China began in 1921, when Guorui Luo utilised an 8 m^3 biogas tank supplied with garbage to generate energy for cooking and lighting (Kougias and Angelidaki 2018). The development of biogas technology was nearly halted towards the end of the 1950s due to fossil fuel oil and gas's low cost. However, interest in biogas resurfaced with the oil crisis of the 1970s, as high oil prices encouraged research into new alternative energy sources (Kougias and Angelidaki 2018; Jørgensen 2009). The rising awareness of climate change and renewable energy has recently boosted interest in biogas technology.

Biogas can be used as an energy and heat source, as well as a transportation fuel. Methane-rich biogas (biomethane) can be stored within the natural gas grid and used as a feedstock to produce chemicals and materials in place of natural gas (Weiland 2010). Biogas production via AD has number of advantages, including reduced greenhouse gas (GHG) emissions in the agricultural sector, reduced odours and pathogens, and improved fertiliser quality (digestate) (Wahid et al. 2015, 2020). In addition, it is a complete waste recycling process, and the biogas usage is regarded as CO_2 neutral as the emission of CO_2 during biogas combustion equals the amount of CO_2 consumed by the plants to create organic matter during photosynthesis (Jørgensen 2009). The utilisation of wastes such as animal manure, crops residues, and industrial waste for biogas generation has contributed to waste minimisation and environmental preservation (Divya et al. 2015).

Wet and dry digestion are the two types of AD processes used to produce biogas. The two digestion systems differ in total solid (TS) concentration, with dry digestion systems having a greater TS content (20–40%) and wet digestion systems having a TS level less than 20%. Wet AD is the most often used method in continuous plants, and it is used to treat liquids or solids in a liquid mix (Angelonidi and Smith 2015; Chiumenti et al. 2018).

4.2.2.2 Syngas

Syngas, or synthesis gas, is a gas mixture mainly composed of hydrogen (H_2) and carbon monoxide (CO) that is generated through gasification or pyrolysis of carbonaceous feedstocks (Woolcock and Brown 2013). Aside from these primary gases, various products, including some solid and liquid compounds as well as additional gaseous components, may be expected during the gasification process. Typically, the solid product consists primarily of ash, whereas the liquid product mainly contains water and tars. Carbon dioxide, nitrogen, and methane are some of the gaseous by-products (Acharya et al. 2014; Hagos et al. 2014).

Syngas is generally derived from natural gas, coal, or refinery waste. However, due to environmental concerns, there is currently a surge in interest in biomass gasification (Rauch et al. 2014; Woolcock and Brown 2013). The composition of syngas is heavily influenced by the feedstock, gasification process, and gasifier types (Hagos et al. 2014; Mondal et al. 2011). For example, raw syngas generated from palm oil wastes typically contains 6–63% H_2, 14–25% CO, 8–19% CO_2, and 1–12% CH_4 (Samiran et al. 2016).

The first commercial use of syngas was in 1812 by the London Gas, Light and Coke Company, and in 1881, the first application of syngas for the internal combustion engine was made (Göransson et al. 2011; Woolcock and Brown 2013). Between 1920 and 1940, small gasifier systems for car applications were developed in Europe due to

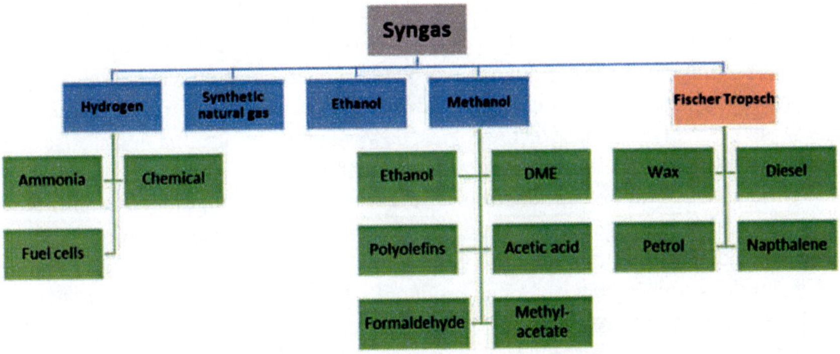

Fig. 4.1 Syngas utilisation in numerous energy-related processes for the manufacture of fuels and chemicals Fischer Tropsch: a process that involves series of chemical reactions that is capable of converting syngas to liquid hydrocarbons; DME: Dimethyl ether

the impending World War II and thus the unpredictable gasoline supply. (Göransson et al. 2011).

Syngas is presently used in a variety of applications, including heat or power usage in integrated gasification combined cycles (IGCCs) for the manufacturing of fuels and chemicals (ammonia, methanol, and higher alcohol), where syngas is used as an intermediate product (Abdul Mujeebu 2016; Woolcock and Brown 2013). The use of syngas for heat, power, and fuels is predicted to expand in the future as global energy consumption rises (Woolcock and Brown 2013). Figure 4.1 summarises the application of syngas and the conversion techniques. Raw syngas contains trace amounts of undesired contaminants such as tars, nitrogen-based chemicals (NH_3, HCN), sulphur-based compounds (H_2S, COS), hydrogen halides (HCl, HF), and trace metals (Na, K). Thus, purification of raw syngas is required before utilising syngas in downstream applications (Acharya et al. 2014; Mondal et al. 2011). Cyclones, water or oil scrubbers, adsorption columns, and filters are some of the most prevalent syngas purification methods. They are used to eliminate contaminants such as char, tar, and particulate matters (Acharya et al. 2014).

4.2.2.3 Biohydrogen

Hydrogen (H_2) is a substantial and prospective energy resource that is expected to play an important role in the future (Das 2009). This is due to its high energy conversion efficiency, significant gravimetric energy density, and eco-friendly oxidation products (Elsharnouby et al. 2013). Sir Henry Cavendish discovered the elemental nature of hydrogen for the first time when dissolving metals in dilute acids (Mona et al. 2020). Cavendish discovered that the action of specific acids on particular metals produced a distinct, peculiar, and highly flammable gas, which he termed

"Inflammable Air". Robert Boyle had previously produced hydrogen, but its properties had not been known; Cavendish reported these in detail, including the density of the gas (West 2014). Conventionally, hydrogen is generated from natural gas through the process of steam reforming. Other industrial processes include coal gasification and water electrolysis. These technologies, however, are not sustainable because they rely on non-renewable energy sources (Manish and Banerjee 2008).

While most hydrogen production is now dependent on fossil fuels, efforts to manufacture biohydrogen from diverse bio-residuals such as wastewater or organic wastes are emerging. Fermentation, biophotolysis, and bio-electrochemical system are examples of biological hydrogen production technologies (Osman et al. 2020a). Biological hydrogen production is a viable, environment-friendly, and sustainable alternative to conventional hydrogen production methods. (Azwar et al. 2014; Levin et al. 2004; Mona et al. 2020). The advantages of biological H_2 production systems include the utilisation of renewable energy resources, energy-intensive, and the capacity to function at ambient temperature and atmospheric pressure (Mona et al. 2020; Kotay and Das 2008). Figure 4.2 illustrates different processes of biological hydrogen production (Osman et al. 2020a; Ghimire et al. 2015).

Biohydrogen production via fermentation is categorised into two techniques based on the necessity for light by bacteria: (a) dark fermentation and (b) photo-fermentation (Osman et al. 2020a). Dark fermentation is a light-independent process that produces biological hydrogen, organic acids, and alcohols from carbohydrate-rich organic wastes employing anaerobic or facultative anaerobic microbes (Osman et al. 2020a). On the other hand, photo-fermentation necessitates light for photo-heterotrophic bacteria to convert organic acids (such as lactic, butyric, and acetic) to CO_2 and hydrogen under anaerobic conditions (Kamaraj et al. 2020).

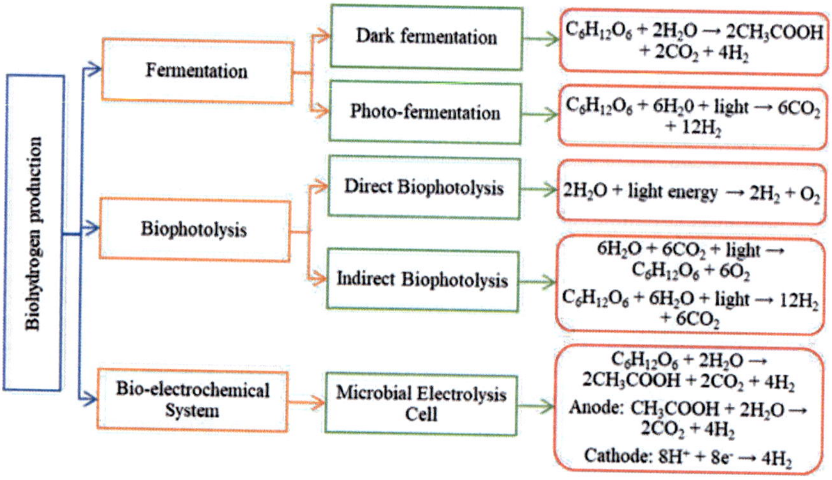

Fig. 4.2 Overview of the routes of biohydrogen production processes

Biophotolysis, also known as water-splitting photosynthesis, is a process whereby oxygenic photosynthetic microorganisms such as cyanobacteria and green microalgae break water molecules to generate hydrogen and oxygen with light as the primary energy source (Shanmugam et al. 2020). There are two methods for producing H_2 via biophotolysis: (a) direct biophotolysis and (b) indirect biophotolysis (Osman et al. 2020a). The direct method involves producing hydrogen under anaerobic conditions at the expense of sunlight through photosynthesis, whereas indirect method requires a two-stage photosynthetic conversion of light energy to carbohydrates as a type of chemical energy (Shanmugam et al. 2020; Mona et al. 2020). The bio-electrochemical system is a new technology that uses microbial electrolysis cells (MEC) to generate H_2 from various substrates (Osman et al. 2020a). MEC is sometimes referred to as electro-fermentation and biocatalysed electrolysis cells. The MEC system consists of two electrodes (cathode and anode) and can be configured as a single-chamber or two-chamber MEC (Osman et al. 2020a). The basic working method is the same in both MEC types. The oxidation of organic compounds in the anode produces electrons that are translocated to the anode. These are then transferred to the cathode, wherein they interact with protons to form H_2 (Osman et al. 2020a).

4.2.3 *Solid Biofuels*

Solid biofuels are solid biomass that can be utilised for heat or electricity production. These include firewood, dry animal manure, charcoal, biochar, and pellets. Direct combustion of solid biofuels such as wood is the most convenient strategy to generate bioenergy in developing countries. However, biomass in its natural form has a low density and poses challenges for modern heating applications. Hence, in the developed nations, solid biofuels, especially those derived from wood or agricultural waste, are promoted to compacted forms with higher energy values (Alatzas et al. 2019; Perea-Moreno et al. 2019). The promoted solid biofuels are easier to operate, transport, storage, and use. They also possess higher energy concentration per volume unit and are free or have reduced toxic compounds such as sulphur and heavy metals.

4.2.3.1 Charcoal

Charcoal is a porous solid biofuel comprised primality of carbon. It is generally produced by pyrolysis, more specifically, through carbonisation, a thermal decomposition process with low oxygen and long residence time. The wood is first dried then pyrolysis at 400 °C in an airless environment (Chandrasekaran et al. 2019). Converting wood into charcoal overcomes the lower energy content of wood when compared to fossil fuels. With an energy content of 28–33 $MJ.kg^{-1}$, charcoal has a comparable/higher energy content than its fossil counterparts, i.e. coal (Jouzani et al. 2020). Charcoal is mainly used as fuel or a reducing agent in the iron and steel industries (Hagemann et al. 2018).

The world's largest charcoal producer and consumer is Brazil (Nabukalu and Gieré 2019). Good charcoal is dark black in colour with a shiny lustre, does not crumble easily, poor conductor, and floats on water (Saravanakumar et al. 2006). The global production of charcoal in 2017 was around 51 Mt (Nabukalu and Gieré 2019).

4.2.3.2 Biochar

Biochar, like charcoal, is also rich in carbon. However, the thermochemical conversion (pyrolysis, hydrothermal carbonisation, and gasification) is conducted in an environment with low oxygen supply (Lehmann et al. 2006). During pyrolysis, biomass is heated in a low or oxygen-free conditions, producing biochar with co-products such as bio-oil and syngas (Manyà 2012). Compared to normal pyrolysis at 400–1200 °C (Rajendran et al. 2019), heating at lower temperatures for longer duration is preferred for biochar production (Tripathi et al. 2016). In addition to production method mentioned/conditions, the feedstock type also impacts biochar properties (Sun et al. 2014). Biochar can be made from lignocellulosic materials (Yuan et al. 2011; Sun et al. 2014), algae (micro and macro) (Yu et al. 2017), animal dung (Cely et al. 2015), sludge from sewage (Yue et al. 2017), and other organic carbonaceous materials. Biochar can be used as an ion exchanger, pH buffer, electron conductor, detoxifier, and as a source of nutrients that support microbial habitation. Hence, the traditional applications of biochar included improving the soil properties (Sun et al. 2020).

4.2.3.3 Biofuel Pellet

Biofuel pellets are produced through extrusion under high pressure and temperatures. This yields small cylinders, which are then cut to the desired lengths (Whittaker and Shield 2017). They are currently made from wood residues such as sawdust, shavings, and wood chips (Picchio et al. 2020; Kuparinen et al. 2014). Among these, sawdust is the ideal substrate for pelleting (Whittaker and Shield 2017). However, alternative sources are being explored as the demand grows such as forest residues (Picchio et al. 2012), spent coffee grounds (Lisowski et al. 2019), and tree leaves from urban areas (Mudryk et al. 2021). Hence, heating and industrial sectors has a high demand for pellets (Proskurina et al. 2019) The utilisation of biomass in the form of pellets (volumetric energy content of around 18 MJ m^{-3}) is more cost-effective and generate less particulate emissions when used for heating and electricity production compared to the direct use of non-modified biomass residues generation (Shen et al. 2012; Shepherd 2000). The USA, Canada, and the EU are the top pellet producers. Among them, the EU, with a consumption of about 20 Mt, leads in terms of importing pellets and its utilisation in energy generation (Thrän et al. 2017).

With the rising energy demands, good quality solid biofuels that are produced in an environment-friendly manner are in high demand (Hartmann 2017). Hence, the International Organization for Standardization (ISO) has developed international pellet quality standards. This includes the EN ISO 17225–2 for the graded wood pellets for industrial and domestic use, the EN ISO 17225-6 for the graded non-woody pellets, and the EN ISO 17225–1 for the general quality requirements, (Picchio et al. 2020).

4.3 Generations of Biofuel Feedstocks

Biofuels are classified into four generations based on the feedstocks used in their production. First generation (G1) is made from food crops such as sugarcane wheat, corn, and vegetable oils. Second generation (G2) is made from non-edible fractions of plants, which includes willow, miscanthus, woods, and straws. Algal biomass, including seaweed and microalgae, is used to produce third generation (G3). Fourth generation (G4), on the other hand, is made from genetically modified microalgae. This section goes over some specifics, such as the benefits and drawbacks of each biofuel generation. Table 4.2 provides a general overview of energy feedstock generations and some examples of biofuels generated by different conversion technologies.

4.3.1 First-Generation (G1) Biofuel Feedstocks

Feedstocks for G1 biofuels are mainly starchy crops (e.g. cereals, cassava, and potato), sugar crops (e.g. sugarcane, sugar beet, and sweet sorghum), and oilseed crops (rapeseed, soybean, and sunflower) which are edible resources (Wang et al. 2012). Crops with starch or simple sugars are used for G1 bioethanol production through fermentation, while oil crops are used for G1 biodiesel production through transesterification. Corn in the USA, wheat in the EU, and sugar cane in Brazil are the most used G1 biofuel feedstock. Almost 99% of biofuels currently used in EU road transport are first generation (Bourguignon 2015). They are associated with concerns such as feasibility in terms of cultivation area (Jiang et al. 2016), direct competition with food (Vermerris 2008), and concerns over negative energy balance (Pimentel and Patzek 2005; Wesseler 2007). G1 biofuels have been produced worldwide in large quantities for which the production process is considered "established technology". However, one of the purposes of producing G1 biofuels, from food crops, was to show that fossil fuels could be replaced. It was clear from the early days that the excessive use of G1 biomass for biofuel production would compromise food security (Smith 2010). The non-edible grade of these crops, such as animal grade wheat and broken or rejected rice, is still rich in sugars or oils and therefore is being used for bioethanol production, biodiesel production or subjected to anaerobic

Table 4.2 Overview of bioenergy feedstock generations

Biofuel Gen	Biomass category	Biomass example	Conversion technology	Biofuel	References
First generation	Sugar crops	Sugarcane	Fermentation	Bioethanol	Raza et al. (2019); Burrett et al. (2009)
		Sugar beet			Dammer et al. (2017)
		Sweet sorghum			Mathur et al. (2017), Whitfield et al. (2012), Ratnavathi et al. (2011)
	Starch crops	Corn	Fermentation after saccharification	Bioethanol	(Mosier and Ileleji (2015), Green et al. (2015)
		Wheat			Chin and Ingledew (1993), Green et al. (2015)
		Cassava			Liu et al. (2013), Ademiluyi and Mepba (2013)
	Oil crops	Palm oil	Transesterification	Biodiesel	Ishola et al. (2020), Singh and Singh (2010)
		Rapeseed			(Rezki et al. (2020), Encinar et al. (2018)
		Sunflower			Boumesbah et al. (2015), Dizge et al. (2009)
		Soybean			Dizge et al. (2009), Singh and Singh (2010)
	Sugar-rich wastes	Molasses	Fermentation	Bioethanol	Zaky et al. (2018a, 2020)

(continued)

Table 4.2 (continued)

Biofuel Gen	Biomass category	Biomass example	Conversion technology	Biofuel	References
	Oil-rich wastes	Fat, oil and grease (FOG)	Transesterification	Biodiesel	Abomohra et al. (2020)
Second generation	Lignocellulosic agriculture residues	Wheat straw	Fermentation after saccharification	Bioethanol	Saha et al. (2005a) Talebnia et al. (2010)
		Rice straw	AD[a]	Biogas	Mothe and Polisetty (2021), Ngan et al. (2020)
		Sugarcane bagasse	Fermentation after saccharification	Bioethanol	Chandel et al. (2012)
		Sugarcane bagasse	Gasification	Syngas	Ahmed and Gupta (2012)
		Wood	Torrefaction	Biocoal	Negi et al. (2020)
		Sawdust	Pyrolysis	Biodiesel	Ahmed et al. (2020)
	Lignocellulosic crops	Miscanthus	Fermentation after saccharification	Bioethanol	Baibakova et al. (2020)
		Willow	Fast pyrolysis	Biodiesel	Miettinen et al. (2017)
		Willow	Steam explosion/fermentation	Bioethanol	Ziegler-Devin et al. (2019)
Third generation	Seaweed	*Laminaria saccharina*	HTL[b]	Biodiesel	Bach et al. (2014)
		Ulva rigida	SSF[c]	Bioethanol	Korzen et al. (2015)
		Ulva intestinalis	Fermentation/transesterification	Bioethanol/ Biodiesel	Osman et al. (2020b)
	Microalgae	*Nannochloropsis Oceanica*	HTL	Biodiesel	Caporgno et al. (2016)
		Chlorella protothecoides	Transesterification	Biodiesel	Xu et al. (2006)

(continued)

Table 4.2 (continued)

Biofuel Gen	Biomass category	Biomass example	Conversion technology	Biofuel	References
Fourth generation	Genetically modified microalgae	*C. reinhardtii*	Dark fermentation	Biohydrogen	Beer et al. (2009)
		Synechocystis sp. PCC6803	Photo-fermentation	Bioethanol	de Farias Silva and Bertucco (2016), Deng and Coleman (1999), Gao et al. (2012)

[a] *AD* Anaerobic digestion, [b] *HTL* Hydrothermal liquefaction, [c] *SSF* Simultaneous saccharification and fermentation

digestion for biogas production. Although it is not edible grade, as of the typical G1 biomass feedstocks, they represent a loss of valuable edible food that affects global food security, and the goal is to reduce this type of biomass.

4.3.1.1 Sugar Crops Used for the G1 Biofuels Production

Sugar crops accumulate high concentrations of fermentable sugars that can be easily extracted then fermented directly by microorganisms, mainly the yeast *S. cerevisiae*, to alcohols such as bioethanol. Sugarcane (*Saccharum officinarum* L.) along with sugar beet (*Beta vulgaris*) is the main sugar crop currently being used for biofuel production at the industrial level. The primary feedstock for bioethanol production in Brazil is sugarcane, while in the UK, sugar beet is one of the main feedstocks for bioethanol production. A ton of sugarcane can produce around 100 kg of sugar and approximately 45 kg of molasses. Molasses contain about 50% fermentable sugars and, therefore, can produce up to 25% ethanol recovery through fermentation (Raza et al. 2019). More than 21 million m^3 of ethanol is produced from sugarcane (Burrett et al. 2009). Sugar beet and sugar cane are very attractive feedstocks for biofuel production because they have high land efficiency compared to other G1 energy crops (Dammer et al. 2017).

In addition to sugar beet and sugarcane, sweet sorghum (*Sorghum bicolor* L. Moench) has also been investigated by many researchers as another promising sugar crop for the G1 bioethanol production at an industrial scale (Mathur et al. 2017). Unlike sugarcane and sugar beet, it has relatively low input requirements with the ability to grow on marginal land. It also has a high tolerance to drought, waterlogging, and salt, so that it could be the best choice as an energy crop for the regions with hot and dry climatic conditions (Mathur et al. 2017). Sweet sorghum refers specifically to genotypes of sorghum that accumulate soluble sugars in their stalks (Whitfield et al. 2012). Its juice contains ~16–18% fermentable sugar that can be directly fermented by yeast into ethanol (Ratnavathi et al. 2011).

4.3.1.2 Starch Crops Used for the G1 Biofuels Production

Starch crops used in G1 biofuel production are grains (e.g. corn and wheat) and tubers (e.g. potatoes and cassava) (Marques et al. 2018). Among these, corn with about 70% starch, from the USA, is the primary source of the world's bioethanol. Globally, around 40.8 billion bushels of corn were produced in 2017 (Mohanty and Swain 2019). Around 2.5–2.9 gallons of bioethanol can be generated from 25 kg of corn (Mosier and Ileleji 2015). As of 2012, more than 40% of the USA corn was utilised for the production of ethanol, though not all ethanol is used as biofuel. Although it has an additional hydrolysis step (compared to sugarcane), corn can use the rest of the plant as G2 feedstock to produce G2 bioethanol (stalk, cob, etc.). In Europe, bioethanol is made mainly from wheat (Chin and Ingledew 1993). More than 713 million metric tons of wheat is grown throughout the world (Shevkani et al.

2017). Compared to corn, cereals such as wheat have a higher potential to produce higher yields of alcohol when the spent grains are also utilised (Green et al. 2015).

With an annual global production of about 550 million metric tons, cassava (*Manihot esculenta* Crantz) is an important drought tolerant tropical food crop (Sivamani et al. 2018). It is one of China's most essential biofuel crops but relies heavily on imports (Liu et al. 2013). The carbohydrate content of cassava has been reported to be higher than other tuber crops such as potato, with one ton of fresh cassava root yielding 150 L of ethanol (Ademiluyi and Mepba 2013). Additionally, ethanol production from cassava could be more profitable than other feedstocks. For example, in Tanzania, the cost of producing cassava ethanol is USD 0.37–0.43/L, which is comparable to production cost in Brazil (USD 0.47), the USA (USD 0.46), and India (USD 0.52) (Arndt et al. 2012). However, its rapid post-harvest deterioration they are generally processed into cassava starch. In the starch form, cassava is an attractive feedstock for microbial fermentation (e.g. ethanol) in China, Thailand, and South Africa (Li et al. 2014).

4.3.1.3 Oil Crops Used for the G1 Biofuel Production

Sugar and starch feedstocks are used to produce bioethanol, while oil crops are used to produce biodiesel. Oil from soybean is the major source of biodiesel in the USA, whereas in Europe and tropical countries, palm oil and rapeseed oil are the main biodiesel feedstocks (Singh and Singh 2010). These oils are easy to convert to biodiesel and can often be used directly in diesel engines with slight modification. Acid, alkaline, and enzymatic catalysts are used in the transesterification process to make biodiesels.

Even with its relatively high price, as a food crop, and a biodiesel yield of only about 70 gallons per acre, soybean (*Glycine max*) is still a major biodiesel feedstock. One ton of soybean can produce 182 kg of oil, 794 kg of soybean meal, and 18 kg of residues (Panichelli et al. 2009). Of the total soy produced, only 6% is used in the form of whole beans (e.g. biofuel and edible oils), while the remaining 94% is used in processed food products (e.g. animal feed, soy milk, and soy sauce) (Oliveira and Schneider 2016). Argentina, the third-largest soybean producer and the world's top exporter of soybean oil and meal, is expected to significantly increase biodiesel production (Panichelli et al. 2009). The USA Energy Information Administration reported that soybean oil is the most common vegetable oil used for biodiesel production in the USA. This represents 30% of the 26 billion pounds of the total U.S. soybean oil supply (Hanson 2019).

Although the oil content of rapeseed is more than 40%, it ranks just below soybeans in terms of production. According to FAOSTAT 2018, the global production of rapeseed oil was around 24.6 million tons in 2018 (FAO 2018). In the EU, rapeseed biodiesel is the more commonly used biodiesel fuel (Torres-Jimenez et al. 2012). Rapeseed oil contains around 98% of triglycerides, more than 60% oleic acid (monounsaturated), and more than 20% linoleic acid (doubly unsaturated), and hence, it has the capability to perform very well at low temperatures (Pinzi et al. 2009).

4.3.1.4 Rich Biomass Waste (Non-lignocellulosic Biological Residues) Used for the G1 Biofuel Production

Many biomasses waste contain high concentrations of starch, fermentable sugars, or oils like those in G1 feedstock. These include molasses, waste vegetable oil (WVO), broken and low-grade starch crops, and fruit waste. In addition, the non-edible grade of crops such as animal grade wheat and broken or rejected grains are still rich in sugars or oils, therefore are being used for bioethanol and biodiesel production, or are being subjected to anaerobic digestion for biogas production. Among them, WVO has been used as a fuel, and some of the earliest diesel engines ran exclusively on vegetable oil. The accumulated free fatty acid in WVO is a viable feedstock for biofuel (Ullah et al. 2015). China produces around 5 million tons of WVO (Chen et al. 2017).

In the USA, biodiesel produced from recycled oil can receive half dollar per gallon as a subsidy, attributed to its near 100% WVO recycling rate in the USA (Zhang et al. 2014). With potassium hydroxide as a catalyst, WVO's transesterification has been reported to yield 92% methyl ester (biodiesel), 1.33% soap materials, and 6% glycerol (Phan and Phan 2008; Lamichhane et al. 2020). Using WVO has also been reported to decrease up to 70% for production costs of biodiesel (Endalew et al. 2011). However, it can decrease engine life if not properly refined. Additionally, collection from the distributed sources, restaurants and homes, is challenging.

Molasses, a thick dark brown liquid generated during sugar production, is also being used for bioethanol production (Zaky et al. 2020). Left after sugar crystallisation, molasses, and the leftover liquid consist of fermentable sugars in high concentration. One ton of sugarcane can produce 40–60 kg of molasses (Wang et al. 2013). According to FAOSTAT 2018, around 63.7 million tons of molasses was produced in 2018 (FAO 2018). Global production of molasses is around 50 million tons, and around 15% of this is beet molasses (Gabra et al. 2019). About 80% of the world's molasses is used for ethanol production through fermentation processing (Raza et al. 2019). In general, sugarcane juice can be mixed with molasses in different percentage to produce bioethanol (Laluce et al. 2016; Zaky et al. 2018b).

4.3.2 Second-Generation (G2) Biofuel Feedstocks

G2 biofuels are produced from non-edible feedstocks comprising lignocellulosic crop residues, wood residues, dedicated energy crops, and municipal solid waste. They are mainly cellulose, hemicellulose, and lignin matrix; they are tough to hydrolyse into fermentable sugars and hence need additional steps such as pre-treatment to disrupt the biomass matrix and make the enzyme accessible.

4.3.2.1 Agricultural Residues Used for the G2 Biofuels Production

Agricultural wastes, such as corn stover (Chang et al. 2001), wheat straw (Hongzhang and Liying 2007), corn cobs (Melekwe et al. 2016; Cao et al. 1996), rice husk (Saha et al. 2005b), and sugarcane bagasse (Neves et al. 2016), have been investigated as potential lignocellulosic biomass for biofuel production. These are produced in billions of tons across the globe each year; however, most of these residues are either discarded or burned resulting in environmental pollution. Wheat straw comprises mainly the plant's stem and leaves after the grain has been harvested. Global wheat production for 2014 was 732 million tons (Taylor and Koo 2015), and assuming a residue/crop ratio of between 1.3 and 1.5:1 (Talebnia et al. 2010), about 1000 million tons of wheat residues are produced annually. This makes wheat straw a promising lignocellulosic biomass for G2 bioethanol production especially in Europe due to its large production and high carbohydrate content (Alvira et al. 2016). In the UK, wheat breeding programs are also focusing on improving the straw's digestibility (Moirangthem et al. 2015; Moirangthem 2017; Calderini et al. 2021).

Another example is rice straw which is main staple food for many Asian countries and had a global production of over 748 million tons in 2016 (FAO 2016). The major residues from rice cultivation are rice straw and rice husk. Due to the large quantities produced, farmers across south-east Asia burn tens of millions metric tons of leftover stalks (Bond et al. 2013; Moirangthem et al. 2021c). The released black carbon or soot forms the second most important contributor to global warming, second only to carbon dioxide (Abraham et al. 2016). The incorporation of the straw into the soil, compared to its removal from the fields, has also been reported to emit greenhouse gases, mainly methane (Bond et al. 2013).

Bagasse is obtained from sugarcane after the juice is extracted from the crushed stem. From one ton of sugarcane, ~280 kg of wet bagasse is generated as residue (Soccol et al. 2010). Annually, more than 279 million metric tons (DM) of bagasse is produced, and about 50% of bagasse is used for power and electricity generation (Chandel et al. 2012). However, they also can be utilised to produce bioethanol, industrial enzymes, xylitol, organic acids, etc. (Parameswaran 2009; Pandey et al. 2000).

4.3.2.2 Dedicated Lignocellulosic Biomass Used for the G2 Biofuels Production

Dedicated lignocellulosic biomass is non-food energy crop that is grown specifically for biofuel production. These crops are generally able to grow on marginal lands and poor soil, e.g. *Miscanthus sinensis* and *M. giganteus* (Lee and Kuan 2015), switch-grass (Chang et al. 2001), giant reed (*Arundo donax*), reed canary grass (*Phalaris arundinacea*), willow, and hybrid poplar (Zabed et al. 2017). Additionally, fibres from green biorefining, such as clover grass and legumes, are also investigated for G2 feedstock in Denmark (Ambye-Jensen et al. 2014).

Perennial grass crops offer enormous growth potential in various regions and climate zones. For example, Miscanthus is a leading bioenergy perennial crop in Europe (Lee and Kuan 2015; Wang et al. 2015). The sterile hybrid *Miscanthus* × *giganteus* is currently being grown as a commercial bioenergy biomass due to its high yield potential (Lewandowski et al. 2000). Willow has many desirable characteristics as a feedstock for biofuel production. With its coppicing ability and vigorous growth, it can produce more than 11 (dried) tones ha^{-1} $year^{-1}$ when grown on marginal land (Volk et al. 2011). *Miscanthus* and short rotation coppice derived from willow have been the most widely planted species in the UK. Productivity of crops considered ranged from 0.1 to 1.75 Mt/ha/year (dry basis) for wheat straw (crop residue), 14 to 16 Mt/ha/year (dry basis) for willow (woody crop), to as high as 44 Mt/ha/ year (dry basis) for *Miscanthus (*herbaceous crop) (Laser and Lynd 2014). Through the thermochemical process, such biomasses are mainly converted to biodiesel and syngas production through the thermal conversion method. However, they have the potential for bioethanol production and other chemicals through fermentation. Hence, perennial grass crops have enormous potential as a feedstock for G2 bioenergy.

4.3.2.3 Forest Biomass Residues Used for the G2 Biofuels Production

Forest biomass is a renewable and sustainable source of feedstock for producing biofuels (Shabani et al. 2013). It is carbon neutral and abundantly available worldwide. These biomasses are generally converted into energy by thermochemical processes, with combustion being the most widely used (Ahmad et al. 2016; Pisupati and Tchapda 2015). The forest bioenergy sector utilises all sorts of wastes and side streams from saw logs and pulpwood (Kumar et al. 2021), bark (Frankó et al. 2015), sawdust (Stoffel et al. 2017), softwood trimmings (pine) (Galbe and Zacchi 2002), hardwood chips (oak) (Perego et al. 1990), and branches (Sasaki et al. 2014). Globally, more than 1.87 billion m^3 of wood fuel are annually consumed (Van den Born et al. 2014). These comprise fuelwood, charcoal, and other wood-based energy sources (Seifert et al. 2014). Around 5–10% of forest residues are used to make wood pellets (2 million tons), accounting for 61% of the total capacity of Canada.

4.3.2.4 Municipal Solid Waste (MSW) Used for the G2 Biofuels Production

MSW is a bioenergy source that usually contains high lignocellulosic content. In 2016, 37% of the 2.01 billion tons of MSW generated was disposed of into landfills (Meng et al. 2021). MSW is an attractive feedstock due to its large availability, low price, and no competition with food production as it is not purposely produced or collected for biofuel production (Farmanbordar et al. 2018). MSW could be converted to a variety of different fuels such as biogas through anaerobic digestion, syngas through gasification, and biochar through pyrolysis (Salman et al. 2019). They are also often used in cogeneration plants, where it is burned to produce heat and electricity (Tsai 2016). In addition, they can also be used to produce bioethanol. With the suitable pre-treatment method, enzymatic hydrolysis, and fermentation process, high bioethanol yields (over 80%) could be achieved from MSW (Barampouti et al. 2019). It was estimated that around 83 billion litres of ethanol can be produced globally from paper MSW (Shi et al. 2009).

4.3.3 Third-Generation (G3) Biofuel Feedstocks

The limitations associated with the G1 and G2 biofuels have encouraged new biomass research that does not compete with the edible crops for freshwater and arable land. Hence, algae were proposed as the biomass feedstock for the G3 biofuel. Algae are photosynthetic organisms that grow in different aquatic environments including oceans, seas, rivers, ponds, lakes, and wastewater. They are able to grow in a wide range of light intensities, pH, salinities, and temperatures. In addition, they can grow independently or in symbiosis with other organisms and tolerate the adverse conditions and inhibitors. Algae are commonly classified according to their colour to three main groups including *Phaeophyta* (brown algae), *Rhodophyta* (red algae), and *Chlorophyta* (green algae). They are also generally classified by their size to macroalgae and microalgae. Macroalgae (known as seaweed) are eukaryotic, large size, multicellular, and visible to the naked eye. On the contrary, microalgae are microscopic single cells that can be prokaryotic (cyanobacteria) or eukaryotic (green algae) (Khan et al. 2018).

Seaweeds are primarily investigated for bioethanol production through fermentation because they contain a high percentage of carbohydrates. However, several studies on seaweed utilisation for biodiesel/syngas production via thermal approaches such as pyrolysis, gasification, and hydrothermal acidification (HTL). On the other hand, microalgae have been mainly investigated for biodiesel production via transesterification as they can accumulate high amounts of lipids. However, some microalgae species accumulate considerable amounts of starch and therefore were investigated for bioethanol production via fermentation. In addition, some studies are looking into biogas production via anaerobic digestion from both microalgae and seaweed (Zaky 2021). Seaweed and microalgae are two distinct types of biomasses in terms of classification, distribution, chemical composition, farming and propagation, and their utilisation for biofuel production.

4.3.3.1 Seaweed

Seaweed is the common term for the marine macroalgae. Seaweeds are among the largest biomass producers in the marine environment (Bhadury and Wright 2004). They have been an important part of the traditional diet and medicine in many Asian countries for so many centuries (Ale and Meyer 2013). Seaweeds are classified under different kingdoms of living organisms and represent an important reservoir of natural bioactive products. They grow rapidly in a wide range of different shapes and colours, including more than 10,000 species of the fastest-growing plants on the planet, constituting one of the most important biomass resources in the marine environment (Irkin and Yayintas 2018). Seaweeds are found attached or floating freely; they lack true root, stem, and leaves. They can grow in deep-sea areas (up to 180 m in depth) as well as in estuaries and in shallow water on solid objects such as rocks, pebbles, dead corals, shells, and plant materials. However, they thrive more in shallow rocky coastal areas, especially those exposed to low tide (Pal et al. 2014).

Seaweeds are grouped into three major groups primarily based on their pigmentation: brown seaweeds (*Phaeophyceae*) mainly contain fucoxanthin, red seaweeds (*Rhodophyceae*) mainly contain phycoerythrin, and green seaweeds (members of the Chlorophyta) mainly contain chlorophyll (Borines et al. 2011). Although they are categorised mainly on their colour, they greatly differ in their chemical composition, geographical distribution, and morphology. Each class is also characterised by the specific carbohydrates they produce. The main storage polysaccharides of seaweeds are ulvan and mannan for green seaweeds; agar and carrageenan for red seaweeds; and fucoidan, alginate, and laminarin for brown seaweeds (Salehi et al. 2019; Jönsson et al. 2020).

Like terrestrial biomass, seaweeds are comprised of a mix of carbohydrates, lipids, proteins, and other components in specific proportions. In general, carbohydrates represent, on average, 50% of the seaweed's dry weight (DW) (Salehi et al. 2019). The total carbohydrate content is vastly different within each class of seaweed depending on the species, location, and season of growth or cultivation. It was reported that the total carbohydrate content (in the DW) of brown seaweed ranges from 12.2 to 56.4%, while in red seaweed ranges from 34.6 to 71.2% and in green seaweed ranges from 29.8 to 58.1% (Salehi et al. 2019). Protein content also varies greatly between seaweed species and is particularly influenced by seasonal variation (Marinho-Soriano et al. 2006). Proteins generally represent 4.3–24.0% of dry weight in brown seaweed, 8.7–32.7% of dry weight in green seaweed, and 8.0–47.0 of dry weight in red seaweed (Salehi et al. 2019). Seaweed lipid content is generally low, accounting for only 0.1–4.5% of dry weight (Salehi et al. 2019). Though there are many lipids, the most abundant are phospholipids and glycoglycerolipids (Salehi et al. 2019; Vieira et al. 2018). Also, polyunsaturated fatty acids (PUFAs), which have shown several health benefits, were found in high concentrations (Kendel et al. 2015; Dawczynski et al. 2007). Other lipids also include pigments, such as carotenoids, sterols, and vitamins—all of which have a commercial value (Salehi et al. 2019).

The remainder of seaweed dry matter is referred to as ash. The total ash content represents around 22% of seaweed's dry weight and dramatically varies between species, ranging from 17.0 to 44.0% in brown seaweed, 7.0 to 37.0% in red seaweed, and 11.0–73.0% in green seaweed (Salehi et al. 2019). Ash is comprised of macrominerals (Na, K, Ca, and Mg) and trace elements (Mn, Fe, Zn, and Cu) (Liu 2017; Rasyid 2017; Lunde 1970). Sodium and calcium are the minerals found most abundantly, whereas, for trace minerals, it is zinc, manganese, and arsenic (Ryan et al. 2012; Rupérez 2002). As they are not biosynthesised, variations in mineral content depend on seaweed's bioabsorption and bioaccumulation capacity and its growth environment. Overall, macroalgae tend to have a higher mineral content than terrestrial plants, making them an easy industrial source. Furthermore, seaweeds contain a variety of bioactive compounds such as ulvan, fucoidan, lectin, taurine, squalene, and others. These compounds have wide range of potential applications in variety of industries, including pharmaceutical, food, feed, and biofuel. Several studies have been conducted to investigate the extensive properties of these chemicals as well as their potential industrial applications. (Salehi et al. 2019; Olatunji 2020; Bayu and Handayani 2018).

Seaweed represents an ideal substrate for the sustainable production of biofuels because, unlike terrestrial plants, it does not require freshwater or arable land for production (Zaky 2021). In addition, seaweed has a high carbohydrate content free of lignin, meaning milder processing conditions compared to lignocellulosic biomass (John et al. 2011; van Hal et al. 2014). Many conversion technologies can be applied to the seaweed biomass for the production of different biofuel vectors such as bioethanol, biogas, syngas, and biodiesel. These technologies include chemical (extraction and transesterification), biochemical (anaerobic digestion and fermentation), and thermochemical (hydrothermal liquefaction, combustion, and pyrolysis) technologies, as are summarised in Fig. 4.3. Some of these conversion technologies, such as gasification, combustion, pyrolysis, and transesterification, require dried seaweed. Whereas hydrothermal liquefaction (HTL), hydrolysis pretreatments, fermentation, and anaerobic digestion (AD) can be performed on wet seaweed biomass (Michalak and Chojnacka 2014; Suutari et al. 2015).

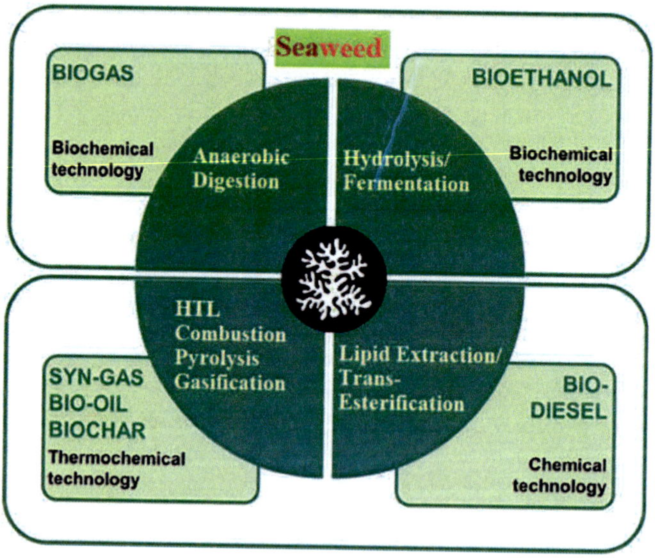

Fig. 4.3 Different conversion technologies and pathways for biofuel production from seaweed

Air-drying of seaweed should be performed under the shadow with occasional turning to preserve the bioactive components from degradation (Michalak 2018). Even for wet processes, partial drying to around 25% of water content is beneficial as it stabilises the biomass and reduces the energy consumption during transportation and processing (Ghadiryanfar et al. 2016). Several studies have investigated the utilisation of seaweed for bioethanol production (Khan and Hussain 2015; Obata et al. 2016; Korzen et al. 2015; Osman et al. 2020b), biogas production (Vanegas and Bartlett 2013; Tabassum et al. 2016; Marquez et al. 2016; Lakshmikandan et al. 2021), and biodiesel production (Bae et al. 2011; Wang et al. 2021; Elshobary et al. 2021). Yet, more research is needed to optimise the process to achieve commercially viable biofuel production from seaweed.

Seaweed was suggested to be used alongside seawater and marine yeast for bioethanol production in a marine biorefinery system that relies solely on marine resources (seawater, seaweed, and marine yeast) (Zaky 2017; Zaky et al. 2021; Zaky 2021). This greatly eliminates the pressures on arable land, freshwater resources, and food crops associated with the use of G1 and G2 biomass. Furthermore, seaweed can serve as a means for marine bioremediation as they have been shown to eliminate heavy metal contaminants in wastewater through bioaccumulation (Henriques et al. 2017; Michalak 2020). Seaweeds can also remove high concentrations of nitrogen and phosphorous from coastal waters (Jiang et al. 2020; Xiao et al. 2017). Therefore, seaweed-based biorefinery would strongly support the sustainability of the marine environment while taking part in global warming mitigation.

4.3.3.2 Microalgae

Microalgae are a broad range of microorganisms that include cyanobacteria (historically), green and red algae, and diatoms (Mata et al. 2010). They are unicellular, primarily photoautotrophic organisms, existing either on their own or in colonies. They are ubiquitous in aquatic environments from cold glacier ice to hot springs and freshwater, seawater and brackish water and feature low down on the food chain as food for higher organisms such as plankton (Norton et al. 1996; Enamala et al. 2018). They are the major driver of the aquatic life cycle (Brodie et al. 2017) because they are the primary producers of the aquatic system. At present, 30,000 species of microalgae have been identified and analysed, but the total number exceeds 50,000 microalgal species (Duong et al. 2015; Behera et al. 2015). Microalgae photosynthetically produce almost 50% of the atmospheric oxygen with great ability for CO_2 sequestrate as they have a high growth rate (up to 50 times quicker than the terrestrial plants) (Chapman 2013; Darwish et al. 2020; Barati et al. 2021). They can fix 183 tons of CO_2 for every 100 tons of microalgal biomass produced and hence have huge potential for global warming mitigation (Chisti 2008).

Most microalgae are autotrophic, so they grow photosynthetically, converting inorganic carbon from the atmosphere to biological macromolecules such as carbohydrates, protein, lipids, and other high-value bioactive chemicals (Khan et al. 2018;

Darwish et al. 2020). Many microalgae species are mixotrophic, and they simultaneously conduct both phototrophic and heterotrophic processes to obtain energy from both organic and inorganic sources of carbon and experience maximum growth. However, some of them are heterotrophs, which only utilises organic carbon such as glucose as the carbon source of energy (Prathima Devi et al. 2013; Darwish et al. 2020).

Microalgae are cultivated in photobioreactors that offer suitable conditions for optimal growth. There are two main types for the photobioreactors including: a) open systems which are usually large open ponds or raceways, and b) closed systems which consisting of usually glass or Perspex tubing or flat plate vessels (Zappi et al. 2019; Ugwu et al. 2008). The main requirements for cultivation are light, a carbon source, and a medium containing inorganic nitrogen, phosphorous, and trace elements. Temperature, agitation, salinity, and pH must also be adjusted in the photobioreactor based on the microalgae strain to sustain a high growth rate (Brennan and Owende 2010). Microalgae grow rapidly with the potential of transforming up to 10% of solar energy they absorb into biomass with theoretical yield biomass of about 77 $g/m^2/day$, corresponding to about 280 ton/ha/year (Formighieri et al. 2012; Melis 2009). However, in practice, microalgae yield is lower in both open and closed culture systems. This is due to the loss of absorbed active radiation (Rodolfi et al. 2009; Béchet et al. 2013). Appropriate shaking and mixing of the culture in the bioreactors are essential to uniform the distribution of light to avail the same energy strength to all cells in order to convert the maximum amount of energy to biomass (Khan et al. 2018).

Studies to explore microalgae as feedstock to produce liquid biofuels started in the 1980s (Behera et al. 2015). Microalgae have been proposed to produce biodiesel as it has a higher lipid content than other biodiesel feedstocks and is not used as a primary food source (Gouveia 2011; Darwish et al. 2020). Most microalgae species are potential biodiesel candidates due to high lipids contents, produced under stress, which usually account for up to 50% of the dry matter and could reach up to 80%, such as in the case of the *B. braunii* and *Schizochytrium* sp. (Chisti 2007; Powell and Hill 2009). Microalgae can produce up to 136,900 L/ha/year algal oil, which can produce up to 121,104 kg^{-1} $ha^{-1}year^{-1}$ biodiesels (Mata et al. 2010). Having a lower land requirement per kilogram of biodiesel produced minimises the land needed to be dedicated to biofuel production, freeing up land for food crop production. This can be improved even further by cultivating marine microalgae on coastal sites, which are usually not suitable for agriculture and near an abundant source of water and minerals (seawater) (Deng et al. 2009; Zaky et al. 2021; Zaky 2021). Table 4.3 provides a comparison between some G1 oil-producing crops and microalgae as candidates for biodiesel production.

Microalgae and cyanobacteria are strong candidates for bioethanol production as they can accumulate 50%, or more, of their dry weight, which can then be hydrolysed and fermented using yeast. Microalgae strains with the potential for bioethanol production are selected primarily per their ability to accumulate carbohydrates, which depends mainly on nutritional and environmental conditions. The main nutritional factors include source, type and concentration of nitrogen, carbon,

Table 4.3 Comparison of various feedstock that are used for the production of biodiesel

Feedstock	Food source	LC (%)	LR (m^2)	WR (kg)	PBP (kg^{-1} ha^{-1} y^{-1})	Reference
Corn	Yes	3–6	172	4015	241 – 438	Sawangkeaw and Ngamprasertsith (2013), Yang et al. (2011), Mata et al. (2010)
Soybean	Yes	18–20	18	13,676	450 – 506	Sawangkeaw and Ngamprasertsith (2013), Yang et al. (2011), Mata et al. (2010)
Canola	Yes	40–45	12	n/a	590 -663	Sawangkeaw and Ngamprasertsith (2013). Mata et al. (2010)
Jatropha	No	30–39	15	n/a	1500	Sawangkeaw and Ngamprasertsith (2013), Lam et al. (2009), Mata et al. (2010)
Palm oil	Yes	45–50	2	n/a	3004 - 5006	Sawangkeaw and Ngamprasertsith (2013), Lam et al. (2009), Mata et al. (2010)
Low lipid microalgae	No	30	0.2	5215 828[a] 745[b]	51,927	Yang et al. (2011), Mata et al. (2010)
Medium lipid microalgae	No	50	0.1	3726 591[a] 532[b]	86,515	Yang et al. (2011), Mata et al. (2010)
High lipid microalgae	No	70	0.1	2235 355[a] 320[b]	121,104	Yang et al. (2011), Mata et al. (2010)

LC lipid content (per dry weight), *LR* land required for kg^{-1} $year^{-1}$ biodiesel, *WR* water required for kg biodiesel, *PBP* potential Biodiesel productivity, *n/a* not available
[a] using recycled harvest water, [b] using seawater

phosphorus, sulphur, and iron, while the main environmental factors are pH, light intensity, temperature, and salinity (Chen et al. 2013; de Farias Silva and Bertucco 2016).

The genera *Chlorella, Chlorococcum, Scenedesmus, Synechococcus,* and *Tetraselmis* have been studied as feedstock for the third generation of bioethanol production. In general, the cultivation under a high light intensity ranging from 150 to 450 μm^{-2} s^{-1} at mesophilic temperature (20–30 °C) and using air supplemented with 2–5% CO_2 achieves around 50% carbohydrate content under nutrient deficiency, mainly nitrogen (de Farias Silva and Bertucco 2016). It was reported that *Scenedesmus obliquus* CNW-N propagated in normal medium accumulated about 50% carbohydrates in their cells (dry weight) after exhaustion of the nitrogen source but the protein content was significantly decreased (Ho et al. 2013b). The growth of *Synechococcus* sp. PCC7002 strain was investigated with various nitrate concentrations. It was found that the depletion of nitrate in the first days of cultivation significantly increased the carbohydrates content to reach about 60% of the dry weight (Möllers et al. 2014).

The carbohydrates in algal biomass are mostly in polymeric form, so they require hydrolysis to fermentable sugars to be readily fermented to bioethanol (Al Abdallah et al. 2016). The yields of fermentable sugars and ethanol differ depending on the species and strains of microalgae. For example, hydrolysis of *Scenedesmus dimorphus* carbohydrate with diluted sulphuric acids yielded 80% fermentable sugars (Chng et al. 2017). Ethanol yield from the fermentation of *Scenedesmus* sp. derived sugars was reported at 93% of the theoretical yield (Sivaramakrishnan and Incharoensakdi 2018). In another study, hydrolysis with 1% sulphuric acid of *C. vulgaris* FSP-E biomass achieved a glucose yield of 93.6% from the microalgal carbohydrates. The fermentation of hydrolysates of this strain reported 11.7 g/L final ethanol concentration which was 87.6% of the theoretical yield (Ho et al. 2013a).

Besides biodiesel and bioethanol, several studies have investigated the use of microalgae as a substrate or co-substrate for biogas production through anaerobic digestion (Varol and Ugurlu 2016; Cavinato et al. 2017; Dębowski et al. 2020). However, despite their great potential, an economically viable commercial scale for biogas production from microalgae is yet to come.

There are many advantages of utilising algal biomass over the conventional G1 and G2 biomass for biofuels production including but not limited to: (a) the ability to grow throughout the year, therefore, algal biomass and oil productivity are much higher than the conventional crops; (b) the growth potential of algal species is very high compared to the conventional crops; (c) the ability to grow under harsh conditions, like saline, brackish water, seawater, which does not affect food production; (e) their water footprint is very low, especially marine algae; (d) they do not require pesticides or herbicides for cultivation; and (f) they have a great ability for bioremediation, and they can even utilise the nitrogen, phosphorus, and the other contaminant in wastewater for their propagation. On the other hand, some limitations are associated with utilising algal biomass as feedstock for biofuel production. The high cultivation cost is one of the main limitations, as harvesting and dewatering of algae biomass require high energy input accounting for about 20–30% of the total cost of production (Behera et al. 2015).

The mitigation of the disadvantages and maximising the advantages of G3 biofuels could be achieved through the co-utilisation of both algae biomass (microalgae and seaweed) in a coastal integrated marine biorefinery system (CIMB). In this system, seaweed is utilised for bioethanol production using seawater and marine yeast. Microalgae are propagated using seawater and the CO_2 produced during bioethanol fermentation and then processed for biodiesel production. High-value chemicals are extracted from the processed algae biomass to offset the production cost and enhance the energy balance (Zaky 2021). More details on the CIMB system are stated in Chap. 20 of this book.

4.3.4 *Fourth-Generation (G4) Biofuel Feedstocks*

Microalgae, the G3 biofuel feedstock, are attractive feedstock for biofuel production due to their high yield, low water and land footprint, ability for bioremediation, and the exceptional ability for CO_2 fixation. However, achieving feasible biofuel production from microalgae requires a major reduction of the operational cost (mainly harvesting and dewatering) and/or a major improvement of the biomass yield and their lipid and carbohydrates content, which can be achieved by genetic engineering. Therefore, genetically modified (GM) microalgae were proposed as a new generation, fourth-generation (G4), feedstock for biofuel production.

Improving photosynthetic efficiency through increasing light penetration and reducing photoinhibition is a common strategy used for the genetic modification of microalgae to increase both biomass productivity and CO_2 sequestration efficiency (Barati et al. 2021; Abdullah et al. 2019). Additionally, the metabolic engineering of microalgae can lead to significant increases in lipid or carbohydrate content. Maximisation of lipids and carbohydrates, as well as increased yield of microalgae biomass, are among the most influential factors in improving bioenergy production efficiency from microalgae. (Naghshbandi et al. 2020; Abdullah et al. 2019). Although there are genome sequences available for more than 30 microalgae, the metabolic pathway is still in the initial stage, but the development of *Chlamydomonas reinhardtii* as a model system for microalgae should facilitate the identification of the genes involved in enzymes and integrate with complex metabolism systems in algae (Banerjee et al. 2016). Nevertheless, the metabolic pathways are very complex, and the main difficulty lies in marking an appropriate pathway capable of producing biofuels (Banerjee et al. 2016).

The cultivation of the GM microalgae can be conducted in closed systems (photobioreactors) or open systems (raceway ponds). The closed systems have tightly controlled conditions, so the contamination and environmental exposure are minimised, but their capital expense is very high. On the other hand, the open systems have lower operating costs than closed systems, but there is an increased risk of the GM algae leaking to the surrounding environment.

Many studies have investigated the potential of GM microalgae for the production of biodiesel, biohydrogen, and bioethanol. Naghshbandi et al. reported more than 25 GM microalgae strains of deferent species and genera that were genetically modified to enhance their lipid content towards enhanced production of G4 biodiesel (Naghshbandi et al. 2020). One of these GM strains, *C. reinhardtii*, showed a 29-fold improvement in triacylglycerides (TAG) content (Naghshbandi et al. 2020; Iwai et al. 2014). Another GM *C. reinhardtii* strain was investigated for the production of G4 biohydrogen which showed enhanced production of up to 13-fold compared to the wild type

(Iwai et al. 2014; Lee et al. 2010). The production of G4 bioethanol was investigated using the hydrolysates of GM microalgae through fermentation by yeast (Chow et al. 2015). In addition, photo-fermentation is an approach for producing G4 bioethanol directly from GM microalgae without hydrolysis and fermentation (de Farias Silva and Bertucco 2016; Deng and Coleman, 1999). It can be also applied to directly produce other G4 biofuels and chemicals such as butanol, acetone, isopropanol, and many others (de Farias Silva and Bertucco 2016; Radakovits et al. 2010).

Although GM microalgae have a great potential as a G4 biomass for biofuel production, there are many concerns and challenges. The main challenge is the difficulty associated with genetically engineer pathways in microalgae because of the complexity of their genome. The main concern with G4 biomass is the risk of contaminating the surrounding environment with GM microalgae, especially when cultivated in open systems. Hence, intensive research is needed to find new technologies and pathways for enhancing the biomass yield, biomass content of lipids and carbohydrates, and the yield of metabolites of the photo-fermentation in GM microalgae. Also, more research to assess the attributed risks to GM microalgae is needed in order to prove their safety on health and the environment. The success of these research could encourage the governments and industry to invest in G4 biofuels.

4.4 Technologies Used for Feedstocks Conversion to Biofuel

4.4.1 Pre-treatment and Hydrolysis

Pre-treatment of lignocellulosic biomass to improve their digestibility has been studied for nearly two centuries (Hassan et al. 2018). The goals have been to remove lignin and structural disruption of cellulose and hemicellulose, facilitating the enzymatic hydrolysis into fermentable sugars (Langsdorf et al. 2021). Lignocellulosic biomass such as wheat and rice straws is highly resistant to hydrolysis due to their rigid structure which is capable of resisting microbial and enzymatic attacks for bioconversion (ElMekawy et al. 2013). This resistance or recalcitrance has been described as the greatest barrier to realising the potential of lignocellulose as an industrial feedstock (Lynd et al. 2008). The major bottleneck in G2 bioethanol production is the high cost associated with the pre-treatment and enzymatic hydrolysis. Pre-treatment is an energy-intensive process and can constitute more than 40% of the total processing cost (Sindhu et al. 2016). There are many pre-treatment methods; however, only a small number are cost-effective at the current time including steam explosion, hydrothermal and dilute acid pre-treatments (Dobbelaere et al. 2014).

Pre-treatment can be physical, chemical, hydrothermal, and enzymatic.

4.4.1.1 Physical Pre-treatment

Physical pre-treatments focus on size reduction of biomass. Depending on the desired final particle size of the material, chipping or grinding or milling process is performed. Particle size of 10–30 mm is usually achieved by chipping, while 0.2–2 mm is often obtained after milling or grinding (Tomás-Pejó et al. 2011). The objective of size reduction is to increase the specific surface area of the biomass while reducing the degree of polymerisation in order to facilitate better enzymatic depolymerisation (Lam et al. 2015; Alvira et al. 2010; Dasari and Eric Berson 2007). In addition, size reduction also allows control of particle size, which also makes the handling of material easier.

To achieve an ethanol concentration of more than 4% (w/w) (the benchmark for an efficient distillation), the sugar concentration in the fermentation medium must be greater than 8% (w/w). To get this sugar concentration from most lignocellulosic biomass, initial solid loading of more than 20% (w/w) is required (Larsen et al. 2008). However, cellulosic slurries become progressively more viscous and difficult to handle at solids concentrations of >15% (w/w). This can be overcome by reducing particle size, which has also been reported to reduce the slurry's viscosity (Viamajala et al. 2009).

As no chemical compounds are needed in this approach, the physical pre-treatments have a smaller environmental impact compared to chemical pre-treatment methods (described below). However, energy consumption is a major limiting factor for fine grinding biomass industrially, compared to most chemical pre-treatment methods (Holtzapple et al. 1991; Licari et al. 2016; Maurya et al. 2015). Energy consumption also differs depending on biomass type. For example, specific energy consumptions for grinding wheat straw and corn stover with a hammer mill under similar parameters conditions were 51.6 and 22.07 kW ht^{-1}, respectively (Mani et al. 2004).

4.4.1.2 Dilute Acid and Alkali Pre-treatment

The most common chemical pre-treatments use diluted acid and alkaline solutions as a catalyst. However, they are generally characterised as corrosive and toxic chemicals that need costly reactors and produce inhibitory compounds as by-products (Jönsson and Martín 2016). Acidic pre-treatments, such as 1–10% sulphuric acid, are the most common pre-treatment for lignocellulosic materials (Badger 2002). Pre-treatment with dilute acid is cheap and effectively hydrolyses hemicelluloses into monomeric sugars, causing transformation in the structure of the lignocellulosic matrix, increasing enzyme accessibility, and improving cellulose conversion (Cao et al. 2012). A downside of the dilute acid pre-treatment method is the formation of by-products that may be inhibitory for the subsequent fermentation process. Sulphuric acid is the acid most often used in this category (Taherzadeh and Karimi 2008). Hydrolysis yields as high as 74% were obtained when the wheat straw is subjected to 0.75% (v/v) of H_2SO_4 for 1 h at 121 °C (Saha et al. 2005a).

Alkaline pre-treatment, on the other hand, is based on the saponification process. Sodium hydroxide, potassium hydroxide, and aqueous ammonia are commonly used in this category. The ammonia and sodium or calcium hydroxides cause the biomass to expand (Rabemanolontsoa and Saka 2016). They act on lignin and hemicelluloses removal and generate fewer inhibitory compounds than acid pre-treatment (Zhang et al. 2016). This mainly is because alkaline pre-treatment could be applied at a lower temperature and pressure which cause less sugar degradation products compared with acid pre-treatment. Pre-treatment with sodium hydroxide has shown to increase the porosity of biomass and surface area (Janu et al. 2011). However, this is at the expense of a longer reaction time (Bali et al. 2015).

4.4.1.3 Hydrothermal Pre-treatment

Hydrothermal pre-treatment includes hot water, steam explosion, and supercritical water pre-treatments. It is mild or less severe compared to dilute acid pre-treatment but results in similar effects on the structure of lignocelluloses (Pu et al. 2013). The reaction medium becomes acidic due to the release of the weak acids, especially acetic acid, from hemicellulose and the auto-dissociation of water at elevated temperatures (hydronium ions) (Gurram et al. 2011; Cantarella et al. 2004). At Above 150 °C, the hemicelluloses followed by lignin begin to solubilise (Bobleter 1994; Garrote et al. 1999). However, the residual lignin still could hinder the enzymatic hydrolysis (Zhuang et al. 2016).

Based on the temperature and pressure used, this treatment can also be divided into: (i) Subcritical water pre-treatment: takes place below the critical point of water, using only water as the catalyst, temperature around 100–375 °C, and sufficient pressure to maintain water in the liquid state, above 50 atm (Allmon and Dorsey 2009; Sánchez and Cardona 2008). ii) Supercritical water pre-treatment: takes place above the critical point of water, using only water as the catalyst, temperature above 374 °C, and corresponding pressure above 218.11 atm (Matsumura et al. 2006; Zhao et al. 2009).

4.4.1.4 Steam Pre-treatment

Steam pre-treatment (or steam explosion) is one of the highly studied pre-treatment methods on a range of lignocellulosic materials (Taherzadeh and Karimi 2007). The biomass is first exposed to saturated steam (typically at 140–260 °C) under high pressure (about 6.81–47.7 atm) for a short time. Here, water penetrates the biomass and facilitates partial hydrolysis of hemicellulose. This is followed by a sudden release of the pressure, which ruptures the fibres. The exposure to steam results in hydrolysis of hemicellulose and activation of lignin, while the pressure drops resulting in disrupting the biomass. In this process, the biomass is exposed to thermal (high temperature), mechanical (sudden vaporisation of water), and chemical (hydrolysis of hemicellulose) processes. The addition of catalyst such as acid catalyst during the treatment or soaking before the treatment has been reported to improve the pre-treatment (Eklund et al. 1995).

4.4.1.5 Microwave Pre-treatment

Hydrothermal pre-treatment can also be conducted assisted by microwave. Microwave radiation heats the biomass in a way that causes disruption of the ligno-cellulosic structures. Microwave pre-treatment has been gradually moved from the laboratory to the pilot scale over the last 30 years (Li et al. 2016). Microwave-assisted pre-treatment technologies of lignocellulosic biomass can be applied using heating temperatures below 200 °C that depolymerise the biomass. Compared to conventional heating, the microwave is more energy-efficient and offers selectivity and uniform heating performance, reducing the formation of degradation products (Dai et al. 2017). This brings about improved enzymatic saccharification via swelling and fragmentation of the fibres (Diaz et al. 2015). Microwave-assisted pre-treatment can also be operated under the thermochemical route for pyrolysis of lignin using high heating temperatures (above 400 °C) to convert biomass to oil and/or gases. This process has been investigated in various biomass sources, including wheat straw, rice straw, willow, miscanthus, and many others (Moirangthem et al. 2021a, b, c).

4.4.1.6 Enzyme Hydrolysis

Enzymatic hydrolysis, especially the conversion of cellulose to fermentable sugars for bioethanol production, is usually an essential step. The rate of cellulose hydrolysis is affected by many factors, including the accessibility of the fibres to enzymes, the structure of cellulose (Balat 2011), the crystallinity of cellulose, and the hemicellulose and lignin content (Sun and Cheng 2002). Typically, two classes of enzymes (cellulases and hemicellulases) are employed together and synergistically to attack cellulose and hemicellulose.

A complex mixture of enzymes termed “cellulases” carries out the degradation of cellulose (Bayer et al. 1998). Cellulases have been conventionally divided into three major groups: (i) Endoglucanases, which cleaves at random amorphous sites in the long chain of the cellulose polysaccharide producing shorter chains of oligosaccharides with different lengths, (ii) Cellobiohydrolases (or exoglucanases) cuts at the reducing and non-reducing ends of the cellulose chains producing glucose or cellobiose. (iii) β-glucosidases, mainly converts soluble cellobiose to glucose (Bai et al. 2017). These three groups of enzymes work synergistically where product of one enzyme becomes the substrate of the another. This reduces the chance of inhibition due to over production of one particular product (Eriksson et al. 2002; Väljamäe et al. 2003). Commercial preparations of cellulolytic cocktails are produced and supplied by several companies such as Novozymes (Denmark), Genecor (Palo Alto, CA), and Iogen (Canada) (De Canio Paola and Patrizia 2011).

The hemicellulosic enzyme system is more complex than cellulases. It involves, for example, glucuronidase, acetylesterase, xylanase, β-xylosidase, galactomannase, and glucomannase (Verardi et al. 2012; Sun and Cheng 2002). Unlike cellulose, a homo-polymer, the hemicellulose is composed of diverse groups of heterogeneous polymers with various side groups. Enzymatic hydrolysis is attractive because it is selective for biomass degradation and can be conducted at mild temperature and pH (Dobbelaere et al. 2014) and does not create a corrosion problem like in acid hydrolysis (Duff and Murray, 1996). However, high enzyme cost is a bottleneck towards its economic feasibility at commercial scale (Ometto et al. 2014; Wei 2016; Hosseini Koupaie et al. 2019).

4.4.2 *Biochemical Conversion (Including the Catalysts)*

4.4.2.1 Fermentation

Fermentation is a well-known process and has been used by humans for thousands of years to make alcoholic beverages and bread, while more recently is also being used to produce biofuels and bio-based chemicals. It was spontaneously carried out long before the biochemical process was thoroughly identified in 1857 by Louis Pasteur, the first scientist to demonstrate that the fermentation process is performed by living cells (Smith 2012).

Fermentation is a biochemical process that occurs in the cells of living organisms to produce energy from sugars in the absence of oxygen. During the fermentation, acids or alcohols and some other compounds are produced as by-products of the process. The industrial biotechnology sector exploited the fermentation phenomena to produce biofuels and high-value chemicals from biomass using microorganisms. The type of products obtained from the fermentation depends mainly on the microorganism, fermentation media, and fermentation conditions used in the fermentation process. Ethanol fermentation by yeast, where glucose is converted to ethanol and CO_2 Fig. 4.4 (Maicas 2020), is one of the most popular types of fermentation. It is being used successfully in the biofuel industry to produce bioethanol commercially from different types of biomasses mainly using the baker's yeast, *S. cerevisiae* (Dev et al. 2019; Zaky et al. 2018a). Ethanol fermentation is conducted commercially in large biorefinery vassals called fermentors or bioreactors with a working volume of up to 250 m^3. There are different types of fermentors that have been designed for different production style. They mainly consist of a cylindrical body with a central rotor from the top equipped with bladed impellers around the shaft used for stirring and a gas sparger for aeration. The fermentor body is surrounded by a water jacket for cooling and contains many ports and probes for different purposes as in Fig. 4.5.

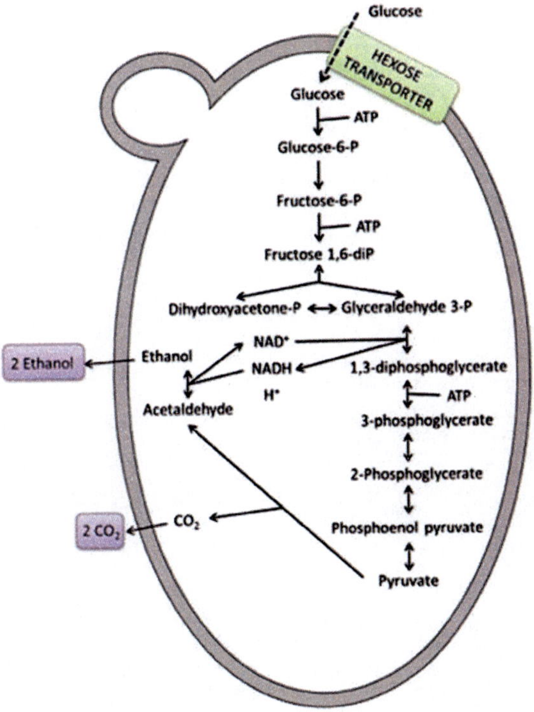

Fig. 4.4 Ethanol fermentation pathway in yeast cell

4.4.2.2 Anaerobic Digestion (AD)

AD is a complex biochemical process that converts organic material to simpler organic compounds in the anaerobic environment by a consortia of anaerobic microorganisms to generate biogas and a semi-solid material (digestate) rich in nutrients that is suitable as a soil fertiliser (Esposito et al. 2012; Lyberatos and Skiadas 1999). Hydrolysis, acidogenesis, acetogenesis, and methanogenesis are the four key processes in AD (Kougias and Angelidaki 2018). Figure 4.6 illustrates the steps involved in anaerobic digestion.

The initial step in AD is hydrolysis. Large molecules such as carbohydrate, protein, and fat are broken down into soluble monomers by hydrolytic bacteria. If a slowly degradable substrate is utilised, hydrolysis is considered as the rate-limiting stage. The hydrolysis of carbohydrates into glucose is shown in Eq. 4.1 (Sattler et al. 2011).

$$C_6H_{10}O_4 + 2H_2O \rightarrow C_6H_{12}O_6 + 2H_2 \tag{4.1}$$

Acidogenic bacteria convert soluble monomers into organic acids such as acetic, propionic, and butyric acids, as well as alcohol, carbon dioxide, and hydrogen during acidogenesis. Some of the reactions involved during this stage are shown in Eq. 4.2 to 4.4 (Sattler et al. 2011).

Fig. 4.5 Diagram of a bioreactor (fermentor) used for bioethanol production. 1—motor; 2—gas outlet; 3—gas/air inlet; 4—probe for inlet of acid; 5—probe for inlet of the base; 6—probe for antifoam; 7—nutrient inlet; 8—ggitator shaft; 9—probe for temperature; 10—impeller; 11—water jacket for cooling; 12—working volume; 13—gas sparger; 14—ingold ports for a pH probe and a dissolved oxygen probe/sensor—15, port for product outlet

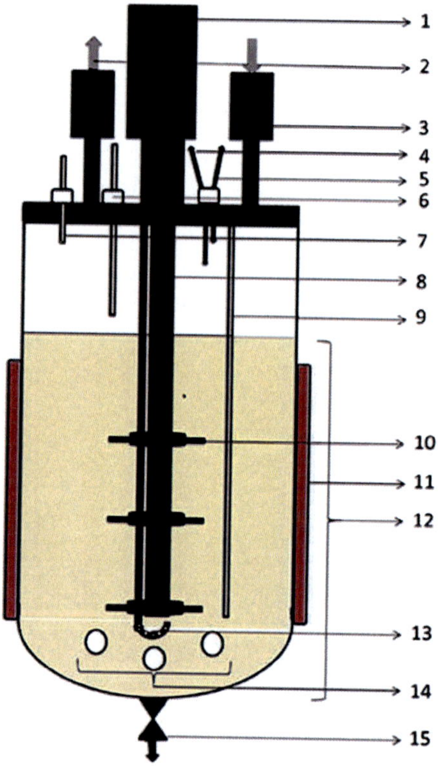

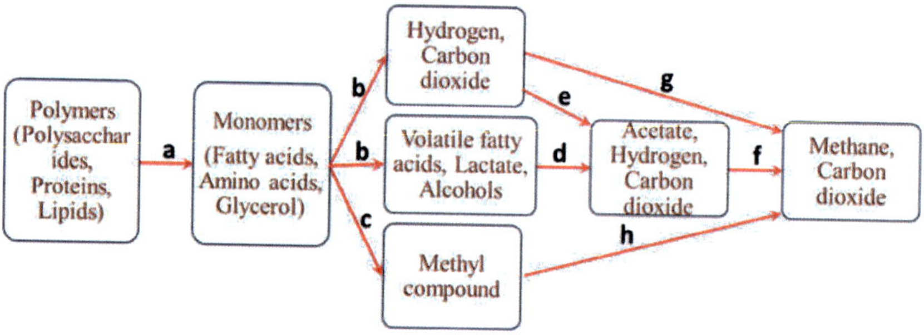

Fig. 4.6 Anaerobic digestion process: a—hydrolysis, b—acidogenesis, c—dimethylation, d—acetogenesis, e—homoacetogenesis, f—acetoclastic methanogenesis, g—hydrogenotrophic methanogenesis, h—methylotrophic methanogenesis

$$C_6H_{12}O_6 \rightarrow 2CH_3CH_2OH + 2CO_2 \tag{4.2}$$

$$C_6H_{12}O_6 + 2H_2 \rightarrow 2CH_3CH_2COOH + 2H_2O \tag{4.3}$$

$$C_6H_{12}O_6 \rightarrow 3CH_3COOH \tag{4.4}$$

Following that, acetogens transform the organic acids and alcohols produced during acidogenesis into acetic acid/acetate, carbon dioxide, and hydrogen. Equations 4.5 and 4.6 show some of the reactions involved during acetogenesis (Sattler et al. 2011).

$$CH_3CH_2COO^- + 3H_2O \rightarrow CH_3COO^- + H^+ + HCO_3^- + 3H_2 \tag{4.5}$$

$$CH_3CH_2OH + H_2O \rightarrow CH_3COO^- + H^+ + 2H_2 \tag{4.6}$$

The final phase in the AD process is methanogenesis. Methanogenic archaea use acetate, simple methylated compounds, alcohols, hydrogen, and carbon dioxide to produce methane, carbon dioxide, and water at this step. Methane can be generated in several ways, including aceticlastic, methylotrophic, and hydrogenotrophic methanogenesis (Lyu et al. 2018). The acetoclastic methanogenesis pathway converts acetate directly into methane and carbon dioxide (Lyu et al. 2018). Aceticlastic methanogens are found to be dominant in environments where hydrogenotrophic methanogens reduce H_2 levels sufficiently to allow for high amounts of acetate production (Lyu et al. 2018). Some acetoclastic methanogens can also use a second pathway called methylotrophic methanogenesis, which uses methanol or methylamines as a substrate to produce methane. The hydrogenotrophic methanogenesis pathway generates methane from hydrogen and carbon dioxide synthesised in preceding steps (Lyu et al. 2018).

4.4.2.3 Lipid Extraction and Transesterification

Lipid extraction is usually being used as the first step in biodiesel production from lipid-rich biomass such as jatropha seeds, palm, jojoba, and microalgae (Sitepu et al. 2020). The methods for extracting lipids are classified as mechanical or chemical. Mechanical extraction processes that do not necessitate chemical aid include bead mills, expeller presses, microwave-assisted pyrolysis extraction, ultrasound-assisted extractions, pulsed electric fields, and hydrothermal liquefaction (Gorgich et al. 2020). Chemical methods are Soxhlet, supercritical fluid, and accelerated solvent extraction (Gorgich et al. 2020).

The typical two-step method (TSP), which comprises a lipid extraction phase of substrate accompanied by lipid transesterification using alcohol, was extensively employed in biodiesel manufacturing (Kim and Yeom 2020). A one-step direct

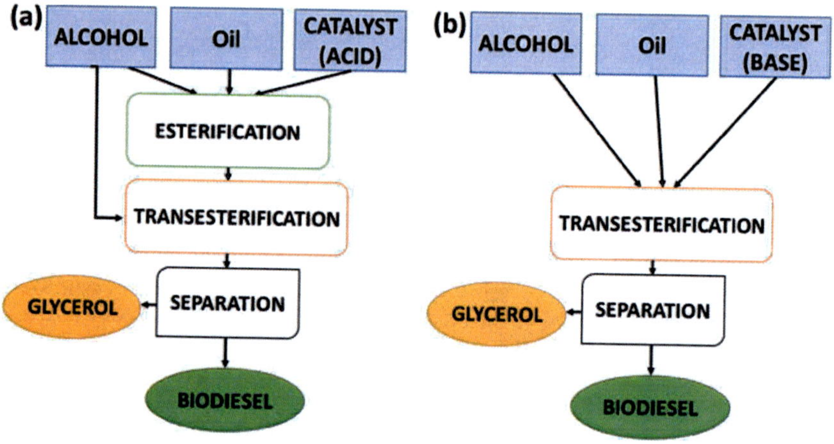

Fig. 4.7 Comparison between **a** two-step versus **b** one-step biodiesel production

process (OSDP) or in-situ transesterification, in which lipid extraction and transesterification happen concurrently in a single reactor, can also be applied to produce biodiesel (Kim and Yeom 2020). Figure 4.7 depicts a schematic diagram for one- and two-step biodiesel production.

Transesterification is a chemical process that converts oil and lipid biomass (fatty acid triglycerides) to fatty acid methyl esters (FAME). It is simply the displacement of an alcohol group with an ester by another alcohol (Azcan and Danisman 2008). Alcohols play a key role in the transesterification process. The most often used alcohols are methanol, ethanol, propanol, butanol, and amyl alcohol, but methanol is widely employed due to its low price and physical advantages (Huang et al. 2010). Transesterification processes are classified into catalytic and non-catalytic. Non-catalytic transesterification is a high-temperature thermochemical process carried out in tubular or bubble column reactors (Tabatabaei and Aghbashlo 2018). Catalytic transesterification, which comprises both homogeneous and heterogeneous catalytic processes, is a popular approach in biodiesel synthesis. Homogeneous catalytic procedures are subdivided into two methods, alkaline and acidic (Mahlia et al. 2020). The application of chemical catalysis for biodiesel production has some drawbacks, including energy-consuming and the production of unwanted by-products, which impede the separation of product from glycerol and di- and mono-acylglycerols, making biocatalysts more appealing (Rizwanul Fattah et al. 2020). Extracellular and intracellular lipases are two types of biocatalysts. Aside from biocatalysts, nanocatalysts have lately gained attention due to their excellent catalytic efficiency under mild operating conditions (Rizwanul Fattah et al. 2020; Chozhavendhan et al. 2020).

4.4.3 Thermochemical Conversion (Including the Catalysts)

Bioenergy can be produced from biomass using two methods: thermochemical and biochemical/biological conversion. Thermochemical conversion is generally more efficient than biochemical/biological processes due to the shorter reaction time and increased capacity to remove most of the organic compounds (Zhang et al. 2010). Thermochemical conversions include hydrothermal liquefaction (HTL), pyrolysis, gasification, combustion, carbonisation, and torrefaction. Table 4.4 summarised the process conditions, benefits, and drawbacks of the thermochemical conversion processes.

4.4.3.1 Hydrothermal Liquefaction (HTL)

HTL is the thermochemical conversion of biomass into liquid products at high temperatures (200–400 °C) and pressures (5–25 MPa) for 20–120 min (Gollakota et al. 2018; Pang 2019). The major products from HTL are bio-oil (with high heating value), water-soluble compounds, biochar, and gases (Toor et al. 2011). Water is commonly utilised as the working medium for HTL technique to improve heat transfer and biomass decomposition (Pang 2019). Water acts as a reactant and a catalyst at high temperatures, causing organic materials to dissolve and reform by introducing hydrogen ions into the hydrocarbons (Patel et al. 2016; Ong et al. 2019). The HTL process is ideal for transforming high-moisture content biomass because no drying is needed. Microalgae, grasses, woody biomass, and manure are examples of common feedstock used in HTL processing (Gollakota et al. 2018; Toor et al. 2011). The key benefits of the HTL method over pyrolysis are improved energy efficiency, lower operating temperature, and reduced tar generation (Ong et al. 2019). The HTL process generates bio-crude with a high heating value, biochar (hydrochar), water-soluble compounds, and gas. Various alkaline catalysts can reduce char formation and boost oil yield and quality (Toor et al. 2011).

4.4.3.2 Pyrolysis

Pyrolysis is the thermochemical decomposition of organic materials using heat in oxygen-free environment to generate liquid bio-oil, solid biochar, and non-condensable gas products (Kan et al. 2016; Venkatachalam et al. 2022). Non-condensable gases include H_2, CH_4, CO, CO_2, and other gaseous hydrocarbons (Ong et al. 2019). The yields of pyrolysis products are determined by process parameters; for instance, when the reaction temperature is below 450 °C, biochar is the main product (Ong et al. 2019). Biomass pyrolysis is classified into three types based on the heating rate and solid residence time: slow pyrolysis, fast pyrolysis, and flash pyrolysis (Venkatachalam et al. 2022; Patel et al. 2020). Slow pyrolysis is applied for biochar production, with relatively low temperatures, a low heating rate, and a very

Table 4.4 Comparison between the different types of thermochemical conversion technologies used for biofuel production

Conversion Method	Reaction Medium	Temp. (°C)	Main Products	Product Gas Compos	Pollutants	Advantages	Disadvantages
Hydrothermal liquefaction	Inert/oxygen-free	250–400	1. Bio-oil 2. Water-soluble products (e.g. organic acids and alcohol) 3. Biochar 4. Syngas	H_2, CO_2, CO, and CH_4	-	1. Does not require preliminary drying 2. High quality of bio-oil 3. Clean process (no or little pollutant)	1. Relatively low oil yield 2. Requires high-pressure equipment, resulting in increased capital costs
Pyrolysis	No oxidant	Slow: 300–700 Fast: 400–800 Flash: 800–1100	1. Bio-oil 2. Biochar 3. Syngas	CO_2, CO, H_2, CH_4, and other hydrocarbons	NO_x, SO_x, NH_3, tar, and ash	1. Direct production of liquid fuels, which can then be directly processed in conventional refineries	1. High energy consumption due to its endothermic nature
Gasification	Partial oxidising atmosphere	>700	1. Syngas	CO, H_2, CO_2, N_2, CH_4, and other hydrocarbons	H_2S, NH_3, tar, and ash	1. Production of various chemical products, such as methanol and other hydrocarbons, provides for flexible market adaption	1. Tar has the potential to clog downstream processes and reduce gasification efficiency

(continued)

Table 4.4 (continued)

Conversion Method	Reaction Medium	Temp. (°C)	Main Products	Product Gas Compos	Pollutants	Advantages	Disadvantages
Combustion	Oxidising atmosphere (oxidant usually higher than the stoichiometric value)	700–1400	1.Syngas 2. Biochar	CO_2, N_2, O_2, and CO	SO_x, NO_x, polycyclic aromatic hydrocarbons (PAHs) and ash	1. Simple technology 2. No changes required for co-combustion of biomass and coal	1. NO_x, SO_x, and particulates are generated. Other potential unburned pollutants, such as CO, PAHs, condensed fumes (tars/oils), soot, and unburned carbon. 2. Elements found in biomass such as Si, K, Na, S, Cl, P, Ca, Mg, and Fe can induce ash fouling and slagging
Carbonisation	Inert/oxygen limited	> 300	1.Biochar 2.Biocoke 3.Bio-oil 4.Syngas	CO, CO_2, and CH_4	Tars, pyroligneous liquors	1. Low cost 2. Large-scale operation	1. Harmful products (CO, tars, and smoke)
Torrefaction	Inert/oxygen limited	200–300	1. Biochar 2. Bio-oil 3. Biogas	CO_2, CO, CH_4, H_2, and other hydrocarbons	Tars	1. Can be applied to biomass with high oxygen and moisture content	1. Flue gas cleaning is needed

high vapour residence time, often 5 to 30 min. Fast pyrolysis is distinguished by a high heating rate with high heat transfer, a short residence time (0.5–10 s), and rapid cooling of aerosols and vapours, resulting in a high bio-oil yield (50–70 wt.%, dry basis). Flash pyrolysis is the rapid degradation of biomass in an inert environment with a high heating rate, high temperatures, and a short vapour residence time (less than 1 s), resulting in extremely high bio-oil yields (75–80 wt.%, dry basis) (Patel et al. 2020; Kan et al. 2016). To boost pyrolysis product yields, catalysts such as AAEMs, zeolites, and quartz sand have been utilised (Ong et al. 2019).

4.4.3.3 Gasification

Gasification is a thermochemical conversion of biomass that generates a large amount of gaseous (CO_2, CH_4, CO, H_2, and gaseous hydrocarbons) and trace amounts of solid and liquid by-products (e.g. ash, chars, tars, and oils) (Pang 2019). The composition of the product gas is influenced by gasification conditions such as temperature, equivalent ratio, and pressure (Ibarra-Gonzalez and Rong 2019; Pereira et al. 2012). Drying, pyrolysis, combustion, and reduction are the four steps in gasification (Ong et al. 2019; Puig-Arnavat et al. 2010). Gasification typically takes place at temperatures above 700 °C. Oxidants or gasifying substances such as O_2, air, and steam are required during the gasification process (Pang 2019; Patel et al. 2016). The advantage of gasification is that it allows for the manufacture of a wide range of chemical products, like methanol and other hydrocarbons, allowing for greater market flexibility. One of the most significant issues encountered during biomass gasification is the development of tar. Tar condenses at low temperatures and polymerises in machinery such as motors and turbines (Pereira et al. 2012). AAEMs and Ni-based catalysts, for example, improve tar conversion into combustion gases (H_2, CO, and CH_4) at reduced temperatures compared to non-catalytic tar conversion. Fixed bed, fluidised bed, and entrained flow are common types of gasifiers (Ong et al. 2019; Pang 2019). Recently, novel gasification technologies have been reported to lower tar content, increase hydrogen fraction in the product gas, and improve the energy efficiency of biomass gasification (Pang 2019).

4.4.3.4 Combustion

Combustion is the most basic and common practice of biomass thermochemical conversion. Heat is generated during the process as a result of the combustion of carbon- and hydrogen-rich feedstock to CO_2 and H_2O (Zhao et al. 2017). Although any sort of biomass can be burned, combustion is only practical for biomass with a moisture content lower than 50% (Goyal et al. 2008). Combustion usually occurs at temperatures ranging from 700 to 1400 °C, with air serving as the reaction medium (Zhao et al. 2017). Combustion is acknowledged as a low-cost and highly reliable technique that is well established and commercially viable. Combustion has the advantage of being a simple technology requiring no modifications for biomass and coal co-combustion. However, this method has several limitations, including the emission of intermediates such as air pollutants as a consequence of incomplete combustion (CH_4, CO, and particulate matter). Additionally, SO_X and NO_X were emitted during the process (Tanger et al. 2013). Also, fouling and slagging of boiler walls and heat exchanger surfaces are common problems with solid biomass combustion (Schneider et al. 2020).

4.4.3.5 Carbonisation

Carbonisation is the thermochemical conversion of biomass into carbon-rich or carbon-containing residues under an inert atmosphere, and it is typically carried out at temperatures above 300 °C (Venkatachalam et al. 2022). It is a complex thermochemical process because several reactions, such as dehydrogenation, hydrogen exchange, condensation, and isomerisation, take place at the same time (Venkatachalam et al. 2022). Carbonisation is accomplished by gradually heating the feedstock at high temperatures over a period of several hours. The primary products of carbonisation include charcoal (fuel), biochar (fertiliser or soil amendment), activated carbon (for adsorption and purification), and biocoke (for metal extraction) (Amer and Elwardany 2020). Carbonisation is performed via the use of various reactors such as kilns, retorts, and converters. A kiln is the classic char-making apparatus, whereas retorts and converters are commercial reactors that can generate char while also capturing some volatiles and bio-oil (Amer and Elwardany 2020).

Torrefaction

Torrefaction is a thermochemical process that, under inert conditions, enhances the energy quality of solid biomass. Moisture and low-weight organic volatiles from biomass are eliminated during the process, resulting in a hydrophobic solid product or biochar with a high heating value and enhanced grindability (Ong et al. 2019). The process is commonly performed at temperatures varying between 200 and 300 °C, with reaction times spanning from 15 to 60 min. (Ong et al. 2019). There are three temperature categories for torrefaction: light (200–235 °C), mild (235–275 °C), and severe (275–300 °C) (Venkatachalam et al. 2022). The main product of torrefaction process is biochar, while the by-products include bio-oil and biogas. Acids, phenols, ketones, alcohols, esters, aldehydes, and water are the constituents of bio-oil. Torrefaction gas products include CO_2, CO, CH_4, H_2, and C_2H_4 (Negi et al. 2020). Torrefaction has the advantage of being able to be applied to biomass that contains significant levels of oxygen and moisture, which other conversion techniques such as combustion, gasification, and pyrolysis cannot. Thus, torrefaction can increase biomass consistency, calorific value, grindability, and biodegradability while decreasing oxygen and moisture content and the biomass's hydrophilic characteristic of the biomass (Venkatachalam et al. 2022; Ong et al. 2019).

References

Abdul Mujeebu M (2016) Hydrogen and syngas production by superadiabatic combustion—a review. Appl Energy 173:210–224. https://doi.org/10.1016/j.apenergy.2016.04.018

Abdullah B, Syed Muhammad SAFa, Shokravi Z, Ismail S, Kassim KA, Mahmood AN, Aziz MMA (2019) Fourth generation biofuel: a review on risks and mitigation strategies. Renew Sustain Energy Rev 107:37–50. https://doi.org/10.1016/j.rser.2019.02.018

Abomohra AE-F, Elsayed M, Esakkimuthu S, El-Sheekh M, Hanelt D (2020) Potential of fat, oil and grease (FOG) for biodiesel production: a critical review on the recent progress and future perspectives. Progr Energy Combus Sci 81:100868. https://doi.org/10.1016/j.pecs.2020.100868

Abraham A, Mathew AK, Sindhu R, Pandey A, Binod P (2016) Potential of rice straw for bio-refining: an overview. Biores Technol 215:29–36. https://doi.org/10.1016/j.biortech.2016.04.011

Acharya B, Roy P, Dutta A (2014) Review of syngas fermentation processes for bioethanol. Biofuels 5(5):551–564

Ademiluyi FT, Mepba HD (2013) Yield and properties of ethanol biofuel produced from different whole cassava flours. ISRN Biotechnol 2013:916481. https://doi.org/10.5402/2013/916481

Aghbashlo M, Peng W, Tabatabaei M, Kalogirou SA, Soltanian S, Hosseinzadeh-Bandbafha H, Mahian O, Lam SS (2021) Machine learning technology in biodiesel research: a review. Progr Energy Combus Sci 85:100904

Ahmad AA, Zawawi NA, Kasim FH, Inayat A, Khasri A (2016) Assessing the gasification performance of biomass: a review on biomass gasification process conditions, optimization and economic evaluation. Renew Sustain Energy Rev 53:1333–1347. https://doi.org/10.1016/j.rser.2015.09.030

Ahmed A, Abu Bakar MS, Sukri RS, Hussain M, Farooq A, Moogi S, Park Y-K (2020) Sawdust pyrolysis from the furniture industry in an auger pyrolysis reactor system for biochar and bio-oil production. Energy Convers Manage 226:113502. https://doi.org/10.1016/j.enconman.2020.113502

Ahmed II, Gupta AK (2012) Sugarcane bagasse gasification: global reaction mechanism of syngas evolution. Appl Energy 91(1):75–81. https://doi.org/10.1016/j.apenergy.2011.07.001

Al Abdallah Q, Nixon BT, Fortwendel JR (2016) The enzymatic conversion of major algal and cyanobacterial carbohydrates to bioethanol. Front Energy Res 4(36). https://doi.org/10.3389/fenrg.2016.00036

Alatzas S, Moustakas K, Malamis D, Vakalis S (2019) Biomass potential from agricultural waste for energetic utilization in Greece. Energies 12(6):1095

Ale MT, Meyer AS (2013) Fucoidans from brown seaweeds: an update on structures, extraction techniques and use of enzymes as tools for structural elucidation. RSC Adv 3(22):8131–8141. https://doi.org/10.1039/C3RA23373A

Allmon SD, Dorsey JG (2009) Retention mechanisms in subcritical water reversed-phase chromatography. J Chromatogr A 1216(26):5106–5111. https://doi.org/10.1016/j.chroma.2009.04.068

Alvira P, Negro MJ, Ballesteros I, González A, Ballesteros M (2016) Steam explosion for wheat straw pretreatment for sugars production. Bioethanol 2(1). https://doi.org/10.1515/bioeth-2016-0003

Alvira P, Tomás-Pejó E, Ballesteros M, Negro MJ (2010) Pretreatment technologies for an efficient bioethanol production process based on enzymatic hydrolysis: a review. Biores Technol 101(13):4851–4861. https://doi.org/10.1016/j.biortech.2009.11.093

Ambye-Jensen M, Johansen KS, Didion T, Kádár Z, Meyer AS (2014) Ensiling and hydrothermal pretreatment of grass: consequences for enzymatic biomass conversion and total monosaccharide yields. Biotechnol Biofuels 7(1):95. https://doi.org/10.1186/1754-6834-7-95

Amer M, Elwardany A (2020) Biomass carbonization. Resour Challenges Appl 211

Angelidaki I, Treu L, Tsapekos P, Luo G, Campanaro S, Wenzel H, Kougias PG (2018) Biogas upgrading and utilization: current status and perspectives. Biotechnol Adv 36(2):452–466. https://doi.org/10.1016/j.biotechadv.2018.01.011

Angelonidi E, Smith SR (2015) A comparison of wet and dry anaerobic digestion processes for the treatment of municipal solid waste and food waste. Water Environ J 29(4):549–557

Arndt C, Pauw K, Thurlow J (2012) Biofuels and economic development: a computable general equilibrium analysis for Tanzania. Energy Economics 34(6):1922–1930. https://doi.org/10.1016/j.eneco.2012.07.020

Azcan N, Danisman A (2008) Microwave assisted transesterification of rapeseed oil. Fuel 87(10):1781–1788. https://doi.org/10.1016/j.fuel.2007.12.004

Azwar M, Hussain M, Abdul-Wahab A (2014) Development of biohydrogen production by photobiological, fermentation and electrochemical processes: a review. Renew Sustain Energy Rev 31:158–173

Bach Q-V, Sillero MV, Tran K-Q, Skjermo J (2014) Fast hydrothermal liquefaction of a Norwegian macro-alga: screening tests. Algal Res 6:271–276. https://doi.org/10.1016/j.algal.2014.05.009

Badger P (2002) Ethanol from cellulose: a general review. Trends New Crops New Uses 14:17–21

Bae YJ, Ryu C, Jeon J-K, Park J, Suh DJ, Suh Y-W, Chang D, Park Y-K (2011) The characteristics of bio-oil produced from the pyrolysis of three marine macroalgae. Biores Technol 102(3):3512–3520. https://doi.org/10.1016/j.biortech.2010.11.023

Bai X, Wang X, Wang S, Ji X, Guan Z, Zhang W, Lu X (2017) Functional studies of β-glucosidases of *Cytophaga hutchinsonii* and their effects on cellulose degradation. Front Microbiol 8(140). https://doi.org/10.3389/fmicb.2017.00140

Baibakova OV, Skiba EA, Budaeva VV, Gismatulina YA, Sakovich GV (2020) Producing bioethanol from Miscanthus: experience of primary scale-up. Catal Ind 12(2):155–161. https://doi.org/10.1134/S2070050420020038

Balat M (2011) Production of bioethanol from lignocellulosic materials via the biochemical pathway: a review. Energy Convers Manage 52(2):858–875. https://doi.org/10.1016/j.enconman.2010.08.013

Balat M, Balat H (2010) Progress in biodiesel processing. Appl Energy 87(6):1815–1835

Bali G, Meng X, Deneff JI, Sun Q, Ragauskas AJ (2015) The effect of alkaline pretreatment methods on cellulose structure and accessibility. Chemsuschem 8(2):275–279. https://doi.org/10.1002/cssc.201402752

Banerjee C, Dubey KK, Shukla P (2016) Metabolic engineering of microalgal based biofuel production: prospects and challenges. Front Microbiol 7(432). https://doi.org/10.3389/fmicb.2016.00432

Barampouti EM, Mai S, Malamis D, Moustakas K, Loizidou M (2019) Liquid biofuels from the organic fraction of municipal solid waste: a review. Renew Sustain Energy Rev 110:298–314. https://doi.org/10.1016/j.rser.2019.04.005

Barati B, Zeng K, Baeyens J, Wang S, Addy M, Gan S-Y, El-Fatah Abomohra A (2021) Recent progress in genetically modified microalgae for enhanced carbon dioxide sequestration. Biomass Bioenergy 145:105927. https://doi.org/10.1016/j.biombioe.2020.105927

Bayer EA, Chanzy H, Lamed R, Shoham Y (1998) Cellulose, cellulases and cellulosomes. Curr Opin Struct Biol 8(5):548–557. https://doi.org/10.1016/S0959-440X(98)80143-7

Bayu A, Handayani T (2018) High-value chemicals from marine macroalgae: opportunities and challenges for marine-based bioenergy development. IOP Conf Ser Earth Environ Sci 209:012046. https://doi.org/10.1088/1755-1315/209/1/012046

Béchet Q, Muñoz R, Shilton A, Guieysse B (2013) Outdoor cultivation of temperature-tolerant *Chlorella sorokiniana* in a column photobioreactor under low power-input. Biotechnol Bioeng 110(1):118–126. https://doi.org/10.1002/bit.24603

Beer LL, Boyd ES, Peters JW, Posewitz MC (2009) Engineering algae for biohydrogen and biofuel production. Curr Opin Biotechnol 20(3):264–271. https://doi.org/10.1016/j.copbio.2009.06.002

Behera S, Singh R, Arora R, Sharma NK, Shukla M, Kumar S (2015) Scope of algae as third generation biofuels. Front Bioeng Biotechnol 2(90). https://doi.org/10.3389/fbioe.2014.00090

Bhadury P, Wright PC (2004) Exploitation of marine algae: biogenic compounds for potential antifouling applications. Planta 219(4):561–578. https://doi.org/10.1007/s00425-004-1307-5

Bobleter O (1994) Hydrothermal degradation of polymers derived from plants. Prog Polym Sci 19(5):797–841. https://doi.org/10.1016/0079-6700(94)90033-7

Bond TC, Doherty SJ, Fahey DW, Forster PM, Berntsen T, DeAngelo BJ, Flanner MG, Ghan S, Kärcher B, Koch D, Kinne S, Kondo Y, Quinn PK, Sarofim MC, Schultz MG, Schulz M, Venkataraman C, Zhang H, Zhang S, Bellouin N, Guttikunda SK, Hopke PK, Jacobson MZ, Kaiser JW, Klimont Z, Lohmann U, Schwarz JP, Shindell D, Storelvmo T, Warren SG, Zender CS (2013) Bounding the role of black carbon in the climate system: a scientific assessment. J Geophys Res Atmos 118(11):5380–5552. https://doi.org/10.1002/jgrd.50171

Borines MG, McHenry MP, de Leon RL (2011) Integrated macroalgae production for sustainable bioethanol, aquaculture and agriculture in Pacific island nations. Biofuels Bioprod Biorefin 5(6):599–608. https://doi.org/10.1002/bbb.310

Boumesbah I, Hachaïchi-Sadouk Z, Ahmia AC (2015) Biofuel production from sunflower oil and determination of fuel properties. In: Progress in clean energy, vol 2. Springer, pp 105–111

Bourguignon D (2015) EU biofuels policy. Dealing with indirect land use change. Members' Research Service

Brennan L, Owende P (2010) Biofuels from microalgae—a review of technologies for production, processing, and extractions of biofuels and co-products. Renew Sustain Energy Rev 14(2):557–577. https://doi.org/10.1016/j.rser.2009.10.009

Brodie J, Ball SG, Bouget F-Y, Chan CX, De Clerck O, Cock JM, Gachon C, Grossman AR, Mock T, Raven JA, Saha M, Smith AG, Vardi A, Yoon HS, Bhattacharya D (2017) Biotic interactions as drivers of algal origin and evolution. 216(3):670-681. https://doi.org/10.1111/nph.14760

Burrett R, Clini C, Dixon R, Eckhart M, El-Ashry M, Gupta D, Haddouche A, Hales D, Hamilton K, House C (2009) Renewable energy policy network for the 21st century

Calderini DF, Castillo FM, Arenas-M A, Molero G, Reynolds MP, Craze M, Bowden S, Milner MJ, Wallington EJ, Dowle A, Gomez LD, McQueen-Mason SJ (2021) Overcoming the trade-off between grain weight and number in wheat by the ectopic expression of expansin in developing seeds leads to increased yield potential. New Phytol 230(2):629–640. https://doi.org/10.1111/nph.17048

Cantarella M, Cantarella L, Gallifuoco A, Spera A, Alfani F (2004) Effect of inhibitors released during steam-explosion treatment of poplar wood on subsequent enzymatic hydrolysis and SSF. Biotechnol Prog 20(1):200–206. https://doi.org/10.1021/bp0257978

Cao NJ, Krishnan MS, Du JX, Gong CS, Ho NWY, Chen ZD, Tsao GT (1996) Ethanol production from corn cob pretreated by the ammonia steeping process using genetically engineered yeast. Biotech Lett 18(9):1013–1018. https://doi.org/10.1007/BF00129723

Cao S, Pu Y, Studer M, Wyman C, Ragauskas AJ (2012) Chemical transformations of *Populus trichocarpa* during dilute acid pretreatment. RSC Adv 2(29):10925–10936. https://doi.org/10.1039/C2RA22045H

Caporgno MP, Pruvost J, Legrand J, Lepine O, Tazerout M, Bengoa C (2016) Hydrothermal liquefaction of *Nannochloropsis oceanica* in different solvents. Biores Technol 214:404–410. https://doi.org/10.1016/j.biortech.2016.04.123

Cavinato C, Ugurlu A, de Godos I, Kendir E, Gonzalez-Fernandez C (2017) Biogas production from microalgae. In: Gonzalez-Fernandez C, Muñoz R (eds) Microalgae-based biofuels and bioproducts. Woodhead Publishing, pp 155–182. https://doi.org/10.1016/B978-0-08-101023-5.00007-8

Cely P, Gascó G, Paz-Ferreiro J, Méndez A (2015) Agronomic properties of biochars from different manure wastes. J Anal Appl Pyrol 111:173–182. https://doi.org/10.1016/j.jaap.2014.11.014

Chandel AK, da Silva SS, Carvalho W, Singh OV (2012) Sugarcane bagasse and leaves: foreseeable biomass of biofuel and bio-products. J Chem Technol Biotechnol 87(1):11–20. https://doi.org/10.1002/jctb.2742

Chandrasekaran A, Subbiah S, Ramachandran S, Narayanasamy S, Bartocci P, Fantozzi F (2019) Natural draft-improved carbonization retort system for biocarbon production from *Prosopis juliflora* Biomass. Energy Fuels 33(11):11113–11124. https://doi.org/10.1021/acs.energyfuels.9b02639

Chang VS, Kaar WE, Burr B, Holtzapple MT (2001) Simultaneous saccharification and fermentation of lime-treated biomass. Biotech Lett 23(16):1327–1333. https://doi.org/10.1023/A:1010594027988

Chapman RL (2013) Algae: the world's most important "plants"—an introduction. Mitig Adapt Strat Glob Change 18(1):5–12. https://doi.org/10.1007/s11027-010-9255-9

Chen C-Y, Zhao X-Q, Yen H-W, Ho S-H, Cheng C-L, Lee D-J, Bai F-W, Chang J-S (2013) Microalgae-based carbohydrates for biofuel production. Biochem Eng J 78:1–10. https://doi.org/10.1016/j.bej.2013.03.006

Chen G, Liu J, Yao J, Qi Y, Yan B (2017) Biodiesel production from waste cooking oil in a magnetically fluidized bed reactor using whole-cell biocatalysts. Energy Convers Manage 138:556–564. https://doi.org/10.1016/j.enconman.2017.02.036

Chin P, Ingledew W (1993) Effect of recycled laboratory backset on fermentation of wheat mashes. J Agric Food Chem 41(7):1158–1163

Chisti Y (2007) Biodiesel from microalgae. Biotechnol Adv 25(3):294–306. https://doi.org/10.1016/j.biotechadv.2007.02.001

Chisti Y (2008) Biodiesel from microalgae beats bioethanol. Trends Biotechnol 26(3):126–131. https://doi.org/10.1016/j.tibtech.2007.12.002

Chiumenti A, da Borso F, Limina S (2018) Dry anaerobic digestion of cow manure and agricultural products in a full-scale plant: efficiency and comparison with wet fermentation. Waste Manage 71:704–710

Chng LM, Lee KT, Chan DCJ (2017) Evaluation on microalgae biomass for bioethanol production. iop conference series: materials science and engineering 206:012018. https://doi.org/10.1088/1757-899x/206/1/012018

Chow T-J, Su H-Y, Tsai T-Y, Chou H-H, Lee T-M, Chang J-S (2015) Using recombinant cyanobacterium (*Synechococcus elongatus*) with increased carbohydrate productivity as feedstock for bioethanol production via separate hydrolysis and fermentation process. Biores Technol 184:33–41. https://doi.org/10.1016/j.biortech.2014.10.065

Chozhavendhan S, Singh MVP, Fransila B, Kumar RP, Devi GK (2020) A review on influencing parameters of biodiesel production and purification processes. Curr Res Green Sustain Chem 1:1–6

Dai L, He C, Wang Y, Liu Y, Yu Z, Zhou Y, Fan L, Duan D, Ruan R (2017) Comparative study on microwave and conventional hydrothermal pretreatment of bamboo sawdust: hydrochar properties and its pyrolysis behaviors. Energy Convers Manage 146:1–7. https://doi.org/10.1016/j.enconman.2017.05.007

Dammer L, Carus M, Piotrowski S, Puente Á, Breitmayer E, Beus Nd, Liptow C (2017) Sustainable first and second-generation bioethanol for Europe: a sustainability assessment in the context of the European Commission's REDII proposal. Ind Biotechnol 13(6):292–300

Darwish R, Gedi MA, Akepach P, Assaye H, Zaky AS, Gray DA (2020) *Chlamydomonas reinhardtii* is a potential food supplement with the capacity to outperform Chlorella and Spirulina. Appl Sci 10(19):6736

Das D (2009) Advances in biohydrogen production processes: an approach towards commercialization. Int J Hydrogen Energy 34(17):7349–7357

Dasari RK, Eric Berson R (2007) The effect of particle size on hydrolysis reaction rates and rheological properties in cellulosic slurries. Appl Biochem Biotechnol 137(1):289–299. https://doi.org/10.1007/s12010-007-9059-x

Dawczynski C, Schubert R, Jahreis G (2007) Amino acids, fatty acids, and dietary fibre in edible seaweed products. Food Chem 103(3):891–899. https://doi.org/10.1016/j.foodchem.2006.09.041

De Canio Paola DBI, Patrizia R (2011) Latest frontiers in the biotechnologies for ethanol production from lignocellulosic biomass. Biofuel Prod Recent Develop Prospects 163

de Farias Silva CE, Bertucco A (2016) Bioethanol from microalgae and cyanobacteria: a review and technological outlook. Process Biochem 51(11):1833–1842. https://doi.org/10.1016/j.procbio.2016.02.016

Dębowski M, Kisielewska M, Kazimierowicz J, Rudnicka A, Dudek M, Romanowska-Duda Z, Zieliński M (2020) The effects of microalgae biomass co-substrate on biogas production from the common agricultural biogas plants feedstock. Energies 13(9):2186

Demirbas A (2007) Importance of biodiesel as transportation fuel. Energy Policy 35(9):4661–4670

Deng M-D, Coleman JR (1999) Ethanol Synthesis by genetic engineering in cyanobacteria. Appl Environ Microbiol 65(2):523–528. https://doi.org/10.1128/AEM.65.2.523-528.1999

Deng X, Li Y, Fei X (2009) Microalgae: a promising feedstock for biodiesel. Af J Microbiol Res 3(13):1008–1014

Dev B, Zaky AS, Jayabalan R (2019) Bioethanol fermentation: the path forward for eco-friendly and sustainable development. In: Technologies for value addition in food products and processes. Apple Academic, pp 233–270

Diaz AB, Moretti MMdS, Bezerra-Bussoli C, Carreira Nunes CdC, Blandino A, da Silva R, Gomes E (2015) Evaluation of microwave-assisted pretreatment of lignocellulosic biomass immersed in alkaline glycerol for fermentable sugars production. Biores Technol 185:316–323. https://doi.org/10.1016/j.biortech.2015.02.112

Divya D, Gopinath L, Christy PM (2015) A review on current aspects and diverse prospects for enhancing biogas production in sustainable means. Renew Sustain Energy Rev 42:690–699

Dizge N, Aydiner C, Imer DY, Bayramoglu M, Tanriseven A, Keskinler B (2009) Biodiesel production from sunflower, soybean, and waste cooking oils by transesterification using lipase immobilized onto a novel microporous polymer. Biores Technol 100(6):1983–1991. https://doi.org/10.1016/j.biortech.2008.10.008

Dobbelaere S, Anthonis T, Soetaert W (2014) Conversion technologies for the production of liquid fuels and biochemicals. In: Cellulosic energy cropping systems, pp 15–30. https://doi.org/10.1002/9781118676332.ch2

Duff SJB, Murray WD (1996) Bioconversion of forest products industry waste cellulosics to fuel ethanol: a review. Biores Technol 55(1):1–33. https://doi.org/10.1016/0960-8524(95)00122-0

Duong VT, Ahmed F, Thomas-Hall SR, Quigley S, Nowak E, Schenk PM (2015) High protein- and high lipid-producing microalgae from northern australia as potential feedstock for animal feed and biodiesel. Front Bioeng Biotechnol 3:53

Eklund R, Galbe M, Zacchi G (1995) The influence of SO_2 and H_2SO_4 impregnation of willow prior to steam pretreatment. Biores Technol 52(3):225–229. https://doi.org/10.1016/0960-8524(95)00042-D

ElMekawy A, Diels L, De Wever H, Pant D (2013) Valorization of cereal based biorefinery byproducts: reality and expectations. Environ Sci Technol 47(16):9014–9027. https://doi.org/10.1021/es402395g

Elsharnouby O, Hafez H, Nakhla G, El Naggar MH (2013) A critical literature review on biohydrogen production by pure cultures. Int J Hydrogen Energy 38(12):4945–4966

Elshobary ME, El-Shenody RA, Abomohra AE-F (2021) Sequential biofuel production from seaweeds enhances the energy recovery: a case study for biodiesel and bioethanol production. Int J Energy Res 45(4):6457–6467. https://doi.org/10.1002/er.6181

Enamala MK, Enamala S, Chavali M, Donepudi J, Yadavalli R, Kolapalli B, Aradhyula TV, Velpuri J, Kuppam C (2018) Production of biofuels from microalgae—a review on cultivation, harvesting, lipid extraction, and numerous applications of microalgae. Renew Sustain Energy Rev 94:49–68. https://doi.org/10.1016/j.rser.2018.05.012

Encinar JM, Pardal A, Sánchez N, Nogales S (2018) Biodiesel by transesterification of rapeseed oil using ultrasound: a kinetic study of base-catalysed reactions. Energies 11(9):2229

Endalew AK, Kiros Y, Zanzi R (2011) Heterogeneous catalysis for biodiesel production from *Jatropha curcas* oil (JCO). Energy 36(5):2693–2700. https://doi.org/10.1016/j.energy.2011.02.010

Eriksson T, Börjesson J, Tjerneld F (2002) Mechanism of surfactant effect in enzymatic hydrolysis of lignocellulose. Enzyme Microb Technol 31(3):353–364. https://doi.org/10.1016/S0141-0229(02)00134-5

Esposito G, Frunzo L, Liotta F, Panico A, Pirozzi F (2012) Bio-methane potential tests to measure the biogas production from the digestion and co-digestion of complex organic substrates. Open Environ Eng J 5(1)

FAO (2016) Rice Market Monitor. vol XIX. Food and Agriculture Organisation of the United Nations

FAOSTAT (2018) http://www.fao.org/faostat/en/#data/QC. Accessed 28 June 2021

Farmanbordar S, Karimi K, Amiri H (2018) Municipal solid waste as a suitable substrate for butanol production as an advanced biofuel. Energy Convers Manage 157:396–408. https://doi.org/10.1016/j.enconman.2017.12.020

Field JL, Richard TL, Smithwick EAH, Cai H, Laser MS, LeBauer DS, Long SP, Paustian K, Qin Z, Sheehan JJ, Smith P, Wang MQ, Lynd LR (2020) Robust paths to net greenhouse gas mitigation and negative emissions via advanced biofuels. Proc Natl Acad Sci 117(36):21968. https://doi.org/10.1073/pnas.1920877117

Formighieri C, Franck F, Bassi R (2012) Regulation of the pigment optical density of an algal cell: filling the gap between photosynthetic productivity in the laboratory and in mass culture. J Biotechnol 162(1):115–123. https://doi.org/10.1016/j.jbiotec.2012.02.021

Frankó B, Galbe M, Wallberg O (2015) Influence of bark on fuel ethanol production from steam-pretreated spruce. Biotechnol Biofuels 8(1):15. https://doi.org/10.1186/s13068-015-0199-x

Gabra FA, Abd-Alla MH, Danial AW, Abdel-Basset R, Abdel-Wahab AM (2019) Production of biofuel from sugarcane molasses by diazotrophic Bacillus and recycle of spent bacterial biomass as biofertilizer inoculants for oil crops. Biocatal Agric Biotechnol 19:101112. https://doi.org/10.1016/j.bcab.2019.101112

Galbe M, Zacchi G (2002) A review of the production of ethanol from softwood. Appl Microbiol Biotechnol 59(6):618–628. https://doi.org/10.1007/s00253-002-1058-9

Gao Z, Zhao H, Li Z, Tan X, Lu X (2012) Photosynthetic production of ethanol from carbon dioxide in genetically engineered cyanobacteria. Energy Environ Sci 5(12):9857–9865. https://doi.org/10.1039/C2EE22675H

Garrote G, Domínguez H, Parajó JC (1999) Hydrothermal processing of lignocellulosic materials. Holz Als Roh- Und Werkstoff 57(3):191–202. https://doi.org/10.1007/s001070050039

Gerpen JV (2005) Biodiesel processing and production. Fuel Process Technol 86(10):1097–1107. https://doi.org/10.1016/j.fuproc.2004.11.005

Ghadiryanfar M, Rosentrater KA, Keyhani A, Omid M (2016) A review of macroalgae production, with potential applications in biofuels and bioenergy. Renew Sustain Energy Rev 54:473–481. https://doi.org/10.1016/j.rser.2015.10.022

Ghimire A, Frunzo L, Pirozzi F, Trably E, Escudie R, Lens PN, Esposito G (2015) A review on dark fermentative biohydrogen production from organic biomass: process parameters and use of by-products. Appl Energy 144:73–95

Gollakota A, Kishore N, Gu S (2018) A review on hydrothermal liquefaction of biomass. Renew Sustain Energy Rev 81:1378–1392

Göransson K, Söderlind U, He J, Zhang W (2011) Review of syngas production via biomass DFBGs. Renew Sustain Energy Rev 15(1):482–492

Gorgich M, Mata T, Martins A, Branco-Vieira M, Caetano N (2020) Comparison of different lipid extraction procedures applied to three microalgal species. Energy Rep 6:477–482

Gouveia L (2011) Microalgae as a feedstock for biofuels. Springer, Heidelberg

Gowlett JAJ (2016) The discovery of fire by humans: a long and convoluted process. Philos Trans R Soc Biol Sci 371(1696):20150164. https://doi.org/10.1098/rstb.2015.0164

Goyal H, Seal D, Saxena R (2008) Bio-fuels from thermochemical conversion of renewable resources: a review. Renew Sustain Energy Rev 12(2):504–517

Green DI, Agu RC, Bringhurst TA, Brosnan JM, Jack FR, Walker GM (2015) Maximizing alcohol yields from wheat and maize and their co-products for distilling or bioethanol production. J Inst Brew 121(3):332–337

Gurram RN, Datta S, Lin YJ, Snyder SW, Menkhaus TJ (2011) Removal of enzymatic and fermentation inhibitory compounds from biomass slurries for enhanced biorefinery process efficiencies. Biores Technol 102(17):7850–7859. https://doi.org/10.1016/j.biortech.2011.05.043

Hagemann N, Spokas K, Schmidt H-P, Kägi R, Böhler MA, Bucheli TD (2018) Activated carbon, biochar and charcoal: linkages and synergies across pyrogenic carbon's ABCs. Water 10(2):182

Hagos FY, Aziz ARA, Sulaiman SA (2014) Trends of syngas as a fuel in internal combustion engines. Adv Mech Eng 6:401587

Hanson S (2019) Soybean oil comprises a larger share of domestic biodiesel production. U.S. Energy Information Administration. https://www.eia.gov/todayinenergy/detail.php?id=39372

Hartmann H (2017) Solid biofuels, fuels, and their characteristics. In: Meyers RA (ed) Encyclopedia of sustainability science and technology. Springer, New York, pp 1–36. https://doi.org/10.1007/978-1-4939-2493-6_245-3

Hassan SS, Williams GA, Jaiswal AK (2018) Emerging technologies for the pretreatment of lignocellulosic biomass. Biores Technol 262:310–318. https://doi.org/10.1016/j.biortech.2018.04.099

Henriques B, Rocha LS, Lopes CB, Figueira P, Duarte AC, Vale C, Pardal MA, Pereira E (2017) A macroalgae-based biotechnology for water remediation: simultaneous removal of Cd, Pb and Hg by living *Ulva lactuca*. J Environ Manage 191:275–289. https://doi.org/10.1016/j.jenvman.2017.01.035

Ho S-H, Huang S-W, Chen C-Y, Hasunuma T, Kondo A, Chang J-S (2013a) Bioethanol production using carbohydrate-rich microalgae biomass as feedstock. Biores Technol 135:191–198. https://doi.org/10.1016/j.biortech.2012.10.015

Ho S-H, Kondo A, Hasunuma T, Chang J-S (2013b) Engineering strategies for improving the CO_2 fixation and carbohydrate productivity of *Scenedesmus obliquus* CNW-N used for bioethanol fermentation. Biores Technol 143:163–171. https://doi.org/10.1016/j.biortech.2013.05.043

Hoekman SK, Broch A, Robbins C, Ceniceros E, Natarajan M (2012) Review of biodiesel composition, properties, and specifications. Renew Sustain Energy Rev 16(1):143–169

Hogarth JR (2017) Evolutionary models of sustainable economic change in Brazil: no-till agriculture, reduced deforestation and ethanol biofuels. Environ Innov Soc Trans 24:130–141. https://doi.org/10.1016/j.eist.2016.08.001

Holtzapple MT, Jun J-H, Ashok G, Patibandla SL, Dale BE (1991) The ammonia freeze explosion (AFEX) process. Appl Biochem Biotechnol 28(1):59–74. https://doi.org/10.1007/BF02922589

Hongzhang C, Liying L (2007) Unpolluted fractionation of wheat straw by steam explosion and ethanol extraction. Biores Technol 98(3):666–676. https://doi.org/10.1016/j.biortech.2006.02.029

Hosseini Koupaie E, Dahadha S, Bazyar Lakeh AA, Azizi A, Elbeshbishy E (2019) Enzymatic pretreatment of lignocellulosic biomass for enhanced biomethane production—a review. J Environ Manage 233:774–784. https://doi.org/10.1016/j.jenvman.2018.09.106

Huang G, Chen F, Wei D, Zhang X, Chen G (2010) Biodiesel production by microalgal biotechnology. Appl Energy 87(1):38–46

Ibarra-Gonzalez P, Rong B-G (2019) A review of the current state of biofuels production from lignocellulosic biomass using thermochemical conversion routes. Chin J Chem Eng 27(7):1523–1535

IEA-ETSAP I (2013) Production of liquid biofuels, technology brief. IEA-ETSAP and IRENA

Irkin L, Yayintas Ö (2018) Pharmacological properties and therapeutic benefits of seaweeds (a review). Int J Trend Sci Res Dev 2:1126–1131

Ishola F, Adelekan D, Mamudu A, Abodunrin T, Aworinde A, Olatunji O, Akinlabi S (2020) Biodiesel production from palm olein: a sustainable bioresource for Nigeria. Heliyon 6(4):e03725. https://doi.org/10.1016/j.heliyon.2020.e03725

Iwai M, Ikeda K, Shimojima M, Ohta H (2014) Enhancement of extraplastidic oil synthesis in *Chlamydomonas reinhardtii* using a type-2 diacylglycerol acyltransferase with a phosphorus starvation-inducible promoter. Plant Biotechnol J 12(6):808–819. https://doi.org/10.1111/pbi.12210

Janu KU, Sindhu R, Binod P, Kuttiraja M, Sukumaran RK, Pandey A (2011) Studies on physico-chemical changes during alkali pretreatment and optimization of hydrolysis conditions to improve sugar yield from bagasse.

Jiang Y, Duarte AV, van den Brink J, Wiebenga A, Zou G, Wang C, de Vries RP, Zhou Z, Benoit I (2016) Enhancing saccharification of wheat straw by mixing enzymes from genetically-modified *Trichoderma reesei* and *Aspergillus niger*. Biotech Lett 38(1):65–70. https://doi.org/10.1007/s10529-015-1951-9

Jiang Z, Liu J, Li S, Chen Y, Du P, Zhu Y, Liao Y, Chen Q, Shou L, Yan X, Zeng J, Chen J (2020) Kelp cultivation effectively improves water quality and regulates phytoplankton community in a turbid, highly eutrophic bay. Sci Total Environ 707:135561. https://doi.org/10.1016/j.scitotenv.2019.135561

John RP, Anisha GS, Nampoothiri KM, Pandey A (2011) Micro and macroalgal biomass: a renewable source for bioethanol. Biores Technol 102(1):186–193. https://doi.org/10.1016/j.biortech.2010.06.139

Jönsson LJ, Martín C (2016) Pretreatment of lignocellulose: formation of inhibitory by-products and strategies for minimizing their effects. Biores Technol 199:103–112. https://doi.org/10.1016/j.biortech.2015.10.009

Jönsson M, Allahgholi L, Sardari RRR, Hreggviðsson GO, Nordberg Karlsson E (2020) Extraction and modification of macroalgal polysaccharides for current and next-generation applications. Molecules 25(4):930

Jørgensen PJ (2009) Biogas-green energy. Aarhus University, Faculty of Agricultural Sciences

Jouzani GS, Aghbashlo M, Tabatabaei M (2020) Biofuels: types, promises, challenges, and role of fungi. In: Fungi in fuel biotechnology. Springer, pp 1–14

Kamaraj M, Ramachandran K, Aravind J (2020) Biohydrogen production from waste materials: benefits and challenges. Int J Environ Sci Technol 17(1):559–576

Kan T, Strezov V, Evans TJ (2016) Lignocellulosic biomass pyrolysis: a review of product properties and effects of pyrolysis parameters. Renew Sustain Energy Rev 57:1126–1140

Kendel M, Wielgosz-Collin G, Bertrand S, Roussakis C, Bourgougnon N, Bedoux G (2015) Lipid composition, fatty acids and sterols in the seaweeds *Ulva armoricana*, and *Solieria chordalis* from Brittany (France): an analysis from nutritional, chemotaxonomic, and antiproliferative activity perspectives. Mar Drugs 13(9):5606–5628. https://doi.org/10.3390/md13095606

Khan AM, Hussain MS (2015) Production of biofuels from marine macroalgae *Melanothamnus afaqhusainii* and *Ulva fasciata*. J Chem Soc Pakistan 37

Khan MI, Shin JH, Kim JD (2018) The promising future of microalgae: current status, challenges, and optimization of a sustainable and renewable industry for biofuels, feed, and other products. Microb Cell Fact 17(1):36. https://doi.org/10.1186/s12934-018-0879-x

Kim JY, Yeom SH (2020) Optimization of biodiesel production from waste coffee grounds by simultaneous lipid extraction and transesterification. Biotechnol Bioprocess Eng 25(2):320–326

Korzen L, Pulidindi IN, Israel A, Abelson A, Gedanken A (2015) Single step production of bioethanol from the seaweed *Ulva rigida* using sonication. RSC Adv 5(21):16223–16229. https://doi.org/10.1039/C4RA14880K

Kotay SM, Das D (2008) Biohydrogen as a renewable energy resource—prospects and potentials. Int J Hydrogen Energy 33(1):258–263

Kougias PG, Angelidaki I (2018) Biogas and its opportunities—a review. Front Environ Sci Eng 12(3):1–12

Kumar A, Adamopoulos S, Jones D, Amiandamhen SO (2021) Forest biomass availability and utilization potential in Sweden: a review. Waste Biomass Valorization 12(1):65–80. https://doi.org/10.1007/s12649-020-00947-0

Kuparinen K, Heinimö J, Vakkilainen E (2014) World's largest biofuel and pellet plants—geographic distribution, capacity share, and feedstock supply. Biofuels Bioprod Biorefin 8(6):747–754. https://doi.org/10.1002/bbb.1516

Lakshmikandan M, Murugesan AG, Wang S, El-Fatah Abomohra A (2021) Optimization of acid hydrolysis on the green seaweed *Valoniopsis pachynema* and approach towards mixotrophic

microalgal biomass and lipid production. Renewable Energy 164:1052–1061. https://doi.org/10.1016/j.renene.2020.10.062
Laluce C, Leite GR, Zavitoski BZ, Zamai TT, Ventura R (2016) Fermentation of sugarcane juice and molasses for ethanol production. In: Sugarcane-based biofuels and bioproducts, pp 53–86. https://doi.org/10.1002/9781118719862.ch3
Lam MK, Lee KT, Mohamed AR (2009) Life cycle assessment for the production of biodiesel: acase study in Malaysia for palm oil versus jatropha oil. Biofuels Bioprod Biorefin 3(6):601–612. https://doi.org/10.1002/bbb.182
Lam WC, Kwan TH, Budarin VL, Mubofu EB, Fan J, Lin CSK (2015) Pretreatment and thermochemical and biological processing of biomass. In: Introduction to chemicals from biomass, pp 53–88. https://doi.org/10.1002/9781118714478.ch3
Lamichhane G, Khadka S, Adhikari S, Koirala N, Poudyal DP (2020) Biofuel production from waste cooking oils and its physicochemical properties in comparison to petrodiesel. Nepal J Biotechnol 8(3):87–94
Langsdorf A, Volkmar M, Holtmann D, Ulber R (2021) Material utilization of green waste: a review on potential valorization methods. Bioresour Bioprocess 8(1):19. https://doi.org/10.1186/s40643-021-00367-5
Larsen J, Østergaard Petersen M, Thirup L, Wen Li H, Krogh Iversen F (2008) The IBUS process—lignocellulosic bioethanol close to a commercial reality. Chem Eng Technol 31(5):765–772. https://doi.org/10.1002/ceat.200800048
Laser M, Lynd L (2014) Introduction to Cellulosic energy crops. In: Cellulosic energy cropping systems. pp 1–14. https://doi.org/10.1002/9781118676332.ch1
Lee HS, Vermaas WF, Rittmann BE (2010) Biological hydrogen production: prospects and challenges. Trends Biotechnol 28(5):262–271. https://doi.org/10.1016/j.tibtech.2010.01.007
Lee W-C, Kuan W-C (2015) Miscanthus as cellulosic biomass for bioethanol production. Biotechnol J 10(6):840–854. https://doi.org/10.1002/biot.201400704
Lehmann J, Gaunt J, Rondon M (2006) Bio-char sequestration in terrestrial ecosystems—a review. Mitig Adapt Strat Glob Change 11(2):403–427. https://doi.org/10.1007/s11027-005-9006-5
Levin DB, Pitt L, Love M (2004) Biohydrogen production: prospects and limitations to practical application. Int J Hydrogen Energy 29(2):173–185. https://doi.org/10.1016/S0360-3199(03)00094-6
Lewandowski I, Clifton-Brown JC, Scurlock JMO, Huisman W (2000) Miscanthus: european experience with a novel energy crop. Biomass Bioenerg 19(4):209–227. https://doi.org/10.1016/S0961-9534(00)00032-5
Li H-g, Luo W, Wang Q, Yu X-b (2014) Direct Fermentation of gelatinized cassava starch to acetone, butanol, and ethanol using *Clostridium acetobutylicum* mutant obtained by atmospheric and room temperature plasma. Appl Biochem Biotechnol 172(7):3330–3341. https://doi.org/10.1007/s12010-014-0765-x
Li H, Qu Y, Yang Y, Chang S, Xu J (2016) Microwave irradiation—a green and efficient way to pretreat biomass. Biores Technol 199:34–41. https://doi.org/10.1016/j.biortech.2015.08.099
Licari A, Monlau F, Solhy A, Buche P, Barakat A (2016) Comparison of various milling modes combined to the enzymatic hydrolysis of lignocellulosic biomass for bioenergy production: glucose yield and energy efficiency. Energy 102:335–342. https://doi.org/10.1016/j.energy.2016.02.083
Lisowski A, Olendzki D, Świętochowski A, Dąbrowska M, Mieszkalski L, Ostrowska-Ligęza E, Stasiak M, Klonowski J, Piątek M (2019) Spent coffee grounds compaction process: its effects on the strength properties of biofuel pellets. Renewable Energy 142:173–183. https://doi.org/10.1016/j.renene.2019.04.114
Liu B, Wang F, Zhang B, Bi J (2013) Energy balance and GHG emissions of cassava-based fuel ethanol using different planting modes in China. Energy Policy 56:210–220. https://doi.org/10.1016/j.enpol.2012.12.052

Liu K (2017) Characterization of ash in algae and other materials by determination of wet acid indigestible ash and microscopic examination. Algal Res 25:307–321. https://doi.org/10.1016/j.algal.2017.04.014

Lunde G (1970) Analysis of trace elements in seaweed. J Sci Food Agric 21(8):416–418. https://doi.org/10.1002/jsfa.2740210806

Lyberatos G, Skiadas I (1999) Modelling of anaerobic digestion–a review. Global Nest Int J 1(2):63–76

Lynd LR, Laser MS, Bransby D, Dale BE, Davison B, Hamilton R, Himmel M, Keller M, McMillan JD, Sheehan J, Wyman CE (2008) How biotech can transform biofuels. Nat Biotechnol 26(2):169–172. https://doi.org/10.1038/nbt0208-169

Lyu Z, Shao N, Akinyemi T, Whitman WB (2018) Methanogenesis. Curr Biol 28(13):R727–R732

Ma F, Hanna MA (1999) Biodiesel production: a review1Journal Series #12109, Agricultural Research Division, Institute of Agriculture and Natural Resources, University of Nebraska–Lincoln.1. Bioresour Technol 70(1):1–15. https://doi.org/10.1016/S0960-8524(99)00025-5

Mahlia TMI, Syazmi ZAHS, Mofijur M, Abas AEP, Bilad MR, Ong HC, Silitonga AS (2020) Patent landscape review on biodiesel production: technology updates. Renew Sustain Energy Rev 118:109526. https://doi.org/10.1016/j.rser.2019.109526

Maicas S (2020) The role of yeasts in fermentation processes. Microorganisms 8(8):1142. https://doi.org/10.3390/microorganisms8081142

Mani S, Tabil LG, Sokhansanj S (2004) Grinding performance and physical properties of wheat and barley straws, corn stover and switchgrass. Biomass Bioenerg 27(4):339–352. https://doi.org/10.1016/j.biombioe.2004.03.007

Manish S, Banerjee R (2008) Comparison of biohydrogen production processes. Int J Hydrogen Energy 33(1):279–286. https://doi.org/10.1016/j.ijhydene.2007.07.026

Manyà JJ (2012) Pyrolysis for biochar purposes: a review to establish current knowledge gaps and research needs. Environ Sci Technol 46(15):7939–7954. https://doi.org/10.1021/es301029g

Marinho-Soriano E, Fonseca PC, Carneiro MAA, Moreira WSC (2006) Seasonal variation in the chemical composition of two tropical seaweeds. Biores Technol 97(18):2402–2406. https://doi.org/10.1016/j.biortech.2005.10.014

Marques S, Moreno AD, Ballesteros M, Gírio F (2018) Starch biomass for biofuels, biomaterials, and chemicals. Biomass Green Chem 69–94

Marquez GPB, Noboribayashi T, Takahashi T, Fujii Y, Takeuchi H, Hasegawa T (2016) Performance of semi-continuous fixed-bed digester for thalassic biogas production of the brown seaweed, *Ecklonia* spp. Energy Sources Part a: Recovery Utilization Environ Effects 38(18):2782–2787. https://doi.org/10.1080/15567036.2015.1120820

Mata TM, Martins AA, Caetano NS (2010) Microalgae for biodiesel production and other applications: a review. Renew Sustain Energy Rev 14(1):217–232. https://doi.org/10.1016/j.rser.2009.07.020

Mathur S, Umakanth AV, Tonapi VA, Sharma R, Sharma MK (2017) Sweet sorghum as biofuel feedstock: recent advances and available resources. Biotechnol Biofuels 10(1):146. https://doi.org/10.1186/s13068-017-0834-9

Matsumura Y, Sasaki M, Okuda K, Takami S, Ohara S, Umetsu M, Adschiri T (2006) Supercritical water treatment of biomass for energy and material recovery. Combust Sci Technol 178(1–3):509–536. https://doi.org/10.1080/00102200500290815

Maurya DP, Singla A, Negi S (2015) An overview of key pretreatment processes for biological conversion of lignocellulosic biomass to bioethanol. 3 Biotech 5(5):597–609. https://doi.org/10.1007/s13205-015-0279-4

Melekwe EI, Lateef SA, Ana GREE (2016) Bioethanol production potentials of corn cob, waste office paper and leaf of *Thaumatococcus daniellii*. Curr J Appl Sci Technol 1–10

Melis A (2009) Solar energy conversion efficiencies in photosynthesis: minimizing the chlorophyll antennae to maximize efficiency. Plant Sci 177(4):272–280. https://doi.org/10.1016/j.plantsci.2009.06.005

Meng F, Dornau A, McQueen Mason SJ, Thomas GH, Conradie A, McKechnie J (2021) Bioethanol from autoclaved municipal solid waste: assessment of environmental and financial viability under policy contexts. Appl Energy 298:117118. https://doi.org/10.1016/j.apenergy.2021.117118

Michalak I (2018) Experimental processing of seaweeds for biofuels. WIREs Energy Environ 7(3):e288. https://doi.org/10.1002/wene.288

Michalak I (2020) The application of seaweeds in environmental biotechnology. In: Bourgougnon N (ed) Advances in botanical research, vol 95. Academic, pp 85–111. https://doi.org/10.1016/bs.abr.2019.11.006

Michalak I, Chojnacka K (2014) Algal extracts: technology and advances. Eng Life Sci 14(6):581–591. https://doi.org/10.1002/elsc.201400139

Miettinen I, Kuittinen S, Paasikallio V, Mäkinen M, Pappinen A, Jänis J (2017) Characterization of fast pyrolysis oil from short-rotation willow by high-resolution Fourier transform ion cyclotron resonance mass spectrometry. Fuel 207:189–197. https://doi.org/10.1016/j.fuel.2017.06.053

Mohanty SK, Swain MR (2019) Bioethanol production from corn and wheat: food, fuel, and future. In: Ray RC, Ramachandran S (eds) Bioethanol production from food crops. Academic, pp 45–59. https://doi.org/10.1016/B978-0-12-813766-6.00003-5

Moirangthem K (2017) Overcoming challenges for the biorefineries of the future. University of Nottingham

Moirangthem K, Baxter D (2016) Alternative fuels for marine and inland waterways. European Commission

Moirangthem K, Waldron P, Tucker G, King I, King J (2015) Dual purpose wheat: for a future that demands more food and energy. Asp Appl Biol 131:163–167

Moirangthem K, Greetham D, Craigon J, Tucker G (2021a) Impact of mixed feedstock of wheat straw, willow and Miscanthus on enzymatic hydrolysis and inhibitor production after microwave hydrothermal pre-treatment in Europe. Biofuels 1–8. https://doi.org/10.1080/17597269.2021.1894782

Moirangthem K, Ramakrishna P, Amer MH, Tucker GA (2021b) Bioactivity and anthocyanin content of microwave-assisted subcritical water extracts of Manipur black rice (Chakhao) bran and straw. Future Foods 3:100030. https://doi.org/10.1016/j.fufo.2021.100030

Moirangthem K, Zaky AS, Tucker GA (2021c) Microwave subcritical water pre-treatment and enzymatic hydrolysis of geographical identification (GI) tag Indian black rice (Chakhao Poireiton) straw for fermentable sugar production. Biofuels 1–8. https://doi.org/10.1080/17597269.2021.1955585

Möllers KB, Cannella D, Jørgensen H, Frigaard N-U (2014) Cyanobacterial biomass as carbohydrate and nutrient feedstock for bioethanol production by yeast fermentation. Biotechnol Biofuels 7(1):64. https://doi.org/10.1186/1754-6834-7-64

Mona S, Kumar SS, Kumar V, Parveen K, Saini N, Deepak B, Pugazhendhi A (2020) Green technology for sustainable biohydrogen production (waste to energy): a review. Sci Total Environ 728:138481. https://doi.org/10.1016/j.scitotenv.2020.138481

Mondal P, Dang GS, Garg MO (2011) Syngas production through gasification and cleanup for downstream applications—recent developments. Fuel Process Technol 92(8):1395–1410. https://doi.org/10.1016/j.fuproc.2011.03.021

Mosier NS, Ileleji KE (2015) How fuel ethanol is made from corn. In: Dahiya A (ed) Bioenergy. Academic, Boston, pp 379–384. https://doi.org/10.1016/B978-0-12-407909-0.00023-7

Mothe S, Polisetty VR (2021) Review on anaerobic digestion of rice straw for biogas production. Environ Sci Pollut Res 28(19):24455–24469. https://doi.org/10.1007/s11356-020-08762-9

Mudryk K, Jewiarz M, Wróbel M, Niemiec M, Dyjakon A (2021) Evaluation of urban tree leaf biomass-potential, physico-mechanical and chemical parameters of raw material and solid biofuel. Energies 14(4):818

Nabukalu C, Gieré R (2019) Charcoal as an Energy resource: global trade, production and socioeconomic practices observed in Uganda. Resources 8(4):183

Naghshbandi MP, Tabatabaei M, Aghbashlo M, Aftab MN, Iqbal I (2020) Metabolic engineering of microalgae for biofuel production. In: Spilling K (ed) Biofuels from algae: methods and protocols. Springer, New York pp 153–172. https://doi.org/10.1007/7651_2018_205

Negi S, Jaswal G, Dass K, Mazumder K, Elumalai S, Roy JK (2020) Torrefaction: a sustainable method for transforming of agri-wastes to high energy density solids (biocoal). Rev Environ Sci Biotechnol 19(2):463–488. https://doi.org/10.1007/s11157-020-09532-2

Neves PV, Pitarelo AP, Ramos LP (2016) Production of cellulosic ethanol from sugarcane bagasse by steam explosion: effect of extractives content, acid catalysis and different fermentation technologies. Biores Technol 208:184–194. https://doi.org/10.1016/j.biortech.2016.02.085

Ngan NVC, Chan FMS, Nam TS, Van Thao H, Maguyon-Detras MC, Hung DV, Cuong DM, Van Hung N (2020) Anaerobic digestion of rice straw for biogas production. In: Gummert M, Hung NV, Chivenge P, Douthwaite B (eds) Sustainable rice straw management. Springer International Publishing, Cham, pp 65–92. https://doi.org/10.1007/978-3-030-32373-8_5

Norton TA, Melkonian M, Andersen RA (1996) Algal biodiversity. Phycologia 35(4):308–326. https://doi.org/10.2216/i0031-8884-35-4-308.1

Obata O, Akunna J, Bockhorn H, Walker G (2016) Ethanol production from brown seaweed using non-conventional yeasts. Bioethanol 2(1):134–145. https://doi.org/10.1515/bioeth-2016-0010

OECD (2016) Biofuels. https://doi.org/10.1787/agr_outlook-2016-13-en

Olatunji O (2020) Aquatic Biopolymers. Springer

Oliveira GdLT, Schneider M (2016) The politics of flexing soybeans: China, Brazil and global agroindustrial restructuring. J Peasant Stud 43(1):167–194. https://doi.org/10.1080/03066150.2014.993625

Ometto F, Quiroga G, Pšenička P, Whitton R, Jefferson B, Villa R (2014) Impacts of microalgae pre-treatments for improved anaerobic digestion: thermal treatment, thermal hydrolysis, ultrasound and enzymatic hydrolysis. Water Res 65:350–361. https://doi.org/10.1016/j.watres.2014.07.040

Ong HC, Chen W-H, Farooq A, Gan YY, Lee KT, Ashokkumar V (2019) Catalytic thermochemical conversion of biomass for biofuel production: a comprehensive review. Renewable and Sustainable Energy Reviews 113:109266. https://doi.org/10.1016/j.rser.2019.109266

Osman AI, Deka TJ, Baruah DC, Rooney DW (2020a) Critical challenges in biohydrogen production processes from the organic feedstocks. Biomass Convers Biorefinery. https://doi.org/10.1007/s13399-020-00965-x

Osman MEH, Abo-Shady AM, Elshobary ME, Abd El-Ghafar MO, Abomohra AE-F (2020b) Screening of seaweeds for sustainable biofuel recovery through sequential biodiesel and bioethanol production. Environ Sci Pollut Res 27(26):32481–32493. https://doi.org/10.1007/s11356-020-09534-1

Pal A, Kamthania MC, Kumar A (2014) Bioactive compounds and properties of seaweeds—a review. Open Access Libr J 1(4):1–17

Pandey A, Soccol CR, Nigam P, Soccol VT (2000) Biotechnological potential of agro-industrial residues. I: sugarcane bagasse. Bioresour Technol 74(1):69–80. https://doi.org/10.1016/S0960-8524(99)00142-X

Pang S (2019) Advances in thermochemical conversion of woody biomass to energy, fuels and chemicals. Biotechnol Adv 37(4):589–597. https://doi.org/10.1016/j.biotechadv.2018.11.004

Panichelli L, Dauriat A, Gnansounou E (2009) Life cycle assessment of soybean-based biodiesel in Argentina for export. The Int J Life Cycle Assessment 14(2):144–159. https://doi.org/10.1007/s11367-008-0050-8

Parameswaran B (2009) Sugarcane bagasse. In: Biotechnology for agroindustrial residues utilization. Springer, Netherlands

Patel A, Agrawal B, Rawal BR (2020) Pyrolysis of biomass for efficient extraction of biofuel. Energy Sources, Part A Recovery Utilization Environ Effects 42(13):1649–1661. https://doi.org/10.1080/15567036.2019.1604875

Patel M, Zhang X, Kumar A (2016) Techno-economic and life cycle assessment on lignocellulosic biomass thermochemical conversion technologies: a review. Renew Sustain Energy Rev 53:1486–1499. https://doi.org/10.1016/j.rser.2015.09.070

Perea-Moreno M-A, Samerón-Manzano E, Perea-Moreno A-J (2019) Biomass as renewable energy: worldwide research trends. Sustainability 11(3):863

Perego P, Converti A, Palazzi E, Del Borghi M, Ferraiolo G (1990) Fermentation of hardwood hemicellulose hydrolysate by *Pachysolen tannophilus*, *Candida shehatae* and *Pichia stipitis*. J Ind Microbiol 6(3):157–164. https://doi.org/10.1007/BF01577690

Pereira EG, da Silva JN, de Oliveira JL, Machado CS (2012) Sustainable energy: a review of gasification technologies. Renew Sustain Energy Rev 16(7):4753–4762. https://doi.org/10.1016/j.rser.2012.04.023

Phan AN, Phan TM (2008) Biodiesel production from waste cooking oils. Fuel 87(17):3490–3496. https://doi.org/10.1016/j.fuel.2008.07.008

Picchio R, Latterini F, Venanzi R, Stefanoni W, Suardi A, Tocci D, Pari L (2020) Pellet production from woody and non-woody feedstocks: a review on biomass quality evaluation. Energies 13(11):2937

Picchio R, Spina R, Sirna A, Monaco AL, Civitarese V, Giudice AD, Suardi A, Pari L (2012) Characterization of woodchips for energy from forestry and agroforestry production. Energies 5(10):3803–3816

Pimentel D, Patzek TW (2005) Ethanol production using corn, switchgrass, and wood; biodiesel production using soybean and sunflower. Nat Resour Res 14(1):65–76. https://doi.org/10.1007/s11053-005-4679-8

Pinzi S, Garcia IL, Lopez-Gimenez FJ, Luque de Castro MD, Dorado G, Dorado MP (2009) The ideal vegetable oil-based biodiesel composition: a review of social, economical and technical implications. Energy Fuels 23(5):2325–2341. https://doi.org/10.1021/ef801098a

Pisupati SV, Tchapda AH (2015) Thermochemical processing of biomass. In: Advances in bioprocess technology. Springer, pp 277–314

Powell EE, Hill GA (2009) Economic assessment of an integrated bioethanol–biodiesel–microbial fuel cell facility utilizing yeast and photosynthetic algae. Chem Eng Res Des 87(9):1340–1348. https://doi.org/10.1016/j.cherd.2009.06.018

Prathima Devi M, Swamy YV, Venkata Mohan S (2013) Nutritional mode influences lipid accumulation in microalgae with the function of carbon sequestration and nutrient supplementation. Biores Technol 142:278–286. https://doi.org/10.1016/j.biortech.2013.05.001

Proskurina S, Junginger M, Heinimö J, Tekinel B, Vakkilainen E (2019) Global biomass trade for energy— Part 2: Production and trade streams of wood pellets, liquid biofuels, charcoal, industrial roundwood and emerging energy biomass. Biofuels, Bioproducts Biorefin 13(2):371–387. https://doi.org/10.1002/bbb.1858

Pu Y, Hu F, Huang F, Davison BH, Ragauskas AJ (2013) Assessing the molecular structure basis for biomass recalcitrance during dilute acid and hydrothermal pretreatments. Biotechnol Biofuels 6(1):15. https://doi.org/10.1186/1754-6834-6-15

Puig-Arnavat M, Bruno JC, Coronas A (2010) Review and analysis of biomass gasification models. Renew Sustain Energy Rev 14(9):2841–2851. https://doi.org/10.1016/j.rser.2010.07.030

Rabemanolontsoa H, Saka S (2016) Various pretreatments of lignocellulosics. Biores Technol 199:83–91. https://doi.org/10.1016/j.biortech.2015.08.029

Radakovits R, Jinkerson RE, Darzins A, Posewitz MC (2010) Genetic engineering of algae for enhanced biofuel production. Eukaryot Cell 9(4):486–501. https://doi.org/10.1128/EC.00364-09

Rajendran K, Lin R, Wall DM, Murphy JD (2019) Influential aspects in waste management practices. In: Taherzadeh MJ, Bolton K, Wong J, Pandey A (eds) Sustainable resource recovery and zero waste approaches. Elsevier, pp 65–78. https://doi.org/10.1016/B978-0-444-64200-4.00005-0

Rasyid A (2017) Evaluation of nutritional composition of the dried seaweed *Ulva lactuca* from Pameungpeuk Waters, Indonesia. Trop Life Sci Res 28(2):119–125. https://doi.org/10.21315/tlsr2017.28.2.9

Ratnavathi CV, Chakravarthy SK, Komala VV, Chavan UD, Patil JV (2011) Sweet sorghum as feedstock for biofuel production: a review. Sugar Tech 13(4):399–407. https://doi.org/10.1007/s12355-011-0112-2

Rauch R, Hrbek J, Hofbauer H (2014) Biomass gasification for synthesis gas production and applications of the syngas. Wires Energy Environ 3(4):343–362. https://doi.org/10.1002/wene.97

Raza G, Ali K, Hassan MA, Ashraf M, Khan MT, Khan IA (2019) Sugarcane as a bioenergy source. In: Sugarcane biofuels. Springer, pp 3–19

Rezki B, Essamlali Y, Aadil M, Semlal N, Zahouily M (2020) Biodiesel production from rapeseed oil and low free fatty acid waste cooking oil using a cesium modified natural phosphate catalyst. RSC Adv 10(67):41065–41077. https://doi.org/10.1039/D0RA07711A

Rico JAP (2007) Programa de Biocombustíveis no Brasil e na Colômbia: uma análise da implantação, resultados e perspectivas. Universidade de São Paulo

Rizwanul Fattah IM, Ong HC, Mahlia TMI, Mofijur M, Silitonga AS, Rahman SMA, Ahmad A (2020) State of the art of catalysts for biodiesel production. Front Energy Res 8(101). https://doi.org/10.3389/fenrg.2020.00101

Röder M, Thiffault E, Martínez-Alonso C, Senez-Gagnon F, Paradis L, Thornley P (2019) Understanding the timing and variation of greenhouse gas emissions of forest bioenergy systems. Biomass Bioenerg 121:99–114. https://doi.org/10.1016/j.biombioe.2018.12.019

Rodolfi L, Chini Zittelli G, Bassi N, Padovani G, Biondi N, Bonini G, Tredici MR (2009) Microalgae for oil: Strain selection, induction of lipid synthesis and outdoor mass cultivation in a low-cost photobioreactor. Biotechnol Bioeng 102(1):100–112. https://doi.org/10.1002/bit.22033

Rupérez P (2002) Mineral content of edible marine seaweeds. Food Chem 79(1):23–26. https://doi.org/10.1016/S0308-8146(02)00171-1

Ryan S, McLoughlin P, O'Donovan O (2012) A comprehensive study of metal distribution in three main classes of seaweed. Environ Pollut 167:171–177. https://doi.org/10.1016/j.envpol.2012.04.006

Saha BC, Iten LB, Cotta MA, Wu YV (2005a) Dilute acid pretreatment, enzymatic saccharification and fermentation of wheat straw to ethanol. Process Biochem 40(12):3693–3700. https://doi.org/10.1016/j.procbio.2005.04.006

Saha BC, Iten LB, Cotta MA, Wu YV (2005b) Dilute acid pretreatment, enzymatic saccharification, and fermentation of rice hulls to ethanol. Biotechnol Prog 21(3):816–822. https://doi.org/10.1021/bp049564n

Salehi B, Sharifi-Rad J, Seca AML, Pinto DCGA, Michalak I, Trincone A, Mishra AP, Nigam M, Zam W, Martins N (2019) Current trends on seaweeds: looking at chemical composition, phytopharmacology, and cosmetic applications. Molecules 24(22):4182

Salman CA, Schwede S, Naqvi M, Thorin E, Yan J (2019) Synergistic combination of pyrolysis, anaerobic digestion, and CHP plants. Energy Procedia 158:1323–1329. https://doi.org/10.1016/j.egypro.2019.01.326

Samiran NA, Jaafar MNM, Ng J-H, Lam SS, Chong CT (2016) Progress in biomass gasification technique – With focus on Malaysian palm biomass for syngas production. Renew Sustain Energy Rev 62:1047–1062. https://doi.org/10.1016/j.rser.2016.04.049

Sánchez ÓJ, Cardona CA (2008) Trends in biotechnological production of fuel ethanol from different feedstocks. Biores Technol 99(13):5270–5295. https://doi.org/10.1016/j.biortech.2007.11.013

Saravanakumar A, Haridasan TM, Bai RK (2006) Technical and feasibility study of conversion of long-stick wood to charcoal in a partial combustion metal kiln. Energy Sustain Dev 10(3):17–25. https://doi.org/10.1016/S0973-0826(08)60540-2

Sasaki C, Okumura R, Asada C, Nakamura Y (2014) Steam explosion treatment for ethanol production from branches pruned from pear trees by simultaneous saccharification and fermentation. Biosci Biotechnol Biochem 78(1):160–166. https://doi.org/10.1080/09168451.2014.877818

Sattler M, Mattingly SP, Sattler M, Chen VC, Afotey BN, Brydia RE, Cazzell M, Theriot S, Upayokin A, Athappan A (2011) Anaerobic processes for waste treatment and energy generation. Integr Waste Manage I:11215

Sawangkeaw R, Ngamprasertsith S (2013) A review of lipid-based biomasses as feedstocks for biofuels production. Renew Sustain Energy Rev 25:97–108. https://doi.org/10.1016/j.rser.2013.04.007

Schneider T, Müller D, Karl J (2020) A review of thermochemical biomass conversion combined with stirling engines for the small-scale cogeneration of heat and power. Renew Sustain Energy Rev 134:110288. https://doi.org/10.1016/j.rser.2020.110288

Seifert T, Ackerman P, Chirwa PW, von Doderer C, du Toit B, Görgens J, Ham C, Kunneke A, Meincken M (2014) Biomass from Wood in the Tropics. In: Seifert T (ed) Bioenergy from wood: sustainable production in the tropics. Springer Netherlands, Dordrecht, pp 1–10. https://doi.org/10.1007/978-94-007-7448-3_1

Shabani N, Akhtari S, Sowlati T (2013) Value chain optimization of forest biomass for bioenergy production: a review. Renew Sustain Energy Rev 23:299–311. https://doi.org/10.1016/j.rser.2013.03.005

Shahid EM, Jamal Y (2011) Production of biodiesel: a technical review. Renew Sustain Energy Rev 15(9):4732–4745. https://doi.org/10.1016/j.rser.2011.07.079

Shanmugam S, Hari A, Pandey A, Mathimani T, Felix L, Pugazhendhi A (2020) Comprehensive review on the application of inorganic and organic nanoparticles for enhancing biohydrogen production. Fuel 270:117453. https://doi.org/10.1016/j.fuel.2020.117453

Sharma YC, Singh B (2009) Development of biodiesel: current scenario. Renew Sustain Energy Rev 13(6):1646–1651. https://doi.org/10.1016/j.rser.2008.08.009

Shen G, Tao S, Wei S, Zhang Y, Wang R, Wang B, Li W, Shen H, Huang Y, Chen Y, Chen H, Yang Y, Wang W, Wei W, Wang X, Liu W, Wang X, Simonich SLM (2012) Reductions in emissions of carbonaceous particulate matter and polycyclic aromatic hydrocarbons from combustion of biomass pellets in comparison with raw fuel burning. Environ Sci Technol 46(11):6409–6416. https://doi.org/10.1021/es300369d

Shepherd P (2000) Biomass co-firing: a renewable alternative for utilities. National Renew Energy Lab., Golden, CO (US)

Shevkani K, Singh N, Bajaj R, Kaur A (2017) Wheat starch production, structure, functionality and applications—a review. Int J Food Sci Technol 52(1):38–58. https://doi.org/10.1111/ijfs.13266

Shi AZ, Koh LP, Tan HTW (2009) The biofuel potential of municipal solid waste. GCB Bioenergy 1(5):317–320. https://doi.org/10.1111/j.1757-1707.2009.01024.x

Siddiquee MN, Rohani S (2011) Lipid extraction and biodiesel production from municipal sewage sludges: a review. Renew Sustain Energy Rev 15(2):1067–1072. https://doi.org/10.1016/j.rser.2010.11.029

Sindhu R, Binod P, Pandey A (2016) Biological pretreatment of lignocellulosic biomass—an overview. Biores Technol 199:76–82. https://doi.org/10.1016/j.biortech.2015.08.030

Singh D, Sharma D, Soni SL, Sharma S, Kumar Sharma P, Jhalani A (2020) A review on feedstocks, production processes, and yield for different generations of biodiesel. Fuel 262:116553. https://doi.org/10.1016/j.fuel.2019.116553

Singh SP, Singh D (2010) Biodiesel production through the use of different sources and characterization of oils and their esters as the substitute of diesel: a review. Renew Sustain Energy Rev 14(1):200–216. https://doi.org/10.1016/j.rser.2009.07.017

Sitepu EK, Heimann K, Raston CL, Zhang W (2020) Critical evaluation of process parameters for direct biodiesel production from diverse feedstock. Renew Sustain Energy Rev 123:109762. https://doi.org/10.1016/j.rser.2020.109762

Sivamani S, Chandrasekaran AP, Balajii M, Shanmugaprakash M, Hosseini-Bandegharaei A, Baskar R (2018) Evaluation of the potential of cassava-based residues for biofuels production. Rev Environ Sci Biotechnol 17(3):553–570. https://doi.org/10.1007/s11157-018-9475-0

Sivaramakrishnan R, Incharoensakdi A (2018) Utilization of microalgae feedstock for concomitant production of bioethanol and biodiesel. Fuel 217:458–466. https://doi.org/10.1016/j.fuel.2017.12.119

Smith KA (2012) Louis pasteur, the father of immunology? Front Immunol 3:68–68. https://doi.org/10.3389/fimmu.2012.00068

Smith WHM (2010) Biofuels. production, application and development, vol 46. Experimental Agriculture, vol 3, 2010/06/25 edn. Cambridge University Press, Wallingford. https://doi.org/10.1017/S0014479710000165

Soccol CR, Vandenberghe LPdS, Medeiros ABP, Karp SG, Buckeridge M, Ramos LP, Pitarelo AP, Ferreira-Leitão V, Gottschalk LMF, Ferrara MA, Silva Bon EPd, Moraes LMPd, Araújo JdA, Torres FAG (2010) Bioethanol from lignocelluloses: status and perspectives in Brazil. Bioresour Technol 101(13):4820–4825. https://doi.org/10.1016/j.biortech.2009.11.067

Stoffel RB, Neves PV, Felissia FE, Ramos LP, Gassa LM, Area MC (2017) Hemicellulose extraction from slash pine sawdust by steam explosion with sulfuric acid. Biomass Bioenerg 107:93–101. https://doi.org/10.1016/j.biombioe.2017.09.019

Sun X, Atiyeh HK, Li M, Chen Y (2020) Biochar facilitated bioprocessing and biorefinery for productions of biofuel and chemicals: a review. Bioresour Technol 295:122252. https://doi.org/10.1016/j.biortech.2019.122252

Sun Y, Cheng J (2002) Hydrolysis of lignocellulosic materials for ethanol production: a review. Biores Technol 83(1):1–11. https://doi.org/10.1016/S0960-8524(01)00212-7

Sun Y, Gao B, Yao Y, Fang J, Zhang M, Zhou Y, Chen H, Yang L (2014) Effects of feedstock type, production method, and pyrolysis temperature on biochar and hydrochar properties. Chem Eng J 240:574–578. https://doi.org/10.1016/j.cej.2013.10.081

Suutari M, Leskinen E, Fagerstedt K, Kuparinen J, Kuuppo P, Blomster J (2015) Macroalgae in biofuel production. Phycol Res 63(1):1–18. https://doi.org/10.1111/pre.12078

Tabassum MR, Wall DM, Murphy JD (2016) Biogas production generated through continuous digestion of natural and cultivated seaweeds with dairy slurry. Biores Technol 219:228–238. https://doi.org/10.1016/j.biortech.2016.07.127

Tabatabaei M, Aghbashlo M (2018) Biodiesel: From Production to Combustion, vol 8. Springer

Taherzadeh MJ, Karimi K (2007) Enzymatic-based hydrolysis processes for ethanol from lignocellulosic materials: a review. BioResources 2(4):707–738

Taherzadeh MJ, Karimi K (2008) Pretreatment of lignocellulosic wastes to improve ethanol and biogas production: a review. Int J Mol Sci 9(9):1621–1651

Talebnia F, Karakashev D, Angelidaki I (2010) Production of bioethanol from wheat straw: an overview on pretreatment, hydrolysis and fermentation. Biores Technol 101(13):4744–4753. https://doi.org/10.1016/j.biortech.2009.11.080

Tanger P, Field J, Jahn C, DeFoort M, Leach J (2013) Biomass for thermochemical conversion: targets and challenges. Front Plant Sci 4(218). https://doi.org/10.3389/fpls.2013.00218

Taylor RD, Koo WW (2015) 2015 Outlook of the US and World Wheat Industries, 2015–2024. Agribusiness & Applied Economics Report. Department of Agribusiness and Applied Economics, North Dakota State University, North Dakota. https://doi.org/10.22004/ag.econ.201310

Thrän D, Peetz D, Schaubach K, Backéus S, Benedetti L, Bruce L (2017) Global wood pellet industry and trade study 2017. IEA Bioenergy Task 40

Tomás-Pejó E, Alvira P, Ballesteros M, Negro M (2011) Pretreatment Technologies for Lignocellulose-to-Bioethanol Conversion in Biofuels. In: Ashok P, Christian L, Steven CR, Claude-Gilles D, Edgard G (eds) Academic, Amsterdam

Toor SS, Rosendahl L, Rudolf A (2011) Hydrothermal liquefaction of biomass: a review of subcritical water technologies. Energy 36(5):2328–2342. https://doi.org/10.1016/j.energy.2011.03.013

Torres-Jimenez E, Kegl M, Dorado R, Kegl B (2012) Numerical injection characteristics analysis of various renewable fuel blends. Fuel 97:832–842. https://doi.org/10.1016/j.fuel.2012.02.069

Tripathi M, Sahu JN, Ganesan P (2016) Effect of process parameters on production of biochar from biomass waste through pyrolysis: a review. Renew Sustain Energy Rev 55:467–481. https://doi.org/10.1016/j.rser.2015.10.122

Tsai W-T (2016) Analysis of municipal solid waste incineration plants for promoting power generation efficiency in Taiwan. J Mater Cycles Waste Manage 18(2):393–398. https://doi.org/10.1007/s10163-014-0345-8

Ugwu CU, Aoyagi H, Uchiyama H (2008) Photobioreactors for mass cultivation of algae. Biores Technol 99(10):4021–4028. https://doi.org/10.1016/j.biortech.2007.01.046

Ullah Z, Bustam MA, Man Z (2015) Biodiesel production from waste cooking oil by acidic ionic liquid as a catalyst. Renewable Energy 77:521–526. https://doi.org/10.1016/j.renene.2014.12.040

Väljamäe P, Kipper K, Pettersson G, Johansson G (2003) Synergistic cellulose hydrolysis can be described in terms of fractal-like kinetics. Biotechnol Bioeng 84(2):254–257. https://doi.org/10.1002/bit.10775

Van den Born G, Van Minnen J, Olivier J, Ros J (2014) Integrated analysis of global biomass flows in search of the sustainable potential for bioenergy production. PBL Netherlands Environmental Assessment Agency, Bilthoven

van Hal JW, Huijgen WJJ, López-Contreras AM (2014) Opportunities and challenges for seaweed in the biobased economy. Trends Biotechnol 32(5):231–233. https://doi.org/10.1016/j.tibtech.2014.02.007

Vanegas CH, Bartlett J (2013) Green energy from marine algae: biogas production and composition from the anaerobic digestion of Irish seaweed species. Environ Technol 34(15):2277–2283. https://doi.org/10.1080/09593330.2013.765922

Varol A, Ugurlu A (2016) Biogas Production from Microalgae (Spirulina platensis) in a Two Stage Anaerobic System. Waste and Biomass Valorization 7(1):193–200. https://doi.org/10.1007/s12649-015-9442-8

Vasudevan PT, Briggs M (2008) Biodiesel production—current state of the art and challenges. J Ind Microbiol Biotechnol 35(5):421. https://doi.org/10.1007/s10295-008-0312-2

Venkatachalam CD, Ravichandran SR, Sengottian M (2022) Lignocellulosic and algal biomass for bio-crude production using hydrothermal liquefaction: conversion techniques, mechanism and process conditions: a review. Environ Eng Res 27(1):200555–200550. https://doi.org/10.4491/eer.2020.555

Verardi A, De Bari I, Ricca E, Calabrò V (2012) Hydrolysis of lignocellulosic biomass: current status of processes and technologies and future perspectives. In: Bioethanol, vol 2012. InTech Rijeka, pp 95–122

Vermerris W (2008) Why bioenergy makes sense. In: Genetic Improvement of Bioenergy Crops. Springer, pp 3–42

Viamajala S, McMillan JD, Schell DJ, Elander RT (2009) Rheology of corn stover slurries at high solids concentrations—effects of saccharification and particle size. Biores Technol 100(2):925–934. https://doi.org/10.1016/j.biortech.2008.06.070

Vieira EF, Soares C, Machado S, Correia M, Ramalhosa MJ, Oliva-teles MT, Paula Carvalho A, Domingues VF, Antunes F, Oliveira TAC, Morais S, Delerue-Matos C (2018) Seaweeds from the Portuguese coast as a source of proteinaceous material: total and free amino acid composition profile. Food Chem 269:264–275. https://doi.org/10.1016/j.foodchem.2018.06.145

Vignesh P, Pradeep Kumar AR, Ganesh NS, Jayaseelan V, Sudhakar K (2021) Biodiesel and green diesel generation: an overview. Oil Gas Sci Technol Rev IFP Energies Nouvelles 76:6

Volk T, Abrahamson L, Cameron K, Castellano PJ, Corbin T, Fabio ES, Johnson G, Kuzovkina-Eischen Y, Labrecque M, Miller RO, Smart L, Staver K, Stanosz G, Rees K (2011) Yields of willow biomass crops across a range of sites in North America. Aspects Appl Biol 67–74

Wahid R, Hjorth M, Kristensen S, Møller HB (2015) Extrusion as pretreatment for boosting methane production: effect of screw configurations. Energy Fuels 29(7):4030–4037

Wahid R, Mulat DG, Gaby JC, Horn SJ (2019) Effects of H 2: CO_2 ratio and H_2 supply fluctuation on methane content and microbial community composition during in-situ biological biogas upgrading. Biotechnol Biofuels 12(1):1–15

Wahid R, Romero-Guiza M, Moset V, Møller HB, Fernández B (2020) Improved anaerobic biodegradability of wheat straw, solid cattle manure and solid slaughterhouse by alkali, ultrasonic and alkali-ultrasonic pre-treatment. Environ Technol 41(8):997–1006

Wang B, Li Y, Ren N (2013) Biohydrogen from molasses with ethanol-type fermentation: effect of hydraulic retention time. Int J Hydrogen Energy 38(11):4361–4367. https://doi.org/10.1016/j.ijhydene.2013.01.120

Wang D, Ikenberry M, Pe L, Hohn K (2012) Acid-functionalized nanoparticles for pretreatment of wheat straw. J Biomater Nanobiotechnol 3(3):242–352. https://doi.org/10.4236/jbnb.2012.33032
Wang S, Liu Q, Lu Z, Xu L, Liu X, Hu Y, Qian L, Xu S, Abomohra AE-F (2021) Visible light-driven conversion of carboxylic acids into esters for enhanced algal bio-crude oil catalyzed by cadmium sulfide quantum dots (CdS-QDs). Fuel Process Technol 216:106778. https://doi.org/10.1016/j.fuproc.2021.106778
Wang Y, Huang J, Li Y, Xiong K, Wang Y, Li F, Liu M, Wu Z, Tu Y, Peng L (2015) Ammonium oxalate-extractable uronic acids positively affect biomass enzymatic digestibility by reducing lignocellulose crystallinity in Miscanthus. Biores Technol 196:391–398. https://doi.org/10.1016/j.biortech.2015.07.099
Wei S (2016) The application of biotechnology on the enhancing of biogas production from lignocellulosic waste. Appl Microbiol Biotechnol 100(23):9821–9836. https://doi.org/10.1007/s00253-016-7926-5
Weiland P (2010) Biogas production: current state and perspectives. Appl Microbiol Biotechnol 85(4):849–860. https://doi.org/10.1007/s00253-009-2246-7
Wesseler J (2007) Opportunities ('costs) matter: a comment on Pimentel and Patzek "Ethanol production using corn, switchgrass, and wood; biodiesel production using soybean and sunflower." Energy Policy 35(2):1414–1416. https://doi.org/10.1016/j.enpol.2006.02.007
West JB (2014) Henry Cavendish (1731–1810): hydrogen, carbon dioxide, water, and weighing the world. Am J Physiol Lung Cellular Mol Physiol 307(1):L1–L6. https://doi.org/10.1152/ajplung.00067.2014
Whitfield MB, Chinn MS, Veal MW (2012) Processing of materials derived from sweet sorghum for biobased products. Ind Crops Prod 37(1):362–375. https://doi.org/10.1016/j.indcrop.2011.12.011
Whittaker C, Shield I (2017) Factors affecting wood, energy grass and straw pellet durability—a review. Renew Sustain Energy Rev 71:1–11. https://doi.org/10.1016/j.rser.2016.12.119
Woolcock PJ, Brown RC (2013) A review of cleaning technologies for biomass-derived syngas. Biomass Bioenerg 52:54–84. https://doi.org/10.1016/j.biombioe.2013.02.036
Xiao X, Agusti S, Lin F, Li K, Pan Y, Yu Y, Zheng Y, Wu J, Duarte CM (2017) Nutrient removal from Chinese coastal waters by large-scale seaweed aquaculture. Sci Rep 7(1):46613. https://doi.org/10.1038/srep46613
Xu H, Miao X, Wu Q (2006) High quality biodiesel production from a microalga Chlorella protothecoides by heterotrophic growth in fermenters. J Biotechnol 126(4):499–507. https://doi.org/10.1016/j.jbiotec.2006.05.002
Yang J, Xu M, Zhang X, Hu Q, Sommerfeld M, Chen Y (2011) Life-cycle analysis on biodiesel production from microalgae: Water footprint and nutrients balance. Biores Technol 102(1):159–165. https://doi.org/10.1016/j.biortech.2010.07.017
Yu KL, Lau BF, Show PL, Ong HC, Ling TC, Chen W-H, Ng EP, Chang J-S (2017) Recent developments on algal biochar production and characterization. Biores Technol 246:2–11. https://doi.org/10.1016/j.biortech.2017.08.009
Yuan J-H, Xu R-K, Zhang H (2011) The forms of alkalis in the biochar produced from crop residues at different temperatures. Biores Technol 102(3):3488–3497. https://doi.org/10.1016/j.biortech.2010.11.018
Yue Y, Cui L, Lin Q, Li G, Zhao X (2017) Efficiency of sewage sludge biochar in improving urban soil properties and promoting grass growth. Chemosphere 173:551–556. https://doi.org/10.1016/j.chemosphere.2017.01.096
Yusoff MNAM, Zulkifli NWM, Masjuki HH, Harith MH, Syahir AZ, Khuong LS, Zaharin MSM, Alabdulkarem A (2018) Comparative assessment of ethanol and isobutanol addition in gasoline on engine performance and exhaust emissions. J Clean Prod 190:483–495. https://doi.org/10.1016/j.jclepro.2018.04.183
Yusuf NNAN, Kamarudin SK, Yaakub Z (2011) Overview on the current trends in biodiesel production. Energy Convers Manage 52(7):2741–2751. https://doi.org/10.1016/j.enconman.2010.12.004

Zabed H, Sahu JN, Suely A, Boyce AN, Faruq G (2017) Bioethanol production from renewable sources: current perspectives and technological progress. Renew Sustain Energy Rev 71:475–501. https://doi.org/10.1016/j.rser.2016.12.076

Zaky AS (2017) Marine fermentation, the sustainable approach for bioethanol production. EC Microbiol 25–27

Zaky AS, Carter CE, Meng F, French CE (2021) A Preliminary life cycle analysis of bioethanol production using seawater in a coastal biorefinery setting. Processes 9(8). https://doi.org/10.3390/pr9081399

Zaky AS, French CE, Tucker GA, Du C (2020) Improving the productivity of bioethanol production using marine yeast and seawater-based media. Biomass Bioenergy 139:105615. https://doi.org/10.1016/j.biombioe.2020.105615

Zaky A, Greetham D, Tucker G, Du C (2018a) The establishment of a marine focused biorefinery for bioethanol production using seawater and a novel marine yeast strain. Sci Rep 8. https://doi.org/10.1038/s41598-018-30660-x

Zaky AS, Greetham D, Tucker GA, Du C (2018b) The establishment of a marine focused biorefinery for bioethanol production using seawater and a novel marine yeast strain. Sci Rep 8 (1):12127. https://doi.org/10.1038/s41598-018-30660-x

Zaky AS (2021) Introducing a marine biorefinery system for the integrated production of biofuels high-value-chemicals and co-products: a path forward to a sustainable future. Processes 9(10):1841. https://doi.org/10.3390/pr9101841

Zappi ME, Bajpai R, Hernandez R, Mikolajczyk A, Fortela DL, Sharp W, Chirdon W, Zappi K, Gang D, Nigam KDP, Revellame ED (2019) Microalgae culturing to produce biobased diesel fuels: an overview of the basics, challenges, and a look toward a true biorefinery future. Ind Eng Chem Res 58(35):15724–15746. https://doi.org/10.1021/acs.iecr.9b01555

Zhang H, Aytun Ozturk U, Wang Q, Zhao Z (2014) Biodiesel produced by waste cooking oil: review of recycling modes in China, the US and Japan. Renew Sustain Energy Rev 38:677–685. https://doi.org/10.1016/j.rser.2014.07.042

Zhang L, Xu C, Champagne P (2010) Overview of recent advances in thermo-chemical conversion of biomass. Energy Convers Manage 51(5):969–982. https://doi.org/10.1016/j.enconman.2009.11.038

Zhang Z, Xie Y, He X, Li X, Hu J, Ruan Z, Zhao S, Peng N, Liang Y (2016) Comparison of high-titer lactic acid fermentation from NaOH- and NH_3-H_2O_2-pretreated corncob by Bacillus coagulans using simultaneous saccharification and fermentation. Sci Rep 6(1):37245. https://doi.org/10.1038/srep37245

Zhao X, Zhou H, Sikarwar VS, Zhao M, Park A-HA, Fennell PS, Shen L, Fan L-S (2017) Biomass-based chemical looping technologies: the good, the bad and the future. Energy Environ Sci 10(9):1885–1910. https://doi.org/10.1039/C6EE03718F

Zhao Y, Lu W-J, Wang H-T, Yang J-L (2009) Fermentable hexose production from corn stalks and wheat straw with combined supercritical and subcritical hydrothermal technology. Biores Technol 100(23):5884–5889. https://doi.org/10.1016/j.biortech.2009.06.079

Zhuang X, Wang W, Yu Q, Qi W, Wang Q, Tan X, Zhou G, Yuan Z (2016) Liquid hot water pretreatment of lignocellulosic biomass for bioethanol production accompanying with high valuable products. Biores Technol 199:68–75. https://doi.org/10.1016/j.biortech.2015.08.051

Ziegler-Devin I, Menana Z, Chrusciel L, Chalot M, Bert V, Brosse N (2019) Steam explosion pretreatment of willow grown on phytomanaged soils for bioethanol production. Ind Crops Prod 140:111722. https://doi.org/10.1016/j.indcrop.2019.111722

Chapter 5
Thermochemical Conversion of Wastes

Sabariswaran Kandasamy, Mathiyazhagan Narayanan, Narayanamoorthy Bhuvanendran, and Zhixia He

Abstract Thermochemical conversion technologies have recently played a significant role in converting energy from waste sources. Thermochemical technologies have promising ways of recycling energy from various waste materials while reducing the environmental impact. This chapter primarily provides the collective information on waste feedstocks used for the thermochemical conversion from the recent review literature. Second, the numerous thermochemical conversion methods are discussed, including direct combustion, pyrolysis, gasification, and hydrothermal liquefaction using various reactors for each technique. It assesses the conversion of multiple wastes to crude bio-oil and the likelihood of converting syngas to bio-oil. Hydrothermal conversions occur at moderate temperatures, but typically at high pressure and in the presence of water. The thermochemical conversion includes the accurate temperature, pressure, and heating rate, which can be accomplished using various reactors. For the large-scale industrialization of biofuels, a greater understanding of the mechanism of conversion, reactors, and feedstock composition is crucial. Moreover, this chapter discusses the various thermochemical conversions of wastes and its bio-oil yield.

Keywords Sources of wastes · Thermochemical conversion · Hydrothermal liquefaction · Biomass conversion · Crude bio-oil

---

S. Kandasamy · N. Bhuvanendran · Z. He (✉)
Institute for Energy Research, Jiangsu University, Zhenjiang 212013, China
e-mail: zxhe@ujs.edu.cn

S. Kandasamy
e-mail: sabari@ujs.edu.cn

Z. He
School of Energy and Power Engineering, Jiangsu University, Zhenjiang 212013, China

M. Narayanan
PG and Research Centre in Biotechnology, Adhiyamaan Educational Research Institute, MGR College, Hosur, Krishnagiri, Tamil Nadu, India

A. E.-F. Abomohra et al. (eds.), *Waste-to-Energy*,
https://doi.org/10.1007/978-3-030-91570-4_5

5.1 Introduction

The demand for energy sources to meet the consumption of human energy is continuing to increase. Fossil fuels are still the dominant source of energy (Awalludin et al. 2015) in the country. Although the amount of fossil fuel is unknown, it is well known that it is depleting and non-renewable. Before using fossil fuels (Purnomo et al. 2021), biomass was the primary energy source for cooking. Fossil fuels like coal, oil, and natural gas have intensified the world's dependence on these sources of fossil fuels. In response to environmental and energy protection challenges, renewable energy becomes more and more relevant. Before the fossil fuel system, wood was the only source of energy for humans. The use of wood or dung (Chan et al. 2019) for heating and cooking is evidence. Today, biomass represents 14% of the world's final energy intake. Examples of biomass sources (Labaki and Jeguirim 2017), either natural or derived from matter, are dedicated energy crops, agricultural residues (Adjin-Tetteh et al. 2018), timber, municipal solid waste (MSW), animal refuse, and sludge (Kandasamy et al. 2021). Waste biomass is easy to obtain, less costly, and does not create market imbalances as food products, unlike energy crops. Biomass plays a key role in renewable waste management and high-quality carbon–neutral biofuels since it is an environmentally efficient energy source with huge potential.

India and China were two leading (Senthil et al. 2019) countries using biodegradation technologies to generate food and human waste biogas in response to the world fossil fuel prices due to the Second World War. Green energy typically existed in this period in more ways than in the second half of the (Uzoejinwa et al. 2018) twentieth century. Biofuels are produced from biomass, but their form and features are determined by the mechanisms of conversion: biological, physical, chemical, or process (Shi et al. 2016). Microbial or enzyme (Long et al. 2019) fermentation can produce biogas, ethanol, and biodiesel, for example, with or without physical and chemical pretreatment procedures. Instead, biomass conversion into bio-oil, biochar, syngas, and other products includes entirely thermochemical processes, including torrefaction (Lü et al. 2018), carbonization, thermal liquefaction, pyrolysis, and gasification. Figure 5.1 shows the schematic diagram of thermochemical conversion and its products.

Because of the current energy shortage, thermochemical research and development are central to a substantial part of the realistic and sustainable energy solution (Ong et al. 2019). A sustainable agricultural activity concept for waste management has also been proposed for producing biofuels from farm manure waste for various reasons, including reduced net greenhouse gas emissions, solid waste reduction, biomass sequestration, soil enrichment, and economic benefits (Li and Jiang 2017). The thermochemical process's physicochemical parameters mainly depend on the raw material and the transformation method utilized (Bora et al. 2020). Moreover, corn/cellulose biomass ethanol, soybean biodiesel, canola, jatropha, animal fats, waste cooking oil, algae, anaerobic manure digestion biogas, and the thermochemical conversion of various biomass into solid, liquid, and gas fuel are examples of commercially accessible biofuels (Parshetti et al. 2015). Thermochemical processes

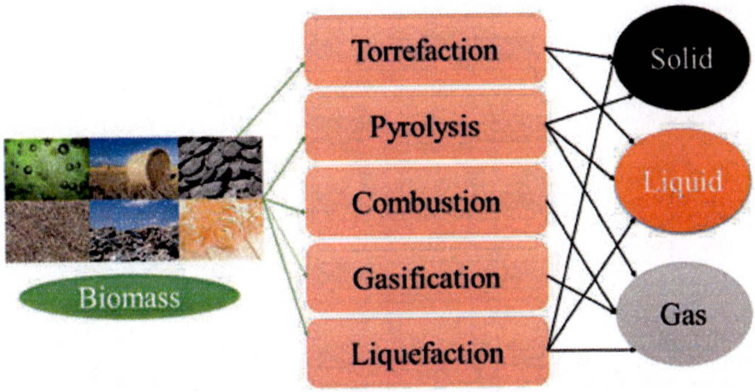

Fig. 5.1 Illustration summarizing the thermochemical conversion pathways and the main products

are recent in terms of trading compared to other alternative energy sources. Still, they have recently come to much attention because these biofuels have various technological and geopolitical advantages (Gracida-Alvarez et al. 2016). Multiple thermochemical conversion methods, such as torrefaction, combustion, pyrolysis, gasification, and hydrothermal liquefaction, will convert wastes into bioenergy (Chan and Wang 2016). Two of the processes successfully transforming biomass into bioenergy were pyrolysis and hydrothermal liquefaction. In this chapter, the different thermochemical processes of waste were discussed briefly. The physical, chemical, and atoms nature of various biomasses used for bio-oil production by thermochemical conversion process is shown in Table 5.1.

5.2 Thermochemical Conversion Technologies

The different thermochemical conversion methods are combustion, torrefaction, pyrolysis, gasification, and hydrothermal liquefaction (HTL) which were involved in the conversion of wastes into bio-oil.

5.2.1 Combustion

Combustion is the most common process (Chan and Wang 2016) for converting biomass. Biomass has a bright future as a source of renewable energy around the world. It represents over 97% of the world's bioenergy supply. The traditional burning of biomass plays an essential (Kumar et al. 2019) role in people's daily lives in many less-developed countries since it is the only energy source for cooking and heating. Combustion is a low-cost, high-reliability technology that is well known and marketable. There are three primary phases during biomass (Zhou et al. 2016) combustion: drying, pyrolysis, and the combustion of volatile gases with solid char.

Table 5.1 Physical, chemicals, and atoms nature of various biomasses used for bio-oil production by thermochemical conversion process

Biomass	Physical, chemicals, and atoms in percentage (%)											Density (kg/m^3)	References
	Lignin	Cellulose	Hemi cellulose	C	FC	H	O_2	N	Ash	VM	M		
Wood	30	50	20–30	51.6	17	6.3	41.5	0.1	1	82	20	1186	(Jahirul et al. 2012)
Wheat straw	20	40	20–25	48.5	16	5.5	3.9	0.3	4	59	16	1233	(Lam et al. 2008)
Barley strew	20	41	21	45.7	18	6.1	38.3	0.4	6	46	30	210	(Fabičovicová et al. 2016)
Switch grass	20	40	10–40	58.04	14.47	3.14	41	0.07	1.01	111	73	266	(Lam et al. 2008)
Sugarcane bagasse	32	19	32–48	46.52	16.9	6.26	46.98	0.18	6.2	72.3	52	120	(Jha and Dass 2020)
Miscanthus	17	24	44	47.73	16	5.85	–	0.28	4.5	83.3	7.97	100	(Jahirul et al. 2012)
Corn stover	21	28	35	43.92	15.23	6.01	40.44	0.42	5.13	75.6	4.01	130	(Przybysz Buzała et al. 2017)
Hazelnut shell	42	28	30.4	51.14	28.30	5.7	41.1	0.42	1.71	70.3	6.4	425	(Monarca et al. 2012)
Cypress	28.2	41.2	41	55	8.12	6.5	38	1.01	0.4	90.8	–	692	(Sudagar et al. 2020)
Switchgrass	28	45	31	44.77	–	5.79	49.13	0.31	5.8	–	15	108	(Tumuluru 2015)
Olive husk/baggage	48	24	23.6	66.9	–	9.2	22	2	–	–	–	323	(Waliszewska et al. 2021)
Corncob	15	50	31	46.46	18.63	6.02	44.29	0.36	2.86	78.5	20	240	(Wang et al. 2020a)
Reed canary grass	10.22	39.29	21.93	45.36	–	5.81	48.49	0.34	5.10	–	23	270	(Zaman et al. 2017)
Rice straw	18	32	24	48.31	11	5.32	34.63	1.23	4.3	79	6	200	(Tumuluru 2015)

(continued)

Table 5.1 (continued)

Biomass	Physical, chemicals, and atoms in percentage (%)											Density (kg/m^3)	References
	Lignin	Cellulose	Hemi cellulose	C	FC	H	O_2	N	Ash	VM	M		
Dactylis glomerata	19.33	37.71	31.48	42.96	–	5.70	49.44	1.90	7.50	–	–	42.96	(Zaman et al. 2017)
Danish pine	28	42	27	45.7	16	7	47	0.1	0.03	–	17	124	(Tumuluru 2015)
Cotton coir	0	95	20	46.52	16.70	5.50	34.64	0.17	13.2	70.2	13.63	709.53	(Lam et al. 2008)
Festuca arundinacea	6.5	33.3	30.4	42.22	–	5.64	50.65	1.50	7.30	–	–	860	(Wang et al. 2020a)
Lolium perenne	13	28.8	21	43.12	–	5.80	49.80	1.28	6.20	–	–	975	(Lam et al. 2008)
Banana waste	14	13	15	44.28	14.0	6.23	37.9	0.80	10.7	75.3	8.57	550	(de Oliveira Maiaa et al. 2014)

Volatile gas combustion accounts for more than 70% of overall heat generation (Cho et al. 2018). It usually occurs over the fuel bed and is characterized by yellow flames. The existence of little blue (Liu et al. 2017) flames indicates that the char in the fuel bed has been combusted (Chan and Wang 2018). Technical contradictions such as biomass fuel characteristics, combustion chamber qualities, and co-fire were challenges for massive biomass combustion.

5.2.2 *Torrefaction*

Torrefaction of biomass is a standard pretreatment method for improving its physical and pyrolysis properties. The thermal activity and developing gas in the torrefaction phase and torrefied biomass have been studied using TG-FTIR. On the other hand, DTG curves can be used to see when most volatiles are reported during torrefaction. FTIR can be used to calculate volatile functional groups correspondingly. Moreover, the lignocellulosic components undergo torrefaction and gas generation under isothermal conditions for 5 h (cellulose, xylan, and lignin) (Vega et al. 2019).

Cellulose was thermally resistant for a short time but subsequently quickly degraded, especially at 280 °C. The volume of CO produced by xylan breakdown increased as the temperature was increased. In the xylan side chains, the 4-O-methyl glucuronic units degraded first (Ong et al. 2020), while in the second degradation, the acetyl groups (Adeniyi et al. 2020) and macromolecule fragments started from 250 °C (Chen et al. 2015). In addition, primarily acetic acid (Ibarra-Gonzalez and Rong 2019) was produced at 280 °C. Lignin decomposition (Adeniyi et al. 2020) reactions at low temperatures (220 and 250 °C) were more critical, leading to the emission of phenol. During the torrefaction process, the chemical content of the biomass is depolymerized (Ganeshan et al. 2016), devolatilized, and carbonated, resulting in biochar (Ibarra-Gonzalez and Rong 2019) and bio-oil (water, organics, and lipids). The gaseous material produced during (Schmitt et al. 2019) the torrefaction phase can be monitored using TG-FTIR. The torrefaction yield and the advanced gas emitted during rice husk torrefaction were evaluated using (Yang et al. 2019b) TG-FTIR. Figure 5.2 shows the general process of torrefaction.

The volatile release at maximum mass (Bundhoo 2018) reduction was seen in FTIR at different torrefaction temperatures. During low torrefaction (Jha and Dass 2020) temperatures, adsorption peaks such as phenolic, carbonyl compounds, aromatic hydrocarbons, and low hydrocarbons were not observed. However, the adsorption peaks (Zhang and Zhang 2019) increased with an increase in the torrefaction temperature. During the initial torrefaction (Chen et al. 2018) or depolymerization phase, the CO_2 peak (Syed-Hassan et al. 2017) was observed. In contrast, the characteristic peaks for aromatic hydrocarbons and carbonyls were relatively lower, which shows that biomass depolymerizes and releases modest amounts (Nicodème et al. 2018) of CO, CO_2, and H_2O. Carbonization is the last step of torrefaction in which the absorption peak has declined slightly, thereby suggesting that the period of torrefaction is over. Torrefaction is a good pretreatment of pyrolysis biomass

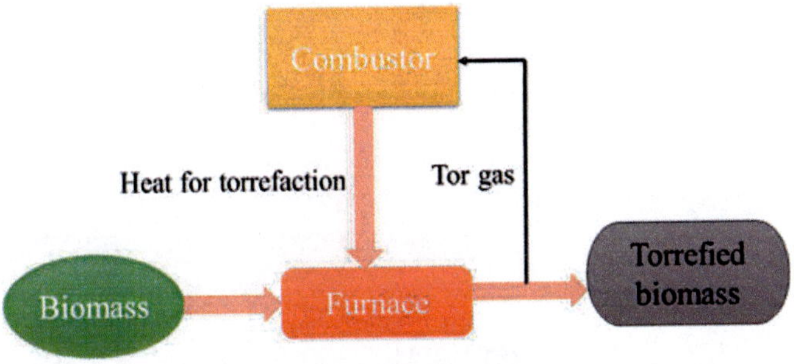

Fig. 5.2 Torrefaction generalized process

and produces high-quality solid fuel. During pyrolysis of herbaceous residues, the gaseous material released was significantly affected by torrefaction. Due to a wide range (Dou et al. 2019) of temperatures (150–600 °C), CO_2 emissions had two peaks during the pyrolysis.

After torrefaction, the CO_2 peak shifted to a higher temperature (Lee et al. 2017) with an amplitude decline. The generation of CO_2 mainly due to hemicellulose and cellulose decomposition led to a substantial reduction of the CO_2 peak at 280 °C in pyrolysis for torrefied biomass. In the meantime, lignin demethoxylation could lead to CO_2 release at high temperatures (more than 500 °C). Two CH_4 peaks were mainly formed during pyrolysis due to hemicellulose de-methylation at low temperatures (less than 400 °C). The total release of CH_4 has changed to 430 °C, after a torrefaction at 280 °C, due to lignin decomposition, as most hemicelluloses have been degraded. The cracking of the oxygen groups also resulted in carbonyl and glycosidic (Patel et al. 2016) (C–O–C) bonds. The torrefaction temperature has an essential effect on torrefied biomass combustion (Pang 2019) behavior. The gas production also decreased as the torrefaction temperature increased, resulting in gas emitted during combustion. During combustion, the SO_2 released was decreased by > 89% when N_2 torrefaction was used at temperatures above 330 °C. Comparing the emissions of combustion gases (CO_2, SO_2, and NH_3) with different torrefaction atmospheres, it was found that CO_2 torrefied biomass released more gas than N_2 torrefied biomass in combustion. Torrefaction processes and torrefied biomass properties can be affected at low temperatures by atmospheric CO_2 torrefaction.

5.2.3 *Pyrolysis*

A large variety of biomasses are used by pyrolysis, a modern thermochemical method for processing characteristics or biocrude (Sani et al. 2017). Pyrolysis, also called dry pyrolysis (Korai et al. 2016), is a solid (biochar), liquid (bio-oil), and gaseous (Bhaskar and Pandey 2015) (syngas) process causing biomass decomposition, with extremely high temperatures, without oxygen or air pressure. Compared to the

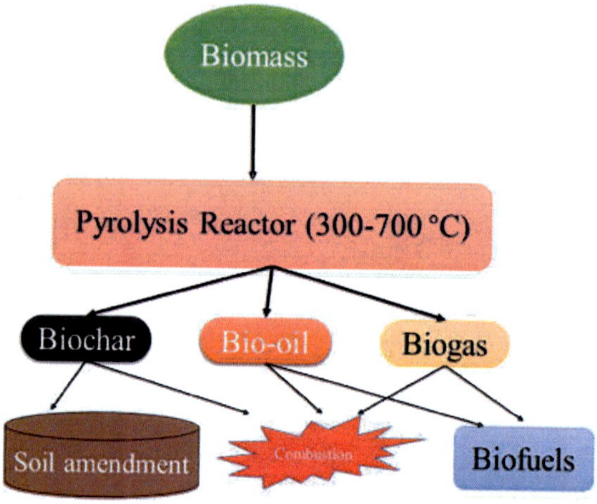

Fig. 5.3 Pyrolysis of biomass and applications

conventional dry pyrolysis process, pyrolysis technology (Matthews et al. 2016) can be used under subcritical or supercritical conditions. Water acts in these circumstances, usually over 374 °C and 22.1 MPa, as the reaction medium at elevated temperatures and pressures. The pyrolysis of biomass and its applications are shown in Fig. 5.3.

As in supercritical conditions, (i) water stays polar and non-polar, and (ii) water is slightly dielectric in a supercritical state, (iii) supercritical pyrolysis is much favored over subcritical (Cocero et al. 2018). Water thus behaves under supercritical conditions as an excellent organic solvent and helps to solubilize non-polar organic compounds. Several technical advances have emerged in bio-oil (Gomaa et al. 2020) production from a range of consistent feedstocks to replace traditional fossil fuels in recent years. Because of characteristics such as the lack of complex polymers such as lignin, cellulose, and hemicellulose (Sawatdeenarunat et al. 2018), furthermore, this allows for (i) low-temperature degradation, (ii) low O/C ratio, resulting in a high calorific value, (iii) low ash level, and (iv) low-energy consumption (Kunwar et al. 2017). Aquatic biomass has now become an essential feedstock for biomass pyrolysis conversion. Pyrolysis generates liquid phenolic ethers (Hossain et al. 2016), alkyl phenolics (Lü et al. 2018), heterocyclic ethers (Ren et al. 2020), and polyaromatic hydrocarbons (Castaldi et al. 2017), while solid char comprises gaseous carbon, hydrogen, and oxygen and gaseous compounds such as CO_2, CO, CH_4, and H_2.

5.2.3.1 Slow Pyrolysis

Slow pyrolysis aims to provide a rich material of carbon, such as activated carbon. Temperature and time of reaction affect the productions. Char yields at a typical 600 °C (Song et al. 2020) are approximately produced 25 wt%. For example, after pyrolysis, the char is triggered by steam to generate activated coal. Coconut shell-powered coal is commercially available. New applications require direct carbon fuel cells (Lopez et al. 2017) that are theoretically more electrically efficient than combustion with turbines.

5.2.3.2 Fast Pyrolysis

The key product is the pyrolysis fluid, which needs a highly high heat of 1000 K s^{-1} and a fast reaction time. Because pyrolysis oil is a (Wang et al. 2020b) intermediate substance, the oil-forming vapor must be refreshed rapidly. Pyrolysis oil (tar) contains hundreds (Matsakas et al. 2017) of different compounds, contributing to gases and chars. To avoid this, fast (Chen et al. 2015) heating rates and quick reaction times are required. Pyrolysis oils contain acids that can induce corrosion and water up to 30%. The heating value of pyrolysis oil (Lee et al. 2019) is around 17 MJ kg^{-1}, equivalent to initial biomass (e.g., birch wood with 7 wt% moisture has a heating value of 17.9 MJ kg^{-1}). Pyrolysis oil can be utilized as a fuel for large engines, such as ships or power plants, however not for chars (Pham et al. 2015). Pyrolysis oil is hydrogenated to make it more appropriate for car fuel, but hydrogen is needed. Fast pyrolysis oils can be used to manufacture resins and used to produce a "barbeque" flavor as liquid smoke.

5.2.4 Gasification

Gasification is performed between 800 and 1000 °C and contains syngas, hydrogen, carbon monoxide (Dabe et al. 2019), and methane mixtures. "Tar-free syngas" can be rendered above 1000 °C. Unless syngas is used as feedstock to produce a catalytic reaction like diesel, methanol, or Oxo-synthesis, and low tar content is beneficial. Catalysts, particularly noble metal catalysts, are disabled by tars, so syngas cleaning, in any case, is needed. Gasification processes are usually (Maya et al. 2016) performed using air or oxygen as the high-temperature gasification agent. The requisite heat is generated when part of the product is combusted. Table 5.2 shows the various biomass utilized in gasification units. One common concern is that such a gasifier (Dabe et al. 2019) should have a high output to minimize the relative cost of production. On the other hand, biomass is widespread and must be transported over long distances. As a first step in the BioliqTM concept, straw is transformed into small plants by simple pyrolysis. For instance, to produce petrol, the slurry is sent to a larger plant (Soares et al. 2020) with a gasifier, gas cleaner, and synthesizer. The definition includes an induced flow gasifier with a cooled wall. The ash condenses on the wall, which protects from corrosion. The ash melt runs down and gathers from the inner surface of the ash crust. The syngas generated here is converted into automotive fuels. Figure 5.4 shows a basic procedure for gasification unit.

Table 5.2 Thermochemical conversion process (gasification) comparison among various biomasses

Type of biomass	Experimental conditions	Name of the process	Outcome	References
Bark waste	100 kW, N: 0.05–2.70%, 850 °C, water: 6.1%, bark waste: 11.9%	Dual fluidized bed gasifier (DFB)	Conversion of fuels with more volume of nitrogen	(Canabarro et al. 2013)
Fruit bunches	50% moisture, 1.0 mm particle size		The decrease of gasification efficiency resulting increases in moisture content A low calorific value yielded higher amount of moisture and oxygen	
Lignite	90 kW, 370 and 510 μm-sized particle 1.3 ratio of steam: carbon, 2.1 ratio of kgH_2O:kg carbon		The reduced particle increases the yield of the gas as around 15.7%	
Mixture of pine and maple oak and waste seed corn	800 kW, gasifying agent: O_2 and steam, 800 °C	Fluidized bed gasifier	At low quantity of N and moisture contents of feedstock, the gasification is more efficient	(Ren et al. 2020)
Indole molecule	About 0.2 mol L^{-1} of indole, 550–700 °C, 3–80 min, 30 MPa pressure	Supercritical water gasification	Increased concentration of indole enhances the yield of CH_4, with 79% and 20% of H and C gasification efficiencies correspondingly	(Patel et al. 2016)
Glycerol	About 10, 20, and 30% of glycerol, 300–430 °C, 5–120 min, 30 MPa pressure		At 350–370 °C with long residence times, the increased coke formation has occurred	(Canabarro et al. 2013)
Raw and torrefied bamboo, high-volatile bituminous coal	Biomass particle size 44–250 μm, O_2 used as gasifying agent, 300 K, 2 MPa pressure	Entrained-flow gasifier	More than 90% of carbon conversions occurred	(Ren et al. 2020)

(continued)

Table 5.2 (continued)

Type of biomass	Experimental conditions	Name of the process	Outcome	References
Sugarcane bagasse	Temperature: 800, 900, and 1000 °C; gasifying agent: 8 g/min of steam; tracer gas: 2.33 g/min of nitrogen; sample: 15 g of sugarcane bagasse	Steam gasification	The increase in reactor temperature resulted in an increase in energy yield and apparent thermal efficiency The enhancement in syngas quality at the 1000 °C case resulted in an increase of energy yield	(Patel et al. 2016)
Raw bamboo; torrefied bamboo; high-volatile bituminous coal	Gasification agent: oxygen; sizes of the particles: 44–250 μm; fuel temperature: 300 K; pressure: 2 MPa	Entrained-flow gasifier	About 90% of carbon conversions occurred	(Canabarro et al. 2013)
Forest residue	Size of the feedstock 20–80 mm with 10–20% of moisture	Atmospheric-type and pressurized-type gasifier	Biomass-based significant yield was noticed than the conventional fuels and chemicals feedstocks	(Ren et al. 2020)
Crude glycerol with olive kernel	750–850 °C with the air ratios of λ = 0.2–0.4	Fixed-bed reactor	Around 19% of H_2 concentration was increased 33% and 19.5% of tar yield was decreased to 2.4% at 850 °C with λ = 0.4 air ratio	(Patel et al. 2016)
Wastes of pine, Red oak, and horse manure	800 °C, 12.2% of pine, 14.8% of red oak, and 18.33% horse manure		About 81.7–84.6% range of thermodynamic efficiencies for the gasifier were recorded	
Blend of polypropylene and poplar sawdust	400–800 °C, 2 mm and 3 mm-sized sawdust and polypropylene, respectively	Packed-bed reactor	High temperature decreases the solid residues fraction with the increased gas yield at the optimum temperature 700 °C	(Canabarro et al. 2013)

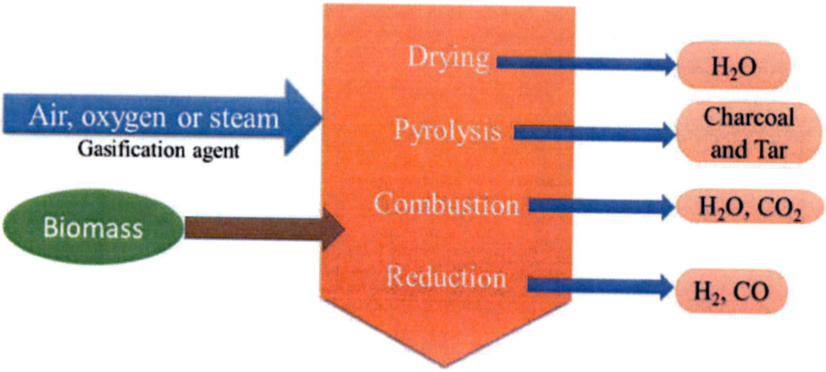

Fig. 5.4 Fundamental procedure for producing syngas using a typical downdraft fix-bed gasifier

5.2.5 *Hydrothermal Liquefaction (HTL)*

HTL is performed at subcritical water temperatures, with water as a profitable, eco-sustainable reaction medium for transforming biomass into fuel. HTL includes bio-oil, stable residues, gases, and aqueous materials in process, at 280–370 °C and a pressure of 10–25 MPa. A main advantage of HTL (He et al. 2020) is the ability to use humid biomass without exposing it to the drying phase. The use of water as a reaction medium has two main advantages in HTL under the above conditions: (i) initiate contact with biomass and help to split the chemical bond between them and (ii) make it easier to distinguish bio-oil (Zhang et al. 2020) from other components. HTL involves a series of structural and chemical water medium transitions with three major steps: depolymerization, decomposition, and recombination. Biocrude or bio-oil is a thick (0.97–1.04 kg L^{-1}), viscous (Kandasamy et al. 2020) (3.27–330 mPas), and dense (0.97–1.04 kg L^{-1}) liquid petroleum made from various biomasses. The bulk of biocrude are phenolics, aromatic hydrocarbons, nitrogen-containing compounds, amides, fatty acids, and esters. Hydrogen, carbon dioxide, carbon monoxide (Kandasamy et al. 2019), and methane are present in HTL-derived gaseous compounds in the stage of 23.7%, 19.1%, 22.8%, and 25.8%, respectively. The most popular experimental reactors or systems used for HTL are load autoclaves (10–100 mL), large batch reactors, and continuous reactors. Depolymerization (hydrolysis), decomposition, and recombination are the three main reaction pathways in HTL (repolymerization). Although most organic compound forms trigger much of the exact HTL reactions, literature studies have shown some of the main paths. The selectivity of such reactions is affected by pH, the severity of the HTL phase (Kantarli et al. 2016) (temperature, stress, ramping, and retention time), solvent form and concentration, and catalyst type and concentration.

For individual products, it reported a biocrude yield of 82 wt% for lipids, 21.1 wt% for proteins, 4.6 wt% for cellulose, 6.6% for xylose, and 1.4% for lignin (Zhang et al. 2018). They observed that protein–carbohydrate and protein–lignin combinations increased yields. However, cellulose–xylose combinations in biocrude processing were ineffective. It was observed that lipid and lignin had an antagonistic relationship. The quinary mix had the highest energy recovery, highlighting the value of a stable substrate for synergistic effects during the HTL phase. The albumin, asparagine, glutamine, soy protein, glucose, carbohydrate starch, and fat are all involved in both HTL pathways. The hydrothermal liquefaction and its products are shown in Fig. 5.5.

The microalgae composition varies according to the organisms. As a consequence, such species will not be suitable for the HTL protocol. On the other hand, municipal sludge has fewer variable characteristics (Ganeshan et al. 2016), and its compression can be slightly tweaked by changing the primary and secondary sludge mixing ratio. There are two main differences in the structure of the two sludge forms. The first is the high protein content of secondary sludge, which can stabilize the substrate composition (Wachter et al. 2021) and benefit from Maillard reactions. The second is the high lipid content of primary sludge, which can be directly converted into biocrude in the HTL phase. Their biocrude yield is also smaller than that of other biomass types. Due to the HTL carbohydrate pathway, excessive organic acids develop during the process and pH reduction (Yang et al. 2019a). HTL reactions favor hydrochar production to biocrude production in acidic environments. In the context of macroalgae and lignocellulosic biomass HTL, an alkaline catalyst must also be used to stabilize the pH. Like microalgae, *Cyanobacteria* sp. have various biological compositions depending on the species. *Cyanobacteria* sp. have higher protein and lower lipid (Buyukada 2017) content than microalgae. As a consequence, Maillard reactions are mostly used in the treatment of cyanobacterial biocrude.

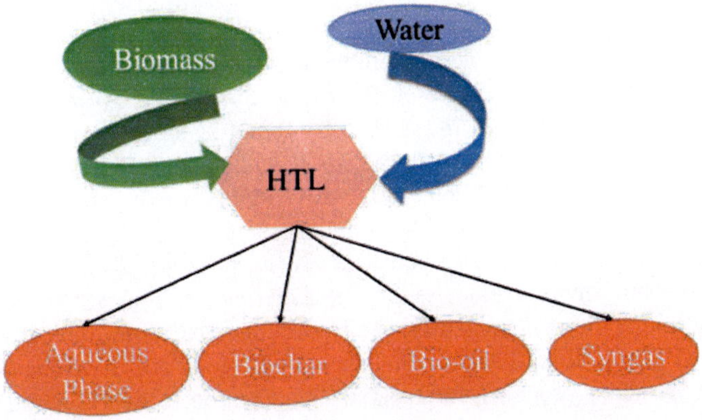

Fig. 5.5 General process of hydrothermal liquefaction and its products

5.3 Thermochemical Conversion of Different Wastes

The various wastes have been involved in the thermochemical conversion process, such as medical wastes, waste rubber seed, sewage, non-lignocellulosic biomass wastes, plastics, waste tires, food, algae, banana leaves, rice husk, sugarcane bagasse, duckweed, and straw waste.

5.3.1 Thermochemical Conversion of Medical Wastes

Medical wastes are known as solid waste from human and animal diagnosis, treatment and immunization, and practices such as biological product studies, tests, and growth. According to the World Health Organization, about 20% of the clinical waste managements (CMW) are infectious, toxic, or radioactive. Indeed, the concept or designation of these CMWs is not standardized in all nations, which results in a wide variety of care and handling alternatives. There are four types of medical waste: hospital, medical, managed health, and infectious waste. The primary ingredient of the bulk of the emulsion polymer isocyanate (EPI) is plastic polymers. Failure to handle this form of waste threatens people and the environment, which is still overburdened with a daily volume of plastic waste (Ishaq and Dincer 2021).

5.3.2 Conversion of Waste Rubber Seed

Rubber seed is the most viable non-edible source since it comprises 52–60% of the kernel and can produce biodiesel. Rubber seed is a solid by-product (Masnadi et al. 2015) of the *hevea brasiliensis*, mostly found in the tropics and commonly grown for its natural rubber latex. In the world, India is ranked fifth in rubber seed production (754,330 t), behind China (928,450 t), Malaysia (Shahabuddin et al. 2020) (1,735,522 t), Thailand (3,172,394 t), and Indonesia (5,367,980 t). Meanwhile, using the rubber seed shell as a biomass feedstock may have greater advantages and power in tackling current environmental issues. The shell represents 40–48% of rubber seed weight. Although the shell contains much less oil than the kernel, direct pyrolysis is beneficial (Falter and Pitz-Paal 2018). Oilseeds and seed shells have been known to provide value-added products. Physical steam activation is used to produce rubber seed shell-activated carbon. Rubber seed shell is a strong precursor for producing high-capacity activated carbon through the pyrolysis process. The average liquid commodity yield (500 °C) in a fixed bed reactor is 46.4 wt% from the palm seed shell. At two pyrolysis (Gaber et al. 2018) temperatures, 500 and 600 °C, the liquid pomegranate yield was estimated up to 54 wt%. According to records, the solid residue is a high heat content and low sulfur carbon-rich fuel.

5.3.3 Conversion of Sewage Sludge

Thermochemical processes have been considered one of the most promising approaches to derive energy and products from waste residues in recent decades. Because of the increased quantity of organic and sufficiently high calorific value, wastewater sludge produced in wastewater treatment plants (WWTPs) may be treated as a special biomass case in its dry form. Sludge has a high humidity level, and desiccation absorbs a significant part of the dry solids' energy content. Despite this, the conversion method for thermochemical sewage sludge is self-sufficient.

5.3.4 Conversion of Non-lignocellulosic Biomass

All biochar production can be carried out using pyrolysis, hydrothermal carbonization, gasification and other thermochemical methods. Non-lignocellulosic biochar (NLBC) is mainly generated by pyrolysis and hydrothermal treatment (Vakalis et al. 2017), depending on precursor physical status and technological application. Owing to the complexities of non-lignocellulosic biomass (NLBM) compounds, the reactions during thermochemical (Zhang et al. 2019a) processes are highly complicated, and NLBC's physicochemical properties have great effects on treatment conditions (e.g., treatment procedure, heating temperature, and rate) and on the existence of NLBM feedstock.

5.3.5 Conversion of Plastic Wastes

Plastics are used to decrease the number of containers, enhance transport safety, keep food fresh and prevent spoilage, make medicines, conserve medicines, and insulate electrical components in many ways of our daily life. Nonetheless, vast amounts of non-biodegradable plastic waste, mostly microplastics, are generated and end up in the environment (Chaturvedi and Dave 2020). Since 2002, global industrial plastics demand has grown by almost 80%. Their degree of recyclability is a division of plastics into seven major classes: polyethylene, terephthalate, polyethylene high density, polyvinyl chloride, polyethylene low density, polypropylene (Shen 2020), polystyrene, and several other materials. Recycling technologies may help reduce plastic waste generation but pollute the environment and cost energy, labor, and services money. Plastics are converted into alternative fuels and chemicals through waste to electricity, such as pyrolysis, liquefaction, and gasification. Thermochemical waste plastics conversion methods are our major focus. This system reduces the volume of plastic that ends up in sites and oceans, reduces greenhouse emissions, and has a high level of conversion and performance. Depending on the conversion process, plastics can be converted to bio-oil, biological-crude oil, synthesis gas, hydrogen, or aromatic charcoal. Certain polymer materials, like polystyrene (PS), can be decomposed thermally into high-yield monomers. That is not the reality with polyethylene (PE) or polypropylene (PP), which accounts for 60–70% of urban solid waste (Materazzi and Holt 2019). Since they bear the expense of collection and pyrolysis, these polyolefins can be converted into valuables rather than monomers. It also appears that a particular form of the catalyst is useful in the pyrolysis process with a specific type of plastic. Sadly, only pure plastic waste can be used and is not recommended for mixed plastic waste. Tertiary disposal of communicated post-consumption plastic waste is a terrifying challenge, since it involves hydrocarbons and blended polymers containing nitrogen and sulfur, as well as specific materials modified.

5.3.6 Waste Tires as a Thermochemical Process Feedstock

Due to their high carbon content, waste tires are a preferred thermochemical process feedstock. Commercial vehicle tires are made from a variety of materials, rubber (45–55 wt%) (Cruz et al. 2021), carbon black (20–35 wt%), metal textiles (15–25 wt%), zinc oxide (1–3 wt%), sulfur (Pashchenko et al. 2020) (1–2 wt%), and chemicals (5–7 wt%). In the car tires, natural rubber (NR) and synthetic (SR) rubbers, namely butyl rubber (BR) and styrene-butadiene rubber (SBR), are used. Due to its appropriate tear resistance, superior bond strength, and rolling strength, NR is used. SBR is also found in pneumatic boxes and sidewalls. In the pneumatic production direction, the vulcanization phase, in which a non-reversible reaction in the different (Nkosi et al. 2021) chemicals and feed materials, connects rubber polymer sequences of strong molecular bridge. The cross-linked elastomers are stiff, insoluble, and inseparable thermosets. Waste pneumatic products have a comparatively high heating value (38–45 MJ kg^{-1}) compared to other waste (Deng et al. 2021) materials such as biomass (22–28.5 MJ kg^{-1}), food waste (3.667–9.920 MJ kg^{-1}), animal manure (12–15 MJ kg^{-1}), plastics (24–44 MJ kg^{-1}) traditional fossil fuels (16.98 MJ kg^{-1}), and subbituminous (33.04 MJ kg^{-1}).

5.3.7 Components of Lignocellulosic Biomass

Lignocellulosic biomass is made up of cellulose (linear homogeneous polysaccharide (Pashchenko 2017) composed of D-glucose units), hemicellulose, and lignin (phenylpropanoid polymer composed of D-xylose, L-arabinose, D-mannose, D-galactose, and D-glucose). Commelinide monocotyledone hemicelluloses contain more xylan and arabinan, which are related to arabinoxylan and glucan. Hemicellulose glucan, xyloglucan, and mannan are more common in no commelinidae dicotyledon hemicellulose.

The chemical composition of the biomass varies as plant biomass matures. The plant's cell walls thicken and lignify throughout the growing season (increase secondary cell walls). The proportion of structural components (Kwon et al. 2019) (cellulose, hemicellulose, and lignin) increased with the continued growth of the collected biomass as the proportion of non-structural components (non-structural sugars, proteins, and mineral compounds) decreased. Moreover, to maximize the utilization of these biomasses in the production of biofuels or green chemicals, it is necessary to have a thorough understanding of the content (Fan et al. 2017), and the molecular structure of the key chemical components will vary depending on the harvesting time. It also relies on the harvesting location, as soil composition varies by location.

5.3.8 *Conversion of Food Wastes*

Food waste is generated daily from various sources, including fruit, vegetables (Aburto et al. 2015), poultry, and meat. It has been developed to produce value-added items, including protein (Shen et al. 2018), gelatin, amino acids, lipids, and produce oil and a renewable energy source. The United Nations, Food and Agriculture Organization reports that close to 1.3–109 t of fresh vegetables, beef, fruit, pastries, and dairy products are missing or absent from the food supply chain due to spoilage, expiry, and insufficient inventory control. Food waste is projected to increase over the next 25 years due to demographic and economic development, particularly in (Dirbeba et al. 2017) Asian countries. Asian countries will increase their annual urban food waste from 278 million tons in 2005 to 416 million tons in 2025. In 2014, the annual production of food waste in Singapore was roughly 790,000 t, and in 2015, it was estimated (Sun et al. 2019) to be 15,000 t/d in Malaysia.

Owing to intrinsic environmental issues and complexities of food waste composition, food waste has become an attractive source of value-added products. In particular, the technology of thermal conversion, particularly, biomass into bio-oil (Hossain et al. 2016), is an innovative recycling process. A various types of thermochemical transformation technologies, including gasification, pyrolysis, and hydrothermal liquefaction (HTL), torrefaction, and direct combustion increasing conversion of biomass into liquids. HTL is a promising strategy to transform biomass into liquefied materials (Pattanayak et al. 2020), breaking the biomass chemical structure into important liquid components utilizing hot pressurized water. HTL is an energy-intensive drying system that is cost-effectively bypassed. According to previous studies, the optimal temperature for biomass liquefaction is from 200 to 450 °C (da Silva Almeida et al. 2016) depending on the feedstock biomass, solvents, catalysts, and other operating parameters. As the reaction temperature increases, biocrude oil yields increase. During liquefaction, decomposing and repolymerizing processes lead to biocrude, aqueous liquid chemical and solid debris, and steam. The high pressure in the reaction causes the water to stay in a liquid state allowing for the further reaction of the main biomass structure. The HTL system benefits from allowing wet biomass with high humidity levels before conversion without the need for a pre-drying (Rodriguez-Sanchez et al. 2019) process. The application of HTL is acceptable for processing biocrude oil from waste algae and wood-based biomass with varying moisture content and reducing production costs, since before the conversion, no drying or dewatering is needed.

5.3.9 *Conversion of Algae Biomass*

Biochemical and thermochemical conversion technologies are used to produce (Kumar et al. 2019) environmentally friendly biofuels such as solid fuel, bio-hydrogen, biodiesel, bioethanol (Das et al. 2021), and microalgal biomass synthetic gas. On the other hand, thermochemical conversion is considered more significant (Nogueira Junior et al. 2018). The feedstock properties are one of the major problems of biochemical conversion. Feedstock enhancements and pretreatment processes have a high capital cost which increases the production costs. Besides, the efficiency of biofuel conversion through biochemical pathways is low. In addition, for each technology, the biochemical pathway has produced just one end product, while, with a single route, the thermochemical route will produce several end products. On the other hand, other thermochemical methods have a clear benefit in product production and consistency. The reaction temperature and the residence time were relatively more important among the different factors influencing process efficiency.

5.3.10 *Conversion of Banana Leaves*

Generally, direct burning uses wood as a primary fuel to produce thermal (Ahmed et al. 2020) and electrical energy. The sulfur and nitrogen levels of biomass are usually low, leading to low concentrations of pollutant gases, such as SO_2, NO_x, and N_2O, during combustion. Though combustion biomass emits CO_2, the volume emitted during photosynthesis during plant growth does not surpass. Global warming is anticipated as the atmosphere's CO_2 concentration increases, mostly due to the combustion of fossil fuels. As a result, clean energy sources that reduce CO_2 emissions need to be identified. Solar, wind, water power, and biomass combustion, are examples of CO_2-neutral energy sources. The features and characteristics of these products (Kumar et al. 2020) are affected by the parameters which determine slow, fast, or flash pyrolysis and the proportions of biomass components. Some studies have shown that the capacity to produce energy for banana waste is focused on the processing of biogas from banana peels, stalks, leaves, and pseudostems, the development of banana peel briquettes, and the generation of banana peel ethanol and banana fruit.

5.3.11 Conversion of Rice Husk

The vast amounts of rice husks discarded after frying are not well handled. If an efficient process is workable, the rice husk can be converted (Weldekidan et al. 2018) into valuable energy to compensate for fossil fuel overuse. Hydrothermal liquefaction is a thermochemical low-temperature (Xu et al. 2019) mechanism that splits biomass into small molecular fragments of water. These volatile and reactive light fragments then convert it into oily compounds of various molecular weights. There was much curiosity in using heat-compressed and sub/supercritical fluids (Unrean et al. 2018) for biomass liquefaction. A few of the solvents used for this treatment method are water, ethanol, methanol, acetone, and 1,4-dioxane. Among, water is the most environmentally conscious solvent.

A moderate volume of water is considered helpful for the processing of oil fractions. However, the carbon content is lower as bio-oil fractions are liquefied in water, the oxygen content is higher, and the heating value is lower in the biocrude. Chemical solvents have been used to increase the production of liquid products containing less oxygen (and thus higher heating value). First of all, the liquefaction reaction is heavily affected by the solvent. There are also some benefits to the physicochemical properties of subcritical ethanol. First, sub/supercritical fluids (Zhang et al. 2019b) have special distribution properties such as gas and liquid density and total miscibility of process liquid/steam materials, which provide a single-phase atmosphere for reactions that may otherwise occur. Second, sub/supercritical fluids will dissolve non-liquid or gaseous solvent-induced materials to encourage reactions in gasification and liquefaction. Third, sub and supercritical ethanol can serve as a hydride donor through the hydride transfer of their α-hydrogen (Makwana et al. 2019) by a process known as "hydrogen shutting." An ethanol–water blended solvent has been selected as the liquefaction solvent to extract bio-oil products from rice husk.

5.3.12 Conversion of Sugarcane Bagasse

For fast pyrolysis, dry biomass (feeds below 10 wt%) is ground to 3 mm thermally decomposed by around 500 °C in an inert atmosphere (David et al. 2018) with a few seconds' hot gas residence result in fast pyrolysis bio-oil (FPBO). The catalyst is also essential in hydrotreatment reactions. As a result, several investigations have been conducted on the development and synthesis of catalysts for hydrotreating FPBO. In particular, nickel-based catalysts have proven successful in transforming model compounds and generating rapid pyrolysis bio-oil from a range of feedstocks. Nickel-based catalysts have previously been identified as having advantages, including low cost, high deoxygenation (Athira et al. 2021), and the ability to use promoters for various selectivity or higher resistance to poisoning and deactivation in the catalyst formulation.

While several groups examined the sugarcane bagasse pyrolysis, only a few examined the liquid fraction hydrotreatment, and none observed at the whole process chain of sugarcane bagasse (SCB). SCB is especially useful since it has gathered focus in the sugar mill instead of other biomasses like sugarcane leaves. In general, sugarcane leaves are left in the field, requiring a simple and cost-effective selection method. In this way, the consolidated sugarcane conversion unit would be economically beneficial, enabling the construction of large units without the need to transport the feedstock or an intermediary component, as is generally proposed for bio-oil-based production. In addition, the high lignin content (17–32 wt%), which is considered an obstacle to carbohydrate hydrolysis, makes the bagasse especially appealing for thermochemical recovery (Dabe et al. 2019), given that desirable targets are not only hydrocarbons but aromatic monomers. This strategy is especially relevant for countries such as Brazil that have had record production of sugarcane ethanol in 2019 and plan to increase production by 2030. The large quantities of agricultural residues generated can be thermochemically processed into products combined with aviation kerosene at the 10% concentration, as stated in the Brazilian National Biofuel Strategy for 2030. Additionally, a new variety of functional chemical products can be produced, such as functional aromatic compounds, expanding the chemical range of the sugarcane refineries (Chan and Wang 2018).

5.3.13 Conversion of Duckweed

Duckweed is rich in protein, has a high starch content, and owing to its deep saccharification, its cell wall material can easily be turned into a biofuel product. The duckweed is commonly used for gaseous or liquid fuel production using biochemical and thermochemical processes such as fermentation, pyrolysis, hydrothermal liquefaction, and gasification. Biofuels such as syngas and alcohol are produced by duckweed fermentation. Since further specifications are required for this process, the approach is inefficient. Pyrolysis and HTL have for some time been known as important ways of manufacturing bio-oil. The crude duckweed bio-oil (DCBO) is a dark brown, oily fluid that cannot be used directly in a combustion engine. There are many inconveniences, including high corrosiveness, a high heteroatomic material, low higher heating value (HHV), and a high acidity index. As a result, substantial upgrades are needed to fulfill the specifications of the program. However, owing to the complexity of DCBO, it is difficult to update and determine the upgrade phase. Duckweed is rich in starch, carbon, and protein, but poor in lipids and the debris of all fragmented bio-macromolecules can be transformed into DCBO during thermochemical conversion (Djandja et al. 2021).

5.3.14 Conversion of Straw Waste

Rice straw is a key component of agro-residues and could be a viable source of feedstock for biofuels production. China and India produce the majority of rice straws. As a result, every year, a large volume of rice straw is produced worldwide (Biswas et al. 2018). The amount of chemical energy stored in raw biomass in each commodity is known as the energy yield. Because of the chemical composition, which mainly depends on carbon and hydrogen, coke, and more dramatically char, energy sinks have essential functions. Although the gas yield was equal to that of carbohydrate in each trial, the chemical energy carriers' output was markedly lower than CO. In particular, H_2 and light hydrocarbons were present at small concentrations. Biomass pretreatment has a significant effect on the molecular composition of the subsequent bio-oils (Titiloye et al. 2013). According to GC–MS analysis, the acid-washed camelina straw showed significantly reduced carboxylic acids, ketones, ethers, and oxygen-coated aromas. On the other hand, CS-ac produced abundant furans and anhydrous sugars, principally levoglucosan. This trend is broadly consistent with previous results from different sources of biomass that are subject to washing. These results show that inorganic matter, in particular temperature, can help convert sugar into smaller molecules such as acids and ketones. The use of HZSM-5 zeolites, on the other hand, induces a higher proportion of oxygenated aromatic substances and especially aromatic hydrocarbons both for CS and CS-ac, largely absent from thermal bio-oil. The liquid fraction of arenas obtained from the CS-ac pyrolysis biomass is more concentrated than in HZSM-5. In the pyrolysis bio-oil, the fraction of oxygenated aromatic material recovered from the CS sample is higher. Catalysts of HZSM-5 produce aromatic agents, and this mechanism is most probably mediated by the Diels–Alders condensation, which is medium to these solid acids by cracking and dehydration. This is an important method to improve bio-oil characteristics as a fuel (Nanda and Berruti 2021).

5.4 Bio-oil Upgrading

Hydrotreating, hydrocracking, catalytic cracking, solvent adding, and emulsion are alternatives to improve bio-oil bio-pyrolysis with an upgraded bio-intended use alternative. Hydrotreating is the most frequent process when the upgraded bio-oil is to be used as a transport fuel. Oxygen is eliminated through hydro-deoxygenation (HDO), and complex bio-oil is converted to hydrocarbons (Shan Ahamed et al. 2021) and water as products from 200 to 400 °C with pressure varying from 2 to 5 MPa. Temperature, pressure, and catalysts all affect the structure of the hydrocarbon component. The upgraded bio-oil (Duan et al. 2015) has dramatically enhanced its properties but requires refining before being used as a liquid fuel for transport. Therefore, refining and extracting bio-oil is difficult and expensive because the target liquid is a transport fuel. However, the one-step supercritical (Pang 2019) upgrade of alcohol is a promising approach when the solvent is used as a burning fuel or a heavy fuel with more excellent stability and heat than the original bio-oil. The HHV of bio-oil can be increased using supercritical ethanol upgrades from about 20–40 MJ kg^{-1} to 36–40 MJ kg^{-1} (Labaki and Jeguirim 2017). As per a different report, the HHV of improved bio-oil is 34.140 MJ kg^{-1} (Long et al. 2019), with a water content of 1.6 wt%. These findings are comparable to those observed for upgraded boil-oil hydrotherapy (40 MJ kg^{-1}).

5.5 Factors Affecting the Bio-oil Composition

Biomass pyrolysis is an absence of oxygen thermochemical breakdown method that produces liquid (bio-oil), solid (char), and coal (Das et al. 2021). Fractions can vary according to the biomass, operating temperature, heating intensity, and duration of the residence. For the liquid target product, fast quenching and adequate operating temperatures of 400–650 °C are required. The chemical composition of bio-oil (Durak 2019) is also affected by the heating rate and the residence time. Different biomass components decompose and produce different products at various temperatures, but the chemical composition of bio-oil often depends on the heating rate and residence time. When cellulose decomposes between 250 and 350 °C, levoglucosan and other anhydrous celluloses (Matthews et al. 2016) are produced. Lignin has the most complex structure and is the most resistant of the three major elements, decomposes to polysensitized phenol oligomers and monomers at temperatures ranging from 280 to 500 °C. There are heavily oxygenated side branches of hemicellulose that are comparatively easy to decompose when heated. Hemicellulose is also the first fraction of the biomass to decompose at temperatures (Kumar et al. 2019) 180 and 320 °C. As hemicelluloses decompose, acetic acid and other organic acids, sugars, and furans are formed.

On the other hand, short-chain acids have decreased with increased ash content and reduced hydrocarbons with increased cellulose content (Patel et al. 2016) in biomass. As a result, biomass pyrolysis-based bio-oil has a complex chemical composition, making it unsuitable for engines or natural chemical feedstock. Recently, several pilots and commercial pyrolysis plants were installed. Biomass Technology Group (BTG) (Chan et al. 2019) has established in Hengelo, the Netherlands, its full-scale commercial plant (Empyro project) that produces 20 million liters of biomass timber oil per year. Bio-oil is produced in all the steam or heat and power generation plants listed for combustion in boilers (Bundhoo 2018). Separation and upgradation are necessary before bio-oil can be used as a fuel or chemical transport.

5.6 Future Perspectives and Concluding Remarks

At present, the only compositional productivity conditions for biomass are sugar content (Sani et al. 2017), organic content, and total ash. In addition, to convert organic matter, other feedstock variables may substantially impact biomass behavior, influencing real yield, product quality, catalyst efficacy (Pang 2019), and reactor cycle life, and these factors effectively affect feedstock yield ability. To date, many of these consistency properties of feedstock, such as ash species and other intrinsic (Pattanayak et al. 2020) inhibitors, have not been considered. Different conversion pathways require feedstock materials with diverse quality requirements to produce biofuels from waste biomass. As such, these feedstock properties should be recognized as essential qualities of consistency, which are taken into account for mitigation by the collection and preparation of feedstock. It is crucial solving feedstock variability problems and the supply quality (Wang et al. 2020b) of controlled, standardized, and feedstocks for biorefineries to achieve (Xu et al. 2019) viable biofuel production. Approaches and policies must be adopted to mitigate the effects of the feedstock content and profitability to ensure a healthy and economic conversion.

While the bulk of the literature concentrates on the nature of manufacturing systems, there is no mention of how to selectively, reliably, and cheaply isolate the intended products from the resulting dynamic matrix. The solvent extraction (Shahabuddin et al. 2020) can be made after conversion or concurrently during catalysis in a biphasic system with reactive (aqueous) and extractive (organic) (Titiloye et al. 2013) layers. For room temperature vacuum distillation and flash separation, low boiling point solvents are safer, while high boiling point solvents are more energy intensive. Green solvents (Weldekidan et al. 2018) from biomass or CO_2 should be checked against traditional synthetic solvents for their environmental benefits. The three-stage product profile (liquid, solid, and gas) of a biorefinery system is a problem in downstream separation and (Wang et al. 2020b) purification processes. Waste by-products such as wastewater effluents, greenhouse gas emissions, and solid residues are infrequently reported (Zhang et al. 2020). When assessing the importance of the recovery of food waste, a life cycle assessment of all sources of products and how much waste is collected and how it is correctly used or disposed of is essential. Future food waste recovery should prioritize green chemistry and low-impact engineering pathways. Right now, it is essential to extend the thermodynamics of a specific reaction from proof-of-concept tests to field-oriented industrial-sized implementations. Additional efficiency, selectivity, mass transit, and throughput rate parameters must be considered when examining the production appeal. A manufacturer-level analysis of total cost and profitability is necessary when best configuration, fixed capital expense, equipment life, materials, energy balance (Yang et al. 2019b), transport and logistics, and product sensitivity analyses are determined. Although more field data will be possible in the future, mathematical models should be built for simulating the corresponding processes and optimizing operating parameters. Green synthesis is essential in this respect of recyclable materials from renewable resources. The innovative technologies explored in the present chapter for food waste transformation into value-added chemicals provide long-term strategies to convert waste from an environmental threat into a lucrative product. These research studies encourage and enhance the (Shen et al. 2018) techno-economic feasibility of renewable biorefineries by maximizing raw materials, method options, energy integration, and demand constraints.

In thermochemical choices, the position of reactor technology is crucial. Operational parameters and reactor setup have a significant influence on product consistency during the process. A reactor may have one positive characteristic most credited. Still, there may also be many other negative aspects, making it less suitable for technological strength and, most notably, market competition (Soares et al. 2020). It could help intensify the process of the value-added product. In the future, more detailed analyses of barriers (high moisture, ash, and heavy metal content) would be required to increase the consistency of the substance (Vakalis et al. 2017) and make it available readily for commercial purposes. The thermochemical conversions of the main components of wastes through torrefaction, combustion, pyrolysis, gasification, and hydrothermal liquefaction are summarized in this chapter. Furthermore, waste characterization methods and thermochemical conversion processes are consistently generalized. Finally, the existing research gap and limitations are highlighted.

5.7 Conclusions

This chapter illustrates the many types of biomass waste and its thermochemical conversions. As the demand for alternative energy sources grows, biomass is used to substitute fossil fuels. In addition, waste and biomass incineration are already widely used in the commercial sector, and methods exist to convert them into biofuels or hydrogen. Even though advanced technologies have proven viable in the laboratory and on pilot scales in many cases, commercialization is still a challenge owing to market competition. Moreover, reactor technologies for thermochemical options play an important role. During the process, operational parameters and reactor design have a significant impact on product quality. However, plenty of other negative factors may be present, making it less appropriate in terms of technological strength and, most importantly, competitiveness.

On the other hand, more extensive investigations on the barriers (high moisture, ash, and heavy metal content) will be required in the future to improve the product's quality and make them widely available for commercial use. Environmental policies and regulations are likely to aid in the widespread adoption of numerous sustainable technologies in the future years. As a result, the thermochemical conversion of waste to biofuel will be a better renewable fuel.

Acknowledgements This work was supported by the National Natural Science Foundation of China (51876083, 51776088), a Project Funded by the Priority Academic Program Development of Jiangsu High Education Institutions and High-tech Research Key laboratory of Zhenjiang (SS2018002).

References

Aburto J, Moran M, Galano A, Torres-García E (2015) Non-isothermal pyrolysis of pectin: a thermochemical and kinetic approach. J Anal Appl Pyrol 112:94–104

Adeniyi AG, Ighalo JO, Onifade DV (2020) Biochar from the thermochemical conversion of orange (*Citrus sinensis*) peel and Albedo: product quality and potential applications. Chem Afr 3(2):439–448

Adjin-Tetteh M, Asiedu N, Dodoo-Arhin D, Karam A, Amaniampong PN (2018) Thermochemical conversion and characterization of cocoa pod husks a potential agricultural waste from Ghana. Ind Crops Prod 119:304–312

Ahmed G, Acharya S, Kawale H, Singh A, Kishore N, Pal S (2020) Thermochemical conversion of Polyalthia longifolia leaves at different temperatures and characterization of their products. Fuel 280:118574

Athira G, Bahurudeen A, Appari S (2021) Thermochemical conversion of sugarcane bagasse: composition, reaction kinetics, and characterisation of by-products. Sugar Tech 23(2):433–452

Awalludin MF, Sulaiman O, Hashim R, Nadhari WNAW (2015) An overview of the oil palm industry in Malaysia and its waste utilization through thermochemical conversion, specifically via liquefaction. Renew Sustain Energy Rev 50:1469–1484

Bhaskar T, Pandey A (2015) Chapter 1—Advances in thermochemical conversion of biomass—introduction. In: Pandey A, Bhaskar T, Stöcker M, Sukumaran RK (eds) Recent advances in thermo-chemical conversion of biomass. Elsevier, Boston, pp 3–30

Biswas B, Singh R, Kumar J, Singh R, Gupta P, Krishna BB, Bhaskar T (2018) Pyrolysis behavior of rice straw under carbon dioxide for production of bio-oil. Renew Energy 129:686–694

Bora RR, Lei M, Tester JW, Lehmann J, You F (2020) Life cycle assessment and technoeconomic analysis of thermochemical conversion technologies applied to poultry litter with energy and nutrient recovery. ACS Sustain Chem Eng 8(22):8436–8447

Bundhoo ZMA (2018) Microwave-assisted conversion of biomass and waste materials to biofuels. Renew Sustain Energy Rev 82:1149–1177

Buyukada M (2017) Uncertainty estimation by Bayesian approach in thermochemical conversion of walnut hull and lignite coal blends. Bioresour Technol 232:87–92

Canabarro N, Soares JF, Anchieta CG, Kelling CS, Mazutti MA (2013) Thermochemical processes for biofuels production from biomass. Sustain Chem Process 1(1):22

Castaldi M, van Deventer J, Lavoie JM, Legrand J, Nzihou A, Pontikes Y, Py X, Vandecasteele C, Vasudevan PT, Verstraete W (2017) Progress and prospects in the field of biomass and waste to energy and added-value materials. Waste Biomass Valoriz 8(6):1875–1884

Chan WP, Wang J-Y (2016) Comprehensive characterisation of sewage sludge for thermochemical conversion processes—based on Singapore survey. Waste Manag 54:131–142

Chan WP, Wang J-Y (2018) Formation of synthetic sludge as a representative tool for thermochemical conversion modelling and performance analysis of sewage sludge—based on a TG-FTIR study. J Anal Appl Pyrol 133:97–106

Chan YH, Cheah KW, How BS, Loy ACM, Shahbaz M, Singh HKG, Yusuf NaR, Shuhaili AFA, Yusup S, Ghani WAWAK, Rambli J, Kansha Y, Lam HL, Hong BH, Ngan SL (2019) An overview of biomass thermochemical conversion technologies in Malaysia. Sci Total Environ 680:105–123

Chaturvedi S, Dave PN (2020) 5—Application of nanotechnology in foods and beverages. In: Grumezescu AM, Holban AM (eds) Nanoengineering in the beverage industry. Academic, pp 137–162

Chen P, Anderson E, Addy M, Zhang R, Cheng Y, Peng P, Ma Y, Fan L, Zhang Y, Lu Q, Liu S, Zhou N, Deng X, Zhou W, Omar M, Griffith R, Kabir F, Lei H, Wang Y, Liu Y, Ruan R (2018) Breakthrough technologies for the biorefining of organic solid and liquid wastes. Engineering 4(4):574–580

Chen Y, Wu Y, Hua D, Li C, Harold MP, Wang J, Yang M (2015) Thermochemical conversion of low-lipid microalgae for the production of liquid fuels: challenges and opportunities. RSC Adv 5(24):18673–18701

Cho D-W, Tsang DCW, Kim S, Kwon EE, Kwon G, Song H (2018) Thermochemical conversion of cobalt-loaded spent coffee grounds for production of energy resource and environmental catalyst. Bioresour Technol 270:346–351

Cocero MJ, Cabeza Á, Abad N, Adamovic T, Vaquerizo L, Martínez CM, Pazo-Cepeda MV (2018) Understanding biomass fractionation in subcritical & supercritical water. J Supercrit Fluids 133:550–565

Cruz G, Rodrigues AdLP, da Silva DF, Gomes WC (2021) Physical–chemical characterization and thermal behavior of cassava harvest waste for application in thermochemical processes. J Therm Anal Calorim 143(5):3611–3622

da Silva Almeida H, Corrêa OA, Eid JG, Ribeiro HJ, de Castro DAR, Pereira MS, Pereira LM, de Andrade Aâncio A, Santos MC, da Mota SAP, da Silva Souza JA, Borges LEP, Mendonça NM, Machado NT (2016) Performance of thermochemical conversion of fat, oils, and grease into kerosene-like hydrocarbons in different production scales. J Anal Appl Pyrol 120:126–143

Dabe SJ, Prasad PJ, Vaidya A, Purohit H (2019) Technological pathways for bioenergy generation from municipal solid waste: renewable energy option. Environ Prog Sustain Energy 38(2):654–671

Das P, Chandramohan VP, Mathimani T, Pugazhendhi A (2021) A comprehensive review on the factors affecting thermochemical conversion efficiency of algal biomass to energy. Sci Total Environ 766:144213

David GF, Justo OR, Perez VH, Garcia-Perez M (2018) Thermochemical conversion of sugarcane bagasse by fast pyrolysis: high yield of levoglucosan production. J Anal Appl Pyrol 133:246–253

de Oliveira Maiaa, B.G, Souzab, O, Marangonib, C, Hotzaa, D, de Oliveiraa, A.P.N, Sellinb, N. 2014. Production and characterization of fuel briquettes from banana leaves waste. Chem Eng 37

Deng Y, Dewil R, Appels L, Ansart R, Baeyens J, Kang Q (2021) Reviewing the thermo-chemical recycling of waste polyurethane foam. J Environ Manag 278:111527

Dirbeba MJ, Brink A, DeMartini N, Zevenhoven M, Hupa M (2017) Potential for thermochemical conversion of biomass residues from the integrated sugar-ethanol process—fate of ash and ash-forming elements. Bioresour Technol 234:188–197

Djandja OS, Yin L, Wang Z, Guo Y, Zhang X, Duan P (2021) Progress in thermochemical conversion of duckweed and upgrading of the bio-oil: a critical review. Sci Total Environ 769:144660

Dou B, Zhang H, Song Y, Zhao L, Jiang B, He M, Ruan C, Chen H, Xu Y (2019) Hydrogen production from the thermochemical conversion of biomass: issues and challenges. Sustain Energy Fuels 3(2):314–342

Duan P, Wang B, Xu Y (2015) Catalytic hydrothermal upgrading of crude bio-oils produced from different thermo-chemical conversion routes of microalgae. Bioresour Technol 186:58–66

Durak H (2019) Characterization of products obtained from hydrothermal liquefaction of biomass (*Anchusa azurea*) compared to other thermochemical conversion methods. Biomass Convers Biorefin 9(2):459–470

Fabičovicová K, Lucas M, Claus P (2016) From barley straw to valuable polyols: a sustainable process using ethanol/water mixtures and hydrogenolysis over ruthenium-tungsten catalyst. Chemsuschem 9(19):2804–2815

Falter C, Pitz-Paal R (2018) Energy analysis of solar thermochemical fuel production pathway with a focus on waste heat recuperation and vacuum generation. Sol Energy 176:230–240

Gaber C, Demuth M, Prieler R, Schluckner C, Hochenauer C (2018) An experimental study of a thermochemical regeneration waste heat recovery process using a reformer unit. Energy 155:381–391

Ganeshan G, Shadangi KP, Mohanty K (2016) Thermo-chemical conversion of mango seed kernel and shell to value added products. J Anal Appl Pyrol 121:403–408

Gomaa MR, Mustafa RJ, Al-Dmour N (2020) Solar thermochemical conversion of carbonaceous materials into syngas by co-gasification. J Clean Prod 248:119185

Gracida-Alvarez UR, Keenan LM, Sacramento-Rivero JC, Shonnard DR (2016) Resource and greenhouse gas assessments of the thermochemical conversion of municipal solid waste in Mexico. ACS Sustain Chem Eng 4(11):5972–5978

He Z, Wang B, Zhang B, Feng H, Kandasamy S, Chen H (2020) Synergistic effect of hydrothermal co-liquefaction of Spirulina platensis and lignin: optimization of operating parameters by response surface methodology. Energy 201:117550

Hossain MA, Jewaratnam J, Ganesan P (2016) Prospect of hydrogen production from oil palm biomass by thermochemical process—a review. Int J Hydrogen Energy 41(38):16637–16655

Ibarra-Gonzalez P, Rong B-G (2019) A review of the current state of biofuels production from lignocellulosic biomass using thermochemical conversion routes. Chin J Chem Eng 27(7):1523–1535

Ishaq H, Dincer I (2021) A new approach in treating industrial hazardous wastes for energy generation and thermochemical hydrogen production. J Clean Prod 290:125303

Jahirul MI, Rasul MG, Chowdhury AA, Ashwath N (2012) Biofuels production through biomass pyrolysis—a technological review. Energies 5(12):4952–5001

Jha P, Dass B (2020) Analysis of biomasses for their thermochemical transformations to biofuels. Int J Energy Prod Manag 5(2):115–124

Kandasamy S, Devarayan K, Bhuvanendran N, Zhang B, He Z, Narayanan M, Mathimani T, Ravichandran S, Pugazhendhi A (2021) Accelerating the production of bio-oil from hydrothermal liquefaction of microalgae via recycled biochar-supported catalysts. J Environ Chem Eng 9(4):105321

Kandasamy S, Zhang B, He Z, Chen H, Feng H, Wang Q, Wang B, Ashokkumar V, Siva S, Bhuvanendran N, Krishnamoorthi M (2020) Effect of low-temperature catalytic hydrothermal liquefaction of Spirulina platensis. Energy 190:116236

Kandasamy S, Zhang B, He Z, Chen H, Feng H, Wang Q, Wang B, Bhuvanendran N, Esakkimuthu S, Ashokkumar V, Krishnamoorthi M (2019) Hydrothermal liquefaction of microalgae using Fe_3O_4 nanostructures as efficient catalyst for the production of bio-oil: optimization of reaction parameters by response surface methodology. Biomass Bioenergy 131:105417

Kantarli IC, Kabadayi A, Ucar S, Yanik J (2016) Conversion of poultry wastes into energy feedstocks. Waste Manag 56:530–539

Korai MS, Mahar RB, Uqaili MA (2016) Optimization of waste to energy routes through biochemical and thermochemical treatment options of municipal solid waste in Hyderabad, Pakistan. Energy Convers Manag 124:333–343

Kumar M, Oyedun AO, Kumar A (2019) A comparative analysis of hydrogen production from the thermochemical conversion of algal biomass. Int J Hydrogen Energy 44(21):10384–10397

Kumar M, Shukla SK, Upadhyay SN, Mishra PK (2020) Analysis of thermal degradation of banana (*Musa balbisiana*) trunk biomass waste using iso-conversional models. Bioresour Technol 310:123393

Kunwar B, Chandrasekaran SR, Moser BR, Deluhery J, Kim P, Rajagopalan N, Sharma BK (2017) Catalytic thermal cracking of postconsumer waste plastics to fuels. 2. Pilot-scale thermochemical conversion. Energy Fuels 31(3):2705–2715

Kwon EE, Kim S, Lee J (2019) Pyrolysis of waste feedstocks in CO_2 for effective energy recovery and waste treatment. J CO_2 Utiliz 31:173–180

Labaki M, Jeguirim M (2017) Thermochemical conversion of waste tyres—a review. Environ Sci Pollut Res 24(11):9962–9992

Lam P, Sokhansanj S, Bi X, Lim C, Naimi L, Hoque M, Mani S, Womac AR, Narayan S, Ye X (2008) Bulk density of wet and dry wheat straw and switchgrass particles. Appl Eng Agric 24(3):351–358

Lee SY, Sankaran R, Chew KW, Tan CH, Krishnamoorthy R, Chu D-T, Show P-L (2019) Waste to bioenergy: a review on the recent conversion technologies. BMC Energy 1(1):4

Lee XJ, Lee LY, Gan S, Thangalazhy-Gopakumar S, Ng HK (2017) Biochar potential evaluation of palm oil wastes through slow pyrolysis: Thermochemical characterization and pyrolytic kinetic studies. Bioresour Technol 236:155–163

Li D-C, Jiang H (2017) The thermochemical conversion of non-lignocellulosic biomass to form biochar: A review on characterizations and mechanism elucidation. Bioresour Technol 246:57–68

Liu X, Zhang Y, Nahil MA, Williams PT, Wu C (2017) Development of Ni- and Fe- based catalysts with different metal particle sizes for the production of carbon nanotubes and hydrogen from thermo-chemical conversion of waste plastics. J Anal Appl Pyrol 125:32–39

Long F, Li F, Zhai Q, Wang F, Xu J (2019) Thermochemical conversion of waste acidic oil into hydrocarbon products over basic composite catalysts. J Clean Prod 234:105–112

Lopez G, Artetxe M, Amutio M, Bilbao J, Olazar M (2017) Thermochemical routes for the valorization of waste polyolefinic plastics to produce fuels and chemicals. A review. Renew Sustain Energy Rev 73:346–368

Lü F, Hua Z, Shao L, He P (2018) Loop bioenergy production and carbon sequestration of polymeric waste by integrating biochemical and thermochemical conversion processes: A conceptual framework and recent advances. Renewable Energy 124:202–211

Makwana JP, Pandey J, Mishra G (2019) Improving the properties of producer gas using high temperature gasification of rice husk in a pilot scale fluidized bed gasifier (FBG). Renewable Energy 130:943–951

Masnadi MS, Grace JR, Bi XT, Lim CJ, Ellis N (2015) From fossil fuels towards renewables: Inhibitory and catalytic effects on carbon thermochemical conversion during co-gasification of biomass with fossil fuels. Appl Energy 140:196–209

Materazzi M, Holt A (2019) Experimental analysis and preliminary assessment of an integrated thermochemical process for production of low-molecular weight biofuels from municipal solid waste (MSW). Renewable Energy 143:663–678

Matsakas L, Gao Q, Jansson S, Rova U, Christakopoulos P (2017) Green conversion of municipal solid wastes into fuels and chemicals. Electron J Biotechnol 26:69–83

Matthews LR, Niziolek AM, Onel O, Pinnaduwage N, Floudas CA (2016) Biomass to liquid transportation fuels via biological and thermochemical conversion: process synthesis and global optimization strategies. Ind Eng Chem Res 55(12):3203–3225

Maya DMY, Sarmiento ALE, Oliveira C, Lora E, Andrade R (2016) Gasification of municipal solid waste for power generation in Brazil, a review of available technologies and their environmental benefits. Journal of Chemistry and Chemical Engineering 10(6):249–255

Monarca, D, Colantoni, A, Cecchini, M, Longo, L, Vecchione, L, Carlini, M, Manzo, A. 2012. Energy Characterization and Gasification of Biomass Derived by Hazelnut Cultivation: Analysis of Produced Syngas by Gas Chromatography. *Mathematical Problems in Engineering*, **2012**, 102914.

Nanda S, Berruti F (2021) Thermochemical conversion of plastic waste to fuels: a review. Environ Chem Lett 19(1):123–148

Nicodème T, Berchem T, Jacquet N, Richel A (2018) Thermochemical conversion of sugar industry by-products to biofuels. Renew Sustain Energy Rev 88:151–159

Nkosi N, Muzenda E, Gorimbo J, Belaid M (2021) Developments in waste tyre thermochemical conversion processes: gasification, pyrolysis and liquefaction. RSC Adv 11(20):11844–11871

Nogueira Junior E, Kumar M, Pankratz S, Oyedun AO, Kumar A (2018) Development of life cycle water footprints for the production of fuels and chemicals from algae biomass. Water Res 140:311–322

Ong, H.C, Chen, W.-H, Farooq, A, Gan, Y.Y, Lee, K.T, Ashokkumar, V. 2019. Catalytic thermochemical conversion of biomass for biofuel production: A comprehensive review. *Renewable and Sustainable Energy Reviews*, **113**, 109266.

Ong, H.C, Chen, W.-H, Singh, Y, Gan, Y.Y, Chen, C.-Y, Show, P.L. 2020. A state-of-the-art review on thermochemical conversion of biomass for biofuel production: A TG-FTIR approach. *Energy Conversion and Management*, **209**, 112634.

Pang S (2019) Advances in thermochemical conversion of woody biomass to energy, fuels and chemicals. Biotechnol Adv 37(4):589–597

Parshetti GK, Suryadharma MS, Pham TPT, Mahmood R, Balasubramanian R (2015) Heterogeneous catalyst-assisted thermochemical conversion of food waste biomass into 5-hydroxymethylfurfural. Bioresour Technol 178:19–27

Pashchenko D (2017) Thermodynamic equilibrium analysis of combined dry and steam reforming of propane for thermochemical waste-heat recuperation. Int J Hydrogen Energy 42(22):14926–14935

Pashchenko D, Gnutikova M, Karpilov I (2020) Comparison study of thermochemical waste-heat recuperation by steam reforming of liquid biofuels. Int J Hydrogen Energy 45(7):4174–4181

Patel M, Zhang X, Kumar A (2016) Techno-economic and life cycle assessment on lignocellulosic biomass thermochemical conversion technologies: A review. Renew Sustain Energy Rev 53:1486–1499

Pattanayak S, Hauchhum L, Loha C, Sailo L (2020) Selection criteria of appropriate bamboo based biomass for thermochemical conversion process. Biomass Conversion and Biorefinery 10(2):401–407

Pham TPT, Kaushik R, Parshetti GK, Mahmood R, Balasubramanian R (2015) Food waste-to-energy conversion technologies: Current status and future directions. Waste Manage 38:399–408

Przybysz Buzała K, Kalinowska H, Przybysz P, Małachowska E (2017) Conversion of various types of lignocellulosic biomass to fermentable sugars using kraft pulping and enzymatic hydrolysis. Wood Sci Technol 51(4):873–885

Purnomo, C.W, Kurniawan, W, Aziz, M. 2021. Technological review on thermochemical conversion of COVID-19-related medical wastes. *Resources, Conservation and Recycling*, **167**, 105429.

Ren J, Liu Y-L, Zhao X-Y, Cao J-P (2020) Biomass thermochemical conversion: A review on tar elimination from biomass catalytic gasification. J Energy Inst 93(3):1083–1098

Rodriguez-Sanchez, S, Ruiz, B.a, Martinez-Blanco, D, Sanchez-Arenillas, M, Diez, M.A, Suarez-Ruiz, I, Marco, J.F, Blanco, J, Fuente, E. 2019. Sustainable thermochemical single-step process to obtain magnetic activated carbons from chestnut industrial wastes. *ACS Sustainable Chemistry & Engineering*, **7**(20), 17293-17305

Sani, J, Sokoto, A.M, Tambuwal, A.D, Garba, N.A. 2017. Effect of NiO/SiO2 on thermo-chemical conversion of waste cooking oil to hydrocarbons. *Heliyon*, **3**(5), e00304.

Sawatdeenarunat C, Nam H, Adhikari S, Sung S, Khanal SK (2018) Decentralized biorefinery for lignocellulosic biomass: integrating anaerobic digestion with thermochemical conversion. Biores Technol 250:140–147

Schmitt N, Apfelbacher A, Jäger N, Daschner R, Stenzel F, Hornung A (2019) Thermo-chemical conversion of biomass and upgrading to biofuel: the thermo-catalytic reforming process—a review. Biofuels Bioprod Biorefin 13(3):822–837

Senthil C, Vediappan K, Nanthagopal M, Seop Kang H, Santhoshkumar P, Gnanamuthu R, Lee CW (2019) Thermochemical conversion of eggshell as biological waste and its application as a functional material for lithium-ion batteries. Chem Eng J 372:765–773

Shahabuddin M, Krishna BB, Bhaskar T, Perkins G (2020) Advances in the thermo-chemical production of hydrogen from biomass and residual wastes: summary of recent techno-economic analyses. Bioresource Technol 299:122557

Shan Ahamed T, Anto S, Mathimani T, Brindhadevi K, Pugazhendhi A (2021) Upgrading of bio-oil from thermochemical conversion of various biomass—mechanism, challenges and opportunities. Fuel 287:119329

Shen Y (2020) A review on hydrothermal carbonization of biomass and plastic wastes to energy products. Biomass Bioenergy 134:105479

Shen Y, Chen X, Ge X, Chen M (2018) Thermochemical treatment of non-metallic residues from waste printed circuit board: pyrolysis vs. combustion. J Clean Prod 176:1045–1053

Shi H, Mahinpey N, Aqsha A, Silbermann R (2016) Characterization, thermochemical conversion studies, and heating value modeling of municipal solid waste. Waste Manag 48:34–47

Soares RB, Martins MF, Gonçalves RF (2020) Thermochemical conversion of wastewater microalgae: the effects of coagulants used in the harvest process. Algal Res 47:101864

Song C, Zhang C, Zhang S, Lin H, Kim Y, Ramakrishnan M, Du Y, Zhang Y, Zheng H, Barceló D (2020) Thermochemical liquefaction of agricultural and forestry wastes into biofuels and chemicals from circular economy perspectives. Sci Total Environ 749:141972

Sudagar S, Adhisegar C, Bernadsha P, Balamurugan R (2020) A comparative study of different biomass properties by using pyrolysis process. AIP conference proceedings. AIP Publishing LLC, p 040001

Sun H, Li Y, Bian Z, Yan X, Wang Z, Liu W (2019) Thermochemical energy storage performances of Ca-based natural and waste materials under high pressure during $CaO/CaCO_3$ cycles. Energy Convers Manag 197:111885

Syed-Hassan SSA, Wang Y, Hu S, Su S, Xiang J (2017) Thermochemical processing of sewage sludge to energy and fuel: fundamentals, challenges and considerations. Renew Sustain Energy Rev 80:888–913

Titiloye JO, Abu Bakar MS, Odetoye TE (2013) Thermochemical characterisation of agricultural wastes from West Africa. Ind Crops Prod 47:199–203

Tumuluru JS (2015) Comparison of chemical composition and energy property of torrefied switchgrass and corn stover. Front Energy Res 3(46)

Unrean P, Lai Fui BC, Rianawati E, Acda M (2018) Comparative techno-economic assessment and environmental impacts of rice husk-to-fuel conversion technologies. Energy 151:581–593

Uzoejinwa BB, He X, Wang S, El-Fatah Abomohra A, Hu Y, Wang Q (2018) Co-pyrolysis of biomass and waste plastics as a thermochemical conversion technology for high-grade biofuel production: recent progress and future directions elsewhere worldwide. Energy Convers Manag 163:468–492

Vakalis S, Sotiropoulos A, Moustakas K, Malamis D, Vekkos K, Baratieri M (2017) Thermochemical valorization and characterization of household biowaste. J Environ Manag 203:648–654

Vega LY, López L, Valdés CF, Chejne F (2019) Assessment of energy potential of wood industry wastes through thermochemical conversions. Waste Manag 87:108–118

Wachter P, Gaber C, Demuth M, Hochenauer C (2021) Towards a recuperative, stationary operated thermochemical reformer: experimental investigations on the methane conversion and waste heat recovery. Appl Therm Eng 183:116121

Waliszewska B, Grzelak M, Gaweł E, Spek-Dźwigała A, Sieradzka A, Czekała W (2021) Chemical characteristics of selected grass species from polish meadows and their potential utilization for energy generation purposes. Energies 14(6):1669

Wang F, Shi D, Han J, Zhang G, Jiang X, Yang M, Wu Z, Fu C, Li Z, Xian M (2020) Comparative study on pretreatment processes for different utilization purposes of switchgrass. ACS Omega 5(35):21999–22007

Wang Z, Burra KG, Zhang M, Li X, Policella M, Lei T, Gupta AK (2020b) Co-pyrolysis of waste tire and pine bark for syngas and char production. Fuel 274:117878

Weldekidan H, Strezov V, Town G, Kan T (2018) Production and analysis of fuels and chemicals obtained from rice husk pyrolysis with concentrated solar radiation. Fuel 233:396–403

Xu S, Hu Y, Wang S, He Z, Qian L, Feng Y, Sun C, Liu X, Wang Q, Hui C, Payne EK (2019) Investigation on the co-pyrolysis mechanism of seaweed and rice husk with multi-method comprehensive study. Renew Energy 132:266–277

Yang X, Tian S, Kan T, Zhu Y, Xu H, Strezov V, Nelson P, Jiang Y (2019a) Sorption-enhanced thermochemical conversion of sewage sludge to syngas with intensified carbon utilization. Appl Energy 254:113663

Yang Z, Wu Y, Zhang Z, Li H, Li X, Egorov RI, Strizhak PA, Gao X (2019) Recent advances in co-thermochemical conversions of biomass with fossil fuels focusing on the synergistic effects. Renew Sustain Energy Rev 103:384–398

Zaman CZ, Pal K, Yehye WA, Sagadevan S, Shah ST, Adebisi GA, Johan R (2017) Pyrolysis: a sustainable way to generate energy from waste. Pyrolysis. IntechOpen, Rijeka, Croatia, p 1

Zhang B, Chen J, Kandasamy S, He Z (2020) Hydrothermal liquefaction of fresh lemon-peel and Spirulina platensis blending-operation parameter and biocrude chemistry investigation. Energy 193:116645

Zhang B, He Z, Chen H, Kandasamy S, Xu Z, Hu X, Guo H (2018) Effect of acidic, neutral and alkaline conditions on product distribution and biocrude oil chemistry from hydrothermal liquefaction of microalgae. Bioresour Technol 270:129–137

Zhang J, Zhang X (2019) 15—The thermochemical conversion of biomass into biofuels. In: Verma D, Fortunati E, Jain S, Zhang X (eds) Biomass, biopolymer-based materials, and bioenergy. Woodhead Publishing, pp 327–368

Zhang L, Wu W, Siqu N, Dekyi T, Zhang Y (2019) Thermochemical catalytic-reforming conversion of municipal solid waste to hydrogen-rich synthesis gas via carbon supported catalysts. Chem Eng J 361:1617–1629

Zhang S, Zhu S, Zhang H, Liu X, Zhang H (2019b) Evaluation of pyrolysis behavior and products properties of rice husk after combined pretreatment of washing and torrefaction. Biomass Bioenergy 127:105293

Zhou X, Broadbelt LJ, Vinu R (2016) Chapter Two—Mechanistic understanding of thermochemical conversion of polymers and lignocellulosic biomass. In: Van Geem KM (ed) Advances in chemical engineering, vol 49. Academic, pp 95–198

Chapter 6
Anaerobic Digestion of Waste for Biogas Production

YanZhao Zhang, Jiafu Lin, Tao Song, and Haifeng Su

Abstract Anaerobic fermentation technology is one of the effective ways to produce biogas energy from biomass waste. Biomass anaerobic fermentation is the effective conversion of organic matter in biomass under the assimilation of anaerobic bacteria, and finally produces methane and part of carbon dioxide with economic value, which can be used for combustion and power generation. Biomass resources such as crops, oil crops, agricultural organic residues, forest trees, and forest industrial residues usually provide energy. This paper reviews the research achievements of biogas production by anaerobic fermentation of biomass waste resources at present, and analyzes the progress of biogas production by means of different organic wastes such as agricultural waste, urban waste, forestry waste, mixed fermentation of different wastes and adding exogenous catalyst. The use of aquatic plants such as duckweed to produce biogas is also highlighted. Based on the above analysis, the economic feasibility of using different kinds of biomass waste was evaluated. This paper provides valuable reference for the production of biogas from material waste by anaerobic fermentation.

Keywords Anaerobic digestion · Biogas production · Anaerobic bacteria · Aquatic plants · Economic feasibility

Y. Zhang · H. Su (✉)
Chongqing Institute of Green and Intelligent Technology, Chinese Academy of Science, 266, Fangzheng Avenue, Shuitu High-Tech Park, Beibei, Chongqing 400714, China
e-mail: suhaifeng@cigit.ac.cn

Y. Zhang
e-mail: zhangyanzhao@cigit.ac.cn

J. Lin · T. Song · H. Su
Chengdu University, Chengdu 61004, China
e-mail: linjiafu@cdu.edu.cn

A. E.-F. Abomohra et al. (eds.), *Waste-to-Energy*,
https://doi.org/10.1007/978-3-030-91570-4_6

6.1 Introduction

Biogas anaerobic fermentation refers to the organic material (such as people and livestock poultry manure, straw and weeds, etc.) in a certain moisture content, temperature, and anaerobic conditions through various kinds and the function of different kinds of microbe catabolism, eventually mixtures, such as the formation of methane and carbon dioxide gas (methane) complex biochemistry process. With the increase in population and economic development, resource scarcity and environmental pollution have become a global issue. China's energy consumption is huge and the structure of energy supply is unreasonable. China's energy consumption reached 4.87 billion tons of standard coal in 2019, increasing by 3.2% over the previous year. Among them, coal consumption was about 2.81 billion tons, increasing by 0.92% over the previous year. Oil consumption reached 920 million tons, up 4.8% on the previous year; Natural gas consumption was 306.7 billion cubic meters, an increase of 7.3% over the previous year (Zhang 2021). China has a high dependence on foreign energy consumption. In 2019, the dependence on foreign oil is 77.8%, and the dependence on foreign gas is 43%. The energy structure is unreasonable, and the proportion of fossil energy is 84.7%, especially the proportion of coal is still as high as 57.7%, which brings greater pressure to China's environmental protection and greenhouse gas emission reduction, and also brings great challenges to the energy transformation and energy revolution (Statistics 2020). In the new complex international situation, it is crucial to build a safe multi-wheel driven energy supply system of coal, oil, gas, nuclear, new energy and renewable energy to meet China's energy needs and development needs (Lingling and Weiquan 2021). Microbial anaerobic fermentation technology is a kind of renewable energy production technology using various wastes, easy to promote and apply, environment friendly, and the production of biogas can be directly used as renewable energy (Kandasamy et al. 2021b). Biogas is a mixed substance primarily composed of methane and carbon dioxide, as shown in Table 6.1 (Bilgen and Sarikaya 2016).

Although many studies have proved that biogas technology using organic waste is an effective way in the field of waste energy and resource recovery. There are also some reviews on anaerobic fermentation mechanism, pretreatment technology of straw agricultural waste biogas fermentation raw materials, adaptability of biogas

Table 6.1 Typical composition of biogas

Main components	Percentage content (%)
CH_4	55–75
CO_2	24–45
CO	0–0.3
N_2	1–5
H_2	0–3
H_2S	0.1–0.5
O_2	Traces

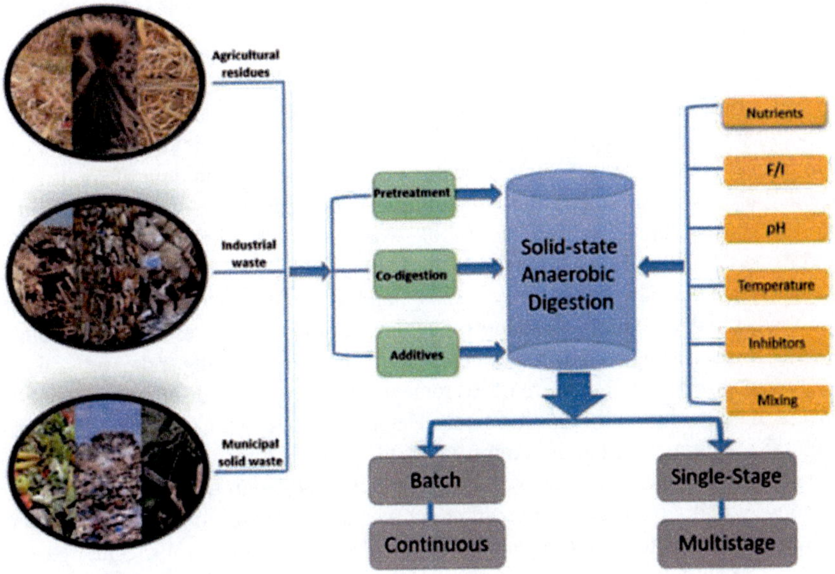

Fig. 6.1 Solid-state anaerobic digestion for biogas production

fermentation of various raw materials, and so on. At the same time, some researchers in the biogas application-related engineering technology research such as solid-state anaerobic digestion Fig. 6.1, combining with the basic theoretical research results and biogas application-related engineering technology research results, have been effectively summarized.

However, there is still a lack of systematic summary on the economic and technical feasibility of using different types of wastes to produce biogas, the deficiency of practical application, the disconnection of relatively mature research status, and the economic feasibility and technical feasibility of using anaerobic fermentation to treat different wastes. In particular, the research on biogas production by anaerobic fermentation of aquatic plants is lacking the corresponding summary. Many research achievements have not been fully utilized due to the active development of biogas production technology by anaerobic fermentation and the extensive application of anaerobic technology to treat different types of wastes in practice. This review will provide a great reference for the rational and effective utilization and treatment of different wastes in the future.

6.2 Methane and Biogas from Waste Materials in Rural Areas

China is the country that produces the most solid waste in the world. The annual solid waste generated by various economic activities and production processes can reach 1.21010 t, with rural waste exceeding 5.3109 t. A large amount of rural waste is piled up at will and discharged, resulting in a massive waste of resources. Rural waste consists primarily of rural household waste, agricultural waste, forestry waste, and livestock and poultry excrement. Rural household waste, mainly includes meal waste, waste plastics, waste paper and ash, and agricultural wastes mainly include crop stalks and agricultural products processing residues, and forestry wastes mainly include forest cutting, wood processing residues and forest pruning and so on. Animal excrement refers to the mixture of excrement, urine and grass discharged by cattle, sheep, pigs, poultry and other livestock (Hetaoli et al. 2017). China is a large agricultural and animal husbandry country with plenty of biomass energy. According to statistics, China has a rich crop of straw. In 2017, the theoretical resource amount of straw was 1.09 billion tons, corn straw was about 251 million tons, straw was about 193 million tons, and wheat straw was about 153 million tons (Shuang et al. 2020). The output of livestock and poultry manure in 2019 is about 3.9 billion tons, and the comprehensive utilization rate is about 75% (Yingtao and Wentao 2020). After resource utilization, the development of green and efficient anaerobic fermentation methane production technology can effectively replace coal and other fossil energy, and the biogas produced can be an important supplement to China's renewable energy, which is of great significance to the adjustment of China's energy structure, the realization of energy conservation and emission reduction, and the development of circular economy.

6.2.1 Methane and Biogas from Straw

In March 2020, the Ministry of Agriculture and Rural Affairs of China formulated the "Key Points of Agricultural and Rural Green Development in 2020", pointing out that we should continue to promote the comprehensive utilization of straw, setting up a technical expert group for the comprehensive utilization of straw, guiding the development of technology demonstration and model integration, and improving the scientific utilization level of crop straw. The No. 1 document of the CPC Central Committee in 2020 emphasizes the promotion of green development of agriculture, the in-depth reduction of pesticides and fertilizers, the strengthening of the treatment of agricultural film pollution, and the promotion of comprehensive utilization of straw. There are more than 20 kinds of main crops in China, among which corn, wheat, and rice produce the most straw. The active components of main stalks are shown in Table 6.2 (Zhibin et al. 2019). Straw contains rich nutrient elements such as C, H, O, N, etc. The main components of straw are cellulose, hemicellulose, and lignin. Lignin

Table 6.2 Effective component of main crop straw (%)

Species	Fat	Protein	Lignin	Cellulose	Hemicellulose
Rice straw	1.40	4.80	12.50	32.00	24.00
The wheat straw	1.50	4.40	18.00	30.50	23.50
Corn stover	16.00	5.70	22.00	34.00	37.50

forms a complex network structure with cellulose and hemicellulose that is difficult to degrade. The protective effect of lignin and hemicellulose on cellulose is the main reason that affects the methane production efficiency of anaerobic fermentation (Kratky and Jirout 2011).

Cellulose is the main component of lignocellulose, a kind of polysaccharide chemical institution as shown in Fig. 6.2. Through the beta 1, 4 glycosidic bond connections of cellobiose (a disaccharide of glucose) of the basic unit of the linear polysaccharide polymer structure for $C_6H_{10}O_5$, molecular cell and orderly arrangement of multiple parallel form filamentous tiny fiber, thus forming the chain of cellulose. With intramolecular and intermolecular hydrogen bonds, the chain-like cellulose is joined to form an insoluble polymeric form (Zhao 2011). The chemical structure

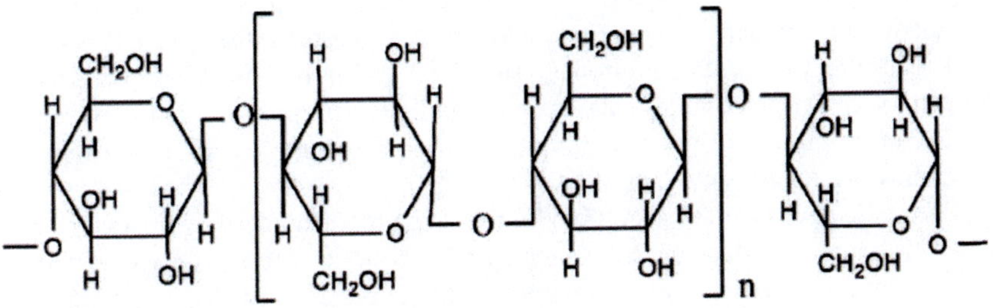

Fig. 6.2 Molecular structure of cellulose

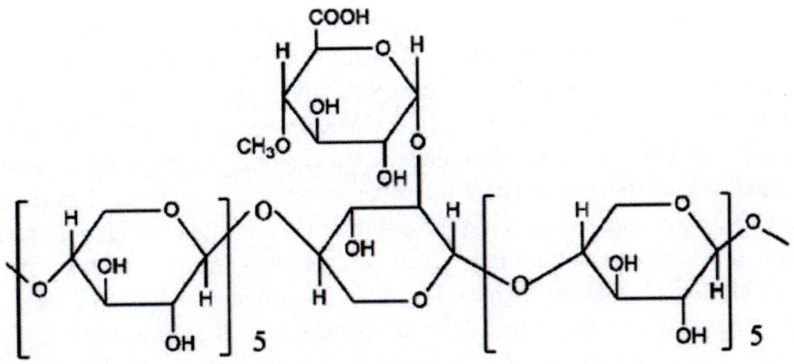

Fig. 6.3 Molecular structure of hemicellulose

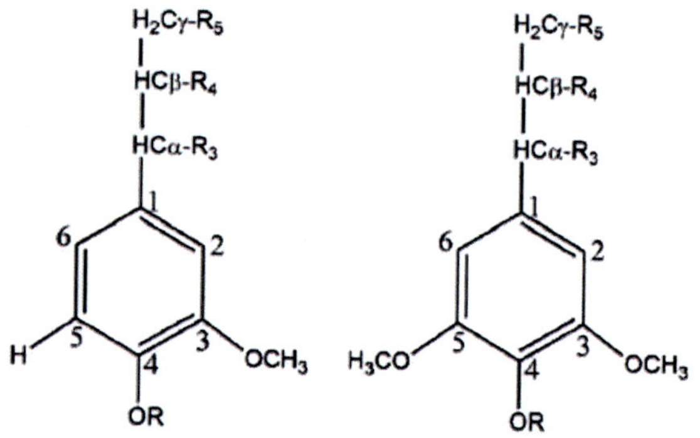

Fig. 6.4 Molecular structure of lignin

of hemicellulose is shown in Fig. 6.3. Hemicellulose is a kind of heteropolysaccharide with a dendritic structure formed by different types of five-carbon sugars and six-carbon sugars, which are the main components of hemicellulose. The basic component unit of hemicellulose is $C_5H_8O_4$, with loose and amorphous structure, so it is easy to be degraded (Peng and Wu 2010). Lignin is a cross-linked phenolic polymer that is irregular and haphazard, rigid, impermeable, and resistant to microbial attack and oxidative stress. (Ponnusamy et al. 2019). Its chemical structure is shown in Fig. 6.4. Lignin is formed by connecting phenylpropane structural units by ether bonds and carbon–carbon double bonds. Its basic unit structure consists of $C_9H_{10}O_3(OCH_3)_{0.9–1.7}$. Phenylpropane structural units are divided into three types: p-hydroxyphenyl propane, syringe propane and guaiacyl propane. Different lignin kinds can be produced by combining the three phenylpropane structural units in different ways (Q 2019).

The main mechanism of anaerobic fermentation is when microorganisms interact with substrates to produce methane and carbon dioxide and other gases while synthesizing substances needed for their own life activities under anaerobic conditions. For straw, cellulose and hemicellulose are the main substrates for anaerobic fermentation. Btyaut proposed in 1979 that anaerobic fermentation could be divided into three stages, and discovered that methanogens and acidogenic bacteria were the main fermentation process. In the same year, Zeichus et al. put forward the four-stage theory of anaerobic fermentation (X 2020) at the First International Anaerobic Fermentations Conference, which is a complex biochemical process. These include Hydrolysis, Acidogenesis, Acetogenesis, and Methanogenesis. The main function of microorganisms is facultative anaerobic bacteria and obligate anaerobe in anaerobic hydrolysis fermentation stage. These microbes secrete large amounts of extracellular enzymes to conduct the solubility of the polymer into simple soluble monomer. For biomass straw, they can convert into sugar, organic matter polypeptide, amino acids and other biological small molecules. It gives off hydrogen and carbon dioxide. The extent of the hydrolysis stage depends on the extent to which cellulose and hemicellulose are encapsulated by lignin. Cellulose and hemicellulose are readily degradable, whereas lignin is difficult to degrade, resulting in low gas production. In the acidification stage, facultative anaerobe and obligate anaerobe transform the small molecular degradable substances produced in the hydrolysis stage into a small amount of volatile fatty acid substances such as acetic acid, propionic acid and alcohol, and release hydrogen and carbon dioxide at the same time, which will significantly reduce the pH of the system in the whole stage. In the phase of hydrogen-acetic acid production, obligate anaerobic hydro-acid-producing bacteria further convert volatile fatty acid substances produced during acidification to acetic acid, which releases hydrogen and carbon dioxide. Because the organic acids decompose at this stage, the pH increases sharply. The main processes of hydrogen production and acetic acid production are as follows (Bi 2020).

$$CH_3CHOHCOO^- + 2H_2O \rightarrow CH_3COO^- + HCO_3^- + H^+ + 2H_2$$
$$CH_3CH_2OH + H_2O \rightarrow CH_3COO^- + H^+ + 2H_2$$
$$CH_3CH_2CH_2COO^- + 2H_2O \rightarrow 2CH_3COO^- + H^+ + 2H_2$$
$$CH_3CH_2COO^- + 3H_2O \rightarrow 2CH_3COO^- + H^+ + 2H_2$$
$$4CH_3OH + 2CO_2 \rightarrow 3CH_3COOH + 2H_2O$$

The last stage is the methanogenic stage, and the microorganisms that play a role in this process are obligate anaerobic methanogens, which decompose acetic acid to form methane and carbon dioxide, and at the same time generate methane through the interaction of hydrogen and carbon dioxide produced in each stage, or generate methane by using formic acid produced by other bacteria (R 2020). The main processes in the methanogenic stage are as follows.

$$CH_3COOH \rightarrow CH_4 + CO_2$$
$$CO_2 + 4H_2 \rightarrow CH_4 + 2H_2O$$
$$HCOOH + 3H_2 \rightarrow CH_4 + 2H_2O$$
$$CH_3OH + H_2 \rightarrow H_2 + CH_4 + 2H_2O$$
$$4CH_3NH_2 + 2H_2O + 4H^+ \rightarrow 3CH_4 + CO_2 + 4NH_4^+$$

6.2.1.1 Pretreatment

Because cellulose, hemicellulose, lignin with tight structures and are bound together by different forces such as hydrogen bonds, mixed crosslinking forms a complex network skeleton, which is difficult to be directly degraded by microorganisms. For example, anaerobic fermentation processing directly for straw, less actual biogas production, lead to too long fermentation time, thus economic efficiency is not high. Therefore, the basic principle of pretreatment is to destroy the crosslinking structure and separate cellulose, hemicellulose and lignin. Straw pretreatment technology can improve the utilization rate of biogas and gas production rate of straw. At present, there are many pretreatment methods for straw, mainly including physical pretreatment, chemical pretreatment, biological pretreatment and combined pretreatment (Zemin et al. 2013).

Physical pretreatment is to reduce the size of raw materials, soften biomass or reduce the degree of polymerization and crystallinity of cellulose. Through physical methods, the cell wall structure of raw materials can be destroyed, increasing the accessibility of enzymes and anaerobic bacteria and organic matter, thus make it easy to digest. Mechanical grinding, grinding, pressure heating, radiation pretreatment, and other physical pretreatment methods are currently in use. Xiong used mechanical pretreatment to treat straw with three particle sizes of 5 cm, 2 cm and 10-mesh sieve respectively for anaerobic fermentation, and the study found that the smaller the particle size, the better the gas production effect (X 2015). Zheng et al. found that pretreatment of corn stalks by ball milling can significantly improve the enzymatic hydrolysis rate of corn stalks (Yuan et al. 2015). Franz et al. found that the degradation rate of wheat straw could be improved by steam explosion pretreatment (Theuretzbacher et al. 2015). Physical pretreatment of straw can increase the rate of anaerobic fermentation and gas production. However, physical pretreatment requires specific machinery and equipment and consumes a large amount of energy in the process, so economic factors and feasibility need to be considered in practical application.

Chemical pretreatment methods remove lignin and part of hemicellulose by acid or alkali, mainly to open the lipid bond between cellulose, hemicellulose and lignin, increase the accessibility of enzymes to cellulose, accelerate the anaerobic fermentation, and obtain more products and yield. The main chemical pretreatments include acid pretreatment, alkali pretreatment, ammonification pretreatment, etc. Kshirsagar et al. pretreated straw in an autoclave with 0.5% sulfuric acid at 121 °C for 60 min to obtain a high reducing sugar yield (Kshirsagar et al. 2015). It was found that straw pretreated with oxalic acid had the greatest effect on enzymatic saccharification, which was 2.68 times higher than that of untreated straw. The optimal treatment conditions were oxalic acid concentration of 5.01%, temperature of 135.91 °C and time of 30.86 min (Amnuaycheewa et al. 2016). Wang et al. pretreated yellow corn stalks with different alkali under hot and humid conditions, and found that the lignin, cellulose and hemicellulose contents of corn stalks were significantly reduced, and using 6% KOH solution had the best degradation ability for lignocellulose, the lignin content decreased as high as 67.04% and hemicellulose content decreased as high as 76.86%. After 4% ammonia solution treatment, anaerobic fermentation has the best gas production effect, and the gas production per unit TS can reach 125.25 mL/g (Yuxiaofei et al. 2020). The chemical pretreatment operation is relatively simple, and the methane production effect of anaerobic fermentation after treatment is also very obvious, but the pretreatment process will produce inhibitory products, which may hinder anaerobic fermentation. Some chemicals' improper treatment will cause environmental pollution problems, so in the practical application to consider a variety of factors, weigh the advantages and disadvantages.

Biological pretreatment refers to the use of some microorganisms with lignin degradation ability to degrade the lignin in straw, destroy the crystal structure of straw, shorten the time of anaerobic fermentation, and improve the gas production rate. The key of biological pretreatment is to find microorganisms with high lignin degradation ability and suitable anaerobic fermentation conditions. Huang Kailin et al. treated corn stalks with a composite microbial strain HK-4 composed of *Aspergillus niger*, *Trichoderma*, *Penicillium oxalate* and white rot fungi at a constant temperature of 28 °C and found that the degradation rates of cellulose, hemicellulose and lignin could reach 64.52%, 51.06%, and 3.89%, respectively (Huang et al. 2018). Studies have shown that the degradation rates of cellulose, hemicellulose and lignin in silage corn stalks pretreated with white-rot fungi for solid-state anaerobic fermentation are 9.9%, 23.2%, and 15.2%, respectively, after 15 days of treatment. The results of anaerobic fermentation experiment showed that the methane yield of VS of pretreatment silage corn stalk fermentation for 21 days was 215.5 mL/g (Shan et al. 2013). It is clear that microbial pretreatment of biomass straw can significantly increase the rate of anaerobic fermentation and gas production, while the overall treatment process is efficient and clean. However, the time of anaerobic fermentation should be strictly controlled. If the biological pretreatment time is too long, it will consume too much substrate to synthesize its own substance, which will affect the gas production performance of subsequent anaerobic fermentation.

Single pretreatment technologies such as acid, alkali and biological have made significant contributions to improving the anaerobic digestibility and gas production rate, but these single pretreatment technologies have a series of problems, and each single pretreatment technology has certain limitations. Therefore, in recent years, scholars at home and abroad have gradually developed the combined pretreatment technology combining various treatment technologies. Wang Xing et al. used steam blasting and calcium oxide combined pretreatment technology to treat rice stalks, and the results showed that the combined pretreatment method destroyed the structure of rice stalks most thoroughly compared with the single pretreatment method. The lignin content in rice stalks after the combined pretreatment decreased from 17.2 to 12.2%, and the removal rate could reach 29.1% (Xing et al. 2017). Dong et al. used dual-frequency ultrasound combined with alkali pretreatment to compare corn stalk with single single-frequency ultrasonic pretreatment and alkali pretreatment, and found that cumulative gas production increased by 11.1% and 28.2%, respectively (Dong et al. 2018). Zhang et al. pretreated straw by extrusion combined with sodium hydroxide, and found that methane production increased by 54% compared with the control group, and energy recovery increased from 38.9% to 59.9% (Zhang et al. 2015). Sabeeh et al. pretreated straw with alkali and photocatalysis, and found that methane production could be increased by 36% (Sabeeh et al. 2020). These combined pretreatment technologies can make up for each other, improve the digestibility of lignocellulose and promote the rate of anaerobic digestion. The combined pretreatment technologies have a wide range of application prospects.

6.2.2 *Methane and Biogas from Livestock and Poultry Manure*

China is the largest livestock farming country in the world, with 3.8 billion t of raw livestock manure per year, and the comprehensive utilization rate is not high (Hemin 2019). The progress of the farming industry drives rural economic development and increases farmers' income. However, with the development of livestock and poultry farming in China toward scale and intensification, fertilizer application has changed from mainly farm organic fertilizer to chemical fertilizer, and livestock and poultry manure has changed from treasure to waste. Livestock manure contains a large amount of organic matter, antibiotics, heavy metals, and carries a large number of pathogenic microorganisms, which will cause pollution to soil, water bodies, and the atmosphere without treatment, and livestock manure has become one of the important sources of environmental pollution in China (Qing et al. 2013). The Central Document No. 1 of 2021 issued by the Central Committee of the Communist Party of China and the State Council clearly pointed out the need to promote the green development of agriculture and strengthen the resource utilization of livestock and poultry manure. As an important biomass energy source, resource utilization of livestock and poultry manure is conducive to saving and replacing primary energy, reducing the dependence on non-renewable energy sources, and realizing the sustainable use of resources.

Due to the different feeding conditions and dietary characteristics of various livestock and poultry, it was found that different livestock and poultry manure differed greatly in composition (Shen et al. 2015). Pigs, sheep, chickens, horses, donkeys, mules, chickens, ducks, geese, and rabbits are the main objects of livestock farming in China. Among them, cattle, pigs, sheep, and chickens are the largest, and the four livestock manures account for 96.5% of the annual national emissions and constitute the main body of livestock manure in China. The component content of the main livestock manure in China is shown in Table 6.3 (Tian et al. 2012).

It can be seen that livestock manure contains a large number of nutrients, which can be used as substrate for anaerobic fermentation and converted into biogas through

Table 6.3 Composition content of main animal manure (kg/t)

Poultry	Feedstock	BOD	COD	NH_3–N	TN	TP
Pig	Manure	57.0	52.0	3.1	5.9	3.4
	Urine	5.0	9.0	1.4	3.3	0.5
Cow	Manure	24.5	31.0	1.7	4.4	1.2
	Urine	4.0	6.0	3.5	8.0	0.4
Sheep	Manure	4.1	4.6	0.8	7.5	2.6
	Urine	–	–	–	14.0	2.0
Chicken	Manure	47.9	45.0	4.8	9.8	5.4

anaerobic fermentation, which can greatly reduce the waste of resources and achieve green economic development.

6.3 Methane and Biogas from Municipal Waste

The rapid development of urbanization, the rapid progress of industrialization, and the increasing population have led to a high concentration and huge amount of various types of waste generated in cities. Municipal waste mainly includes domestic waste, kitchen waste, construction waste, urban sludge, pharmaceutical factory wastewater, etc. (Pattnaik and Reddy 2010). According to China Statistical Yearbook 2020, the amount of general solid waste generated in large and medium-sized cities in China was 1.55 billion t in 2018, an increase of 18.32% year-on-year. Terms such as "garbage siege" are increasingly appearing in major media, and urban waste without treatment not only causes a serious waste of land resources, but also seriously pollutes the soil, groundwater, and atmosphere, posing a great threat to the living environment of residents (Yi 2016). The generation of municipal waste has become an important factor limiting the development of cities, and municipal waste is receiving more and more attention. Solid waste is considered a misplaced resource, and "trituration" treatment of municipal waste is an effective way to treat and dispose of municipal waste. Chinese municipal waste is rich in organic matter, so it can be treated by microbial anaerobic fermentation, which converts the organic matter in municipal waste through the action of microorganisms and releases methane in the process. This can not only effectively treat and dispose of municipal waste, but also generate clean energy, which is important for China to relieve energy pressure.

6.3.1 Methane and Biogas Production from Domestic Waste

With the growth of population and economy, the amount of urban domestic waste generated in China is increasing. The massive surge of urban domestic waste, piled up around the city, seriously restricts the green and healthy development of the urban economy. According to China Statistical Yearbook 2020, China produced 242.062 million t of domestic waste in 2019, with a harmless treatment rate of 99.2%. As a major component of solid waste, urban domestic waste mainly includes residential domestic waste, garden waste, office waste discharged from institutions, street sweeping waste, and waste generated from public places (e.g., parks, stations, airports, docks, etc.) (Minglong 2015). Urban domestic waste in China mainly contains food residues, wood fertilizers, paper, textiles, plastics, and rubber (Zhou et al. 2015). A comparison of the composition of municipal domestic waste in China and other major countries is shown in Table 6.4 (Yunpan 2016). It can be seen that the composition of urban domestic waste in China is complex and has different forms.

Table 6.4 Composition of municipal solid waste in different countries (%)

Country	Food waste	paper	Plastic	Glass	Metal	Others
China	52.6	6.9	7.3	1.6	0.5	31.1
Australia	47.0	23.0	4.0	7.0	5.0	14.0
Japan	34.0	33.0	13.0	5.0	3.0	12.0
France	32.0	20.0	9.0	10.0	3.0	26.0
Switzerland	29.0	20.0	15.0	4.0	3.0	29.0
Korea	28.0	24.0	8.0	5.0	7.0	28.0
American	25.0	34.0	12.0	5.0	8.0	16.0
Canada	24.0	47.0	3.0	6.0	13.0	7.0
Germany	14.0	34.0	22.0	12.0	5.0	13.0
Singapore	9.5	21.2	11.5	1.0	14.6	42.2

It is mainly kitchen waste with high organic matter content, high water content, and low calorific value.

The harmless treatment methods of general urban domestic waste are sanitary landfill and incineration. Sanitary landfill is an important final treatment means of municipal domestic waste. It is also the main way of municipal domestic waste treatment in China at this stage, mainly characterized by mature technology, simple operation and management, large treatment capacity, high flexibility, wide application, and relatively low investment and operation costs. However, landfills take up a lot of land and have poor capacity reduction effect. If the landfilled waste is not treated harmlessly, a large number of bacteria and viruses will remain, heavy metal pollution potential, and waste leachate will also pollute to groundwater. Incineration is a comprehensive treatment process of high-temperature decomposition and deep oxidation of municipal waste. Incineration technology is characterized by large treatment capacity, good capacity reduction, complete harmlessness, and thermal energy recovery, so this method is a waste treatment technology commonly used in China and most countries in the world (Zhang and Xu 2012). However, the waste incineration process produces dioxins, large amounts of acidic gases and incompletely burned organic matter, such as dioxins and other highly toxic organochlorine compounds, which are directly discharged into the environment and can pollute the atmosphere and cause harm to human health.

Kitchen waste accounts for more than 50% of the household waste in major cities in China, and is the most important component of household waste (Dong et al. 2016). Due to the high-water content of kitchen waste, this poses a series of problems for its subsequent treatment. On the one hand, the high water content leads to a lower level of calorific value of the waste. On the other hand, high water content wastes space in sanitary landfills and leads to a large amount of leachate generation. Therefore, a high proportion of kitchen waste should be collected separately and treated separately for resource utilization.

6.3.2 Methane and Biogas Production from Kitchen Waste

Kitchen waste is an important part of domestic waste, and the main sources are food waste and meal waste from households, school canteens, and the catering industry. Since China is a large population country, it generates a huge amount of kitchen waste. Based on a monthly generation of 1 t/million people, China's national population of 1.4 billion will produce 1.68 million tons of kitchen waste per year (Qiang et al. 2013). The increasing amount of kitchen waste emissions, especially for large and medium-sized cities, has become a constraint to economic development. Kitchen waste presents a solid–liquid mixed state with complex composition, mainly a mixture of oil, water, fruit peels, vegetables, rice and noodles, fish, meat, bones, and fee tableware, plastic, paper towels and other substances (Wang et al. 2019). Analyzed in terms of chemical composition, it mainly has inorganic salts, organic acids, various biological macromolecular organic compounds such as proteins, fats, starch, cellulose, etc., and is rich in many trace elements such as potassium, calcium, magnesium, iron and phosphorus (Qiaoling 2012). Due to the high water content of kitchen waste, it contains a large number of pathogenic microorganisms such as *Salmonella* and *Staphylococcus aureus*, in addition to breeding flies, mosquitoes, and other pests. If not treated in time, it may also produce toxic and harmful substances such as dioxins, which endanger people's lives and the environment.

The main hazards of kitchen waste are reflected in the following aspects. (1) The high organic content led to very easy to deteriorate, decay, ferment, which will produce a large number of toxins. (2) At the same time, due to the complex source, and containing a variety of bacteria and pathogenic bacteria, kitchen waste will be easy to become a vector for infectious diseases (Li et al. 2016b). The high-water content of kitchen waste increase difficulties of the collection, transportation and treatment of kitchen waste, increase the cost of sewage treatment operations. At the same time, the waste leachate generated by the kitchen waste pile can cause water pollution through surface runoff and infiltration. (3) Endangering human health, the meat protein and animal fatty substances in kitchen waste, the main source of these substances for the provision of meat food livestock and poultry, livestock in direct food without effective treatment of kitchen waste, prone to "like eating like" homologous pollution. (4) The higher organic matter and moisture in kitchen waste also can quickly decay and deteriorate by the role of microorganisms. Especially in summer, when the temperature is higher, the decay and deterioration rate is faster. It is easy to produce a large amount of leachate as well as foul-smelling gas, breeding mosquitoes and insects, causing a bad impact on environmental health, and causing cross-contamination of epidemics between humans and animals, endangering human health and promoting the spread of certain fatal diseases to some extent (Zixu and Jing 2015).

Traditional food waste treatment and disposal methods mainly include crushing direct discharge, feed disposal, landfill method, incineration method, composting method, etc. However, these traditional treatment methods have their own disadvantages. Crushing direct discharge of water consumption increases the amount of urban sewage generation, increasing the difficulty of treatment in sewage treatment plants, not suitable for large-scale food waste treatment and disposal. In the past, kitchen waste was collected and used directly to feed pigs, which affects food safety and makes human consumption of such pork susceptible to various zoonotic diseases and cancer, and has been banned by the government. Landfill method usually occupies a lot of land because water, organic content of the kitchen waste is so high. The use of landfill method is also easy to produce malodorous gas and a lot of leachate, spread disease contamination of soil and groundwater, so landfill method treatment is increasingly restricted. Incineration method of domestic waste treatment, domestic wastes are prone to inadequate combustion, producing dioxins and other toxic and harmful gases (Zhang et al. 2011). Due to the complex composition of domestic kitchen waste, many harmful substances will enter the soil with composted products causing secondary pollution, so the application of the composting method is also limited to some extent (Cunsheng 2013). The anaerobic fermentation of kitchen waste can produce a large amount of biogas, which is a clean renewable energy that can be used for power generation and as fuel. The fermentation residue can be used to make organic fertilizer. The fermentation system is completely closed without generating odor. With the decreasing fossil fuel reserves, anaerobic digestion for methane production is not a better choice as a mature technology (Kondusamy and Kalamdhad, 2014). Therefore, anaerobic fermentation to produce biogas is the future direction of kitchen waste treatment.

Studies have found that the anaerobic fermentation of restaurant kitchen waste can produce gas at a rate of 435 mL/g, that is, the anaerobic fermentation of 1 t restaurant kitchen waste can produce 87 m^3 of biogas. Complete combustion of 1 m^3 biogas can produce heat equivalent to 0.7 kg of anthracite or 0.7 kg of gasoline or 0.8 kg of incomplete combustion (Longsheng et al. 2012). Therefore, the anaerobic fermentation of kitchen waste to produce biogas is an effective way to realize “three transformations” and turn waste into treasure. Many factors are affecting anaerobic fermentation, including temperature, pH, alkalinity, organic load, C/N, inoculation amount, hydraulic retention time, toxic substances, and so on. Meng et al. took kitchen waste as substrate to study the influence of organic loading on single-phase anaerobic fermentation of kitchen waste. The results showed that pH, ammonia nitrogen, and soluble chemical oxygen demand remained relatively stable in the fermentation process under organic loading of 0.75–1.25 g/L day. The increase and loading were conducive to the increase of biogas yield and yield. When the organic load reached a certain height of 1.5 g/L day, the pH of the fermentation process decreased and the gas production efficiency decreased (Xianwu et al. 2011). Zheng et al. studied the influence of dry anaerobic fermentation of kitchen waste without inoculation ratio. The results showed that when inoculation ratio was 0.90, pH gradually increased. The system operated normally, COD removal rate was as high as 90.29%. Methane content was as high as 255.4 L with methane production higher than 60% (Xiaowei et al. 2014). Therefore, it is necessary to control proper conditions to maximize the methane yield and yield for the anaerobic fermentation of kitchen waste. At present, the most commonly used is the mesothermal anaerobic fermentation of kitchen waste. However, with the further development of waste classification and technological progress, the high-temperature anaerobic fermentation technology of kitchen waste has attracted more and more attention due to its high gas production efficiency and more efficient killing of pathogenic bacteria.

Due to the complex composition and high water content of kitchen waste, anaerobic fermentation without treatment will affect the gas production efficiency and gas production, so it is necessary to pretreat kitchen waste before anaerobic fermentation. Pretreatment methods can be categorized as physical, chemical, biological and combined pretreatment. There are various pretreatment methods, and the effects of different pretreatment methods on the gas production performance of anaerobic fermentation vary greatly. Chao et al. carried out anaerobic fermentation after crushing, squeezing and filtering pretreatment of kitchen waste respectively. The results showed that the gas production effect of crushing > and squeezing > filtration was 145.9% and 82.83% higher than that of filtration and squeezing respectively (Hui et al. 2017). Yuan et al. explored the influence of different pretreatment methods on the anaerobic methane production of kitchen waste through acid pretreatment, alkali pretreatment and hot pretreatment. The results showed that the cumulative load methane production volume of kitchen waste treated with acid pretreatment was the highest, which could reach 669.26 mL/g(VS) (Lingli et al. 2015). Sun found that the main mechanism of acid, alkali and heat pretreatment was to change the microbial community structure of the fermentation system, and the dominant strains after three different pretreatment methods were consistent, among which the dominant strains after acid pretreatment had the highest dominance, with Simpson index of 0.92 (Yanbin 2013). Ariubaatar et al. studied the thermal pretreatment and ozonation pretreatment before Chinese anaerobic fermentation and found that the optimal methane production was 647.5 ± 10.6 mLCH_4/g(VS), which was about 52% higher than that of the control group without pretreatment (Ariunbaatar et al. 2014). It can be seen that pretreatment can change the properties of kitchen waste, improve the performance of anaerobic fermentation, and increase the gas production efficiency of kitchen waste anaerobic fermentation.

6.3.3 Methane and Biogas from Municipal Sludge

With the increasing number of urban population and urbanization in China, urban sewage production has been increasing. The number of sewage treatment plants, treatment capacity and sludge production have been increasing. According to the statistics of the Ministry of Housing and Urban–Rural Development, by the end of 2019, China's urban sewage discharge was 5,546,474 m^3, and there were 2471 urban sewage treatment plants. The generation of dry sludge discharged from urban wastewater treatment plants is more than 11 million t, and the annual growth rate is greater than 10% (Xiaomei et al. 2008). Therefore, the residual sludge generated from urban wastewater treatment has become a serious problem. The types of sludge from municipal wastewater treatment plants mainly include primary sludge, secondary sludge, activated sludge, residual sludge, and digested sludge. The content of various

Table 6.5 Content of main components of different sludge

Type	VS%	OM%	ASH%
Primary sludge	60–90	60–80	20–40
Activated sludge	60–88	61–75	25–39
Digested sludge	30–60	30–60	40–70

dry weight components of different sludge is shown in Table 6.5 (Wanqin et al. 2014a).

Traditional modes of sludge disposal include landfill, incineration, and land use. Municipal sludge contains a large number of nutrients, such as nitrogen, phosphorus and various trace elements, as well as organic matter that is difficult to degrade, heavy metals and harmful substances such as pathogenic microorganisms and parasite eggs (Cao 2020). The organic composition of sludge is shown in Table 6.6, and in sludge, the organic matter content accounts for more than 50% of the sludge mass. Improper disposal of municipal sludge can cause secondary pollution, causing groundwater, surface water, and air pollution. The large volume of sludge will consume a large number of land resources and even cause ecological damage in an area. Because sludge generally has a water content higher than 90%, it has a colloidal structure, is very difficult to dehydrate, has unstable organic matter, is prone to rot and odor, and is prone to release toxic and harmful pollutants into the environment. With the advent of energy and resource crisis, the organic matter and nutrients contained in urban sludge are important resources for the sustainable development of modern society, and the most appropriate direction is to realize the reduction, harmlessness and resource nation of sludge through reasonable technology. Anaerobic fermentation can improve sludge characteristics, increase dewatering characteristics and remove organic matter, and the biogas produced is a high-quality clean energy source (Ting and Lee 2007). It has the advantages of protecting the environment and saving resources. The residue after fermentation can still be used for subsequent applications such as the synthesis of agricultural fertilizers, construction materials, etc.

Many scholars have done a lot of research work on methane production by anaerobic fermentation of municipal sludge. Ge uses mixed fermentation of municipal sludge and kitchen waste to improve the degradation rate of organic matter and greatly improve the stability of fermentation system and gas production characteristics (Yuan 2019). Duan et al. studied solid-phase anaerobic fermentation of municipal sludge under medium temperature conditions, and the results proved that high solid anaerobic digestion treatment of dehydrated sludge was completely feasible in a fully mixed semi-continuous reactor (Duan et al. 2012). However, the traditional

Table 6.6 Organic composition of sludge (%)

Sludge composition	Ash	Protein	Carbohydrate	Fat	Hemicellulose	Cellulose	lignin
Mass fraction	40	20	11	6	3	1	19

sludge fermentation has many disadvantages, such as slow reaction, long sludge retention time, low methane yield, and low gas production efficiency, which seriously affect the popularization and application of sludge anaerobic fermentation technology. Researchers at home and abroad have conducted numerous experiments and engineering applications to address the issues that extracellular polymer (ESP) is difficult to degrade and cell wall is difficult to break during sludge anaerobic fermentation. The results show that pretreatment can promote the degradation of ESP, improve the hydrolysis of macromolecular organic matter in sludge, and break the cell wall of sludge bacteria. Effectively improving the performance of sludge anaerobic fermentation, shorten hydraulic retention time, increase methane production and gas production rate.

Li et al. explored the influence of different biological treatments on sludge anaerobic digestion, and found that directly adding fungus-*Paecilomonas vannici* as pretreatment for anaerobic fermentation had the best methane production effect with the net cumulative gas production and methane production increased by 85.79% and 42.76%, respectively, comparing that 1 kg sludge can produce 12.69 L of methane using pure sludge, comparing that it is increased by 42.74% with untreated sludge (Xue et al. 2015). The results show that the anaerobic fermentation technology of dewatering sludge by microorganism pretreatment has good feasibility and application value. Gulsen Akbay et al. found that hot water pretreatment of sludge could improve the rate of sludge anaerobic fermentation, and the optimal reaction conditions were 180 °C and 76 min pretreatment time. At this time, the relative abundance of methanogens in the microbial community, such as methanogens *Methanosarcina kluyver* and *van Niel*, *methanogens* and *methanogens hydrogen-producing*, was higher than that in the unpretreated area (Gulsen Akbay et al. 2021). Due to the low biodegradability of cell wall and the existence of extracellular biopolymer, it is a strong limiting factor for anaerobic fermentation. Yeneneh et al. found that the use of ultrasonic wave, microwave, and pretreatment of municipal sludge can significantly improve the methane production and solid removal rate and dehydration rate. Microwave-ultrasonic combined pretreatment can enhance sludge decomposition, flocculate destruction, cell wall rupture and release of soluble organic matter (Yeneneh et al. 2013). Luo et al. found that pretreatment of activated sludge by iron activation with persulfate can improve sludge dewatering properties and anaerobic fermentation performance of sludge (Luo et al. 2018). It can be seen that pretreatment of municipal sludge prior to anaerobic fermentation can improve sludge properties as well as the gas production rate and gas production performance of sludge anaerobic fermentation.

Due to the characteristics of low calorific value and low C/N ratio of municipal sludge, the anaerobic fermentation process is often combined with different biomass for anaerobic fermentation, which can change the C/N ratio in the fermentation process and increase the gas production efficiency. Ajeej et al. believed that methanogens were particularly sensitive to the substrate and C/N ratio in the process of anaerobic fermentation. Therefore, adding some carbon-rich wastes to the sludge could significantly improve the activity of methanogens, thus improving methane production and gas production rate (Ajeej et al. 2015). Yuan et al. mixed primary settling sludge, surplus sludge, dehydrated sludge and wheat straw for anaerobic fermentation, and used wheat straw as a carbon source to adjust the sludge C/N ratio (Demirbas et al. 2016). The results showed that when the ratio of carbon to nitrogen was 25:1, the cumulative methane production of different sludge was the largest. The cumulative methane production of primary sludge, residual sludge, and dehydrated sludge were 19,600 mL, 18,790 mL, and 16,300 mL, respectively, which were 43.4%, 49.4%, and 54.4% higher than the cumulative gas production of single sludge, respectively (Rong et al. 2015). Wang et al. It was assumed that the organic matter content of food waste leachate was high, as was the carbon-to-nitrogen ratio. The problem of low anaerobic digestion efficiency of sludge could be improved by mixing food waste leachate with municipal sludge for anaerobic fermentation. The research found that the cumulative methane production of cooked landfill leachate and sludge by anaerobic fermentation could reach 2102 mL, which was 4.6 times of that without adding leachate (Dandan et al. 2014). Zhang et al. studied the co-digestion characteristics of pig manure and dewatered sludge, and found that under the optimal ratio, the cumulative methane yield was 315.8 mL/g, which was 82.4% higher than that of single digestion (Wanqin et al. 2014b). Sebastian et al. used pig manure and chicken manure for anaerobic fermentation of municipal sludge and found that the maximum methane production could be obtained from sewage sludge mixtures with 30% pig manure addition (Borowski et al. 2014). Therefore, the co-anaerobic fermentation of municipal sludge and biomass with high carbon content can greatly increase methane production, and is an effective way to realize the stabilization, harmless and resource utilization of sludge and other organic wastes.

6.3.4 *Methane and Biogas Production from Pharmaceutical Wastewater*

The pharmaceutical industry plays a very important role in the development of China's national economy. With the increasing demand for pharmaceutical products and the ongoing development of the pharmaceutical industry, there has also been a steady discharge of various pharmaceutical wastewater in recent years. According to statistics, as of 2016, China has a total of 7622 pharmaceutical manufacturing enterprises with a wastewater discharge of over 600 million tons, accounting for about 3% of the total amount of industrial waste water (Yan 2018). Pharmaceutical wastewater mainly includes fermentation drug production wastewater, chemical synthesis production wastewater, extraction production wastewater, traditional Chinese medicine production wastewater, biological engineering production wastewater, as well as washing water and flushing wastewater in the production process of various preparations (Yafeng and Ying). General pharmaceutical factory industrial wastewater characteristics are mainly, high concentration of organic pollutants in

Table 6.7 Physicochemical properties of common biopharmaceutical wastewater

Item	Concentration	Item	Concentration
COD (mg/L)	2000–10,000	SS (mg/L)	200–500
Chromaticity	500–1000	BOD5 (mg/L)	1000–2500
Temp. (°C)	25–80	TP (mg/L)	50–250
pH	4–9	TN (mg/L)	500–1500

wastewater of high concentration of organic matter, COD can be as high as 5000–20,000 mg/L, BOD can reach 2000–10,000 mg/L, SS concentration can reach 5000–23,000 mg/L, the TN reaches 600–1000 mg/L, and the biodegradability is poor. The water quality and quantity change greatly. The physical and chemical properties of the common biopharmaceutical wastewater are shown in Table 6.7 (Li and Li 2015). The pharmaceutical process is very complex, in which a large number of intermediates and metabolites are produced, as well as a large number of highly concentrated, highly toxic, and difficult to degrade wastewater. Pharmaceutical wastewater has become one of the most difficult wastewaters in the world. All countries are actively exploring the efficient treatment and disposal methods of pharmaceutical wastewater.

At present, the main treatment methods of pharmaceutical wastewater include biological treatment technology, physicochemical treatment technology, chemical treatment technology, etc. (Qianxing 2021). Biological treatment technology mainly includes aerobic biological treatment technology, anaerobic biological treatment technology and aerobic-anaerobic combination treatment technology. The long-term practice has proved that the biological treatment technology is the most economical treatment way at present. Anaerobic biological treatment has attracted a lot of attention because of its high organic load, low sludge yield, low energy consumption, less need for nutrients, and the ability to produce biogas for energy recovery. The commonly used anaerobic treatment processes for pharmaceutical wastewater include anaerobic membrane bioreactor (ANMBR), up-flow anaerobic sludge bed (UASB), anaerobic sequencing batch reactor (ANSBR), moving biofilm reactor (MBBR), etc. (Shi et al. 2017).

Because pharmaceutical wastewater contains a high concentration of organic matter, this organic matter can be used as a substrate for anaerobic fermentation to produce clean energy biogas. Chen et al. studied the treatment of pharmaceutical wastewater containing m-cresol and isopropanol by anaerobic biofilm reactor under hydraulic retention time of 48, 36, 24, 18 and 12 h, and explored the characteristics of methane fermentation under different hydraulic retention time. The results showed that as the hydraulic retention time increased, so did the efflux rate of organic matter. Biogas production increased significantly when optimal hydraulic retention time was 36 h, and methane content maintained a stable trend in this stage (Chen et al. 2018). Sun et al. used two anaerobic systems composed of up-flow anaerobic sludge bed reactor (UASB) and internal circulation reactor to treat Chinese herbal medicine wastewater, and found when the hydraulic retention time was 21 h, the energy conversion efficiency could be increased from $9.6 \pm 2.5\%$ for hydrogen production alone to

72.4 ± 2.5% for hydrogen and methane combined production. The removal rate of COD can reach 90.1 ± 2.1% (Sun et al. 2019). Pharmaceutical wastewater contains complex components, some of which may inhibit the anaerobic fermentation process. Li et al. used up-flow anaerobic sludge bed (UASB) to treat the synthetic pharmaceutical wastewater containing organic sulfides and sulfate compounds, and found that when the hydraulic retention time was 1.3 days, it had a good removal effect of organic compounds. It was also found that there is a symbiotic and competitive relationship between sulfate-reducing bacteria and methanogens, and methanogens may be inhibited when the sulfide concentration is too high (Li et al. 2015). Aydin et al. found that erythromycin, tetracycline, and sulfamethoxazole antibiotics had a certain inhibitory effect on the anaerobic treatment of pharmaceutical wastewater (Aydin et al. 2015).

Pharmaceutical wastewater has a negative impact on anaerobic fermentation due to its high concentration of organic matter, complex water quality, and strong biological inhibition. Most pharmaceutical wastewater needs to be treated by physical and chemical treatment to change the organic composition, reduce the load of organic matter, improve the biodegradability of wastewater, and then be treated by biological method (Shon et al. 2006). At present, there are many new technologies and different processes for the biological treatment of pharmaceutical wastewater, but most of them remain in the laboratory model stage, and there are few researches on the methane production performance of anaerobic fermentation (Lefebvre et al. 2014).

6.4 Forest Waste Products for Methane and Biogas

China is a country rich in forest biomass resources. According to statistics, about 2.1 billion tons of forest wastes and agricultural and forestry by-products are produced in China every year, and a large amount of forest wastes in China have not been effectively utilized (Weihong and Xiaoxu 2017). Forest biomass resources mainly include forestry waste and forestry by-product waste. Forestry waste mainly includes bark, branches and leaves, and forestry by-product waste mainly refers to some fruit skins (Tingting et al. 2016). The production of clean energy methane from forest wastes via anaerobic fermentation is critical for waste resource utilization.

Branches and leaves contain more lignocellulose which is difficult to degrade, so pretreatment is needed to destroy the structure of lignocellulose, while leaves are easier to degrade. Ge et al. found that when anaerobic digestion was used to treat the leaves and sawdust of albizia biomass, the maximum methane yield of anaerobic fermentation could reach 161 and 113 L^{-1}kgVS (Ge et al. 2014). The common pretreatment methods include physical, chemical, and biological pretreatment. Leonidas et al. used birch for anaerobic fermentation, and methane production could reach 40 $mLCH_4/g$ (VS). After hydrothermal pretreatment, methane production increased by 6.7 times to as high as 254 $mLCH_4/g$ (VS) (Matsakas et al. 2015). Yao et al. produced methane by anaerobic digestion of poplar processing residues pretreated with sodium hydroxide, and found that the methane yield was 271.9 L/kg under the optimal ratio condition, which was 113.8% higher than that of the samples without sodium hydroxide pretreatment. The results of scanning electron microscopy, infrared spectroscopy, and crystallinity tests showed that the lignocellulosic structure of poplar processing residues was destroyed by sodium hydroxide (Yiqing et al. 2013). Maria et al. pretreated lignocellulosic substrates with ligninolytic fungus *Phanerochaete flavido-alba*, and found that lignin content decreased and biogas production increased (López et al. 2013).

Forest waste is high in carbon content. In order to obtain good activity of microorganisms in the anaerobic fermentation process, methane is usually produced by combined fermentation with other organic matter. He studied the gas production performance of the mixed fermentation of paper mulberry leaves with cow dung, kitchen waste and rice straw, and found that when the C/N ratio was 20, the best gas production effect of the mixed anaerobic fermentation of paper mulberry leaves with kitchen waste could reach 411.71 mL/g (VS), which was about 46% higher than that of paper mulberry leaves alone by fermentation of 283.06 mL/g (VS) (He 2019). Li et al. mixed poplar leaves and pig dung in different proportions at 35 °C to investigate the biogas yield. It was proved that gas production was greatly increased after adding pig manure in different proportions. Especially when the ratio of leaves to feces was 1:2, the total gas production reached 7338 mL (Li et al. 2016a). Prabhu et al. found that the co-anaerobic fermentation of cow dung and dry leaves could improve methane production and gas production rate (Prabhu et al. 2020). As a kind of biomass material, forestry waste has a high potential for methane production by anaerobic fermentation. Due to its own characteristics, biogas production can be improved by combining with other biomass with high nitrogen content. Looking for fermentation technology and process to improve gas production efficiency and gas production is the research focus and direction in the future.

6.5 Aquatic Plants Produce Methane and Biogas

Energy is the most important resource and driving force for human survival and development. With the development of economic globalization and the continuous improvement of human living standards, the contradiction between energy supply and demand is becoming increasingly fierce. Traditional fossil fuels, such as coal, oil and natural gas, are still dominant, but these are non-renewable energy, will be exhausted in the future, and the burning will release a large amount of greenhouse gases (Kandasamy et al. 2021a). The dense and complex cell walls of lignocellulosic plants, which are high in lignin, limit the use of anaerobic digestion on a large scale. An alternative approach is to use aquatic plants as substrates for anaerobic fermentation. Aquatic plants, which are rich in organic matter, are an important biomass resource, and the anaerobic fermentation of hydrophytes to produce energy has attracted more and more attention from all countries (Yin et al. 2016). Biogas is a clean and renewable energy and will become an important breakthrough point of energy development in the future. Anaerobic fermentation technology, which uses organic matter in aquatic plants to produce biogas, can not only solve secondary pollution of aquatic plants but also realize resource utilization, which is something that more and more countries around the world are paying attention to.

Different aquatic plants have different performance in producing methane by anaerobic fermentation. Song et al. took 7 different wetland aquatic plants (*Hydrocotyle vulgaris*, *Thalia dealbata* Fraser, *calamus*, *purple taro* and *Pontederia cordata* L) as fermentation substrates and investigated their gas production effects under the condition of 37 °C. The gas production rates of the seven aquatic plants were 513.23 mL/g (VS), 539.09 mL/g (VS), 577.96 mL/g (VS), 508.95 mL/g (VS), 555.05 mL/g (VS) and 629.41, respectively ML/g (VS), 473.09 mL/g (VS), indicating that aquatic plants have good methane production potential, and purple taro has the best gas production effect (Xuehong et al. 2012). Barua et al. carried out anaerobic fermentation experiment on aquatic plant water hyacinth by hot air oven pretreatment, and the results showed that the maximum methane production of unpretreated water hyacinth was 143 ± 14 mLCH_4/g (VS) on the 32nd day, while that of pretreated water hyacinth with hot air oven was 193 ± 22 mL CH_4/g (VS) on the 14th day (Barua and Kalamdhad 2017). It can be seen that aquatic plants have good methane production potential. Pretreatment can change the properties of substrate, reduce the fermentation time and increase the methane production rate.

Duckweed is a floating aquatic plant belonging to Duckweed subfamily of Araceae, consisting of 36 species in 5 genera. Duckweed is small in size (about 0.55–1.5 cm), simple in structure (only composed of phylloids and pseudoroots), fast in asexual reproduction (its biomass doubles in about 24 h), and has strong adaptability to the environment. Duckweed is widely distributed in all kinds of freshwater environments in the world (Jingjing et al. 2021). Duckweed is rich in starch and high protein, with a volatile substance content of 84.24 ± 0.2% and a lignin content of 12.2%, which has full potential for the production of biogas. The composition analysis of Duckweed is shown in Table 6.8 (Yadav et al. 2017). Compared with other aquatic plants, duckweed grows faster and can cover the entire surface of water in a very short time, forming a layer of plant cover, affecting the gas exchange of water and damaging the aquatic ecosystem. At present, duckweed has been used in pollution remediation, livestock and poultry feed, energy processing and pharmaceutical industries. Duckweed is regarded as a promising biomass energy alternative to lignocellulosic plants due to the complex cell wall structure of lignocellulosic plants, which limits the widespread application of methane production in anaerobic digestion.

Some studies found that dry duckweed was used for anaerobic fermentation, and the results showed that the maximum methane production was 390 ± 0.1 mL/g(VS), indicating that duckweed had an excellent ability to produce methane by anaerobic fermentation (Calicioglu and Brennan 2018). Kaur et al. carried out anaerobic fermentation of duckweed through three different combination processes of acid production fermentation (HAF), electro hydrogenation (HMEC) and methane production (MAD) to explore an effective comprehensive method. The results showed that the three-stage system combined with the three processes had the best gas production effect, with a total energy yield of 38.77 mol Biogas/kg TOCr, and the

Table 6.8 Composition analysis of duckweed

Proximate analysis (%)				Ultimate analysis (%)				Lignocellulosic content (%)		
Volatile matter	Moisture	Ash	Fixed carbon	C	H	N	C/N ratio	Cellulose	Hemicellulose	Lignin
84.24 ± 0.2	11.76 ± 0.2	3.79 ± 0.1	0.3 ± 0.4	47.44 ± 0.2	5.84 ± 0.1	5	9.5	55.2	32.6	12.2

hydraulic retention time and organic load could also be reduced by using the three-stage process (Kaur et al. 2019). In order to improve the C/N ratio in the fermentation process, duckweed and other organic matter co-fermentation have better methane production rate. Zahid et al. carried out anaerobic fermentation of duckweed and activated sludge at different degrees of hydrolysis under medium temperature (35 °C), and the results showed that the maximum methane production was 468 mL/g (VS) (Gaur et al. 2017). Henderson et al. studied the combined anaerobic digestion of duckweed and cow dung, and the results showed that the addition of 2–3% duckweed in the reactor could significantly increase the methane production. Finally, 2% was determined to be the optimal incorporation ratio, and the methane production under the optimal ratio was 1.4–1.9 times that of the control group without duckweed (Henderson et al. 2012). Gu et al. also studied the mixed anaerobic fermentation of duckweed and cow dung, and carried out the anaerobic dry fermentation under medium temperature conditions. The results showed that the optimal ratio of dung to duckweed was 1.00:0.67 when the hydraulic retention time was 30 days, volume gas production rate and feedstock gas production rate (measured by volatile solids VS) were 0.50 L/L day and 0.33 L/g, respectively (Gu et al. 2014). Therefore, it is feasible to use duckweed as the main body of anaerobic fermentation. In order to increase the biogas yield of duckweed anaerobic digestion, Ren et al. mixed duckweed and biological sludge to conduct anaerobic fermentation. It was found that when the two substrates were mixed, the delay stage of anaerobic fermentation was shortened and methane production was higher than that of anaerobic fermentation alone. The optimal treatment condition is the mixture of duckweed and residual sludge at 1:1 and the heat-alkali pretreatment before fermentation, at which the maximum biogas production is 3309 mL and methane production is 1883 (Ren et al. 2018). Jiang also studied the biogas production by anaerobic digestion of duckweed and residual sludge, and found that when duckweed and biological sludge were mixed for anaerobic digestion, the two substrates had complementary advantages, which could shorten the time of acidification stage, and the cumulative gas production was 11% higher than the calculated value (Nan et al. 2017). Zhang studied the mixed anaerobic digestion of water hyacinth and pig manure to produce methane, and carried out anaerobic fermentation of the mixture through up-flow anaerobic fermentation sludge bed reactor (UASB). It was found that COD concentration at about 4000 mg/L had the maximum gas production efficiency, and the average COD removal rate at this concentration was 78.4%. Methane accounted for 57% of the produced biogas, and methane production rate was 210.7 mL/g (Zhang 2011).

China is rich in biomass energy, and its resource utilization can effectively relieve the pressure of fossil energy supply. Anaerobic digestion is an effective way to produce clean energy biogas. The typical rural solid wastes such as straw, animal manure, forestry waste and urban waste such as domestic garbage, municipal sludge and pharmaceutical factory waste water can be used as resources by anaerobic fermentation. Duckweed, as an aquatic plant with fast growth rate and large biomass, is rich in organic matter and has excellent methane production potential. The combined anaerobic fermentation of duckweed and other wastes and the corresponding process are the research hotspots in the future.

6.6 Conclusions

Today, when fossil energy is increasingly exhausted, it is necessary to develop renewable energy to solve the energy problem. Resource utilization can effectively relieve the pressure of fossil energy supply. Anaerobic digestion is an effective way to produce clean energy biogas. Anaerobic fermentation can be used to recover resources from common rural solid wastes such as straw, animal manure, and forestry waste, as well as urban wastes such as domestic garbage, municipal sludge, and pharmaceutical factory wastewater. It not only solves the pollution problem of these wastes, but also realizes the resource utilization of wastes. The biogas residue produced in the fermentation process can also be used for subsequent use. However, the traditional anaerobic fermentation has disadvantages such as low gas production rate and long fermentation time, which limit its wide application. Compared with the medium temperature anaerobic fermentation, the high temperature anaerobic fermentation technology has higher gas production efficiency and can kill a large number of pathogenic microorganisms. As a result, it is a hot direction to investigate the highly efficient and cost-effective high temperature anaerobic fermentation process. Duckweed has excellent methane production potential and can replace lignocellulosic plants in anaerobic fermentation to produce methane. The combined anaerobic fermentation of duckweed and other wastes and the corresponding process are the research hotspots in the future.

References

Ajeej A, Thanikal JV, Narayanan CM, Kumar RS (2015) An overview of bio augmentation of methane by anaerobic co-digestion of municipal sludge along with microalgae and waste paper. Renew Sustain Energy Rev 50:270–276

Amnuaycheewa P, Hengaroonprasan R, Rattanaporn K, Kirdponpattara S, Cheenkachorn K, Sriariyanun M (2016) Enhancing enzymatic hydrolysis and biogas production from rice straw by pretreatment with organic acids. Ind Crops Prod 87:247–254

Ariunbaatar J, Panico A, Frunzo L, Esposito G, Lens PNL, Pirozzi F (2014) Enhanced anaerobic digestion of food waste by thermal and ozonation pretreatment methods. J Environ Manage 146:142–149

Aydin S, Ince B, Ince O (2015) Inhibitory effect of erythromycin, tetracycline and sulfamethoxazole antibiotics on anaerobic treatment of a pharmaceutical wastewater. Water Sci Technol 71(11):1620–1628

Barua VB, Kalamdhad AS (2017) Biochemical methane potential test of untreated and hot air oven pretreated water hyacinth: a comparative study. J Cleaner Prod

Bi S (2020) Anaerobic fermentation of corn straw pretreated through yellow silage for biogas production. Master, NanyangNormal University (in Chinese)

Bilgen S, Sarikaya I (2016) Utilization of forestry and agricultural wastes. Energy Sources Part A-Recovery Utilization and Environmental Effects 38(23):3484–3490

Borowski S, Domanski J, Weatherley L (2014) Anaerobic co-digestion of swine and poultry manure with municipal sewage sludge. Waste Manage 34(2):513–521

Calicioglu O, Brennan RA (2018) Sequential ethanol fermentation and anaerobic digestion increases bioenergy yields from duckweed. Biores Technol 257:344–348

Cao B (2020) Research progress on the characteristics of municipal sludge and its resource utilization technology. Environ Dev 32(08):104+106 (in Chinese)
Chen ZB, Li X, Hu DX, Cui YB, Gu FG, Jia FQ, Xiao TT, Su HY, Xu J, Wang HX, Wu P, Zhang Y, Jiang N (2018) Performance and methane fermentation characteristics of a pilot scale anaerobic membrane bioreactor (AnMBR) for treating pharmaceutical wastewater containing m-cresol (MC) and iso-propyl alcohol (IPA). Chemosphere 206:750–758
Cunsheng Z (2013) Study on biogas production from kitchen waste by anaerobic fermentation technology. Beijing University of chemical technology (in Chinese)
Dandan W, Xianqiang G, Li G, Qiang h, Hainan A, Lili Y (2014) Enhanced anaerobic digestion of urban sludge by adding restaurant garbage leachate. J Environ Sci 34(10):2566–2572 (in Chinese)
Demirbas A, Taylan O, Kaya D (2016) Biogas production from municipal sewage sludge (MSS). Recovery Utilization, and Environmental Effects 38(20):3027–3033
Dong Y, Zou D, Liu Y, Ye B (2016) Study on characteristics of domestic waste in China and its comprehensive treatment technologies: taking Hangzhou City as an example. Acta Agriculturae Zhejiangensis 28(06):1055–1060 (in Chinese)
Dong C, Chen J, Guan R, Li X, Xin Y (2018) Dual-frequency ultrasound combined with alkali pretreatment of corn stalk for enhanced biogas production. Renew Energy 127:444
Duan NN, Dong B, Wu B, Dai XH (2012) High-solid anaerobic digestion of sewage sludge under mesophilic conditions: feasibility study. Biores Technol 104:150–156
Gaur RZ, Khan AA, Suthar S (2017) Effect of thermal pre-treatment on co-digestion of duckweed (*Lemna gibba*) and waste activated sludge on biogas production. Chemosphere 174:754–763
Ge XM, Matsumoto T, Keith L, Li YB (2014) Biogas energy production from tropical biomass wastes by anaerobic digestion. Biores Technol 169:38–44
Gu X, Wang W, Zhu Q, Tang X, Hu Q (2014) Experimental study on dry anaerobic digestion using duckweed mixed with cow dung. Environ Pollut Prev 36(07):18–21 (in Chinese)
Gulsen Akbay HE, Dizge N, Kumbur H (2021) Enhancing biogas production of anaerobic co-digestion of industrial waste and municipal sewage sludge with mechanical, chemical, thermal, and hybrid pretreatment. Bioresource Technol 340:125688
He Y (2019) Research on methane production process:co-digestion of *Broussonetia papyrifera* leaves and organic waste. Lanzhou University of Technology (in Chinese)
Hemin Z (2019) Progress in anaerobic digestion of livestock and poultry manure. Guangdong Chem Ind 46(19):117–118
Henderson SL, Triscari PA, Reinhold DM (2012) Enhancement of methane production by codigestion of dairy manure with aquatic plant biomass. Biol Eng Trans 5(3):147–157
Hetaoli H, Haoran Y, Xiaofeng L, Hanping C, Tingzhou L, Yong C (2017) Study on the strategy of rural waste classification and resource utilization in China. China Eng Sci 19(04):103–108
Huang K, Zhao L, Feng J, Yao Z, Yu J, Luo J (2018) Pretreatment of corn stalk by composite microbial strain improving its methane production performance by anaerobic digestion. Trans Chin Soc Agric Eng 34(16):184–189
Hui C, Peng P, Diyong J, Yeqing L, Hongjun Z (2017) Effects of three pretreatment methods on liquid phase anaerobic fermentation of food waste. Green Technol 12:1–3 (in Chinese)
Jingjing Y, Xuyao Z, Gaojie L, Shiqi H, Yan C, Zuoliang S, Hongwei H (2021) Research and application in duckweeds: a review. Sci Bull 66(09):1026–1045 (in Chinese)
Kandasamy S, Devarayan K, Bhuvanendran N, Zhang B, He Z, Narayanan M, Mathimani T, Ravichandran S, Pugazhendhi A (2021a) Accelerating the production of bio-oil from hydrothermal liquefaction of microalgae via recycled biochar-supported catalysts. J Environ Chem Eng 9(4):105321
Kandasamy S, Narayanan M, He Z, Liu G, Ramakrishnan M, Thangavel P, Pugazhendhi A, Raja R, Carvalho IS (2021b) Current strategies and prospects in algae for remediation and biofuels: an overview. Biocatalysis Agric Biotechnol 35:102045
Kaur M, Srikanth S, Kumar M, Sachdeva S, Puri SK (2019) An integrated approach for efficient conversion of *Lemna minor* to biogas. Energy Convers Manage 180:25–35

Kondusamy D, Kalamdhad AS (2014) Pre-treatment and anaerobic digestion of food waste for high rate methane production—a review. J Environ Chem Eng 2(3):1821–1830

Kratky L, Jirout T (2011) Biomass size reduction machines for enhancing biogas production. Chem Eng Technol 34(3):391–399

Kshirsagar SD, Waghmare PR, Loni PC, Patil SA, Govindwar SP (2015) Dilute acid pretreatment of rice straw, structural characterization and optimization of enzymatic hydrolysis conditions by response surface methodology. RSC Adv 5(58):46525–46533

Lefebvre O, Shi X, Wu CH, Ng HY (2014) Biological treatment of pharmaceutical wastewater from the antibiotics industry. Water Sci Technol 69(4):855–861

Li WC, Niu QG, Zhang H, Tian Z, Zhang Y, Gao YX, Li YY, Nishimura O, Yang M (2015) UASB treatment of chemical synthesis-based pharmaceutical wastewater containing rich organic sulfur compounds and sulfate and associated microbial characteristics. Chem Eng J 260:55–63

Li Y, Jin Y, Li J, Chen Y, Gong Y, Li Y, Zhang J (2016) Current situation and development of kitchen waste treatment in China. Procedia Environ Sci 31:40–49

Li X, Li GY (2015) A review: pharmaceutical wastewater treatment technology and research in China. In: Carl J (ed) Proceedings of the 2015 Asia-Pacific energy equipment engineering research Conference, vol 9. AER-Advances in Engineering Research, pp 345–348

Li A, Liu Y, Wang J, Liu H (2016a) Biogas production by co-fermentation of poplar leaves and swine feces. J S Agric 47(05):705–709 (in Chinese)

Lingli Y, Yanbin S, Xue W, Xiujin L, Hairong Y, Dexun Z, Yanping L (2015) Effect of different pretreatment on anaerobic co production of hydrogen and methane from kitchen waste. Biogas of China 33(02):13–18 (in Chinese)

Lingling J, Weiquan W (2021) A study on collaborative governance of energy. Environ Climate Change China 05:100–103

López MJ, Suárez-Estrella F, Vargas-García MC, López-González JA, Verstichel S, Debeer L, Wierinck I, Moreno J (2013) Biodelignification of agricultural and forest wastes: effect on anaerobic digestion. Biomass Bioenergy 58:343–349

Longsheng Y, Linghua R, Xin W, Hao W (2012) Analysis of physicochemical properties and anaerobic fermentation potential of food waste. J Central S Univ (Sci Technol) 43(04):1584–1588 (in Chinese)

Luo JY, Zhang Q, Wu LJ, Feng Q, Fang F, Xue ZX, Li C, Cao JS (2018) Improving anaerobic fermentation of waste activated sludge using iron activated persulfate treatment. Biores Technol 268:68–76

Matsakas L, Rova U, Christakopoulos P (2015) Sequential parametric optimization of methane production from different sources of forest raw material. Fronti Microbiol 6

Minglong Z (2015) Experimental study on classification scheme and anaerobic digestion of typical organic components of municipal solid waste in Lanzhou. Lanzhou Jiaotong University (in Chinese)

Nan J, Hongyan R, Wenquan R, Jialin L (2017) Anaerobic digestion experiments about duckweed and excess sludge to produce biogas. J Environ Eng 11(08):4757–4765 (in Chinese)

Pattnaik S, Reddy MV (2010) Assessment of municipal solid waste management in Puducherry (Pondicherry), India. Resour Conserv Recycl 54(8):512–520

Peng Y, Wu S (2010) The structural and thermal characteristics of wheat straw hemicellulose. J Anal Appl Pyrol 88(2):134–139

Ponnusamy VK, Nguyen DD, Dharmaraja J, Shobana S, Banu JR, Saratale RG, Chang SW, Kumar G (2019) A review on lignin structure, pretreatments, fermentation reactions and biorefinery potential. Bioresour Technol 271:462–472

Prabhu AV, Manimaran R, Raja SA, Jeba P (2020) Biogas production from anaerobic co-digestion of *Prosopis juliflora* pods with water hyacinth, dry leaves, and cow manure. Energy Sources Part A-Recovery Utilization and Environmental Effects 42(3):375–386

Q WY (2019) Study on pretreatment of corn straw by anaerobic fermentation. Heilongjiang 81 Agricultural University (in Chinese)

Qiang Z, Ye J, Wei J (2013) Research progress of energy production from kitchen garbage. Progr Chem Ind 32(03):558–562 + 574 (in Chinese)
Qianxing L (2021) Analysis of wastewater treatment technology in biopharmaceutical factory. Sci Technol Wind 07:112–113 (in Chinese)
Qiaoling W (2012) Study on Influencing Factors of anaerobic fermentation process of kitchen waste. Nanjing University (in Chinese)
Qing L, Guangpo W, Zepu J, Ying X, Dongliang H, Yangrui L (2013) Research progress on resource utilization of livestock and poultry manure. J S Agric 44(02):338–343
R L (2020) Study on solid state pretreatment of corn straw by anaerobic fermentation. Jiangnan University
Ren H, Jiang N, Wang T, Omar MM, Ruan W, Ghafoor A (2018) Enhanced biogas production in the duckweed anaerobic digestion process. J Energy Resour Technol Trans ASME 140(4)
Rong YH, Chao Z, Fei LR, Fang WY, Xin Z, Jin LX, Xun Z (2015) Investigation on anaerobic co-digestion performance of sewage sludge and wheat straw. China Biogas 33(03):38–44 (in Chinese)
Sabeeh M, Zeshan, Liaquat R, Maryam A (2020) Effect of alkaline and alkaline-photocatalytic pretreatment on characteristics and biogas production of rice straw. Bioresour Technol 309
Shan L, Shubiao W, Wanqin Z, Changle P, Yu D, Renjie D (2013) Effects of white rot fungi pretreatment on methane production by anaerobic fermentation of corn straw. Trans Chin Soc Agric Mach 44(S2):124–129+142
Shen X, Huang G, Yang Z, Han L (2015) Compositional characteristics and energy potential of Chinese animal manure by type and as a whole. Appl Energy 160:108–119
Shi XQ, Leong KY, Ng HY (2017) Anaerobic treatment of pharmaceutical wastewater: a critical review. Biores Technol 245:1238–1244
Shon HK, Vigneswaran S, Snyder SA (2006) Effluent organic matter (EfOM) in wastewater: constituents, effects, and treatment. Crit Rev Environ Sci Technol 36(4):327–374
Shuang L, Wenzheng G, Zhiwei S, Yi Y, Wei Q (2020) Methane Production of Different Crop Strawby Anaerobic Digestion. 48(11):105–109
Statistics NBo (2020) China statistical yearbook
Sun CY, Liu F, Song ZW, Li LX, Pan Y, Sheng T, Ren GM (2019) Continuous hydrogen and methane production from the treatment of herbal medicines wastewater in the two-phase ‘UASB(H)-ICM’ system. Water Sci Technol 80(6):1134–1144
Theuretzbacher F, Lizasoain J, Lefever C, Saylor MK, Enguidanos R, Weran N, Gronauer A, Bauer A (2015) Steam explosion pretreatment of wheat straw to improve methane yields: Investigation of the degradation kinetics of structural compounds during anaerobic digestion. Biores Technol 179:299–305
Tian Z, Meidong B, Wei G (2012) Current situation of livestock manure pollution and biogas potential in China. Chin J Ecol 31(05):1241–1249 (in Chinese)
Ting CH, Lee DJ (2007) Production of hydrogen and methane from wastewater sludge using anaerobic fermentation. Int J Hydrogen Energy 32(6):677–682
Tingting D, Sining Y, Jiang Z, Xinlei H, XianMei Z, Hanhou Z (2016) Research progress of anaerobic fermentation of different biomass waste. Biogas of China 34(02):46–52 (in Chinese)
Wang H, Xu J, Sheng L (2019) Study on the comprehensive utilization of city kitchen waste as a resource in China. Energy 173:263–277
Wanqin Z, Quanyuan W, Shubiao W, Dandan Q, Wei L, Zhuang Z, Renjie D (2014a) Batch anaerobic co-digestion of pig manure with dewatered sewage sludge under mesophilic conditions. Appl Energy 128:175–183
Wanqin Z, Dandan Q, Shubiao W, Renjie D, Wansheng Z (2014b) Effect of different pretreatment methods on sewage sludge anaerobic digestion. J Agric Mach 45(09):187–198 (in Chinese)
Weihong C, Xiaoxu S (2017) Application and research status of agricultural and forestry waste in China. Mod Agric Technol 18:148–149 (in Chinese)
X X (2015) Study on the effect of pulverization pretreatment on the formation of scum in biogas fermentation. Chinese Academy of Agricultural Sciences

X Z (2020) Study on acid-producing methane-producing microorganisms in anaerobic fermentation of livestock manure. Henan Agriculture (11) :51–52+60

Xianwu M, Xiaohui X, Zhiman Y, Meng D, Rongbo G (2011) Effect of organic loading on single phase anaerobic fermentation of food waste. Anhui Agric Sci 39(25):15567–15569 (in Chinese)

Xiaomei P, Zhiya Y, Hongguang Z (2008) Research development of anaerobic fermentation treatment of excess sludge from municipal wastewater in China. China Biogas (01):25–29 (in Chinese)

Xiaowei Z, Bing L, Yi L, Liping C (2014) Effect of inoculation ratio on start-up of dry anaerobic fermentation of food waste. J Environ Eng 8(03):1157–1162 (in Chinese)

Xing W, Qiang L, Zheng Z, Jing H, Yayue D, Min Z, Xiaobo Y (2017) Effects of steam explosion and calcium oxide pretreatment on the anaerobic dry fermentation of rice straw. J Agro-Environ Sci 36(02):394–400

Xue L, Cong L, Jundong S, Xiang L, Fei W (2015) Effect of different biopretreatments on anaerobic digestion process performance of sludge. J Agric Mach 46(08):186–191 (in Chinese)

Xuehong S, Xinyuan J, Mengjie Z (2012) Trail on anaerobic fermentation of 7 kinds of wetland aquatic plants for biogas production. Guangdong Agric Sci 39(03):91–94 (in Chinese)

Yiqing Y, Mulan H, Yubing R, Liying M, Yang L, Hongmei S, Yun X, Hua Z, Qien L, Lizhe A (2013) Anaerobic digestion of poplar processing residues for methane production after alkaline treatment. Biores Technol 134:347–352

Yadav D, Barbora L, Bora D, Mitra S, Rangan L, Mahanta P (2017) An assessment of duckweed as a potential lignocellulosic feedstock for biogas production. Int Biodeterior Biodegradation 119:253–259

Yafeng L, Ying G, Research progress of pharmaceutical wastewater treatment technology. Water Treatment Technol 40(05):1–4 + 9 (in Chinese)

Yan Z (2018) Research progress in the treatment technologies of pharmaceutical wastewater. Ind Water Treatment 38(05):5–9 (in Chinese)

Yanbin S (2013) Effect and mechanism of different pretreatment on anaerobic co production of hydrogen and methane from kitchen waste. Beijing University of Chemical Technology (in Chinese)

Yeneneh AM, Chong S, Sen TK, Ang HM, Kayaalp A (2013) Effect of ultrasonic, microwave and combined microwave-ultrasonic pretreatment of municipal sludge on anaerobic digester performance. Water Air Soil Pollut 224(5)

Yi Z (2016) Contemporary thinking on urban waste treatment. Jiangsu Constr 04:120–126 (in Chinese)

Yin Z, Dan X, Enrong X, Zhenbin W (2016) Progress on the technology and methods of the energy utilization of aquatic plants. Environ Sci Technol 39(11):118–126 (in Chinese)

Yingtao D, Wentao L (2020) Research on key technologies of biogas purification from agricultural waste by anaerobic fermentation. Sci Technol Innov 34:31–32

Yuan Z, Long J, Wang T, Shu R, Zhang Q, Ma L (2015) Process intensification effect of ball milling on the hydrothermal pretreatment for corn straw enzymolysis. Energy Convers Manage 101:481–488

Yuan G (2019) Study on mixed anaerobic fermentation of food waste and municipal sludge. Northwest A & F University (in Chinese)

Yunpan D (2016) Research on high pressure extrusion dry and wet separation and anaerobic digestion of wet components of municipal solid waste. Shenzhen University (in Chinese)

Yuxiaofei W, Yunbo S, Wei Q (2020) Effects of alkaline pretreatment on the anaerobic fermentation of corn straw under mild humidity and heat conditions Jiangsu. Agric Sci 48(05):236–243

Zemin X, Guoming Z, Li M (2013) Research and discussion on pretreatment technology of straw biogas. J Agric Mech Res 35(09):227–230

Zhang Y, Chen X, Gu Y, Zhou X (2015) A physicochemical method for increasing methane production from rice straw: extrusion combined with alkali pretreatment. Appl Energy 160:39–48

Zhang Q, Xu H (2012) Status and devel development suggestion of incineration technology of domsetic garbages. Environ Eng 30(02):79–81+89 (in Chinese)

Zhang Y, Shang X, Li K, Zhang C, Zhang K, Rong H (2011) Technologies status and management strategies of municipal solid waste disposal in China. Ecol Environ J 20(02):389–396 (in Chinese)
Zhang D (2011) The anaerobic digestion study on aquatic plants mixtured with swine manure and process design. University of Science and Technology of China (in Chinese)
Zhang T (2021) Discussion on the current situation and development trend of energy utilization in china. (05):76–78 (in Chinese)
Zhao L (2011) Study on the characteristics of corn stalk dry anaerobic fermentation based on bio-pretreatment and the utilization of residue media. Shenyang Agricultural University, Doctor
Zhibin Z, Hong Z, Dongmei G (2019) Study on pretreatment of straw anaerobic fermentation for biogas production. Renew Resour Circ Econ 12(08):39–41
Zhou H, Long Y, Meng A, Li Q, Zhang Y (2015) Classification of municipal solid waste components for thermal conversion in waste-to-energy research. Fuel 145:151–157
Zixu L, Jing P (2015) Study on characteristics and treatment technology of food waste. Environ Sci Manage 40(07):102–104 (in Chinese)

Chapter 7
Waste Fermentation for Energy Recovery

Jiafu Lin, YanZhao Zhang, Tao Song, and Haifeng Su

Abstract Biological fermentation engineering is an important part of biological engineering. Carbohydrates are used to produce various industrial solvents and chemical raw materials by microorganisms, among which energy production through biotransformation is a very important research field. In this chapter, the common organic wastes, including typical rural solid wastes, forestry solid wastes, and naturally-grown aquatic plants such as duckweed, as well as urban waste are discussed as potential feedstock for liquid bioenergy including bioethanol, biobutanol, and bio-olefin. The pretreatment technology and fermentation modes of these wastes as well as the microbial species used are compared and discussed in depth. Duckweed was used as a typical example to evaluate the potential of producing ethanol, butanol, higher alcohols, and biodiesel via fermentation pathways. Moreover, the economic feasibility of producing liquid biofuel through fermentation from different waste feedstocks is evaluated. This chapter provides an important reference and insight for future research on organic waste fermentation.

Keywords Organic wastes · Bioenergy · Fermentation · Duckweed · Bioethanol

7.1 Introduction

The oil and natural gas resources suitable for extraction are only enough for 30 years, at most 50 years, and coal reserves are only enough for 300 years (Chen and Qiu 2007). Bioenergy from organic wastes is the only energy product that can replace

Y. Zhang · H. Su (✉)
Chongqing Institute of Green and Intelligent Technology, Chinese Academy of Science, 266, Fangzheng Avenue, Shuitu High-tech Park, Beibei, Chongqing 400714, China
e-mail: suhaifeng@cigit.ac.cn

Y. Zhang
e-mail: zhangyanzhao@cigit.ac.cn

J. Lin · T. Song · H. Su
Chengdu University, Chengdu 61004, China
e-mail: linjiafu@cdu.edu.cn

A. E.-F. Abomohra et al. (eds.), *Waste-to-Energy*,
https://doi.org/10.1007/978-3-030-91570-4_7

petroleum fuels on a large scale. Hydro, wind, solar, nuclear, and other new sources of energy apply only to power and heat. Bioenergy products are diverse via fermentation. Energy products include liquid bioethanol and diesel, solid prototype and molding fuels, gaseous biogas, and other energy products. It can replace oil, coal, and natural gas, as well as heat and electricity generation. Bioenergy has diversity in raw materials. Biofuel can use crop straw, forest processing residues, livestock and poultry manure, organic wastewater residue of food processing industry, municipal waste, but also can use low-quality land to grow a variety of energy plants. Bioenergy has a variety of "material" applications, like oil and coal to produce plastic, fiber, and other materials as well as chemical raw materials and other material products, forming a huge biochemical production system. Bioenergy is "recyclable" and "environmentally friendly". Biofuels are produced in the process of harmless and resource recovery of agricultural and rural organic wastes. All the life materials of biofuel can enter the biological cycle of the earth, and even the carbon dioxide released will be absorbed by plants again and participate in the cycle of the earth, so as to achieve zero emissions. The sustainability of material and the circulability of resources are a modern and advanced production mode. Biofuels can expand agricultural production, promote rural economic development and increase farmers' income; It can also boost manufacturing, construction, automobile, and other industries. The development of biofuels in China and other countries can also promote agricultural industrialization and the development of small and medium-sized towns and narrow the gap between workers and farmers, which is of great political, economic and social significance.

Biofuel will increase the number of "crude oil" producing countries from 20 to 200. By producing fuel independently, it will suppress the price of imported oil and reduce the cost of imported oil, so that more funds can be used to improve people's lives and fundamentally solve the food crisis. Bioenergy can create jobs and build domestic markets. Brazil's experience shows that for every job in petrochemicals, 152 jobs are created in ethanol. The petrochemical industry invested $220,000 per job, while the fuel industry invested only $11,000. "The renewable energy industry will create 20.4 million jobs by 2030, including 12 million in biofuels," according to the United Nations Environment Program's "Green Jobs" report.

However, bioenergy exists in physical form and is the only renewable energy source that can be stored and transported. In addition, it is the most widely distributed, not limited by weather and natural conditions, wherever there is life there is biomass. The carrier source of bioenergy is usually organic waste. Organic waste is a misplaced resource for rural waste such as biomass straw and livestock manure. Forestry solid waste such as biomass waste from forestry. Municipal waste such as domestic waste, kitchen waste, urban sludge, pharmaceutical factory wastewater, etc. is used for fermentation to produce ethanol, butanol, and other renewable energy in different countries. For example, China is a major energy consumer, facing the pressure of traditional fossil energy depletion and carbon emission reduction, the development of renewable energy has become urgent. It has an important role and significance for China to adjust the energy structure and realize the resource utilization of waste.

Despite numerous biomass energy production, collection, and utilization research, many studies reported on the biomass, biomass utilization technology, including biomass production, collection, and processing technology, biomass energy and the environment, the biomass of anaerobic biogas, biomass to ethanol, and diesel oil producing technology, biomass direct combustion technology. However, fermentation methods, modes, and technologies from waste to energy generation include pretreatment, microbial types of energy production including bacteria and fungi, and energy types produced by fermentation including ethanol, butanol, olefin, and other energy sources. Of particular importance is the lack of systematic treatment of wastes (plant species) such as duckweed in the development and application of bioenergy. The purpose of this paper is to introduce systematically the research status of biomass energy fermentation technology, the types of energy, and the microbial species of waste energy fermentation, and the application of duckweed in biomass energy.

7.2 Fermentation Methods, Modes, and Techniques

7.2.1 Solid-State Fermentation

Solid-state fermentation is a microbial fermentation process in which the medium is in a solid state and the fermentation system is carried out in the absence of almost absence of free water present (Xu et al. 2002). Solid-state fermentation substrates are water-insoluble polymers, and the substrate is not only a site for microbial growth and development, but also provides the carbon source, nitrogen source, inorganic salts, water, and other nutrients needed for microbial growth. Compared with other culture methods, solid-state fermentation has the following advantages: (1) the medium is simple and widely available, mostly inexpensive natural substrates such as biomass straw, etc. (2) Low investment and low energy consumption. The technology is simple. (3) The yield of the product is high. (4) Low substrate water content, small bioreactor size, no subsequent wastewater treatment, environmentally friendly, post-treatment processing is relatively simple. (5) The fermentation process generally does not require an aseptic operation. (6) Continuous aeration is not required, and the air is generally not strictly sterile (Huang et al. 2003).

Solid-state fermentation is a multiphase system of gas, liquid, and solid phases as well as microorganisms (Thomas et al. 2013). There is almost no flowing free water in the solid-state fermentation substrate, the content of bound water is about 12–80% (mostly maintained at about 60%), and the oxygen required for microbial growth and development, and metabolism comes mainly from the gas phase of the continuous phase, with relatively low energy consumption. The history of solid-state fermentation can be traced back to thousands of years ago, and with the development of solid-phase fermentation technology, a variety of substances can be used as substrates for fermentation. Biomass feedstocks with more applications at this stage are mainly starchy feedstocks, sugar feedstocks, and lignocellulosic feedstocks. Starchy raw materials such as potatoes and grain grains, sugar raw materials such as sugar cane and sweet sorghum. But the use of these raw materials there is "competition with the people for food, and food for land" problem (Wang et al. 2018). Biomass feedstocks such as biomass straw do not have these problems and are the focus of global research on biological fermentation.

There are many factors that affect the solid-state fermentation process, mainly depending on the type of substrate, the type of microorganism, and the size of the production scale, based on the above conditions, which can be divided into biochemical, physicochemical, and environmental factors, all of which are closely related and cannot be viewed individually. (1) Microbial influence: fungi and bacteria are the microorganisms used more in solid-state fermentation, and relatively speaking inoculation of filamentous fungi is preferable because solid-state fermentation simulates the living environment of filamentous fungi (Soccol et al. 2017). Inoculation of fungal spores is more convenient and flexible, and easy to keep for a longer period of time, but also has the same disadvantages of long-term lag and large spore inoculum. (2) The effect of water and water activity: water is the main medium of solid-state fermentation, and changes in the water content of the substrate have an important impact on the growth and metabolic capacity of microorganisms. The high-water content will reduce the volume of gas in the substrate and the intensity of gas exchange, making it difficult to cool down and ventilate, and increasing the risk of contamination with miscellaneous bacteria. The low water content will inhibit the growth and metabolism of microorganisms and also cause the substrate to swell. The range of water content during solid-state fermentation should be controlled between 30 and 80%, and the amount of water control should be different for inoculating different microorganisms for fermentation. The ability of microorganisms to grow and metabolize on a substrate depends on the water activity (Aw) of that substrate.

For different microbial species, water activity is generally different for bacteria (0.90–0.99), most yeasts (0.80–0.90), and fungi (0.60–0.70). During solid-state fermentation, contamination by trash bacteria can be avoided because of the low water activity requirement of fungi (Li et al. 2011). (3) Substrate and particle size: the general substrate for solid-state fermentation is usually agricultural by-products such as lignocellulose, and these substrates have a macromolecular structure that will wrap the carbon and nitrogen sources, which is not conducive to fermentation, so some pretreatment should be carried out before fermentation. Mainly through physical, chemical, and biological pretreatment means to reduce particle size, release degradable substances and improve the efficiency of fermentation. (4) Temperature and pH: Microorganisms release large amounts of heat during growth and metabolism. A number of microbial life activities such as metabolism, protein synthesis, reproduction, etc. are sensitive to temperature, so this heat should be removed in time to avoid the impact on the growth and metabolism of the bacterium. pH is also one of the key factors affecting the growth and metabolism of microorganisms, and different species of microorganisms have different adaptation ranges for pH. The optimum pH range of fungi is 3.8–6.0, and the optimum pH range of yeast is 4.0–5.0. But the excellent buffering property of some materials in solid-state fermentation helps to reduce the need for pH control. Therefore, in solid-state fermentation, as long as the initial pH is adjusted to the desired value, the fermentation process usually does not need to detect and control pH. (5) Fermentation time: It is very important to control the fermentation time to improve the yield of target products. There is an optimal time phase for microbial fermentation. Too short a time is not enough to obtain the required yield, too long, because the environment has been unfavorable to the growth of bacteria, often resulting in autolysis of bacteria, yield decline. Therefore, the fermentation time should be determined according to the fermentation strain, fermentation process, fermentation products to conduct the corresponding experiments.

7.2.2 *Liquid Fermentation*

Liquid fermentation is the process of preparing the substrate into a liquid state and then inoculating the strain into the liquid substrate for biological reaction to prepare the product. The uniformity of heat and mass transfer in liquid fermentation is more advantageous than that in solid-state fermentation. Liquid fermentation has the advantages of high density, precise control, and high degree of automation, so industrial mass production is still dominated by liquid fermentation Fig. 7.1.

A large number of studies have been conducted by domestic and foreign scholars on the production of energy substances by liquid fermentation. Wen et al. explored the optimal fermentation conditions for in situ enzymatic saccharifications of rice straw by liquid fermentation using rice straw as substrate and *Trichoderma reesei*

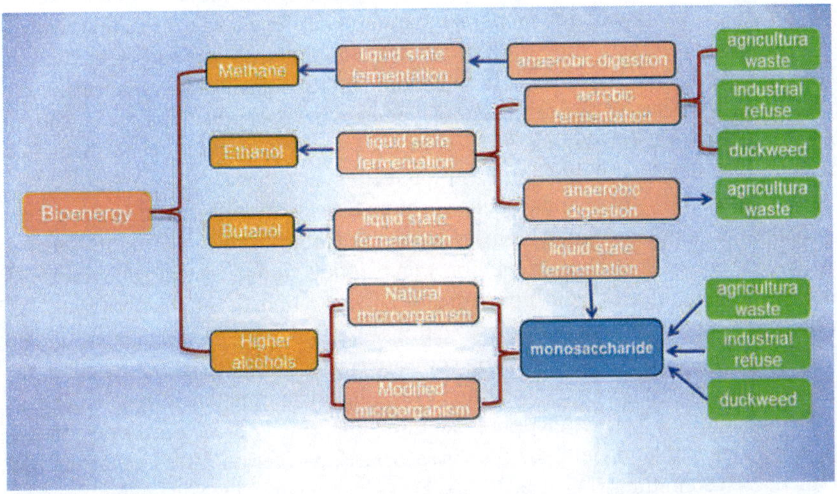

Fig. 7.1 Fermentation patterns for different bioenergy sources including methane, butanol, and higher alcohols: Isobutanol of C5, 1-methyl-butanol, amyl alcohol, hexanol of C6, etc.

as enzyme-producing microorganism. The results showed that the optimal enzyme-producing fermentation conditions were at fermentation temperature of 30 °C, initial pH 6.5, fermentation time of 24 h, and rice straw addition of 30 g/L. The specific sugar yield of straw at pH 4.8, enzymatic digestion temperature of 50 °C, and enzymatic digestion time of 24 h can be as high as 0.350 g/g (Wen et al. 2014). It indicates that liquid enzymatic saccharification is an effective way to achieve resource utilization of biomass straw. Si et al. produced 2,3-butanediol by fermenting corn stover hydrolysate with Klebsiella Oxytoca ZU-03 at an initial pH point of 6.0 and fermentation temperature of 60 °C for 64 h. The sugar utilization reached 99.36% and the yield of 2,3-butanediol reached the theoretical 94% of the maximum yield of 0.468 g/g (Si and Xia 2010). Song et al. conducted a study on the production of cellulosic ethanol by fermentation of corn stover saccharate by *Pachysolen tannophilus* (P-01), and the volume fraction of ethanol was 2.05% under optimal reaction conditions, which was 33.17% higher than that of the control (Song et al. 2008). Sasaki et al. used xylose-assimilating *Saccharomyces cerevisiae* to ferment rice straw hydrolysate after membrane concentration, and ethanol yield of 5.34–6.44 g/L could be achieved after dilute acid pretreatment (Sasaki et al. 2013).

7.2.3 Simultaneous Saccharification and Fermentation

The general fermentation process of ethanol, butanol, and other biofuels can be divided into four steps: (1) physical, chemical, or biological pretreatment, (2) enzymatic hydrolysis, (3) microbial fermentation, (4) separation and concentration. Due to the different conditions of action of microorganisms and enzymes in the hydrolysis process and fermentation process, the traditional stepwise fermentation hydrolysis process and fermentation process are divided into two vessels of fermentation. The synchronous saccharification fermentation process synchronizes hydrolysis and fermentation process in one fermentation vessel, which occupies less space, shorter fermentation time, and higher yield compared with the traditional fermentation method. It is the most studied fermentation method at present.

Liu considered synchronous saccharification fermentation as a promising process for biotransformation of lignin biomass and studied the effect of synchronous saccharification fermentation of corn stover for ethanol production after steam blast pretreatment, with ethanol yield and final ethanol concentration of 77.2% and 59.8 g/L, respectively, under optimal reaction conditions (Liu et al. 2016). Du et al. investigated high temperature brewer's yeast using corn stover as a raw material for synchronous saccharification. The optimal conditions for ethanol production were obtained as 7.4% inoculum, 34.2 °C temperature, 5.0 initial pH, and 49.36 U/g enzyme concentration, and the ethanol yield was 59.88% at 150.12 h of fermentation under optimal fermentation conditions (Du et al. 2016). The effect of conditions on ethanol production by simultaneous saccharification fermentation of wheat straw was investigated and found that the concentration of ethanol reached a maximum of 38.32 g/L after 120 h of fermentation at 38 °C, 16.0% solids content, 35 FPU/g cellulase dosing, and

8 g/L yeast concentration, with a yield of 71.71% of the theoretical differential rate (Zhang et al. 2012). Saha et al. used wheat straw as raw material and The maximum ethanol yield was 36.0 g/L and 0.43 g/L at pH 6.0 and fermentation temperature 35 °C for 83 h (Saha et al. 2015). Lin et al. investigated the simultaneous saccharification fermentation of wheat straw at an initial pH of 4.6, enzyme addition of 30 FPU/g and temperature of 37.5 °C, and the maximum ethanol. The maximum ethanol yield could reach 70.76% at an initial pH of 4.6, enzyme addition of 30 FPU/g and temperature of 37.5 °C (Zhang et al. 2013). Synchronous saccharification fermentation can be applied not only to the production of alcohols from biomass straw but also to the production of biofuels from other wastes. Zhang et al. conducted research work on the production of ethanol from synchronous saccharification fermentation of kitchen waste. The results showed that the optimal reaction conditions were glycosylase addition concentration of 100 U/g, protease addition of 150 U/g, cellulase of 100 U/g, pH 5.3, and the concentration of ethanol could reach 54.6 g/L after 100 h of fermentation time (Zhang et al. 2015). Yan et al. investigated the induction of cellulase by cow manure and the possibility of converting cow manure feedstock into bioethanol by simultaneous saccharification and fermentation of ethanol yield of 25.65 g/L (Yan et al. 2018).

7.2.4 Pretreatment Techniques

Pretreatment technologies can be classified according to their nature as physical, chemical, combined physical and chemical methods, and biological methods. Lignocellulosic raw materials such as biomass straw, forestry waste, municipal waste, and other wastes such as livestock manure, municipal sludge, and pharmaceutical plant wastewater can be used to ferment biomass for fuel production due to their high organic matter content.

China is a largely agricultural country that produces a large amount of biomass straw every year, and the national production of biomass straw was 816 million t in 2016, of which corn straw was the largest, accounting for 36.88% of the total (Zhang et al. 2018). The use of biomass straw to produce biofuels such as ethanol and butanol has attracted increasing interest from researchers due to its wide source and high economic efficiency. The lignocellulosic feedstock has a complex structure, with hemicellulose and lignin intertwined and covering the wood surface to form a dense structure, making it impossible for cellulase to act directly with the cellulose. The component content of the main biomass straws is shown in Table 7.1 (Wang 2015). Direct fermentation yield is low, so pretreatment is performed to destroy the dense structure formed by cellulose, lignin, and hemicellulose. For lignocellulosic materials, the main physical pretreatment techniques are mechanical crushing, ultrasonic pretreatment, microwave method, high energy radiation, and liquid hydrothermal method. Chemical pretreatment technology mainly includes acid hydrolysis method, alkali treatment method, organic solvent method, wet oxidation method, etc. Physicochemical methods mainly include steam blasting method,

Table 7.1 Lignocellulose raw materials

Lignocellulose feedstock	Cellulose%	Hemicellulose%	Lignin%
Corn stalk	35–39.6	16.8–35	7–18.4
Wheat stalk	32.9–50	24–35.5	8.9–17.3
Rice straw	36.2–47	19–24.5	9.9–24

ammonia fiber blasting method, etc. The biological method mainly uses certain microorganisms for pretreatment.

Chen et al. investigated the effect of irradiation pretreatment of rice straw on the production of ethanol by enzymatic saccharification and fermentation, and the results showed that there was an increase in cellulose conversion and ethanol conversion after irradiation pretreatment compared to the control group (Chen et al. 2015). Sun et al. conducted a study on hydrothermal pretreatment of corn straw and found that 56.08% hemicellulose was leached after hydrothermal pretreatment, which showed that hydrothermal pretreatment destroyed the dense structure of lignocellulosic (Sun et al. 2019). Lian et al. determined the optimal reaction conditions of enzymatic digestion time of 9 h, fermentation time of 7 d, and fermentation temperature of 34 °C by pretreating corn stover with hydrogen peroxide to produce fuel ethanol using semi-synchronous saccharification fermentation. Under the optimal conditions, ethanol conversion reached 76.54% and ethanol concentration reached 23.64 g/L (Lian et al. 2018). Kim et al. explored the effect on ethanol production using nitric acid pretreatment of rice straw. The results showed that 0.65% nitric acid at 158.8 °C and 5.86 min reaction time enzymatic digestion was best with a digestibility rate of 83.0% and ethanol yield increased from 10.92 to 14.50 g/L (Kim et al. 2014). Amiri et al. conducted a study on the production of ethanol, butanol, and acetone by pretreatment of rice straw with organic solvents and found that pretreatment at 180 °C gave a glucose yield of 46.2%, after fermentation, 22.5 g of ethanol, 22.1 g of propanol, and 80.3 g of butanol could be obtained (Amiri et al. 2014). Kuamr et al. used green solvent (GS) pretreatment of rice straw as raw material to produce cellulosic ethanol, and the results showed that the saccharification efficiency after green solvent pretreatment was as high as 87.1%, and the maximum yield of reducing sugar was 226.7 g/L. After 36 h of fermentation, 36.7 g/L of ethanol could be obtained yield, with a conversion rate of 90.1% (Kumar et al. 2016). Arora considered biological pretreatment as a green pretreatment method and it used white-rot fungus, *Trametes hirsuta* pretreatment of rice straw and compared it with steam blast pretreatment to investigate the effect on the production of bioethanol. It was found that the lignin removal rate of the biological pretreatment was higher than that of the steam blasting pretreatment, and the ethanol yields after biological pretreatment and steam blasting pretreatment were 0.86 g/L and 1.13 g/L, respectively, which showed that the ethanol yield of the steam blasting pretreatment was better than that of the biological pretreatment (Arora et al. 2016). It is evident that lignocellulosic feedstocks, after pretreatment, are able to increase the ethanol yield and production rate.

Lignocellulosic material is considered an effective feedstock for bioethanol and butanol production, but other wastes also have some potential for biofuel production. The production of biofuels from other biomass wastes and the pretreatment methods have been studied by domestic and foreign scholars. Liu et al. used campus kitchen waste as raw material and fermented it with brewer's yeast after pretreatment using Rhizobium and Bacillus subtilis for ethanol production. It was found that the maximum yield of ethanol could reach 6.67% under optimal fermentation conditions with 22.79% substrate water content, 15% brewer's yeast inoculum, 30 °C, and 5 d fermentation time (Liu et al. 2013). Woldesenbet et al. used different concentrations of dilute sulfuric acid (0.2, 0.4, 0.6, 0.8, and 1.0 mol/L) to pretreat livestock manure to investigate livestock manure production The potential of bioethanol was investigated and it was shown that pretreatment of chicken manure with 0.8 mol/L released more reducing sugars with a maximum ethanol yield of 50 g/L (Woldesenbet et al. 2013). Ji conducted a study on the preparation of fuel ethanol from MSW and the ethanol yield per gram of dry kitchen waste was 0.255 mL/g (Ji 2014).

7.3 Types of Energy from Fermentation

In recent years, with the rapid growth of the global economy, the world's energy consumption has increased significantly. In the face of the increasing depletion of traditional fossil energy sources such as coal, oil, and natural gas, the increasingly serious environmental pollution, and the greenhouse effect caused by the massive emission of greenhouse gases, the energy issue has become one of the most important problems plaguing countries around the world. The development of new renewable energy sources such as ethanol, butanol, olefins, etc. has received increasing attention from all over the world Fig. 7.2.

7.3.1 Ethanol

Ethanol with molecular formula C_2H_6O, commonly known as alcohol, is a flammable, volatile, colorless, and transparent liquid at room temperature and pressure. Fuel ethanol generally refers to anhydrous ethanol with a volume fraction of 99.5% or more, which is a good octane blending component and gasoline oxygenating agent. Ethanol gasoline can effectively reduce the emission of PM2.5 and CO in automobile exhaust and can supplement fossil fuel resources, which has important research significance and value for reducing the foreign dependence on petroleum resources, reducing greenhouse gas emissions, and environmental pollution.

Since the 1970s, biofuel ethanol has been extensively researched around the world as a vehicle fuel, and ethanol is considered one of the most important renewable fuels of the future. After decades of development fuel ethanol is now globally recognized as the most mature alternative fuel to gasoline. Several countries around the world have

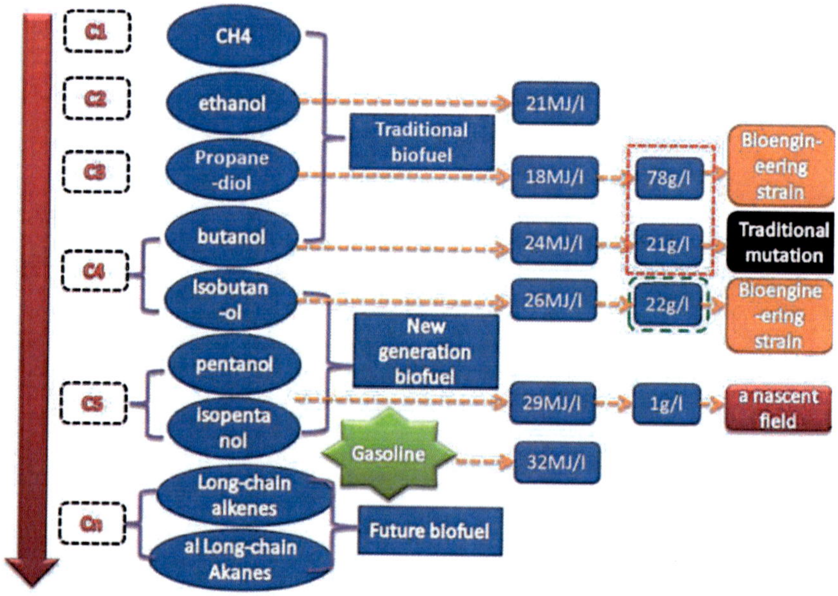

Fig. 7.2 With the utilization of biomass as the object, it is mainly engaged in the research of biomass energy production by biosynthesis, involving the biomass energy of C1–C6, including methane, butanol, Isobutanol of C5, 1-methyl-butanol, amyl alcohol, hexanol of C6, etc.

spared no effort to develop bioethanol and use ethanol as an additive or fuel alternative to petroleum, and have developed a series of supportive policies. The United States and Brazil have the largest fuel ethanol applications in the world, accounting for more than 10% and 50% of their gasoline fuel consumption, respectively. China's current fuel ethanol usage accounts for more than 2.1% of gasoline usage, with huge room for development. The development of renewable bio-liquid fuels has a very important role in reducing the dependence on fossil fuels, reducing greenhouse gas emissions, and activating the rural economy in China (Guo et al. 2016). The current first generation fuel ethanol, which uses food as the main raw material, has been gradually abandoned due to high cost and endangering land and food security. Second-generation fuel ethanol technologies based on non-food feedstocks such as waste, straw, and algae are being actively developed (Li et al. 2013).

Lignocellulose is the most abundant renewable resource on earth, and the production of bioethanol from lignocellulosic feedstocks has good prospects for development. The development of bioenergy from lignocellulosic feedstock by modern biotechnological means has become an important part of the energy development strategy of major countries in the world. Khaleghian et al. studied the production of bioethanol from rice straw by simultaneous saccharification and fermentation of the straw after pretreatment with sodium carbonate, and the results showed that the enzymatic hydrolysis yield was up to 100% and the ethanol yield was increased by more than 40% (Khaleghian et al. 2015). Phitsuwan et al. conducted a study on the production of ethanol from rice straw, which showed that rice straw is a promising biomass feedstock for ethanol production (Phitsuwan et al. 2016). Suriyachai et al. optimized the process of ethanol production from rice straw using Simultaneous saccharification and co-fermentation (SSCF) fermentation, and the optimized ethanol yield could reach a maximum of 15.2 g/L and ethanol concentration up to 28.6 g/L (Suriyachai et al. 2013). Wu et al. conducted a study on ethanol production from biomass straw treated with *Trichoderma reesei* Aq-5b and *Trichoderma viride* NSW-XM and found that pretreatment greatly improved the ethanol production efficiency and shortened the fermentation cycle, and the ethanol yield was as high as 2.17 g/(L h) after pretreatment (Wu et al. 2016).

Although the cellulose ethanol production technology is maturing and entering the industrial demonstration process, there is still room for improvement in pretreatment, enzyme preparation and enzymatic process, pentose/hexose co-fermentation strains and process, and equipment and equipment.

7.3.2 *Butanol*

Butanol is a colorless and transparent liquid with the molecular formula C_4H_9OH, has a special odor, is slightly soluble in water, and is miscible in any ratio with various organic solvents such as ethanol and ether. Although bioethanol is generally favored as a gasoline blending component, the development of its application is somewhat limited due to its low energy density, high vapor pressure, and corrosive transport pipelines. The main properties of ethanol and butanol are shown in Table 7.2 (Liu et al. 2008). Compared with bioethanol, biobutanol has the advantages of high energy density and fuel calorific value, miscibility with gasoline in any ratio, and low corrosiveness for pipeline transportation (Gao et al. 2018). After bioethanol, biobutanol has become a hot spot for research and development of a new generation of renewable energy.

Many countries around the world are vigorously developing biobutanol technology in the face of the energy and resource revolution. The U.S. biobutanol fuel project has been put into operation in 2009. The UK has already accounted for about 10% of the market share of biofuels in 2015. China's research on biobutanol has also made substantial progress. Wang conducted a study on the production of biobutanol from corn stover, using 2% NaOH, 1.5% H_2SO_4, steam blasting, and steam

Table 7.2 Main properties of ethanol and butanol

Fuel	Density (kg/L)	Heat of gasification (kJ/kg)	RON	CN	Molar calorific value (MJ/mol)
Ethanol	0.7893	854	110	8	1233.6
Butanol	0.8109	430	96	25	2601.9

blasting in combination with 2% NaOH and 1.5% H_2SO_4 to pretreat corn stover to investigate the effect of enzymatic hydrolysis and the effect of butanol yield. The results showed that 2% NaOH pretreatment was the most effective, with a high lignin removal rate of 81.7% and enzymatic hydrolysis rate of 70.5% for corn stover. After ABE fermentation, the butanol concentration could reach 212.0–232.0 g/L (Wang 2016). Tian conducted a study on the production of butanol by hydrolysis and fermentation of corn stover and studied the intention of citric acid-sodium citrate instead of traditional acetic acid-sodium acetate as a buffer for enzymatic hydrolysis of corn stover pretreatment, and the experiment proved that the concentration of summary reducing sugar in the hydrolysate was 14% higher, and through UV mutagenesis, a mutant strain CM20 was used to ferment the hydrolysate with a butanol yield of 10.8 g/L (Tian 2015). Boonsombuti et al. studied the effects of ionic liquids, acids, and bases on the production of butanol after pretreatment of rice straw, respectively, and the results showed that the alkali treatment was more effective and that isobutanol appeared in the NaOH pretreated fermentation broth, indicating that isobutanol can be produced under suitable conditions (Boonsombuti et al. 2020). Moradi et al. similarly studied the effect of NaOH and concentrated phosphoric acid pretreatment of rice straw for butanol production and the results showed that 163.5 and 44.2 g of butanol per kg of substrate was hydrolyzed after alkali pretreatment with glucose. After concentrated phosphoric acid pretreatment, 192.3 g of glucose, as well as 44.2 g of butanol per kg of substrate, was hydrolyzed (Moradi et al. 2013).

Lignocellulosic feedstock is a hot spot for research on the production of biobutanol, and other wastes such as kitchen waste have received some attention and research due to their high organic matter content as a potentially cheap and high-quality biomass resource. Zhang et al. conducted a study on the direct use of strain an amylolytic Clostridium sp. strain BOH3 for the production of butanol from kitchen waste without pretreatment. The strain BOH3 was found to have the gene encoding amylase and could produce 14.1 g/L of butanol from food waste (Zhang et al. 2020). The butanol production process is affected by various factors such as pH. Shi et al. conducted a study on the fermentation of kitchen waste by *Clostridium beijerinckii* NCIMB 8052 for the production of biobutanol and showed that the butanol yield was only 5.96 g/L without the addition of any nutrients, and the addition of pH buffer increased the butanol yield and Wang et al. investigated the feasibility of direct butanol fermentation from kitchen waste without saccharification and achieved a butanol yield of 12.1 g/L and a maximum production rate of 0.725 g/(L h) at a solid-to-liquid ratio of 1:1 (Wang et al. 2019).

7.4 Plants

The source and cost of biomass feedstock are one of the major constraints on the development of bioenergy, so many scientists are focusing on lignocellulosic feedstock, which is cheaper and more widely available. Han et al. determined the fiber composition and the potential for ethanol production in reed, kelp, and shore moss

3. Ethanol fermentation experiments were conducted on reed with high cellulose content and controlled with straw, and the ethanol content was 0.43–0.47% higher than that of straw at 0.29–0.31% under the same conditions of fermentation for 20 h (Han et al. 2019).

It is evident that reed, kelp, and shorea feedstocks have great potential for ethanol production. Li explored the feasibility of kelp for biofuel production using enzymatic saccharification technology and fermentation of ethanol and butanol, and the results showed that ethanol and butanol of 0.58% (V/V) by volume and 0.34% (V/V) by volume, respectively, could be obtained after fermentation (Li 2013). Zhou et al. investigated the feasibility of ethanol production by fructose-based fermentation of the energy plant inulin as the sole carbon source and showed that bioethanol production from inulin could be well converted using *Saccharomyces cerevisiae* strain BY47442, with no significant difference compared to ethanol production using glucose or fructose as substrates (Zhou et al. 2008). Yang likewise conducted a study on the production of ethanol by simultaneous saccharification fermentation of inulin and showed that ethanol fermentation using *Saccharomyces cerevisiae* strain Y05 resulted in ethanol concentrations up to 71.18 g/L (Yang 2010). Sornvoraweat et al. conducted a study on the production of bioethanol using water hyacinth as a raw material for isolated hydrolytic fermentation (SHF) and obtained an ethanol yield of 0.25 g/g at a concentration of 3.39 g/L under optimal fermentation conditions (Sornvoraweat et al. 2010).

7.4.1 Biodiversity of Floating Duckweed

Duckweed, a floating aquatic plant, consists of about 40 species in five genera, namely, *Spirodela, Landoltia, Lemna, Wolffiella, Wolffia,* among which Duckweed is the smallest flowering plant in the world. Duckweed is widely distributed in various freshwater environments, such as lakes, ponds, and rice paddies. The duckweed plant has a simple structure, which is composed of phylloids and pseudoroots (Spirodeka, Landoltia, Lemna) or only phylloids (Wolffiella, Wolffia). It is easy to culture and fast to grow asexually (the biomass doubles about 24 h) (Yang et al. 2021). Studies have shown that the starch content in duckweed can be controlled by controlling the growth conditions, such as pH of culture medium and phosphate concentration (Cui and Cheng 2015). Due to the high biomass accumulation, short reproduction cycle, and rich organic matter of duckweed, duckweed is a high-quality raw material for the development of biomass energy such as ethanol, butanol, and biogas Fig. 7.3.

7.4.2 Floating Ethanol

With the increasing scarcity of fossil energy and the increasing ecological load, countries around the world are successively taking alternative energy sources as an

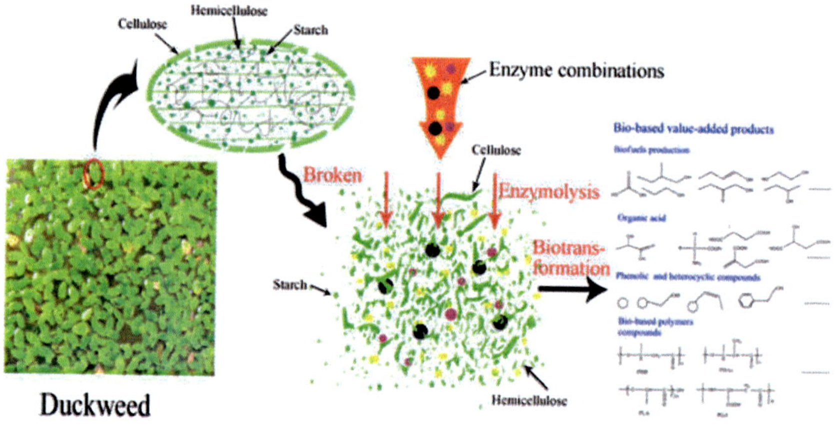

Fig. 7.3 Transformation process of duckweed in biomass energy field

important energy policy to achieve sustainable development. Lignocellulosic fermentation for ethanol production is considered as one of the most mature processes for biofuel production, but the dense structure of lignin and hemicellulose composition of lignocellulosic feedstock and pretreatment technology become one of the main factors limiting its development, leading to high costs.

Aquatic plants grow without taking up land, grow faster compared to terrestrial plants, and have a high content of starch as well as cellulose components that can be converted into fermentable sugars, making them potential feedstocks for raw fuel ethanol (Xue et al. 2013). Aquatic puffballs are fast-growing plants, rich in starch and cellulose, and have become a research hotspot for the fermentation of bioethanol production due to their excellent biochemical properties. Xu et al. concluded that floating duckweed has a starch content of 31–45.8% (dry weight) and after fermentation, can convert up to 94.7% of starch to ethanol, which is much higher than the ethanol yield of most other potential crops (Xu et al. 2012). Lee et al. conducted a comparative study of ethanol production from floating duckweed and corn starch and showed that floating duckweed plants have a lower ethanol yield than corn starch, but a higher ethanol conversion rate (Lee et al. 2016). Xu et al. conducted a study on the conversion of high amylose floating duckweed to bioethanol and found that after enzymatic fermentation by yeast, the ethanol conversion was 94.7% of the theoretical yield, which was higher than about 50% of the ethanol produced with corn as substrate (Xu et al. 2011). Calicioglu and Brennan conducted a study on continuous fermentation of floating duckweed for ethanol and methane production and found that the combined bioethanol-biomethane fermentation process obtained 70.4% more bioenergy from floating duckweed compared to fermentation alone (Calicioglu and Brennan 2018).

As floating duckweed contains lignocellulose, pretreatment is required to destroy the structure of lignocellulose, increase the accessibility of enzymes, and improve the yield and efficiency of ethanol. Zhao et al. pretreated floating duckweed by steam blasting and later produced ethanol using simultaneous saccharification fermentation technique, and found that the ethanol yield was 70% higher than that of the unpretreated control (Zhao et al. 2015). Fontinelle Souto conducted a hydrothermal pretreatment followed by simultaneous saccharification fermentation after hydrothermal pretreatment and found that ethanol yield reached 88.81% of the theoretical yield under optimal conditions pretreatment conditions (200 °C, 10 min) (Fontinelle Souto et al. 2019). Chen et al. investigated the effect of pectinase pretreatment of floating duckweed on bioethanol yield and found that the maximum glucose yield after pretreatment was 218.64 ± 3.10 mg/g and ethanol concentration was 30.8 ± 0.8 g/L with a yield of 2.20 g/(L h) (Chen et al. 2012).

7.4.3 Duckweed Butanol

Butanol is a clean and efficient new energy source. The production of butanol by biological fermentation is a research hotspot in recent years. At present, the key factor restricting the development of biobutanol industry is its economy, so it is very important to choose raw materials with wide sources and good economy easily. Duckweed is a non-food aquatic plant with high organic matter content and its source has been widely concerned.

Long carried out a study on high-starch duckweed butanol fermentation and found that the duckweed fermentation with an initial sugar concentration of 60 g/L could yield 11.65 g/L butanol, which was 99.57% of corn butanol fermentation yield and 98.9% of cassava butanol fermentation yield (Long, 2012). It can be seen that duckweed is an effective substitute for food crops to produce butanol by fermentation of aquatic plants.

7.4.4 Duckweed Advanced Alcohol

Facing the energy crisis, it is necessary to develop new alternative energy sources. Ethanol is considered an ideal biofuel to replace petroleum, but bioethanol has some shortcomings such as low calorific value and certain corrosivity. Ethanol renewable alcohol obtained by biological fermentation can be used as an effective substitute for ethanol. Su et al. conducted a study on the production of biofuel advanced alcohols by duckweed fermentation Fig. 7.4 and found that 12.03 g/L butanol could be produced by *C. acetobutylicum* CPCC 8012. In addition, 16.24 mg/L butanol, 4.68 mg/L isoamyl alcohol, and 198.85 mg/L amyl alcohol can be produced by fermentation of *Escherichia coli* bioengineered strains (Su et al. 2014).

SC: Saccharomyces cerevisiae AH109; EC: Escherichia coli; CA: Clostridium acetobutylicum CICC 8012. PFM: fermentation substrate (duckweed) for acid hydrolysate was filtered with a 0.22-micron membrane. NFM: the fermentation substrate without filtration. EP: Pretreatment with enzymatic hydrolysis. AP:

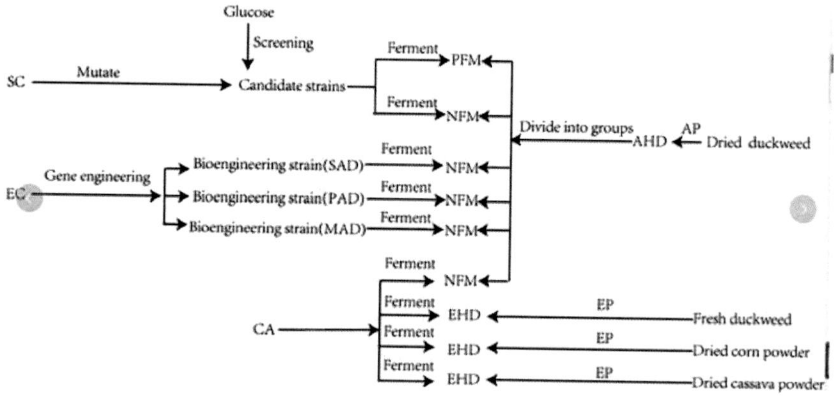

Fig. 7.4 Flowsheet of the experimental design for fermentation processes

Pretreatment with acid hydrolysis. AHD: acid hydrolysate of duckweed as the fermentation substrate, meaning the products resulting from pretreatment of duckweed with acid hydrolysis. EHD: enzymatic hydrolysate as the fermentation substrate, meaning the products resulting from pretreatment of duckweed with enzymatic hydrolysis.

7.5 Conclusions

There are various technical methods for fermentation to produce energy, and simultaneous saccharification fermentation technology is the most applied fermentation technology, and the screening of efficient strains of homosaccharide fermentation and the optimization of the process is the focus of future research. The fermentation of biomass straw to produce biomass energy is an effective way to alleviate the depletion of fossil energy and realize the resource utilization of waste, but because biomass straw is rich in lignin and cellulose, pretreatment is required for fermentation. Therefore, research on new efficient pretreatment means and development of new fermentation processes are hot spots for future research.

With its rich biodiversity and fast growth rate rich in starch and cellulose, duckweed is a good potential feedstock for biofuel production. The application of biotechnology and a deeper understanding of the microscopic fermentation process will bring a technological revolution in the production of biofuels from floating duckweed. In addition, the development of an efficient and clean production process of floating pondweed as well as pretreatment technology is a hot spot for future research on floating pondweed produced fuels.

References

Amiri H, Karimi K, Zilouei H (2014) Organosolv pretreatment of rice straw for efficient acetone, butanol, and ethanol production. Biores Technol 152:450–456

Arora A, Priya S, Sharma P, Sharma S, Nain L (2016) Evaluating biological pretreatment as a feasible methodology for ethanol production from paddy straw. Biocatal Agric Biotechnol 8:66–72

Boonsombuti A, Trisinsub O, Luengnaruemitchai A (2020) Comparative study of three chemical pretreatments and their effects on the structural changes of rice straw and butanol production. Waste Biomass Valorization 11:2771–2781

Calicioglu O, Brennan RA (2018) Sequential ethanol fermentation and anaerobic digestion increases bioenergy yields from duckweed. Biores Technol 257:344–348

Chen H, Qiu W (2007) The crucial problems and recent advance on producing fuel alcohol by fermentation of straw. Prog Chem Z2:1116–1121 (in Chinese)

Chen L, Su X, Xiong X, Chen J, Wu X, Hu Q, Zhang Y, Wang K (2015) Enzymatic saccharification and fermentation of irradiated rice straw for ethanol production. Acta Nucifera Sin 29(04):704–709 (in Chinese)

Chen Q, Jin Y, Zhang G, Fang Y, Xiao Y, Zhao H (2012) Improving production of bioethanol from duckweed (*Landoltia punctata*) by pectinase pretreatment. Energies 5:3019–3032

Cui W, Cheng JJ (2015) Growing duckweed for biofuel production: a review. Plant Biol 17:16–23

Du J, Bu W, Hu X, Wang Y (2016) Experimental study on ethanol production by simultaneous saccharization of corn straw. Acta Energiae Solaris Sin 37(09):2251–2256 (in Chinese)

Fontinelle Souto LR, da Silva IF, Ninow JL, Collins SRA, Elliston A, Waldron KW (2019) Effect of hydrothermal pre-treatment on duckweed (*Landoltia punctata*) biomass for simultaneous saccharification and fermentation process. Biomass Bioenergy 127

Gao Y, Guo X, Yang Y, Zhang M, Li W, Lu D (2018) Research progress of biobutanol fermentation. Biotechnol Bull 34(08):27–34 (in Chinese)

Guo X, Luo H, Deng L (2016) Progress of global fuel ethanol industry. Contem Chem Ind 45(09):2244–2248 (in Chinese)

Han W, Ming K, Ma Y, Lv J, Zhang B (2019) Study on the possibility of several coastal plants as biomass materials of producing ethanol. Anhui Agric Sci 47(05):77–79 (in Chinese)

Huang D, Wu Q, Lu J, Guan G (2003) Current research progress of solid state fermentation technology and equipment. Food Ferment Ind 06:87–91(in Chinese)

Ji L (2014) Study on fuel ethanol production from municiple solid wastes. Beijing Univ Chem Technol (in Chinese)

Khaleghian H, Karimi K, Behzad T (2015) Ethanol production from rice straw by sodium carbonate pretreatment and *Mucor hiemalis* fermentation. Ind Crops Prod 76:1079–1085

Kim I, Lee B, Park JY, Choi S-A, Han J-I (2014) Effect of nitric acid on pretreatment and fermentation for enhancing ethanol production of rice straw. Carbohyd Polym 99:563–567

Kumar AK, Parikh BS, Shah E, Liu LZ, Cotta MA (2016) Cellulosic ethanol production from green solvent-pretreated rice straw. Biocatal Agric Biotechnol 7:14–23

Lee CJ, Yangcheng H, Cheng JJ, Jane JL (2016) Starch characterization and ethanol production of duckweed and corn kernel (vol 67, pp 348, 2015). Starch-Starke 68:581–581

Li F (2013) The explored research of fuel ethanol and butanol production from *Laminaria japonica* fermentation. Guangxi Univ Sci Technol (in Chinese)

Li L, Yang X, Xue Y (2011) Research progress of modern solid state fermentation technology, equipment and application. J Henan Univ Technol (Nat Sci Edn) 32(01):89–94 (in Chinese)

Li Z, Li D, Huang G, Wei H (2013) Insights on current development of fuel ethanol. Prog Chem Indus 32(07):1457–1467 (in Chinese)

Lian Z, LV Z, Liu B, Liu X, Li J, Tian B, Zhuang Q, Liu T (2018) Study on alkaline pretreatment and semi synchronous saccharification and fermentation of high concentration corn straw to produce ethanol. Food Ferment Indus 44(10):124–129 (in Chinese)

Liu A, Xu S, Cai X, Peng P, Lu C (2013) Fuel ethanol production from food waste by being mixed microbical fermentation. Chin J Environ Eng 7(02):727–731 (in Chinese)

Liu Y, Liu H, Zhang J, Cheng K, Chen Z (2008) Research progress in new biofuel butanol. Mod Chem Indus 06:28–31 + 33 (in Chinese)
Liu Z, Chen H (2016) Simultaneous saccharification and co-fermentation for improving the xylose utilization of team exploded corn stover at high solid loading. Bioresour Technol 201:15–26
Liu ZH, Qin L, Zhu JQ et al (2016) Simultaneous saccharification and fermentation of steam-exploded corn stover at high glucan loading and high temperature. Biotechnol Biofuels 7
Long F (2012) Study on butanol conversion technology of Canna edulis and duckweed. Univ Chin Acad Sci (in Chinese)
Moradi F, Amiri H, Soleimanian-Zad S, Ehsani MR, Karimi K (2013) Improvement of acetone, butanol and ethanol production from rice straw by acid and alkaline pretreatments. Fuel 112:8–13
Phitsuwan P, Permsriburasuk C, Waeonukul R, Pason P, Tachaapaikoon C, Ratanakhanokchai K (2016) Evaluation of fuel ethanol production from aqueous ammonia-treated rice straw via simultaneous saccharification and fermentation. Biomass Bioenerg 93:150–157
Saha BC, Nichols NN, Qureshi N, Kennedy GJ, Iten LB, Cotta MA (2015) Pilot scale conversion of wheat straw to ethanol via simultaneous saccharification and fermentation. Biores Technol 175:17–22
Sasaki K, Sasaki D, Sakihama Y, Teramura H, Yamada R, Hasunuma T, Ogino C, Kondo A (2013) Ethanol fermentation by xylose-assimilating *Saccharomyces cerevisiae* using sugars in a rice straw liquid hydrolysate concentrated by nanofiltration. Biores Technol 147:84–88
Si Y, Xia L (2010) 2, 3-butanediol was produced by fermentation of corn straw hydrolysate. Food Ferment Indus 36(02):26–29 (in Chinese)
Soccol CR, da Costa ESF, Letti LAJ, Karp SG, Woiciechowski AL, de Vandenberghe LPS (2017) Recent developments and innovations in solid state fermentation. Biotechnol Res Innov 1(1):52–71
Song A, Liu Y, Dong Y, Du F (2008) Optimization of fiber fuel system for corn straw saccharification liquid fermentation. Food Ferment Ind 34(12):65–68 (in Chinese)
Sornvoraweat B, Kongkiattikajorn J (2010) Separated hydrolysis and fermentation of water hyacinth leaves for ethanol production. Asia-Pacific J Sci Technol 15(9):794–802
Sornvoraweat B, Kongkiattikajorn J, Kasetsart U (2010) Development of separate hydrolysis and fermentation of water hyacinth for ethanol production, 48th Kasetsart University Annual Conference, Thailand, pp 268–275
Su H, Zhao Y, Jiang J, Lu Q, Li Q, Luo Y, Zhao H, Wang M (2014) Use of duckweed (*Landoltia punctata*) as a fermentation substrate for the production of higher alcohols as biofuels. Energy Fuels 28:3206–3216
Sun M, Wang G, Han X, Zhang C (2019) Effect of hydrothermal pretreatment on characteristics of corn straw hydrolysate. Acta Solar Energy Sin 40(06):1656–1663 (in Chinese)
Suriyachai N, Weerasaia K, Laosiripojana N, Champreda V, Unrean P (2013) Optimized simultaneous saccharification and co-fermentation of rice straw for ethanol production by *Saccharomyces cerevisiae* and *Scheffersomyces stipitis* co-culture using design of experiments. Biores Technol 142:171–178
Thomas L, Larroche C, Pandey A (2013) Current developments in solid-state fermentation. Biochem Eng J 146–161
Tian L (2015) Study on butanol fermentation using corn straw hydrolysate as substrate. Qingdao University (in Chinese)
Wang L (2015) Techno-economical ananlysis of cellulosic ethanol production process by different pretreatments. Dalian University of technology (in Chinese)
Wang L, Zhao Q, Chen H (2018) Research progress in solid state fermentation of bio-energy. Biotechnology 03:41–44 (in Chinese)
Wang S (2016) Biobutanol production from corn stover and product separation. Dalian Univ Technol (in Chinese)
Wang Y, Gao M, Bodjui Olivier A, Wang F, Wang Q (2019) Bio-Butanol production from direct fermenation of kitchen wastes. Environ Eng 37(04):137–142 (in Chinese)

Wen X, Bai G, Li T, Ma Y, Zou W (2014) Optimization of in situ enzymatic hydrolysis for saccharification of rice straw by liquid fermentation. Food Ferment Ind 47(04):166–172+181 (in Chinese)

Woldesenbet AG, Gizachew S, Bhagwan SC (2013) Bio-ethanol production from poultry manure at Bonga Poultry Farm in Ethiopia. Afr J Environ Sci Technol 7(6):435–440

Wu X, Zhang J, Xu E, Liu Y, Cheng Y, Addy M, Zhou W, Griffith R, Chen P, Ruan R (2016) Microbial hydrolysis and fermentation of rice straw for ethanol production. Fuel 180:679–686

Xu F, Chen H, Li Z (2002) New developments in engineering aspects of solid-state fermentation. Prog Biotechnol 01:44–48 (in Chinese)

Xu J, Cui W, Cheng JJ, Stomp A-M (2011) Production of high-starch duckweed and its conversion to bioethanol. Biosys Eng 110:67–72

Xu J, Zhao H, Stomp A-M, Cheng J (2012) The production of duckweed as a source of biofuels. Biofuels 3(5):589–601

Xue H, Dong Z, Fang Y, Jin Y, Zhao H (2013) Producing fuel ethanol from energy hygrophyte duckweed. Renew Energy 31(07):55–59 (in Chinese)

Yan Q, Liu X, Wang Y, Li H, Li Z, Zhou L, Qu Y, Li Z, Bao X Cow manure as a lignocellulosic substrate for fungal cellulase expression and bioethanol production. Amb Express

Yang J, Zhao X, Li G, Hu S, Chen Y, Sun Z, Hou H (2021) Research and application in duckweeds: a review. Sci Bull 66(09):1026–1045 (in Chinese)

Yang L (2010) (in Chinese) Study on simultaneous saccharification and fermentation of Jerusalem artichoke to produce fuel ethanol. Hebei Univ Sci Technol

Zhang C, Li T, Su G, HE J (2020) Enhanced direct fermentation from food waste to butanol and hydrogen by an amylolytic Clostridium. Renew Energy 153:522–529

Zhang W, Li W, Zhao J, Lin Y, Peng Z, Wang X (2012) Fuel ethanol was prepared by simultaneous saccharification and fermentation of wheat straw. Food Ferment Ind 38(12):50–54 (in Chinese)

Zhang W, Lin Y, Zhang Q, Wang X, Wu D, Kong H (2013) Optimisation of simultaneous saccharification and fermentation of wheat straw for ethanol production. Fuel 112:331–337

Zhang Y, Chen X, Gu Y, Zhou X (2015) A physicochemical method for increasing methane production from rice straw: extrusion combined with alkali pretreatment. Appl Energy 160:39–48

Zhang Y, Guo D, Wang K, Wu X, Qi H, Deng M, Chen L (2018) Fermentation for ethanol production with γ ray irradiation-pretreated corn straw. Brewed China 37(09):58–61 (in Chinese)

Zhao X, Moates GK, Elliston A, Wilson DR, Coleman MJ, Waldron KW (2015) Simultaneous saccharification and fermentation of steam exploded duckweed: improvement of the ethanol yield by increasing yeast titre. Biores Technol 194:263–269

Zhou Z, Cao H, Zhu Y, Li S, Bai X, Zhao X, Du Y (2008) The primary study of ethanol production by fermenting of jerusalem artichoke instead of corn. J Northwest Agric 04:297–301 + 305 (in Chinese)

Chapter 8
Esterification/Transesterification of Lipidic Wastes for Biodiesel Production

Mohammadhosein Rahimi, Mina Tajmirriahi, Fateme Saadatinavaz, and Su Shiung Lam

Abstract Biodiesel, a promising alternative fuel to diesel, has recently attracted worldwide attention due to its low toxicity, biodegradability, and eco-friendly nature. This fuel can solely be used in diesel engines or blended with diesel without complicated modifications to the engine. Generally, biodiesel is produced from edible and non-edible oil crops, lipidic wastes, and algae. Previous reports indicate that more than 75% of the overall cost of biodiesel production is related to feedstock. In this regard, feedstock selection for biodiesel production is of significant importance. Therefore, this chapter aims to provide an overview of biodiesel production through esterification and/or transesterification from diverse lipid-rich resources, especially the low-cost wastes. The potential of biodiesel production from various feedstocks and the impact of fatty acids profiles on biodiesel quality are discussed. The effect of different process conditions such as pretreatment and downstream processing for enhanced production of biodiesel is reviewed. Finally, the economic feasibility of biodiesel production from wastes is estimated and discussed.

Keywords Biodiesel · Esterification · Fatty acids composition · Lipidic wastes · Transesterification

M. Rahimi (✉) · M. Tajmirriahi · F. Saadatinavaz
Department of Chemical Engineering, Isfahan University of Technology, 84156-83111 Isfahan, Iran
e-mail: mh.rahimi@ce.iut.ac.ir

M. Tajmirriahi
e-mail: m.tajmirriahi@ce.iut.ac.ir

F. Saadatinavaz
e-mail: f.saadatinavaz@ce.iut.ac.ir

S. S. Lam
Pyrolysis Technology Research Group, Higher Institution Centre of Excellence (HICoE), Institute of Tropical Aquaculture and Fisheries (AKUATROP), Universiti Malaysia Terengganu, 21030 Kuala Nerus, Terengganu, Malaysia
e-mail: lam@umt.edu.my

A. E.-F. Abomohra et al. (eds.), *Waste-to-Energy*,
https://doi.org/10.1007/978-3-030-91570-4_8

8.1 Introduction

In today's world, with the rapid growth of population, renewable and sustainable fuel utilization represents a suitable strategy for reducing fossil fuel overconsumption. The global energy demand has risen dramatically during the recent decades due to industrialization and urbanization, and fossil fuels have been considered the primary energy source to meet these needs (Abomohra et al. 2020a). As a consequence of the environmental concerns, limited resources of conventional fossil fuels, and fluctuations in their price, scientists have investigated the possibility of replacing fossil fuels with renewable energies (Mahlia et al. 2020). Besides, the existence of fossil fuel resources in limited parts of the world has intensified the nation's interest in biofuels (Bhuiya et al. 2016a). In this regard, biofuels utilization is increasing, aiming to serve as an alternative to fossil fuels in the future (Sitepu et al. 2020), providing beneficial features such as being easily available, economically feasible, and less environmentally hazardous.

Due to the high energy conversion, diesel fuels are extensively used by heavy trucks, buses, and electric generators. Despite its wide range of applications, diesel exhaust emission and particulate matter were found to cause different diseases such as genotoxicity, carcinogenicity, respiratory infections, adverse pulmonary, and cardiovascular health effects (Mohd Ali et al. 2020). Besides, nitrogen oxide (NO_x) emissions of diesel and biodiesel are the reason for different unfavorable impacts on human health, such as headaches, eye irritation, and breathing difficulties (Mohd Ali et al. 2020). In this regard, biodiesel is suggested to be advantageous over conventional diesel with respect to lower emission of particulate matter and NO_x emissions than diesel. Besides, biodiesel is a renewable, sustainable, technically feasible, economically competitive, non-toxic, non-flammable, and biodegradable fuel. It is defined as fatty acid methyl esters (FAMEs) derived from the free fatty acids (FFAs) and triacylglycerols (TAGs) of edible/non-edible oils and animal fats through the esterification and/or transesterification process, either in the presence or absence of a catalyst (Bhuiya et al. 2016a). According to the American standard ASTM D6751, biodiesel is "a fuel comprising monoalkyl esters of long-chain fatty acids (FAs) derived from vegetable oils or animal fats, designated as B100" (Andreo-Martínez et al. 2020). Biodiesel is a sulfur-free fuel composed of about 10% oxygen and trace amounts of nitrogen (Mahlia et al. 2020). Compared to conventional diesel, the net CO_2 emissions could be reduced by 78% (Andreo-Martínez et al. 2020). The reduction in the emissions of polycyclic aromatic hydrocarbons and unburned hydrocarbons occurs up to 90% when biodiesel is combusted (Mahlia et al. 2020). Biodiesel can be blended with diesel at different blending ratios. Blends with up to 20% biodiesel are commonly adopted and reported to be beneficial in terms of performance and cost (Jeon and Park 2018; Wong et al. 2019). Further, they could be used in internal combustion engines without engine modifications, based on the engine manufacturer's guidelines (Wong et al. 2019).

Accordingly, in this chapter, an overview of biodiesel production through various lipidic wastes is provided. In this regard, a summary of different feedstocks for the production of biodiesel is first presented. Afterward, the processes of esterification and/or transesterification of the lipid-rich feedstocks are summarized. In addition, the quality and characteristics of biodiesel derived from different feedstocks are presented as well as the effect of FAs of various resources on the quality of biodiesel. Finally, the methods of biodiesel production enhancement and the economic feasibility of its production are highlighted.

8.2 Different Feedstocks for Biodiesel Production

The most critical problems that concern civilizations are hunger, worldwide energy shortage, and environmental degradation. It seems that the sustainable production of biofuels, especially biodiesel, can fight these issues and somehow mitigate the overall energy needs. Some factors that affect the availability of biodiesel feedstock are regional climate, geographical location, field conditions, and farming methods (Arshad et al. 2018). Since feedstock value stands for the major portion of biodiesel production expenses, about 70–95%, many studies have been carried out to develop different feedstocks for biodiesel production (Bhuiya et al. 2016a). These attempts aimed to increase the economic justification of biodiesel and improve its sustainable supply. A broad range of feedstocks is available for the sustainable production of biodiesel. According to the feedstock type, biodiesel is categorized into three distinct generations (Arshad et al. 2018).

First generation biodiesel is produced using edible plant oils, such as soybean, corn, coconut, and olive. Second generation biodiesel is made from non-edible plant oils, waste cooking oils, and animal waste fats. Third generation biodiesel is obtained from algal oils such as *Chlorella sp.*, *Isochrysis sp.*, and *Nannochloris sp.* (Arshad et al. 2018). These feedstocks have specific advantages and disadvantages, as presented and discussed in the following:

8.2.1 *Edible Plant Oils*

Edible plant oils include soybean, corn, palm, rapeseed, mustard, coconut, sunflower, canola, and olive. These oils are currently the commercial feedstock for biodiesel production (Pinzi and Pilar 2012). Edible oils are a renewable and biodegradable source of biodiesel production. There are many controversies about exploiting these feedstocks considering that the use of these feedstocks can increase food prices worldwide. These materials are costly that can in turn influence the economic viability of biodiesel (Pinzi and Pilar 2012). Besides, their cultivation needs fertile lands, and the allocation of these lands for vegetable oil crops can contribute to deforestation. These plants are also inflexible to climate conditions (Pikula et al. 2020).

8.2.2 Non-edible Plant Oils

Non-edible plant oils include Jatropha, jojoba, castor, linseed, and cottonseed. The oil obtained from these oilseeds contains high FFAs content and toxic materials, so it is not suitable for human or animal consumption. Therefore, they have no conflicts with food sources (Adewale et al. 2015). Using these materials can decrease the consumption of edible plant oils for biodiesel production (Arshad et al. 2018). These plants are globally available at a reasonable price and can grow in contaminated areas with low fertility. They have good adaptability to different climatic factors and are not vulnerable to disease. They are also potent to produce valuable by-products. Moreover, their cultivation promotes the reforestation and restoration of degraded lands (Pikula et al. 2020). However, they have a lower oil yield compared to edible plant oils. Hence, extensive land requirements with low productivity keep these feedstocks far from sustainable replacement (Pikula et al. 2020). Although they do not directly compete with food sources, the competition of edible and non-edible plants for farmlands could become a contentious issue (Adewale et al. 2015). They also require intricate downstream process technologies (Ewunie et al. 2021).

8.2.3 Waste Cooking Oils

Large amounts of these wastes are produced in restaurants, food processing factories, and households (Bhuiya et al. 2016a). According to the reports, 15 million tons of these wastes are generated annually all around the world. Hence, vast volumes of these materials drain into rivers and landfills that pollute the environment. Only one liter of these wastes is capable of contaminating 1000 L of water. Therefore, conversion of these wastes for biodiesel production is a helpful strategy to diminish environmental pollution (Pinzi and Pilar 2012; Casallas et al. 2018).

Waste cooking oil has no competition with food security and land use. In addition, they are far cheaper than plant oils. Also, under optimum operating conditions, the quality of biodiesel produced from these wastes and plant oils is similar (Adewale et al. 2015). However, impurities such as FFAs and water can result in saponification and hydrolysis, respectively (Bhuiya et al. 2016a). Besides, the lack of a centralized collection system is a severe drawback for the large-scale production of biodiesel (Pikula et al. 2020).

8.2.4 Fat, Oil, and Grease (FOG)

FOG is a by-product from food industries and restaurants, which is rich in lipid. FOG is composed of TAGs, esters, phospholipids, sterols, and high proportions of FFAs. The FOG disposal is a serious challenge since it clogs the pipes and blocks the sewer. There are several routes for FOG management, including incineration, recycling, landfill, and biofuel (biogas and biodiesel) production. FOG is an inexpensive feedstock for biodiesel production. However, the high content of FFAs inhibits the transesterification process. Thus, a pretreatment step is required to remove FFAs or convert them to TAGs or FAMEs (Abomohra et al. 2020b). According to the FFAs content, FOG is classified into yellow grease and brown grease. Yellow grease contains less than 15% of FFAs, while brown grease contains water and more than 15% of FFAs. A high concentration of FFAs negatively affects biodiesel yield. Therefore, yellow grease is a popular feedstock for biodiesel production (Abomohra et al. 2020b).

8.2.5 Dairy Wastes

The dairy industry is related to producing various products such as butter, cheese, cream, and milk. During the production process, various wastes such as whey, dairy sludges, and wastewater are generated that are hazardous for the environment. Therefore, its disposal to the environment is not recommended, and a suitable method should be performed to reduce the unfavorable contents (e.g., organic compounds, fats, sulfates, etc.) (Ahmad et al. 2019). One suitable method is direct or indirect biodiesel production using dairy wastes. In the direct method, the oil of dairy wastes should be separated and subjected to transesterification. In contrast, the indirect method uses the oil content of microorganisms cultivated on dairy wastes, especially dairy wastewater. For instance, direct utilization of dairy waste scum for biodiesel production using eggshell wastes as catalyst was investigated previously (Kavitha et al. 2019). In this study, a maximum biodiesel yield of 96% was obtained. In another investigation, the microalga *Acutodesmus dimorphus* was cultivated on raw dairy wastewater, resulted in considerable enhancement of biomass production in which the biomass contained 25% of lipids (Chokshi et al. 2016). The lipids were then converted to biodiesel (78% conversion).

8.2.6 Animal Waste Fats

Animal fat is a by-product of meat processing in slaughterhouses and fish factories, such as tallow, chicken fat, pork lard, mutton fat, leather industry solid waste, and fish oil. This fat comprises TAGs, water, protein, and various minerals. Some of these animal fats are currently utilized on an industrial scale (Adewale et al. 2015). This fat has low levels of FFAs and water content. The produced biodiesel from this substance has a high cetane number and the conversion of these fats to biodiesel is inexpensive (Pikula et al. 2020). The NO_x emission of produced biodiesel is lower or equal compared to plant oils. The produced biodiesel has good oxidative stability because of the presence of saturated FAs (Adewale et al. 2015). However, the produced biodiesel has a low pour point, high viscosity, and high flash point. Besides, the absence of an efficient collection system hinders the large-scale production of biodiesel (Pikula et al. 2020).

8.2.7 Algal Oils

Microalgal oil is an attractive lipid source for biodiesel production. It has the potential to fulfill the global need for transport fuels (Bhuiya et al. 2016a). It is easy to cultivate algae, as it does not require fertile lands and can grow in saltwater, sewage, or marginal lands. So, it does not compete with agricultural lands (Bhuiya et al. 2016a). Algae has a fast growth rate with high lipid content compared to edible and non-edible substrates. Its growth rate can increase via the addition of specific nutrients and optimum aeration (Pikula et al. 2020). Microalgae are cultivated in open ponds or closed (photobioreactors) systems, and they can grow using sunlight and CO_2 as their energy and carbon source. Using microalgal oil bestows environmental benefits, including consumption and elimination of CO_2 content of industrial flue gases, removal of nutrients constituents of wastewaters (i.e., nitrogen, phosphorus, and carbon), utilization of biomass residual after oil extraction for methane or ethanol production, or as fertilizers (Pikula et al. 2020). However, algal oil contains many unsaturated lipids that result in low stability of produced biodiesel. Algae cultivation demands a high amount of nutrients (nitrogen and phosphorus) and requires enormous investment. Besides, there are some challenges for scaling up the cultivation and oil extraction process. Some specific algae species are capable of generating harmful toxins (Singh et al. 2020).

8.2.8 Potential of Different Feedstocks for Biodiesel Production

Life cycle assessment (LCA) is a useful strategy for analyzing the potential environmental impacts of a process and its products. For comparison and evaluation of different feedstocks for biodiesel production using the LCA method, a few parameters should be considered, including the availability of farm and water, fertility of the soil, cultivation practice, fertilizers and pesticides requirement, the cost of feedstock, its transportation and storage, its greenhouse gases emissions, and possible impacts on air quality (Bhuiya et al. 2016a). To date, several LCA studies have been done on feedstocks producing biodiesel of different generations.

Guo and Hanaki studied the LCA of biodiesel production from four energy crops, namely *Jatropha*, *Pistacia*, *Swida*, and *Xanthoceras* (Guo and Hanaki 2010). They reported that by replacing 1 t of fossil fuels with biodiesel, the CO_2 emission decreased from 2.08 to 1.13 t, equivalent to a 54% reduction. They also declared that the overall cost of biodiesel production could compete with fossil fuel in terms of price (Guo and Hanaki 2010). Chatterjee et al. examined the LCA of biodiesel production from Jatropha (Chatterjee et al. 2012). They introduced some ways to improve the LCA of this process, including employing novel oil extraction technologies to improve the oil output, production of value-added by-products to enhance the economy of the process, and the development of machinery for harvesting and dehulling of seeds to decrease the production cost (Chatterjee et al. 2012).

Chung et al. had a comparative study on the LCA of biodiesel production from Jatropha oil and waste cooking oil, using KOH and CaO as catalysts (Chung et al. 2019). They claimed that biodiesel from waste cooking oil has fewer environmental impacts since it does not need the cultivation or fertilizing stage. In biodiesel production from Jatropha oil, KOH was used as a homogeneous catalyst with many ecological impacts (Chung et al. 2019). The KOH production requires chemical substances and additives, and the biodiesel production stage needs a complicated process of purification and neutralization. The biodiesel produced from waste cooking oil using waste chicken eggshell-derived CaO as a heterogeneous catalyst was claimed to have fewer environmental effects (Chung et al. 2019).

Saranya and Ramachandra studied six distinct scenarios for biodiesel production from estuarine microalgae to analyze the LCA (Saranya and Ramachandra 2020). They used three different nutrient input methods (no nutrient, wastewater as nutrient, and fertilizer as nutrient) and two microalgal oil conversion methods (acid catalyst and biocatalyst). They reported that the wastewater-biocatalyst scenario is the best because it has the least greenhouse gas footprint, the highest net energy ratio, and inexpensive wastewater remediation (Saranya and Ramachandra 2020). Dufour and Iribarren performed a comparative study on biodiesel production from first generation (soybean and rapeseed) and second generation (waste cooking oil, beef tallow, and poultry fat) feedstocks (Dufour and Iribarren 2012). They reported that among the mentioned feedstocks, waste cooking oils are beneficial since it has negligible environmental impacts (Dufour and Iribarren 2012).

Siqueira et al. compared different scenarios for producing third generation biodiesel from *Phormidium autumnale* to optimize the process parameters (Siqueira et al. 2018). They reported that a carbon/nitrogen ratio of 40 had the largest amounts of produced biomass and oil, the highest energy output, and the optimum CO_2 emission (18.09 kg CO_2-eq/kg). They also demonstrated the advantages of biodiesel production from *P. autumnale* over soybean (Siqueira et al. 2018). Soybean requires pesticides, insecticides, and larger arable lands for plantation, but microalgae do not need arable lands. The energy of the produced biodiesel from microalgae is twice as much as that made from soybean. The highest CO_2 emission was 27.12 kg CO_2-eq/kg, which is lower than that emitted from soybean biodiesel (70 kg CO_2-eq/kg). The amount of water used during different stages of the biodiesel production process is called water footprint (WF). WF of the biodiesel production from microalgae was 28.38 m^3 per kg of biodiesel produced, which is far lower than the WF of biodiesel produced from soybean (42×10^5 m^3/kg) (Siqueira et al. 2018).

By considering the pros and cons of the three distinct generations of biodiesel feedstocks, and the results of some LCA studies, it can be revealed that second- and third generation feedstocks are more advantageous than the first generation.

The FAs composition varies in different oil sources that influence the production and quality of biodiesel. These FAs have different carbon chain lengths and different numbers of unsaturated bonds (Singh and Singh 2010). The oil content and composition of the FAs of some oil sources are given in Table 8.1.

8.3 Overview of Biodiesel Production

Due to the high viscosity, high FFAs content, and low volatility of raw oils (e.g., vegetable oils and animal fats), their direct use as fuel in diesel engines possess some disadvantages like ignition delay variation and gum formation, which may damage the engine (Wong et al. 2019; Athar and Zaidi 2020). Therefore, an appropriate method is required for the conversion of such oils into a convenient fuel. Up to now, biodiesel is produced through different processes like pyrolysis, microemulsion, and transesterification. Among, transesterification has attracted more attention for converting various raw oils to biodiesel as this method is cost effective and its conversion efficiency is high (Yesilyurt et al. 2020). Besides, transesterification has been reported as a promising method for reducing viscosity and enhancing the eventual fluidity of the produced biodiesel (Behçet et al. 2015; Sahani et al. 2020). The reaction between the TAGs of different oils/fats with short-chain alcohol (e.g., ethanol and methanol) and a catalyst (e.g., acidic, basic, and enzymatic) for the production of FAMEs and glycerol is called transesterification. Figure 8.1 shows the general process of biodiesel production from lipidic wastes as well as the transesterification process. Transesterification is classified into catalytic and non-catalytic transesterification. In terms of catalytic transesterification, a suitable catalyst selection is vital to enhance the product yield. Up to now, scientists have studied many different catalysts' properties and applications to determine their ability in the transesterification of different feedstocks. Generally, catalysts are divided into acid (homogenous and heterogeneous), base (homogenous and heterogeneous), and enzyme (lipases). On the other hand, non-catalytic transesterification contains reactions under supercritical conditions. In this section, different methods of transesterification are discussed.

8.3.1 *Catalytic Transesterification*

Nowadays, biodiesel is generally produced by homogenous base catalysts due to some advantages. NaOH and KOH are widely used among different homogenous base catalysts as they are abundant and cheap. Their efficiency in the whole product yield is high in shorter durations, and the reaction requires low temperature and pressure when these catalysts are used (Verma et al. 2016). Generally, base catalysts are

Table 8.1 Oil content and fatty acids composition of various feedstocks

Feedstock	Oil (%dw)	Fatty acids									References
		14:0	16:0	16:1	18:0	18:1	18:2	18:3	20:0	Others[a]	
Edible oil											
Canola	40–45	0.05	6.23	0.34	2.49	61.46	22.12	5.11	1.43	0.37	(Yaşar 2020; Singh et al. 2020)
Coconut	63–65	19.9	7.8	0.1	3	4.4	0.8	–	–	63.9	(Singh and Singh 2010; Singh et al. 2020)
Corn	45–55	0.02	12.23	0.13	2.62	31.4	51.21	0.85	0.32	0.13	(Yaşar 2020; Karmakar et al. 2010)
Mustard	28–32	–	3.22	–	1.19	18.69	14.38	5.84	4.02	51.35	(Sharif et al. 2017)
Olive oil	45–70	–	13.27	0.86	3.69	68	12.48	0.76	0.48	0.24	(Yaşar 2020; Karmakar et al. 2010)
Palm	30–60	–	42.6	0.3	4.4	40.5	10.1	0.2	–	1.1	(Singh and Singh 2010; Singh et al. 2020)
Peanut	45–55	–	11	–	2	48	32	1	1	3	(Singh and Singh 2010; Singh et al. 2020)

(continued)

Table 8.1 (continued)

Feedstock	Oil (%dw)	Fatty acids									References
		14:0	16:0	16:1	18:0	18:1	18:2	18:3	20:0	Others[a]	
Rapeseed	38–46	–	3.5	–	0.9	64.1	22.3	8.2	–	–	(Singh and Singh 2010; Singh et al. 2020)
Soybean	15–20	0.07	11.43	0.07	4.03	24.85	55.33	3.34	0.25	0.57	(Yaşar 2020; Singh et al. 2020)
Sunflower	25–35	0.08	5.93	0.14	3.44	36.22	52.95	0.38	0.23	0.46	(Yaşar 2020; Singh et al. 2020)
Non-edible oil											
Castor	48–60	–	1.1	–	3.1	4.9	1.3	–	–	89.6	(Singh and Singh 2010)
Cottonseed	18–25	0.69	21.47	0.56	2.61	18.21	55.45	0.15	0.06	0.14	(Yaşar 2020; Karmakar et al. 2010)
Drumstick	30–40	–	9.1	–	2.7	79.4	0.7	0.2	5.8	2.1	(Singh and Singh 2010) (Demirbas et al. 2016)
Jojoba	45–50	–	–	–	–	0.55–0.77	–	–	28–31	14.45	(Karmakar et al. 2010)

(continued)

Table 8.1 (continued)

Feedstock	Oil (%dw)	Fatty acids									References
		14:0	16:0	16:1	18:0	18:1	18:2	18:3	20:0	Others[a]	
Karanja	27–39	–	3.7–7.9	–	2.4–8.9	44.5–71.3	10.8–18.3	–	2.2–4.7	–	(Karmakar et al. 2010)
Linseed	35–45	–	5.1	0.3	2.5	18.9	18.1	55.1	–	–	(Singh and Singh 2010)
Mahua	35–42	–	20–25	–	20–25	41–51	10–14	–	0–3.3	–	(Karmakar et al. 2010)
Neem	40–50	0.2–2.6	13.6–16.2	–	14.4–24	49–62	2.3–15.8	–	0.8–3.4	–	(Karmakar et al. 2010)
Passiflora	18–29		8.8–11	–	2.2–3.1	13.6–16.9	67.8–74.3	0.3–0.4	–	–	(Karmakar et al. 2010)
Polanga	60–75	0.75	14.4	0.25	15.57	34.41	28.34	0.15	–	2.67	(Chavan et al. 2013; Demirbas et al. 2016)
Rice bran	15–23	0.4–1	12–18	–	1–3	40–50	29–42	0.5–1	–	0.2–0.4	(Karmakar et al. 2010)
Safflower	–	0.1	6.7	0.08	2.4	11.5	79	0.15	–	–	(Yaşar 2020)
Sal	10–12	–	4.5–8.6	–	34.2–44.8	34.2–44.8	2.7	–	6.3–12.2	–	(Singh and Singh 2010)
Sesame	50–60	–	13.1	–	3.9	52.8	30.2	–	–	–	(Singh and Singh 2010)
Soapnut	42.7	–	3.8	–	1.1	55.2	8.3	2.4	5.3	23.9	(Chen et al. 2013)

(continued)

Table 8.1 (continued)

Feedstock	Oil (%dw)	Fatty acids									References
		14:0	16:0	16:1	18:0	18:1	18:2	18:3	20:0	Others[a]	
Syringa	10–45	–	9.7–10.1	–	3.31–3.67	21.69–22.31	63.27–64.89	0.34–0.43	0.19–0.21	0–0.36	(Stavarache et al. 2008; Demirbas et al. 2016)
Tobacco	30–43	0.09	10.96	0.2	3.34	14.54	69.49	0.69	–	0.69	(Singh and Singh 2010; Karmakar et al. 2010)
Waste oil											
Beef tallow	–	1.6	21.6	17.7	2.5	31.5	3.3	1.3	–	–	(Toldrá-Reig et al. 2020)
Camel fat	–	9.82	34.6	9.3	11.07	25.44	–	–	0.2	7.38	(Sbihi et al. 2014)
Chicken fat	–	0.74	21.33	3.25	5.67	37.02	28.91	1.5	–	–	(Behçet et al. 2015)
Date pit oil	–	11.85	9.82	0.09	2.96	44.92	8.47	0.21	0.43	21.28	(Phoenix 2012)
Fish oil	–	6.71	20.2	6.59	4.2	19.71	2.63	1.64	0.79	–	(Behçet et al. 2015)
Leather waste fat	–	2.99	25.5	7.23	11.6	1.8	43.44	0.66	0.12	4.91	(Keskin et al. 2020)
Mutton tallow	–	2.2	21.1	11.6	2.1	38.7	10.2	0.6	–	–	(Toldrá-Reig et al. 2020)

(continued)

Table 8.1 (continued)

Feedstock	Oil (%dw)	Fatty acids									References
		14:0	16:0	16:1	18:0	18:1	18:2	18:3	20:0	Others[a]	
Pig tallow	–	–	21.6	4.1	14.6	52.3	4.1	–	1.03	2.27	(Suresh et al. 2021)
Poultry fat	–	0.4	21.6	6.3	3.2	30	28.4	2.4	–	4.5	(Toldrá-Reig et al. 2020)
Pork lard	–	1.6	25.1	12.6	2.8	36.5	16.5	1.1	–	0.5	(Toldrá-Reig et al. 2020)
Waste cooking oil	–	–	1.4	34.9	1.8	4	42.5	13.5	0.6	1.3	(Farid et al. 2020)
Microbial oil											
Chaetoceros muelleri	43.4	16.50	28.84	39.48	1.92	0.96	0.79	0.51	–	11.0	(Wang et al. 2014)
Chlorella vulgaris	34.4	–	13.98	–	2	70.84	10.43	0.8	–	1.95	(Wang et al. 2013)
Chlorococcum pamirum	64.9	1.55	25.77	10.29	28.84	13.86	10.28	7.4	–	1.95	(Feng et al. 2014)
Rhodosporidium kratochvilovae	53.2	0.78	43.06	–	28.74	17.34	0.48	–	–	–	(Patel et al. 2015)
Lipomyces starkeyi	34.8–58.7	1	25.8	–	7.4	53.1	10.6	2.1	–	–	(Tsakona et al. 2014)
Nannochloropsis oceanica	54.28	4.57	33.08	26.71	3.67	18.16	1.73	–	–	12.08	(Xiao et al. 2015)
Cryptococcus curvatus	61	–	13.2	–	22.62	50.72	11.2	0.89	–	–	(Xu et al. 2014)

[a]Sum of other reported fatty acids

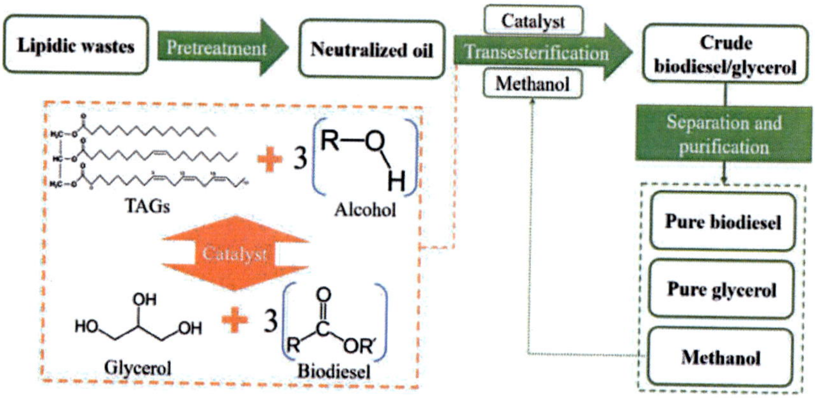

Fig. 8.1 General process of biodiesel production from lipidic wastes and transesterification reaction

cheap and highly effective in mild conditions during short reaction times. However, these kinds of catalysts are sensitive to the FFAs and water content of the feedstock. The high FFAs content of the feedstock results in soap and water formation, which decreases the biodiesel yield. In addition, the water inhibits the transesterification by hydrolyzing TAGs to diacylglycerols, which results in more FFAs formation (Pikula et al. 2020). Besides, the loss of the ester produced is a disadvantage when there is soap formation, resulting in the reduction of the final yield and increasing the costs. Further, purification is expensive and time consuming, and the disposal of highly basic waste streams may cause environmental problems (Sitepu et al. 2020). Generally, feedstock with FFAs content of less than 2 wt% is recommended when a base catalyst is used (Wong et al. 2019; Yesilyurt et al. 2020). However, as provided in Table 8.2, homogenous base catalysts were used in the production of biodiesel from lipidic wastes such as beef tallow, camel fat, chicken fat, and waste cooking oil, and resulted in a considerable yield of biodiesel. For instance, performing KOH as the catalyst for transesterification of waste vegetable oil and fish oil resulted in 89 and 96% yield of biodiesel, respectively (Fadhil and Ali 2013; Chavan et al. 2017).

In contrast, using acid catalysts like HCl and H_2SO_4 during the transesterification solves the problem of high FFAs content as their performance is higher when the FFAs content is more than 1 wt% (Farobie and Matsumura 2017). Thus, there is no need for pretreatment for cheaper feedstocks that contain high FFAs content (e.g., tallow and waste cooking oil). They are also efficient in catalyzing esterification, transesterification, and simultaneous esterification and transesterification (Bhatti et al. 2008; Madhu et al. 2016; Wong et al. 2019). As provided in Table 8.2, a mixture of sunflower and soybean oil with 8.5% FFAs content was converted to biodiesel through direct esterification (simultaneous esterification and transesterification) using 2.5 wt% H_2SO_4, resulting in 96.6% conversion to biodiesel (Farag et al. 2011). In another investigation, transesterification of chicken fat and mutton fat resulted in a considerable yield of 99 and 93.2%, respectively, when H_2SO_4 was used

Table 8.2 Summary of different feedstocks employed in biodiesel production through different methods

Feedstock	Methods	Catalyst (amount)	Alcohol	Alcohol:oil	Time	Temp (°C)	Conversion (C) or yield (Y)	References
Edible								
Canola	Transesterification	MgO (3 wt%)	Methanol	20:3	2 h	190	Y = 82.8%	(Jeon et al. 2013)
		t-MgO (3 wt%)	Methanol	20:3	2 h	190	Y = 98.2%	
Palm	Transesterification	ZnO (250 mg)	Methanol	6:1	5 h	60	Y = 83.2%	(Madhuvilakku and Piraman 2013)
		TiO_2-ZnO (200 mg)	Methanol	6:1	5 h	60	Y = 92.2%	
Rapeseed	Transesterification	KOH (1 wt%)	Methanol	6:1	2 h	65	Y = 95–96%	(Rashid and Anwar 2008)
		Na/FAP (6 wt%)	Methanol	10:1	8 h	120	Y = 98%	(Essamlali et al. 2019)
Soybean	Transesterification	NaOH (0.4 wt%)	Methanol	6:1	1 h	60	Y = ~97%	(Dias et al. 2008)
		K/BC-Fe_2O_3 (2.5 wt%)	Methanol	8:1	1 h	60	Y = 98%	(Liu et al. 2018)
Sunflower	Transesterification	NaOH (0.4 wt%)	Methanol	6:1	1 h	60	Y = ~96%	(Dias et al. 2008)
Sunflower and soybean mixture	Direct esterification	H_2SO_4 (2.5 wt%)	Methanol	6:1	1 h	60	C = 96.6%	(Farag et al. 2011)
Non-edible								
Castor	Microwave-assisted transesterification	H_2SO_4/C (5 wt%)	Methanol	12:1	1 h	65	Y = 94%	(Yuan et al. 2009)
Desert date	Transesterification	KOH (1.7 wt%)	Methanol	6:1	1 h	Room temp	Y = ~90%	(Chapagain et al. 2009)
Jatropha	Microwave-assisted two-step transesterification	KOH-CaO (3.17 wt%)	Methanol	8.42:1	67.9 min	60	C = 97.1%	(Liao and Chung 2013)

(continued)

Table 8.2 (continued)

Feedstock	Methods	Catalyst (amount)	Alcohol	Alcohol:oil	Time	Temp (°C)	Conversion (C) or yield (Y)	References
Kapok seed	Microwave-assisted esterification and transesterification	KOH (2.15 wt%)	Methanol	9.85:1	3.29 min	57.09	C = 98.9%	(Bokhari et al. 2015)
Karanja	Two-step transesterification	CaO (2.5 wt%)	Methanol	8:1	2 h	65	Y = 94%	(Madhu et al. 2016)
		$BaCeO_3$ (1.2 wt%)	Methanol	19:1	100 min	65	C = 98.41%	(Sahani et al. 2019)
Mesua ferrea	Two-step transesterification	Co doped ZnO (2.5 wt%)	Methanol	9:1	3 h	60	C = 98.03%	(Borah et al. 2019)
Rubber seed	Transesterification-assisted hydrodynamic cavitation	KOH (1 wt%)	Methanol	6:1	20 min	55	C = 97.1%	(Bokhari et al. 2016)
Tobacco	Transesterification	NaOH (0.75 wt%)	Methanol	6:1	2 h	55	Y = ~88%	(Usta et al. 2011)
Wastes								
Beef tallow	Transesterification	KOH (0.8 wt%)	Methanol	6:1	2 h	60	Y = 90.8%	(Mata et al. 2011)
		Immobilized lipase (20 wt%)	Ethanol	12:1	48 h	50	Y = 89.7%	(Da Rós et al. 2010)
Camel fat	Transesterification	NaOH (1 wt%)	Methanol	6:1	2 h	65	Y = 98.6%	(Sbihi et al. 2014)
Chicken fat	Transesterification	KOH (0.8 wt%)	Methanol	6:1	2 h	60	Y = 76.8%	(Mata et al. 2011)
	Two-step transesterification	KOH (0.8 wt%)	Methanol	8:1	1 h	60	Y = 82%	(Chavan et al. 2017)
	Transesterification	H_2SO_4 (1.25 g)	Methanol	30:1	24 h	50	Y = ~99%	(Bhatti et al. 2008)
	Two-step transesterification	NaOMe (1%)	Methanol	1:1	3 h	70	C = 98.1%	(Shi et al. 2013)
	Transesterification	$CaO/CuFe_2O_4$ (3 wt%)	Methanol	15:1	4 h	70	Y = 94.5%	(Seffati et al. 2019)

(continued)

Table 8.2 (continued)

Feedstock	Methods	Catalyst (amount)	Alcohol	Alcohol:oil	Time	Temp (°C)	Conversion (C) or yield (Y)	References
Fish oil	Transesterification	KOH (0.5 wt%)	Methanol	6:1	1 h	32	Y = 96%	(Fadhil and Ali 2013)
Leather waste fat	Transesterification	KOH (0.8 wt%)	Methanol	6:1	1 h	63	Y = 86.8%	(Keskin et al. 2020)
Mutton fat	Transesterification	H_2SO_4 (2.5 g)	Methanol	30:1	24 h	60	Y = ~93.2%	(Bhatti et al. 2008)
		MgO-KOH (4 wt%)	Methanol	22:1	20 min	65	C = 98%	(Mutreja et al. 2011)
Pig tallow	Ultrasound-assisted two-step transesterification	CuO (2.07 wt%)	Methanol	29.87:1	35.36 min	60	Y = 97.82%	(Suresh et al. 2021)
Pork lard	Transesterification	KOH (0.8 wt%)	Methanol	6:1	2 h	60	Y = 91.4%	(Mata et al. 2011)
Swine fat	Ultrasound-assisted two-step transesterification	KOH (1.11 wt%)	Methanol	7.42:1	176.14 min	62.3	Y = 98%	(He et al. 2020)
Used vegetable oil	Simultaneous microwave and ultrasound-assisted transesterification	BaO (0.75 wt%)	Methanol	6:1	2 min	–	Y = ~93.5%	(Martinez-Guerra and Gude 2014)
Waste cooking oil	Transesterification	KOH (1.4 wt%)	Methanol	10:1	4 h	60	Y = 96.9%	(Farid et al. 2020)
		$NaOCH_3$ (3 wt%)	Methanol	12:1–18:1	5 min	60	Y = 98%	(Eze et al. 2018)
		SO_4/Fe-Al-TiO_2 (3 wt%) nano	Methanol	10:1	2.5 h	90	Y = 96%	(Gardy et al. 2018)
Waste frying oil	Transesterification	CH_3ONa (0.4 wt%)	Methanol	6:1	1 h	60	Y = ~92%	(Dias et al. 2008)
Waste lard	Microwave-assisted two-step transesterification	CaO/zeolite (8% w/v)	Methanol	30:1	1.25 h	65	Y = 90.89%	(Lawan et al. 2020)
	Ultrasound-assisted transesterification	Immobilized lipase (6 wt%)	Methanol	4:1	30 min	–	Y = 96.8%	(Adewale et al. 2016)

(continued)

Table 8.2 (continued)

Feedstock	Methods	Catalyst (amount)	Alcohol	Alcohol:oil	Time	Temp (°C)	Conversion (C) or yield (Y)	References
Waste vegetable oil	Two-step transesterification	KOH (0.8 wt%)	Methanol	8:1	1 h	60	Y = 89%	(Chavan et al. 2017)
Algae								
Caulerpa peltata	Ultrasound-assisted two-step transesterification	NaOH (1 wt%)	Methanol	12:1	16 min	60	C = 98.11%	(Suganya and Renganathan 2014)
Enteromorpha compressa	Two-step transesterification	NaOH (1 wt%)	Methanol	9:1	70 min	60	Y = 90.6%	(Suganya et al. 2013)
Nannochloropsis oculata	Transesterification	$Ca(OCH_3)_2$ (3 wt%)	Methanol	30:1	3 h	80	Y = 99%	(Teo et al. 2016)
Spirulina maxima	Two-step transesterification	KOH (0.75 wt%)	Methanol	9:1	20 h	65	Y = 86.1%	(Rahman et al. 2017)

during the transesterification in optimum conditions in which methanol-to-oil ratio was 30:1 and the reaction time was 24 h (Bhatti et al. 2008). Despite the advantages of homogenous acid catalysts, they are not applicable for biodiesel production in the industry because they corrode equipment and materials and cause environmental problems. Besides, the influence of these catalysts on the transesterification rate is about 4000 times lower than the homogenous base catalysts under comparable conditions (Farobie and Matsumura 2017). Further, the separation of homogenous acid catalysts from the product is difficult (Pikula et al. 2020). Also, during the reaction, high temperature and a high molar ratio of alcohol-to-oil are required (Lam et al. 2010).

Besides the mentioned benefits of homogenous catalysts, a more sustainable catalyst with a convenient separation process and reusability potential could be a more profitable choice during transesterification. In this regard, heterogeneous catalysts could be an appropriate option to replace the homogenous ones (Wong et al. 2019). They could be reused in the transesterification process, and their separation is not as complicated as the homogenous catalysts. Besides, saponification does not happen when heterogeneous catalysts are used in the process (Farobie and Matsumura 2017). Recently, base earth metal oxides, basic zeolites, and hydrotalcite were used as heterogeneous base catalysts (Yesilyurt et al. 2020). They could be synthesized from cheap feedstocks like calcium hydroxide and limestone. This type of catalyst produces lower concentrations of basic wastewater, and their catalytic activity under the same conditions is almost the same as the homogenous basic catalysts (Farobie and Matsumura 2017). The base earth metals (e.g., Ba, Ca, Mg, and Sr) are the source of solid base catalysts such as BaO, CaO, MgO, and SrO. CaO is the most attractive heterogeneous base catalyst due to its strong basic activity and poor solubility in methanol (Zabeti et al. 2009). As an example, a CaO catalyst derived from crab shell was used as the catalyst for biodiesel production from karanja seed oil (Madhu et al. 2016), which resulted in 94% of biodiesel yield. In terms of using lipidic wastes as feedstock for biodiesel production, performing used vegetable oil resulted in a production yield of 93.5% when 0.75 wt% BaO was used as catalyst (Martinez-Guerra and Gude 2014).

The heterogeneous acid catalysts are more efficient than base catalysts and homogenous acid catalysts in terms of ease of separation, resulting in the better reusability of the catalyst. Thus, they are used on industrial scale (Pikula et al. 2020). Besides, these catalysts are less corrosive than homogenous acid catalysts. They also possess some advantages like low toxicity, no need for biodiesel washing step, and low product contamination (Farobie and Matsumura 2017). Different heterogeneous acid catalysts such as ZrO_2, TiO_2, and SnO_2 have been utilized in transesterification. However, it is noteworthy that transesterification reaction in the presence of heterogeneous acid catalysts may result in unwanted side reactions. The reaction rate is slow, and the reaction usually requires high temperature and pressure (Farobie and Matsumura 2017; Yesilyurt et al. 2020). In an investigation, Borah and colleagues prepared TiO_2 based nanocomposites for the production of biodiesel from waste cooking oil (Borah et al. 2018). In this study, transesterification under optimum conditions of 65 °C, methanol-to-oil ratio of 12:1, catalyst concentration of 1.5 wt%, and 3 h of reaction time, the highest biodiesel conversion of 98% was achieved.

According to the mentioned characteristics of chemical catalysts, several concerns are associated with their utilization, comprising the need for wastewater treatment and catalyst removal from biodiesel and/or glycerol. Nowadays, in comparison with all mentioned catalysts, biocatalysts (enzymes) have attracted significant interest in overcoming the chemical catalysts problems. Generally, when enzymes are used in the transesterification at moderate temperatures (35–45 °C), low alcohol-to-oil ratio is required. The separation and recovery of the product are easy, and the FFAs and water content of the feedstock are not important. Besides, saponification and hydrolysis will not happen in the reaction, and FFAs and TAGs could be simultaneously converted to biodiesel (Farobie and Matsumura 2017). The main enzyme involved in transesterification is lipase isolated from different sources (e.g., animals, bacteria, fungi, plants, and yeasts) (Kant Bhatia et al. 2021). Further, lipase properties are highly dependent on their source. Accordingly, lipases are efficient in catalyzing both esterification and transesterification, which results in the production of high-grade glycerol. In addition, the enzymes could be recycled, and they do not require high reaction temperatures (Sitepu et al. 2020). However, there are some demerits in enzymatic catalysts, as they are costly and their reaction rate is slow. Besides, methanol has an inhibitory effect on them and can deactivate them in the reaction (Farobie and Matsumura 2017; Yesilyurt et al. 2020). Immobilization methods could overcome the performance of the enzymes in terms of reusability and instability, but their cost is still challenging. For instance, immobilized lipase was used for the production of biodiesel from beef tallow and waste lard, resulted in considerable production yields of biodiesel (Table 8.2).

In addition to the catalyst type, the amount of catalyst affects the final yield. When the catalyst loading is higher than the necessary loading, saponification happens that reduces the biodiesel yield. On the other hand, when the catalyst amount is lower than the essential loading, the reaction cannot be completed and lowers the product yield (Yesilyurt et al. 2020). Besides, the separation of biodiesel and glycerol is difficult because of the soap formation.

The alcohol type and amount are significant parameters in the production of biodiesel. Different alcohols such as butanol, ethanol, methanol, octanol, pentanol, and propanol have been utilized in the transesterification of different feedstocks. The two most commonly used alcohols in the reaction are methanol and ethanol. As provided in Table 8.2, the widely available methanol is usually used because of its low cost, chemical, and physical benefits. When ethanol is used, the cloud and pour points of the produced biodiesel are slightly lower. Simultaneously, its viscosity is somewhat higher compared to those biodiesel produced in the presence of methanol (Athar and Zaidi 2020). Besides, ethanol removal is hard, and biodiesel yield is low when ethanol is used as alcohol (Yesilyurt et al. 2020). Other alcohols could be utilized in the reaction, but they cost more. The amount of alcohol in the process is usually reported as the ratio of the alcohol-to-oil. This amount is highly dependent on the feedstock, type of alcohol, and catalyst.

Reaction time and temperature are two crucial parameters influencing the yield and rate of transesterification. According to the reversible nature of transesterification, the reaction duration should be optimum. Generally, it takes 5 min for the reaction to accelerate, and in the first 30 min, the transesterification conversion is 80%. If the reaction continues more than the optimum duration, the product yield will be reduced due to product loss (reversible reaction) and soap formation. In terms of the reaction temperature, the optimum temperature is lower than the boiling point of the alcohol, which prevents evaporation. Further, the viscosity of the oil may reduce in high temperatures. As an example, it has been indicated that when methanol is used as alcohol, the temperatures close to the boiling point of methanol are suitable (Yesilyurt et al. 2020).

8.3.2 Conversion of Wastes to Catalysts for Biodiesel Production

As discussed above, due to the limitations of using homogenous acid and base catalysts for transesterification, solid heterogeneous catalysts have attracted researchers' attention. In this regard, many efforts have been made to develop these catalysts, such as using various wastes (e.g., industrial and bio-based wastes) for catalysts production. The utilization of wastes as catalysts enhances the sustainability of biodiesel production and is a cost-effective and environmentally friendly approach. These types of catalysts, as depicted in Fig. 8.2, included but are not limited to various waste

Fig. 8.2 Potential resources for utilization as catalysts for biodiesel production

shells (e.g., chicken/duck/quail eggshells, and mussel/cockle/snail/oyster shells, etc.), animal bones (e.g., bovine, chicken, fish, sheep, etc.), waste fish scales, rocks, industrial slag (e.g., carbide slag, gasified slag, blast furnace slag, etc.), industrial waste ash (e.g., coal fly ash, etc.), and biomass ash (e.g., banana peel ash, coconut husk ash, rice husk ash, etc.) (Marwaha et al. 2018; Shan et al. 2018; Alagumalai et al. 2021).

In this regard, waste chicken eggshells were used as catalysts for biodiesel production through the transesterification of waste cooking oil, resulted in the 96.07% yield of biodiesel under optimum conditions (Gupta and Rathod 2018). Also, river snail shell ash was used in the transesterification process of used cooking oil for biodiesel production, in which 92.5% of conversion was achieved when 3 wt% of catalyst was used in the process (Kaewdaeng et al. 2017). In terms of biodiesel production using animal bones waste as catalyst, the yield of 89.5% obtained when used cooking oil subjected to transesterification using the mixture of waste chicken and fish bones as catalyst (Tan et al. 2019b). In another investigation, montmorillonite clay K-30 used as catalyst during the transesterification of waste frying oil resulted in 93.7% biodiesel yield when 5 wt% catalyst was used (Inayat et al. 2019). According to the mentioned studies and literature, considerable biodiesel production yields could be achieved when waste catalysts are used during the process. Therefore, this could be a promising way to overcome the problems of other catalysts in the production of biodiesel.

8.3.3 Non-catalytic Transesterification

The production of biodiesel under supercritical conditions has been proposed as an alternative for the catalytic production of biodiesel. It has been stated that many problems of catalytic biodiesel production have been overcome in supercritical conditions. This process does not require any catalyst, has a high rate in shorter reaction times, has fewer process steps compared to the catalytic methods, and supports simultaneous FFAs esterification and TAGs transesterification. Besides it is not sensitive to FFAs and water content, the purification and separation phases are easier and are efficient in the conversion of different feedstocks (Farobie and Matsumura 2017; Andreo-Martínez et al. 2020). Despite the advantages of biodiesel production under supercritical conditions, there are some drawbacks such as high oil-to-alcohol ratio, high pressure (35–60 MPa), temperature (250–400 °C) requirements, and low energy efficiency (Zabeti et al. 2009; Farobie and Matsumura 2017; Pikula et al. 2020). Therefore, some methods are proposed to overcome the bottlenecks of supercritical conditions. In order to produce biodiesel in milder conditions in the presence of methanol, co-solvents (e.g., acetone, carbon dioxide, ethane, n-hexane, and propane) are used to increase the mutual solubility of TAGs and methanol for biodiesel production under more moderate conditions (Pikula et al. 2020). Also, two-step supercritical procedure (7 MPa, 270 °C) consists of hydrolyzing TAGs in subcritical water, followed by the supercritical conversion of produced FFAs (during the first step) in lower oil-to-methanol ratio was also efficient in reducing the energy consumption comparing to the one-step process (20–50 MPa, 350 °C) (Farobie and Matsumura 2017). Besides, heterogeneous catalysts utilization could improve the conversion yields of supercritical conditions (Andreo-Martínez et al. 2020).

8.3.4 Microwave and Ultrasound-Assisted Transesterification

Recently, microwave and ultrasound-assisted procedures have been proposed to improve the transesterification of different feedstocks for biodiesel production (Bokhari et al. 2015; He et al. 2020; Suresh et al. 2021). Microwave-assisted transesterification via the use of electromagnetic waves is advantageous due to their lower reaction durations, less energy consumption, higher biodiesel yields, easier separation process, and environmentally friendly nature compared to the conventional transesterification procedures (Lam et al. 2010; Lawan et al. 2020). During microwave-assisted transesterification, the energy is transferred in electromagnetic form and eliminates the disadvantages of conventional heating (e.g., uneven heating, heat, and energy losses) (Pikula et al. 2020). Microwave irradiations could be applied in both catalytic and non-catalytic transesterification (Nayak et al. 2019). However, this method is faster and more efficient than conventional techniques; the major bottleneck in this method is the limitation for large-scale application partly because of the low penetration depth of the microwaves (Nayak et al. 2019). To overcome this issue, researchers proposed some solutions such as effective stirring of the reaction mixture, performing film-type reactors or continuous processes, using a higher source of microwave frequency, and a higher power reactor (Nayak et al. 2019; Pikula et al. 2020). Another concern about this method is its potential harmful effects on the human body (Nayak et al. 2019).

However, performing ultrasound-assisted transesterification results in considerable amounts of biodiesel yield (He et al. 2020; Suresh et al. 2021). This technology is still relatively new and requires more investigations to overcome the challenges. Based on frequency, ultrasound technology is categorized as low (20–100 kHz) and high (2–10 MHz) frequency which respectively are applicable in the mass transfer enhancement and chemical reactions (e.g., chemical synthesis and wastewater treatment). In terms of transesterification, low frequencies either in pulsed or continuous modes are suggested for mass transfer enhancement through the immiscible reactants (Tan et al. 2019a). Compared to conventional methods, this method is more efficient in biodiesel production since the required amount of catalyst, the reaction duration, alcohol-to-oil ratio, and energy consumption are reduced (Lam et al. 2010; Pikula et al. 2020).

To sum up, the biodiesel production procedure is highly dependent on the feedstock type and its FFAs content. In terms of using lipidic wastes as feedstock, the high FFAs content (5–30%) is the biggest challenge. Besides, the high FFAs contents of waste oils cause a reduction in the biodiesel yield. Also, the separation and purification of the product are difficult, which increases the production cost

(Toldrá-Reig et al. 2020). Therefore, to obtain an efficient product yield from lipidic wastes, reducing the FFAs content to below 0.5–2% or the acid value to lower than 2 mg KOH/g of oil (Idowu et al. 2019) is recommended. Some researchers proposed the two-step transesterification to reduce FFAs content and/or acid value of the feedstock through an esterification reaction (first step) in the presence of an acid catalyst followed by the transesterification of TAGs using a base catalyst (second step). For example, in a two-step process for biodiesel production from microalga *Spirulina maxima*, the acid value of the algal oil was reduced from 10.66 to 0.51 mg KOH/g through acid esterification using H_2SO_4 (1 wt%) (Rahman et al. 2017). In another study, the non-edible oil, *Mesua ferrea*, was used as feedstock for biodiesel production (Borah et al. 2019). The acid value of extracted oil from *M. ferrea* was 14.8 mg KOH/g, and it was reduced to less than 1 mg KOH/g using H_2SO_4.

8.4 Biodiesel Quality, Performance, and Exhaust Emissions Characteristics

The quality of biodiesel refers to its physical and chemical properties, such as acid value, ash content, cetane number, calorific value, density, flash point, oxidation stability, viscosity, and water content. Since the quality of biodiesel influences engine performance and emissions, various standards have been set for biodiesel, and the physical and chemical characteristics of produced biodiesel must comply with them. Many factors, such as the feedstock, its FAs composition, and the biodiesel production method, affect the biodiesel quality. Different characteristics of the standards may vary depending on the region as different countries, including Austria (ON), Czech Republic (CSN), Germany (DIN), India (BIS), Italy (UNI), and the USA (ASTM), have their own standards. Among, the most important standards are the American Standards for Testing Materials (ASTM 6751) and the European Union (EN 14214) standard. This section provides an overview of biodiesel characteristics and quality, engine performance, and emissions. Further, the properties of biodiesel are briefly explained. Table 8.3 provides the quality of biodiesel derived from different feedstocks.

8.4.1 Biodiesel Characteristics

The fuel solidification results in the clogging of the fuel lines and filters and may damage the engine. The possibility of biodiesel utilization in low-temperature conditions is investigated according to cloud point (CP), pour point (PP), and cold filter plugging point (CFPP) (Yaşar 2020). The CP is the lowest temperature at which the fuel efficiently ignites. It also could be defined as the temperature at which the biodiesel crystals become visible when the fuel is cooled. It is noteworthy that

Table 8.3 Properties of biodiesel derived from different feedstocks

Feedstock	Biodiesel characteristics (unit)												References
	FP (°C)	CN (–)	D (kg/m^3)	AV (mg KOH/g)	IN (g I_2/100 g)	CP (°C)	CFPP (°C)	PP (°C)	CV (MJ/kg)	V (40 °C mm^2/s)	OS (h)	Others	
Edible oil													
Canola	137–186.5	54.3–55	866–884.4	0.25–0.48	122	− 3	− 8 to − 13	− 9	40.195	4.40–4.526	2.6–7.9	(97.2–98.3)[a], (0.83–5.33)[b]	(Atabani et al. 2013; Öztürk 2015; Yaşar 2020)
Coconut	118.5	–	860.5	–	–	1	− 1	− 4	38.300	3.1435	8.01	0.94[b]	(Atabani et al. 2013)
Corn	172	56	878	0.15	–	–	− 13	–	–	4.42	1.3	98.4[a], 5.42[b]	(Yaşar 2020)
Olive	177	58	882	0.32	–	–	− 12	–	–	4.53	2.3	98.6[a], 5.22[b]	(Yaşar 2020)
Olive pomace	138	58.7	876	0.1	78	− 2	− 9	− 6	39.96	4.53	15.5	97.72[a]	(López et al. 2014)
Palm	214.5	–	859.1	–	–	13	12	15	40.009	4.6889	23.56	1.81[b]	(Atabani et al. 2013)
Rapeseed	170–172	54	880	0.22–0.30	108	− 4 to − 8	− 11	− 12 to − 15	37.00	4.48–4.55	2.2–6.11	98.5[a], (0.9–4.60)[b]	(Ramadhas et al. 2005; Mihaela et al. 2013; Yaşar 2020)
Soybean	170–202.5	51	869–882	0.16–0.18	123	− 2 to 1	− 2 to − 6	− 3 to 1	39.76–39.976	4.08–4.42	1.5–5.80	97[a], 0.86[b]	(Ramadhas et al. 2005; Atabani et al. 2013; Mihaela et al. 2013; Yaşar 2020)
Sunflower	178	50	882	0.12	–	–	− 4	–	–	4.30	0.9	97.2[a], 5.54[b]	(Yaşar 2020)
Non-edible oil													
Castor	149	–	913	0.996–1.008	–	− 13.4	7	–	38.7–39.16	15.25	1.1	–	(Panwar et al. 2010; Bhuiya et al. 2016b)
Cottonseed	70–175	56	883	0.21	–	–	− 12	− 15	40.32	4.0–4.33	1.88	96.8[a], 5.46[b]	(Ramadhas et al. 2005; Yaşar 2020)

(continued)

Table 8.3 (continued)

Feedstock	Biodiesel characteristics (unit)												References
	FP (°C)	CN (–)	D (kg/m^3)	AV (mg KOH/g)	IN (g I_2/100 g)	CP (°C)	CFPP (°C)	PP (°C)	CV (MJ/kg)	V (40 °C mm^2/s)	OS (h)	Others	
Croton megalocarpus	164	–	870.4	–	–	− 3	− 4	− 2	39.786	4.0707	0.71	–	(Atabani et al. 2013)
Desert date	122–131	53.56	870–890	–	97–100	3–7	1–3	–	–	3.7–4.2	–	(2–5)^b	(Chapagain et al. 2009)
Jatropha	161–186.5	–	864.2–882	–	–	10	10	10	39.44–39.738	4.9476	3.05–4.84	3.84^b	(Atabani et al. 2013)
Karanja	142–180	58	871–890	0.42	89	–	− 7	–	35.56–39.56	4.85	1.82–6	–	(Agarwal et al. 2015; Bhuiya et al. 2016b)
Neem	127–168	56.78	887–899.1	0.22	46.84	20	–	12	39.87	5.50	1.39	–	(Betiku et al. 2014; Agarwal et al. 2015)
Rubber seed	130–183	53	–	0.118–0.12	–	3–4	–	− 6 to − 8	36.50–42.37	5.642–5.81	–	–	(Ramadhas et al. 2005; Dhawane et al. 2015)
Safflower	180	53	880–885	0.22–0.33	145	− 7	− 7 to − 9	− 13	–	4.23–4.60	0.57–2.4	98^a, (1.8–6.32)^b	(Mihaela et al. 2013; Yaşar 2020)
Sesame seed	170	50.48	867.2	–	80.32	− 6	–	− 14	40.4	4.2	–	0^b	(Saydut et al. 2008)
Tobacco	165.4	51–51.6	886.8–888.5	0.3–0.45	112–136	–	− 5	–	39.811	3.5–4.23	0.8	(98.15–98.6)^a, (6–8)^b	(Usta 2005; Usta et al. 2011)
Waste oil													
Animal fat	–	63.88	875	0.38	83.02	–	3	–	36.73	4.25	13.03	99.99^a	(Ballesteros et al. 2011)

(continued)

Table 8.3 (continued)

Feedstock	Biodiesel characteristics (unit)												References
	FP (°C)	CN (–)	D (kg/m³)	AV (mg KOH/g)	IN (g I_2/100 g)	CP (°C)	CFPP (°C)	PP (°C)	CV (MJ/kg)	V (40 °C mm²/s)	OS (h)	Others	
Beef tallow	160–171	–	870	0.147–0.20	44.4	16	10–14	10	–	4.82–5.35	–	–	(Toldrá-Reig et al. 2020)
Trout oil	–	51.3	885	–	–	–	–	–	37.8	4.25	–	–	(Buyukkaya et al. 2013)
Mutton fat	–	59	856	0.65	126	− 4	–	− 5	–	–	–	244.5[c]	(Bhatti et al. 2008)
Camel fat	158	58.7	871	–	65.3	12.7	–	15.5	39.52	3.39	–	(0.031 wt%)[b], 202.3[c]	(Sbihi et al. 2014)
Chicken fat	177	54.8–61	867	0.25	130	− 5	–	− 6	39.934	4.35	–	251.23[c]	(Bhatti et al. 2008; Behçet et al. 2015)
Leather waste fat	173.5	58.8	875.5	0.42	–	–	–	9	39.7	4.636	–	–	(Keskin et al. 2020)
Pig tallow	178	67	–	0.26	–	2	–	6	41.3	4.53	–	99.2[a], 0.87[b], 162.3[c]	(Suresh et al. 2021)
Pork lard	147–175	56.9	870–873	0.22–0.23	66.7–75.6	–	− 20 to 5	–	–	4.74–5.08	–	–	(Toldrá-Reig et al. 2020)
Poultry fat	160–172	–	877	0.044–0.55	78.8	6.1	2–3	–	–	4.50–6.86	–	–	(Toldrá-Reig et al. 2020)
Waste cooking oil	175	–	889.1	0.4	63.8	–	–	–	–	4.8	–	97[a]	(Farid et al. 2020)
Fish oil	114–177	47–52.6	881–885	–	–	–	–	–	40.057–40.54	4.45–4.74	–	–	(Gnanasekaran et al. 2016; Singh et al. 2019)
Waste fish oil	–	–	875.64	0.22	83.29	–	–	–	39.76	4.53	5.8	–	(Mota et al. 2019)

(continued)

Table 8.3 (continued)

Feedstock	Biodiesel characteristics (unit)												References
	FP (°C)	CN (–)	D (kg/m^3)	AV (mg KOH/g)	IN (g I_2/100 g)	CP (°C)	CFPP (°C)	PP (°C)	CV (MJ/kg)	V (40 °C mm^2/s)	OS (h)	Others	
Algae													
Caulerpa peltata	178	57.9	885.47	0.40	–	3	–	− 1	–	4.251	–	–	(Suganya and Renganathan 2014)
Enteromorpha compressa mac	166	58.5	878.47	0.43	–	3	–	− 2	–	4.35	–	–	(Suganya et al. 2013)
Spirulina maxima	178	48	864	0.475	–	–	–	–	38.43	4.47	–	–	(Rahman et al. 2017)
Chlorella vulgaris	–	54.1–56.7	880	–	65–110.4	2.44	4.5	–	40.8	4.38	–	99.8[a], 217.8[c]	(Francisco et al. 2010; Song et al. 2013)
Chlorella minutissima	–	84	880	–	74	–	− 6	–	44	3.8	12	145[c]	(Arora et al. 2018)
Crypthecodinium cohnii	95	46.5	912	0.14	–	16.1	–	–	37.42–39.86	5.06	–	7.5[b]	(Islam et al. 2015)

[a]Ester content
[b]Sulfur content
[c]Saponification value

different FAs profiles of different feedstocks result in variations of CP. CFPP is another indicator for the operability of fuels in low temperatures. It is defined as the lowest temperature at which the fuel components start to gel or crystalize, and the fuel is no longer able to flow through test filters. PP refers to the lowest temperature at which the liquid fuel is still liquid and becomes solid and cannot flow or pump anymore at lower temperatures (Singh et al. 2019). In general, the CP and PP of biodiesel are higher than diesel. Fuels containing fewer concentrations of saturated FAs have better cold-flow properties and suitable viscosities comparing to that of high saturated FAs concentrations (Yaşar 2020). In addition, the mono and polyunsaturated FAs result in high CFPP values (Deshmukh et al. 2019). The standard limits of CP and PP for biodiesel according to the ASTM D6751 are − 3 to − 12 °C and − 15 to − 16 °C, respectively, while the CFPP standard is a maximum of + 5 °C (Bhuiya et al. 2016b).

Flash point (FP) defines as the temperature at which biodiesel forms vapor near its surface and ignites when exposed to a fire source (Atabani et al. 2013). The approximate FP of conventional diesel is 55–65 °C, while that of biodiesel is more than 150 °C. Due to the higher FP of biodiesel, it is supposed to be a safer fuel (than diesel) in terms of storage, handling, and transit. The FP standards according to ASTM D6751 and EN 14214 are the minimum of 130 and 101 °C, respectively (Bhuiya et al. 2016b). As provided in Table 8.3, the FP of approximately most of the biodiesel derived from waste oils is in line with the provided standards. For instance, the FP of biodiesel from camel fat and pig tallow are 158 and 178 °C, respectively (Sbihi et al. 2014; Suresh et al. 2021).

Calorific value (CV) is also known as heating value (HV) and is one of the most important characteristics of biodiesel as it is related to the fuel's energy content. The quantity of released energy by the combustion of a unit amount of fuel is called CV (Singh et al. 2019). Because of the higher oxygen content of biodiesel, the CV of biodiesel is about 10% lower than diesel. In this regard, the CV of biodiesel blends (with diesel) reduces by the increment of blend level. It is noteworthy that the moisture content of the feedstock highly affects the CV of biodiesel. However, the CV standard limitation is not specified in ASTM D6751 and EN 14214 standards, but the minimum value of this parameter is 35 MJ/kg based on EN 14,213 (Bhuiya et al. 2016b). In this regard, utilization of lipidic wastes for biodiesel production is efficient in terms of CV as they possess considerable amounts more than the mentioned standard (Table 8.3). For instance, the CV of biodiesel derived from chicken fat and waste fish oil is approximately 40 MJ/kg (Behçet et al. 2015; Mota et al. 2019).

The biodiesel quality in terms of self-ignition is determined by the cetane number (CN), and this feature directly affects the ignition delay of the fuel. This content is highly dependent on the composition of the feedstock. The CN increases by incrementing chain length and saturation of FAs (Deshmukh et al. 2019). On the other hand, the CN is negatively affected by the number of double bonds (Yesilyurt et al. 2020). Accordingly, one of the most important characteristics in the feedstock selection for biodiesel production is CN. Low CN is not advantageous since it causes knocking and increases gaseous emissions. Besides, the higher oxygen content of biodiesel results in a higher CN compared to conventional diesel. The ignition delay decreased as a result of the CN increment (Yaşar 2020). A minimum of 47 and 51 are required for the CN standards, according to ASTM D6751 and EN 14214 standards, respectively (Yesilyurt et al. 2020). The CN of biodiesel produced from various feedstocks, including edible/non-edible oils, waste oils, and algae provided in Table 8.3, is in line with the standards, suggesting the feasibility of these feedstocks for biodiesel production. Regarding lipidic wastes, the CN of mutton fat and leather waste fat was 59 and 58.8, respectively (Bhatti et al. 2008; Keskin et al. 2020).

The value of FFAs in the biodiesel sample is determined by the acid value (AV) (Singh et al. 2019). It is also defined as the required amount of KOH (mg) to neutralize one gram of FAMEs. High concentrations of FFAs content indicate the high AV, which causes severe problems, such as the corrosion of the supply system of the engine. The maximum AV of 0.5 mg KOH/g has been approved for both ASTM D6751 and EN 14214 standards (Yesilyurt et al. 2020). Approximately the AV of all reported biodiesel (e.g., biodiesel from beef tallow, chicken fat, pork lard, etc.) in Table 8.3 is in line with the provided standards.

The flow capability of fuel is an important parameter indicated by viscosity (V) (Dhawane et al. 2015). The atomization of fuel spray and injection equipment operation are affected by V, especially at low temperatures. Compared to fossil fuel, the high molecular weight and large chemical structure of biodiesel result in a higher V (10–15 times higher than fossil fuel) (Bhuiya et al. 2016b). Insufficient atomization of the fuel happens when the V is higher, resulting in the deposition of dirt and the decrement of thermal efficiency. In contrast, the lower the V, the smaller the size of the droplet, which makes the fuel delivery into the combustion chamber easier. Previous studies showed that during transesterification, the V of methyl esters decreases. This property limitation based on ASTM D6751 and EN 14214 standards is 1.9–6.0 and 3.5–5.0 mm^2/s at 40 °C, respectively (Yesilyurt et al. 2020). According to the literature, the V of produced biodiesel from various feedstocks (Table 8.3) are in line with ASTM D6751 and EN 14214 standards. The V of waste cooking oil and waste fish oil was 4.8 and 4.53 mm^2/s (Mota et al. 2019; Farid et al. 2020).

Density (D) is a physical property defined as the weight per unit volume used to investigate the approximate volume of required fuel to supply adequate fuel combustion. The FAs composition of feedstock and the purity of fuel affect the D, and biodiesel has a higher D than conventional diesel. The D of biodiesel should be between 860 and 900 kg/m^3 according to the EN 14214 standard and, it is 880 kg/m^3 according to the ASTM D6751 (Yesilyurt et al. 2020). D is a significant factor for the injector nozzle because it affects the engine operation and fuel atomization (Singh et al. 2019). Fuel properties like CV and CN are correlated to D. The increasing of FAs chain length and decreasing the double bonds result in the density decrement (Yesilyurt et al. 2020). Using lipidic wastes such as leather waste fat, pork lard, poultry fat, and waste cooking oil for biodiesel production led to the D of 875.5, 870–873, 877, and 889.1 kg/m^3, respectively (Farid et al. 2020; Keskin et al. 2020; Toldrá-Reig et al. 2020), that is in the range of standards.

Another critical characteristic to investigate the quality and performance of biodiesel is oxidation stability (OS) which measures the level of the fuel reacting with air, the level of oxidation, and determines the need for antioxidants. A minimum of 3 and 8 h is required according to ASTM D6751 and EN 14214 standards for biodiesel OS (Yesilyurt et al. 2020). Different factors such as light, heat, metal traces, peroxides, unsaturated FAs, and double bonds cause oxidation (Bhuiya et al. 2016b). The fuels derived from the feedstock containing high amounts of polyunsaturated FAs are more sensitive to oxidation comparing to those that contain saturated or monounsaturated FAs (Deshmukh et al. 2019; Yaşar 2020).

The iodine number (IN) is an indicator of double bonds amount in biodiesel which evaluates the biodiesel degree of unsaturation and oxidation tendency when the fuel contacts with air. The amount of IN varies according to the feedstock and should not exceed 120 g I_2/100 g according to the EN 14214 standard (Yesilyurt et al. 2020). The IN of biodiesel reported in previous studies is in line with the EN 14214 standard (Table 8.3); however, some (e.g., mutton fat and chicken fat) are more than 120 g I_2/100 g. Previous investigations indicated that IN is highly correlated to CN, V, and CFPP (Bhuiya et al. 2016b).

The value of sulfur oxides emitted from fuel combustion depends on the amount of sulfur content (SC) of the fuel. The biodiesel fuels derived from animal fats (e.g., pig tallow) and vegetable oils (e.g., safflower, Jatropha, palm, etc.) contain fewer sulfur quantities (Bhuiya et al. 2016b). The SC is at a maximum of 10 mg/kg based on the EN 14214 standard (Yesilyurt et al. 2020).

The temperature at which the vapor pressure equals the surrounding pressure is known as the boiling point (BP). It indicates the fuel volatility so that the lower contents of BP show the higher volatility (Singh et al. 2019). According to ASTM D6751, the range of 100–615 °C is standard for BP (Yesilyurt et al. 2020).

The amount of water and sediment in the biodiesel indicates its purity. The presence of water in biodiesel is either in the form of dissolved water or suspended droplets of water. The presence of water in the fuel may result in some problems like the corrosion of different parts of the engine such as fuel pumps and tubes and injector pumps, as well as the calorific value decrement. On the other hand, the presence of sediments (suspended rust and dirt particles) in the fuel may cause clogging in fuel lines. Water and sediment content should not exceed 0.05 vol% and 500 mg/kg based on ASTM D6751 and EN 14214, respectively (Atabani et al. 2012; Yesilyurt et al. 2020).

The biodiesel quality, which highly depends on the feedstock, is determined by ester content (EC). The inappropriate conditions of transesterification and impurities of feedstock oil may reduce the EC. The EC should be at least 96.5% based on EN 14214 standard (Yesilyurt et al. 2020). The EC of pig tallow and waste cooking oil was 99.2 and 97%, respectively, which are more than the standard limit (Farid et al. 2020; Suresh et al. 2021).

8.4.2 Engine Performance

Many different parameters influence the engine performance when biodiesel or its blend is used, such as biodiesel quality and its origin, engine operating parameters (e.g., load and speed), the actual start of combustion, air turbulence, air–fuel mixture quality, fuel injection pressure, spray pattern, and ignition delay. Different indicators for engines performance like brake power and torque, brake specific fuel consumption (BSFC), brake thermal efficiency (BTE), and exhaust gas temperature (EGT) have been studied when the engine is fueled with different types of biodiesel (Wan Ghazali et al. 2015; Bhuiya et al. 2016b).

Generally, in investigations about the performance of biodiesel and its blends, a reduction in brake power torque has been reported by many researchers. The main reason is reported to be the lower heating value of biodiesel compared to that of diesel. Further, biodiesel's lower brake power and torque are due to its high viscosity and density, which results in poor spray characteristics in the injector (Yesilyurt et al. 2020). Other researchers also stated that the main reason for the brake power reductions of fuel blends might be their lower heating values (Ilkiliç et al. 2011; Wan Ghazali et al. 2015). On the other hand, other researchers indicated that the lower viscosity of diesel results in higher engine torque values (Kul et al. 2016).

The ratio of the mass flow rate of the fuel to the engine power is known as BSFC. In other words, it is defined as the consumed fuel value by a vehicle for each unit of power output (Işik and Aydin 2019). Compared with conventional diesel, the BSFC value of biodiesel fuels is higher according to the inverse relationship between the calorific value and BSFC. Generally, since the biodiesel heating value is lower, and its density and viscosity are higher than diesel, the biodiesel BSFC increases compared to diesel (Datta and Mandal 2016; Yesilyurt et al. 2020).

BTE is defined as the brake power of an engine as a function of the thermal input from the fuel. It is the most crucial characteristic of an engine in terms of energy, which shows how well the heat from the fuel is converted to mechanical energy. As previously reported, biodiesel utilization may decrease the efficiency according to its heating value and viscosity, which results in the BSFC increment and further the fuel efficiency decrement. This decrease in efficiency could be enhanced by increasing the injection pressure, controlling the injection time and compression ratio (Işik and Aydin 2019). According to the literature, the BTE in diesel–biodiesel blends is a little lower or the same as diesel (Yesilyurt et al. 2020).

EGT is significant due to the exhaust emissions clarification. The previous studies state that the EGT of either biodiesel or its blends is higher than conventional diesel. The reason, as depicted in the earlier sections, is the oxygen content of biodiesel. The biodiesel combustion improves according to the high content of its oxygen, which increases EGT values (Yesilyurt et al. 2020).

8.4.3 Exhaust Emissions

The emissions of different fuels (diesel, biodiesel, and biodiesel blends) in compression ignition (CI) engines vary according to the fuel quality, engine design, and operating conditions. In other words, the emission level differs from one fuel to another as well as from one engine to another. These emissions are mainly particulate matter, nitrogen oxides, hydrocarbons, carbon monoxide, and carbon dioxide (Reşitoğlu et al. 2015). The adverse effects of these emissions on the environment and public health have been extensively studied.

The hydrocarbons are released due to inefficient combustion according to the reduction of combustion temperature, lack of oxygen, and incomplete combustion (Yesilyurt et al. 2020). When the oxygen content in fuel is low, the combustion is deficient, resulting in the rapid increment of hydrocarbons emissions. According to the higher oxygen content of biodiesel in comparison with diesel, previous studies show that hydrocarbon emissions are reduced when biodiesel is used instead of diesel. Besides, the high cetane number of biodiesel, which results in the combustion delay reduction, is another reason for declining of hydrocarbon emissions (Ilkiliç et al. 2011).

Particulate matter emissions present in the exhaust gas may be produced from the agglomeration of tiny particles of partly burned fuel and lube oil, ash content of fuel oil, and cylinder lube oil or sulfates and water. However, most of these particulate matters are released as a result of incomplete combustion of a hydrocarbon. Generally, particulate matters in diesel consist of soot, soluble inorganic fraction, and an inorganic fraction (Reşitoğlu et al. 2015). When using biodiesel, the particulate matter emissions are reduced. The oxygen content of biodiesel improves the combustion and provides the oxidation of already formed soot. In addition, the sulfur content of biodiesel that is lower than that of diesel prevents sulfate formation, resulting in lower particulate matter emissions (Ilkiliç et al. 2011; Yesilyurt et al. 2020).

The sum of NO, NO_2, N_2O, N_2O_3, and N_2O_5 is defined as nitrogen oxide or NO_x emissions. Generally, the NO_x emissions of diesel are lower than that of biodiesel or its blends. The NO_x emissions of fuel are dependent on different parameters such as oxygen content, cetane number, and iodine number. The combustion of biodiesel is better according to its higher oxygen content and cetane number, but resulting in higher NO_x emissions. In addition, the higher iodine value of biodiesel causes higher emissions of nitrogen oxides. The engine load and speed may also alter the NO_x emissions of fuel (Ilkiliç et al. 2011; Yesilyurt et al. 2020).

CO_2 is an important parameter that represents a sign of complete combustion and is emitted as a result of burning the fuels' hydrocarbons. CO_2 is well known for its greenhouse effect, which results in global warming. The emissions of CO_2 and the engine load are directly related to each other. When the engine load is increased, the mass fuel consumption increases and results in the increment of CO_2 emissions. Comparing to diesel, biodiesel is a lower carbon fuel and has lower carbon-to-hydrogen ratio. Therefore, biodiesel combustion results in lower CO_2 emissions (Wan Ghazali et al. 2015). According to the results of some previous studies, biodiesel has no negative effect on global warming, as the emitted CO_2 to the atmosphere was absorbed by the oil plants in the photosynthesis cycle (Bhuiya et al. 2016b).

As a result of incomplete combustion of the fuel, when the oxidation does not happen completely, the carbon monoxide is emitted in the exhaust gases. The air-to-fuel ratio of the mixture is the reason for CO emissions. In other words, the oxygen content inadequacy in the combustion chamber is the most important cause of CO emissions formation. As previously indicated by scientists, these emissions are minimum when biodiesel is used as fuel. As stated in the previous sections, the higher oxygen content and cetane number of biodiesel result in the fuel's better combustion efficiency, which prevents the formation of CO (Reşitoğlu et al. 2015).

8.5 Pretreatments, Downstream Processing, and By-Products Manipulation

8.5.1 Pretreatment

As previously mentioned, feedstock is the most expensive contributor to biodiesel production. Therefore, the utilization of low-cost materials such as waste oils or animal fats is an appropriate strategy to decrease the production cost. However, these inexpensive substances usually contain considerable quantities of undesirable materials, such as water, gums, proteins, suspended particles, oxidative products, phospholipids, and mainly FFAs, which hinder the direct conversion of oil to biodiesel. Pretreatment is a way to decrease or remove undesirable substances. Various types of pretreatments are employed in the industrial production of biodiesel (Šánek et al. 2015). In this section, different pretreatments in the biodiesel production process are summarized.

Degumming is the first step of pretreatment, which is used for phospholipid removal. Phospholipids can cause water formation in the oil, and they make the oil appearance muddy. Degumming is thus conducted using water, acid, enzyme, or semi-permeable membranes (Šánek et al. 2015). Water and acid degumming is used to eliminate soluble and insoluble phospholipids, respectively. It is worth noting that after degumming, other refining approaches are required (Ayoub et al. 2021).

Distillation or steam stripping is a method performed to remove the content of FFAs. Steam or nitrogen gas is used as the stripping gas. Nitrogen gas is more beneficial for stripping since it decreases the oil losses and produces a more stable deacidified oil (Šánek et al. 2015). This method is carried out for the feedstocks with less than 10% of FFAs (Tafesh and Basheer 2013). Solvent extraction is another way of pretreatment for FFAs content reduction using different solvents, such as dimethyl ether, methanol, polyethylene glycol, and ethane. This method is useful for animal fat wastes refining, which has high FFAs content (Šánek et al. 2015).

Liquid acid treatment requires strong acids such as sulfuric acid and an excess amount of methanol. These hazardous materials have negative environmental impacts. This method is suitable for feedstocks with more than 10% of FFAs. However, it needs a neutralization step that undesirably produces a large quantity of salt, and it also generates water. Therefore, the disadvantage of this method is the production of a low-quality solution of methanol, water, and salts (Tafesh and Basheer 2013). Esterification of FFAs in the presence of resins is another beneficial method for FFAs reduction. Resin beads are heterogeneous, so their separation from the solution is easier than homogeneous catalysts. In addition, this method does not need a neutralization step, so it produces no water or salts (Tafesh and Basheer 2013).

The reaction of glycerol with FFAs is called glycerolysis or glycerol esterification. This method is capable of reducing the FFAs content of low-quality feedstocks. The products of this treatment are water and glycerides. It is not common in the biodiesel production processes since it requires expensive metallic catalysts and high temperatures (Elgharbawy et al. 2021).

Lipases are among the enzymes that are utilized in the esterification of FFAs. In the presence of methanol, these enzymes convert FFAs to biodiesel. This procedure is cost-effective and eco-friendly since it can reduce the adverse effects of yellow and brown grease on the environment. This method is suitable for feedstocks with FFAs contents ranging from 0 to 100%. Besides, using this method eliminates the need for the degumming process (Tafesh and Basheer 2013).

Pretreatment of feedstocks with adsorbents, such as bentonite, magnesium silicates, or zeolite, is an inexpensive, effective way for FFAs removal in commercial scales. These materials adsorb impurities and bind them to a solid particulate with weak bonds. After the reaction, they are filtered and separated from the feedstock. Because of the chemical bond of adsorbents and impurities, it is not worthy of recycling and reusing the adsorbents (Tafesh and Basheer 2013).

8.5.2 Downstream Processing and By-Products Manipulation

After the transesterification, purification of the crude biodiesel and glycerol is necessary to enhance the quality of the products. The crude biodiesel contains different concentrations of impurities such as catalysts, water, and FFAs, which negatively affect the engine performance. Therefore, purification is needed to enhance the biodiesel quality to meet ASTM D6751 and EN 14214 standards. Several methods such as distillation, evaporation, washing with water, and gravitational settling have been proposed for biodiesel separation and purification. Crude glycerol, the major by-product of the process, is known as a high-value by-product and has several critical applications; therefore, its purification and utilization in different industrial sectors could reduce the overall cost of biodiesel production. For instance, if the glycerol is purified to the pharmaceutical grade, biodiesel production is more economical. During the biodiesel production process, crude glycerol and other by-products are produced, which have different applications. This section discusses various by-products of biodiesel and their applications in the industry.

When different feedstocks of edible/non-edible crops and algae are subjected to oil extraction, the residue after the extraction is the first by-product. Its type is dependent on the type of feedstock and the extraction procedure. When the oil is extracted by the mechanical press, the solid residue is known as oil cake. If the oil is extracted by a solvent, the residue is called an oil meal. In addition to the feedstock type, the oil cake/meal composition is dependent on the feedstock preparation conditions and the oil extraction methods (Kolesárová et al. 2011). The oil cake/meal contains high concentrations of protein, nitrogen, potassium, and phosphorus and is divided into edible and non-edible types. Therefore, it could be used as animal feed in poultry, ruminants, and fish industries and also as a suitable alternative for chemical fertilizers as they contain considerable concentrations of nitrogen, potassium, and phosphorus. Besides, this by-product has many biotechnological applications in producing amino acids, antibiotics, bio-pesticides, enzymes, ethanol, and other biochemicals (Ramachandran et al. 2007). High digestible components of some oil cakes/meals make them a suitable substrate for biogas production through anaerobic digestion. It is noteworthy that using these oil cakes/meals for different applications is extremely dependent on their type. For instance, the oil cakes of Jatropha and karanja contain toxic materials and could not be used in feed or fertilizer industries; however, they could be useful in bio-pesticides production (Rastegari et al. 2019).

Glycerol is well known as the main by-product of transesterification, while approximately about 10% (w/w) glycerol is generated during the process. Crude glycerol contains different concentrations of glycerol, methanol, organic and inorganic salts, FAMEs, FFAs, and soap depending on the biodiesel production methods (Xiao et al. 2013). The crude glycerol should be subjected to a convenient process to remove the impurities. Its purification is expensive and is not economical in small-scale plants. Therefore, to overcome the problems of crude glycerol, it is subjected to anaerobic digestion or transferred to industrial-scale plants for purification (Rastegari et al. 2019). Purified glycerol has numerous applications in the food, pharmaceutical, and cosmetic industries. It has been used as the initial substrate to produce various chemicals such as carboxylic acids (e.g., glyceric, tartronic, and mesoxalic acids), acrolein, and 3-hydroxypropionaldehyde through different reactions such as dehydration, etherification, oxidation, and pyrolysis. Another exciting application of glycerol is hydrogen production. Generally, glycerol is converted to hydrogen through aqueous-phase reforming, auto-thermal reforming, partial oxidation gasification, steam reforming, and supercritical water reforming processes (Avasthi et al. 2013). In some cases, glycerol is transformed into fuel additives (e.g., alcohol, ethers) used to enhance fuel properties (Anitha et al. 2016). In addition to the mentioned characteristics of glycerol, it has been regarded as a suitable carbon source for different bioprocesses. For instance, glycerol has been extensively used in the mixotrophic cultivation of various microorganisms such as microalgae for enhanced biomass and oil content of the microorganism (Xu et al. 2019). In the anaerobic digestion of sewage sludge, crude glycerol is regarded as a potential co-substrate. Besides, glycerol is a suitable substrate in fermentation as it could be fermented to generate products like butanol, ethanol, and lactate. Many researchers have studied the application of glycerol in wastewater treatment. It has been proposed as a hole scavenger as well as a potential carbon source in the removal of nitrates, nitrite, and phosphorus (Anitha et al. 2016).

As mentioned in the previous sections, methanol is widely used in the biodiesel production process. Accordingly, the overall consumption of methanol could be reduced when recycled and reused in the process. In this regard, the excess amount of methanol is separated through distillation and reused in the next transesterification processes (Rastegari et al. 2019).

When the transesterification is finished, the crude biodiesel and glycerol are separated and undergo purification. The crude glycerol impurities were discussed in the previous sections. Crude biodiesel also contains some impurities (e.g., acylglycerols, catalyst, FFAs, free glycerol, residual alcohol, salts, and soap) that should be removed for biodiesel quality improvement. Generally, the biodiesel is purified using a washing process, which results in the generation of wastewater (Berrios and Skelton 2008). Obviously, this wastewater contains biodiesel impurities that could be used as feedstock for some high-value products production. Besides, biodiesel wastewater could be treated using several methods, such as physicochemical and electrochemical treatments, advanced oxidation technologies, or reused in the transesterification process (Rastegari et al. 2019).

8.6 Economic Feasibility

Generally, the utilization of renewable resources for biofuel production is advantageous in terms of environmental protection. Their combustion will reduce the net greenhouse gas compared to conventional fuels (e.g., crude oil, coal, and gas) (Xu et al. 2019). However, the biofuel production and utilization in terms of economic feasibility are doubtful; their environmental benefits strengthen the global tendency to use them. In line with the contents of this chapter, the most crucial challenge about biodiesel is its higher cost of production compared to that of conventional diesel, which hinders the widespread use of biodiesel. In this context, lowering the biodiesel production cost may be possible by considering some modifications through the process, such as production technologies improvement for productivity and/or yield enhancement (Mahlia et al. 2020). Regarding the production of biodiesel, a suitable method should be considered, which requires low energy and chemicals, produces low toxic by-products, and could be easily scaled up (Abomohra et al. 2020b).

It has been well studied that the high cost of biodiesel production is related to the type of raw material used as feedstock, in which between 60 and 75% of the total cost of biodiesel fuel is related to the feedstock (Atabani et al. 2012; Rekhate and Prajapati 2019). Accordingly, in terms of large-scale biodiesel production, the main issue is employing a suitable sustainable feedstock. Currently, most of the world's biodiesel is produced using edible oil crops. Therefore, as these sources are not sustainable, finding more reliable, economically feasible, and sustainable sources is needed. In this regard, using inexpensive feedstocks like non-edible vegetable oils and waste oils/fats should reduce the overall cost. Regarding the cost of feedstock, it has been reported that vegetable oils cost about 2.5–3 times higher than waste cooking oil (Rezania et al. 2019). However, the costs of biodiesel production could be comparable with conventional diesel when government policies are in line with collecting waste cooking oil for biodiesel production at no cost (Rekhate and Prajapati 2019). The economic feasibility of biodiesel production using virgin vegetable oil and waste cooking oil was investigated through various scenarios as follows

(Zhang et al. 2003); using base catalyst in the transesterification of virgin vegetable oil (1), esterification followed by base catalyst in the transesterification of waste cooking oil (2), using acid catalyst in the transesterification of waste cooking oil (3), and using acid catalyst in the transesterification of waste cooking oil and hexane as an extraction solvent (4). In this study, utilization of waste cooking oil in the presence of acid catalyst was more economically feasible.

Moreover, there are some bottlenecks in using wastes as feedstock. They commonly contain higher FFAs and water in their structure, which directly affects the transesterification yield and the quality of the products. For instance, it has been reported by researchers that the biodiesel conversion rate reduces by increasing the FFAs level (Toldrá-Reig et al. 2020). Accordingly, as mentioned in previous sections, additional processes are required to reduce the FFAs and water content and an alternative approach for the production, leading to additional costs for biodiesel production. Therefore, when low-cost feedstocks are used, performing a suitable configuration, either upstream or downstream processing, is required in order to prevent the additional costs to the whole process. For example, increasing the FFAs level from 5 to 33% by the addition of palmitic acid to soybean oil resulted in a significant drop of biodiesel conversion from 90.5 to 58.8%, in which the reaction optimization caused a high conversion of 98% in the high-level FFAs samples (Abomohra et al. 2020b).

Regarding process intensification, three different processes were studied to produce biodiesel from spent oil containing 5% of FFAs (Marchetti et al. 2008). The processes are homogenous acid catalyst with acid pre-esterification, homogenous acid catalyst, and heterogenous solid catalyst. As they reported, the investment cost of the first process (homogenous acid catalyst with acid pre-esterification) was the highest, but the operation costs for biodiesel production were lower than the other two processes. In another study, the economic feasibility of biodiesel production from waste cooking oil utilizing homogenous acid, base, and enzymatic catalysts was investigated (Karmee et al. 2015), in which acid catalyst transesterification was reported to be the most cost-effective. The reason is fewer steps for production and purification.

In addition to the production technologies enhancement, some strategies like catalysts, water, and methanol reusability are efficient in reducing the costs of the whole process. Regarding the catalyst type, utilization of waste catalysts (as discussed in Sect. 8.3) could enhance the process's economic feasibility. For instance, waste freshwater mussel shell was used as catalyst for transesterification of Chinese tallow oil (Hu et al. 2011). In this study, the prepared catalyst was active for seven reaction cycles with more than 90% of biodiesel yield, which could reduce the costs related to the catalyst utilization. In terms of process intensification, performing waste catalysts in simultaneous esterification and transesterification for biodiesel production could be beneficial. In this regard, rice husk was used for the production of solid acid catalysts in order to produce biodiesel from waste cooking oil via simultaneous esterification and transesterification (Li et al. 2014). According to the results, the proposed procedure was promising as the 98.17% conversion of FFAs was obtained after 3 h, as well as the FAME yield of 87.57% after 15 h.

8.7 Conclusions and Perspectives

This chapter provides extensive information on different methods for biodiesel production as well as biodiesel characteristics and by-products. There are many different methods for biodiesel production that can be performed according to the feedstock. While there might be different promising feedstocks for biodiesel production, biodiesel production from renewable, sustainable resources attracts worldwide attention to overcome the problems derived from the use of current feedstocks. Accordingly, because of the cheap and abundant nature of lipidic wastes, biodiesel production using these feedstocks could be an alternative for the current biodiesel feedstocks. However, there are many drawbacks to biodiesel production using these wastes, which encourages the researchers to investigate suitable methods to make better use of these wastes in biodiesel production.

References

Abomohra AE-F, El-Naggar AH, Alaswad SO et al (2020a) Enhancement of biodiesel yield from a halophilic green microalga isolated under extreme hypersaline conditions through stepwise salinity adaptation strategy. Bioresour Technol 310:123462. https://doi.org/10.1016/j.biortech.2020.123462

Abomohra AE-F, Elsayed M, Esakkimuthu S et al (2020b) Potential of fat, oil and grease (FOG) for biodiesel production: a critical review on the recent progress and future perspectives. Prog Energy Combust Sci 81:100868. https://doi.org/10.1016/j.pecs.2020.100868

Adewale P, Dumont MJ, Ngadi M (2015) Recent trends of biodiesel production from animal fat wastes and associated production techniques. Renew Sustain Energy Rev 45:574–588. https://doi.org/10.1016/j.rser.2015.02.039

Adewale P, Dumont MJ, Ngadi M (2016) Enzyme-catalyzed synthesis and kinetics of ultrasonic assisted methanolysis of waste lard for biodiesel production. Chem Eng J 284:158–165. https://doi.org/10.1016/j.cej.2015.08.053

Agarwal AK, Khurana D, Dhar A (2015) Improving oxidation stability of biodiesels derived from karanja, neem and Jatropha: step forward in the direction of commercialisation. J Clean Prod 107:646–652. https://doi.org/10.1016/j.jclepro.2015.05.055

Ahmad T, Aadil RM, Ahmed H et al (2019) Treatment and utilization of dairy industrial waste: a review. Trends Food Sci Technol 88:361–372. https://doi.org/10.1016/j.tifs.2019.04.003

Alagumalai A, Mahian O, Hollmann F, Zhang W (2021) Environmentally benign solid catalysts for sustainable biodiesel production: a critical review. Sci Total Environ 768:144856. https://doi.org/10.1016/j.scitotenv.2020.144856

Andreo-Martínez P, Ortiz-Martínez VM, García-Martínez N et al (2020) Production of biodiesel under supercritical conditions: state of the art and bibliometric analysis. Appl Energy 264:114753. https://doi.org/10.1016/j.apenergy.2020.114753

Anitha M, Kamarudin SK, Kofli NT (2016) The potential of glycerol as a value-added commodity. Chem Eng J 295:119–130. https://doi.org/10.1016/j.cej.2016.03.012

Arora N, Patel A, Pruthi PA et al (2018) Utilization of stagnant non-potable pond water for cultivating oleaginous microalga *Chlorella minutissima* for biodiesel production. Renew Energy 126:30–37. https://doi.org/10.1016/j.renene.2018.03.033

Arshad M, Zia MA, Shah FA, Ahmad M (2018) An overview of biofuel. In: Perspectives on water usage for biofuels production, pp 1–37

Atabani AE, Silitonga AS, Badruddin IA et al (2012) A comprehensive review on biodiesel as an alternative energy resource and its characteristics. Renew Sustain Energy Rev 16:2070–2093. https://doi.org/10.1016/j.rser.2012.01.003

Atabani AE, Mahlia TMI, Masjuki HH et al (2013) A comparative evaluation of physical and chemical properties of biodiesel synthesized from edible and non-edible oils and study on the effect of biodiesel blending. Energy 58:296–304. https://doi.org/10.1016/j.energy.2013.05.040

Athar M, Zaidi S (2020) A review of the feedstocks, catalysts, and intensification techniques for sustainable biodiesel production. J Environ Chem Eng 8:104523. https://doi.org/10.1016/j.jece.2020.104523

Avasthi KS, Reddy RN, Patel S (2013) Challenges in the production of hydrogen from glycerol—a biodiesel byproduct via steam reforming process. Procedia Eng 51:423–429. https://doi.org/10.1016/j.proeng.2013.01.059

Ayoub M, Hizami M, Yusoff M et al (2021) A comprehensive review on oil extraction and biodiesel production technologies. Sustainability 13:788. https://doi.org/10.3390/su13020788

Ballesteros R, Monedero E, Guillén-Flores J (2011) Determination of aldehydes and ketones with high atmospheric reactivity on diesel exhaust using a biofuel from animal fats. Atmos Environ 45:2690–2698. https://doi.org/10.1016/j.atmosenv.2011.02.051

Behçet R, Oktay H, Çakmak A, Aydin H (2015) Comparison of exhaust emissions of biodiesel–diesel fuel blends produced from animal fats. Renew Sustain Energy Rev 46:157–165. https://doi.org/10.1016/j.rser.2015.02.015

Berrios M, Skelton RL (2008) Comparison of purification methods for biodiesel. Chem Eng J 144:459–465. https://doi.org/10.1016/j.cej.2008.07.019

Betiku E, Omilakin OR, Ajala SO et al (2014) Mathematical modeling and process parameters optimization studies by artificial neural network and response surface methodology: a case of non-edible neem (*Azadirachta indica*) seed oil biodiesel synthesis. Energy 72:266–273. https://doi.org/10.1016/j.energy.2014.05.033

Bhatti HN, Hanif MA, Qasim M, Ata-ur-Rehman (2008) Biodiesel production from waste tallow. Fuel 87:2961–2966. https://doi.org/10.1016/j.fuel.2008.04.016

Bhuiya MMK, Rasul MG, Khan MMK et al (2016a) Prospects of 2nd generation biodiesel as a sustainable fuel—Part 1: selection of feedstocks, oil extraction techniques and conversion technologies. Renew Sustain Energy Rev 55:1109–1128. https://doi.org/10.1016/j.rser.2015.04.163

Bhuiya MMK, Rasul MG, Khan MMK et al (2016b) Prospects of 2nd generation biodiesel as a sustainable fuel—Part 2: properties, performance and emission characteristics. Renew Sustain Energy Rev 55:1129–1146. https://doi.org/10.1016/j.rser.2015.09.086

Bokhari A, Chuah LF, Yusup S et al (2015) Microwave-assisted methyl esters synthesis of Kapok (*Ceiba pentandra*) seed oil: parametric and optimization study. Biofuel Res J 2:281–287. https://doi.org/10.18331/BRJ2015.2.3.6

Bokhari A, Chuah LF, Yusup S et al (2016) Cleaner production of rubber seed oil methyl ester using a hydrodynamic cavitation: optimisation and parametric study. J Clean Prod 136:31–41. https://doi.org/10.1016/j.jclepro.2016.04.091

Borah MJ, Devi A, Saikia RA, Deka D (2018) Biodiesel production from waste cooking oil catalyzed by in-situ decorated TiO2 on reduced graphene oxide nanocomposite. Energy 158:881–889. https://doi.org/10.1016/j.energy.2018.06.079

Borah MJ, Devi A, Borah R, Deka D (2019) Synthesis and application of Co doped ZnO as heterogeneous nanocatalyst for biodiesel production from non-edible oil. Renew Energy 133:512–519. https://doi.org/10.1016/j.renene.2018.10.069

Buyukkaya E, Benli S, Karaaslan S, Guru M (2013) Effects of trout-oil methyl ester on a diesel engine performance and emission characteristics. Energy Convers Manag 69:41–48. https://doi.org/10.1016/j.enconman.2013.01.017

Casallas ID, Carvajal E, Mahecha E et al (2018) Pre-treatment of waste cooking oils for biodiesel production. Chem Eng Trans 65:385–390. https://doi.org/10.3303/CET1865065

Chapagain BP, Yehoshua Y, Wiesman Z (2009) Desert date (*Balanites aegyptiaca*) as an arid lands sustainable bioresource for biodiesel. Bioresour Technol 100:1221–1226. https://doi.org/10.1016/j.biortech.2008.09.005

Chatterjee R, Sharma V, Kumar S et al (2012) Life cycle assessment of energy performance of biodiesel produced from Jatropha curcas Life cycle assessment of energy performance of biodiesel produced from Jatropha curcas. J Renew Sustain Energy 4:053110. https://doi.org/10.1063/1.4754156

Chavan SB, Kumbhar RR, Deshmukh RB (2013) *Callophyllum Inophyllum* Linn ("honne") oil, a source for biodiesel production. Res J Chem Sci 3:24–31

Chavan SB, Yadav M, Singh R et al (2017) Production of biodiesel from three indigenous feedstock: optimization of process parameters and assessment of various fuel properties. Environ Prog Sustain Energy 36:788–795. https://doi.org/10.1002/ep.12606

Chen Y, Chiang T, Chen J (2013) Properties of soapnut (*Sapindus mukorossi*) oil biodiesel and its blends with diesel. Biomass Bioenerg 52:15–21. https://doi.org/10.1016/j.biombioe.2013.02.025

Chokshi K, Pancha I, Ghosh A, Mishra S (2016) Microalgal biomass generation by phycoremediation of dairy industry wastewater: an integrated approach towards sustainable biofuel production. Bioresour Technol 221:455–460. https://doi.org/10.1016/j.biortech.2016.09.070

Chung ZL, Tan YH, Chan YS et al (2019) Biocatalysis and Agricultural Biotechnology Life cycle assessment of waste cooking oil for biodiesel production using waste chicken eggshell derived CaO as catalyst via transesterification. Biocatal Agric Biotechnol 21:101317.https://doi.org/10.1016/j.bcab.2019.101317

Da Rós PCM, Silva GAM, Mendes AA et al (2010) Evaluation of the catalytic properties of *Burkholderia cepacia* lipase immobilized on non-commercial matrices to be used in biodiesel synthesis from different feedstocks. Bioresour Technol 101:5508–5516. https://doi.org/10.1016/J.BIORTECH.2010.02.061

Datta A, Mandal BK (2016) A comprehensive review of biodiesel as an alternative fuel for compression ignition engine. Renew Sustain Energy Rev 57:799–821. https://doi.org/10.1016/j.rser.2015.12.170

Demirbas A, Bafail A, Ahmad W, Sheikh M (2016) Biodiesel production from non-edible plant oils. Energy Explor Exploit 34:290–318. https://doi.org/10.1177/0144598716630166

Deshmukh S, Kumar R, Bala K (2019) Microalgae biodiesel: a review on oil extraction, fatty acid composition, properties and effect on engine performance and emissions. Fuel Process Technol 191:232–247. https://doi.org/10.1016/j.fuproc.2019.03.013

Dhawane SH, Kumar T, Halder G (2015) Central composite design approach towards optimization of flamboyant pods derived steam activated carbon for its use as heterogeneous catalyst in transesterification of *Hevea brasiliensis* oil. Energy Convers Manag 100:277–287. https://doi.org/10.1016/j.enconman.2015.04.083

Dias JM, Alvim-Ferraz MCM, Almeida MF (2008) Comparison of the performance of different homogeneous alkali catalysts during transesterification of waste and virgin oils and evaluation of biodiesel quality. Fuel 87:3572–3578. https://doi.org/10.1016/j.fuel.2008.06.014

Dufour J, Iribarren D (2012) Life cycle assessment of biodiesel production from free fatty acid-rich wastes. Renew Energy 38:155–162. https://doi.org/10.1016/j.renene.2011.07.016

Elgharbawy AS, Sadik WA, Sadek OM et al (2021) A review on biodiesel feedstocks and production technologies. J Chil Chem Soc 66:5098–5109. https://doi.org/10.4067/S0717-97072021000105098

Essamlali Y, Amadine O, Fihri A, Zahouily M (2019) Sodium modified fluorapatite as a sustainable solid bi-functional catalyst for biodiesel production from rapeseed oil. Renew Energy 133:1295–1307. https://doi.org/10.1016/j.renene.2018.08.103

Ewunie GA, Morken J, Lekang OI, Yigezu ZD (2021) Factors affecting the potential of Jatropha curcas for sustainable biodiesel production: a critical review. Renew Sustain Energy Rev 137:110500. https://doi.org/10.1016/j.rser.2020.110500

Eze VC, Phan AN, Harvey AP (2018) Intensified one-step biodiesel production from high water and free fatty acid waste cooking oils. Fuel 220:567–574. https://doi.org/10.1016/j.fuel.2018.02.050

Fadhil AB, Ali LH (2013) Alkaline-catalyzed transesterification of *Silurus triostegus* Heckel fish oil: optimization of transesterification parameters. Renew Energy 60:481–488. https://doi.org/10.1016/j.renene.2013.04.018

Farag HA, El-Maghraby A, Taha NA (2011) Optimization of factors affecting esterification of mixed oil with high percentage of free fatty acid. Fuel Process Technol 92:507–510. https://doi.org/10.1016/j.fuproc.2010.11.004

Farid MAA, Roslan AM, Hassan MA et al (2020) Net energy and techno-economic assessment of biodiesel production from waste cooking oil using a semi-industrial plant: a Malaysia perspective. Sustain Energy Technol Assess 39:100700. https://doi.org/10.1016/j.seta.2020.100700

Farobie O, Matsumura Y (2017) State of the art of biodiesel production under supercritical conditions. Prog Energy Combust Sci 63:173–203. https://doi.org/10.1016/j.pecs.2017.08.001

Feng P, Deng Z, Hu Z et al (2014) Characterization of *Chlorococcum pamirum* as a potential biodiesel feedstock. Bioresour Technol 162:115–122. https://doi.org/10.1016/j.biortech.2014.03.076

Francisco EC, Neves DB et al (2010) Microalgae as feedstock for biodiesel production: carbon dioxide sequestration, lipid production and biofuel quality. J Chem Technol Biotechnol 85:395–403. https://doi.org/10.1002/jctb.2338

Gardy J, Osatiashtiani A, Céspedes O et al (2018) A magnetically separable SO_4/Fe-Al-TiO_2 solid acid catalyst for biodiesel production from waste cooking oil. Appl Catal B Environ 234:268–278. https://doi.org/10.1016/j.apcatb.2018.04.046

Gnanasekaran S, Saravanan N, Ilangkumaran M (2016) Influence of injection timing on performance, emission and combustion characteristics of a DI diesel engine running on fish oil biodiesel. Energy 116:1218–1229. https://doi.org/10.1016/j.energy.2016.10.039

Guo R, Hanaki K (2010) Potential and life cycle assessment of biodiesel production in China potential and life cycle assessment of biodiesel production in China. J Renew Sustain Energy 2:033107. https://doi.org/10.1063/1.3449298

Gupta AR, Rathod VK (2018) Waste cooking oil and waste chicken eggshells derived solid base catalyst for the biodiesel production: optimization and kinetics. Waste Manag 79:169–178. https://doi.org/10.1016/j.wasman.2018.07.022

He C, Mei Y, Zhang Y et al (2020) Enhanced biodiesel production from diseased swine fat by ultrasound-assisted two-step catalyzed process. Bioresour Technol 304:123017. https://doi.org/10.1016/j.biortech.2020.123017

Hu S, Wang Y, Han H (2011) Utilization of waste freshwater mussel shell as an economic catalyst for biodiesel production. Biomass Bioenergy 35:3627–3635. https://doi.org/10.1016/j.biombioe.2011.05.009

Idowu I, Pedrola MO, Wylie S et al (2019) Improving biodiesel yield of animal waste fats by combination of a pre-treatment technique and microwave technology. Renew Energy 142:535–542. https://doi.org/10.1016/j.renene.2019.04.103

Ilkiliç C, Aydin S, Behcet R, Aydin H (2011) Biodiesel from safflower oil and its application in a diesel engine. Fuel Process Technol 92:356–362. https://doi.org/10.1016/j.fuproc.2010.09.028

Inayat A, Nassef AM, Rezk H et al (2019) Fuzzy modeling and parameters optimization for the enhancement of biodiesel production from waste frying oil over montmorillonite clay K-30. Sci Total Environ 666:821–827. https://doi.org/10.1016/j.scitotenv.2019.02.321

Işik MZ, Aydin H (2019) Investigation on the effects of gasoline reactivity controlled compression ignition application in a diesel generator in high loads using safflower biodiesel blends. Renew Energy 133:177–189. https://doi.org/10.1016/j.renene.2018.10.025

Islam MA, Rahman MM, Heimann K et al (2015) Combustion analysis of microalgae methyl ester in a common rail direct injection diesel engine. Fuel 143:351–360. https://doi.org/10.1016/j.fuel.2014.11.063

Jeon J, Park S (2018) Effect of injection pressure on soot formation/oxidation characteristics using a two-color photometric method in a compression-ignition engine fueled with biodiesel blend (B20). Appl Therm Eng 131:284–294. https://doi.org/10.1016/j.applthermaleng.2017.12.005

Jeon H, Kim DJ, Kim SJ, Kim JH (2013) Synthesis of mesoporous MgO catalyst templated by a PDMS-PEO comb-like copolymer for biodiesel production. Fuel Process Technol 116:325–331. https://doi.org/10.1016/j.fuproc.2013.07.013

Kaewdaeng S, Sintuya P, Nirunsin R (2017) Biodiesel production using calcium oxide from river snail shell ash as catalyst. Energy Procedia 138:937–942. https://doi.org/10.1016/j.egypro.2017.10.057

Kant Bhatia S, Kant Bhatia R, Jeon JM et al (2021) An overview on advancements in biobased transesterification methods for biodiesel production: oil resources, extraction, biocatalysts, and process intensification technologies. Fuel 285:119117. https://doi.org/10.1016/j.fuel.2020.119117

Karmakar A, Karmakar S, Mukherjee S (2010) Properties of various plants and animals feedstocks for biodiesel production. Bioresour Technol 101:7201–7210. https://doi.org/10.1016/j.biortech.2010.04.079

Karmee SK, Patria RD, Lin CSK (2015) Techno-economic evaluation of biodiesel production from waste cooking oil—a case study of Hong Kong. Int J Mol Sci 16:4362–4371. https://doi.org/10.3390/ijms16034362

Kavitha V, Geetha V, Jacqueline PJ (2019) Production of biodiesel from dairy waste scum using eggshell waste. Process Saf Environ Prot 125:279–287. https://doi.org/10.1016/j.psep.2019.03.021

Keskin A, Mehmet Ş, Osman A (2020) Experimental studies on biodiesel production from leather industry waste fat and its effect on diesel engine characteristics. Fuel 276:118000. https://doi.org/10.1016/j.fuel.2020.118000

Kolesárová N, Hutňan M, Hutňan H et al (2011) Utilization of biodiesel by-products for biogas production. J Biomed Biotechnol. https://doi.org/10.1155/2011/126798

Kul BS, Kahraman A, Knuth KH (2016) Energy and exergy analyses of a diesel engine fuelled with biodiesel-diesel blends containing 5% bioethanol. Entropy 18:387. https://doi.org/10.3390/e18110387

Lam MK, Lee KT, Mohamed AR (2010) Homogeneous, heterogeneous and enzymatic catalysis for transesterification of high free fatty acid oil (waste cooking oil) to biodiesel: a review. Biotechnol Adv 28:500–518. https://doi.org/10.1016/j.biotechadv.2010.03.002

Lawan I, Garba ZN, Zhou W et al (2020) Synergies between the microwave reactor and CaO/zeolite catalyst in waste lard biodiesel production. Renew Energy 145:2550–2560. https://doi.org/10.1016/j.renene.2019.08.008

Li M, Zheng Y, Chen Y, Zhu X (2014) Biodiesel production from waste cooking oil using a heterogeneous catalyst from pyrolyzed rice husk. Bioresour Technol 154:345–348. https://doi.org/10.1016/j.biortech.2013.12.070

Liao CC, Chung TW (2013) Optimization of process conditions using response surface methodology for the microwave-assisted transesterification of Jatropha oil with KOH impregnated CaO as catalyst. Chem Eng Res Des 91:2457–2464. https://doi.org/10.1016/j.cherd.2013.04.009

Liu K, Wang R, Yu M (2018) An efficient, recoverable solid base catalyst of magnetic bamboo charcoal: preparation, characterization, and performance in biodiesel production. Renew Energy 127:531–538. https://doi.org/10.1016/j.renene.2018.04.092

López I, Quintana CE, Ruiz JJ et al (2014) Effect of the use of olive-pomace oil biodiesel/diesel fuel blends in a compression ignition engine: preliminary exergy analysis. Energy Convers Manag 85:227–233. https://doi.org/10.1016/j.enconman.2014.05.084

Mata TM, Cardoso N, Ornelas M et al (2011) Evaluation of two purification methods of biodiesel from beef tallow, pork lard, and chicken fat. Energy Fuels 25:4756–4762. https://doi.org/10.1021/ef2010207

Madhu D, Chavan SB, Singh V et al (2016) An economically viable synthesis of biodiesel from a crude *Millettia pinnata* oil of Jharkhand, India as feedstock and crab shell derived catalyst. Bioresour Technol 214:210–217. https://doi.org/10.1016/j.biortech.2016.04.055

Madhuvilakku R, Piraman S (2013) Biodiesel synthesis by TiO_2-ZnO mixed oxide nanocatalyst catalyzed palm oil transesterification process. Bioresour Technol 150:55–59. https://doi.org/10.1016/j.biortech.2013.09.087

Mahlia TMI, Syazmi ZAHS, Mofijur M et al (2020) Patent landscape review on biodiesel production: technology updates. Renew Sustain Energy Rev 118:109526. https://doi.org/10.1016/j.rser.2019.109526

Marchetti JM, Miguel VU, Errazu AF (2008) Techno-economic study of different alternatives for biodiesel production. Fuel Process Technol 89:740–748. https://doi.org/10.1016/j.fuproc.2008.01.007

Martinez-Guerra E, Gude VG (2014) Transesterification of used vegetable oil catalyzed by barium oxide under simultaneous microwave and ultrasound irradiations. Energy Convers Manag 88:633–640. https://doi.org/10.1016/j.enconman.2014.08.060

Marwaha A, Rosha P, Mohapatra SK et al (2018) Waste materials as potential catalysts for biodiesel production: current state and future scope. Fuel Process Technol 181:175–186. https://doi.org/10.1016/j.fuproc.2018.09.011

Mihaela P, Josef R, Monica N, Rudolf Z (2013) Perspectives of safflower oil as biodiesel source for South Eastern Europe (comparative study: safflower, soybean and rapeseed). Fuel 111:114–119. https://doi.org/10.1016/j.fuel.2013.04.012

Mohd Ali MA, Gimbun J, Lau KL et al (2020) Biodiesel synthesized from waste cooking oil in a continuous microwave assisted reactor reduced PM and NO_x emissions. Environ Res 185:109452. https://doi.org/10.1016/j.envres.2020.109452

Mota FAS, Costa Filho JT, Barreto GA (2019) The Nile tilapia viscera oil extraction for biodiesel production in Brazil: an economic analysis. Renew Sustain Energy Rev 108:1–10. https://doi.org/10.1016/j.rser.2019.03.035

Mutreja V, Singh S, Ali A (2011) Biodiesel from mutton fat using KOH impregnated MgO as heterogeneous catalysts. Renew Energy 36:2253–2258. https://doi.org/10.1016/j.renene.2011.01.019

Nayak SN, Bhasin CP, Nayak MG (2019) A review on microwave-assisted transesterification processes using various catalytic and non-catalytic systems. Renew Energy 143:1366–1387. https://doi.org/10.1016/j.renene.2019.05.056

Öztürk E (2015) Performance, emissions, combustion and injection characteristics of a diesel engine fuelled with canola oil-hazelnut soapstock biodiesel mixture. Fuel Process Technol 129:183–191. https://doi.org/10.1016/j.fuproc.2014.09.016

Panwar NL, Shrirame HY, Rathore NS et al (2010) Performance evaluation of a diesel engine fueled with methyl ester of castor seed oil. Appl Therm Eng 30:245–249. https://doi.org/10.1016/j.applthermaleng.2009.07.007

Patel A, Sindhu DK, Arora N et al (2015) Bioresource Technology Biodiesel production from non-edible lignocellulosic biomass of *Cassia fistula* L. fruit pulp using oleaginous yeast *Rhodosporidium kratochvilovae*. Bioresour Technol 197:91–98. https://doi.org/10.1016/j.biortech.2015.08.039

Phoenix L (2012) 4:56–61

Pikula K, Zakharenko A, Stratidakis A et al (2020) The advances and limitations in biodiesel production: feedstocks, oil extraction methods, production, and environmental life cycle assessment. Green Chem Lett Rev 13:11–30. https://doi.org/10.1080/17518253.2020.1829099

Pinzi S, Pilar M (2012) Feedstocks for advanced biodiesel production. In: Advances in biodiesel production: processes and technologies. Woodhead Publishing Limited, pp 69–90

Rahman M, Aziz M, Ali Al-khulaidi R et al (2017) Biodiesel production from microalgae *Spirulina maxima* by two step process: optimization of process variable. J Radiat Res Appl Sci 10:140–147. https://doi.org/10.1016/j.jrras.2017.02.004

Ramachandran S, Singh SK, Larroche C et al (2007) Oil cakes and their biotechnological applications—a review. Bioresour Technol 98:2000–2009. https://doi.org/10.1016/j.biortech.2006.08.002

Ramadhas AS, Jayaraj S, Muraleedharan C (2005) Biodiesel production from high FFA rubber seed oil. Fuel 84:335–340. https://doi.org/10.1016/j.fuel.2004.09.016

Rashid U, Anwar F (2008) Production of biodiesel through optimized alkaline-catalyzed transesterification of rapeseed oil. Fuel 87:265–273. https://doi.org/10.1016/j.fuel.2007.05.003

Rastegari H, Jazini H, Ghaziaskar HS, Yalpani M (2019) Applications of biodiesel by-products. In: Tabatabaei M, Aghbashlo M (eds) Biodiesel: from production to combustion. Springer, Cham, pp 101–125

Rekhate C, Prajapati AK (2019) Production, engine performance, combustion, emission characteristics and economic feasibility of biodiesel from waste cooking oil: a review. Environ Qual Manag 29:7–35. https://doi.org/10.1002/tqem.21645

Reşitoğlu İA, Altinişik K, Keskin A (2015) The pollutant emissions from diesel-engine vehicles and exhaust aftertreatment systems. Clean Technol Environ Policy 17:15–27. https://doi.org/10.1007/s10098-014-0793-9

Rezania S, Oryani B, Park J et al (2019) Review on transesterification of non-edible sources for biodiesel production with a focus on economic aspects, fuel properties and by-product applications. Energy Convers Manag 201:112155. https://doi.org/10.1016/j.enconman.2019.112155

Sahani S, Roy T, Chandra Sharma Y (2019) Clean and efficient production of biodiesel using barium cerate as a heterogeneous catalyst for the biodiesel production; kinetics and thermodynamic study. J Clean Prod 237:117699. https://doi.org/10.1016/j.jclepro.2019.117699

Sahani S, Roy T, Sharma YC (2020) Smart waste management of waste cooking oil for large scale high quality biodiesel production using Sr-Ti mixed metal oxide as solid catalyst: optimization and E-metrics studies. Waste Manag 108:189–201. https://doi.org/10.1016/j.wasman.2020.04.036

Šánek L, Pecha J, Kolomazník K, Bařinová M (2015) Biodiesel production from tannery fleshings: feedstock pretreatment and process modeling. Fuel 148:16–24. https://doi.org/10.1016/j.fuel.2015.01.084

Saranya G, Ramachandra TV (2020) Energy Conversion and Management: X Life cycle assessment of biodiesel from estuarine microalgae. Energy Convers Manag X 8:100065. https://doi.org/10.1016/j.ecmx.2020.100065

Saydut A, Duz MZ, Kaya C et al (2008) Transesterified sesame (*Sesamum indicum* L.) seed oil as a biodiesel fuel. Bioresour Technol 99:6656–6660. https://doi.org/10.1016/j.biortech.2007.11.063

Sbihi HM, Nehdi IA, Tan CP, Al-Resayes SI (2014) Production and characterization of biodiesel from *Camelus dromedarius* (Hachi) fat. Energy Convers Manag 78:50–57. https://doi.org/10.1016/j.enconman.2013.10.036

Seffati K, Honarvar B, Esmaeili H, Esfandiari N (2019) Enhanced biodiesel production from chicken fat using $CaO/CuFe_2O_4$ nanocatalyst and its combination with diesel to improve fuel properties. Fuel 235:1238–1244. https://doi.org/10.1016/j.fuel.2018.08.118

Shan R, Lu L, Shi Y et al (2018) Catalysts from renewable resources for biodiesel production. Energy Convers Manag 178:277–289. https://doi.org/10.1016/j.enconman.2018.10.032

Sharif RH, Paul RK, Bhattacharjya DK, Ahmed KU (2017) Physicochemical characters of oilseeds from selected mustard genotypes. J Bangladesh Agric Univ 15:27–40. https://doi.org/10.3329/jbau.v15i1.33527

Shi W, Li J, He B et al (2013) Biodiesel production from waste chicken fat with low free fatty acids by an integrated catalytic process of composite membrane and sodium methoxide. Bioresour Technol 139:316–322. https://doi.org/10.1016/j.biortech.2013.04.040

Singh SP, Singh D (2010) Biodiesel production through the use of different sources and characterization of oils and their esters as the substitute of diesel: a review. Renew Sustain Energy Rev 14:200–216. https://doi.org/10.1016/j.rser.2009.07.017

Singh D, Sharma D, Soni SL et al (2019) Chemical compositions, properties, and standards for different generation biodiesels: a review. Fuel 253:60–71. https://doi.org/10.1016/j.fuel.2019.04.174

Singh D, Sharma D, Soni SL et al (2020) A review on feedstocks, production processes, and yield for different generations of biodiesel. Fuel 262:116553. https://doi.org/10.1016/j.fuel.2019.116553

Siqueira SF, Deprá MC, Zepka LQ, Jacob-lopes E (2018) Life Cycle Assessment (LCA) of third-generation biodiesel produced heterotrophically by *Phormidium autumnale* Open Biotechnol J 12:270–281. https://doi.org/10.2174/1874070701812010270

Sitepu EK, Heimann K, Raston CL, Zhang W (2020) Critical evaluation of process parameters for direct biodiesel production from diverse feedstock. Renew Sustain Energy Rev 123:109762. https://doi.org/10.1016/j.rser.2020.109762

Song M, Pei H, Hu W, Ma G (2013) Evaluation of the potential of 10 microalgal strains for biodiesel production. Bioresour Technol 141:245–251. https://doi.org/10.1016/j.biortech.2013.02.024

Stavarache CE, Morris J, Maeda Y et al (2008) Syringa (*Melia azedarach* L.) Berries oil: a potential source for biodiesel fuel. Rev Chim 59:672–677. https://doi.org/10.37358/RC.08.6.1853

Suganya T, Renganathan S (2014) Ultrasonic assisted acid base transesterification of algal oil from marine macroalgae *Caulerpa peltata*: optimization and characterization studies. Fuel 128:347–355. https://doi.org/10.1016/j.fuel.2014.03.037

Suganya T, Nagendra Gandhi N, Renganathan S (2013) Production of algal biodiesel from marine macroalgae *Enteromorpha compressa* by two step process: optimization and kinetic study. Bioresour Technol 128:392–400. https://doi.org/10.1016/j.biortech.2012.10.068

Suresh T, Sivarajasekar N, Balasubramani K (2021) Enhanced ultrasonic assisted biodiesel production from meat industry waste (pig tallow) using green copper oxide nanocatalyst: comparison of response surface and neural network modelling. Renew Energy 164:897–907. https://doi.org/10.1016/j.renene.2020.09.112

Tafesh A, Basheer S (2013) Pretreatment methods in biodiesel production processes. In: Fang Z (ed) Pretreatment techniques for biofuels and biorefineries. Springer, Berlin, Heidelberg, pp 417–434

Tan SX, Lim S, Ong HC, Pang YL (2019a) State of the art review on development of ultrasound-assisted catalytic transesterification process for biodiesel production. Fuel 235:886–907. https://doi.org/10.1016/j.fuel.2018.08.021

Tan YH, Abdullah MO, Kansedo J et al (2019b) Biodiesel production from used cooking oil using green solid catalyst derived from calcined fusion waste chicken and fish bones. Renew Energy 139:696–706. https://doi.org/10.1016/j.renene.2019.02.110

Teo SH, Islam A, Taufiq-Yap YH (2016) Algae derived biodiesel using nanocatalytic transesterification process. Chem Eng Res Des 111:362–370. https://doi.org/10.1016/j.cherd.2016.04.012

Toldrá-Reig F, Mora L, Toldrá F (2020) Trends in biodiesel production from animal fat waste. Appl Sci 10:1–17. https://doi.org/10.3390/app10103644

Tsakona S, Kopsahelis N, Chatzifragkou A et al (2014) Formulation of fermentation media from flour-rich waste streams for microbial lipid production by *Lipomyces starkeyi*. J Biotechnol 189:36–45. https://doi.org/10.1016/j.jbiotec.2014.08.011

Usta N (2005) Use of tobacco seed oil methyl ester in a turbocharged indirect injection diesel engine. Biomass Bioenergy 28:77–86. https://doi.org/10.1016/j.biombioe.2004.06.004

Usta N, Aydoğan B, On AH et al (2011) Properties and quality verification of biodiesel produced from tobacco seed oil. Energy Convers Manag 52:2031–2039. https://doi.org/10.1016/j.enconman.2010.12.021

Verma P, Sharma MP, Dwivedi G (2016) Impact of alcohol on biodiesel production and properties. Renew Sustain Energy Rev 56:319–333. https://doi.org/10.1016/j.rser.2015.11.048

Wan Ghazali WNM, Mamat R, Masjuki HH, Najafi G (2015) Effects of biodiesel from different feedstocks on engine performance and emissions: a review. Renew Sustain Energy Rev 51:585–602. https://doi.org/10.1016/j.rser.2015.06.031

Wang W, Zhou W, Liu J et al (2013) Biodiesel production from hydrolysate of *Cyperus esculentus* waste by *Chlorella vulgaris*. Bioresour Technol 136:24–29. https://doi.org/10.1016/j.biortech.2013.03.075

Wang X, Liang J, Luo C et al (2014) Biomass, total lipid production, and fatty acid composition of the marine diatom *Chaetoceros muelleri* in response to different CO_2 levels. Bioresour Technol 161:124–130. https://doi.org/10.1016/j.biortech.2014.03.012

Wong KY, Ng JH, Chong CT et al (2019) Biodiesel process intensification through catalytic enhancement and emerging reactor designs: a critical review. Renew Sustain Energy Rev 116:109399. https://doi.org/10.1016/j.rser.2019.109399

Xiao Y, Xiao G, Varma A (2013) A universal procedure for crude glycerol purification from different feedstocks in biodiesel production: experimental and simulation study. Ind Eng Chem Res 52:14291–14296. https://doi.org/10.1021/ie402003u

Xiao Y, Zhang J, Cui J et al (2015) Simultaneous accumulation of neutral lipids and biomass in *Nannochloropsis oceanica* IMET1 under high light intensity and nitrogen replete conditions. Algal Res 11:55–62. https://doi.org/10.1016/j.algal.2015.05.019

Xu X, Kim JY, Cho HU et al (2014) Bioconversion of volatile fatty acids from macroalgae fermentation into microbial lipids by oleaginous yeast. Chem Eng J 264:735–743. https://doi.org/10.1016/j.cej.2014.12.011

Xu S, Elsayed M, Ismail GA et al (2019) Evaluation of bioethanol and biodiesel production from *Scenedesmus obliquus* grown in biodiesel waste glycerol: a sequential integrated route for enhanced energy recovery. Energy Convers Manag 197:111907. https://doi.org/10.1016/J.ENCONMAN.2019.111907

Yaşar F (2020) Comparison of fuel properties of biodiesel fuels produced from different oils to determine the most suitable feedstock type. Fuel 264:116817. https://doi.org/10.1016/j.fuel.2019.116817

Yesilyurt MK, Cesur C, Aslan V, Yilbasi Z (2020) The production of biodiesel from safflower (*Carthamus tinctorius* L.) oil as a potential feedstock and its usage in compression ignition engine: a comprehensive review. Renew Sustain Energy Rev 119:109574. https://doi.org/10.1016/j.rser.2019.109574

Yuan H, Yang BL, Zhu GL (2009) Synthesis of biodiesel using microwave absorption catalysts. Energy Fuels 23:548–552. https://doi.org/10.1021/ef800577j

Zabeti M, Wan Daud WMA, Aroua MK (2009) Activity of solid catalysts for biodiesel production: a review. Fuel Process Technol 90:770–777. https://doi.org/10.1016/j.fuproc.2009.03.010

Zhang Y, Dubé MA, McLean DD, Kates M (2003) Biodiesel production from waste cooking oil: 2. Economic assessment and sensitivity analysis. Bioresour Technol 90:229–240. https://doi.org/10.1016/S0960-8524(03)00150-0

Chapter 9
Microbial Fuel Cells (MFCs) for Waste Recycling and Energy Production

Qian Zhao, Taotao Gao, Xiaoqin Li, and Dan Xiao

Abstract Microbial fuel cell (MFC) is a kind of device that uses electricity-producing microorganisms as anode catalysts to directly convert chemical energy in organic matter into electrical energy. It has broad application prospects in the field of wastewater treatment and new energy development. This chapter aims to discuss the concept, reaction mechanism, and application prospect of bioelectrochemical systems using wastes for electricity generation. The types, properties, and activity parameters of microbe for MFC are discussed. In addition, the design and optimization of MFCs devices for enhanced efficiency are evaluated. Particularly, critical discussions are provided on application of MFCs technology for different types of wastewater as well as corresponding key bottlenecks. Besides, the recent progress in wastewater treatment and seawater desalination, as well as the possibility of the combination between wastewater treatment and seawater desalination integrated with biomass production is evaluated. This chapter also discusses the limitations and challenges of MFCs industrialization and large-scale applications. The advantages and inadequacies of MFCs compared with other technical solutions for waste utilization are analyzed in detail. Moreover, the economic feasibility, future research perspectives in order to enhance the MFCs performance in large-scales are presented.

Keywords Microbial fuel cell · Waste utilization · Biomass · New energy · Biofuel

Q. Zhao
College of Mechanical Engineering, Chengdu University, Chengdu 610106, PR China
e-mail: zhao_qian@cdu.edu.cn

Q. Zhao · T. Gao · X. Li · D. Xiao (✉)
Nano Energy and Catalytic Materials Research Center, Institute for Advanced Study, Chengdu University, Chengdu 610106, PR China
e-mail: xiaodan@scu.edu.cn

X. Li
e-mail: lixiaoqin0321@163.com

A. E.-F. Abomohra et al. (eds.), *Waste-to-Energy*,
https://doi.org/10.1007/978-3-030-91570-4_9

9.1 Introduction

In the recent decades, consumption of energy within the world has had a prosperous trend. Energy sources are classified into three batches: fossil fuels, renewable sources, and nuclear sources, in which non-renewable sources of energy, which include an enormous portion of energy consumption, could be categorized into two major classifications: nuclear and fossil energy (Rahimnejad et al. 2009). The consumption of fossil fuels to release hazardous gases has severely threatened human life through its drastic aftermaths, such as global warming and atmospheric pollution (Rahimnejad et al. 2012). Therefore, the search for clean and renewable energy is going on. This process will continue until there is a stable solution available as an alternative to fossil fuels.

However, countries around the world have made remarkable efforts to find a piece of cogent solution for energy crisis by turning the eyes into renewable energy sources such as solar energy, wind energy and hydroenergy. As an upshot of these efforts, one type of the latterly proposed alternative energy sources is the fuel cells (FCs) which generate energy using high-value metal catalysts. In actual fact, FC is of plethora advantages over other kinds of energy generators, e.g., no emissions of environmental polluting gases (such as SO_x, NO_x, CO_2, and CO), higher efficiency, no existence of mobile parts, as a result, lack of sonic pollution, and so forth (Peighambardoust et al. 2010). In contrast, the new energy source of FCs has the disadvantages of the high cost and high mass generation.

One type of FCs is microbial fuel cells (MFCs) that use an active microorganism as a biocatalyst in an anaerobic anode compartment for the production of bioelectricity, which represents a clean and renewable energy resource (Zhao et al. 2009; Elshobary et al. 2021). Although electrical current produced by bacteria was observed by Potter in 1911 (Potter 1911), limited feasible results were acquired in this area by the next 50 years (Lewis 1966). The concept of using microorganisms as catalysts in fuel cells was explored from the 1970s (Roller et al. 1984; Suzuki 1976). In the early 1990s, FCs became far more appealing devices and MFCs were considered as promising technology which was presented to treat domestic wastewater (Allen and Bennetto 1993; Arun et al. 2020). Furthermore, research related to MFCs turned booming from 1999 once it was discovered that mediator was not a necessary component within MFCs (Nader et al. 2012). However, it is only in the past one decade, MFCs with an enhanced power output (Schröder et al. 2003; Rabaey et al. 2004; Liu et al. 2004; Serra and Espírito-Santo 2021) have been developed which provides new opportunities for the sustainable production of energy from biodegradable, reduced compounds Fig. 9.1.

MFCs function on different carbohydrates but also on complex substrates present in wastewaters which have operational and functional advantages over the technologies currently used for generating energy from organic matter. First, the direct conversion of substrate energy to electricity enables high conversion efficiency. Second, MFCs operate efficiently at ambient, and even at low, temperatures distinguishing

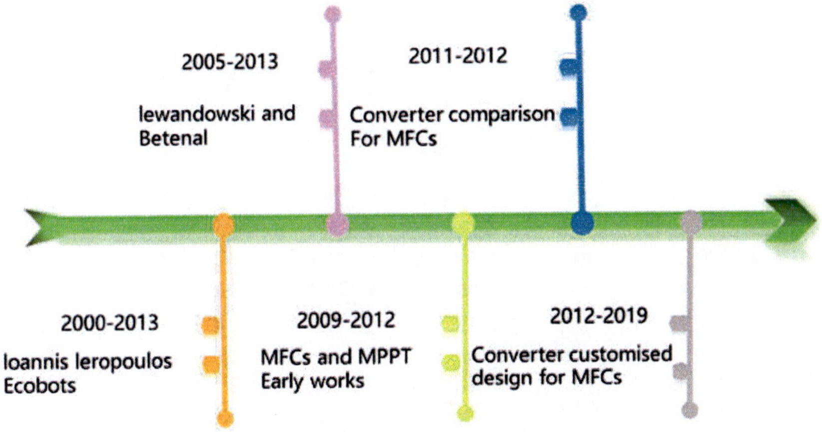

Fig. 9.1 Microbial fuel cell and energy-harvesting research highlights from 2000 until 2019

them from all current bio-energy processes. Third, an MFC does not require gas treatment because the off-gases of MFCs are enriched in carbon dioxide and normally have no useful energy content. Fourth, MFCs do not need energy input for aeration provided the cathode is passively aerated (Liu et al. 2004). Fifth, MFCs have potential for widespread application in locations lacking electrical infrastructures and also to expand the diversity of fuels used to satisfy the energy requirements (Cai et al. 2020). MFCs are considered to be a very promising device for the treatment of the contaminants in wastewater (Arun et al. 2020) and soil (Abbas et al. 2021) and achieve the bioelectricity generation simultaneously. However, the power output of MFCs remains to be improved and the key parameters affecting on bioelectricity generation of MFCs should also be explored sustainably in promoting the application of MFCs for the waste recycling and the energy production.

9.2 Principles for MFCs

Three basic theories, called electrochemistry, thermodynamics, and reaction kinetics, are commonly recognized in MFCs. Currently, the low power output (< 6 W m^{-2}; ≤ 500 W m^{-3}) of MFCs is caused by several factors related to the cathode, the anode, the chemical species, the ion-exchange, the microbial species and activity, microbial concentrations, substrate, metabolites, operational conditions, and fuel cell configuration (Zhao et al. 2009). Generally, the current generation of MFCs can be affected by the transfer of substrate, protons, and metabolites between solution and electrode surface. Studies reveal that the performance of MFCs can be improved by decreasing the polarization voltage of MFCs which is related to the surface area of electrode, the thickness of membrane, and the conductivity of electrolyte (Reimers et al. 2007).

The voltage is an essential parameter for batteries, for MFCs, the maximum operating voltages are 0.3–0.7 V. The voltage (E) is a function of the external resistance (R_{ex}) and the current (I), which can be denoted as Eq. 9.1:

$$E = IR_{\text{ex}} \tag{9.1}$$

The E also can be calculated from the redox potential or electrode potential of the half-cell. The MFCs includes two half-cell chambers, and the relationship between the voltage and the electrode potential can be presented as Eq. 9.2:

$$E = \varphi_{\text{r}} - \varphi_{\text{l}} \tag{9.2}$$

where φ_r and φ_l are the electrode potentials of the half-cell at the anodic and cathodic chambers, respectively. The biological redox reactions also follow the rule of the change of Gibbs free energy (ΔG):

$$\Delta G = -nFE \tag{9.3}$$

where F is Faraday constant (F = 96,493 C); n is the number of the transferred electrons during reaction.

For example, the MFCs based on glucose, the reaction is as follows:

Anodic reaction:

$$C_6E_{12}O_6 + 24OH^- \leftrightarrow 6CO_2 + 18H_2O + 24e^-$$

Cathode reaction:

$$6O_2 + 12H_2O + 24e^- \leftrightarrow 24OH^-$$

Battery reaction:

$$C_6H_{12}O_6 + 6O_2 \leftrightarrow 6CO_2 + 6H_2O$$

In this battery reactor, n = 24. At 25 °C with atmospheric pressure (0.1 MPa), $\Delta G = -2870\,\text{kJ/mol}$, the conversion potential of glucose: E = 1.24 V.

As well known that the redox reaction with H^+ can be described as:

$$O_{\text{ox}} + \text{m}H^+ + ne^- \leftrightarrow R_{\text{red}}$$

where O_{ox} and R_{red} are the oxidized species and reduced form, respectively. The electrode potential can be calculated by:

$$E = E^0 - \frac{2.303RT}{nF} \lg \frac{a_{\text{red}}}{a_{\text{ox}}} + \frac{2.303RT}{nF} \lg \alpha_{H^+}^{\text{m}} \tag{9.4}$$

When $T = 298$ K

$$E = E^0 - \frac{0.05916\,\text{V}}{n} \lg \frac{a_{\text{red}}}{a_{\text{ox}}} - \frac{0.05916}{n} \text{mpH} \tag{9.5}$$

where E^0 is the standard electrode potential, a_{ox} is the activity of the oxide in the electrolyte, and a_{red} is the activity of the reduced species. If the electrode reaction is carried out under a fixed pH, two constant values (pH and E^0) can be combined as $E^{\oplus}$, thus

$$E = E^{\oplus} - \frac{0.05916\,\text{V}}{n} \lg \frac{a_{\text{red}}}{a_{\text{ox}}} \tag{9.6}$$

$E^{\oplus}$ is called the biochemical standard electrode potential, representing the potential of electrode reaction under a fixed pH when the activities of oxide and reduced species equal 1.

For example, in CH_3CHO MCF,

Battery reaction:

$$CH_3CHO + NADH + H^+ \leftrightarrow CH_3CH_2OH + NAD^+$$

Anodic reaction:

$$NADH + H^+ \leftrightarrow NAD^+ + 2H^+ + 2e^- \quad E^{\oplus} = -0.324\,\text{V}$$

Cathodic reaction:

$$CH_3CHO + 2H^+ + 2e^- \leftrightarrow CH_3CHO \quad E^{\oplus} = -0.197\,\text{V}$$

$$E^{\oplus} = E_+^{\oplus} - E_-^{\oplus} = -0.197 - (-0.324) = 0.127\,\text{V}$$

$$\Delta_r G_m^{\oplus} = -nFE = -2 \times 96485 \times 0.127 = -24.51\,\text{kJ/mol}$$

Thermodynamics is also an important theoretical basis for the study of MFCs. According to chemical thermodynamics, the relationship between ΔG and the change of enthalpy (ΔH) and entropy (ΔS):

$$\Delta G = \Delta H - T\Delta S \tag{9.7}$$

Or

$$-nFE = \Delta H - T\Delta S \tag{9.8}$$

In the research of MFCs, the charge/discharge of battery is proceeded under the equilibrium condition, and the microbial chemical reactions involved in the battery show the characteristic of reversibility under a certain condition, which is one of the features of biochemical catalytic process (Jan 2014). The reversibility of MFCs reveals that the catalytic reaction process in two electrodes can shift from equilibrium condition to positive or opposite direction. From the thermodynamic perspective, MFCs also include two types: reversibility and irreversibility. Based on the concept of thermodynamics, the reversibility refers to the reversible electron transfer in biological reaction, which is the basis of thermodynamics. In MFCs, the reversibility means biochemical reversibility and thermodynamic reversibility. The biochemical reversibility is the materials reversibility that is the reactions in two electrodes are reversible. The thermodynamic reversibility is the energy reversibility, where the released energy in discharge process is coincident equal to the expended energy in charge process. In reversibility system, the cell can retune to the initial state after charge/discharge.

The kinetics of MFCs is directly linked to the reactions occurring at the electrode. The recent studies show that the key parameter for MFCs' research is to produce electrons in microbial reaction and promote the directional migration of electrons. The metabolism for most of the microorganisms can proceed through respiration. In the process of metabolism and growth, the microorganisms with electrochemical activity can produce electrons by feeding glucose or other carbohydrates. The produced electrons can transfer to anode by various paths and then migrate to cathode via external circuit. Finally, the cathodic electrons react with oxygen and protons to form water. The directional migrated electrons can generate current (Renslowa et al. 2011).

Here, the glucose is further used as the reacted substrate of MFCs to discuss and analyze the various influence factors for MFCs' reaction kinetics as well as the corresponding improved strategies. The current output of MFCs is related to the consumption rate of glucose. It is well known that the generated current by MFCs based on glucose can be expressed by the following Equation:

$$i = nF\frac{\mathrm{d}N}{\mathrm{d}t} \tag{9.9}$$

where $\frac{\mathrm{d}N}{\mathrm{d}t}$ is the consumption rate of glucose, mol s^{-1}. It can be seen that the current output of MFCs is proportional to the consumption rate of glucose. The current is the direct measurement of the glucose's consumption rate; in other words, aiming to increase the current output, it is prerequisite to accelerate the consumption rate of glucose.

The electrochemical reactions of MFCs are heterogeneous and can only occur at the interface between electrode and solution. Thus, to improve the performance of MFCs, it is important to accelerate the reaction rate on the electrode surface. The voltage loss caused by reaction kinetics is a significant part of the capacity loss of MFCs. The voltage loss of MFCs mainly includes charge transfer overpotential, ohmic overpotential, and mass transport overpotentials. Among them, the charge transfer overpotential can be improved by reducing the temperature of solution and boosting exchange current density. The ohmic overpotential can be overcome through the utilization of solutions (anodic and cathodic) with high conductivity, shortening the distance of electrodes, and increasing the sectional area of reactor. Similar to charge transfer overpotential, the mass transport overpotentials also can be restricted by reducing the temperature of solution as well as increasing the ultimate current density.

The charge transfer overpotential is always driven from the irreversibility of the reactions occurring on the surface of electrode and appears at low current density. The charge transfer overpotential can be expressed by a simplified formula:

$$j = j_0 \frac{nF\eta_{\text{act}}}{RT} \tag{9.10}$$

where j is the current density, j_0 is the exchange current density, η_{act} is the charge transfer overpotential. It is obvious that the charge transfer overpotential is proportional to current density, and the changed range is related to the magnitude of the exchange current density. In order to obtain larger j, the η_{act} should be improved as well as increase the j_0. As well known that j_0 can be calculated as follows:

$$j_0 = nFc_R^* f_1 e^{-\Delta G/(RT)} \tag{9.11}$$

where c_R^* is the concentration of glucose at the reaction interface, f_1 is the decayed rate from reactant to product. Thus, the following strategies can be used to improve j_0, that is (1) increase the concentration of glucose, and decrease the effect of the heat and mass transfer, (2) reducing the activated energy barrier, (3) increase the temperature of anodic cell, (4) increase the surface roughness of the electrode.

The internal resistance is an important index to evaluate the performance of MFCs, which can affect the power output of battery. The internal resistance is also one of the key research points in MFCs, which includes the connection resistance between the electrode and wire, contact resistance between the electrodes and solution, the ion transport resistance, the membrane resistance of the proton exchange membrane. The ohmic overpotential located at intermediate currents, which can be presented as follows:

$$\eta_{\text{ohmic}} = iR = i(R_{\text{elec}} + R_{\text{ionic}}) \tag{9.12}$$

where R_{elec} is the resistance of electron transport and R_{ionic} is the resistance of ion transport. As the ion transport is much more difficult than electron transport, the resistance caused by ion transport plays a major role. The ion transport resistance can be presented as follows:

$$R_{\text{ionic}} = \frac{L}{A\sigma} \tag{9.13}$$

where A is the cross-sectional area,σ is the conductivity of the solution, and L is the transmission distance. To reduce the resistance of MFCs, the following strategies can be carried out: (1) increase A and σ, decrease L, and optimized the structure of electrodes to reduce the R_{ionic}, (2) improve the connection between electrodes to reduce the connection resistance, (3) reduce the thickness of the proton exchange membrane to reduce the membrane resistance.

During the operation of MFCs, there is a concentration difference of reactants between the reaction interface and the solution due to the consumption of interfacial reactants. This difference is related to the current density of MFCs, as follows:

$$j = -nFD^{\text{eff}}\frac{c_R^* - c_R^0}{\delta} \tag{9.14}$$

where c_R^* is the concentration of reactants at the reaction interface, c_R^0 is the concentration of reactants in solution, δ is the thickness of diffusion layer, and D^{eff} is the effective diffusion coefficient of reactants.

The concentration loss is called mass transport overpotential, which is caused by the concentration change of reactants or products between the interface of electrode and electrolyte due to the depletion of reactants and accumulation of products. When the current output of MFC is large, the concentration loss plays a major role in the voltage loss; thus, it is beneficial to increase the voltage output of MFC by restraining concentration loss. The concentration loss can be expressed by the following Equation:

$$\eta_{\text{conc}} = \frac{RT}{nF}\left(1 + \frac{1}{a}\right)\ln\frac{j_{\text{L}}}{j_{\text{L}} - j} \tag{9.15}$$

where j_{L} is the limiting current density, $j_{\text{L}} = nFD^{\text{eff}}\frac{c_R^0}{\delta}$, which can be improved by the following methods: (1) increase the concentration of glucose in the anodic cell, (2) boosting the flow of anodic solution, and (3) optimizing the structure of motor.

9.3 MFCs Microbiology

9.3.1 *Mechanism of Electron Transfer*

The transfer of electrons produced by microbes from intracellular to extracellular electron acceptors is an important step in MFC electricity generation, and also one of the key factors is limiting the electricity production performance. Microbes are so far known to transfer electrons to extracellular via three mechanisms: nanowires, cell contact transfer, and mediator transfer.

9.3.1.1 Nanowires

Gorby et al. (2006) found that the cell surface of some Geobacter and Shewanella species could produce a kind of conductive appendage—nanoscale pili, which were called microbial "nanowires." Here the pilis are only 3–5 nm wide, but 1000 times longer than the width. Reguear et al. (2005) measured the electrical conductivity of the pili on the surface of G. Sulfurreducens by conductive-probe atomic force microscope (AFM) and found that the pili would produce a strong electrical response while the probe was applied with external voltage. The result confirmed that the pili possessed good electrical conductivity. In the process of electron transfer, the two ends of nanowires were connected with the outer membrane and the surface of the electrode, respectively, so as to realize the rapid transfer of electrons from the cell to the electrode.

9.3.1.2 Cell Contact Transfer

The cell membranes of some electricity-producing microorganisms (Geobacter, Shewanella, Desuffuromonas, etc.,) directly contact the anode surface, and the electrons generated during the metabolism process can be transferred to the electrode through the outer membrane redox protein. Holmes et al. (2006) had proved that two outer membrane cytochromes, OmcS and OmcE, both play an important role in the reduction process, and OmcS participates in the direct transfer of electrons from microbial cells to the electrode. In cell contact transfer of electrons, only a single layer of microorganism close to the surface of the electrode has electrochemical activity for that can transfer electrons to the surface of anode. Here the density of the single layer of microorganisms on the anode surface limits the electricity generation performance of MFCs (Schröder et al. 2003).

9.3.1.3 Mediator Transfer

Chemical mediators or electron mediators have often been added to the MFCs to allow bacteria and even yeast to transfer electrons. The mediators can be divided into two categories, specifically exogenous and endogenous mediators. Here the exogenous mediator mainly includes neutral red, 2,6-anthracene, disulfonic acid (AQDS), thionine, potassium ferricyanide, methyl viologen, etc. The use of exogenous mediator can significantly improve the electron transfer rate from intracellular to extracellular, but this not only increases the cost, but also poisons microorganisms, so its application is greatly limited. The endogenous mediators are primarily self-produced or endogenous chemical intermediaries, such as pseudomonas aeruginosa and the related compounds produced by its metabolism (Rabaey et al. 2005).

9.3.2 Electricigens

Electricigens are a kind of microorganisms that can transfer electrons generated during the oxidation of organic matter to electrodes through the electron transport chain to generate electric current. Here the microorganisms can maintain their growth by using the energy obtained in the process of electron transfer. Electricigens are often used as biocatalysts of MFCs and that play an irreplaceable role in the electricity production. At present, electricigens are mainly bacteria and fungi. Most of the bacteria belong to Proteobacteria or Firmicutes, while the fungi are mostly yeast and algae.

9.3.2.1 Bacteria

1. Proteobacteria

   (1) Alpha-proteobacteria

   Rhodopseudomonas palustris: It is gram-negative facultative anaerobic bacteria with flagella on its surface. The best growth mode is to use light and organic carbon sources for photo-energy heterotrophic. Under hypoxia, the substrate can be fermented and autotrophic with hydrogen, sodium sulfate, hydrogen sulfide, etc., as electron donors.

   Ochrobactrum anthropic: A gram-negative aerobic bacterium and which can utilize a variety of complex organics and simple organic acids such as acetic acid, butyric acid, grapes, glycerin to generate electricity.

   Acidiphilium cryptum: *Acidiphilium cryptum* is one of the gram-negative facultative anaerobic iron-reducing bacteria that has a certain degree of acid resistance of the presence of oxygen. Under anaerobic conditions, Fe (III) can be used as an electron donor for electricity generation.

   (2) Beta-proteobacteria

   Rhodoferax ferrireducens: It is a gram-negative facultative anaerobic bacterium. The electrons can be directly transferred to the electrode without artificially adding influence factors, which can completely to oxidize glucose, xylose, fructose, sucrose, etc. to form CO_2.

   Comamonas denitrificans: A gram-negative facultative anaerobic non-iron-reducing bacterium, and it mainly adopts acetic acid, lactic acid, malonic acid, pyruvic acid, and fumaric acid as electron acceptors to produce electricity, while nitrate can also be used to generate nitrogen. Based on this feature, nitrate can be added to generate nitrogen for maintaining anaerobic environment during the electricity generation process.

(3) Gamma-proteobacteria

Pseudomonas aeruginosa: A gram-negative facultative anaerobic bacterium. *Pseudomonas aeruginosa* is the earliest reported microorganism that can produce electron mediators by itself. The bacteria will produce pyocyanin during the electricity generation process, and which can be used as an electron transfer mediator between itself as well as other species and the electrode, thereby increasing the ability to generate electricity.

Klebsiella pneumoniae: It is a gram-negative facultative anaerobic bacterium with a ticker capsule, which can generate a biofilm that catalyzed and oxidizes a variety of organic substances to produce electricity on the anode in an oxygen-free environment.

(4) Delta-proteobacteria

Geobacteraceae: It is a type of gram-negative bacteria. Among them, *Geobacter sulfurreducens* is obligate anaerobic bacteria that can only use acetic acid and ammonia as electron donors, and Fe(III), S, Co-EDTA, fumaric acid, and malic acid as electron acceptors. *Geobacter metallireducens* is strict anaerobe bacteria which able to oxidize aromatic compounds and has a strong electron conversion rate.

(5) Epsilon-proteobacteria

Arcobacter butzleri: It can generate electricity in an acidic environment and is the first electricity-producing microorganism found in the Epsilon-proteobacteria.

2. Phylum Firmicutes

(1) Clostridium

Clostridium butyricum: It is a gram-positive strict anaerobic iron-reducing bacterium with poor electricity production capacity. The cell wall of gram-positive bacteria is much thicker than those of negative bacteria, making it harder for the electrons to pass through. Clostridium butyricum can reduce high valent iron and hydrolyze sucrose, cellobiose, starch and other complex polysaccharides.

(2) Enterococcus

Enterococcus is a type of gram-positive aerobic or facultative anaerobe bacterium. *Enterococcus gallinarum* is the first Fe(III) reducing bacteria isolated from underwater soil.

(3) Bacillus

Bacillus subtilis is a widely used bacteria that can grow and stably produce electricity in an anaerobic environment.

9.3.2.2 Microalgae and Fungi

1. *Microalgae*

Microalgae can produce a variety of nutrients in the metabolic process, so it is widely used in various fields. Microalgae biological MFCs can be divided into anode type and cathode type. Microalgae biological anode MFCs use the microalgae in the anode chamber to directly generate electricity or indirectly generates electricity through the synergistic effect of microalgae and electricity-generating bacteria while the microalgae biological cathode MFCs adopts the microalgae photosynthesis to absorb CO_2 produced by anaerobic bacteria metabolism and produce O_2 which can be used as electron acceptor.

2. *Saccharomyces*

Saccharomyces cerevisiae: It can use maltose, glucose, and trehalose for aerobic respiration and decompose complex organic matter under anaerobic conditions. *Saccharomyces cerevisiae* has a simple metabolic process contributing to study the power generation reaction mechanism in MFCs.

Hansenula anomala: It can use glucose as an electron donor. Studies have shown that there are electrochemically active enzymes on the outer membrane surface.

9.4 MFC Structure

At present, MFC can be divided into many types from the perspective of battery structure. The single-chamber MFC reactor consists of a closed cylindrical glass barrel containing eight graphite rod anodes. At the center of the glass cylinder is a carbon/platinum air cathode supported by a perforated plastic tube. The proton exchange membrane is hot pressed on the cathode. The anode of the reactor has a larger membrane-hanging area, while avoiding the diffusion of dissolved oxygen as much as possible, and improving the electric energy output of the microbial fuel cell. In the experiment, the simulated domestic sewage with a COD of 50–220 mg/L was used as the substrate, and the maximum power density obtained during the hydraulic retention time of 333 h was 262 mW/m^2, and the COD removal rate could reach 80%. The advantage of the single-chamber MFC is that the anode and cathode are closer, the cathode mass transfer rate is improved, the operating cost is reduced because no aeration is required, the area is small, the structure is simple, and the power of MFC can be further increased by removing the proton exchange membrane. It can be seen that low energy consumption, low cost, and large output power are the goals that people are pursuing. Therefore, research and development of MFC with direct air cathode system will have certain competitiveness.

However, if the distance between the anode and cathode is too small, O_2 will easily pass through the proton exchange membrane to reach the anode, which will also have a certain impact on the generating microorganisms and reduce the coulomb efficiency of the battery.

Double-chamber MFCs are commonly used in laboratories. A typical double-chamber MFC as shown in Fig. 9.2 generally has components such as an anode chamber, a cathode chamber, and a proton exchange membrane (Rabaey et al. 2005; Logan et al. 2005). The biggest feature of the dual-chamber MFC is the use of a membrane (PEM) between the anode and the cathode. Therefore, based on this principle and feature, various types of MFC reactors can be designed. Double-chamber MFC is divided into rectangular type, double-cylinder type, flat plate type, and upflow type. The reactor of the rectangular microbial fuel cell is composed of a rectangular cathode chamber and an anode chamber, and the two chambers are separated by a proton exchange membrane. Similar to the structure of a rectangular reactor, the cathode and anode chambers of the double-bottle microbial fuel cell (also known as H-type MFC) are connected by a cylindrical glass bridge at a certain distance from the bottom of the bottle. The protons between the two bridges rubber pads the exchange membrane separate the two chambers.

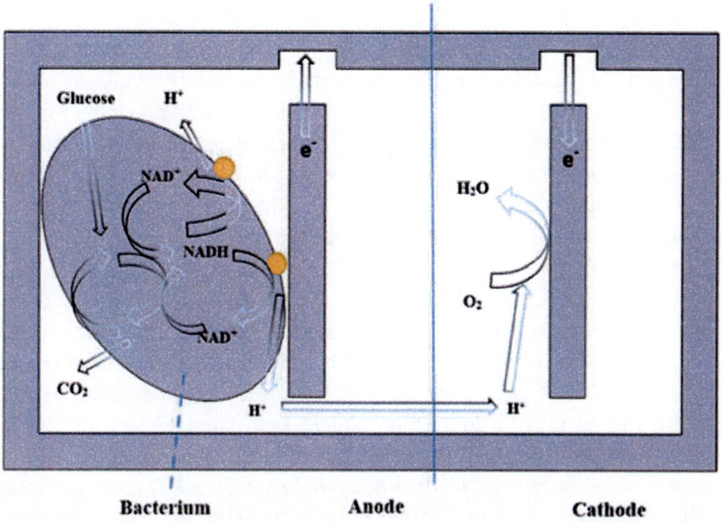

Fig. 9.2 Operating principles of an MFC. A bacterium in the anode compartment transfers electrons obtained from an electron donor (glucose) to the anode electrode. This occurs either through direct contact, nanowires, or mobile electron shuttles (small spheres represent the final membrane-associated shuttle). During electron production, protons are also produced in excess. These protons migrate through the cation exchange membrane (CEM) into the cathode chamber. The electrons flow from the anode through an external resistance (or load) to the cathode where they react with the final electron acceptor (oxygen) and protons

"Three-in-one" MFC is a new type of microbial fuel cell that combines anode, proton exchange membrane, and cathode. It can reduce the internal resistance of MFC to a large extent and increase the output power of MFC. Research and experiment results show that the internal resistance of the "three-in-one" MFC is only 10–302, which is much lower than other forms of MFC, and the maximum output power density can reach 300 mW/m^2. In addition, there are many other types of microbial fuel cells, such as multi-cathode MFCs and other configurations, which are also favored by scientists.

From the perspective of battery structure, the existing microbial fuel cells can be roughly divided into five categories. Microbial fuel cells assembled with different structures have their own advantages and disadvantages, and novel battery structures are also favored by scientists. With the acceleration of urbanization, non-renewable energy is increasingly exhausted; all kinds of wastewater discharge and its discharge standards have become larger and higher, people pay more and more attention to microbial fuel cells. As an innovative technology, microbial fuel cell (MFC) cannot only reduce the cost of wastewater treatment, but also provide a new way for human to obtain energy.

9.4.1 Up-Flow MFC

The upflow MFC is based on UASB reactor and has the advantages of both UASB and MFC. Upflow MFC is characterized by simple structure and large load, and the correlation with wastewater treatment process is increased by fully mixing medium and microorganisms. In the same tank, the anode and cathode are filled with carbon felt separated by fiberglass and glass beads, and the anode treats the wastewater in a bottom-up sequence. When the cathode was replaced by porous platinum-loaded graphite and the bipolar separator was replaced by a polyacrylic plate, the power density was increased to 560 mW/m^2.

Compared with traditional MFC, up-flow MFC is more applicable in practice. The advantages of UAMFC are mainly reflected in the following aspects: (1) Using activated carbon particles as the anode, not only increases the adhesion area of biofilm, increases the biomass, but also greatly reduces the material cost; (2) The cathode area is large, and the reaction overpotential is reduced; (3) The anode and cathode are separated by a barrier, and the cathode covers the anode. The distance between the positive electrode and the negative electrode is the shortest, and the internal resistance of the battery is the least; (4) In the process of operation, continuous upflow operation is more suitable for sewage treatment.

9.4.2 Double-Chamber H-type MFC

H-type MFC is the most commonly used form in current research. Because this kind of reactor is mostly composed of two glass bottles with one arm sandwiched with cation exchange membrane in the middle, it looks like the letter "H" in appearance, so it is also called H-type MFC. Double-chamber H-type MFC consists of anode chamber and cathode chamber, separated by cation exchange membrane in the middle, which ensures the spatial independence of anode electron donor and cathode electron acceptor. Because the anode and cathode are in different spaces, it can ensure that the two chambers do not affect each other.

A two-compartment microbial fuel cell (MFC) was developed with F026 as anode and Cr(VI) wastewater as cathode. The results showed that when the pH of wastewater was 2 and the temperature of MFC was 303 K, MFC had the best treatment effect on Cr(VI) wastewater. Under these conditions, 92.1% of Cr (VI) can be effectively reduced to Cr(VI) after MFC treatment for 80 h α-Cr_2O_3. The results show that MFC is a promising process for the treatment of wastewater containing Cr(VI).

9.4.3 Flat MFC

In terms of the cell structure, the current microbial fuel cell generally has two anode and cathode chambers separated by proton exchange membrane, but this structure is not conducive to the amplification of the cell. The PEM structure, which is wrapped around the cathode rod and placed in the anode chamber, facilitates cell amplification and has been used in large-scale wastewater treatment.

The flat plate MFC is an improvement of the double-cavity MFC system. The anode and cathode are flattened together with the proton exchange membrane, and the bacteria are enriched on the anode due to the action of gravity, thus reducing the internal resistance and increasing the output power. In order to improve the mass transfer on the electrode surface, different flow fields were arranged on the anode and cathode sides of the plate MFC instead of the liquid storage tank.

9.4.4 Double Tube Microbial Fuel Cell

The researchers developed a new type of cylindrical double-cavity MFC that can be viewed as a variation of the square MFC. The MFC consists of a cylindrical diaphragm that tightly surrounds the anode and an external cathode chamber. This structure greatly reduces the distance between the poles and increases the area of the proton exchange membrane, so that the internal resistance is only 4 Ω. The filling MFC can improve the power generation capacity of MFC, while the tubular proton membrane can effectively reduce the internal resistance of MFC by increasing the current channel in MFC. Therefore, a dual tubular microbiofuel cell was constructed based on a tubular proton membrane. The device integrates a cathode chamber and an anode chamber. The tubular proton membrane was used to increase the current channel in cells, reduce the internal resistance of MFC, improve its power generation capacity, and strengthen the purification of sewage.

Dual tube MFC increases the proton membrane area per unit volume and reduces the internal resistance of MFC. The area power density of the two MFCs is similar. However, due to the more compact structure of the dual-cylinder MFC, the membrane area of the dual-cylinder MFC is approximately 10 times that of the filled MFC for the same equipment volume. Therefore, the volumetric power density of the two-cylinder MFC is also about 10 times that of the filled MFC. The results show that the double tube MFC can increase the proton membrane density and the electric density of MFC per unit volume.

9.4.5 Series MFC

Given existing research, the amount of energy produced by a single fuel cell is very small, so some researchers are trying to improve power generation by connecting multiple independent fuel cells in series. The anode and cathode consist of graphite rods embedded with granular graphite, the glucose used for continuous power generation. It is found that the maximum power density (258 mW/m^3), the open-circuit voltage is 4.16 V, and the internal resistance is 49.1 Ω. The parallel short-circuit current is 425 mA, and the resistance of the series microbial fuel cell is 1.3 Ω. Coulomb efficiency is only 12% in series operation and 78% in parallel operation.

In order to realize the industrial application of microbial fuel cells, single cells must be in series or parallel. At present, the microbial fuel cell stack is still in the experimental research stage. The University of Queensland has done a lot of research work on the application of air cathode microbial fuel cells in sewage treatment engineering, but there are still some problems to be solved.

9.5 Electrode Materials for MFCs

9.5.1 Anode Materials

Microbial fuel cells (MFCs) are a novel bioelectrochemical device with dual functions of decontamination and productivity. Anode material, as an important part of microbial growth and electron collection, is an important factor affecting the performance of MFCs. Therefore, the development of high-performance anode materials is an important way to break through the bottleneck of MFCs output power.

9.5.1.1 Traditional Anode Materials

(1) Carbon materials

Carbon materials are widely used in MFCs due to their excellent electrical conductivity and biocompatibility. Traditional carbon materials include carbon rod, carbon plate, carbon paper, carbon cloth, carbon felt, activated carbon, and carbon fiber brush. Because the smooth surface of graphite rod is not conducive to the attachment of microorganisms, and there are few electrochemically active sites, the output power of MFCs is low.

(2) Metal materials

Metal materials have significant advantages in reducing MFCs internal resistance and reaction activation energy due to their excellent conductor properties and strong oxidation. Therefore, platinum, stainless steel, nickel, and titanium are also widely used in the field of MFCs. However, metallic materials are easily passivated and corroded in complex bioelectrochemical systems, and their long-term discharge behavior needs further observation.

9.5.1.2 Modified Anode Materials

When traditional carbon materials and metal materials are directly applied to MFCs, their electrical performance is often low, and a large number of studies began to focus on the modification of anode materials. The biocompatibility, electrical conductivity, and specific surface area of anode materials can be improved by surface treatment, modification of nano-carbon materials, introduction of transition state metal oxides, and modification of conductive polymers, so as to improve the electrical performance of MFCs.

(1) Surface treatment

Surface treatment can improve the surface properties of the anode material. A large number of oxygen-containing functional groups are introduced through surface treatment to improve the wettability of the material surface, which is conducive to the attachment of microorganisms. At the same time, the anode material after surface treatment is rougher, which increases the specific surface area of the anode material and provides more attachment sites for electrogenic microorganisms, so as to improve the electrical performance of MFCs.

(2) Modification of nano-carbon materials

Compared with traditional carbon materials, nano-carbon materials have more abundant pore structure, better electrical conductivity, and better biocompatibility. The introduction of nano-carbon material not only greatly improves the surface roughness of the substrate material, but also significantly enhances the kinetics of bioelectric catalysis and improves the electron transfer efficiency of extracellular. Electrochemical reduction of GO provides a large number of sites for bacteria to attach to, while greatly improving the electrical conductivity of the material.

(3) Conducting polymer modification

Conductive polymers are favored in the field of MFCs anode modification due to their high conductivity and abundant redox sites. The main purpose of using the conductive polymer as the anode material for MFCs is to improve the surface activity and electron transfer rate of the anode, so as to obtain better electrical performance.

9.5.2 *Cathode Materials*

In MFCs, the use of simple organic substrates, such as glucose, sodium acetate, etc., can produce a higher power density. However, if the organic substrate of the anode is replaced by actual organic wastewater, this value will be greatly reduced, mainly because there are some substances in the actual waste that are difficult to biodegrade and cannot be oxidized into current. MFCs cathode for fuel wastewater treatment, mainly through the MFCs anode substrate is degraded by microorganisms generated electrons are transferred to the cathode. On the one hand, the electrons are directly accepted by the cathode pollutants, and part of the pollutants are directly reduced. On the other hand, after the electrons are accepted by oxygen, H_2O_2 is produced. H_2O_2 itself has a certain oxidation and has a small amount of oxidation and degradation effect on organic matter.

The reaction characteristics of the cathode limit the overall power output of MFCs. In order to improve the power output and overall efficiency of MFCs, some measures should be taken to reduce one, two or three losses of the cathode as far as possible. In addition, in addition to improving the performance of the cathode, the more important factor limiting the engineering and commercial application of MFCs is the structure of the battery, which greatly increases the capital investment of MFCs.

At present, the cathode electron acceptor of MFCs is mainly divided into liquid cathode and air cathode. The most commonly used electron acceptor is O_2, and it is divided into dissolved oxygen in water and gaseous oxygen. Liquid cathode MFCs reaction is carried out in solution with low mass transfer and reaction resistance, so this type of battery can usually get a higher power output. However, its biggest disadvantage is that it needs to constantly supplement electrolyte to the cathode, and the cathode products may bring secondary pollution, so it is not suitable for practical engineering applications. Air cathode MFCs use oxygen in the air as electron acceptor, but the three-phase reduction reaction rate of oxygen molecules on the cathode surface is very slow, and precious metal platinum is needed as catalyst to reduce the overpotential loss of the reaction, which greatly increases the cost of fuel cell.

As an electron acceptor, O_2 has the advantages of high oxidation voltage, cheap and easy to obtain, and the reaction product is water and no pollution. For microbial fuel cells with dissolved oxygen as acceptor, oxygen concentration is one of the main limiting factors when dissolved oxygen does not reach saturation. At present, many researches directly expose the platinum-loaded cathode to air to form an air cathode single-cell microbial fuel cell. This design can reduce the energy consumption caused by aeration and can effectively solve the problem of O_2 transmission, thus improving the O_2 reduction rate and increasing the power output.

Due to the double-chamber microbial fuel cell cathode chamber is separated from the reaction of the anode chamber, and so reduce the interaction between the polar chamber, which is beneficial to microbial fuel cell anode chamber microbial conductive mechanism and basic parameters of microbial fuel cell structure research, as well as to the fixed anode chamber basic conditions, to study the effect of cathode for battery power output. At present, the cathode oxidants of two-compartment microbial fuel cells mainly include dissolved oxygen, ferricyanide, potassium permanganate, manganese dioxide, and so on.

9.5.3 *Membrane*

In recent years, a large number of scholars have applied MFCs to the pollution treatment of dye wastewater, coking wastewater, landfill leachate, and other industrial wastewater. Studies have reported that MFCs can effectively treat coking wastewater (COD = 2000 mg/L), and the maximum voltage can be stabilized at about 370 mV. It can realize the degradation of pollutants containing sulfur, nitrogen, phenols, rings, and alkanes. As a barrier separating the anode from the cathode, the membrane is critical in the composition of MFCs. An ideal membrane can effectively improve the power density and coulomb efficiency of the cell. The performance of the membrane is mainly affected by ion concentration, buffer type, temperature, and other factors related to substrate and ion transfer. At present, the diaphragm used more mainly includes proton exchange membrane, cation exchange membrane, anion exchange membrane, bipolar membrane, and ultrafiltration membrane.

9.5.3.1 Cation Exchange Membrane

The most commonly used cation exchange membrane is Nafion perfluorinated sulfonic acid membrane, which has a negative sulfonate group and can conduct a variety of cations. The Solvay-Solexis Hyflon polymer membrane consists of short side chains containing fluorine, which provides higher electrical conductivity and chemical stability, but also higher internal resistance. Some researchers prepared a polysulfone organic mineral membrane containing 85% ZrO_2, which was modified to have a lower internal resistance than Nafion membrane, but a higher oxygen pass rate, resulting in an increase in anode potential, causing concentration polarization, and reducing the overall power output. For the problem of concentration polarization, some studies have reported that the use of phosphate buffer solution can alleviate the concentration polarization, but still cannot eliminate the concentration polarization near the biofilm, cannot provide an appropriate microenvironment for microorganisms.

9.5.3.2 Anion Exchange Membrane

Considering the limitations of cation exchange membrane on the proton transfer efficiency and hydroxyl compared with proton through a cation exchange membrane more easily through the anion exchange membrane, the researchers to look to the anion exchange membrane, such as ZuO using phosphate or carbonate as the carrier of the proton transfer and pH buffer will anion exchange membranes used in membrane air cathode, The power density is 13.1 W/m^3, while the CEM film is only 8.3 W/m^3. However, with the extension of use time, the anion exchange membrane will deform and bend to produce voids, which will make the substrate diffuse and increase the internal resistance. Meanwhile, the use of buffer solution will also increase the operating cost.

9.5.3.3 Forward Osmosis Membrane

In recent years, in order to overcome the shortage of traditional diaphragm, forward osmosis (FO) technology has been applied in the fields of seawater desalination, water purification, sewage treatment, and reuse by virtue of its outstanding advantages such as good effluent quality and small membrane pollution trend. In the field of wastewater treatment and reuse, FO membrane and its combination technology have the characteristics of high retention, low tendency of membrane pollution, and low energy consumption, which has become a new research hotspot. However, as a new type of membrane, the application of FO membrane in MFCs is still restricted by concentration polarization and membrane fouling. Related theories are being studied.

9.5.4 Electrolyte

The electrolyte is the most traditional electrolyte. The electrolyte is obtained by heating the organic solvent of CAMMA butylactone and the weak acid salt capacitor. The cathode of the aluminum electrolytic capacitor in the ordinary sense is this electrolyte. Electrolyte is the medium used in chemical batteries, electrolytic capacitors, etc. (with a certain degree of corrosivity), which can provide ions for the normal work of the battery, and ensure that the chemical reactions occurring in the work are reversible. There are many advantages to using an electrolyte as a cathode: the liquid has a large contact area with the medium, which helps to increase the electrical capacity. In addition, the electrolytic capacitor uses the electrolyte as the cathode, when the medium is broken down, as long as the breakdown current does not continue, then the capacitor can heal itself. But electrolyte also has its shortcomings: in the high-temperature environment easy to leakage and volatilization, on the life and stability of great impact, in high-temperature and high-pressure electrolyte may be instantaneous vaporization, resulting in volume increase caused by explosion (i.e., we often say the blast).

9.6 Applications of MFCs

9.6.1 MFCs for Wastewater Treatment

The water pollution, especially the organic wastewater generated in various aspects such as household water and industrial production, is the major issue in environmental governance. The treatment of organic wastewater is mainly to remove the toxic organic matter, heavy metals, nitrogen, sulfide, suspended matter, and other harmful substances in the wastewater. In order to effectively control water pollution, various water treatment methods have been proposed. At present, the common water treatment technologies include physical non-destructive adsorption methods and coagulation methods, which can convert pollutants from the liquid phase to the solid phase and realize the water purification. But it is difficult to realize large-scale application due to the high cost. Although the treatment methods based on chemical, photochemical, biological processes can destroy the structure of the contaminants and have remarkable effect, they need a long processing cycle. Therefore, the microbial treatment technologies with the characteristics of fast, high efficiency and low-cost are the currently main methods, especially the MFC technologies have aroused wide concern due to the efficient capability to purify the organic wastewaters while generating electrical energy.

At present, MFCs display a low power, which is far lower than the hydrogen–oxygen fuel cell (low three orders of magnitude), but given the application for efficient purification and reuse of organic wastewater, it is of great significance to open a breakthrough in wastewater treatment. These treatment technologies use MFCs as the reactor, bacteria as biocatalysts, and the wastewater treatment processes can realize the transformation of the chemical energy in available biodegradable organic pollutants into electrical energy Fig. 9.3, which relates microbial metabolism to electrochemical reactions (Nikhil et al. 2018; Palanisamy et al. 2019).

Due to the wide range of sources and low treatment cost, microbial wastewater treatment has become the most promising wastewater treatment technology. In essence, all microbial treatment technologies can be divided into three metabolic types based on the content of dissolved oxygen in the water environment, including aerobic process, anaerobic process, and the combination of the two treatment processes. The MFC wastewater technologies have significant technical advantages Table 9.1 compared with traditional biological treatment technologies (Palanisamy et al. 2019; He et al. 2017), such as (1) the organic pollutants are directly converted

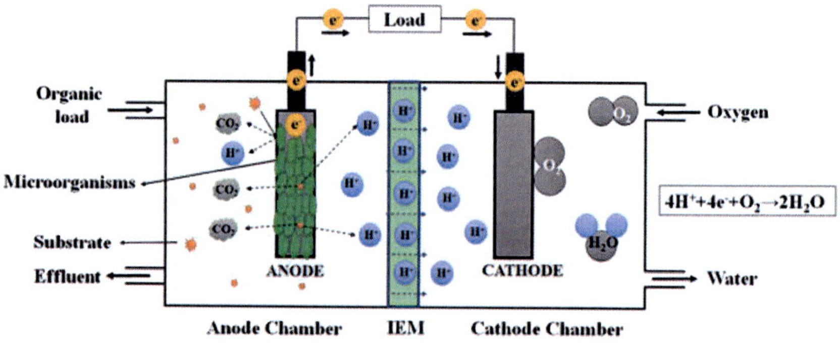

Fig. 9.3 Working principle of the MFCs for electricity and wastewater treatment

Table 9.1 Comparison between microbial fuel cell technologies and traditional biological treatment technologies for wastewater treatment

Methods	Aerobic treatment technology	Anaerobic treatment technology	MFC treatment technologies
Key microorganism	Aerobes	Methanogens	Electricigens
Condition demand	Medium condition	Harsh condition	Normal pressure and temperature
Mode of energy generation	Consuming energy	Methane	Electric power
Energy use complexity	–	Complex	Simple
The sludge quantity	High	High	Low
Scope of application	Low and medium and concentration	Medium and high concentration	Low, medium, and high concentration

into electric energy via microbial degradation, (2) the universal working conditions (various pH, work temperatures, and diverse microorganisms), (3) the low production of sludge (biomass), (4) without external energy input. Therefore, the MFC wastewater treatment technologies have received extensive attention and become the focus of researches.

9.6.1.1 Removal of Organic Pollutants

According to the mechanism of MFCs, the microorganisms in the battery need degrading organic matters to produce electrons and protons, which is the foundation for electricity production. Therefore, the wastewaters treated by microbial fuel cells are mostly organic wastewaters. In recent years, microbial fuel cells have been used more and more widely in the treatment of various organic wastewaters, such as the wastewaters from food-processing, livestock, and petrochemical industries. Lu et al. (2017) designed a MFC system with a volume of 20 L to treat the organic pollutants in brewery effluent. The maximum COD removal efficiency was about 94.6%, and the flow rate and hydraulic retention time were 1 mL min^{-1} and 313 h, respectively. Firdous et al. (2018) reported that the high COD removal percentage (80–90%) and high voltage about 5839 mV could be realized for the vegetable oil wastewater in a double-chamber MFC system at the temperature of 35 °C. Besides, the MFC wastewater treatment efficiency of this system could be enhanced by increasing the treatment temperature and hydraulic retention time. Therefore, it is of importance to adopt the optimal working conditions to realize the improvement of organic pollutant removal efficiency.

9.6.1.2 Removal of Nitrogen and Sulfide

In addition to the treatment of organic wastewater, MFCs also have great potential for the treatment of other types of pollutants, such as nitrogen and sulfide-based hazardous substances in wastewater. It is necessary to remove nitrogen and sulfide from wastewater to prevent the accumulated pollution and the occurrence of eutrophication of water body. The nitrogen exists in wastewater as the ammonium and the conventional removal processes mainly include nitrification and denitrification processes. But these processes need large oxygen supply and result in high amount of sludge (Arredondo et al. 2015). On the contrary, the MFC-based treatment technologies can overcome these shortcomings. The MFCs can transform the existence form of nitrogen and sulfur in wastewater through oxidation–reduction reaction Fig. 9.4 (Palanisamy et al. 2019). By adjusting the voltage, the sulfide in pollutants can be transformed to zero-valent sulfur, sulfites, and sulfates. The ammonia nitrogen with

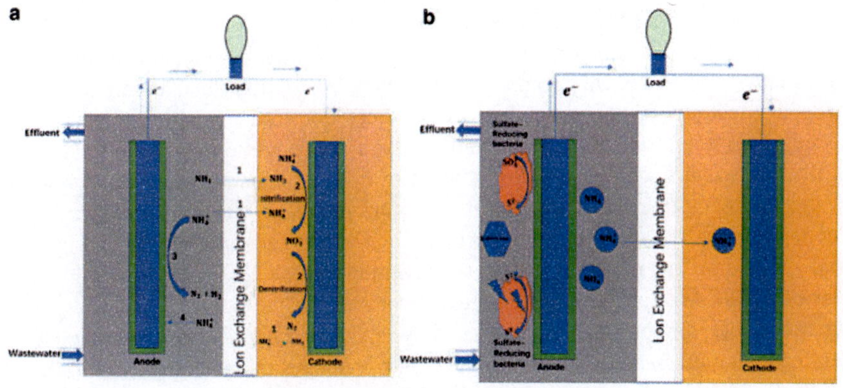

Fig. 9.4 MFCs for the removal of **a** nitrogen and **b** sulfide in wastewaters

a low redox potential can be used as an electron donor to provide energy for microbial growth during aerobic or anaerobic ammonia oxidation. Therefore, the MFCs can effectively achieve the effects of desulfurization and denitrification. Recently, a novel membrane-less MFC to purify the ammonium in wastewater, in which the ion-exchange membrane is replaced by baffles (Palanisamy et al. 2019). This novel MFC system showed high pollutant removal efficiency, and the ammonium nitrogen, total nitrogen and COD removal efficiencies are about 91.76%, 87.66%, and 97.07%, respectively. The sulfide as the by-product is discharged in large quantities from various industries, particularly the tanneries, petrochemical plants, and viscose rayon factories. It is extremely urgent to remove sulfide from wastewater due to its harm to human body and the environment. The traditional treatment methods, including physicochemical and biological processes, often require high cost and large energy input. Besides, the characteristic of the polyvalent state of sulfide also increases the difficulty of removal efficiency via traditional treatment methods. Therefore, the MFC systems have been explored to realize the sulfide removal with low cost while electricity generation. Wang et al. (2018) designed a potentiostatic three-chambered MFC system and studied the electrode potential interactive effect on the removal efficiency of nitrate and sulfide in wastewater. A suitable anode potential from 188 to 146 mV benefited for the sulfide removal and sulfur recovery. In addition to the nitrate and sulfide removal, the MFCs can also be applied to remove many harmful metal ions in wastewater (Li et al. 2021). Therefore, the MFC-based wastewater treatment technologies have a bright application prospect.

9.6.1.3 Combination of MFCs with Other Wastewater Treatment Technologies

The MFCs are often coupled with traditional microbial wastewater treatment technologies to promote the improvement of pollutant removal efficiency and speed, and increase the production efficiency of electricity and organic fuels (Kumar et al. 2018). At present, the MFCs still have some drawbacks, such as uneven distribution of microorganisms in the stirring process at anode and the high technical difficulty in scale-up application. On the contrary, the anaerobic fluidized bed has a good mass transfer effect and fast biochemical reaction rate. Due to the small particles of the carrier, the large total surface area and the high concentration of microorganisms in the biological fluidized bed, the number of microorganisms in per unit volume of the reactor is increased. Besides, there is no blockage during reaction process due to the fluidized state. More importantly, the advantages in scale-up and industrial application can make up for the shortcomings of the MFCs. The combination between MFCs and anaerobic biological treatment technology also provides new research ideas for the industrialization of wastewater treatment of microbial fuel cell technologies. In addition, due to the dual advantages of anaerobic filtration and anaerobic activated sludge processes, the MFCs based on upflow anaerobic sludge bed (UASB) reactor display higher COD removal capacity and lower treatment cost than other biochemical treatment processes and the unitary MFCs (Papaharalabos et al. 2015). Therefore, this system is widely used in the treatment of the wastewaters with high concentration of COD. For example, the brewery wastewater can obtain a maximum COD reduction of ~ 87% (Dong et al. 2015). The MFCs based on the UASB adopt a continuous liquid feeding method, which can overcome the limitation of unstable discharge voltage caused by batch feeding. This coupled system possesses the advantages of MFC and UASB, so the stability of electricity production is improved. Domestic wastewater and industrial wastewater containing a large number of organic matters can be used as fuel to obtain electricity. Therefore, the research and development of MFCs combined with the UASB process have become a research hotspot in current pollution control and development of new energy sources. With the increase of wastewater discharge standards, the anaerobic biological treatment process is introduced into the urban wastewater treatment process in order to meet the discharge requirements of nitrogen and phosphorus. The microbial fuel cells coupled with the anaerobic–anoxic–oxic (AA/O) treatment technologies are widely used (Liu et al. 2017). Anaerobic–anoxic (AA) is the anaerobic–anoxic stage, which can achieve the purpose of denitrification through the denitrification of denitrifying bacteria; O (Oxic) is the aerobic stage, which removes phosphorus-containing organic matter and other organic matter in the water through the metabolism of aerobic microorganisms. When the AA/O treatment process is used to treat urban wastewater, the COD in the wastewater can be removed while also obtaining an ideal nitrogen and phosphorus removal effect. In response to this changing trend of wastewater treatment technology, it has become feasible to apply microbial fuel cells with AA/O treatment process to the actual treatment process of urban wastewater. Besides, adding aerobic sludge to the cathode cannot only improve the power generation performance, but also greatly improve the COD removal efficiency in the urban wastewater treatment process. The high COD value of industrial wastewater and high salinity provide favorable conditions for the application of the MFCs with AA/O treatment process in industrial wastewater. The azo dyes with azo groups (N–N) in the molecules have the characteristics of high chroma, complex structure, low biodegradability, and great environmental hazard, so they are recognized as the difficult-to-treat organic wastewaters. The MFCs can be combined with traditional electrochemical degradation as well as the technologies of biodegradation to realize the efficient and low-cost degradation of azo dyes.

The development tendencies of the MFC researches are to increase the power output while reducing cost and to realize the large-scale practical application in wastewater treatment and the environmental bioremediation and pollution control. With the continuous development of science and technology and the deepening of cross-scientific research, especially relying on the research progresses in biosensors and bioelectrochemistry, modified electrodes and nanoscience, the MFCs for wastewater treatment have very broad prospects. Especially, the tremendous development of biotechnology in recent years has also provided huge material, knowledge, and technical reserves for the development of MFCs for wastewater treatment.

9.6.2 Application of MFCs for Desalination

With the explosive growth of population, freshwater resources are increasingly strained. Besides, the large discharge of wastewater with high salt content from factories results in the salinization of soil and the pollution of underground freshwater resources, which seriously damage the water environment and further exacerbate the freshwater crisis. Therefore, high-efficiency and low-cost desalination and wastewater treatment technologies have become the focus of the development in the field of environmental engineering. Seawater desalination also become an efficient way to increase the available water resources. The microbial desalination cells (MDCs) are developed from microbial fuel cells (MFCs) and can realize the removal of pollutants as well as desalination and then producing electricity based on the principle of bioelectrochemistry (Yang et al. 2019). Compared to the traditional MFCs, an additional chamber between the anode and cathode chambers is added to the MDCs reactor to desalinate saline water.

Since the advent of the microbial desalination fuel cells, they have received wide attention as a new method for desalination. Cao et al. (2009) firstly used the MDCs to treat the solution with different initial salt concentrations (5, 20, and 35 g/L) with acetate used as the substrate for the bacteria, which effectively reduce the operating cost and realize the practical application of MDCs for desalination. The microbial desalination fuel cells are a new energy self-sufficient and green desalination technology. In recent years, the researches on the structure and function of microbial desalination fuel cell reactors have developed rapidly, by coupling with other technologies to form a variety of configurations, such as stacking type, anode and cathode reflow type, resin-filled type, upflow type (Sayed et al. 2020). There are some differences in the desalting performance, but they all have the promoting effect on the development of seawater desalination. However, the microbial desalination fuel cells have been facing many problems, such as unstable pH of the anode and cathode chamber liquid and low dissolved oxygen concentration at the cathode, which severely limit the performance of the reactor and the scale-up of the device (Salehmin et al. 2020). Therefore, the realization of the practical large-scale application of MDCs for desalination remains needing great efforts to overcome these challenges.

9.6.3 The Application of MFCs for Biosensors

It is necessary to design the online water-monitoring system to real-time feedback the pollution level of wastewater, which benefits for the proper discharge and usage of wastewaters from industries or municipal department, and then achieve the goal of protecting the public health as well as water resources. The MFCs based on the microbial redox reaction possess the promising prospect in biosensors and have been confirmed as the ideal biosensor to detect the organic pollutants in the domestic and industrial wastewaters (Din et al. 2021). With regard to the traditional biosensors, the transducer is essential. But the MFCs can be used as the transducer in themself, so they are the cost-effective biosensors.

In the MFC-based biosensors, the biological recognition element or signal generator is the exoelectrogens in the anode chamber, and the electrodes serve as the transducer. The good long-term stability is the prominent advantage of the MFC-based biosensors due to that the exoelectrogenic biofilms can improve the lifespan of sensing element. The basic principle of MFC-based biosensor is the signal of microbial electrochemical redox reaction with the existence of organic substrates. To be specific, when the pollutants touch the anode chamber and the output voltage will change according to the type and concentration of the pollutants. The sudden change of the voltage is used as the signal for pollutants detection. At present, the MFC-based biosensor has been used in the detection of some metal ions, such as copper (II) and Cr(IV) (Jiang et al. 2016; Liu et al. 2014).

In addition to the wastewater treatment, desalination, and biosensors, the MFCs with the microbial degradation and electrochemical processes can also be applied in the recycling of sludge and biological hydrogen production (Dai et al. 2016). With the continuous progress in nanometer materials and microbial technologies, the new and valuable application directions of the MFCs will further emerge to promote the green and sustainable development of society.

9.7 Prospects of MFCs

Currently, the research on MFCs mainly focuses on two aspects: The first priority is to analyze and identify the interaction between cells and electrodes through molecular biology and genetic engineering technology, improve and regulate electroactivity of microbial cells, and reduce or eliminate the shielding effect or resistance of the electron transfer process. It is urgent to upgrade the reactor configuration for improving the power generation of MFCs and the utilization of biomass energy. The second is to accelerate the industrialization of MFCs. Therefore, MFCs have practical application value from the perspective of waste utilization.

As can be seen from the patent distribution of MFCs-related application domains that about 800 patents associated with sewage and sludge treatments are ranked second, accounting for one-third of the total number of patents. In other word, sewage and sludge treatments are the most important application fields of MFCs apart from power generation. In recent years, the application of MFCs has increasingly shown a diversified trend. The implantable medical device MFCs constructed in human bloodstream can employ glucose as energy source to provide stable and continuous power supply for electronic devices in the human body, including pacemakers, biochips, and drug delivery systems in the body, which broaden application prospects of MFCs in biomedicine. As for marine MFCs, the organic matters and minerals extracted from the seabed sediments are mainly used as fuels to provide stable and inexhaustible electrical energy for marine monitoring equipment that serves in the deep sea and the open sea. Unfortunately, the electricity generation capacity of the current MFCs is very limited and awaits further optimized and improved. In short, benefiting from its many advantages such as negligible biotoxicity, outstanding operability, stability, and sustainability, MFCs will play an important role in more fields.

Nowadays, MFCs are difficult to be applied on a large scale due to its high cost, low energy recovery efficiency, and undesirable power generation should be optimized. Thus, the small-scale experiment will be the mainstream for a long time. Many efforts need to be made to realize the commercialization of MFCs.

9.7.1 *Improving the Output Power of MFCs*

The unsatisfactory output power is one of the restrictions on the large-scale application of MFCs. Over the past several decades, although the battery output power of MFCs has been increased from milliwatts to watts, it still cannot meet the practical commercial requirements. In the future, the following aspects should be considered for the relevant researches:

(1) Improving the power density of MFCs. Attributed to the low power generation, MFCs are still incapable of using as an independent power source in large-scale industrial production so far. Therefore, how to combine MFCs, as a renewable energy source and pollution treatment device, with other energy sources to increase the power density and achieve the lower emission simultaneously is the key to facilitating the development of MFCs.

(2) Enhancing the energy harvesting of MFCs. It is universally accepted that the selection of electricity-producing bacteria, the optimization of proton exchange membranes and electrode materials, the alteration of the size of anode and cathode chambers, and the combination of conductive cables will cause the large fluctuations in the power output of MFCs system. Currently, a myriad of scholars has conducted numerous studies to improve the power generation of MFCs, but it is still a considerable problem that the energy generated by MFCs can be used efficiently under the premise of ensuring the output of the system. Under the precondition that the power generation cannot be significantly increased in the short term, to efficiently collect and reserve energy by designing the external circuits and energy-harvesting systems and reasonably use the low-output energy has gradually become the crucial approach for large-scale applications of MFCs.

(3) Constructing the state-space model of control-oriented MFCs. At present, a large number of models for different types of MFCs have been proposed, but the reaction mechanisms and parameters have not been described in most of the models. It is difficult to utilize classic algorithms of control theory to perform control research and develop the state space of MFCs with surface control. Predictably, the model established with the linear and nonlinear control methods is of great significance to the theoretical research of MFCs.

(4) Solving the control problem of MFCs. As a system with complex models, strong nonlinearity, and serious time-lag, MFCs system is a complement for MFCs modeling, which aims to obtain a more stable output by using more advanced control methods. Imaginably, based on the ideal model, using the advanced control methods to obtain the optimal controller will provide accurate guidance for MFCs.

9.7.2 Improving the Power Generation Capacity of MFCs

According to the analysis of the recognition results of inter-process communication (IPC) code and latent Dirichlet allocation (LDA) subject, improving power generation performance is still the focus and hotspot of MFC's basic research. The means for improving the capacity of electricity production is mainly in the following aspects:

(1) Battery configurations. From laboratory to practical application, the power density of MFCs will be reduced on a large scale, which is widely found in the single-chamber MFCs, double-chamber MFCs, "three-in-one" MFCs, multi-chamber MFCs, multi-cathode MFCs, and other configurations. The improvement of battery configuration is the crucial segment to alleviate this problem.

(2) Electrode materials. Electrode material and electrode structure will affect the properties of the electrode, thereby determining the entire performance of MFCs. High cost and poor stability are the main drawbacks of the current electrodes. Experiments are conducted to hunt for new and inexpensive electrode materials, optimize the electrode structure, reduce the cost and improve the electrocatalytic activity of electrodes, further enhancing the performance of MFCs. Thereinto, the development of conductive materials with good electrical conductivity, favorable corrosion resistance, large specific surface area, and low price is the key to reduce the electrical resistance of electrodes for improving the electricity generation capacity of MFCs.

(3) Proton exchange membranes. As an important part of MFC, the main function of the proton exchange membrane is to isolate reaction environment of the anode and cathode chamber, maintain the potential difference in-between, and efficiently transfer protons from anode to cathode. The nature of the proton exchange membrane directly affects the working efficiency of MFCs. Moreover, the high cost is also the vital factor restricting the industrialization of MFCs.

(4) The electron transfer efficiency of electricigens. The electron transport mechanism of electricigens is not fully understood up to now. It is expected to achieve a breakthrough in the research and development of MFCs by further disclosing the extracellular electron transport process. However, the electricigens used at

this stage mainly include strains such as "Shewanella oneidensis" and "Pseudomonas aeruginosa," which only account for a small fraction of the microbial species in nature. Therefore, the selection of high-efficiency electricity-producing strains, the ecological optimization and regulation of the flora, the genetic modification and the isolation and cultivation of traditional strains are the indispensable part for improving the electricity production performance of MFCs. With the continuous and intensive research, the practical marketization of MFCs is just around the corner.

9.7.3 Increasing the Use of Biomass Energy

In recent years, the tremendous progress of microbiology, nanotechnology, electrochemistry, and environmental engineering have provided favorable material, knowledge and technical reserves for the research of MFCs. Therefore, in the near future, MFCs are expected to make important progress in the efficient conversion and utilization of biomass energy. In terms of biomass sources, the wastewater-based biomass can be effectively used by MFCs. For fibrous biomass, especially plant fibers, eliminating or mitigating the effects of inhibitors such as phenols and furans originated from pretreatment or degradation on the biological activity is a primary issue that needs to be addressed in the resource utilization of MFCs.

(1) In terms of catalysts: High cost, unstable catalytic activity, and lower efficiency of the current catalysts will be hopefully overcome by exploring the cheap raw materials with the effective catalytic property. For instance, the research on three-dimensional composites, ceramic materials, and non-platinum catalysts is expected to reduce the cost of MFCs in practical applications. Moreover, the catalytic activity of catalysts can be enhanced by introducing filler to increase its pore size and specific surface area.
(2) In terms of biofilms: The growth of biofilm has a great influence on the electrochemical properties of MFCs. The electrochemical activity of MFCs is progressively strengthened with the gradual maturity of biofilm. However, the permeability and stability of biofilm will attenuate with the increase of thickness, which extremely affects the transport of extracellular electrons, directly leading to the deteriorative performance of MFCs. Fortunately, the permeability of the biofilm can be promoted by blending fillers (such as nanofibers and sophorolipids) or using surfactants to expand the pore size of the biofilm. Hence, the performance of MFCs will be improved effectively.
(3) In terms of substrates: Due to the ever-increasing cost of simple substrates such as glucose over the years, it is proposed to find some complex substrates, such as wastewater (food-processing wastewater, slaughterhouse wastewater, breeding wastewater, beer wastewater, etc.). Once a large amount of biomass in degraded biomass wastewater or waste semisolid is used, the cost of MFCs and environmental pollution can be controlled simultaneously, and the electricity generation capacity of MFCs also can be increased. However, the complex

substrates may bring about the generation of mixed potentials, which disrupts the flow of internal current, further affecting the properties of the electrodes and reducing the power production of MFCs.

(4) Coupling the plants and MFCs: Some plants (such as rice and sweetgrass) have abundant rhizosphere microorganisms in their roots. By placing electrodes near the roots of plants to construct a coupling system, the electrochemical catalysis of rhizosphere microorganisms can be used to utilize and degrade plants. Using organic matter near the roots is another exploration to promote the conversion and utilization of biomass energy for MFCs.

Considering the high cost of the traditional MFCs, the membrane-less MFCs should be the focus of future development. Furthermore, the treatment of toxic and harmful substances and sewage also will be a significant development direction of MFCs. On the basis of existing research results, the effectiveness of environmental control should be further improved.

9.7.4 Research on the Combination of Multiple MFCs

In future research, in addition to investigating the impact factors of a single MFCs, multiple MFCs and coupling with other biotechnologies should also be developed. The major research directions include as follows:

(1) The development of new-type cathode oxidants: the current being used cathode oxidants can improve the operational effect of MFCs, but it is non-renewable and the price is relatively high. In single-chamber MFCs, the transfer of oxygen has not been reported as a limiting factor, but it is foreseeable that the water film formed by the cathode reduction will seriously hinder the oxidation in the large-scale applications of MFCs. According to the working principle of MFCs, oxidizing wastewater (such as metal oxide wastewater) can be added to the cathode, which can completely get rid of the limitation of oxygen.

(2) The process simulation of MFCs: Without convection, many electricity-producing bacteria cannot be accumulated on electrodes with large porosity so that the current density of MFCs also cannot be improved. In addition, the permeability of the membrane to hydrogen ions has a greater impact on pH changes than the buffer system. Finally, the future direction for the in-depth study of MFCs can be pointed out by the simulation results.

(3) The Coupling of anaerobic fermentation technology and MFCs: Due to the high concentration of complex organic matter and particulate matter contained in the actual sewage, the shortcoming that only the dissolved small molecular organic matter can be utilized by MFCs, which can be amended by coupling of anaerobic fermentation technology and MFCs. After coupling, the complex organic matter can be degraded into simple organic matter by anaerobic fermentation section; further, the biomass energy can be generated by the system simultaneously. Subsequently, the simple organic matter served as fuel of MFCs can

be used to produce electricity. The entire system will achieve the degradation of higher concentrations of complex organic matter, while generating a large amount of energy.

(4) The performance of multiple MFCs: Under ideal conditions, the theoretical voltage of a single MFCs is only 1.1 V, but the rated voltage of the actual load far exceeds this value. Therefore, it is necessary to consider connecting multiple MFCs in series, parallel, or mixed in future practical applications. Moreover, the battery reversal phenomenon is more obvious in a number of MFCs studies, which will lead to the reduced microbial activity. The regulation of substrate is helpful for evaluating the performance of multiple MFCs.

(5) The efficiency of anode materials. In the future, the interdisciplinary integration of chemistry, microbiology, materials science, energy, and environmental engineering should be emphasized on the research of anode materials. Therefore, the anode materials can be studied from the following three aspects: using naturally available materials to prepare green, pollution-free, inexpensive anode materials with excellent performance; using 3D printing technology to construct anode materials with fine pore structure, increased attachment sites of microorganisms, and electrochemically active sites; after obtaining a new anode, its durability and operational stability should be further evaluated, so that the development of high-performance anode materials for MFCs will be more in-depth and systematic.

9.8 Conclusions

As a clean energy technology with broad prospects that MFCs can realize sewage treatment and electric energy recovery simultaneously, representing the future development direction of the utilization of wastewater resources. Whereas, the current electricity generation capacity of MFCs is still far from the realization of engineering applications. With the fruitful advance in battery structure, continuous optimization of electrode materials, and the exploration of high-efficiency electricity-generating microorganisms, the electricity generation capacity of MFCs will be improved significantly. Finally, the MFCs will play a vital role in actual production, eventually resulting in immeasurable economic, environmental, and social benefits.

References

Abbas SZ, Rafatullah M, Chen M-Y et al (2021) Recent advances in soil microbial fuel cells for soil contaminants remediation. Chemosphere 272:129691

Allen RM, Bennetto HP (1993) Microbial fuel-cells. Appl Biochem Biotechnol 39(1):27–40

Arredondo MR, Kuntke P, Jeremiasse AW et al (2015) Bioelectrochemical systems for nitrogen removal and recovery from wastewater. Environ Sci Water Res Technol 1(1):22–33

Arun S, Sinharoy A, Pakshirajan K et al (2020) Algae based microbial fuel cells for wastewater treatment and recovery of value-added products. Renew Sustain Energy Rev 132:110041
Cai T, Meng L, Chen G, Xi Y et al (2020) Application of advanced anodes in microbial fuel cells for power generation: a review. Chemosphere 248:125985
Cao XX, Huang X, Liang P et al (2009) A new method for water desalination using microbial desalination cells. Environ Sci Technol 43(18):7148–7152
Dai H, Yang H, Liu X et al (2016) Electrochemical evaluation of nano-$Mg(OH)_2$/graphene as a catalyst for hydrogen evolution in microbial electrolysis cell. Fuel 174(15):251–256
Din M, Nabi A, Hussain Z et al (2021) Microbial fuel cells—a preferred technology to prevail energy crisis. Int J Energy Res 45(6):8370–8388
Dong Y, Qu Y, He W et al (2015) A 90-liter stackable baffled microbial fuel cell for brewery wastewater treatment based on energy self-sufficient mode. Biores Technol 195:66–72
Elshobary ME, Zabed HM, Yun J et al (2021) Recent insights into microalgae-assisted microbial fuel cells for generating sustainable bioelectricity. Int J Hydrogen Energy 46(4):3135–3159
Firdous S, Jin W, Shahid N et al (2018) The performance of microbial fuel cells treating vegetable oil industrial wastewater. Environ Technol Innov 10:143–151
Gorby YA, Yanina S, Mclean JS et al (2006) Electrically conductive bacterial nanowires produced by *Shewanella oneidensis* strain MR-1 and other microorganisms. Proc Natl Acad Sci USA 103(30):11358–11363
He L, Du P, Chen Y et al (2017) Advances in microbial fuel cells for wastewater treatment. Renew Sustain Energy Rev 71:388–403
Holmes D, Chaudhuri S, Nevin K et al (2006) Microarray and genetic analysis of electron transfer to electrodes in *Geobacter sulfurreducens*. Environ Microbiol 8(10):1805–1815
Jan D (2014) Syntrophy in microbial fuel cells. ISME J 8(1):4–5
Jiang Y, Liang P, Liu P et al (2016) A cathode-shared microbial fuel cell sensor array for water alert system. Int J Hydrogen Energy 42(7):4342–4348
Kumar R, Singh L, Zularisam A et al (2018) Microbial fuel cell is emerging as a versatile technology: a review on its possible applications, challenges and strategies to improve the performances. Int J Energy Res 42:369–394
Lewis K (1966) Symposium on bioelectrochemistry of microorganisms IV. Biochemical fuel cells. Bacteriological Rev 30(1):101
Li M, Zhou J, Bi YG et al (2021) Polypyrrole/sewage sludge carbon as low-cost and high-effective catalyst for enhancing hexavalent chromium reduction and bio-power generation in dual chamber microbial fuel cells. Sep Purif Technol 256:117805
Liu H, Ramnarayanan R, Logan BE (2004) Production of electricity during wastewater treatment using a single chamber microbial fuel cell. Environ Sci Technol 38(7):2281–2285
Liu B, Lei Y, Li B (2014) A batch-mode cube microbial fuel cell based 'shock' biosensor for wastewater quality monitoring. Biosens Bioelectron 62(15):308–314
Liu R et al (2017) Microbial community dynamics in a pilot-scale MFC-AA/O system treating domestic wastewater. Bioresour Technol 241:439–447
Logan BE, Hamelers B, Rozendal R, Schröder U, Keller J, Freguia S, Aelterman P, Verstraete W, Rabaey K (2005) Microbial fuel cells: methodology and technology. Environ Sci Technol 40(17):5181–5192
Lu M, Chen S, Babanova S et al (2017) Long-term performance of a 20-L continuous flow microbial fuel cell for treatment of brewery wastewater. J Power Sources 356(15):274–287
Nader M, Wan R, Rahimnejad M et al (2012) Bioelectricity generation in biological fuel cell with and without mediators. World Appl Sci J 18(4):559–567
Nikhil GN, Chaitanya DK, Srikanth S (2018) Applied resistance for power generation and energy distribution in microbial fuel cells with rationale for maximum power point. Chem Eng J 335(1):267–274
Palanisamy G, Jung H-Y et al (2019) A comprehensive review on microbial fuel cell technologies: Processes, utilization, and advanced developments in electrodes and membranes. J Clean Prod 221(1):598–621

Papaharalabos G, Greenman J, Melhuish C et al (2015) A novel small scale microbial fuel cell design for increased electricity generation and waste water treatment. Int J Hydrogen Energy 40(11):4263–4268

Peighambardoust SJ, Rowshanzamir S, Amjadi M (2010) Review of the proton exchange membranes for fuel cell applications. Int J Hydrogen Energy 35(17):9349–9384

Potter MC (1911) Electrical effects accompanying the decomposition of organic compounds. In: Proceedings of the royal society of London. Series B, containing papers of a biological character 84(571):260–276

Rabaey K, Boon N, Siciliano SD et al (2004) Biofuel cells select for microbial consortia that self-mediate electron transfer. Appl Environ Microbiol 70(9):5373–5382

Rabaey K, Boon N, Höfte M et al (2005) Microbial phenazine production enhances electron transfer in biofuel cells. Environ Sci Technol 39(9):3401–3408

Rahimnejad M, Ghoreyshi A, Najafpour G et al (2012) A novel microbial fuel cell stack for continuous production of clean energy. Int J Hydrogen Energy 37(7):5992–6000

Rahimnejad M, Mokhtarian N, Najafpour G et al (2009) Effective parameters on performance of microbial fuel cell. In: 2009 Second international conference on environmental and computer science 23:411–415

Reguera G, Mccarthy KD, Mehta T et al (2005) Extracellular electron transfer via microbial nanowires. Nature 435(7045):1098–1101

Reimers CE, Stecher HA, John C et al (2007) Substrate degradation kinetics, microbial diversity, and current efficiency of microbial fuel cells supplied with marine plankton. Appl Environ Microbiol 73(21):7029–7040

Renslowa R, Donovana C, Shima M et al (2011) Oxygen reduction kinetics on graphite cathodes in sediment microbial fuel cells. Phys Chem Chem Phys 13(48):21573–21584

Roller SD, Bennetto HP, Delaney GM et al (1984) Electron-transfer coupling in microbial fuel cells: 1. Comparison of redox-mediator reduction rates and respiratory rates of bacteria. J Chem Technol Biotechnol 34(1):3–12

Salehmin M, Lim SS, Satar I, et al (2020) Pushing microbial desalination cells towards field application: prevailing challenges, potential mitigation strategies, and future prospects. Sci Total Environ 759(12):143485

Sayed ET, Shehata N, Abdelkareem MA et al (2020) Recent progress in environmentally friendly bio-electrochemical devices for simultaneous water desalination and wastewater treatment. Sci Total Environ 748(15):141046

Schröder U, Nießen J, Scholz F (2003) A generation of microbial fuel cells with current outputs boosted by more than one order of magnitude. Angew Chem Int Ed 115(25):2986–2989

Serra P, Espírito-Santo A (2021) Sourcing power with microbial fuel cells: a timeline. J Power Sour 482(15):228921

Suzuki S (1976) Fuel cells with hydrogen-forming bacteria. Hosp Hyg Gesundheitswes Desinfekt 159

Wang K, Zhang S, Chen Z et al (2018) Interactive effect of electrode potential on pollutants conversion in denitrifying sulfide removal microbial fuel cells. Chem Eng J 339(1):442–449

Yang E, Chae KJ, Choi MJ et al (2019) Critical review of bioelectrochemical systems integrated with membrane based technologies for desalination, energy self-sufficiency, and high efficiency water and wastewater treatment. Desalination 452(15):40–67

Zhao F, Robert CT, Varco JR (2009) Techniques for the study and development of microbial fuel cells: an electrochemical perspective. Chem Soc Rev 38(7):1926–1939

Chapter 10
Energy Recovery from Fat, Oil and Grease (FOG)

Mahdy Elsayed, Ahmed Tawfik, and Abd El-Fatah Abomohra

Abstract The high percentage of fats, oils and grease (FOG) in wastewater discharges from kitchen waste streams is increasing rapidly due to the increasing demand for food and modern lifestyle. Direct discharge of FOG into sewer system results in many environmental and technical problems. It is energy-rich waste, while effective management is required to recover its energy. FOG collection from wastewater lines prior to discharge into the sewer networks is essential, and further conversion into biofuel could generate additional revenue. Therefore, recent research is focusing on different routes of FOG conversion into usable biofuel. Deep energetic and environmental analysis of FOG conversion into bioenergy concluded that FOG conversion is a very promising route for various biofuel production. This chapter presents an overview of engineered challenges related to various technologies used for energy recovery from FOG wastes and biofuel production. The different routes of biofuels production (e.g., biohydrogen, biomethane and biodiesel) through new integrated routes for sustainable biofuel industry are evaluated. Herein, this chapter provides a successive high throughput of the full conversion of FOG wastes into biofuel toward a *zero-waste* system.

Keywords FOG wastes · Bioenergy · Biomethane · Biohydrogen · Biodiesel

M. Elsayed (✉)
Department of Agricultural Engineering, Faculty of Agriculture, Cairo University, Giza 12613, Egypt
e-mail: mahdy.mahdy@agr.cu.edu.eg

A. Tawfik (✉)
Water Pollution Research Department, National Research Centre, P.O. Box 12622, Giza, Egypt

A. E.-F. Abomohra
Department of Environmental Engineering, School of Architecture and Civil Engineering, Chengdu University, Chengdu 610106, China
e-mail: abomohra@cdu.edu.cn

A. E.-F. Abomohra et al. (eds.), *Waste-to-Energy*,
https://doi.org/10.1007/978-3-030-91570-4_10

10.1 Introduction

The global energy transition is being driven by the need to mitigate climate change and ensuring sustainable growth in an era of rapid change (Elsayed et al. 2020). The world's high energy security and the negative environmental reflections of using fossil fuels have led to a pressing need to investigate novel feedstocks of biofuels (Xu et al. 2019). As a result, additional studies are needed to ensure the energy security while also reducing the pollutants. Among many biofuel feedstocks, fats, oils and grease (FOG) from kitchen waste streams, slaughterhouses, restaurants and food processing plants proved to be the most cost-effective source of biofuels (Abomohra et al. 2020). Traditionally, dumping of FOG-rich wastewater directly into the sewer system would certainly accumulate FOG inside the networks, creating serious environmental problems (Abomohra et al. 2020). In that context, about 47% of sewer lines are blocked, resulting in lessen pipes' cross-section due to FOG deposition (Elsheikh et al. 2013; He et al. 2011). In addition, FOG is energy-rich waste that requires effective management to recover energy. The classical technologies used to treat FOG-rich wastewater, including chemical coagulation, membrane bioreactors, attached growth system and suspended growth system, are very expensive and consume a huge amount of energy and/or chemicals, with simultaneous production of vast quantities of sludge. Therefore, recent research is focusing on biological wastewater treatment (Abomohra et al. 2018) and FOG conversion for bioenergy (Alqaralleh et al. 2018; Del Mundo and Sutheerawattananonda 2017; Solé-Bundó et al. 2020). Deep economic and environmental analysis of FOG conversion into biodiesel concluded that FOG conversion is a promising route if dual-fuel potential was proposed (Abomohra et al. 2020). In addition, anaerobic co-digestion of different substrates showed higher conversion efficiency (Alqaralleh et al. 2018; Hagos et al. 2017; Long et al. 2012; Venturin et al. 2018). However, FOG conversion technology should be economically feasible with minimum chemical consumption, low energy requirements, minimum toxic by-products and applicability on large scales (Tabatabaei et al. 2019). Hydrolysis and methanogenesis of lipids rich in long chain fatty acids (LCFAs) are the rate-limiting step during anaerobic digestion (AD) process (Salama et al. 2019). In addition, due to the high water and free fatty acids (FFAs) content of FOG, pretreatment is required as an essential step to eliminate the water and FFAs for efficient biodiesel production. In that context, various pretreatment/conversion methods, such as biological conversion (Badoei-dalfard et al. 2019), steam stripping (Usseglio et al. 2019), acid esterification (Suwannakarn et al. 2009), nanocatalytic technology (Gardy et al. 2016), supercritical esterification (Ghoreishi and Moein 2013) and glycerolysis (Costa et al. 2015), have been suggested. Direct conversion of FOG into biodiesel at supercritical conditions without prior pretreatment was recently recommended as a promising approach (Abomohra et al. 2020). At the applied critical conditions, LCFAs will be simultaneously degraded (Charuwat et al. 2018), which further enhance the residue's anaerobic digestion.

Hence, this chapter presents an overview of FOG as a potential raw material for bioenergy production. Thus, the necessity for FOG as a feedstock for bioenergy with a strong emphasis on the engineered challenges associated with the present scenario of various technologies are discussed for energy recovery from FOG wastes and biofuel production. The different routes of biofuels production (e.g., biohydrogen, biomethane and biodiesel) through new integrated routes for sustainable biofuel industry are also evaluated. Herein, this chapter provides a successive high throughput of the full conversion of FOG wastes into biofuel toward a *zero-waste* system.

10.2 FOG Wastes (Composition and Technical Challenges)

The issues of FOG discharge and eventual clogging range from residential properties to global sewage blockages (Abomohra et al. 2020). The significant repercussions include sewage network blockages, which result in sanitary sewage overflows (SSO), which further deteriorate the environment and hygiene. The major issues are becoming more visible these days, particularly in industrialized nations' crowded cities (50 kg per capita year) (Williams et al. 2012). As a result, FOG deposition is a preceding warning internationally and necessitating proper management. Although FOG deposition and blockage were decades old, the appropriate management was given no attention. Grease traps or grease interceptors are basic management strategies that use the idea of gravity separation by allowing suspended items to settle.

The chemical composition of FOG is crucial for understanding and elucidating the mechanism of its deposition, as well as possible use and management. Furthermore, weight, density, moisture content, yield strength and melting profile are some of the physical features of FOG that have been extensively studied in previous research (Gross et al. 2017). Aside from them, FOG seems colorless, odorless, tasteless and water-insoluble (Sincero and Sincero 2002). The overall mass of fatty acids has the greatest influence on the weight and density of FOG. It is mostly composed of free fatty acids, triacylglycerols, esters waxes, phospholipids, sterols and sterol esters (Husain et al. 2014). FFAs are particularly important due to their high concentration and chemical reactivity. Usually, FOG obtained from restaurants contains about 15% of FFAs, and it is entirely dependent on the source for FOG generation. Based on FFAs content, FOG can be classified into two groups, namely yellow grease (less than 15%) and brown grease (above15%) (Canakci 2007). However, the high water and free fatty acids (FFAs) contents result in saponification leading to lower conversion rate, with high amounts of FFAs in the residue. Figure 10.1 summarizes the challenges of FOG conversion to energy. Therefore, the present chapter suggests a new route for high bioenergy recovery from FOG through integration of transesterification and anaerobic co-digestion as discussed in Sect. 3.

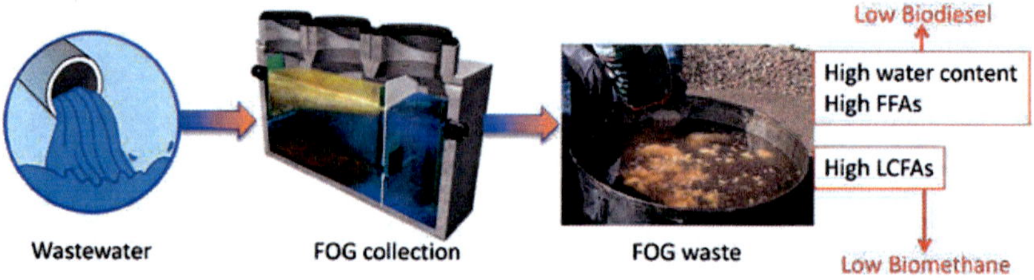

Fig. 10.1 Challenges associated to fats, oils and grease (FOG) for bioenergy (biomethane and biodiesel) production

10.3 Types of Pretreatments of FOG Wastes

The growing urbanization and development results in an obvious establishment that generates FOG; hence, optimal utilization is critical for effective management. To prevent sewage overflows, FOG is often segregated on-site or from wastewater streams. Food service companies install grease interceptors at the source to eliminate FOG as "trap grease."

Because of the high FFA and LCFAs levels, pretreatment is required to remove or convert the majority of the FFAs into TAGs as intermediates or directly to FAMEs. Several pretreatment strategies have been used to improve biofuel production from FOG. The ideal approach should be economically practical, with low energy needs, low chemical usage, low harmful by-product effluent and ease of application on a wide scale (Tabatabaei et al. 2019). Many pretreatment procedures were proposed to lower the high FFAs and LCFAs in FOG to acceptable levels, including acid esterification, steam stripping, biological pretreatment, glycerolysis and supercritical esterification of FOG. Figure 10.2 shows the different methods of pretreatment and processing of FOG wastes, which can be explained in detail below:

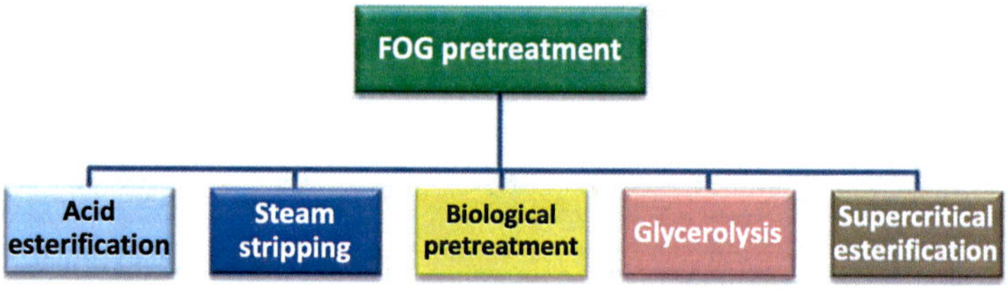

Fig. 10.2 Different methods used for the pretreatment of FOG wastes

10.3.1 Acid Esterification

The traditional technique of pretreatment of FFA-rich feedstocks is acid esterification, often known as "*alcoholysis.*" The interaction of methanol with FFAs in the presence of acid catalysts such as sulfuric acid, hydrochloric acid and phosphoric acid at 65 °C producing esters has been intensively researched for pretreatment of FOG. This investigation revealed that the best conditions for esterification pretreatment were 20:1 methanol-to-FFA ratio and 10% H_2SO_4 at reaction duration, temperature and mixing speed of 24 h, 50 °C and 200 rpm, respectively, resulting in a decrease of acid value to 1 mg KOH/g (Hasuntree et al. 2011). In addition, a complete economic feasibility analysis of acid esterification utilizing various catalysts and reaction conditions will aid in providing a fair appraisal of the best parameters.

10.3.2 Steam Stripping

As a physical refining procedure for FOG, steam stripping can be utilized. FFAs have higher volatility than triacylglycerols (TGs) in FOG. Usseglio et al. (2019) suggested utilizing a non-equilibrium model to analyze stripping processes using a structured packed column and two distinct types of vegetable oils (palm oil and soybean oil). The study demonstrated a final FFAs concentration as low as 0.005%, confirming its technical viability for FOG treatments. The fundamental benefit of steam stripping is that no soap is formed, the effluent water is cleaned, and the pollution by-products are limited. The primary disadvantage of steam stripping is its high energy consumption as well as its high operational and investment expenses (Anderson 2014).

10.3.3 Biological Pretreatments

Biopretreatments have generally been related to the employment of microorganisms (e.g., bacteria, fungi, microalgae, and, insect larvae) capable of using FFAs for biomass production or enzymatically converting FFAs to FAMEs. The benefits of employing microorganisms include low energy consumption, decreased chemical use and decreased pollutant by-products, as well as important biomass production that may be used as a biofuel feedstock in the future.

10.3.4 Glycerolysis

Glycerolysis or glycerin re-esterification is another pretreatment for FOG containing more than 10% FFAs that helps minimize the acidity of biodiesel while saving cost by eliminating the need for acid, methanol or vacuum stripping (Tu et al. 2017). This reaction may be carried out by adding glycerin to FOG at 250 °C to produce MAGs. Furthermore, DAGs and TAGs can be produced by additional interaction with glycerol. After 3 h of reaction at 200 °C, glycerolysis reduces the FFA level of soapstocks from 50 to 5%. Besides, it is effective in lowering FFA levels without the need for any catalyst (Felizardo et al. 2011). Glycerolysis pretreatment is a viable method for the pretreatment of brown grease (>15% FFAs). It demonstrated a high efficiency of 85% reduction in FFAs of brown grease from wastewater scum (70% FFAs) after 120 min at 200 °C.

In this context, Tu et al. (2017) verified that biodiesel synthesis through glycerolysis-treated FOG needed less energy input (0.251 MJ kg^{-1} biodiesel) than the traditional approach (0.534 MJ kg^{-1} biodiesel). Further reductions might improve the process economy in energy input or effective heat recycling. An integrated strategy that recycles heat from glycerolysis in the acid esterification distillation process for menthol recovery might enhance overall energy output while lowering costs.

10.3.5 Supercritical Esterification

Supercritical esterification is a catalytic process that employs alcohol (typically methanol) under supercritical pressure and temperature conditions. The increased solubility of methanol results in the catalyst acting in the place of a catalyst while performing the reaction because of its high solubility/dispersion; this causes the reaction of transition into a rate-limiting reaction instead of a solubility-limiting esterification reaction (Saka and Kusdiana 2001). Because FFAs and glycerides react readily with alcohol under supercritical conditions, there is no need to separate or pretreat these components. At the critical temperatures and pressures for methanol (450 °C and 200 bar, respectively), methanol transforms into a non-polar solvent capable of evenly dissolving an oil sample (Saka and Kusdiana 2001). However, various reaction conditions such as residence time, methanol:FFAs ratio and mixing speed influence the efficiency of FFAs conversion under supercritical conditions. As a result, producing biodiesel with a novel supercritical reagent will reduce total production costs because biodiesel and a valuable by-product can be produced in a single step. This innovative and cost-effective process will be able to generate new income for the biodiesel industry, resulting in a cost-effective biodiesel way.

10.4 Different Technologies of Bioenergy Production from FOG

10.4.1 *Biodiesel*

The fatty acid compositions of different types of FOG are compared with other biodiesel feedstocks. It is evident that the biodiesel significant fatty acids such as palmitic, stearic and oleic acids are comparatively high in FOG than on edible feedstocks except for oleic acid in rapeseed and corn. In a similar way, edible feedstock displayed higher proportions than FOG. Even within different types of FOG, such as yellow and brown greases, potential differences in linoleic acid proportions were observed. These potential differences in fatty acid composition obviously reflect the quality of biodiesel. The comparison of FAME yield from FOG to other feedstocks is shown in Fig. 10.3. The calculated average FAME yield from FOG computed yield revealed the lowest value among other feedstock. Canakci and Van Gerpen (1999) examined the link between FFA levels and transesterification efficiency during acid-catalyzed biodiesel production. Palmitic acid was added to soybean oil in proportions of 5% and 33% to create sample mixes. As the FFAs level climbed from 5 to 33%, the conversion rate of oil mixtures to biodiesel declined from 90.54 to 58.77%. However, optimizing the reaction conditions raises the output by up to 98%, which is equivalent to the yield achieved from other commonly used feedstocks such as soybean and rapeseed oil as shown in Fig. 10.3. Exploring novel eco-friendly technologies for

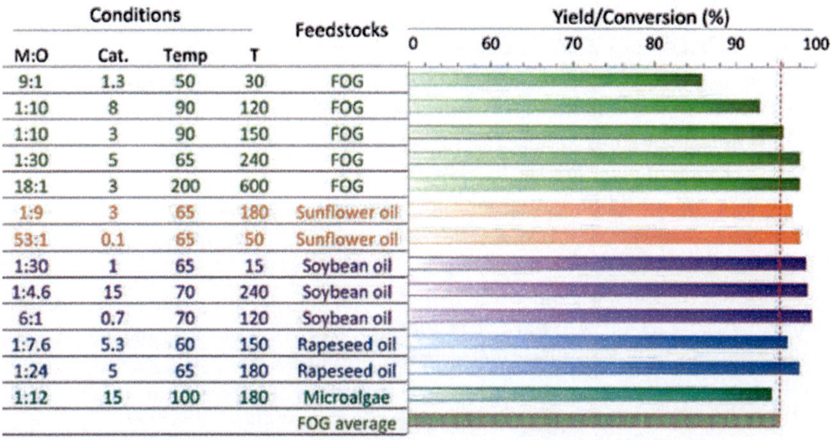

Fig. 10.3 FAME yield or conversion efficiency of FOG in comparison with other feedstocks and the corresponding reaction conditions of methanol:oil ratio (M:O), catalyst amount (Cat, wt%), temperature (Temp, °C) and reaction time (T, min). The graph was adopted from Ref. (Abomohra et al. 2020) with permission number 5158670835706

successful pretreatment and increasing biodiesel output is, thus, critical to commercializing FOG biodiesel production. Furthermore, constructing a cost-effective transesterification reactor suited for the conversion process is a critical parameter. Various reactor geometries and structures, such as spinning reactors, tubular/plug-flow reactors, microwave reactors, cavitation reactors and simultaneous reaction-separation reactors, have lately been considered by Tabatabaei et al. (2019).

10.4.2 Anaerobic Technologies

Anaerobic technologies are effective for not only producing and effluent quality but also producing energy in the form of hydrogen, methane and hythane (Arora et al. 2021; Mostafa et al. 2017). Jeganathan et al. (2006) attempted continuous treatment of wastewater industry rich lipids by an up-flow anaerobic sludge blanket reactor at hydraulic retention time of 5 days and 1.25 days. A packed bed reactor and up-flow anaerobic sludge blanket reactor connected in series were continuously operated at an HRT of 1.25 days. All treatment processes achieved FOG and COD removal efficiencies of 80%, and biomethane yield was varied from 0.32 to 0.37 $mLCH_4$/g CODt removed at an organic loading rate of 3 kg COD/m^3/day. However, the efficiency of reactors highly deteriorated at higher organic loading rates resulting in a severe active sludge flotation and the biomethane production was sharply dropped to 0.18–0.24 $mLCH_4$/gCODt removed. The accumulation of LCFAs occurred at a high loading rate which covered the surface of anaerobes and limited the substrate transfer onto the bacterial cells. Moreover, intentional excess sludge withdrawn from the anaerobic reactors is necessary to avoid and minimize the accumulation of LCFAs in the digesters (Elreedy et al. 2015; Farghaly et al. 2017). The anaerobic co-digestion process is an excellent option where a balanced diet is saved for anaerobes to accomplish their biological activities and various metabolites pathway (Grosser and Neczaj 2016). A continuous stirred tank reactor was employed for biomethanization of co-substrate of slaughter wastewater (SWW) industry and olive mill wastewater (OMW) (Fountoulakis et al. 2008). Partial inhibition occurred at high organic loading conditions yielding methane of 0.298 ± 0.081 $m^3CH_4/kgCOD_{initial}$ for SWW and 0.1707 ± 0.018 m^3 $CH_4/kgCOD_{initial}$ for the co-digestion of SWW with OMW. A substantial reduction in biomethane yield from 70.1 mL CH_4 at a loading rate of 1.5 gCOD-lipids/L/day to 27.1 mL CH_4 at 6.1 gCOD-lipids/L/day was taken place in UASB module treating lipid-rich diary wastewater (Kim and Shin 2010). Based on these results, it is preferred to apply a two-stage anaerobic digesters connected in series to avoid the failure of the process due to the accumulation of LCFAs. Two-stage hyper-thermophilic (70 ± 1 °C)/thermophilic (55 ± 1 °C) anaerobic co-digestion was utilized for anaerobic co-digestion of waste activated sludge and fat, oil and grease and improved the methane yield by 112.7% (Alqaralleh et al. 2016). The sludge residence is the key parameter for degradation of waste-rich lipids. The technologies enjoyed a long SRT is preferred for biomethanization of biowaste containing high lipids concentrations.

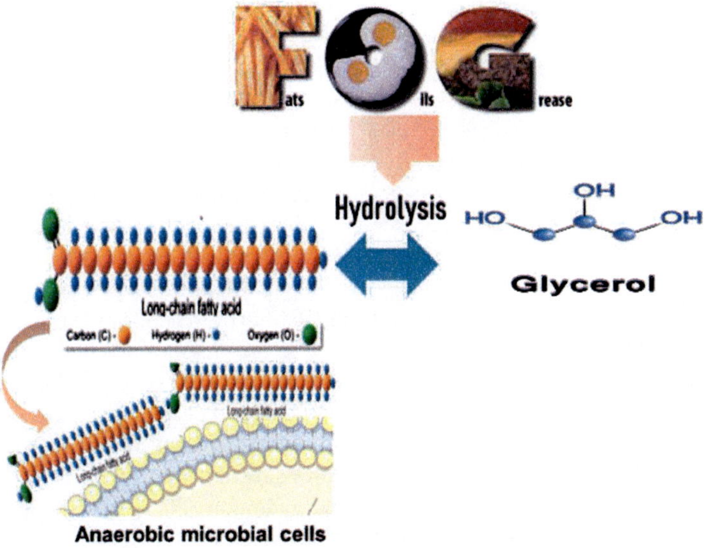

Fig. 10.4 Accumulation of LCFA on the surface of anaerobes

10.4.3 Barriers for Biomethane and Degradation of FOGs by Anaerobic Community

The anaerobic degradation of FOGs generates mainly LCFA, which accumulates on the surface of anaerobe cells preventing the substrate diffusion onto the active bacterial cells causes a failure of biomenthanation process as reported earlier by (Alves et al. 2009; Plante et al. 2019) Fig. 10.4. The aceticlastic methanogens are the main anaerobes negatively affected by the accumulation of the LCFA in the digesters leading to toxicity of the microbes (Ahmad et al. 2011; Alves et al. 2009). Hydrogenotrophic methanogens and syntrophic bacteria as well are inhibited in the presence of high levels of LCFA in the digesters, which should be avoided (Silvestre et al. 2014).

10.4.4 Mitigation of the Inhibition Effect of LCFA Accumulation During Biomethanation

LCFAs accumulation in the anaerobic digesters could be overcome by anaerobic co-digestion process (Hao et al. 2020; Silvestre et al. 2014). Stepwise increase in FOG loading provides a great chance for anaerobes for adaptation process, and subsequently, higher LCFA β-oxidation and methanogenic activities are taken place in the digesters. (Alves et al. 2001) found that LCFA biodegradation by anaerobes was increased with long-term exposure in an anaerobic fixed-bed bioreactor. Moreover, the toxicity of LCFA accumulation is eliminated at long sludge residence time (SRT) of microbes degrading lipids (Mahmoud et al. 2004; Tawfik and Elsamadony 2018). Long-term adaptation of anaerobes degrading fats and oils was identified as the main factor influencing the resilience of the toxicity of LCFA in a digester fed with skim milk and oleaterich wastewaters (Silva et al. 2014). The bacterial changes in LCFA degrading microbial community structure caused by the gradual increases in FOG loading rate were comprehensively reported by (Ziels et al. 2016) during anaerobic co-digestion of fats, oils and grease with excess sewage sludge. The addition of calcium chloride to the anaerobic digester fed with lipid-rich waste highly reduced the inhibitory effect of the accumulation of LCFAs in the reaction medium. However, the inhibition effect of LCFAs was quite high in the presence of calcium carbonate due to its insolubility (Hanaki et al. 1981). The calcium (0.1–1% w/v) addition during anaerobic digestion of FOGs at high loadings (2% v/v) of fat, oil and grease (FOG) increased the biomethane yield by sixfold due to the high growth of *Firmicutes* from the Clostridium, *Syntrophomonas* and Sedimentibacter genera. Furthermore, the genus Methanosaeta was increased with the addition of 0.5% calcium and subsequently increased the biomethane yield (Salama et al. 2020). However, biomethane production decreased with calcium (>0.5%), where anaerobic microflora growth was substantially reduced, and the calcium ions were precipitated on the surface of anaerobes, limiting the substrate diffusion. (Alqaralleh et al. 2019) found that microwave pretreatment of thickened waste-activated sludge and fat, oil and grease at 175 °C improved the methane yield by 137% higher than the control for samples due to an increase of the substrate solubility (Eq. 10.1).

$$\text{Solubilization } (\%) = 100 \times \frac{\text{SCOD}_{\text{final}} - \text{SCOD}_{\text{initial}}}{\text{SCOD}_{\text{initial}}} \tag{10.1}$$

where $\text{SCOD}_{\text{final}}$ is the SCOD of the wastewater after pretreatment and SCO $_{\text{initial}}$ is the SCOD of the wastewater before pretreatment.

Semi-continuous mesophilic reactors were operated for mono-digestion of sewage sludge and co-digestion sewage sludge with fat, oil and grease (FOG) over 180 days (Yang et al. 2016). The biomethane yield higher in the co-digestion process due to the release of extracellular polymeric substances (EPS) was increased by 40%. Thermophilic (55 °C) co-digestion of primary raw sludge and FOG in semi-continuous flow digesters exhibited a high biogas production rate of 17.4 ± 0.86 L/day and methane content 67.9 ± 1.46% at HRT of 24 days and OLR of 2.43 ± 0.15 g TVS/L/day (Kabouris et al. 2008; Li et al. 2013).

Hydrolysis of FOG is the first step in the anaerobic digestion process by acidogenic bacteria (extracellular lipase enzymes) resulted in glycerol and long-chain fatty acids (LCFAs), which are subsequently converted into short-chain volatile fatty acids (acetate, propionate, butyrate etc.) and H_2 by acetogenic anaerobes. Methanogenesis converts the acetate into methane as well as syntrophic bacteria produce methane from carbon dioxide and hydrogen (Fig. 10.5). Co-digestion of fats, oils and greases (FOG) with excess sludge from wastewater treatment plant substantially

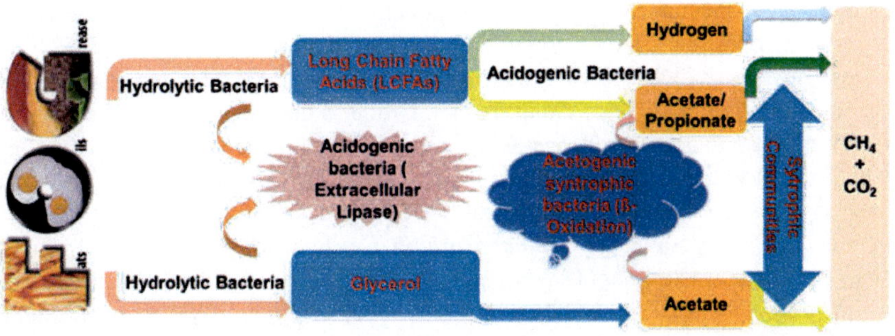

Fig. 10.5 Hydrolysis of FOG (glycerol and long-chain fatty acids (LCFAs)) under anaerobic digestion process

improved biomethane recovery due to minimization of the accumulation of LCFA in the digester (Ziels et al. 2016). Anaerobic co-digestion of the excess sludge and fat deposit (FD) of the sewer system at the substrate-to-inoculum (S/I) ratio of 0.25–1.2, producing a maximum methane production of 685.7 ± 24.1 mL/gVS$_{added}$, at a S/I ratio of 0.5, was four times higher than those obtained from solely anaerobic digestion of excess sludge (Hao et al. 2020). Moreover, the methane yield from anaerobic digestion of fate deposit (FD) was higher than those from FOG due to its physical and chemical properties where the saponification between LCFAs and calcium in the FDs was relatively high, facilitating the methanogenic process (He et al. 2013, 2017). The saponification between metals, i.e., calcium and lipids containing waste/wastewater improved anaerobic biodegradability and minimized the inhibition effect of LCFAs (Battimelli et al. 2010; Salama et al. 2020). A Gompertz model (Eq. (10.2) (Elsayed et al. 2019) was modified to calculate the methane production during the anaerobic digestion process,

$$B_t = B_0 \exp\left\{-\exp\left[\frac{R_m^e}{B_0}(\lambda - t) + 1\right]\right\},\quad t \geq 0 \tag{10.2}$$

where B_t is the specific methane yield at a time (mL/gVS$_{added}$); B_0 is the maximum methane potential (mL/gVS$_{added}$); t is the digestion time (day); R_m is the maximum daily methane production rate (mL/gVS$_{added}$·d); λ is the lag phase (day); e is 2.718.

The methane yield (MY) from anaerobic digestion of FOG was the highest (563 ml CH_4/g VS) as compared to harvested microalgae (140 ml CH_4/gVS) and excess sludge (299 ml CH_4/g VS) (Solé-Bundó et al. 2020). However, co-digestion of FOG with microalgae and sludge augmented the MY by a value of 42%. The MY in the anaerobic digester was increased 2.9 times with the addition of 48% volatile solids (VS) of FOG onto excess sludge from sewage (Kabouris et al. 2009). This was mainly due to the sufficient nutrients, buffering capacity and saving a biodegradable organic source for anaerobes. Moreover, the concentration of LCFAs is not much to accumulate on the bacterial cells due to the dilution of feedstock with co-substrates. (Grosser and Neczaj 2016) added a portion of FOG (10–18% VS) to the sewage sludge for enhancement of biomethane production in the semi-continuous anaerobic digester; (Silvestre et al. 2011) pointed out the gradual supplementation of FOG to excess sewage sludge (23% VS) successfully improving the MY and reducing the risk of inhibition of LCFAs. This was mainly due to the adapted anaerobes, particularly syntrophic acetogenic, acetoclastic methanogenic and ß-oxidation activities, which increase the methane production rate. Similarly, (Kurade et al. 2019) stated that co-digestion of excess sludge with FOG enhanced the MY due to the dominant acetoclastic pathway by *Methanosarcina*. Co-digestion of FOG (1–10%) with slaughterhouse wastewaters (SWW) enhanced biomethane generation up to a level of 562 and 777 ml CH_4/g CODs removed compared to sole SWW (Agabo-García et al. 2020). The addition of FOG improved the biodegradability of the feedstock up to 66–70% and enhanced the acetate utilizing methanogens.

10.4.5 Microbial Activity Responsible for Biomethanization of FOG

Bacterial degradation of FOG is the most important step for the conversion of such substrate onto methane. The hydrolytic anaerobes initially convert the fats and oils into glycerol and LCFA (Deaver et al. 2020). Syntrophic partnership of proton-reducing acetogenic anaerobes (aceticlastic and hydrogen otrophicmethanogenic archaea) is responsible for biomethane production, which utilizes the β-oxidation pathway to transform LCFA into acetate and/or formate and biogenic hydrogen (Ali et al. 2017; Siddique et al. 2011). β-oxidize LCFA bacterial species syntrophically mainly belong to two families of *Syntrophomonadaceae* and *Syntrophaceae* (Ziels et al. 2016). The microbial conversion of LCFA into biomethane is the rate-limiting step due to slowing of lipids biodegradation in anaerobic biodigesters (Deaver et al. 2020). The anaerobic co-digestion of FOG with a co-substrate would mitigate the inhibition effect of LCFA accumulation and enhance the biomethanization process due to the existence of various anaerobes anaerobic functional communities (Cirne et al. 2007). However, the anaerobes are mainly affected by increasing the FOG loading rate, where some anaerobes could be adapted for and others would be inhibited. (Ziels et al. 2016) found that co-digestion of FOG with excess sludge from wastewater treatment plant increased the 16S rRNA gene concentration of the syntrophic ß-oxidizing genus *Syntrophomonas*by 12% of the bacteria community. Moreover, *Methanosaeta* and *Methanospirillum* were the most dominant methanogenesis-producing biomethane from the co-digestion process and represented 80% of the archaea community. However, the loading rate of FOG was the most parameter that affects the structure of microbial functioning activities in the digesters. A higher methane yield was recorded by the anaerobic co-digestion of fats withdrawn from sewers with excess sludge where higher levels of the microbial community of *Geobacter* were detected, suggesting a key role for the direct interspecies electron transfer between *Methanosaeta* and *Geobacter* (Hao et al.

2020). Increasing the FOG addition from 1 to 10% to the slaughterhouse wastewater industry during the co-digestion process resulted in a reduction in *Eubacteria species* and enhanced the a*cetate-utilizing methanogens* (Agabo-García et al. 2020). Both bacterial and archaeal consortia underwent an apparent succession with a gradual FOG addition where acetoclastic genera, methanosaeta Methanosaeta and N09 were dominant, resulting in a high methane yield (Yang et al. 2016). Salama et al. (2019) reviewed the obstacles of anaerobic co-digestion of FOG and confirmed that inhibition of anaerobes by long-chain fatty acids (LCFAs, C14-C24) produced by FOG hydrolysis is a major operational challenge in anaerobic systems, particularly when the digester has a high lipid content. Palatsi et al. (2009) reported that accumulation of LCFAs reduces cell permeability, changes cellular morphology and affects mass transport in the best scenarios where low concentrations of LCFAs can be degraded into short-chain fatty acids within 15 days (Angelidaki and Ahring 1995), which increases the hydraulic retention time of the process. As a result, researching solutions to improve anaerobic digester performance stability at high FOG loading rates is a critical idea for increased biomethane output. Furthermore, as compared to biodiesel synthesis, biogas generation was identified as a less efficient technique of using the energy content of the FOG (Tu 2015).

Park and Li (2012) achieved biomethane productivity of 0.54 L CH_4 g VS/day from algal biomass residue co-digested with FOG waste. In addition, it was observed that co-digestion of *Chlorella* sp. with waste sludge increases the biogas yields by 73–79% and improves the dewaterability of digestate effluents (Wang et al. 2013). Solé-Bundó et al., (2020) examined the effect of thermal pretreatment of microalgae on co-digestion with primary sludge or FOG. The biomethane yield was increased by 25% and 42% by adding FOG in a ratio of 10 and 20% VS, respectively. Therefore, pretreatment of microalgae and co-digestion with other wastes is a promising approach to improve biomethane production from algae (Abomohra and Almutairi 2020).

10.5 Dual-Fuel Integrated Approach

10.5.1 Biohydrogen and Bioethanol

Discharge of wastewater-rich FOG into water streams causes severe environmental problems and clogging the sewer network systems (Ali et al. 2017; Long et al. 2012). Nevertheless, wastewater containing FOG represents a sustainable and attractive low-cost substrate source for biogenic hydrogen and bioethanol production via the fermentation process. Biohydrogen and bioethanol production from wastewater-rich FOG using up-flow anaerobic staged reactor was investigated at different hydraulic retention times and organic loading rates (Ali et al. 2017). Biohydrogen and bioethanol production was highly increased from 44.6 to 80.3 L/day and from 107.6 to 195.2 mg/L at reducing the HRT from 22 to 12 h. Likely, bioH_2 was reduced from 120 to 26.2 L/day at decreasing the organic loading rate from 45 to 12.7 gCOD/L/day. The oleate, palmitate and stearate were significantly augmented from 67 to 1187 mg/L, from 100 to 1249 mg/L and from 122 to 1213 mg/L at increasing the concentration of the influent lipid from 2 to 10 g/L, respectively. The authors detected two *Clostridium* sp. (EBR-02E0045 and EBR-02E-0046) in the sludge, indicating the successful process for treating wastewater-rich FOG with two biofuels production simultaneously at an organic loading rate of 45 g COD/L/day and hydraulic retention time of 12 h.

10.5.2 Sequential Biodiesel and Biomethane

The integrated strategy of sequential biodiesel and biomethane may be employed by producing biodiesel from the lipid component of organic waste and then using the non-lipid material for AD biomethane synthesis. Thus, integrated conversion of FOG into biodiesel should not replace AD, but rather complement it by offering a quicker and much greater energy output via the "FOG biorefinery" concept. According to recent research, Kobayashi et al. (2014) examined the "*dual-fuel production*" from restaurant GTW by transesterification of the FOG layer and anaerobic co-digestion of the dewatered food waste layer. This study showed that the total energy produced from 1 L of GTW in a dual-fuel process (13.4 MJ L^{-1} GTW) was 39.6% higher than that produced from co-digestion only. In addition, (Tu and Mcdonnell 2016) reported that AD of solids from GTW reduced both energy consumption and GHG emissions. This appears to be the most successful strategy for GTW utilization, due to the lowest layer is separated and produces biogas, and the higher-grade FOG layer is pretreated and utilized for biodiesel. Overall, sequential bioenergy production reduces the physical and carbon impact of the feedstock while increasing energy yield compared to individual conversion into biodiesel or biomethane.

10.6 Conclusions and Future Perspectives

The major emphasis of field research should be biofuel feedstocks that do not compete with food or arable soil and which deliver considerable GHG reductions. This chapter provides an overview of the conversion from FOG waste to biodiesel, as a biowaste with various environmental and economic concerns. Although high progress was carried out in the FOG handling and management, the fact that FOG-related problems continue to plague cities is a proof that the current processes are inadequate and need further support from individuals, industry and governments. In order to decrease FOG from dumping the sewage system and to provide the ideal techniques for effective collection and treatment, social media efforts, public education campaigns and boosting awareness at the school level are very necessary. Although much progress has been made in FOG management, it remains evident that the present systems are unsatisfactory and that greater assistance from individuals, businesses and governments is required for problems linked with FOG to continue plaguing communities. Social media activities, public education camps and school awareness promotion play a key part in limiting the number of FOG from being introduced into the sewage system and/or in providing optimal collection and treatment procedures. In addition, sequential conversion of FOG improved the gross bioenergy output, and studying the environmental impact and life cycle assessment of the different suggested conversion routes is of great importance for the future application. In addition, the engineered challenges compilations related to sequential biofuel production from FOG need to extensively studied in order to achieve a "zero-waste" system.

Acknowledgements The second author is very grateful to the Science, Technology & Innovation Funding Authority (STIFA) for funding this research under the grant of basic science (ID: 26271) and pollution control (ID: 41591), Academy of Scientific Research and Technology (ASRT) under the call no. 2/2019/ASRT-Nexus), Imhotep project for partially financially support the research and the National Research Centre for partially supporting the research (Project ID: 12030202).

References

Abomohra A, Eladel H, El-Esawi M, Wang S, Wang Q, He Z, Feng Y, Shang H, Hanelt D (2018) Effect of lipid-free microalgal biomass and waste glycerol on growth and lipid production of Scenedesmus obliquus: innovative waste recycling for extraordinary lipid production. Bioresour Technol 249:992–999

Abomohra AE-F, Almutairi AW (2020) A close-loop integrated approach for microalgae cultivation and efficient utilization of agar-free seaweed residues for enhanced biofuel recovery. Bioresour Technol, 124027

Abomohra AE-F, Elsayed M, Esakkimuthu S, El-Sheekh M, Hanelt D (2020) Potential of fat, oil and grease (FOG) for biodiesel production: a critical review on the recent progress and future perspectives. Prog. Energy Combust. Sci. 81:100868

Agabo-García C, Solera R, Pérez M (2020) First approaches to valorizate fat, oil and grease (FOG) as anaerobic co-substrate with slaughterhouse wastewater: biomethane potential, settling capacity and microbial dynamics. Chemosphere 259:127474. https://doi.org/10.1016/j.chemosphere.2020.127474

Ahmad A, Ghufran R, Wahid ZA (2011) Bioenergy from anaerobic degradation of lipids in palm oil mill effluent. Rev Environ Sci Biotechnol 10:353–376. https://doi.org/10.1007/s11157-011-9253-8

Ali M, Danial A, Tawfik A (2017) Self-dark fermentation of lipids rich wastewater for 2-biofuels (H_2 and Et-OH) production. Process Saf Environ Prot 109:257–267. https://doi.org/10.1016/j.psep.2017.04.007

Alqaralleh RM, Kennedy K, Delatolla R (2019) Microwave vs. alkaline-microwave pretreatment for enhancing thickened waste activated sludge and fat, oil, and grease solubilization, degradation and biogas production. J Environ Manage 233:378–392. https://doi.org/10.1016/j.jenvman.2018.12.046

Alqaralleh RM, Kennedy K, Delatolla R (2018) Improving biogas production from anaerobic co-digestion of Thickened Waste Activated Sludge (TWAS) and fat, oil and grease (FOG) using a dual-stage hyper-thermophilic/thermophilic semi-continuous reactor. J Environ Manage 217:416–428

Alqaralleh RM, Kennedy K, Delatolla R, Sartaj M (2016) Thermophilic and hyper-thermophilic co-digestion of waste activated sludge and fat, oil and grease: evaluating and modeling methane production. J Environ Manage 183:551–561. https://doi.org/10.1016/j.jenvman.2016.09.003

Alves MM, Mota Vieira JA, Alvares Pereira RM, Pereira MA, Mota M (2001) Effect of lipids and oleic acid on biomass development in anaerobic fixed-bed reactors. Part I: Biofilm growth and activity. Water Res 35:255–263. https://doi.org/10.1016/S0043-1354(00)00241-4

Alves MM, Pereira MA, Sousa DZ, Cavaleiro AJ, Picavet M, Smidt H, Stams AJM (2009) Waste lipids to energy: how to optimize methane production from long-chain fatty acids (LCFA). Microb Biotechnol 2:538–550. https://doi.org/10.1111/j.1751-7915.2009.00100.x

Anderson BE (2014) Tech topics glycerolysis for lowering free fatty acid levels. Superior Process Technol: Render Mag, 34–35

Angelidaki I, Ahring BK (1995) Establishment and characterization of an anaerobic thermophilic (55 °C) enrichment culture degrading long-chain fatty acids. Appl Environ Microbiol 61:2442–2445

Arora AS, Nawaz A, Qyyum MA, Ismail S, Aslam M, Tawfik A, Yun CM, Lee M (2021) Energy saving anammox technology-based nitrogen removal and bioenergy recovery from wastewater: inhibition mechanisms, state-of-the-art control strategies, and prospects. Renew Sustain Energy Rev 135.https://doi.org/10.1016/j.rser.2020.110126

Badoei-dalfard A, Malekabadi S, Karami Z, Sargazi G (2019) Magnetic cross-linked enzyme aggregates of Km12 lipase: a stable nanobiocatalyst for biodiesel synthesis from waste cooking oil. Renew Energy 141:874–882

Battimelli A, Torrijos M, Moletta R, Delgenès JP (2010) Slaughterhouse fatty waste saponification to increase biogas yield. Bioresour Technol 101:3388–3393. https://doi.org/10.1016/j.biortech.2009.12.043

Canakci M (2007) The potential of restaurant waste lipids as biodiesel feedstocks. Bioresour Technol 98:183–190. https://doi.org/10.1016/j.biortech.2005.11.022

Canakci M, Van Gerpen J (1999) Biodiesel production viaacid catalysis. Trans ASAE 42:1203

Charuwat P, Boardman G, Bott C, Novak JT (2018) Thermal degradation of long chain fatty acids. Water Environ Res 90:278–287

Cirne DG, Paloumet X, Björnsson L, Alves MM, Mattiasson B (2007) Anaerobic digestion of lipid-rich waste-Effects of lipid concentration. Renew Energy 32:965–975. https://doi.org/10.1016/j.renene.2006.04.003

Costa ET, Almeida MF, Dias JM, Matos A (2015) Glycerolysis of two high free fatty acid waste materials for biodiesel production. In: WASTES 2015—Solutions, treatments and opportunities: selected papers from the 3rd edition of the international conference on wastes: solutions, treatments and opportunities, Viana Do Castelo, Portugal, 14–16 Sept 2015. CRC Press, p 55

Deaver JA, Diviesti KI, Soni MN, Campbell BJ, Finneran KT, Popat SC (2020) Palmitic acid accumulation limits methane production in anaerobic co-digestion of fats, oils and grease with municipal wastewater sludge. Chem Eng J 396:125235. https://doi.org/10.1016/j.cej.2020.125235

Del Mundo DMN, Sutheerawattananonda M (2017) Influence of fat and oil type on the yield, physico-chemical properties, and microstructure of fat, oil, and grease (FOG) deposits. Water Res 124:308–319

Eladel H, Abomohra AE-F, Battah M, Mohmmed S, Radwan A, Abdelrahim H (2019) Evaluation of *Chlorella sorokiniana* isolated from local municipal wastewater for dual application in nutrient removal and biodiesel production. Bioprocess Biosyst Eng 42:425–433

Elreedy A, Tawfik A, Kubota K, Shimada Y, Harada H (2015) Hythane ($H_2 + CH_4$) production from petrochemical wastewater containing mono-ethylene glycol via stepped anaerobic baffled reactor. Int Biodeterior Biodegradation 105:252–261. https://doi.org/10.1016/j.ibiod.2015.09.015

Elsayed M, Abomohra AEF, Ai P, Jin K, Fan Q, Zhang Y (2019) Acetogenesis and methanogenesis liquid digestates for pretreatment of rice straw: a holistic approach for efficient biomethane production and nutrient recycling. Energy Convers Manag 195:447–456. https://doi.org/10.1016/j.enconman.2019.05.011

Elsayed M, Ran Y, Ai P, Azab M, Mansour A, Jin K, Zhang Y, Abomohra AE-F (2020) Innovative integrated approach of biofuel production from agricultural wastes by anaerobic digestion and black soldier fly larvae. J Clean Prod 263:121495. https://doi.org/10.1016/j.jclepro.2020.121495

Elsheikh MA, Saleh HI, Rashwan IM, El-samadoni MM (2013) Hydraulic modelling of water supply distribution for improving its quantity and quality. Sustain Environ Resour 23:403–411

Farghaly A, Elsamadony M, Ookawara S, Tawfik A (2017) Bioethanol production from paperboard mill sludge using acid-catalyzed bio-derived choline acetate ionic liquid pretreatment followed by fermentation process. Energy Convers Manag 145.https://doi.org/10.1016/j.enconman.2017.05.004

Felizardo P, Machado J, Vergueiro D, Correia MJN, Pereira J (2011) Study on the glycerolysis reaction of high free fatty acid oils for use as biodiesel feedstock Fatty acid. Fuel Process Technol 92:1225–1229. https://doi.org/10.1016/j.fuproc.2011.01.020

Fountoulakis MS, Drakopoulou S, Terzakis S, Georgaki E, Manios T (2008) Potential for methane production from typical Mediterranean agro-industrial by-products. Biomass Bioenerg 32:155–161. https://doi.org/10.1016/j.biombioe.2007.09.002

Gardy J, Hassanpour A, Lai X, Ahmed MH (2016) Synthesis of $Ti(SO_4)O$ solid acid nano-catalyst and its application for biodiesel production from used cooking oil. Appl Catal A Gen 527:81–95. https://doi.org/10.1016/j.apcata.2016.08.031

Ghoreishi SM, Moein P (2013) Biodiesel synthesis from waste vegetable oil via transesterification reaction in supercritical methanol. J Supercrit Fluids 76:24–31. https://doi.org/10.1016/j.supflu.2013.01.011

Gross MA, Jensen JL, Gracz HS, Dancer J, Keener KM (2017) Evaluation of physical and chemical properties and their interactions in fat, oil, and grease (FOG) deposits. Water Res 123:173–182. https://doi.org/10.1016/j.watres.2017.06.072

Grosser A, Neczaj E (2016) Enhancement of biogas production from sewage sludge by addition of grease trap sludge. Energy Convers Manag 125:301–308. https://doi.org/10.1016/j.enconman.2016.05.089

Hagos K, Zong J, Li D, Liu C, Lu X (2017) Anaerobic co-digestion process for biogas production: progress, challenges and perspectives. Renew Sustain Energy Rev 76:1485–1496

Hanaki K, Matsuo T, Nagase M (1981) Mechanism of inhibition caused by long-chain fatty acids in anaerobic digestion process. Biotechnol Bioeng 23:1591–1610. https://doi.org/10.1002/bit.260230717

Hao J, de los Reyes FL, He X (2020) Fat, oil, and grease (FOG) deposits yield higher methane than FOG in anaerobic co-digestion with waste activated sludge. J Environ Manage 268:110708. https://doi.org/10.1016/j.jenvman.2020.110708

Hasuntree P, Toomthong V, Yoschoch S, Thawornchaisit U (2011) The potential of restaurant trap grease as biodiesel feedstock. Songklanakarin J Sci Technol 33:525–530

He X, de los Reyes FL, Ducoste JJ (2017) A critical review of fat, oil, and grease (FOG) in sewer collection systems: challenges and control. Crit Rev Environ Sci Technol 47:1191–1217.https://doi.org/10.1080/10643389.2017.1382282

He X, de los Reyes FL, Leming ML, Dean LO, Lappi SE, Ducoste JJ (2013) Mechanisms of fat, oil and grease (FOG) deposit formation in sewer lines. Water Res 47:4451–4459.https://doi.org/10.1016/j.watres.2013.05.002

He X, Iasmin M, Dean LO, Lappi SE, Ducoste JJ, de los Reyes FL (2011) Evidence for Fat, oil, and grease (FOG) deposit formation mechanisms in sewer lines. Environ Sci Technol 45:4385–4391.https://doi.org/10.1021/es2001997

Husain IAF, Alkhatib F, Jammi MS (2014) Problems, control, and treatment of fat, oil, and grease (FOG): a review. J Oleo Sci. https://doi.org/10.5650/jos.ess13182

Jeganathan J, Nakhla G, Bassi A (2006) Long-term performance of high-rate anaerobic reactors for the treatment of oily wastewater. Environ Sci Technol 40:6466–6472

Kabouris JC, Tezel U, Pavlostathis SG, Engelmann M, Dulaney J, Gillette RA, Todd AC (2009) Methane recovery from the anaerobic codigestion of municipal sludge and FOG. Bioresour Technol 100:3701–3705. https://doi.org/10.1016/j.biortech.2009.02.024

Kabouris JC, Tezel U, Pavlostathis SG, Engelmann M, Todd AC, Gillette RA (2008) The anaerobic biodegradability of municipal sludge and fat, oil, and grease at mesophilic conditions. Water Environ Res 80:212–221. https://doi.org/10.2175/106143007x220699

Kim S-H, Shin H-S (2010) Enhanced lipid degradation in an upflow anaerobic sludge blanket reactor by integration with an acidogenic reactor. Water Environ Res 82:267–272. https://doi.org/10.2175/106143009x442899

Kobayashi T, Kuramochi H, Maeda K, Tsuji T, Xu K (2014) Dual-fuel production from restaurant grease trap waste: bio-fuel oil extraction and anaerobic methane production from the post-extracted residue. Bioresour Technol 169:134–142

Kurade MB, Saha S, Salama ES, Patil SM, Govindwar SP, Jeon BH (2019) Acetoclastic methanogenesis led by Methanosarcina in anaerobic co-digestion of fats, oil and grease for enhanced production of methane. Bioresour Technol 272:351–359. https://doi.org/10.1016/j.biortech.2018.10.047

Li C, Champagne P, Anderson BC (2013) Biogas production performance of mesophilic and thermophilic anaerobic co-digestion with fat, oil, and grease in semi-continuous flow digesters: effects of temperature, hydraulic retention time, and organic loading rate. Environ Technol (United Kingdom) 34:2125–2133. https://doi.org/10.1080/09593330.2013.824010

Long JH, Aziz TN, Francis L III, Ducoste JJ (2012) Anaerobic co-digestion of fat, oil, and grease (FOG): a review of gas production and process limitations. Process Saf Environ Prot 90:231–245

Mostafa A, Elsamadony M, El-Dissouky A, Elhusseiny A, Tawfik A (2017) Biological H_2 potential harvested from complex gelatinaceous wastewater via attached versus suspended growth culture anaerobes. Bioresour Technol 231.https://doi.org/10.1016/j.biortech.2017.01.062

Mahmoud N, Zeeman G, Gijzen H, Lettinga G (2004) Anaerobic stabilization and conversion of biopolymers in primary sludge. Effect of temperature and sludge residence time. Water Res 38:983–991

Palatsi J, Laureni M, Andrés MV, Flotats X, Nielsen HB, Angelidaki I (2009) Strategies for recovering inhibition caused by long chain fatty acids on anaerobic thermophilic biogas reactors. Bioresour Technol 100:4588–4596

Park S, Li Y (2012) Evaluation of methane production and macronutrient degradation in the anaerobic co-digestion of algae biomass residue and lipid waste. Bioresour Technol 111:42–48

Plante L, Sheehan NP, Bier P, Murray K, Quell K, Ouellette C, Martinez E (2019) Bioenergy from biofuel residues and waste. Water Environ Res 91:1199–1204. https://doi.org/10.1002/wer.1214

Saka S, Kusdiana D (2001) Biodiesel fuel from rapeseed oil as prepared in supercritical methanol. Fuel 80:225–231

Salama E-S, Saha S, Kurade MB, Dev S, Chang SW, Jeon B-H (2019) Recent trends in anaerobic co-digestion: fat, oil, and grease (FOG) for enhanced biomethanation. Prog Energy Combust Sci 70:22–42

Salama ES, Jeon BH, Kurade MB, Patil SM, Usman M, Li X, Lim H (2020) Enhanced anaerobic co-digestion of fat, oil, and grease by calcium addition: boost of biomethane production and microbial community shift. Bioresour Technol 296:122353. https://doi.org/10.1016/j.biortech.2019.122353

Siddique T, Penner T, Semple K, Foght JM (2011) Anaerobic biodegradation of longer-chain n-alkanes coupled to methane production in oil sands tailings. Environ Sci Technol 45:5892–5899. https://doi.org/10.1021/es200649t

Silva SA, Cavaleiro AJ, Pereira MA, Stams AJM, Alves MM, Sousa DZ (2014) Long-term acclimation of anaerobic sludges for high-rate methanogenesis from LCFA. Biomass Bioenerg 67:297–303. https://doi.org/10.1016/j.biombioe.2014.05.012

Silvestre G, Illa J, Fernández B, Bonmatí A (2014) Thermophilic anaerobic co-digestion of sewage sludge with grease waste: effect of long chain fatty acids in the methane yield and its dewatering properties. Appl Energy 117:87–94. https://doi.org/10.1016/j.apenergy.2013.11.075

Silvestre G, Rodríguez-Abalde A, Fernández B, Flotats X, Bonmatí A (2011) Biomass adaptation over anaerobic co-digestion of sewage sludge and trapped grease waste. Bioresour Technol 102:6830–6836. https://doi.org/10.1016/j.biortech.2011.04.019

Sincero AP, Sincero GA (2002) Physical-chemical treatment of water and wastewater. CRC Press

Solé-Bundó M, Garfí M, Ferrer I (2020) Pretreatment and co-digestion of microalgae, sludge and fat oil and grease (FOG) from microalgae-based wastewater treatment plants. Bioresour Technol 298:122563

Suwannakarn K, Lotero E, Ngaosuwan K, Goodwin JG (2009) Simultaneous free fatty acid esterification and triglyceride transesterification using a solid acid catalyst with in situ removal of water and unreacted methanol. Ind Eng Chem Res, 2810–2818

Tabatabaei M, Aghbashlo M, Dehhaghi M, Panahi HKS, Mollahosseini A, Hosseini M, Soufiyan MM (2019) Reactor technologies for biodiesel production and processing: a review. Prog Energy Combust Sci 74:239–303

Tawfik A, Elsamadony M (2018) Development of dry anaerobic technologies of bio-waste and unlock the barriers for valorization. Optim Appl Bioprocesses. https://doi.org/10.1007/978-981-10-6863-8_13

Tu Q (2015) Fats, oils and greases to biodiesel: technology development and sustainability assessment. University of Cincinnati. http://rave.ohiolink.edu/etdc/view?acc_num=ucin1448037796

Tu Q, Lu M, Knothe G (2017) Glycerolysis with crude glycerin as an alternative pretreatment for biodiesel production from grease trap waste: parametric study and energy analysis. J Clean Prod 162:504–511

Tu Q, Mcdonnell BE (2016) Monte Carlo analysis of life cycle energy consumption and greenhouse gas (GHG) emission for biodiesel production from trap grease. J Clean Prod 112:2674–2683. https://doi.org/10.1016/j.jclepro.2015.10.028

Usseglio M, Salvadori V, Siri G (2019) Advanced modeling of vegetable oils steam stripping with structured packing columns. Comput Chem Eng 121:654–669. https://doi.org/10.1016/j.compchemeng.2018.12.007

Venturin B, Frumi Camargo A, Scapini T, Mulinari J, Bonatto C, Bazoti S, Pereira Siqueira D, Maria Colla L, Alves SL, Paulo Bender J, Luís Radis Steinmetz R, Kunz A, Fongaro G, Treichel H (2018) Effect of pretreatments on corn stalk chemical properties for biogas production purposes. Bioresour Technol 266:116–124. https://doi.org/10.1016/j.biortech.2018.06.069

Wang M, Sahu AK, Rusten B, Park C (2013) Anaerobic co-digestion of microalgae *Chlorella* sp. and waste activated sludge. Bioresour Technol 142:585–590

Williams JB, Clarkson C, Mant C, Drinkwater A, May E (2012) Fat, oil and grease deposits in sewers: characterisation of deposits and formation mechanisms. Water Res 46:6319–6328

Xu S, Elsayed M, Ismail GA, Li C, Wang S, Abomohra AE-F (2019) Evaluation of bioethanol and biodiesel production from *Scenedesmus obliquus* grown in biodiesel waste glycerol: a sequential integrated route for enhanced energy recovery. Energy Convers Manag 197:111907. https://doi.org/10.1016/j.enconman.2019.111907

Yang ZH, Xu R, Zheng Y, Chen T, Zhao LJ, Li M (2016) Characterization of extracellular polymeric substances and microbial diversity in anaerobic co-digestion reactor treated sewage sludge with fat, oil, grease. Bioresour Technol 212:164–173. https://doi.org/10.1016/j.biortech.2016.04.046

Ziels RM, Karlsson A, Beck DAC, Ejlertsson J, Yekta SS, Bjorn A, Stensel HD, Svensson BH (2016) Microbial community adaptation influences long-chain fatty acid conversion during anaerobic codigestion of fats, oils, and grease with municipal sludge. Water Res 103:372–382. https://doi.org/10.1016/j.watres.2016.07.043

Chapter 11
Energy Recovery from Nuisance Algae Blooms and Residues

Gilvana Scoculi de Lira, José V. C. Vargas, André B. Mariano, and Ihana A. Severo

Abstract The nuisance of algae blooms can cause serious ecological and socio-economic damage and can incur a huge cost in their proper disposal. Therefore, it is imperative to research technological solutions that can value algal biomass and minimize its harmful effects on the environment. In this sense, the purpose of this chapter is to compile information about energy recovery from nuisance algal blooms and residues. Here, general issues about algae biology are described. The harmful effect of excessive algal proliferation and the prevention, control, and mitigation strategies are outlined. This chapter also presents an overview of environmental issues, including aspects of eutrophication and life cycle assessment. Finally, the potential applications of biomass as biofuels, bioenergy, and other uses are presented and discussed.

Keywords Algae biomass · Nutrient recovery · Eutrophication · Biofuels · Life cycle assessment · Circular bioeconomy

11.1 Introduction

The earth is covered by large extensions of water, whether freshwater or marine. However, scientific and political discussions regarding anthropic impacts on global water resources have been strengthened, especially concerning their quality and availability. In this wide aquatic environment, problems related to eutrophication are increasingly frequent around the world. This process consists of the increase of effluents with nutrient-enriched organic matter in a water body, which is inducing

G. S. de Lira · J. V. C. Vargas · A. B. Mariano · I. A. Severo (✉)
Technology Sector, Postgraduate Program in Materials Science and Engineering (PIPE), Sustainable Energy Research & Development Center (NPDEAS), Federal University of Paraná (UFPR), P.O. Box 19011, Curitiba, PR 81531-980, Brazil

J. V. C. Vargas
e-mail: viriato@ufpr.br

A. B. Mariano
e-mail: andrebmariano@ufpr.br

A. E.-F. Abomohra et al. (eds.), *Waste-to-Energy*,
https://doi.org/10.1007/978-3-030-91570-4_11

changes in this entire ecosystem. As a consequence, aquatic environments, mainly coastal and lake, remain contaminated, which triggers the excessive multiplication of algae. Eutrophication has been a longstanding problem, and to date, practically no solution has been found. The absence of strict regulations that penalize polluting sites is why toxic compounds proceed to enter watercourses and feed these devastating blooms (Bernauer and Böhmelt 2020; Du Preez and Van Huyssteen 2020).

The factors that favor this phenomenon, in addition to the excess of nutrients, such as phosphorus (P) and nitrogen (N), include water temperature, poor circulation of the medium (water stagnation), and physical dimensions. Moreover, the eutrophication process may cause many negative impacts to the entire ecosystem, such as modification of biogeochemical cycling, full oxygen deprivation (anoxia), aquatic flora and fauna disintegration, nuisance algal blooms, biodiversity decrease, and ecosystem damage (Pandhal et al. 2018; Ghosh and Bakshi 2019; Preisner et al. 2021; Paredes et al. 2021).

Given this scenario, algae, as heterogeneous organisms, are capable to thrive and grow in extreme environmental conditions, as reported above. Although they are known as bioremediation agents for polluting compounds' removal (Laurens 2021; Singh et al. 2021), excessive algal blooms can be extremely harmful to aquatic environments. Blooms and their toxins are also a serious threat to human health, animals, property values, recreational activities, tourism, commercial fishing, drinking water supply, and the local economy, which may incur additional costs related to cleaning, monitoring, and management of the generated biomass (Pan et al. 2011). In Southern Florida (USA), algal blooms spread annually throughout the coastal region. According to the Environmental Protection Agency, toxicity levels have been 100 times higher than the standard considered safe. Estimated economic losses are in the range of USD 100 million and USD 2.2 billion per year for algal blooms in coastal waters and freshwater, respectively (EPA 2021).

The undesirable bloom includes different species of algae, which can be harmful to health and the environment. For example, *Microcystis aeruginosa*, a cyanobacterium that produces harmful cyanotoxins to the human body (called microcystins), either by direct contact, ingestion, or inhalation of contaminated water droplets, causes diseases in the nervous system, intestines, and liver. Increasing populations of this species are a response to nutrient overload, causing environmental problems due to its high capacity to form large blooms. In Brazil, many coastal areas support large populations of *M. aeruginosa* due to the highly favorable environment (Sellner et al. 2003; Sant'Anna et al. 2011; Ali et al. 2020). Therefore, it is crucial to find solutions to minimize nuisance algae, since their proliferation has become a major problem and is increasingly out of control. Despite that, algae blooms, which are usually disposed of as waste in landfills for incineration, can be used as value-added raw materials (Renita and Kumar 2020).

Algae generate biomass containing considerable proteins, carbohydrates, lipids, and other biomolecules of commercial interest. In general, biomass is produced in open or closed cultivation systems, which allow control of operational variables. Under these conditions, it is possible to obtain products with applications in the food, feed, cosmetics, and pharmaceutical industries (Severo et al. 2019). However, as previously reported, biomass from harmful algae blooms is generated in contaminated environments without any control. Therefore, it could be intended for use as a bulk commodity, such as energy and chemicals products (Kim et al. 2015).

Considering that the harmful algae blooms have become a major challenge for sustainable management authorities, it is essential to seek alternatives to capitalize on this biomass and incorporating it into the production chain. Within this framework, the environmental impacts caused by nuisance algae, whether by anthropic actions or by natural occurrences, also lack mitigating actions (Ma et al. 2018; Ali et al. 2020). In this sense, this chapter aims to subtly present issues about energy recovery from nuisance algal blooms and residues. The work covers general aspects of algae biology, nuisance algal blooms, environmental issues, and potential applications.

11.2 Algae: An Overview

Linnaeus (1753) proposed the term "algae" to name some fungi, lichens, and organisms that are nowadays understood as algae. Afterward, the term "algae" became popular to characterize some heterogeneous organisms, such as any plants without roots, stem, and leaves, of a brownish color, commonly called seaweed (Bicudo and Menezes 2010). Algae are considered in three different kingdoms: Monera, Protista, and Plantae. Macroalgae (seaweeds) are classified within Kingdom Plantae, which is composed of more complex organisms. Microalgae are part of the Protista Kingdom; they are considered simpler living organisms. Cyanobacteria (blue-green algae) are organisms belonging to Monera Kingdom (Vidotti and Rollemberg 2004).

Algae have approximately thirty thousand species already identified, but less than 10% have been formally described. They are organisms belonging to different phylogenetic groups, macroalgae (multicellular), and microalgae (unicellular), which are microscopic. Both can be found in a wide range of habitats, such as marine, freshwater, and minority in sediments and soils (Lee et al. 2020). These two groups are photoautotrophic and have chlorophyll, a pigment responsible for photosynthesis. In general terms, the photosynthetic mechanism occurs when the light energy is captured in chemical form by the cells, together with the CO_2 fixation, to synthesize several organic compounds. Photosynthesis is primarily responsible for the entry of energy into the biosphere and is performed in parts by algae (Raven et al. 2010).

Seaweeds can be called by an informal term "macroalgae." Macroalgae are the major primary producers of the sea. Seaweeds are a large and diverse group of aquatic plants (Sudhakar et al. 2018); they are distinguished according to the photosynthetic pigments of their cells and are classified into three phyla: *Chlorophyta* (green macroalgae), *Phaeophyta* (brown macroalgae), and *Rhodophyta* (red algae) (Ortega et al. 2019). These aquatic organisms are becoming essential in world aquaculture due to their non-competition with other crops for land and freshwater. Furthermore, macroalgae have fast growth rates, high biomass production, and high polysaccharide content. Most species have been found in the marine environment (Sudhakar et al. 2018).

Microalgae are microorganisms that constitute the first occurrence group of mitochondria, chloroplasts, endoplasmic reticulum, and well-developed Golgi complex. Besides, generally presenting a single nucleus (Vidotti and Rollemberg 2004). The term microalgae is purely commercial and has no taxonomic value. Usually, they refer to eukaryotic photosynthetic microorganisms, whose screening comprises thousands of species already cataloged (Guiry and Guiry 2021). The classification of microalgae has undergone many changes over the years. The taxonomic division patterns rely upon their morphological, physiological, and structural characteristics, dividing these organisms within the taxa Chlorophyta, Rhodophyta, Glaucocytophyta, Euglenophyta, Chlorarachiniophyta, Heterokonta, Haptophyta, Cryptophyta, and Alveolata (Brasil et al. 2017).

Cyanobacteria are prokaryotes (absence of nuclear envelope) and autotrophs. They are considered the progenitors of the higher plant chloroplast. The characteristic of this group color (blue-green) is due to the presence of phycobiliproteins (phycocyanin and phycoerythrin) and chlorophyll a. They present different cellular strategies, physiological capacities, and adaptations related to the colonization of very extreme environmental conditions. The great importance of cyanobacteria is related to their nitrogen fixing forms; in addition, some species are used as a protein source, such as Spirulina (Knoot et al. 2018; Demay et al. 2019).

11.3 Nuisance Algal Blooms

The rapid and excessive algae growth under favorable conditions leads to a large formation of visible biomass, called an algae bloom. It is believed that this phenomenon has increased considerably in the recent years with a huge impact around the world, mainly in Europe, North America, Asia, Australia, and Central and South America. In India, for example, the first reports of algal blooms were in 1908 along the Malabar Coast to the Laccadive Islands. Harmful species have varied, and there are several different related aspects, such as frequency of occurrences, location, and intensity of bloom development, including upwelling, currents, winds, vertical mixing, surface water temperature, and nutrition (D'Silva et al. 2012; Townhill et al. 2018).

Algae abundance is triggered by the high availability of nutrients within a water body, which is generally attributed to pollution with agricultural effluents, domestic and industrial sewage, fertilizers, groundwater inputs, among others (Miyawaki et al. 2021). Such activities discharge a large number of residues rich in N and P, which are assimilated by algae. In addition, this causes eutrophication (which will be discussed in the next section). On the other hand, if other primary producers such as phytoplankton are present to compete for nutrients and light, the same abundance of nutrients in this alternating stable state may not be able to support a bloom, and the ecosystem remains in balance (Brookfield et al. 2021).

Algae associated with harmful blooms are toxic and can have serious consequences. In aquatic ecosystems, the problem is due to the release of toxins (hepatotoxins and neurotoxins) through the physical structure of the algal cells or accumulated biomass, affecting human health, aquatic biota, and, consequently, causing changes to the food chain activities, as well as economic implications (Al-Ghelani et al. 2005; Grattan et al. 2016). In addition, the contaminated water body becomes unsuitable for domestic or industrial supply (Sun et al. 2018). Submarine groundwater discharge can also lead to eutrophication in water bodies from groundwater-derived nutrient and solute loads, such as N, P, Fe, and Si. Thus, groundwater can also influence algal blooms in surface waters (Brookfield et al. 2021).

In terms of species, cyanobacteria *Microcystis aeruginosa* is one of the most widespread that form harmful algal blooms. It is the dominant genus and is present in all regions of the world, and due to toxicity, has harmful effects at different trophic levels in an aquatic food chain, e.g., phytoplankton, zooplankton, fish, and mollusks (Smith et al. 2019; Mirasbekov et al. 2021). The combination of abiotic conditions, water temperature, low lighting, and water with little turbulence may benefit the growth of this species (Paerl and Otten 2013; Nolan and Cardinale 2019; McPartlin et al. 2017). Species such as *Ulva rigida*, *Chaetomorpha linum*, and *Gracilaria confervoides* have been observed in the Venice Lagoon (Italy) (Barbot et al. 2016; Joniver et al. 2021). In China, in addition to *Microcystis*, the *Aphanizomenon flos-aquae* is proliferated in the Dianchi Lake. In the USA, this same species of cyanobacteria has high cell densities in Upper Klamath Lake. The bloom of the brown macroalgae *Fucus vesiculosus* is dominant in Germany and Denmark (Paerl et al 2018; Renita and Kumar 2020).

Given this scenario, what would be the solution to so many problems? The negative impacts of nuisance algal blooms have boosted many management actions of different forms and levels, as recently reported by Corcoran and Hunt (2021) and summarized in Fig. 11.1.

Prevention, control, and mitigation strategies are under the scope of managing the harmful algae blooms. They are based on the multiphase life cycle of algal blooms. First, prevention is related to methods that minimize the propensity or severity of blooms. On the other hand, the control is based on measures that target eliminating blooms, including mechanical or manual harvesting methods, such as skimming the algae on the water surface, or biochemical treatment. Mitigation is attributed to measures that reduce the negative impacts caused, such as water treatment. More details can be found in the report by Sengco (2009). Importantly, once the prevention or control process fails, the mitigation requirements are greater.

Today, there is a great demand in mitigating failures linked to algal blooms. Legislations and biotoxin monitoring programs in aquatic biota are being implemented around the world in an attempt to resolve these reported problems. According to the New Jersey Department of Environmental Protection, occurrences and information on state response strategies for algae blooms can be routinely reported (https://www.state.nj.us/dep/hab/). Besides, since 1985, reports on these harmful phenomena with great social repercussions have been compiled in the Harmful Algae Event Database,

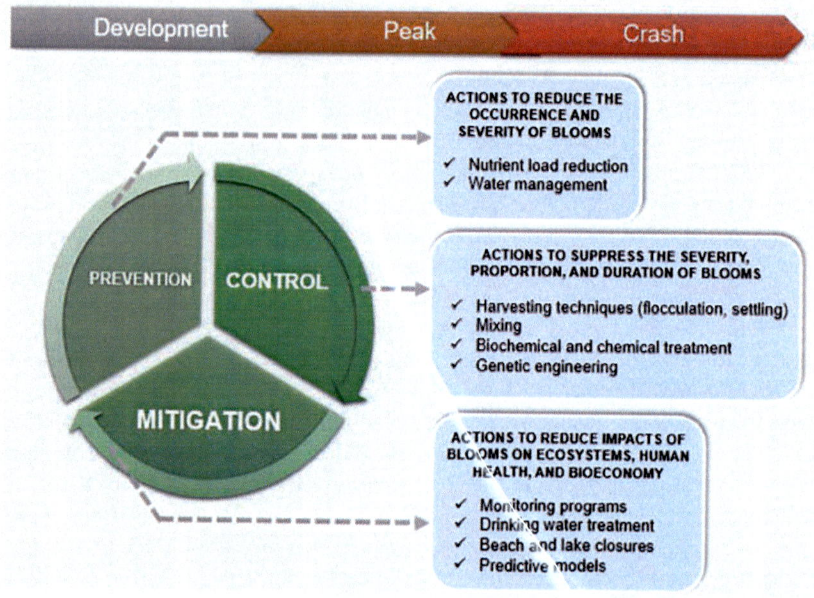

Fig. 11.1 Schematic summary of measures to prevent, control, and mitigate harmful algae blooms

HAEDAT (http://haedat.iode.org). It is a component of the Harmful Algal Information System (HAIS) within the "International Oceanographic Data and Information Exchange" (IODE) of the "Intergovernmental Oceanographic Commission" (IOC) of UNESCO.

11.4 Environmental Issues

Over the past few years, many environmental complications have intensified associated with urban, agricultural, and industrial development. Water pollution is among these issues, which is excessively enriched with nutrients, mainly N and P, promoting accelerated rates of primary production. This process causes eutrophication, which favors the periodic proliferation and dominance of nuisance algae. The damages of eutrophication are alarming worldwide, either to marine and freshwater ecosystems, which are experiencing impacts such as warming, acidification, and oxygen depletion (Griffith and Gobler 2020).

The environmental issues can vary from one region to another with local modifications. With the global increase in algal blooms, all maritime countries must understand the mechanism of occurrence in their territories. In the Indian context, the increasing frequency of algal blooms is a principal concern because of their detrimental effects on ecosystems and the economic, social, and ecological services they provide. India's marine resources are responsible for the income of more than 3.5 million people, estimated to be worth around USD 7 billion a year (Oyeku and Mandal 2021). The Gulf of Mannar (Southeast India), for instance, is an area that encompasses a large number of islands, is responsible for the livelihood of thousands of fishermen in the region and is also one of the four major coral reefs in India. In it, a significant increase in the species *Noctiluca scintillans* has been observed; the blooms of this algae have disturbed the surrounding coral reefs and caused the death of thousands of fish and other marine organisms (Raj et al. 2020). In addition, the species *Hemidiscus hardmanianus* has also proliferated in this area, impacting water quality. However, these phenomena are not attributed to any particular parameter but to a set of them that have caused these environmental issues (Sathishkumar et al. 2021).

Additionally, the problems arising from algal blooms are very diverse; the causes are infinite and are continually increasing. One of the possible reasons for these events is related to long-term climate change and eutrophication. These environmental hotspots can be evaluated and quantified using the life cycle assessment (LCA) tool (see Box 1).

Box 1 Life cycle assessment: Eutrophication potential

Life cycle assessment (LCA) is a tool to analyze the potential environmental impacts of a given process or product and supports sustainable decision-making in a more reliable and comprehensive way. LCA is based on the guidelines of the International Organization for Standardization (ISO) 14,000 series, which is defined in four phases: (i) goal and scope definition, (ii) inventory analysis, (iii) impact assessment, and (iv) interpretation. In addition, it is based on different metrics and sustainability indicators attributed to categories such as water footprint, energy demand, global warming potential, land use, eutrophication, acidification, abiotic depletion, toxicity, photochemical ozone formation, and ozone depletion (Deprá et al. 2020).

The eutrophication category refers to the potential of water to become excessively enriched with nutrients that trigger the algae blooms and reduction of oxygen, and, consequently, a decrease of biodiversity. LCA studies on eutrophication and the harmful algal blooms are scarce in the literature and normally target to evaluate a cause-effect chain for freshwater and marine ecosystems. Applicable LCA models are used to quantify the impacts associated with the release of substances to the environment and the use of natural resources. Considering eutrophication, the LCA tool estimates the relative severity of emissions at various levels in the ecosystem. In addition, the LCA is applied at two levels: (1) midpoint, which estimates the relative contribution of impacts (e.g., N and P concentrations in marine and freshwater). Already the (2) endpoint refers to the final impacts of interest (e.g., effects on human health measured in years of life or ecological impacts measured through the loss of aquatic species over time). Eutrophication potential in LCA is described according to the equation:

$$I_i = \sum F^i_{xmn} P^i_{xn} M_{xm}$$

where I_i is the potential impact of all chemicals (x) released to all media compartments (m) with all modeled exposure routes (n) for a given impact category (here, eutrophication potential). For a given chemical, x, F_{xmn} represents the fate and transport pathway; P_{xn} represents the potency; and M_{xm} represents the mass. LCA uses characterization factors (CF) to translate substance flows into potential environmental impacts (I_i) (Morelli et al. 2018).

Eutrophication is a very slow natural occurrence process; however, anthropic activities have accelerated the rate and extended impacts of this process. This phenomenon results in dense phytoplankton blooms with an unpleasant smell and reduces water clarity, affecting their quality. The eutrophication process is able to destroy aquatic ecosystems by increasing levels of toxins in the waterbody Fig. 11.2.

The high photosynthetic rates of algae can deplete dissolved inorganic carbon and lower the pH, making the water more acidic, which impairs the survival of other organisms. One of the main harmful effects of algal blooms is observed when these microorganisms die, and their decomposition results in severe depletion of oxygen

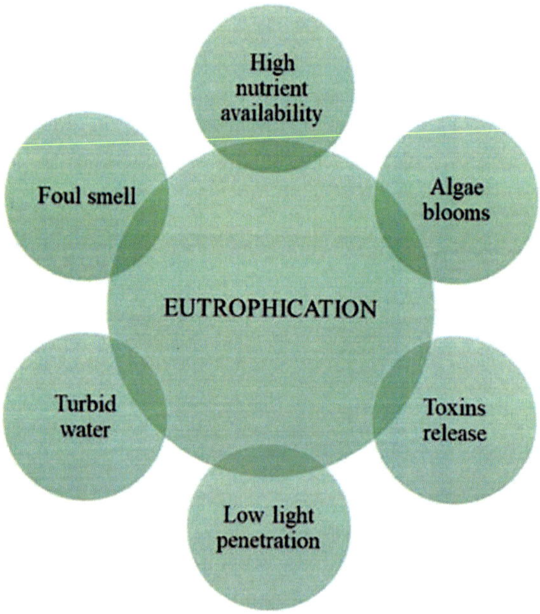

Fig. 11.2 Elements that influence eutrophication

levels (anoxia), leading to a decline in phytoplankton and zooplankton. Eutrophication is an environmental issue that has drawn attention because there may be no recovery of the ecosystem or it may remain eutrophicated for long periods when its regeneration is slow. Algae are able to assimilate large amounts of nutrients, such as C, N, and P, for their growth and development; these inputs into aquatic bodies generally come from residues, such as wastewater, sewage, industrial effluents, and waste from agriculture as well as urban areas (Ghosh and Bakshi 2019; Singh et al. 2021). Figure 11.3 outlines the eutrophication process and the formation of algal blooms.

Some developed countries are succeeding in reducing eutrophication. However, progress has been slow in many other countries due to some factors such as lack of legislation and scientific disagreements about nutrient action on the eutrophication process. On the other hand, it is known that algae cultivation can also be a profitable business with regard to the biotechnological industry. Therefore, these algal blooms can be useful and be an entry point to support the circular bioeconomy.

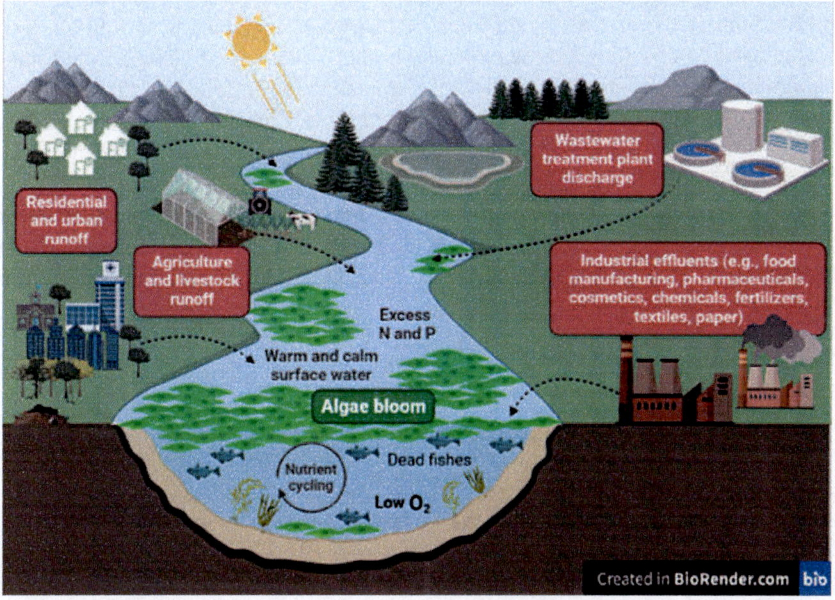

Fig. 11.3 Eutrophication and formation of algae blooms. *Note* Anthropogenic nutrient over-enrichment, associated with the elevation of the ambient temperature and recurrent hydrologic events, such as droughts and storms, boost eutrophication and trigger the increase of algae blooms

11.5 Potential Applications

The frequency and intensity of algal blooms around the world are unlikely to change in the near future. With the depletion of natural resources and to avoid or reduce the risks associated with the occurrence of nuisance algae blooms, many biomass management and recovery approaches have been researched and developed to allow the industrial exploitation of these events. Although algae biomass has been recognized as a bulk commodity that can be marketed, its commercial production is still very restricted (Corcoran and Hunt 2021). This usually is due to the high costs of cultivation and biomass downstream processing steps, which include harvesting, dewatering, drying, and separation of the desired target compounds. However, direct recovery of bloom biomass in water bodies offers a profitable business model. In the context of the circular bioeconomy, the use of this biomass, supported by atmospheric CO_2 and N and P from residues, represents a unique strategy in transforming pollution streams into value-added raw material (Amador-Castro et al. 2021). Table 11.1 presents some comparisons of algal biomass collected directly from blooms for industrial purposes.

The biomass from algal blooms can be recovered and fractionated. In general, its chemical composition is predominantly protein (40–70%), followed by substantial amounts of carbohydrates (10–25%), lipids (5–30%), minerals (5–25%), and pigments (1–5%). These values are generalizations of an immense range of algae species distributed in the most diverse ecosystems in the biosphere. Furthermore, the composition of the biomass of the same species is strongly influenced by the growth conditions, which can induce the production of a certain cell constituent to detriment of another (Matos et al. 2016; Seghiri et al. 2019). Thus, compounds derived from algal biomass create building blocks for the manufacture of multiple bioproducts

Table 11.1 Advantages and disadvantages of the algae to the industry

Parameter	Advantages	Disadvantages
Cultivation	High cell growth rates, do not require large areas for cultivation, and can be harvested all-year-round	Environmental, ecological, social, and human health damage
Biomass	Produces more biomass when compared to soil-based agriculture	Need for cheaper and more efficient harvesting, cleaning, control, and management techniques
Pollution	Can convert greenhouse gasses and wastewater	Produces high levels of toxins
Adaptability	Great versatility to thrive and develop in adverse conditions environments	This great adaptability can cause harm to other species, such as habitat competition
Exploitation	Capable of being used in different industrial sectors	Need to implement its cost-effectiveness

Source Adapted from Tang et al. (2020) and Amador-Castro et al. (2021)

from various industrial sectors, such as energy, food, agriculture, cosmetics, pharmaceuticals, materials, and chemicals. These potentialities must be able to meet the growing need for algal extracts, such as fractions or pure compounds, for economic purposes (Severo et al. 2020, 2021).

11.5.1 Algae Blooms as a Source of Biofuels and Bioenergy

The supply of sufficient amounts of energy with minimal environmental impact is among the main challenges of the macroenergy world. The search for fossil energy substitutes that meet the requirements of sustainability and renewability drives the development of biofuels (Oncel 2013). For this reason, there is a broad expectation to convert the high volume of biomass from algal blooms primarily into biodiesel, bioethanol, biomethane, and biohydrogen through physical, chemical, and/or biological operations (Chia et al. 2022).

According to Hu et al. (2013), about 6,800 tons per year of bio-oil can be obtained from the nuisance blue-green algae blooms in Dianchi Lake (China). This demonstrates the importance of this feedstock for biofuels manufacture. In the study by Maddi et al. (2011), nuisance algae from natural blooms can be applied as liquid fuels through thermochemical processes. Recently, Amador-Castro et al. (2021) reported that the brown alga *Sargassum* may have application in biofuels production. However, factors related to its composition can restrict its potential. From another point of view, these macroalgae have a high moisture content, requiring drying. The energy consumed to dry the biomass with 88% moisture was estimated at around 0.7 MJ/kg, representing a great economic impact. Biofuel production methods where the biomass does not require drying are preferable. Chia et al. (2022) demonstrated that biomass from algal blooms can be blended with coal for energy generation via combustion, gasification, or pyrolysis, as well as from microbial fuel cells (MFCs). In addition, they reported that microalgae species such as *Chlorella sorokiniana* and *Selenastrum gracile* present in blooms are potential candidates for biodiesel output. They can yield 1.3 kg of lipids and 8 kg of biomass in a water volume of 10,000 L (Pandey et al. 2019).

The study by Kuo (2011) was evaluated the net energy balance and the economics of harvesting and converting biomass from algal blooms into biofuels. They modeled different harvesting methods (plankton net trawling, traveling screen, and screw pump filtration) as well as different conversion techniques, such as hydrothermal liquefaction (HTL), anaerobic digestion, transesterification, and fermentation. The net trawl was the most favorable method to harvest the biomass, assuming a speed of 0.5 m/s, a depth of 0.5 m, and a highly eutrophicated area (40 mg-chlorophyll/m^3). Under this scenario, it was possible to collect around 100 kg of dried algae and spend 1 GJ of energy. In terms of energetics, HTL and anaerobic digestion were the most suitable methods for energy production, yielding between 17–29 and 11–19 GJ per km^2 of the estuary, respectively, with a chlorophyll concentration in the field of 50 mg/m^3. In contrast, energy production from the net conversion of transesterification (biodiesel) is unfavorable due to the low lipid content (~16–18%). Similar behavior was observed for fermentation due to low yield.

It is noteworthy that the energy production from algae blooms depends mainly on the harvesting method used and biomass productivity. There are multiple approaches to this purpose, including sedimentation, centrifugation, flocculation, biofilms, dissolved air flotation, and filtration. Direct harvesting, however, is considered the cheapest method, but it has technical, logistical, and regulatory bottlenecks that can make the use of biomass unfeasible. Thus, applying some of these methods without valuing the harvested biomass can incur high costs (Corcoran and Hunt 2021). Another limitation unrelated to the process of obtaining biofuels itself is the fact that algal blooms are seasonal, which can hinder the continued production of this input (Joniver et al. 2021).

Regardless of these aspects, with recent technological advances, researchers are focusing on improving the productivities of cellular compounds through genetic engineering, such as synthetic biology, where algae-blooming strains can be bioengineered into desirable traits to have wide adoption. Targets include the modification of genes related to (i) lipid metabolism to directly secrete some molecules of triacylglycerols, alkanes, alcohols, and isoprenoid derivatives; (ii) carbohydrate metabolism; (iii) increased nutrient conversion; (iv) improved photosynthetic efficiency; and (v) better cell disruption for oil extraction (Henley et al. 2013).

11.5.2 Food, Feed, Health, Agricultural, and Other Uses

Algae have been widely used as a source of nutraceuticals since their exploration began in the 1960s in Asia and North America. Among its produced metabolites, pigments, lipids, proteins, polysaccharides, and phenolic compounds can be sources of functional food ingredients due to the presence of many compounds with bioactive properties (Dias et al. 2020).

Global pigment trade (β-carotene, astaxanthin, fucoxanthin, phycocyanin, phycoerythrin, lutein, and chlorophyll) is estimated to be worth around USD 452.4 million, at a CAGR of 4% during the forecast period to 2025 (Research and Markets 2021). Carotenoids, such as β-carotene and astaxanthin, have a broad market share to be applied as dyes, food additives and supplements, beverages, and animal feed. There are several facilities in operation worldwide in countries like China, Japan, Australia, Israel, Germany, USA, Spain, and Brazil, which are economically viable for this purpose. Furthermore, phycobiliproteins also represent an important source of natural dyes. Phycocyanin and phycoerythrin can be applied in drinks and sweets in general (Maroneze et al. 2020). Another way of applying algae biomass is in its whole form, added to food supplements, as the cyanobacterium *Aphanizomenon flos-aquae* (Nuzzo et al. 2018). As a feed, algae are useful for seafood species, playing a relevant role in aquaculture. However, applications for both human and animal food are often restricted due to the accumulation of toxins in many strains. Biomass is usually generated under the input of contaminated waste streams, causing quality problems (Chen et al. 2021).

In terms of health-related products, algal bloom biomass, as a natural input, is essential as medicines, nutraceuticals, and cosmetics. For example, fucoxanthin, a pigment present in brown algae and diatoms, has application as an antioxidant and anti-inflammatory agent, helping in the process against diabetes, obesity, and cancer. Phycocyanin and phycoerythrin can also be used as therapeutic compounds in diseases induced by oxidative stress. Chlorophylls have anti-proliferative action on cancer cells, antioxidant, anti-mutagenic, anti-inflammatory, antimicrobial, and antiviral. Several species of algae produce substantial amounts of fatty acids such as linolenic, eicosapentaenoic (EPA), and docosahexaenoic (DHA), which exhibit hypolipidemic activity by reducing cardiovascular problems (Borowitzka 2013; Kim et al. 2015; Dias et al. 2020). Polymers derived from algal blooms, such as alginate, have also been explored for use as food, in fiber materials, bioremediation, and in the biomedical field. Despite these potentials, the major limitation for application in the health area lies in collecting the biomass and extracting the desired compounds before being released into the surrounding environment (Joniver et al. 2021).

Other applications include the algae as raw material in modern agriculture as biofertilizers, pesticides, and soil conditioners, which can induce plant growth and improve health land. However, the bottleneck is also the presence of toxic compounds in some species of algae. The use of algal blooming biomass has been consistently reviewed and indicated for some bioproducts, according to Kim et al. (2015), Chen et al. (2021), and Amador-Castro et al. (2021). Progress in algal biotechnology research, including genetic engineering to modify organisms, has also led to increased use of new metabolites (Borowitzka 2013).

11.6 Conclusions

The harmful algae bloom causes eutrophication problems and incur significant economic losses for their removal from the aquatic environment and waste proper disposal. The most prominent issue so far is the enormous amount of naturally generated biomass. So, useful applications of this biomass are essential, but few actions have been carried out aiming to take advantage of this resource as a business model. Furthermore, efficient and scalable harvesting must be considered, not only as a bioremediation approach to algal blooms but also due to technical, logistical, and regulatory challenges. Finally, if these issues are addressed holistically, that is, in the integrated biorefinery context, biomass recovery from nuisance algae blooms has the potential to be used as energy input and contribute to a more circular bio-based economy.

References

Ali J, Wang L, Waseem H, Song B, Djellabi R, Pan G (2020) Turning harmful algal biomass to electricity by microbial fuel cell: a sustainable approach for waste management. Environ Pollut 266(115373):1–10

Al-Ghelani HM, Alkindi AYA, Amer S, Al-Akhzami YK (2005) Harmful algal blooms: physiology, behavior, population dynamics and global impacts—a review. Sultan Qaboos Univ J Sci 10:1–30

Amador-Castro F, García-Cayuela T, Alper HS, Rodriguez-Martinez V, Carrillo-Nieves D (2021) Valorization of pelagic *sargassum* biomass into sustainable applications: current trends and challenges. J Environ Manag 283:112013

Barbot YN, Al-Ghaili H, Benz R (2016) A review on the valorization of macroalgal wastes for biomethane production. Mar Drugs 14(6):120

Bernauer T, Böhmelt T (2020) International conflict and cooperation over freshwater resources. Nat Sustain 3:350–356

Bicudo CEM, Menezes M (2010) Introdução as Algas do Brasil. In: Forzza RC (ed) Instituto de Pesquisas Jardim Botânico do Rio de Janeiro. Catálogo de plantas e fungos do Brasil [online], vol 1. Andrea Jakobsson Estúdio: Instituto de Pesquisa Jardim Botânico do Rio de Janeiro, Rio de Janeiro, pp 49–60. ISBN 978-85-8874-242-0

Borowitzka MA (2013) High-value products from microalgae—their development and commercialisation. J Appl Phycol 25:743–756

Brasil BSAF, Siqueira FG, Salum TFC, Zanette CM, Spier MR (2017) Microalgae and cyanobacteria as enzyme biofactories. Algal Res 25:76–89

Brookfield AE, Hansen AT, Sullivan PL, Czuba JA, Kirk MF, Li L, Newcomer ME, Wilkinson G (2021) Predicting algal blooms: are we overlooking groundwater? Sci Total Environ 769(144442):1–15

Chen L, Giesy JP, Adamovsky O, Svirčev Z, Meriluoto J, Codd GA, Mijovic B, Shi T, Tuo X, Li S-C, Pan B-Z, Chen J, Xie P (2021) Challenges of using blooms of *Microcystis* spp. in animal feeds: a comprehensive review of nutritional, toxicological and microbial health evaluation. Sci Total Environ 764:142319

Chia SR, Nomanbhay SBHJM, Chew KW, Munawaroh HSH, Shamsuddin AH, Show PL (2022) Algae as potential feedstock for various bioenergy production. Chemosphere 287:131944

Corcoran AA, Hunt RW (2021) Capitalizing on harmful algal blooms: from problems to products. Algal Res 55:1–6

Demay J, Bernard C, Reinhardt A, Marie B (2019) Natural products from cyanobacteria: focus on beneficial activities. Mar Drugs 17(6):320

Deprá MC, Severo IA, dos Santos AM, Zepka LQ, Jacob-Lopes E (2020) Environmental impacts on commercial microalgae-based products: Sustainability metrics and indicators. Algal Res 51:102056

Dias RR, Severo IA, Deprá MC, Maroneze MM, Zepka LQ, Jacob-Lopes E (2020) Chapter 2: The next-generation of microalgae-based products. In: Shukla P (ed) Microbial enzymes and biotechniques. Springer, Singapore, pp 15–42

D'Silva MS, Anil AC, Naik RK, D'Costa PM (2012) Algal blooms: a perspective from the coasts of India. Nat Hazards 63:1225–1253

Du Preez CC, Van Huyssteen CW (2020) Threats to soil and water resources in South Africa. Environ Res 183(109015):1–16

Environmental Protection Agency (EPA) (2021). Available at: https://www.epa.gov/nutrientpollution/harmful-algal-blooms#act. Accessed 14 May 2021

Ghosh T, Bakshi BR (2019) Designing biofuel supply chains while mitigating harmful algal blooms with treatment wetlands. Comput Chem Eng 126:113–127

Grattan LM, Holobaught S, Glenn Morris Jr J (2016) Harmful algal blooms and public health. Harmful Algae 57:2–8

Griffith A, Gobler CJ (2020) Harmful algal blooms: a climate change co-stressor in marine and freshwater ecosystems. Harmful Algae 91:1–12

Guiry MD, Guiry GM (2021) AlgaeBase. World-wide electronic publication, National University of Ireland, Galway. Available at: https://www.algaebase.org/about/?tc=accept&undefined. Accessed 18 May 2021

Henley WJ, Litaker RW, Novoveská L, Duke CS, Quemada HD, Sayre RT (2013) Initial risk assessment of genetically modified (GM) microalgae for commodity-scale biofuel cultivation. Algal Res 2(1):66–77

Hu Z, Zheng Y, Yan F, Xiao B, Liu S (2013) Bio-oil production through pyrolysis of blue-green algae blooms (BGAB): product distribution and bio-oil characterization. Energy 52:119–125

Joniver CFH, Photiades A, Moore PJ, Winters AL., Woolmer A, Adams JMM (2021) The global problem of nuisance macroalgal blooms and pathways to its use in the circular economy. Algal Res 58:102407

Kim JK, Kottuparambil S, Moh SH, Lee TK, Kim YJ, Rhee JS, Choi EM, Kim BH, Yu YJ, Yarish C, Han T (2015) Potential applications of nuisance microalgae blooms. J Appl Phycol 27:1223–1234

Knoot CJ, Ungerer J, Wangikar PP, Pakrasi H (2018) Cyanobacteria promising biocatalysts for sustainable chemical production. J Biol Chem 293(14):5044–5052

Kuo C-T (2011) Harvesting natural algal blooms for concurrent biofuel production and hypoxia mitigation. MS Thesis, University of Illinois at Urbana-Champaign

Laurens LML (2021) A new algae technical standards focus group: summarizing and guiding the algae state of art. Algal Res 53(102141):1–4

Lee XJ, Ong HC, Gan YY, Chen WH, Mahlia TMI (2020) State of art review on conventional and advanced pyrolysis of macroalgae and microalgae for biochar, bio-oil and bio-syngas production. Energy Convers Manag 210:1–34

Linnaeus C (1753) Species plantarum. Holmiae: Laurentii Salvii, vol 1, 200 p

Ma H, Guo Y, Qin Y, Li YY (2018) Nutrient recovery technologies integrated with energy recovery by waste biomass anaerobic digestion. Bioresour Technol 269:520–531

Maddi B, Viamajala S, Varanasi S (2011) Comparative study of pyrolysis of algal biomass from natural lake blooms with lignocellulosic biomass. Bioresour Technol 102:11018–11026

Maroneze MM, Dias RR, Severo IA, Queiroz MI (2020) Chapter 11: Microalgae-based processes for pigments production. In: Jacob-Lopes E, Queiroz MI, Zepka LQ (eds) Pigments from microalgae handbook. Springer, Cham, pp 241–264

Matos AP, Feller R, Moecke EHS, de Oliveira JV, Junior AF, Derner RB, Sant'Anna ES (2016) Chemical characterization of six microalgae with potential utility for food application. J Am Oil Chem Soc 93(7):963–972

McPartlin DA, Loftus JH, Crawley AS, Silke A, Murphy CS, O'Kennedy RJ (2017) Biosensors for the monitoring of harmful algal blooms. Curr Opin Biotechnol 45:164–169

Mirasbekov Y, Zhumakhanova A, Zhantuyakova A, Sarkybayev K, Malashenkov DV, Baishulakova A, Dashkova V, Davidson TA, Vorobjev IA, Jeppesen E, Barteneva NS (2021) Semi-automated classification of colonial *Microcystis* by FlowCAM imaging flow cytometry in mesocosm experiment reveals high heterogeneity during seasonal bloom. Sci Rep 11(9377):1–14

Miyawaki B, Mariano AB, Vargas JVC et al (2021) Microalgae derived biomass and bioenergy production enhancement through biogas purification and wastewater treatment. Renew Energy 163:1153–1165

Morelli B, Hawkins TR, Niblick B, Henderson AD, Golden HE, Compton JE, Cooter EJ, Bare JC (2018) Critical review of eutrophication models for life cycle assessment. Environ Sci Technol 52:9562–9578

Nolan MP, Cardinale BJ (2019) Species diversity of resident green algae slows the establishment and proliferation of the cyanobacterium *Microcystis aeruginosa*. Limnologica 74:23–27

Nuzzo D, Presti G, Picone P, Galizzi G, Gulotta E, Guiliano S, Mannino C, Gambino V, Scoglio S, Di Carlo M (2018) Effects of the *Aphanizomenon flos-aquae* extract (Klamin®) on a neurodegeneration cellular model. Oxid Med Cell Longev 9089016:14

Oncel SS (2013) Microalgae for a macroenergy world. Renew Sustain Energy Rev 26:241–264

Ortega A, Geraldi NR, Alam I, Kamau A, Acinas SG, Logares R, Gasol JM, Massana R, Krause-Jensen D, Duarte CM (2019) Important contribution of macroalgae to oceanic carbon sequestration. Nat Geosci 12:748–754

Oyeku OG, Mandal SK (2021) Historical occurrences of marine microalgal blooms in Indian peninsula: probable causes and implications. Oceanologia 63:51–70

Paerl HW, Otten TG (2013) Harmful cyanobacterial blooms: causes, consequences, and controls. Microb Ecol 65:995–1010

Paerl HW, Otten TG, Kudela R (2018) Mitigating the expansion of harmful algal blooms across the freshwater-to-marine continuum. Environ Sci Technol 52:5519–5529

Pan G, Chen J, Anderson DM (2011) Modified local sands for the mitigation of harmful algal blooms. Harmful Algae 10(4):381–387

Pandey MK, Dasgupta CN, Mishra S, Srivastava M, Gupta VK, Suseela MR, Ramteke PW (2019) Bioprospecting microalgae from natural algal bloom for sustainable biomass and biodiesel production. Appl Microbiol Biotechnol 103:5447–5458

Pandhal J, Choon WL, Kapoore RV, Russo DA, Hanotu J, Wilson IAG, Desai P, Bailey M, Zimmerman WJ, Ferguson AS (2018) Harvesting environment microalgal blooms for remediation and resources recovery: a laboratory scale investigation with economic and microbial community impact assessment. Biology 7(4):1–16

Paredes I, Ramírez F, Aragonés D, Bravo MÁ, Forero MG, Green AJ (2021) Ongoing anthropogenic eutrophication of the catchment area threatens the Doñana World Site (South-west Spain). Wetlands Ecol Manag 29:41–65

Preisner M, Neverova-Dziopak E, Kowalewski Z (2021) Mitigation of eutrophication caused by wastewater discharge: a simulation-based approach. Ambio 50:413–424

Raj KD, Mathews G, Obura DO, Laju RL, Bharath MS, Kumar PD, Arasamuthu A, Kumar TKA, Edward JKP (2020) Low oxygen levels caused by *Noctiluca scintillans* bloom kills corals in Gulf of Mannar, India. Sci Rep 10(22133):1–17

Raven PH, Evert RF, Eichhorn SE (2010) Biologia Vegetal, 8ª. Editora Guanabara Koogan, Rio de Janeiro, p 856

Renita AA, Kumar PS (2020) Valorization of waste algal boom for value-added products. In: Jerold M, Arockiasamy S, Sivasubramanian V (eds) Bioprocess engineering for bioremediation: valorization and management techniques, vol 104, pp 129–138

Research and Markets (2021) The world's largest market research store. Available at: https://www.researchandmarkets.com/. Accessed 5 Sept 2021

Sant'Anna CL, Carvalho LR, Fiore MF, Silva-Stenico ME, Lorenzi AS, Rios FR, Konno K, Garcia C, Lagos N (2011) High toxic *Microcystis aeruginosa* strain, isolated from São Paulo—Brazil, produce hepatotoxins and paralytic shellfish poison neurotoxins. Neurotox Res 19:389-402

Sathishkumar RS, Sundaramanickam A, Sahu G, Mohanty AK, Ramesh T, Khan AS (2021) Intense bloom of the diatom *Hemidiscus hardmanianus* (Greville) in relation to water quality and plankton communities in Tuticorin coast, Gulf of Mannar, India. Mar Pollut Bull 163(111757):1–8

Seghiri R, Kharbach M (2019) Essamri A (2019) Functional composition, nutritional properties, and biological activities of Moroccan *Spirulina* microalga. J Food Qual 3707219:11

Sellner KG, Doucette GJ, Kirkpatrick GJ (2003) Harmful algal blooms: causes, impacts and detection. J Microbiol Biotechnol 30(7):383–406

Sengco M (2009) Mitigation of effects of harmful algal blooms. In: Shumway SE, Rodrick GE (eds) Shellfish safety and quality. A volume in Woodhead Publishing series in food science, technology and nutrition

Severo IA, Siqueira SF, Deprá MC, Maroneze MM, Zepka LQ, Jacob-Lopes E (2019) Biodiesel facilities: what can we address to make biorefineries commercially competitive? Renew Sustain Energy Rev 112:686–705

Severo IA, Deprá MC, Dias RR, Jacob-Lopes E (2020) Chapter 26: Process integration applied to microalgae-based systems. In: Jacob-Lopes E, Manzoni MM, Queiroz MI, Zepka LQ (eds) Handbook of microalgae-based processes and products, 1st edn. Elsevier, Cambridge, pp 709–735

Severo IA, Dias RR, Sartori RB, Maroneze MM, Zepka LQ, Jacob-Lopes E (2021) Chapter 4. Technological bottlenecks in establishing microalgal biorefineries, In: Shekh A, Schenk P, Sarada R (eds) Microalgal biotechnology, 1st edn. Royal Society of Chemistry, Londres, pp 118–134
Singh DV, Bhat RA, Upadhyay AK, Singh R, Singh DP (2021) Microalgae in aquatic environs: a sustainable approach for remediation of heavy metals and emerging contaminants. Environ Technol Innov 21(101340):1–19
Smith RB, Bass B, Sawyer D, Depew D, Watson SB (2019) Estimating the economic costs of algal blooms in the Canadian Lake Erie Basin. Harmful Algae 87(101624):1–17
Sudhakar K, Mamat R, Samykano M, Azmi WH, Ishak WFW, Yusaf T (2018) An overview of marine macroalgae as bioresource. Renew Sustain Energy Rev 91:165–179
Sun R, Sun P, Zhang J, Esquivel-Elizondo S, Wu Y (2018) Microorganisms-based methods for harmful algal blooms control: a review. Bioresour Technol 248(Part B):12–20
Tang DYY, Khoo KS, Chew KW, Tao Y, Ho SH, Show PL (2020) Potential utilization of bioproducts from microalgae for the quality enhancement of natural products. Bioresour Technol 304(122997):1–11
Townhill B, Tinker J, Jones M, Pitois S, Creach V, Simpson SD, Dye S, Bear E, Pinnegar JK (2018) Harmful algal blooms and climate change: exploring future distribution changes. ICES J Mar Sci 75(6):1882–1893
Vidotti EC, Rollemberg MCE (2004) Algas: da economia nos ambientes aquáticos à biorremediação e à química analítica. Quim Nova 27(1):139–154

Chapter 12
Organic Rankine Cycles (ORCs) for Waste Heat Utilization

Yong-qiang Feng and Tzu-Chen Hung

Abstract Considering the thermodynamic performance, economic performance, and environmental performance simultaneously, the thermo-environmental optimization for a series dual-pressure organic Rankine cycle (STORC) and a cascaded organic Rankine cycle (CORC) has been investigated. The effects of key operating parameters on the thermal and economic performance of the system were analyzed. Based on a 3 kW ORC experimental prototype, the heat exchanger performance of different mixtures has been studied and compared. Five different proportions of working fluids were selected, respectively, R123, 0.33R123/0.67R245fa, 0.5R123/0.5R245fa, 0.67R123/0.33R245fa, and R245fa. The comparison between the experimental test and simulation result without considering the pressure drop was addressed. Results indicate that R245fa/R152a is the best candidate working fluids for CORC LT-Loop. The evaporator heat transfer coefficients of R245fa are highest, followed by the R245fa/R123 mixture and while R123 shows the lowest evaporator heat coefficients. The measured thermal efficiency of 0.67R123/0.33R245fa is the highest, which is 7.33% and the maximum simulated thermal efficiency is14.55%.

Keywords Organic Rankine cycle (ORC) · Thermo-environmental optimization · Evaporator heat coefficient · Pressure drop

12.1 Introduction

At the current energy consumption rate and population growth rate, the world's proven reserves of coal can last up to 200 years, and oil and natural gas can last up to 50 years. Building a clean, safe, low-carbon, and efficient modern energy system is an important way to achieve high-quality urban energy development. With

Y. Feng (✉)
School of Energy and Power Engineering, Jiangsu University, Zhenjiang, China
e-mail: yqfeng@ujs.edu.cn

T.-C. Hung
Department of Mechanical Engineering, National Taipei University of Technology, Taipei, Taiwan
e-mail: tchung@ntut.edu.tw

A. E.-F. Abomohra et al. (eds.), *Waste-to-Energy*,
https://doi.org/10.1007/978-3-030-91570-4_12

the development of economy, urbanization rate increase, the production and living energy consumption is also growing, but the total energy is limited, environmental capacity is limited, and how to maintain the continuous growth of the industrial economy and the continuous improvement of people's living standards and to control the total energy consumption and improve energy utilization efficiency become the problem that needs to solve. So based on the present situation of the traditional fossil energy at a time and the burning of fossil fuels at a time using great pollution to the environment after the two main aspects, the world health organization and various countries work together to develop a variety of the constraints of energy-saving and emission reduction policy, such as developing renewable energy, upgrading system equipment and product innovation, and the recycle of energy strategy. However, the recycling and utilization of energy have been paid more and more attention by human beings, because the utilization efficiency of heat of combustion is limited. In general, the utilization efficiency of heat of combustion does not exceed 50% of the heat generated by combustion, and the part of heat not used by combustion is lost and wasted in vain. The global energy efficiency is low, an important reason is that more than a large amount of waste heat pressure is not fully achieve cascade utilization of resources, the key factors to restrict energy efficiency of energy-saving technology is backward, and there are many types of use of waste heat recovery technology; main technologies are heat exchange technology, cooling, heating, and thermal power generation technology, and because the structure of the heat exchange and thermal conversion technology is relatively single, and the application range is smaller, the recovery efficiency is low, so we chose thermal power generation technology for key research. Thermal power generation technology is mainly based on the thermal power generation system of Rankine cycle, like the organic Rankine cycle (ORC), Kalina cycle, water vapor expansion cycle, etc. However, ORC is the most widely studied and applied. ORC is a circulation system with low boiling point organic matter as the working medium, which can recover low-temperature thermal energy of different temperature gradients and has a high efficiency in sensible heat recovery (Zhou et al. 2013; Yamada et al. 2014).

Under one atmosphere, some organic working fluids have lower boiling points, so the lower heat absorption can be evaporated to form a higher pressure. According to these characteristics of organic working fluid with a lower boiling point, the L–T (low temperature) heat source can be used to heat it to produce high-pressure (H-P) gas, and the expander can be pushed to do work to realize the recovery and utilization of L–T heat energy. Compared with water, organic working fluid has strong superiority, an ORC system the working fluid of choice is proper, can improve the system efficiency and output power, choose the ORC system cycle working substance to the heat source temperature, evaporation temperature, evaporation pressure, condensation temperature and the pinch point temperature difference and so on carries on the comprehensive consideration, for the characteristics of different fluid to domestic and foreign scholars have conducted a long-term research. Tian et al. (2012, 2014, 2017; Song and Gu 2015) used R134a and R245fa as working medium and used experimental methods to compare the organic Rankine cycle system with flue gas as heat source at 400 °C. The comprehensive comparison found that R134a had better performance in the L–T section, while R245fa had better performance in the H-T (high-temperature) section. Shu et al. (2016) adopted regenerative organic Rankine cycle system and selected five working fluids, R134a, R245fa, pentane, MDM and toluene, to analyze the influence of key parameters on the thermodynamic performance of the system. It is found that the cycle efficiency can be improved by increasing evaporation pressure, evaporation temperature, and expansion ratio, but the superheat has little effect on the cycle performance. Dong et al. (2017) taken R245fa as working fluid, although the calculation model can be simplified for only one type of working fluid, due to the limitations of each type of working fluid, the pure working fluid may not achieve the best effect. For some circulatory systems, mixed working media are more advantageous. Wang et al. (2018) compared the coupling ORC system with single stage across critical ORC system performance and probe into the critical temperature and the working medium combination working medium influence on coupling ORC system performance. Uusitalo et al. (2014) studied a coupling across the critical and subcritical organic Rankine's circulatory system and performance analysis of the coupling ORC system in detail. Javanshir et al. (2017) proposed the ejector organic Rankine cycle and carried out thermodynamic studies on the ejector organic Rankine cycle under subcritical, transcritical, and supercritical conditions. Lim et al. (2021) evaluated the system with nine groups of mixed refrigerants, and the results showed that R123-R245fa had the highest net power. Zhang et al. (2021a) used 36 kinds of working fluids and determined the type of pump suitable for working fluids, which provided reference for practical operation.

In addition to changing the choice of working medium, the thermal efficiency of the system can be improved by optimizing the structural parameters of the system. Wang et al. (2020a) optimized and compared the single-stage evaporation and two-stage evaporation organic Rankine cycles and found that the latter has the smallest generation cost and investment payback period. Kim et al. (2020) proposed a system combining phosphoric acid fuel cell (PAFC) and organic Rankine cycle, and the thermal efficiency increased by 11.31%. Esra and Kiliç (2020) combined ORC with VCRC, and the exergy efficiency of the system was increased by 19.5%. Renno et al. (2020) proposed a new ORC system (CPV/T). Compared with the original ORC, the system efficiency was improved by 2.6%. Zhang et al. (2021b) proposed a new ICE-ORC system for different rated power of ICE, and compared with the traditional system, the cycle efficiency was improved by 17%. Kang et al. (2020) analyzed the performance of ORC systems combined with other circulatory systems and found that the power generation could be increased by up to 20%. Huang et al. (2019) added ORC equipment on the basis of CCHP. Through data comparison, they found that all indexes of CCHP-ORC system were superior to the original system. Shu et al. (2014a) conducted a CORC to recover various waste heat and found that R1234yf was the best working fluid.

In addition, many teams have presented interests in the mixed working fluid. Shu et al. (2020) investigated the dynamic performance of cascade system by dynamic simulation. Results show that the cascade systems are better than other systems in terms of waste heat utilization. Shu et al. (2014b) investigated the thermal efficiency for ORC system used two retardants. The results revealed that in comparison with pure working fluids, zeotropic mixtures presented a more satisfactory thermodynamic performance. The similar conclusion can be found by Heberle et al. (2012) and Zhang et al. (2014). Lecompte et al. (2014) examined the exergy efficiency on ORC system used mixed working fluid. The results showed that the thermal efficiency of mixed medium is higher than that of pure medium. Dong et al. (2014) studied the first law thermal efficiency on ORC system used mixed working fluid. Feng et al. (2015a) compared the thermodynamic properties of pure working medium and mixed working medium, and found that the mixed working fluid had low thermodynamic properties. In their subsequent study (Feng et al. 2015b), a thermo-economical compassion considering exergy efficiency and leveled energy cost (LEC) simultaneously was examined. The results indicated that the zeotropic working fluids do not always have more advantageous than pure working fluids. The similar conclusion can be found in Xiao et al. (2015), and studies have shown that the composite properties of mixed working fluids are sometimes not as good as those of pure working fluids. The above reference demonstrates that the mixture working fluids represented better thermodynamic performance and economic performance than the pure working fluid at certain heat source condition and mass fraction. Tian et al. (Wang et al. 2020b) proposed dual-loop ORC system which uses zeotropic mixtures based on siloxanes. Results indicated that D4/R123 (0.3/0.7) has best performance with net-power output of 21.66 kW and the thermal efficiency of 22.84%. Aziz et al. (2018) optimized the H-T organic Rankine cycle by genetic algorithm, and the UA value of decane increased by 45.5%. Wang et al. (2020c) used genetic algorithm for multi-objective optimization of ORC and sorted the Pareto optimal solution to obtain the optimal operation parameters. Imran et al. (2016) evaluated three different orcs and compared them by genetic algorithm. It was found that the thermal efficiency and exergy efficiency of RORC were higher than ORC, but the average investment cost was higher. Li and Li (2018a) carried out two objective optimization on SORC, and further analyzed the influence of different parameter combinations in Pareto boundary and SORC scale on economic performance.

In this chapter, the thermo-environmental optimization for a series dual-pressure organic Rankine cycle (STORC) and a cascaded organic Rankine cycle (CORC) has been researched. The effects of key parameters on the thermal and economic performance of the system were analyzed. Based on a 3 kW ORC experimental prototype, the heat exchanger performance of different mixtures has been studied and compared. Five different proportions of working fluids were selected, respectively, R123, 0.33R123/0.67R245fa, 0.5R123/0.5R245fa, 0.67R123/0.33R245fa, and R245fa. The comparison between the experimental test and simulation result without considering the pressure drop was addressed.

12.2 Thermo-environmental Optimization of a Novel Supercritical–Subcritical Organic Rankine Cycle

12.2.1 System Description of STORC

Figure 12.1 shows the STORC system flowchart, which mainly includes a H-P evaporator, a L-P (low pressure) evaporator, a preheater, a condenser, and two expanders. The T-s diagram of the STORC system is shown in Fig. 12.2. The working fluid is preheated by the preheater (process 2–3) and divided into two parts, and the first part flows into the H-P cycle, while the remaining working fluid flows into the L-P cycle. The working fluid is further pressurized by a H-P pump to a higher evaporative pressure (process 3–5), and then vaporized to saturation or overheating by absorbing heat from the heat source (process 5–8). The working fluid with H-T and H-P flows into the H-P expander, and output work is done (process 8–9). The remaining working fluid is heated by the L-P evaporator until it is saturated or overheated (process 3–10). The working fluid from the H-P expander (point 9) mixes with the one from the L-P evaporator (point 10) and then flows into the L-P expander to generate work output (process 11–12). The working fluid is then condensed through the condenser (process 12–1) and then fed into the working pump for the next cycle. It should be noted that the H-P and L-P loop are subcritical ORC cycle, while the mixture working fluid is adopted in this study.

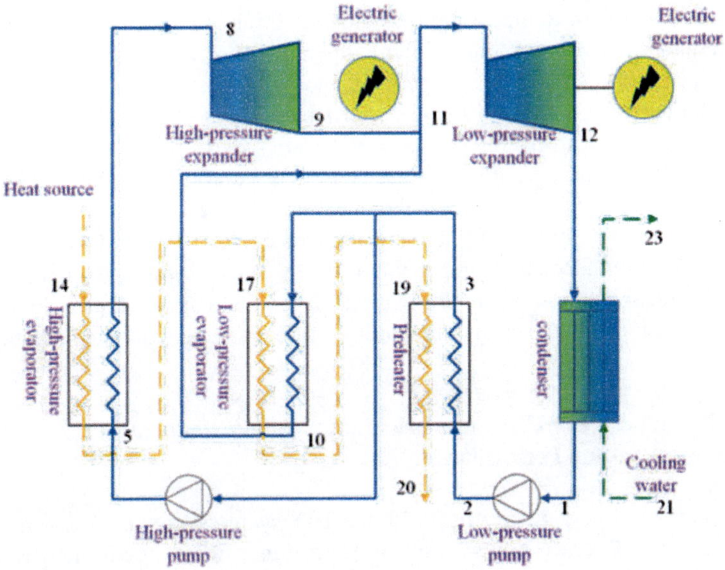

Fig. 12.1 Schematic of the STORC system

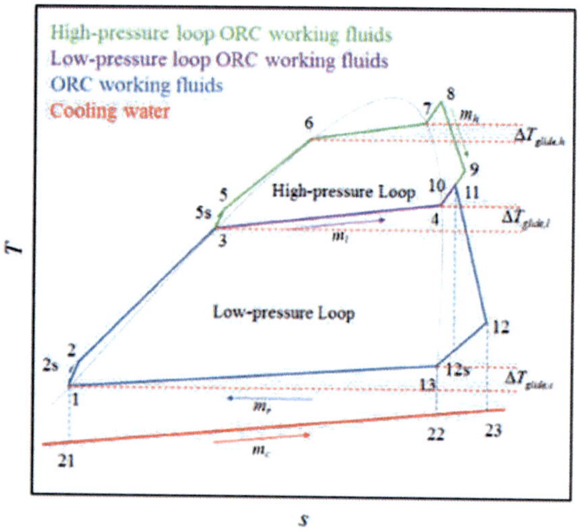

Fig. 12.2 T-s diagram of the STORC system

12.2.2 *Analysis of STORC Operating Parameters*

12.2.2.1 Parametric Analysis

The hot and cold sources in this article remain the same as in the second chapter. For related parameters, please refer to the second chapter. From the above STORC system's flowchart and temperature entropy chart, it can be found that mixed working fluid components, H-P evaporator inlet temperature, L-P evaporator inlet temperature, condensation temperature, superheat, and pinch point temperature difference have important effects on the system's performance. This section analyzes the impact of these six operating parameters on the system's performance. In this section, R245fa, pentane, and their mixtures are chosen as working fluids for the STORC. R245fa and pentane have good thermal properties. Table 12.1 shows the basic parameters of the STORC system. The bubble and dew point positions of the evaporator and condenser influence the comparison of the system's performance under mixed working fluid and pure working fluid. In this paper, the saturated liquid phase point is set to correspond to the evaporation temperature or condensation temperature.

12.2.2.2 Influence of H-P Circulation Evaporation Temperature on System Performance

The change of system's efficiency with the H-T cycle evaporation temperature is shown in Fig. 12.3. The results showed that the tritium efficiency of the three working media showed the same trend. The specific analysis shows that when the H-T circulation evaporation temperature rises, the pinch point of the H-P evaporator moves

Table 12.1 Initial operating parameter of the STORC system

Parameters	Symbol	Value
Heat mass flow (kg/s)	$\dot{m}_g$	1
Heat inlet temperature (K)	$T_{g,\text{in}}$	423
Heat stress (kPa)	P_g	101.325
Cooling water inlet temperature (K)	$T_{\text{c,in}}$	288
Cooling water pressure (kPa)	P_c	101.325
The narrow temperature difference in H-P evaporator (K)	$\Delta T_{\text{pp,h}}$	3–10
The pinch point temperature difference between L-P evaporator (K)	$\Delta T_{\text{pp,1}}$	5
The narrow temperature difference of preheater (K)	$\Delta T_{\text{pp,p}}$	5
The narrow temperature difference of condenser (K)	$\Delta T_{\text{pp,1}}$	5
H-P circulating superheat (K)	$\Delta T_{\text{sup},l}$	1–10
L-P circulating superheat (K)	$\Delta T_{\text{sup},l}$	5
H-P cycle supercoiling (K)	$\Delta T_{\text{sup},h}$	0
L-P cycle undercooling (K)	$\Delta T_{\text{sup},l}$	0
Isentropic efficiency of expander (%)	η_{i}	85
Mechanical efficiency of expander (%)	η_{e}	95
Pump isentropic efficiency (%)	η_{p}	85
The environment temperature (K)	T_0	288

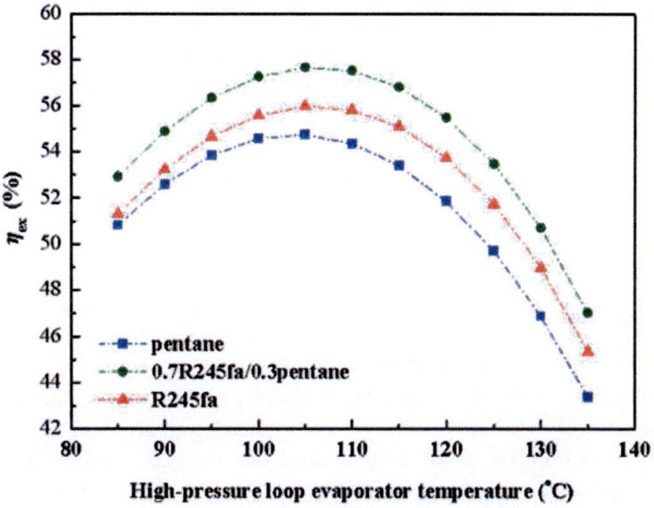

Fig. 12.3 Variations of exergy efficiency with HT-Loop evaporation temperature

to the right, which reduces the mass flow of the H-P circulating working fluid and increases the specific enthalpy of the H-P expander. The overall enthalpy difference of the H-P expander appears first. When the increase is lower, the maximum value of the work done can be found. At the same time, the mass flow of the working fluid in the L-P cycle gradually decreases with the increase of the evaporation temperature. The total working fluid flow decreases first and then increases, and the work of the L-P expander decreases and then increases, but the change is less than for the H-P expander. In theoretical analysis, the power loss of the working fluid pump is negligible relative to the work performed by the expander, so the net output work increases first and then decreases with the rise of H-P evaporation temperature. On the other hand, as the evaporation temperature of the H-P cycle increases, the temperature of the heat source at the outlet of the H-P evaporator rises, and the radon loss of the H-P evaporator decreases, while the radon loss of the L-P evaporator decreases with the mass flow of the L-P refrigerant and the inlet temperature. The temperature of the system gradually increases and gradually becomes the dominant factor of radon loss. Therefore, the total radon loss of the system decreases first and then increases, and finally, the radon efficiency increases first and then decreases. It can also be found in the figure that the radon efficiency of the mixed working fluid is better than that of the pure working fluid, and the radon efficiency under the pentane is the lowest. It is because the existence of the slip temperature that reduces the temperature difference between the cold and heat sources in the heat exchanger. When the H-T cycle evaporation temperature reaches 105 °C, the radon efficiencies of 0.7R245fa/0.3pentane, R245fa, and pentane are 57.67%, 55.94%, and 54.73%, respectively. From the perspective of tritium analysis, it can be known that the STORC system can have better thermal performance using mixed working fluids.

Figure 12.4 shows the relationship between LEC and evaporation temperature under different working media. LEC shows a trend of decreasing first and then increased gradually, and there is a minimum point. This is mainly because when the H-T circulation evaporation temperature rises, the logarithmic average temperature difference of the H-P evaporator decreases and the investment cost increases accordingly. The investment cost of the generator is positively related to the respective output work. The investment cost of the H-P evaporator is increasingly affected by the evaporation temperature. In the end, the LEC decreases first and then gradually increases. The LEC under R245fa is always lower than the other two working fluids, indicating that the economic performance under this working condition which is best followed by 0.7 R245fa/0.3 pentane. When the H-T circulation evaporation temperature is 105 °C, the LEC of R245fa can reach a minimum of 20.26×10^{-2} \$/kWh, and at the same evaporation temperature, the minimum LEC of pentane is 21.63×10^{-2} \$/kWh; and the H-P circulation evaporation temperature is 110 °C, the LEC of R245fa/pentane can reach a minimum of 21.18×10^{-2} \$/kWh. It can be found that the economic performance of the system under pure working medium is better than that under mixed working medium.

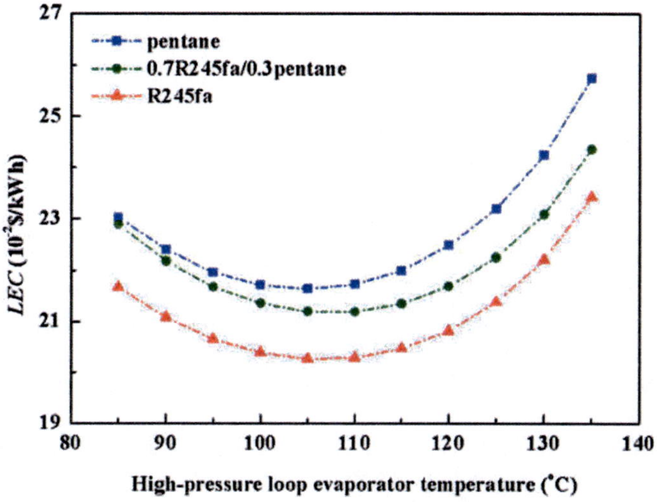

Fig. 12.4 Variations of LEC with HT-Loop evaporation temperature

12.2.2.3 Effect of H-T Cycle Superheat and Pinch Point Temperature Difference on System Performance

Figure 12.5 shows the changes in the efficiency of three kinds of working fluids with the H-T cycle superheat and the pinch point temperature difference. The radon efficiencies, superheating degree, and pinch point temperature difference of the three working fluids maintained opposite trends. This is because when the superheat degree rises, the mass flow of the working fluid decreases, so that the work decreases, and

Fig. 12.5 Relationship between exergy efficiency with HT-Loop superheating degree and pinch point temperature difference

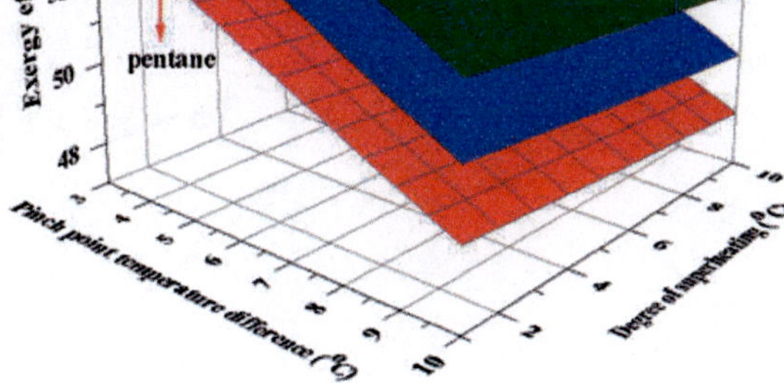

due to the power consumption of the pump which is ignored, so the net output work is continuously reduced. At the same time, we can find that the total loss of the H-P evaporator increases. Eventually, the efficiency of the loss of the H-P evaporator decreases. On the other hand, as the temperature difference at the pinch point increases, the mass flow of the working fluid decreases, resulting in a reduction in the net output power. At the same time, the total loss of the system is increasing, which is mainly due to the increase of the loss caused by the rise in the temperature of the heat source at the inlet of the L-P evaporator, so the efficiency of the loss is decreasing. Under the same superheat degree and pinch point temperature difference, the radon efficiency of R245fa/pentane is better than that of R245fa and pentane, indicating that the thermal performance of pentane is the worst. When the superheat degree is 1 K and the pinch point temperature difference is 3 K, the pentane, 0.7R245fa/0.3pentane, and R245fa can reach the maximum efficiency of 53.62%, 57.63%, and 55.75%.

The change of the LEC of the system with the superheat degree of the H-P cycle and the pinch point temperature difference is shown in Fig. 12.6. It can be seen that the LEC and the H-T cycle superheat degree and the pinch point temperature difference, all have a positive correlation. When H-P cycle superheat is increased, the work done by the high and L-P expanders decreases, which reduces the investment cost. The investment cost of the L-P evaporator also reduces the heat exchange area, although the investment cost of the H-P evaporator has increased. However, the total investment cost has been reduced, and under the combined effect of investment costs and output power, the LEC and superheat degree remain the same trend. The LEC increases with the increase of the temperature difference at the pinch point. The LEC under R245fa/pentane is between two pure working fluids, which means that the best economic performance may not be obtained with mixed working fluids. At the same time, the LEC under R245fa is the lowest and the economic performance is the best,

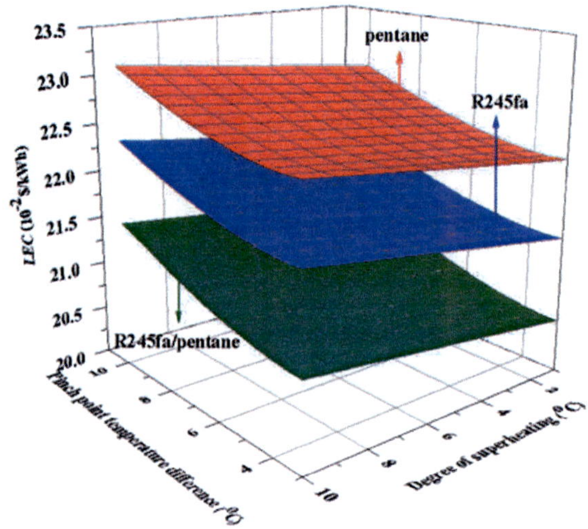

Fig. 12.6 Variations of LEC with HT-Loop superheating degree and pinch point temperature difference

while the pentane has the highest LEC and the economic performance worst. When the superheat is 1 °C and the pinch point temperature difference is 3 °C, the LECs of pentane, 0.7R245fa/0.3pentane and R245fa can reach the lowest values of 22.26 $\times 10^{-2}$ \$/kWh, 21.44 $\times 10^{-2}$ \$/kWh, and 15.50 $\times 10^{-2}$ \$/kWh, respectively.

12.2.3 Thermo-economical Optimization of STORC

In this section, the non-dominated genetic algorithm (NSGA-II) is used to optimize the STORC system for two goals. In order to consider the thermal performance and economic performance of the system, efficiency and LEC were chosen as the objective functions. At the same time, five mass fractions of R245fa, evaporation temperature of H-P cycle, pinch point temperature difference, superheat, evaporation temperature, and condensation temperature of the L-P cycle were selected as the decision variables, and the operating parameters are given in Table 12.2. Genetic algorithm parameter settings are that genetic scale is 100, genetic generation is 200, crossover factor is 0.8, and mutation factor is 0.2. Its Pareto optimal front is shown in Fig. 12.7. Efficiency and LEC maintain the opposite trend. The LEC at point A is 20.10 $\times 10^{-2}$ \$/kWh, and the efficiency is 59.01%, which indicates the best economic performance under point A, and the thermal performance is the worst. The efficiency at point B is 61.39%, and the LEC is 21.91 $\times 10^{-2}$ \$/kWh, which means that the system has the best thermal performance under the condition of point B. The system cannot reach the ideal state in which the highest chirp efficiency and the lowest LEC coexist. On the other hand, the range of variation of the unitary efficiency and LEC is small, which indicates that the system's optimal unitary efficiency and the corresponding operating parameters under the minimum LEC are close.

The solutions of the Pareto optimal front can be used as final results, and the relative order of the Pareto optimal front solutions can be determined according to different decision-making methods. Commonly used decision methods are TOPSIS

Table 12.2 Two-objective optimization results for mixture working fluid

Design parameters	A	B	TOPSIS	LINMAP	Shannon
$E_{sys}(\%)$	59.01	61.39	60.10	60.32	59.07
LEC (10^{-2} \$/kWh)	21.91	20.10	20.41	20.48	20.11
$T_{eva,h}$ (°C)	116.70	114.24	114.58	114.25	116.69
$T_{eva,l}$ (°C)	76.06	76.79	76.85	76.67	76.14
T_{con} (°C)	30.00	30.00	30.00	30.01	30.00
$\Delta T_{sup,h}$ (°C)	3.74	4.26	4.10	4.00	3.74
$\Delta T_{pp,h}$ (°C)	5.07	4.07	4.24	4.08	4.96
r_{245fa}	0.85	0.62	0.77	0.76	0.85

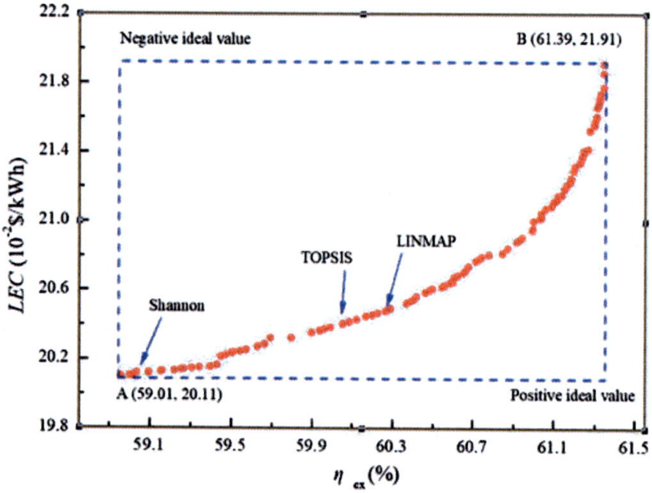

Fig. 12.7 Pareto frontier solution for two-objective optimization

(distance decision of pros and cons), LINMAP (multidimensional preference analysis linear programming decision), and Shannon (entropy decision). Among them, TOPSIS and LINMAP determine the order of the frontier solutions by determining the distance between the frontier solution and the ideal solution by different rules, and Shannon decision can be used to evaluate the system by combining the gray correlation method (GRA) to calculate the gray correlation between the frontier solution and the ideal solution. This section uses these three decisions to solve the optimal solution.

The TOPSIS decision first calculates the distance between each sample and the positive ideal solution and the negative ideal solution and then finds the relative closeness to the positive ideal solution. The greater the relative closeness, the higher the degree of matching with the ideal solution, the better the evaluation result. In the TOPSIS decision, the relative closeness of the individuals in the Pareto front solution is 0.733, 60.10% and 20.41×10^{-2} $/kWh.

After calculating the distance of ideal solution from Euclidean distance, first, calculate the distance between each sample and the positive ideal solution and the negative ideal solution. And then find the distance to the positive ideal solution. The boundary solution with the smallest distance can be used as the final solution. The distance between the individual and the ideal solution in the Pareto front solution set in the LINMAP decision that the best advantage appears at point 53 with a distance of 0.0025. The corresponding Pareto solution has a martingale efficiency and a LEC of 60.32% and 20.48×10^{-2} $/kWh, respectively.

Shannon and GRA decision making first calculate the weighting factors of the optimization objects according to the Shannon method, which intuitively represents the relative importance of the optimization objects. Then through GRA, the correlation coefficient between each Pareto frontal solution and the ideal solution is obtained to obtain the gray correlation value. The larger the gray correlation value, the more the solution and the ideal solution match, so the final solution can be determined based on the gray correlation value. The gray correlation value of the frontier solution and the ideal solution in the Shannon and GRA decision that the best advantage is at point 81 and the gray correlation is 0.761. The Pareto solution corresponds to a martingale efficiency and a LEC of 59.07% and 20.11 × 10^{-2} \$/kWh.

12.2.4 Thermo-environmental Optimization of STORC

By efficiency and LEC double objective optimization, which can be concluded that the thermal performance and economic performance of operation parameters, including working fluid the best mix proportion is 0.77/0.23, because of the working fluid under different mixing proportion of equivalent coefficient of carbon dioxide emissions to the environment impact of indicators of ECE difference. This section will be used only for 0.23 pentane/0.77 R245fa as working fluid, STORC system further optimized, in search of both systems operating conditions of thermal environment performance. Exergic efficiency, LEC, and environmental ECE were selected as the objective functions. H-P cyclic evaporation temperature, superheat and pinch temperature difference, L-P cyclic evaporation temperature, and cold and warm temperature were selected as the decision variables. Genetic algorithm parameter settings are that the genetic scale is 120, the genetic algebra is 300, the crossover factor is 0.8, and the variance factor is 0.2. The optimal Pareto front is shown in Fig. 12.8.

The two-dimensional diagram of Pareto is shown in Fig. 12.9 to better reveal the relationship of the objective function. It can be found that the lowest ECE at point C (57.98%, 20.81 × 10^{-2} \$/kWh, 9.25 kgCO_2eq/kWh) indicates the optimal system environmental performance under this working condition, while exergic efficiency and thermodynamic performance are the worst. Meanwhile, the low LEC indicates optimal economic performance. Exergic efficiency was the highest at E point (62.11%, 21.06 × 10^{-2} \$/kWh, 9.54 kgCO_2eq/kWh), while both LEC and ECE reached the maximum value, indicating that economic performance and environmental performance were decreasing while thermal performance was improved. The LEC at point D (59.17%, 20.66 × 10^{-2} \$/kWh, 9.27 kgCO_2eq/kWh) was the lowest, indicating that the economic performance was the best under the working condition of this point. Both ECE and exergic efficiency were in low positions, indicating that the environmental performance was better, while the thermal performance was poor. Table 12.3 lists the points C, D, and F, as well as the values of the objective function and running parameters under the optimal solution after the decision by TOPSIS, LINMAP, and Shannon. Further calculations show that the migration factors of the optimal solution under the three decisions are 0.283, 0.288, and 0.541, respectively. TOPSIS decision points are selected as the final solution.

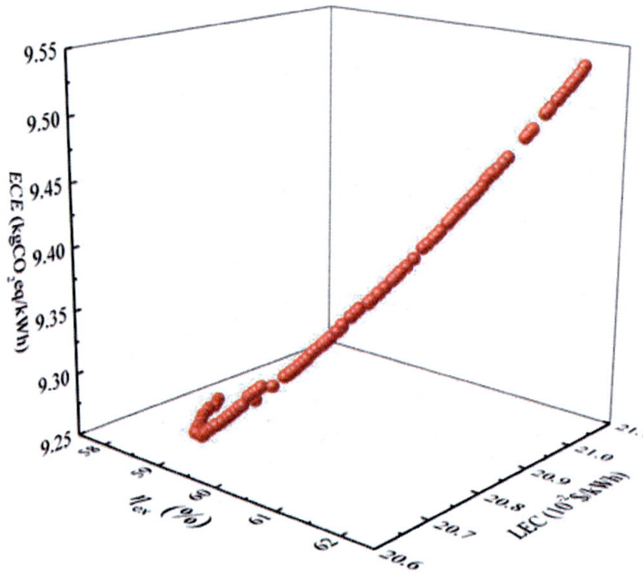

Fig. 12.8 Pareto frontier solution for three-objective optimization

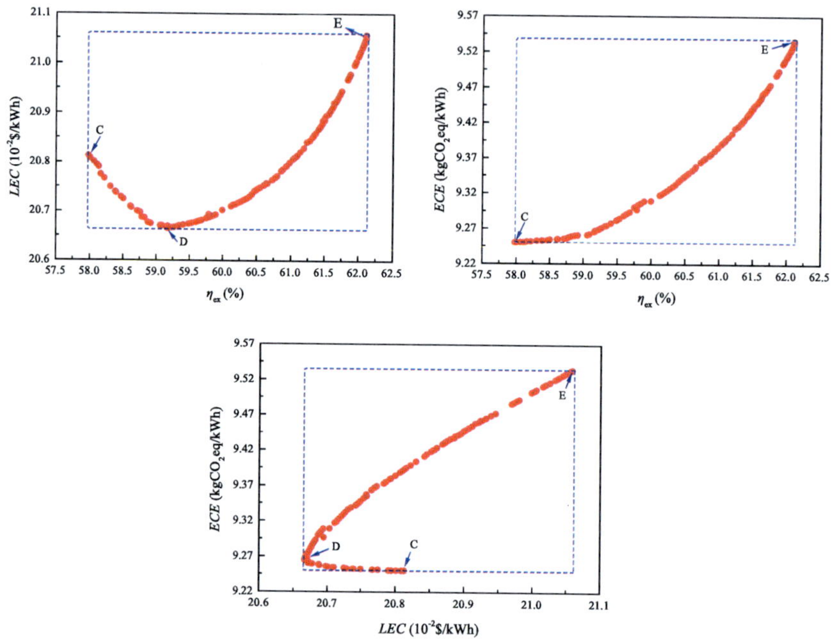

Fig. 12.9 2D graph of Pareto frontier solution for three-objective optimization

Table 12.3 Three-objective optimization results

Parameters	C	D	F	TOPSIS	LINMAP	Shannon
E_{sys} (%)	57.98	59.17	62.11	61.37	61.62	59.16
LEC (10^{-2} \$/kWh)	20.81	20.66	21.06	20.86	20.91	20.67
ECE (kgCO_2eq/kWh)	9.25	9.27	9.54	9.43	9.46	9.27
$T_{eva,\ h}$ (°C)	122.45	113.94	116.89	117.48	117.52	114.08
$T_{eva,l}$ (°C)	78.40	79.34	80.00	80.00	79.99	79.31
T_{con} (°C)	30.00	30.00	30.00	30.00	30.00	30.00
$\Delta T_{sup,h}$ (°C)	1.01	1.00	1.00	1.00	1.00	1.00
$\Delta T_{pp,h}$ (°C)	10.00	10.00	3.00	4.66	4.10	10.00

12.3 Thermo-environmental Optimization of a Cascaded Organic Rankine Cycle (CORC) Using Mixture Working Fluids

12.3.1 System Description

As illustrated in Fig. 12.10, the CORC includes a HT-Loop ORC and a LT-Loop (L–T loop) ORC. The HT-Loop is composed of a pump (pump-1), two evaporators

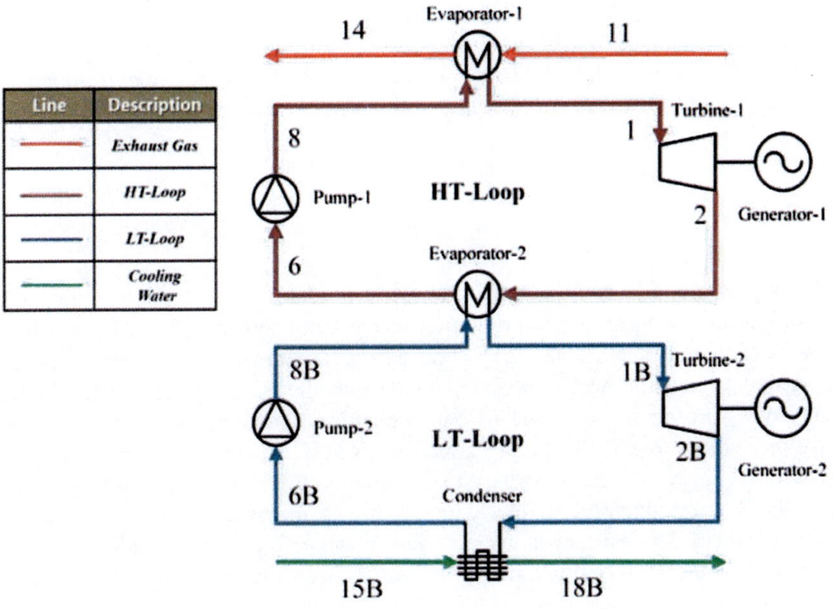

Fig. 12.10 Schematic diagram of the CORC system for L–T waste heat

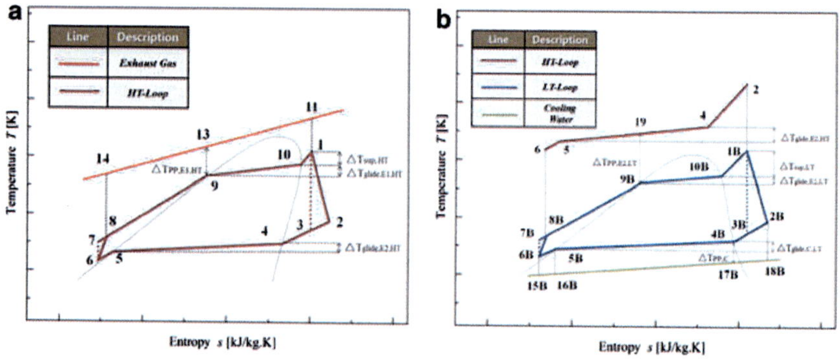

Fig. 12.11 **a** *T-s* diagram of the HT-Loop. **b** *T-s* diagram of the LT-Loop using mixture working fluid

(evparotor-1 and evaporator-2), a turbine (turbine-1), and a pump (pump-1), while the LT-Loop owns a turbine (turbine-2), a pump (pump-2), and a condenser (condenser). It should be noted that the evaporator (evaporator-2) for LT-Loop acts as condenser for HT-Loop.

In HT-Loop, the exhaust gas heated up the working fluid and that working fluid is evaporated in the evaporator-1 and then enters the expander-1 for work. After doing work, the waste steam enters the evaporator-2 and exchanges heat with the LT-Loop working fluid-2. Finally, working fluid is entered in the evaporator-1 by pump for the next HT-Loop. LT-Loop performs a cycle like the HT-Loop, but the heat from HT-Loop.

The key advantage of using mixture working fluid is that pure working fluids can better match the heat sink and heat source. Figure 12.11 shows the corresponding *T-s* diagram of CORC system by using working fluid mixture.

12.3.2 Selection of Working Fluids

The primary criteria of selection of working fluid for the HT-Loop is the high decomposition temperature, representing that it can exert in H-T working condition. Meanwhile, it presents a high boiling temperature at atmospheric pressure to easily convert the heat in working fluids into the LT-Loop with only regular condensing pressures. As to the LT-Loop, the critical temperatures of working fluids take precedence. Meanwhile, R245fa/R227ea, R245fa/R600a, R245fa/pentane, and R245fa/R152a were commonly used to compare with the pure working fluids. R11 and R123 were suggested as the flame retardants for waste heat recovery using mixing with cyclohexane, cyclopentane, and benzene. Therefore, Cyclohexane/R11 is chosen as the HT-Loop working fluid, while those four working fluid mixtures

Table 12.4 Physical properties of working medium

No.	Working fluids	M (kg kmol^{-1})	T_{cr} (K)	P_{cr} (MPa)	$T_{boiling}$ (K)
1	Cyclohexane	84.161	553.64	4.07	353.89
2	R11	137.37	471.11	4.41	296.86
3	R245fa	134.05	427.05	3.65	288.29
4	R227ea	170.03	374.90	2.93	256.81
5	R600a	58.122	407.81	3.63	261.40
6	R152a	66.051	386.41	4.51	249.13
7	Pentane	86.175	507.82	3.03	341.86

(R245fa/R600a, R245fa/R227ea, R245fa/pentane, and R245fa/R152a) are considered as research object for LT-Loop. The detailed thermophysical properties of five nominated mixture working fluids are illustrated in Table 12.4.

12.3.3 *Multi-objective Optimization*

12.3.3.1 Objective Functions and Decision Variables

This study explores thermal economy optimization and thermal optimization for exergy efficiency maximization and LEC minimization. At the same time, the seven system parameters, which includes outlet temperature of evaporator of HT-Loop, condenser temperature of HT-Loop, degree of superheat of HT-Loop, pinch point temperature difference, outlet temperature of evaporator, condenser temperature of LT-Loop, and the mass fraction of R245fa are chose as the decision variables. The list of these constraints is presented in Table 12.5.

In order to examine the effects of operation parameters on system performance, 0.75 Cyclohexane/0.25 R11 is selected for HT-Loop, while four mixture working fluids are chosen for LT-Loop. The effects of six operation parameters (outlet temperature of evaporator of HT-Loop, condenser temperature, pinch point temperature difference of HT-Loop, degree of superheat, outlet temperature of evaporator of LT-Loop, and condenser temperature).

The operational parameters of HT-Loop and LT-Loop have a noteworthy effect on the system's behavior. Meanwhile, the mass fraction of mixture working fluid affects the temperature glide, thereby influencing the match between the heat source/heat sink and the working fluid. Therefore, the parametric analysis on the exergy efficiency and LEC is conducted at first. It should be renowned that the working fluid mixtures with a constant mass fraction are adopted for parametric analysis due to the similar phenomenon which can be obtained for the working fluid mixtures with the other mass fractions.

Table 12.5 Main assumptions for the CORC system

Item	Unit	Value
Heat sources temperature	K	573
Expander isentropic efficiency	[%]	80
Generator efficiency	[%]	90
Pump isentropic efficiency	[%]	80
Pump efficiency	[%]	90
Mass flow of heat sources	kg s^{-1}	0.33
Cooling water temperature	K	283
HT-Loop evaporator outlet temperature	K	513
LT-Loop evaporator outlet temperature	K	333
HT-Loop degree of superheat	K	5
LT-Loop degree of superheat	K	5
Pinch point temperature difference in evaporator-1	K	6
Pinch point temperature difference in condenser	K	5
HT-Loop condenser temperature	K	413
LT-Loop condenser temperature	K	308
Environmental temperature	K	293

12.3.3.2 Effect of Parameters of HT-Loop on Exergy Efficiency and LEC

The relationship between exergy-efficiency, outlet temperature of evaporator of HT-Loop, and condenser temperature of HT-Loop using working fluid mixtures is shown in Fig. 12.12. The HT-Loop evaporator outlet temperature increases from 463 to 513 K, while the HT-Loop condenser temperature increases from 403 to 413 K. Obviously, for those four different working fluid mixtures, a better exergy efficiency is yielded with a higher evaporator outlet temperature or a higher HT-Loop condenser temperature. What should be emphasized that the condenser of HT-Loop acts as an evaporator for the LT-Loop.

As can be seen from the picture, when the outlet temperature of evaporator of HT-Loop keeps rising, the difference of specific enthalpy goes up, whereas the mass flow rate yields a reverse trend. The increasing difference of specific enthalpy is predominant than that of the decreasing mass flow rate, which results the net-power output increase. Meanwhile, with the increase of the evaporator outlet temperature (exergy), failure shows the same trend. Moreover, rising outlet temperature of evaporator of HT-Loop causes a decline in net-power output, and so the exergy efficiency goes down. Among those four working fluid mixtures, 0.5R245fa/0.5R227ea obtains the highest value of exergy efficiency.

The variation of LEC with outlet temperature of evaporator of HT-Loop and condenser temperature of HT-Loop by using working fluid mixtures is illustrated in

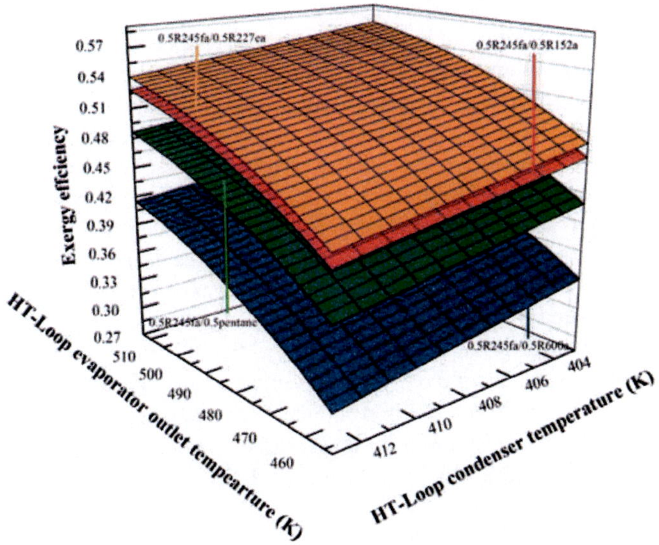

Fig. 12.12 Relationship between exergy efficiency, HT-Loop evaporator outlet temperature, and condenser temperature using mixture working fluids

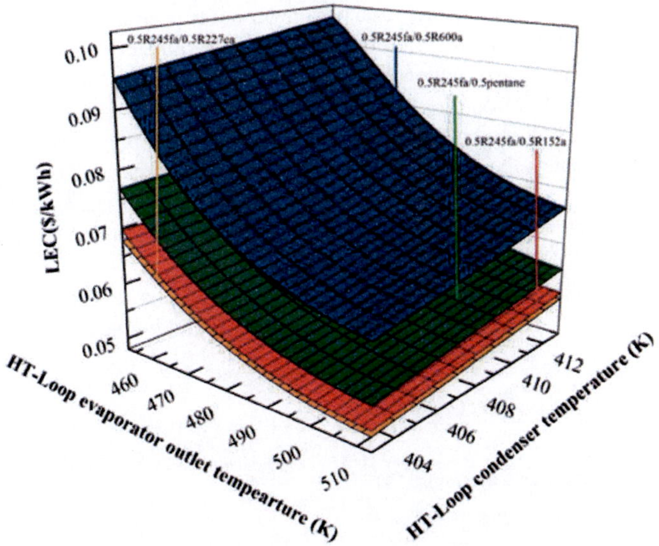

Fig. 12.13 Variation of LEC with HT-Loop evaporator outlet temperature and condenser temperature using mixture working fluids

Fig. 12.13. The LEC of the mixture of the four working media increases with the increase of the temperature of the condenser in the H-T circuit.

The rising outlet temperature of evaporator of HT-Loop decreases the area of heat transfer, resulting in a decline of investment cost in the system. The influence of the descending investment cost of the system and the ascending net-power output comprehensively leads to the descending in LEC at first. With the further rise of outlet temperature of evaporator of HT-Loop, the descending net-power output turns out slowly significant, and thus, the decline of LEC is gradually slowing. Moreover, the increasing condenser temperature of HT-Loop keeps the difference of mean heat-transfer temperature in condensing section of condenser to ascends, causing a decrease in the area of heat transfer of condenser and a decline in the investment cost of the system. However, the influence of the decline of net-power output would outperform that of the decline of system investment cost. Therefore, a rising trend of LEC can be seen with the HT-Loop condenser temperature. 0.5R245fa/0.5R600a yields the highest LEC, followed by 0.5R245fa/0.5pentane, 0.5R245fa/0.5R152a, and 0.5R245fa/0.5R227ea which has the lowest LEC.

12.3.3.3 Thermo-economical Optimization of Maximizing Exergy Efficiency and Minimizing LEC

In order to obtain the optimal operation parameters, it is necessary to carry out dual-objective optimization. Meanwhile, seven operation parameters (outlet temperature of evaporator of HT-Loop, condenser temperature of HT-Loop, pinch point temperature difference of HT-Loop, degree of superheat of HT-Loop, outlet temperature of evaporator of LT-Loop, condenser temperature of LT-Loop, and mass fraction of R245fa) are chosen as the decision variables. The lower bound and upper bound of those seven decision variables are listed in Table 12.6. The parameters of genetic algorithm including generation size, population size, crossover fraction, and migration fraction are set to 100, 60, 0.8, and 0.2, respectively.

The Pareto optimal frontiers of LEC for working fluid mixtures are demonstrated in Fig. 12.14. Obviously, a distinct trade-off can be found between thermodynamic

Table 12.6 Data of the parameter optimization

Decision variables	Lower bound	Upper bound
HT-Loop evaporator outlet temperature (K)	453	513
LT-Loop evaporator outlet temperature (K)	333	363
HT-Loop condenser temperature (K)	403	413
LT-Loop condenser temperature (K)	303	313
HT-Loop pinch point temperature difference (K)	6	12
HT-Loop degree of superheat (K)	2	5
Mass fraction of R245fa	0.1	0.9

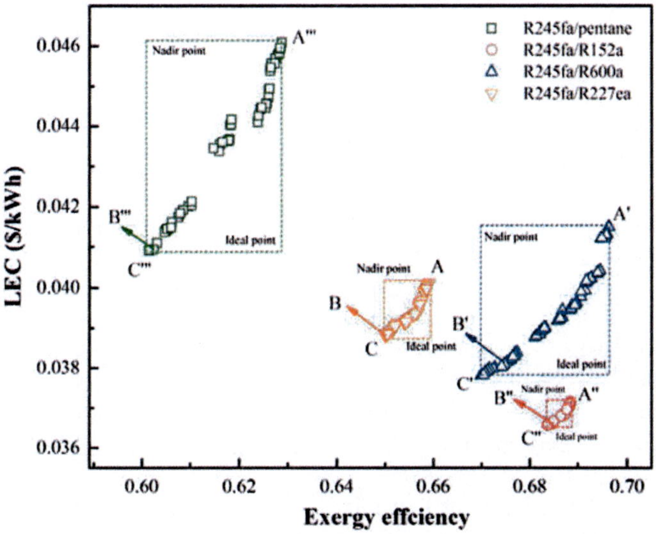

Fig. 12.14 Pareto optimal frontiers of LEC with exergy efficiency using mixture working fluid

and economic performance. As exergy efficiency increases, the LEC starts to ascend slightly at first and then a sharp increase. Taking R245fa/R227ea as an example, the highest thermodynamic performance, which is represented by point A, is 65.91% in exergy-efficiency, but holds the highest economic performance at the same time (0.04 $/kWh for LEC). Meanwhile, the optimal point cannot satisfy both of exergy efficiency and LEC which are optimal, so the TOPSIS decision making is used to get the Pareto optimal solution. The optimal point is illustrated in Fig. 12.14 with the point of B for R245fa/R227ea. It should be mentioned that point A presents the optimum thermodynamic performance, which is obtained by single-objective optimization of maximizing exergy efficiency, while point C presents the optimum economic factors, owning by single-objective optimization of minimizing LEC.

A hypothetical point, which denotes the ideal point, is considered as the best solution to get the optimal point. The ideal point is situated at lower right. From the viewpoint of thermo-economic optimization, a working fluid with Pareto optimal frontier solution located at lower right should take precedence. As observed in Fig. 12.14, R245fa/R152a is the nearest working fluid to the ideal point, while R245fa/pentane is the nearest working fluid to the nadir point. It demonstrates that R245fa/R152a is the best candidate working fluids for CORC LT-Loop.

According to the objective optimization of exergy efficiency, R245fa/R600a has the highest exergy efficiency of 69.63%, which is 5.64% higher than R245fa/R227ea, 1.15% higher than R245fa/R152a, and 10.72% higher than R245fa/pentane. Through the single-objective optimization of LEC, the lowest LEC of 0.036 $/kW h is yielded for R245fa/R152a, which is 5.26% lower than R245fa/R227ea, 5.26% lower than

R245fa/R600a, and 12.19% lower than R245fa/pentane. Meanwhile, the optimum mass fraction is ranging from 38.93 to 69.92%.

12.3.3.4 Thermo-environmental Optimization of Exergy Efficiency, LEC, and Minimizing SEI

Although thermal economy optimization improves system performance, but it not considers the environmental factor. Taking the SEI into account, the tri-objective optimization of maximizing exergy efficiency, minimizing LEC, and minimizing SEI is addressed. It should be noted that those seven operation parameters are selected as decision variables, while maximizing exergy efficiency, minimizing LEC, and minimizing SEI are considered simultaneously. The Pareto optimal frontiers for thermo-environmental optimization mixture using working fluids are displayed in Fig. 12.15. The Pareto optimal solution can be obtained based on the TOPSIS decision making.

Pareto optimal solution of four mixed working media are the points A, B, C, and D, respectively. Pareto optimal solutions for R245fa/R227ea, R245fa/R600a, R245fa/R152a, and R245fa/pentane in pairs of (LEC ($/kWh), exergy efficiency (%), and SEI (kg CO_2eq/m^2)) are (0.038, 65.24, 0.498), (0.036, 69.75, 0.415), (0.037,

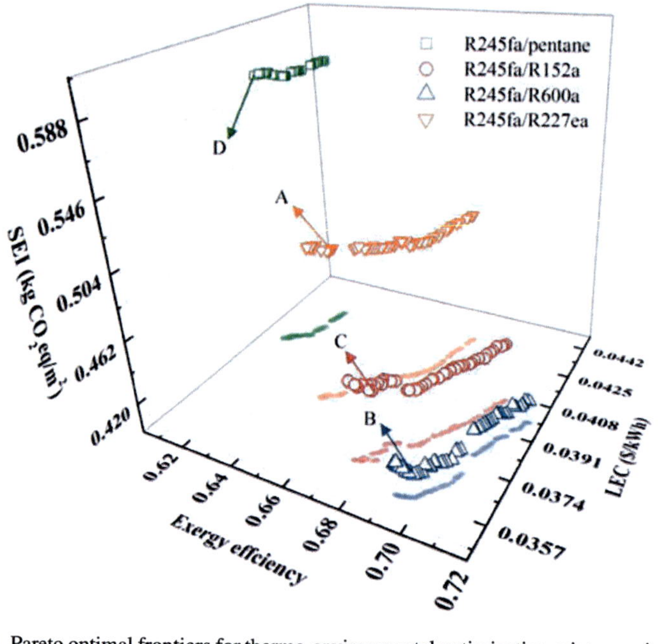

Fig. 12.15 Pareto optimal frontiers for thermo-environmental optimization mixture using working fluids

67.98, 0.446), and (0.041, 61.28, 0.584), respectively. The mass fraction is ranging from 41.64 to 89.11%.

12.4 Experimental Investigation of Heat Exchanger Characteristics on a 3 kW ORC

12.4.1 Experimental Setup Description

As shown in Fig. 12.16, a 3 kW ORC experimental unit is build. The process includes three sub-cycles. The first cycle is the ORC loop, which is the main loop. The cycle consists of five main components, evaporator, expander, condenser, working pump, and generator, as well as valves, pipelines, measuring and control instruments, etc. The low boiling working fluids are heated to a superheated vapor state in the evaporator. This high-pressure vapor is fed to the expander to generate work. The exhaust is cooled in the condenser. The working pump extracts the working fluid from the storage tank and pressurizes it into the evaporator, and the working fluid absorbs heat again in the evaporator. Thereby, working fluid completes the cycle of "heat absorption-work-condensation-pressurization" and simultaneously realizes the

Fig. 12.16 Photograph of experimental setup

conversion process from low-grade thermal energy to high-grade electrical energy. A plunger pump is utilized as the working pump because of its high-volume ratio and small flow rate, and a scroll expander is selected. The second cycle is the heat source loop. In this experiment, the heat source is simulated by means of electric heating conductive oil, and the conductive oil enters the evaporator and transfers heat to the working fluids. The third cycle is the cooling water loop, which exchanges heat with the working fluid by cooling water.

12.4.2 Comparison of Heat Transfer Performance with Various Mass Fraction

Based on the literature review, the comparison of heat transfer coefficients between mixtures and pure working fluid were not well reported. To provide a better guide for the heat exchanger design using mixture working fluids, a detailed discussion on heat transfer performance with various mass fraction is examined in this section.

Figure 12.17 reveals the variation of the evaporator heat transfer coefficients at different R245fa mass fraction. It can be seen from the figure that there is a positive correlation between the two. This shows that R245fa is better for the heat transfer of the evaporator than the R123. R245fa has higher latent heat of evaporation than R123, so R245fa unit mass heat absorption is greater than R123. When the mass fraction of R245fa increases, the heat transfer rate with the heat source will gradually increase, and thus, the heat transfer performance of evaporator is also improved.

Meanwhile, the evaporator heat transfer coefficients are in range of 211.6–273.3 W/m^2 °C for R123, 220.8–334.3 W/m^2 °C for R24fa/R123 mixtures, and 290.6–446.4 W/m^2 °C for R245fa. The evaporator heat transfer coefficients for

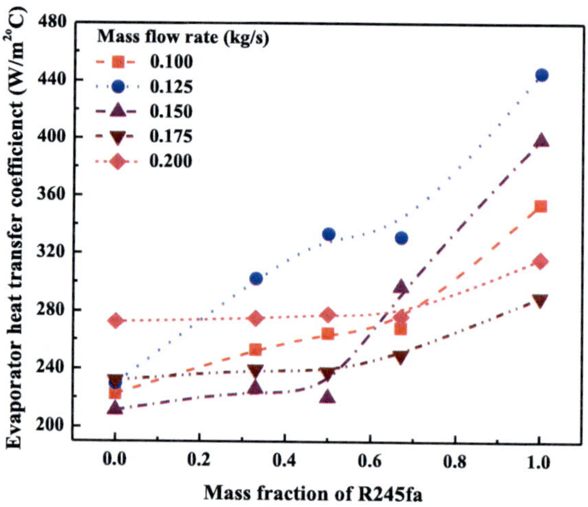

Fig. 12.17 Variation of evaporator heat transfer coefficients with mass fraction of R245fa

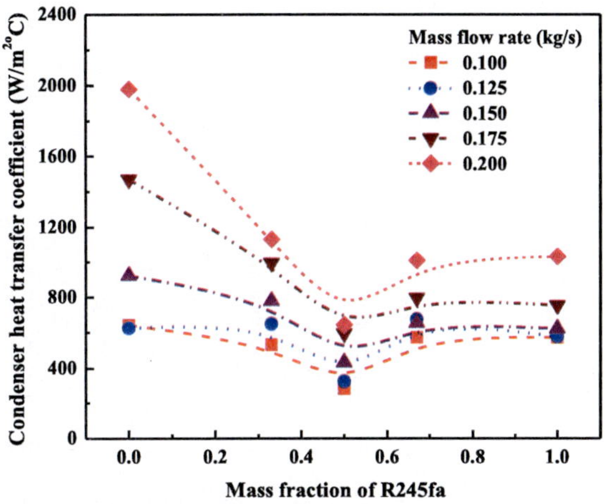

Fig. 12.18 Variation of condenser heat transfer coefficients with mass fraction of R245fa

R245fa/R123 mixtures are 1–45% higher than that for R123, but 12–45% lower than that for R245fa. Meanwhile, R245fa has 16–94% higher evaporator heat transfer coefficients than R123. It can be also found that the mass flow rate has a great effect on evaporator overall heat transfer coefficient for R245fa and R245fa/R123 mixtures, but no obvious effect for R123. The evaporator overall heat transfer coefficient presents a "N" shape trend with mass flow rate.

The variation of the condenser heat transfer coefficients is illustrated in Fig. 12.18. As can be seen from the figure, 0.5R123/0.5R245fa obtains the lowest condenser heat transfer coefficient. The condenser heat transfer coefficients are in the range of 625.3–1978.1 W/m^2 °C for R123, 277.6–1127.1 W/m^2 °C for R24fa/R123 mixtures, and 569.6–1030.0 W/m^2 °C for R245fa. The condenser heat transfer coefficients of R245fa/R123 mixtures (only for part specified mass fraction) are higher than that of R245fa, but lower than that of R123. Taking a mass flow rate of 0.200 kg/s as an example, the condenser heat transfer coefficients decrease from 1978.1 to 639.1 W/m^2 °C and then increase to 1030.0 W/m^2 °C. 0.5R123/0.5R245fa owns a worse condenser heat transfer coefficient, which is 38% lower than that of R245fa and 68% lower than that of R123.

12.4.3 Comparison Between the Experimental Test and Simulation Result Without Considering the Pressure Drop

The ORC theoretical calculation has a larger deviation with the experimental results. One reason is that the simulation ignored the pressure drop and the components' characteristics, resulting in the theoretical analysis cannot better guide the experiment prototype design. In order to ascertain the effect of pressure drop, a detailed

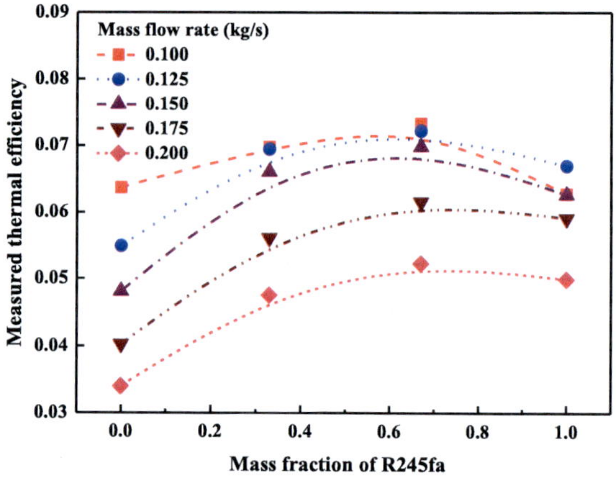

Fig. 12.19 Variation of measured thermal efficiency with mass fraction of R245fa

comparison on the thermal efficiency between the experimental test and simulation result without considering the pressure drop is addressed in this section.

The measured thermal efficiency with R245fa mass fraction is shown in Fig. 12.19. The trend in thermal efficiency indicating that an optimal mass fraction is existed to ensure the maximum thermal efficiency. The measured thermal efficiency is in range of 3.41–6.38% for R123, 4.76–7.33% for R123/R245fa mixtures, and 5.00–6.28% for R245fa. 0.67R123/0.33R245fa yields the highest measured thermal efficiency of 7.33%, which is 15% higher than R123 and 10% higher than R245fa. Meanwhile, a higher mass flow rate owns a smaller measured thermal efficiency. The reason for this is that a higher mass flow rate causes a higher system heat input and pump consumption. The mass flow rate presents insignificant effect on the measured thermal efficiency for R245fa and R123/R245fa mixtures for a mass flow rate smaller than 0.150 kg/s. However, the measured thermal efficiency for R123 has a sharp decrease from 6.38% to 3.41%. The maximum measured thermal efficiency of 0.67R123/0.33R245fa are 7.33%, 7.22%, 6.99%, 6.16%, and 5.24% while the mass flow rate varies from 0.1 to 0.2 kg/s, respectively.

The simulated thermal efficiency without considering the heat exchanger pressure drop is plotted in Fig. 12.20. As can be seen from the figure, the thermal efficiency presents a quadratic curve law with the increase of the mass fraction of R245fa. The simulated thermal efficiency is in range of 11.95–13.02% for R123, 12.86–14.55% for R123/R245fa mixtures, and 10.49–12.64% for R245fa. When the flow rate in the heat exchanger reaches 0.200 kg/s, 0.67R123/0.33R245fa yields the thermal efficiency of 14.55%, which is 13% higher than R123 and 15% higher than R245fa. Meanwhile, lower mass flow rate represents a positive effect on the simulated thermal efficiency. For a mass flow rate higher than 0.175 kg/s, it has a small influence on the simulated thermal efficiency, whereas a significant effect when it is less than 0.150 kg/s. The maximum simulated thermal efficiency of 0.67R123/0.33R245fa are 12.93%, 14.02%, 14.39%, 14.38%, and 14.55%, respectively.

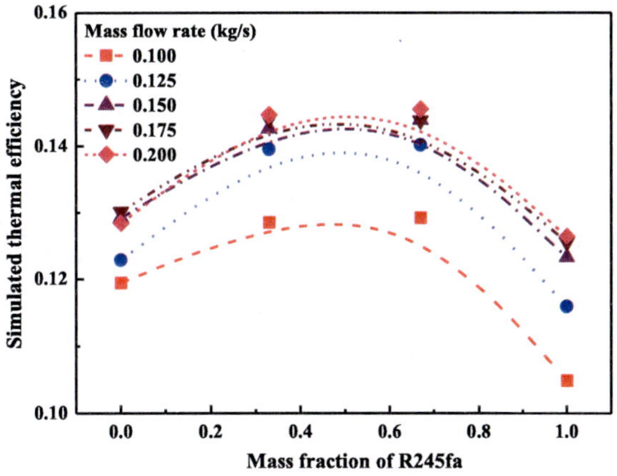

Fig. 12.20 Variation of simulated thermal efficiency with mass fraction of R245fa

To better understand the deviation between the theoretical and experimental results, the measured and simulated thermal efficiency for mass flow rate of 0.100 and 0.200 kg/s are addressed in Fig. 12.21. Obviously, with the increase of the mass fraction of R245fa, both the measured and simulated values of thermal efficiency firstly increase and then decrease. However, the increase of mass flow rate will lead to the decrease of measured thermal efficiency, whereas a reverse trend for the simulated thermal efficiency. Meanwhile, the simulated thermal efficiency is 67–277% higher than the measured one. The simulated thermal efficiency for 0.67R123/0.33R245fa has a 76% higher for mass flow rate of 0.100 kg/s and 178% higher for mass flow rate of 0.200 kg/s than the measured thermal efficiency. It demonstrates that the pressure drop of heat exchanger is significantly related to the system performance and the pressure drop of heat exchanger should be fully considered in theoretical analysis.

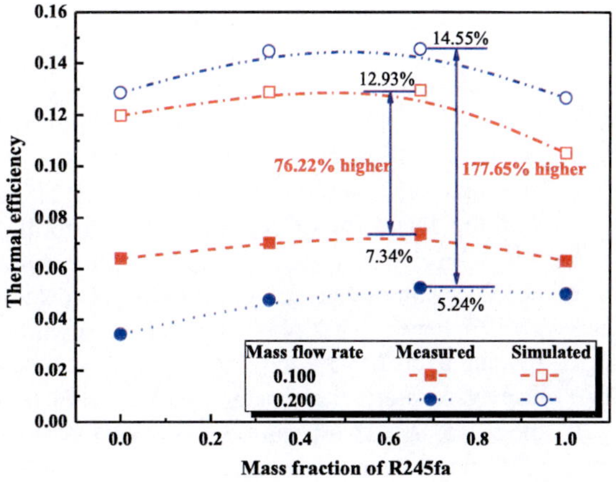

Fig. 12.21 Comparison between the measured and simulated thermal efficiency

12.5 Conclusions

Based on the thermal model and economic model of STORC system, the thermal performance of mixed working fluid R245fa/pentane and pure working fluid R245fa and pentane are understood. The effects of the six parameters including the mass ratio of the mixed working fluid, the H-P circulating evaporation temperature, the superheat, pinch point temperature difference, the L-P circulating evaporation temperature, and the condensation temperature were analyzed. Considering the thermodynamic performance, economic performance, and environmental performance simultaneously, parametric analysis and thermo-environmental optimization of a cascaded organic Rankine cycle (CORC) using mixture working fluids have been investigated. Cyclohexane/R11 is selected as working fluids for HT-Loop. The following four working fluids (R245fa/R227ea, R245fa/R600a, R245fa/R152a, and R245fa/pentane) are the candidate working fluids of the LT-Loop. The heat exchanger performance of different mixtures has been studied and compared in a 3 kW ORC experimental prototype. Five different proportions of working fluids are selected, respectively, R123, 0.33R123/0.67R245fa, 0.5R123/0.5R245fa, 0.67R123/0.33R245fa, and R245fa. The heat transfer characteristics of heat exchangers under different working fluids are studied and compared. The comparison between the experimental test and simulation result without considering the pressure drop is addressed. The main conclusions are as follows:

(1) Considering the thermodynamic performance and economic performance of the system comprehensively, selected exergy efficiency and LEC as the dual-objective optimization goals. Finally, the optimal solution under the TOPSIS decision was selected. Exergy efficiency and LEC were 60.10% and 0.2041 \$/kWh, respectively. The exergy efficiency of Pareto front solution of mixed working fluid ranges from 59.01% to 61.39%, and the maximum exergy efficiency is 1.87% and 1.16% higher than Pentane and R245fa, respectively. ECE was selected as an environmental evaluation index, and the exergy efficiency, LEC, and ECE of STORC system under mixed working fluid were optimized. Finally, TOPSIS decision point (61.37%, 0.2086 \$/kWh, 9.43 kgCO_2eq/kWh) was selected as the optimal solution.

(2) Compared with different working medium from the perspective of thermo-economic optimization, a working fluid with Pareto frontier solution located at lower right should take precedence. R245fa/R152a is the best candidate working fluids for CORC LT-Loop. According to the minimum LEC, the lowest leveled energy cost of 0.036 \$/kW h is yielded for R245fa/R152a, which is 5.26% lower than R245fa/R227ea, 5.26% lower than R245fa/R600a, and 12.19% lower than R245fa/pentane. Meanwhile, the optimum mass fraction is ranging from 38.93% to 69.92%.

(3) The evaporator heat transfer coefficients of R245fa are highest, followed by the R245fa/R123 mixture and while R123 shows the lowest evaporator heat coefficients. And the condenser heat transfer coefficients of R123 are highest and that of 0.5R245fa/0.5R123 are lowest. The measured thermal efficiency

of 0.67R123/ 0.33R245fa is the highest, which is 7.33% and the maximum simulated thermal efficiency is14.55%.

Acknowledgements This work has been supported by the National Natural Science Foundation of China (51806081), the Key Research and Development Program of Jiangsu Province, China (BE2019009-4), the Natural Science Foundation of Jiangsu Province (BK20180882), the Key Research and Development Program of Zhenjiang City, China (SH2019008), the Key Project of Taizhou New Energy Research Institute, China (2018-20), and the Open Foundation Program of Key Laboratory of Wind Energy and Solar Energy Technology (Inner Mongolia University of Technology), Ministry of Education of China (2020ZD02).

References

Aziz F, Roshaan M, Man-Hoe K (2018) Exergetic and heat load optimization of H-T organic Rankine cycle. Energy Convers Manag

Campana C, Cioccolanti L, Renzi M, Caresana F (2019) Experimental analysis of a small-scale scroll expander for L-T waste heat recovery in Organic Rankine Cycle. Energy 187:115929

Dong BS, Xu GQ, Cai Y, Li HW (2014) Analysis of zeotropic mixtures used in hightemperature organic Rankine cycle. Energy Convers Manag 84:253–260

Dong BS, Xu GQ, Li TT et al (2017) Parametric analysis of organic Rankine cycle based on a radial turbine for low-grade waste heat recovery. Appli Therm Eng 126:470–479

Esra OK, Kiliç M (2020) Comparative performance analysis of ORC–VCRC combined systems based on refrigerant selection. Sci Technol Built Environ

Feng YQ, Hung TC, Zhang YN, Li BX, Yang JF, Shi Y (2015a) Performance comparison of low-grade ORCs (organic Rankine cycles) using R245fa, pentane and their mixtures based on the thermoeconomic multi-objective optimization and decision makings. Energy 93:2018–2029

Feng YF, Hung TC, Greg K, Zhang YN, Li BX, Yang JF (2015b) Thermoeconomic comparison between pure and mixture working fluids of organic Rankine cycles (ORCs) for L-T waste heat recovery. Energy Convers Manag 106:859–872

Heberle F, Preißinger M, Brüggemann D (2012) Zeotropic mixtures as working fluids in organic Rankine cycles for low-enthalpy geothermal resources. Renew Energy 37(1):364–370

Huang WC, Chang YF, Yuan YX (2019) Complementary configuration and optimal energy flow of CCHP-ORC systems using a matrix modeling approach. Complexity

Imran M, Muhammad U, Byung-Sik P, Youngmin Y (2016) Comparative assessment of organic Rankine cycle integration for L-T geothermal heat source applications. Energy

Javanshir A, Sarunac N (2017) Thermodynamic analysis of a simple organic Rankine cycle. Energy 118:85–96

Kang HS, Kim MH, Shin YH (2020) Thermodynamic modeling and performance analysis of a combined power generation system based on HT-PEMFC and ORC. Energies

Kim HR, Lee JH, Kim TS (2020) Optimal thermo-economic design of a pafc-orc combined power system. J Mech Sci Technol

Lecompte S, Ameel B, Ziviani D, van den Broek M, De Paepe M (2014) Exergy analysis of zeotropic mixtures as working fluids in organic Rankine cycles. Energy Convers Manag 85:727–739

Li S, Li WY (2018) Thermo-economic optimization of solar organic Rankine cycle based on typical solar radiation year. Energy Convers Manag

Lim TW, Choi YS, Hwang DH (2021) Optimal working fluids and economic estimation for both double stage organic Rankine cycle and added double stage organic Rankine cycle used for waste heat recovery from liquefied natural gas fueled ships. Energy Convers Manag
Meng FX, Zhang HG, Yang FB, Hou XC, Lei B, Zhang L (2017) Study of efficiency of a multistage centrifugal pump used in engine waste heat recovery application. Appl Therm Eng 110:779–786
Renno C, Petito F, Diana D (2020) Modeling of a CPV/T-ORC combined system adopted for an industrial user. Energies
Shu G, Liu L, Tian H, Wei H, Yu G (2014a) Parametric and working fluid analysis of a dual-loop organic Rankine cycle (DORC) used in engine waste heat recovery. Appl Therm Eng 113(1):1188–1198
Shu GQ, Gao YU, Tian H, Wei HQ, Liang XY (2014b) Study of mixtures based on hydrocarbons used in ORC (organic Rankine cycle) for engine waste heat recovery. Energy 74:428–438
Shu GQ, Wang X, Tian H (2016) Theoretical analysis and comparison of Rankine cycle and different organic Rankine cycles as waste heat recovery system for a large gaseous fuel internal combustion engine. Appl Therm Eng 108:525–537
Song J, Gu CW (2015) Performance analysis of a dual-loop organic Rankine cycle (ORC) system with wet steam expansion for engine waste heat recovery. Appl Eng 156:280–289
Sun QX, Wang YX, Cheng ZY, Wang JF, Zhao P, Dai YP (2020) Thermodynamic and economic optimization of a double-pressure organic Rankine cycle driven by L-T heat source. Renewable Energy 147:2822–2832
Tian H, Shu GQ, Wei HQ et al (2012) Fluids and parameters optimization for the organic Rankine cycles (ORCs) used in exhaust heat recovery of Internal Combustion Engine (ICE). Energy 47(1):125–136
Tian H, Liu LN, Shu GQ et al (2014) Theoretical research on working fluid selection for a H-T regenerative transcritical dual-loop engine organic Rankine cycle. Energy Convers Manag 86:764–773
Tian H, Chang LW, Gao YY et al (2017) Thermo-economic analysis of zeotropic mixtures based on siloxanes for engine waste heat recovery using a dual-loop organic Rankine cycle (DORC). Energy Convers Manag 136:11–26
Uusitalo A, Honkatukia J, Turunen-Saaresti T et al (2014) A thermodynamic analysis of waste heat recovery from reciprocating engine power plants by means of organic Rankine cycles. Appl Therm Eng 70(1):33–41
Wang DB, Ma YZ, Tian R et al (2018) Thermodynamic evaluation of an ORC system with a L-P Saturated Steam heat source. Energy 149:375–385
Wang Q, Wang J, Li T, Meng N (2020) Techno-economic performance of two-stage series evaporation organic Rankine cycle with dual-level heat sources. Appl Therm Eng 171, 115078 (2020)
Wang X, Shu GQ, Tian H, Wang R, Cai JW (2020) Dynamic performance comparison of different cascade waste heat recovery systems for internal combustion engine in combined cooling, heating and power. Appl Therm Eng 260:114245
Wang LB, Bu XB, Li HS (2020) Multi-objective optimization and off-design evaluation of organic Rankine cycle (ORC) for low-grade waste heat recovery. Energy
Xiao L, Wu SY, Yi TT, Liu C, Li YR (2015) Multi-objective optimization of evaporation and condensation temperatures for subcritical organic Rankine cycle. Energy 83:723–733
Xu W, Zhang JY, Zhao L, Deng S, Zhang Y (2017) Novel experimental research on the compression process in organic Rankine cycle (ORC). Energy Convers Manag 137:1–11
Yamada N, Tominaga Y, Yoshiba T (2014) Demonstration of 10-Wp micro organic Rankine cycle generator for low-grade heat recovery. Energy 78:806–813
Zhang J, Zhang HG, Yang K, Yang FB, Wang Z, Zhao GY et al (2014) Performance analysis of regenerative organic Rankine cycle (RORC) using the pure working fluid and the zeotropic mixture over the whole operating range of a diesel engine. Energy Convers Manag 84(1):282–294

Zhang XX, Zhang Y, Wang JF (2021) Evaluation and selection of dry and isentropic working fluids based on their pump performance in small-scale organic Rankine cycle. Appl Therm Eng
Zhang T, Li XP, Liu LC (2021) Thermo-economic analysis and optimization of ICE-ORC systems based on a splitter regulation. Energy
Zhou NJ, Wang XY, Chen Z et al (2013) Experimental study on organic Rankine cycle for waste heat recovery from L-T flue gas. Energy 55:216–225

Chapter 13
CO_2-Mediated Energy Conversion and Recycling

Sivakumar Esakkimuthu, Shuang Wang, and Abd El-Fatah Abomohra

Abstract Carbon-di-oxide (CO_2) is inevitable in most of the anthropogenic activities of wider range from household to energy sector. Global CO_2 emissions are always monitored and stringent regulations are constantly being made by different countries to mitigate its level, as the exceedance is linked to severe health problems. Carbon dioxide capture and utilization (CCU) is an effective method to mitigate and balance the ever-raising CO_2 levels in the atmosphere. For instance, CO_2 utilization by microalgal biomass is shown to be significantly effective, which serves the dual purpose of reducing atmospheric CO_2 and effective biomass production. On the other hand, synthetic fuel production by forming isobutanol through sequential utilization of CO_2 along with hydrogen from electrolysis of water is a significant emerging approach. Carbon from waste CO_2 emission can also be effectively utilized to produce polymers like plastic with various innovative technologies in recent years. Thus, the utilization of waste carbon for energy production is vast and scattered. This chapter aims to consolidate the basics, advancements, carbon footprint and promising directions in this field to provide greater insights for cutting down the proportion of CO_2 accumulation in greenhouse emissions.

Keywords CO_2 utilization · Greenhouse gases · Microalgae biomass · Carbon footprint

S. Esakkimuthu · S. Wang (✉)
New Energy Department, School of Energy and Power Engineering, Jiangsu University, Jiangsu 212013, China

S. Esakkimuthu
Algal Technology Program, Center for Sustainable Developments,s College of Arts and Sciences, Qatar University, Doha, Qatar

A. E.-F. Abomohra
Department of Environmental Engineering, School of Architecture and Civil Engineering, Chengdu University, Chengdu 610500, China

A. E.-F. Abomohra et al. (eds.), *Waste-to-Energy*,
https://doi.org/10.1007/978-3-030-91570-4_13

13.1 Introduction

Carbon-di-oxide (CO_2) is a carbon atom covalently linked to two oxygen atoms and is a colourless and acidic gas. Although the level of CO_2 in the earth atmosphere is traced, its constantly raised by anthropogenic activities. Though, autotrophs such as plants and algae require CO_2 for synthesizing their sugars, the CO_2 levels in urbanized environment is dramatically increasing (Claassens et al. 2018). CO_2 is one among the greenhouse gases which potentially restricts the escape of heat from the environment and leads to global warming (Robert and Kummert 2012). CO_2 emission contributes to more than 50% of the total greenhouse gases and the emission is raised by 1.9% annually for the past three decades (Parry et al. 2007). Such huge raise is mainly attributed to the extended usage of fossil fuels due to economic growth and energy consumption increase in last three decades. To be specific, the major sources of CO_2 emission are industries such as power plants, iron and steel industries, oil industries, cement industries, burning of fossil fuels, some chemical production processes and deforestation and others (Sharifi 2020).

According to IPCC, the global CO_2 emission will have a steep increase of up to 90% by 2030, among which emissions due to energy production will increase by 110% which is relatively huge, requiring stringent control (Pachauri et al. 2014). Such raise in emission and consequential effects on climatic change and human health result in many policy changes and regulations across the globe to devise controlling measures. Kyoto protocol in 1997 devised the goal for reducing the greenhouse gas (GHG) emission to 5.2% from 2008 to 2012, later extended to 2020. However, the countries signed in this protocol contributed to only 15% of the total GHG emissions. In contrast, developing countries such as India, China, Brazil, Mexico Philippines, South Africa, etc., contribute more than 40% of the total global GHG emissions (IEA 2011). The major issue with mitigation of CO_2 in developing countries is the inevitability of growing urbanization (Nejat et al. 2015). However, the lack of control over CO_2 emission leads to environmental and economic disaster and hence the countries are urged to devise policies for balancing the growth along with mitigation. Nejat et al. 2015 demonstrated the energy consumption from buildings perspectives such as residential and industrial buildings in which energy consumption is about three quarters in residential buildings as compared to industrial buildings. In last 40 years, energy consumption by both building sectors were increased by about 1.8% every year due to rapid urbanization (Nejat et al. 2015). It was also suggested that such energy consumption from residential sector could effectively be solved and energy efficiency improvisation could be achieved. As linked to energy consumption, these building sectors are also responsible for one third of global GHG emission with CO_2 as major contributor. Hence, effective management of residential energy consumption could potentially reduce the CO_2 emission. For effective management CO_2 emitted need to be captured and converted to other energy forms which can lead to the environmentally and economically lucrative process of "Waste to energy". Wastes are examined worldwide for energy production to make the overall process lucrative and one such recent example is utilizing Fat, Oil and grease (FOG) for biodiesel production (Abomohra et al. 2020).

There are various ways to utilize waste carbon into potential energy forms which include the production of useful products such as plastics, microbial biomass production and relevant microbial products (fuels, nutraceuticals, pharmaceuticals, etc.). To achieve this, CO_2 capture is vital and is achieved in two ways (1) Carbon capture sequestration (CCS) and (2) carbon capture utilization (CCU), both differing with the fate of captured CO_2. For instance, CCS is storing the captured CO_2 for long term use whereas CCU is utilizing carbon dioxide directly for producing useful products. Among these, CCU is found to be useful and helpful for circular economy and helps in mitigating the CO_2 levels in the atmosphere (van Heek et al. 2017). On the other hand, conversion of CO_2 is energy intensive as it is thermodynamically stable, however, the production of useful chemicals and fuels is advantageous. Hence, this chapter aims in elaborating on the concept of CO_2 utilization and subsequent production of energy or products.

13.2 Basics for CO_2 Utilization

The global emission of CO_2 is more than 13,000 Mt y^{-1} (million tonnes per year) whereas the utilization is only about 200 Mt y^{-1} which inevitably demand larger utilization of CO_2 for its considerable abatement (Rafiee et al. 2019). CO_2 utilization can be broadly classified into two which are direct utilization of CO_2 and conversion of CO_2 into chemicals or fuels (Huang and Tan 2014). In direct utilization, CO_2 is directly used as reacting solvent whereas in conversion usage, CO_2 is converted into fuels and chemicals Fig. 13.1. For instance, dry ice, carbonated drinks, fire extinguisher and welding medium are used as small-scale direct applications of CO_2 which leads to lesser mitigation of CO_2. There are some other potential direct applications of CO_2 as an enhanced geothermal system (EGS), enhanced oil recovery (EOR), enhanced gas recovery (EGR) and enhanced coal bed methane recovery (ECBM) which are large-scale direct applications of CO_2. In direct or physical application, CO_2 is directly involved and does not undergo any cracking, whereas, in conversion, it gets reacted in chemical process and forms Carbon monoxide and carbon by which it enhances the respective process (Muradov 2014). CO_2 can act as the primary constituent for many of C_1 containing compounds and hence the application could be vast. The conversion methodologies require energy intensive process by employing reactants such as hydrogen and peroxides along with efficient catalysts for cracking this thermodynamically stable CO_2. The conversion of CO_2 is usually done using mineralization or reduction processes as it has high oxidation state (Nocito and Dibenedetto 2020). Urea is one great example to produce from CO_2 which lead to greater consumption of about 120 Mt y^{-1} followed by methanol production (2–3 Mt y^{-1}) respectively (Zangeneh et al. 2011). In conversion processes, CO_2 can be directly fixed in organic substrate which requires a lesser amount of heat (less than 150 °C) whereas, in fuel production processes, CO_2 is reduced to C_1

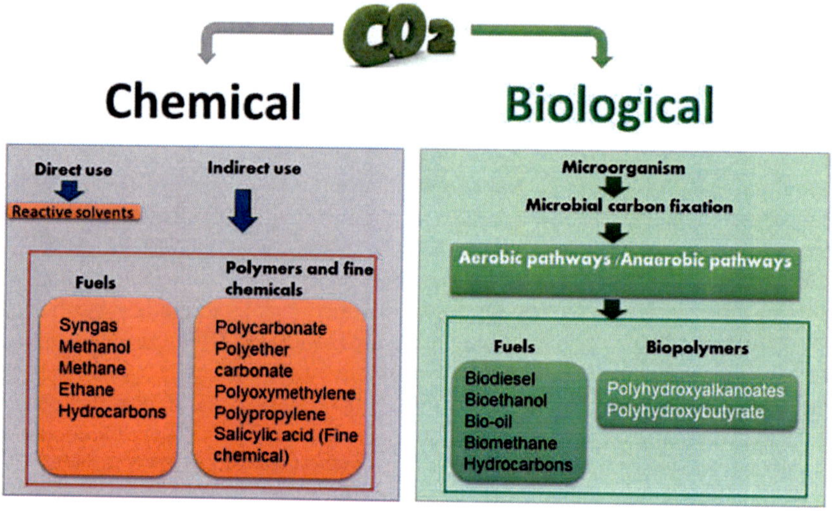

Fig. 13.1 Routes of CO_2 utilization and conversion

which is facilitated at higher temperatures of up to 700 °C (Aresta and Dibenedetto 2003). CO_2 can be utilized in two ways that are direct incorporation under low temperature for the production of carbamates, carbonates, isocyanates, carboxylates, etc. whereas, at higher temperatures, CO_2 can be reduced for the production of carbon monoxide, methanol, hydrocarbon, oxylates, formaldehyde, hydrocarbons, etc. (Aresta and Dibenedetto 2003). Hence, the choice of reaction mode depends on the product synthesis and methane is considered to be the familiar production because of its higher CO_2 consumption rate.

13.3 CO_2 Based Fuel Conversion

13.3.1 Syngas Production

Syngas production is one of the major examples of indirect conversion of CO_2, produced by dry reforming process. It's a mixture of varying proportions of hydrogen, carbon monoxide or carbon-di-oxide. The main reaction mechanism of syngas formation includes dry reforming of methane followed by adsorption and dissociation reactions. The reaction steps include the adsorption of methane over the metal part of the catalyst in dissociated form and lead to the production of H_2 and hydrocarbons. The hydrocarbons and hydrogen species are attached to the metal support and hence hydrogen species are recombined to form hydrogen molecules resulting in desorption of gas phase. Consequently, dissociation of CO_2 occurs as a result of spillover of hydrogen molecules. As a result of CO_2 dissociation, oxidation of hydrocarbon species occurs and further dissociation leads to the production of carbon monoxide and hydrogen molecules (Cheng et al. 2001). Hence the metal support and catalyst play critical roles in the overall reaction and so many catalysts have been reported. The common catalyst used are metal catalysts (Ir, Pt, Pd, Ni, Co, etc.) with oxide supports (MgO, TiO_2, SiO_2, Al_2O_3, etc.) (Wittich et al. 2020). Noble catalysts are efficient whereas Ni-based catalysts are on par with noble catalysts and relatively cost effective. Similarly, Co-based catalysts are also exhibiting better activity than Ni catalysts and are preferred widely (Bouarab et al. 2004). Both acidic and basic metal oxides are employed, among which basic metal oxides are desirable because of effective CO_2 deposition over catalyst. The major challenge with this noble metal acid catalyst is formation of coke (carbonaceous deposition) which results in catalyst inactivation (Bradford and Vannice 1999). Hence, basic catalyst supports are better for adsorption and dissociation of CO_2 and thereby prevents coke formation. CeO_2 is one of the excellent catalysts which generates oxygen species and prevent coke formation (Stroud et al. 2018). Apart from such challenges, this dry reforming process has greater advantages such as utilization of CO_2 along with natural gas for syngas production, production of syngas with high ratio of CO/H_2 and has the ability to utilize the natural gas with CO_2 impurities (Ravanchi and Sahebdelfar 2020). CO_2 mediated energy conversions, especially into fuels are concentrated globally at large scale and few of them are displayed in Table 13.1.

13.3.2 Methanol

Methanol has multiple applications such as fuel, fuel blend and for constituents of methanol conversion to olefins and gasoline. It is also used for the production of formaldehyde and energy storage material. Compared to petrol, its density is half with superior combustion characteristics and is therefore attractive for fuel applications. However, the major consumption of methanol is for formaldehyde and olefin production. The practice of methanol production from CO_2 was first initiated by Lurgi through setting a plant in 1994 (Ashok et al. 2019). Methanol can be produced from CO_2 in two ways, (1) direct hydrogenation of CO_2 to methanol and (2) reduction of CO2 to CO followed by hydrogenation to methanol. In addition, both processes generate larger amount of water resulting in catalyst inactivation. In general, the two-step process was efficient and give higher yield than the one-step process. There are several catalysts identified for the hydrogenation of CO_2 mainly involving copper and zinc with various modifiers (Ce, Ti, Si, Ga, Si, Al, etc.) (Arena et al. 2007). Economically, the cost of CO_2 conversion seems to be a prime factor and hence efforts towards applying waste CO_2 are highly appreciable and pave the way for sustainable fuel production. In Iceland, methanol production plant was set in 2010 by carbon recycling international which produced about 3000 ty^{-1}. This plant utilized 18 tons of CO_2 to produce 10 tons of methanol and the source for carbon is from aluminum industries and H_2 from renewable electricity through water electrolysis. This sets an example for recycling CO_2 and more than 5000 ty^{-1}

Table 13.1 Industrial conversion of CO_2 into fuels

Fuel product	Reaction process	Reactant source	Scale	Industry/institute	Country	References
Methanol	Direct hydrogenation	H2 was supplied by photoelectrolysis using solar energy	Pilot plant 100 ty^{-1}	Mitsui Chemicals Inc. and Research Institute of Innovative Technology for the Earth	Japan	Samiee and Gandzha (2019)
Methanol	Direct hydrogenation	CO_2 obtained from aluminum steel plant; H_2 electrolysis of alkaline water from renewable grid electricity	3000 ty^{-1}	Carbon Recycling International (CRI)	Iceland	CRI—Carbon Recycling International (n.d.)
Methanol	–	H_2 produced from alkaline electrolyzer and renewable electricity was used	–	BSE engineering	Denmark	Schweitzer (2017)
Methanol	Reverse water gas shift	–	75 mty^{-1}	Korean Institute of Science and Technology (KIST)	Korea	Centi et al. (2013)
Methane	Sabatier reaction	H_2 was produced from electrolysis using sun or wind energy	Pilot scale	PtG alpha plants	Germany	Rönsch et al. (2016)
Methane	CO_2 methanation	H2 was supplied by electrolysis	Commercial scale 1000 ty^{-1}	Etogas Company	Germany	Rönsch et al. (2016)

(continued)

Table 13.1 (continued)

Fuel product	Reaction process	Reactant source	Scale	Industry/institute	Country	References
Methane	CO_2 methanation	Heat produced in the reaction used for electrolysis for H_2 production	–	Haldor Topsøe	Denmark	Rönsch et al. (2016)
Methane	CO_2 methanation	H_2 production by electrolysis using solar energy	–	Sendai Technology Institute	Japan	Habazaki et al. (1998)
Methane	CO_2 methanation	CO_2 obtained from flue gas and renewable hydrogen was used	–	RCO_2AS	Norway	Centi et al. (2013)
Hydrocarbons C_{2+}	Fisher-Tropsch reaction	–	1 million ty^{-1}	Shenhua group	China	Amouroux et al. (2014)
Hydrocarbons C_{2+}	Fisher-Tropsch reaction	Generated heat used for electrolysis	–	Sunfire	Germany	Zhu (2019)
Hydrocarbons C_{2+}	Fisher-Tropsch reaction	Direct air capture followed by electrolysis by clean electricity	–	Carbon Engineering	Canada	Carbon Engineering|Direct Air Capture of CO_2|Home (n.d.)

of CO_2 was recycled. In this system, CO_2 obtained from aluminum industries was subjected to a conditioning section where impurities are removed and pure CO_2 for methanol production was obtained. CO_2 feedstock and H_2 were mixed at a ratio of 1:3 followed by reaction under compression. Such methanol production is highly exothermic and the resulting heat was subjected for distillation in downstream unit. This downstream unit is involved in producing the methanol with high fuel grade eligible to blend with gasoline. Such efficient production of methanol from waste CO_2 was patented as Emission to liquids (ETL) (Stefansson 2015). Another methodology of CO_2 mediated methanol production was demonstrated by BSE engineering. Renewable electricity and alkaline electrolyzer were used in this process to generate hydrogen production to which the captured carbon after purification was mixed in specific ratio. The reaction was carried out in methanol synthesis reactor and the heat generated was used for purifying the methanol produced. About 64% of methanol and 36% of hydrogen was obtained in the resulting product which was further purified in distillation section to produce high purity (99.85 wt%) methanol. This production process demonstrated by BSE engineering is highly flexible and the concept of power plant to methanol was envisioned last year (2019–2020) (Schweitzer 2017).

Another strategy of CO_2 mediated methanol production was developed by Korea institute of science and technology, Korean Institute of Energy and Research and Korea gas corporation. The process involves the production of methanol by hydrogenation through reverse gas shift which is named CAMERE (Choi and Cho 2008). The process involves reverse water gas shift reaction on catalyst ($ZnAl_2O_4$) followed by water removal and methanol synthesis on $Cu/ZnO/ZrO_2/Ga_2O_3$ catalyst. Requirement of two different catalysts and reactors was the main disadvantage but higher productivity was achieved because of intermediate water removal (Park et al. 2004). There are potential challenges in methanol production from CO_2 due to its low reactivity. Such lower reactivity results in a lesser yield compared to CO mediated methanol production. In addition, water generation was higher in CO_2 mediated methanol production which results in inhibiting the reaction kinetics. In addition, methanol production from pure CO_2 reduced the production of number of biproducts such as ketones, ether and higher alcohols. However, such limited number of biproducts helps in keeping the purification process simpler (Ravanchi and Sahebdelfar 2020).

13.3.3 Methane

Production of methane from CO_2 is accomplished by CO_2 methanation in two ways. The first method involves, associative adsorption of CO_2 followed by reaction with adsorbed hydrogen resulting in the formation of oxygenates. Hydrogenation results in methane formation and the process does not have any CO intermediates. In a second way, dissociation of CO_2 into CO and O followed by sequential hydrogenation of CO results in methanation (Aldana et al. 2013). The reaction of methane production from CO_2 is almost completing the equilibrium conversions. For instance, more than 90%

of conversions are possible at highly optimized conditions. As the reaction process is exothermic, lower temperature is preferred for CO_2 methanation. To attain more than 90% of conversion, temperature below 300 °C is preferred along with ambient pressure (Sahebdelfar and Ravanchi 2015). Also, a suitable catalyst is required for methane production with higher selectivity towards methane is preferable. Most of the metal groups are the preferred catalysts with nickel and cobalt as good choices because of its cost effectiveness (Li and Wang 2004). Nickel catalysts are proven to exhibit high selectivity towards methane and high activity at lower temperatures. Here also, certain mechanisms (poisoning, fouling) results in catalyst deactivation which can be prevented by using additional metal in catalyst preparation (Rönsch et al. 2016).

The process of CO_2 methanation started in 1970s and Germany was found to be the pioneer in CO_2 methanation process since the country aimed to make 100% renewable energy systems. Plenty of reactors for CO_2 methanation were established and are in use. Germany developed an attractive process called power-to-gas (PTG) generation in which renewable electricity and hydrogen generation was done and used to convert CO_2 thermodynamically to produce methane (Rönsch et al. 2016). Similarly, a project named E-gas was started where hydrogen was produced by electrolysis and used for methane production. The project produced methane by utilizing CO_2 obtained from biogas production and power for running the whole plant obtained from wind energy (EtoGas Power-to-Gas—Hitachi Zosen Inova AG n.d.). Another joint project was initiated between Zosen Corp and Daiki Ataka Engineering Corp, PTTEP and Hitachi, where solar energy was used for running the methane production plant in which hydrogen was obtained from electrolysis of sea water (Habazaki et al. 1998). Similarly, another renewable attempt was made in Norway, where CO_2 was obtained from flue gas along with renewable hydrogen for producing methane which in turn was used for chemical industries to keep lower carbon footprint (Centi et al. 2013). One major challenge with CO_2 methanation is to manage the generated heat and hence fixed bed reactors with gas recycling and cooling systems are used. In a project initiated by TREMP in Denmark, the heat generated from methanation reaction was used in electrolysis unit (Zhu 2019).

13.3.4 Hydrocarbons (C_{2+})

There are two ways of producing hydrocarbons from CO_2 in which cleavage of carbon-hydrogen bond results in forming CO which can be further converted to methane along with an intermediate of adsorbed carbon. In another route, CO_2 is converted to carbonates followed by hydrogenation to produce hydrocarbons (Aldana et al. 2013). The production of hydrocarbons (C_{2+}) and methanation differ in having coupling barrier for C–C for generating hydrocarbons (C_{2+}). Another important aspect in hydrocarbon production is maintaining the appropriate ratio of carbon and hydrogen in the medium as excess hydrogenation results in methane production which in turn affects the further CO_2 conversion process (Gnanamani et al. 2016).

Fischer–Tropsch (FT) reaction is one common mode for hydrocarbon production which is an exothermic process and conventionally involves CO as carbon feedstock. In FT process, involvement of both CO and CO_2 together as feedstock results in faster rate of CO hydrogenation due to competitive adsorption. This shows CO hydrogenation is the primary route and CO_2 hydrogenation is the secondary route (Visconti et al. 2009). The catalysts employed for hydrocarbon production should be capable of catalyzing both Fischer–Tropsch reactions and reverse water gas shift reactions and it was observed that high rate of Fischer–Tropsch reaction could lead to higher conversion of CO_2. Fe based catalysts are preferred for this process because it enhances the carbon- carbon bond formation which is one of the rate limiting step of hydrocarbon production (Li and Wang 2004). The major challenges with identifying catalysts for hydrocarbon production includes presence of water as biproduct that can cause catalyst poisoning and presence of CO_2 that affects CO hydrogenation (Pan et al. 2010). A comparative study showed that Fe based catalysts are efficient than CO catalysts (Dorner et al. 2010). Hence, finding suitable catalysts is one emerging area in related research. In Germany, Sunfire process used Fischer–Tropsch for the production of hydrocarbons from syngas which was produced from co-electrolysis of water and CO_2. As it's an exothermic reaction, heat generation was further used in vaporization of water in electrolysis with an efficiency of around 70% (Renewables Everywhere—Sunfire n.d.). In Canada, a company (Carbon Engineering) demonstrated a project of "air to liquid" in which CO_2 is separated from atmospheric air which is allowed to react with hydrogen atom. Renewable electricity was used in electrolysis of water for producing the hydrogen atom. (Carbon Engineering|Direct Air Capture of CO_2|Home n.d.). This thermodynamical reaction with appropriate catalyst resulted in syngas production which is then converted into hydrocarbons. Similarly, air to liquid projects were carried out widely and were able to produce hydrocarbons with good efficiency but the cost of overall production seems to be disadvantageous.

13.4 Polymers—CO_2 Based Plastics

Alike above-mentioned fuels, CO_2 plays exceptional role in polymer production too and some widely produced polymers in industries are displayed in Table 13.2. There are two ways to conversion of CO_2 into polymers. In direct mode, CO_2 can be used as comonomer for the production of polymers such as polyurea, polycarbonates, etc. In another route, CO_2 is converted into monomer like urea from which polymers are produced (Langanke et al. 2015). Application of epoxide such as epichlorohydrin, Polypropylene carbonates, etc. in copolymerization of CO_2 for polymer production is common (Klaus et al. 2011). The exact process of copolymerization includes opening of ring in epoxide to give metal alkoxide followed by carbon-di-oxide insertion in which metal halide facilitates the ring opening and process is termed as consecutive coordination and insertion. The ring opening is achieved by the presence of metal complex with epoxides attacking the nucleophile group such as halide

Table 13.2 Industrial CO_2 conversion into polymers

Polymer product	Scale	Industry/institute	Country	References
Polycarbonates	1.07 Mt y^{-1}	Asahi Kasei Corporation	Taiwan	Zhu (2019)
Polycarbonates	150 kt y^{-1}	Chimei Asai plant	Taiwan	Aresta and Dibenedetto (2003)
Polyether carbonates polyols	6 Mt y^{-1}	Bayer Material Science and Bayer Technology Service, the RWTH University of Applied Science	Germany	Quadrelli et al. (2011)
Polyetherpolycarbonate polyols	5000 t y^{-1}	Covestro	Germany	Solution Center Covestro|Covestro AG (n.d.)
Polypropylene carbonate	25 kt y^{-1}	Nanyang Zhongju Tianguan Low Carbon Technology Co	China	Quadrelli et al. (2011)
Polycarbonate	150 t y^{-1}	Chimei-Asahi Corp	Taiwan	Brunner (2005)
polypropylene carbonate	3 kt y^{-1}	National Offshore Oil Corporation	China	Quadrelli et al. (2011)
Polyols	10 kt y^{-1}	Huasheng Polymer Co	China	Liu and Wang (2017)

or carbonate and the reaction is called coordination and the reaction is initiated. This metal alkoxide gets inserted with CO_2 resulting in the formation of metal carbonate which in turn acts as place for another coordinated epoxide and successive metal oxide formation. The above-mentioned process repeats for the entire process of polymerization (Trott et al. 2016). Metal-porphyrin and Zinc Phenolate are the familiar homogenous catalysts involved in polymer production from CO_2. Among heterogenous catalysts, metals and metal oxides are the preferred catalysts due to its cost effectiveness and higher activity. CeO_2 based catalysts are very highly active and efficient in producing polycarbonates from carbon-di-oxide and polyols (Arbeláez et al. 2019). For producing polymers that are macroporous, supercritical CO_2 was used in which change in CO_2 density could alter the porosity of the polymer whereas, the change in temperature and pressure helps in removal of CO_2. Asahi Kasei Corporation, an industry in Japan, produces polycarbonate by using ethylene oxide, CO_2 and bisphenol. This process was found to be more efficient as the polymers were produced with higher purity and nullified the cost of purification. In 2007, 150 kt of polycarbonate was produced by the co-operation of Asahi Kasei and Chi Mei industries in Taiwan was achieved (Fukuoka et al. 2007). Another industry in Germany produced polyether carbonates polyols by utilizing about 20% of CO_2 as feedstock in which the catalyst preparation was more significant with nullifying byproducts

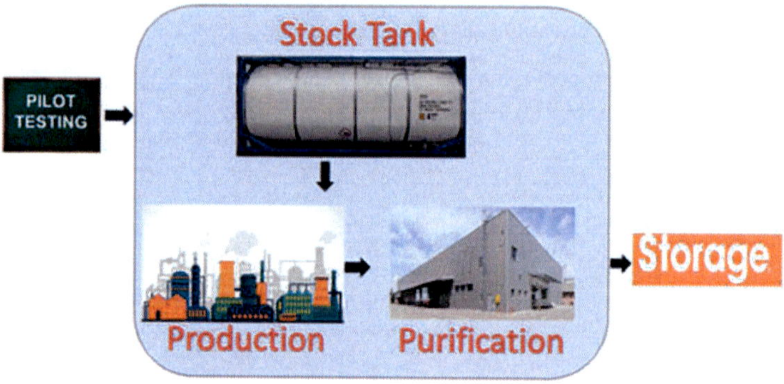

Fig. 13.2 Contour of polymer producing company in Henan (Xu et al. 2018)

and the company was able to produce 5000 ty^{-1} products (Solution Center Cove-stro|Covestro AG n.d.). In the USA, polypropylene carbonate polyols were produced with 50 wt% of CO_2 by a company called Novomer in 2004. CO_2 was supplied from ethanol fermentation industries and proven to be more economical. For instance, the above produced polyols were cheaper than the conventionally produced polyols and were more ecofriendly (Home Page|Novomer n.d.).

In China, polypropylene carbonate was produced from the copolymerization of propylene oxide and CO_2 in the presence of zinc glutarate catalyst by Nanyang Zhongju Tianguan Low Carbon Technology. About 10 kt y^{-1} production capacity for polyol production was constructed in Nantong, China by Huasheng Polymer (Quadrelli et al. 2011). Thus, worldwide production of polymers from CO_2 is emerging in a greater fashion for a couple of decades and contour of polymer producing industries are illustrated in Fig. 13.2. The major bottleneck with this consecutive coordination and insertion using epoxide is lesser availability of epoxides and ecotoxic production process.

13.4.1 Major Advantages of CO_2 Based Plastics

The synthetic plastics such as polystyrene, Polyvinyl chloride, Polyethylene, polypropylene, etc., are produced from fossil-based petroleum products which result in the carbon footprint formation of about 4–6 tons of CO_2 for the production of one ton of plastics (Dormer et al. 2013). Such carbon footprint can be significantly reduced with CO_2 based plastics. For instance, production of polypropylene carbonate using CO2 and propylene oxide results in 40% lesser consumption of petrol-based feedstock and is relatively cheaper as propylene oxide is cost effective chemical (Lee and Cyriac 2011). Now, even better strategies are devised with usage of complete renewable reactants for plastic production.

The resistance and durability of CO_2 based plastics are superior which includes mechanical, thermal and scratch-resistance. Aliphatic CO_2 based plastics is one of the best examples of such kind which has wider potential applications in biomedical and electrochemical industries (Feng et al. 2012).

Biodegradable polymers used for biomedical applications nullify the procedure of removal after implanting with its biodegradable nature which in turn avoids another round of surgical procedures (Nair and Laurencin 2007).

On the other hand, electrolytes made of CO_2 based polymers show some magnificent interfacial stability for ionic passage with lithium metals. As it contains higher proportions of oxygen, it helps in enhancing salt dissociation and exhibits higher level of thermal decomposition temperature (Yu et al. 2010).

13.4.2 Applications of CO_2 Based Plastics

CO_2 based polymers produce greater barrier properties, as it is because of the fact that the poly propylene carbonate (PPC) contains hydrophobic and polar carbonate linkage resulting in strong cohesion which produces tight molecular packing density resulting in higher gas barrier properties (Lee et al. 2013; Seo et al. 2011). Similarly, PPC based composite film shows reduced rates of oxygen transmission, water absorption and vaporization. To enhance the barrier properties, clays are added to PPC along with plasticizers (Guan et al. 2007; Zhai et al. 2015). Similarly, foam is another application that has wider advantages such as good soundproof, low density and conductivity with higher specific strength. PPC with advantages such as impact resistance and co-friendly property makes it more suitable for preparing foam materials (Guan et al. 2007, 2006). Similar to conventional electrolytes (organic solvents) used in batteries, PPC also showed excellent properties such as good interfacial contact and compatibility with lithium ions (Zhou et al. 2013). One challenge with PPC material is that it loses the mechanical strength soon and hence butadiene rubbers are mixed to give proper mechanical strength through crosslinking polymeric framework (Huang et al. 2015). Aliphatic CO_2 based copolymers with glycidyl ethers are now preferred as it provides greater heat resistance and ionic conductivity (Konieczynska et al. 2015). Similarly, CO_2 based polymers have wider potential applications in biomedical applications. For instance, polycarbonates showed exceptional application potential in controlled delivery of drugs, protein and stent (Bege et al. 2012, 2011; Chu et al. 2013a, b). These biomedical applications need greater in vivo compatibility which is achieved by poly ethylene carbonate (PEC). For instance, implantation of PEC made products showed very lesser inflammation and wound healing responses. It was also observed that in vivo degradation of implanted products is achieved by surface erosion and not by any form of enzymatic hydrolysis. The duration of degradation purely depends upon the molecular weight and is roughly between 15 and 30 days (Acemoglu et al. 1997). The surface hydrophilicity of PEC based nanoparticles is advantageous in in vitro experiments performed with macrophages. The biocompatibility and biodegradability of aliphatic polycarbonates are attributed for its supreme applications such as organ implant, 3D scaffolds, bone repair, etc. (Bege et al. 2011). The improvisation needed in hydrophobicity of PPC for the betterment of cytocompatibility is required. The cell vitality rate was considerably high with scaffolds made of PPC suggesting it as prominent material for scaffold in tissue engineering (Zhong et al. 2012).

13.4.3 Recycling of Plastics

Plastics are used worldwide with many potential applications and most of it is made of synthetic or partially synthetic organic substances. Plastics with its flexible nature can be converted to many forms ranging from commodities to containers (Thompson et al. 2009). However, the non-biodegradable nature of plastics is highly worrying and effective management or recycling is highly essential. To date, the effective management of plastics is not implemented fully throughout the world. Inappropriate handling of plastic wastes and disposal led to many devastating effects on the environment and lives. Inappropriate waste management and recycling results in blockage of sewage pipes, fresh water bodies and non-degradable nature of plastics keeps it in the soil for a larger while and affects the soil quality and eventually agriculture too (Chae and An 2018). Plastics are heavily consumed and consequent heavy metal accumulation was observed in invertebrates. Since plastics can enter multiple ecosystems including land, water and sea, it adversely affects the food web (Lwanga et al. 2017). Hence, appropriate management of plastic wastes is the need of this ever developing and plastic producing world. Global attention for this plastic management are provided timely and many policies were devised especially from developed countries. However, developing countries found it difficult to manage plastic waste as growth and developments are their sincere goals and are reluctant in waste management. In developing nation like India, plastics below 50 microns were strictly prohibited in 2016 and manufacturing of those were abandoned by a stringent regulation on plastic waste management by the country. However, it was not fully implemented as many manufacturers urged about the amendment and the ban was not successful because of the amendment in 2018. Some provincial governments of India came with a rule of avoiding plastics that are having single use in 2019. At present, it was reported that about 60% of reusable plastics are recycled in India and comprehensive management for plastics is yet to be implemented (Gopinath et al. 2020). Avoiding plastics pose greater challenges as finding alternates is even challenging as it has wide spread applications. Hence, effective management of plastics is the vital option which is recycling that minimizes the newer synthesis and devising strategies for converting it into value-added products. Among plastic management technologies, wood plastic composite is vital recycling in which plastics are converted to a form with good characteristics and wider acceptance (Winandy et al. 2004). The other mode of plastic waste management includes pyrolysis, hydrothermal liquefaction and blending with cement and road which is illustrated in Fig. 13.3.

Fig. 13.3 Recycling of plastics into various useful products

Pyrolysis is the process of application of high temperature without air which facilitates the thermal degradation of organic substances resulting in bio-oil, biochar and biogas production. This is one predominant mode of fuel production from a variety of substances and plastics found to be an attractive feedstock for the same. Pyrolysis of plastics is considered a tertiary mode of recycling which is proven to be an effective method for waste plastic management. Alike pyrolyzing other feedstock, type of reactors plays critical roles in pyrolyzing plastics. Similarly, the role of catalysts is also immense in determining the product qualities of the pyrolysis of plastics. The products obtained on pyrolyzing plastic solid wastes entirely depends upon the characteristic of the feedstock, choice of catalyst and types of reactors. The products are majorly sectioned as oil, char and biogas (Till et al. 2018). The oil obtained can be used as fuel or fuel blend for gasoline which shows properties on par with standards. The liquid fraction contains aromatics, olefins, paraffins, propane and isoparaffins (Miandad et al. 2017). Char obtained from pyrolysis of plastics majorly contains fixed carbon and volatile carbon. However, some characteristic observations were made that surface has pore formation when co-pyrolyzed with biomass (Jamradloedluk and Lertsatitthanakorn 2014). There were contrasting reports on gaseous yield of pyrolyzing plastic wastes that in some cases it has lower yield than liquid whereas its vice versa in some reports (Basu 2010). The major proportions of gaseous product from pyrolyzing plastic wastes are methane, ethane, butane and propane (Çepelioğullar and Pütün 2013).

In addition, liquefaction is another important strategy used to produce energy and plastics are attractively employed. Plastics are also combined with biomass as a process of co-liquefaction which produces bio-oil, biogas and biochar with better quality. Hydrothermal liquefaction without biomass can also be done but is less common compared to co-liquefaction with biomass. The liquid product predominantly contains carbons which give better properties to the fuel (Yang et al. 2016). One challenge with hydrothermal liquefaction is excess polymerization which results in coke formation, hence co-liquefaction with plastics tend to reduce the formation of coke which in turn increases the bio-oil content (Christensen 2014). Apart from these two modes of fuel conversions, plastics are effectively managed by utilizing it for making roads, concrete, etc. The conversion of plastics to liquid tar is achieved by the process of gasification in the presence of catalyst (Kato et al. 2003). The process of conversion does not need to produce tar with greater fuel characteristics (Saha et al. 2018). Hence the process is so efficient with optimized reaction conditions serving dual purpose of plastic waste management and useful product synthesis. Such strategy of using plastics in making roads provide dual advantages of effective management of plastics and quality road with good integrity. It was practised in many countries (Australia, USA, India, Indonesia and Britain) and using plastics reduced the proportion of sand in plastic mixed constructions. The compressive properties of plastic mixed concrete were excellent and on par with conventional concrete (Nkwachukwu et al. 2013).

The above mentioned are the major management techniques that are prevalently in use across the world and the associated environmental concerns are very important and require appropriate attention. The greenhouse gas emission such as CO_2 and CO are majorly produced during handling of these waste plastics. For instance, pyrolyzed plastic produces fuel which on combustion emits CO_2 which is of great concern. Hence carbon capturing is highly essential to further reduce the carbon levels in the atmosphere. In liquefaction, requirement of water is the major concern as the reaction requires water in higher quantities (Gopinath et al. 2020). The gases such as CO, CO_2 and methane produced from plastic waste management practices severely increased the pollution burden which in turn requires appropriate management. The requirement of land and the reusability of the land after plastic management is a major concern. Similarly, the quantity of plastics used is important, as large-scale management could potentially manage larger amounts of plastic wastes that are dramatically increasing in today's world.

13.5 Microbial CO_2 Fixation and Conversion

Carbon fixation by microorganisms is an effective process that helps to control the CO_2 levels in the atmosphere. In addition, these microorganisms are well exploited for utilizing CO_2 and production of value-added products and which seems more efficient as compared to the energy intensive process of catalytic conversion of CO_2 (Claassens et al. 2018). Microorganisms such as microalgae and cyanobacteria are the autotrophic groups that potentially use CO_2 for photosynthesis. The most commonly exploited microalgal and cyanobacterial species for CO_2 fixation includes *Scenedesmus obliquus, Chlorella pyrenoidosa, Chlorella vulgaris, Thermosynechococcus longates, Rhodovulum viride*, etc. (Du et al. 2019; García-Cubero et al. 2018; Liang et al. 2019; Ma et al. 2019). Ample of research has been carried out and reported in enhancing the microalgal CO_2 fixation potential and achieved relevant insights. The critical factors for enhancing the Carbon fixation includes engineering the cultivation conditions such as temperature, pH, nutrient concentrations, CO_2 feeding rate, reactors, etc. (Morales et al. 2018). In addition to photosynthetic microorganisms, there are non-photosynthetic microorganisms which are bacterial species under the phyla Thermodesulfobacteria, actinobacter, proteobacteria, Chloroflexi, aquificae, Chlorobi and firmicutes. The bacteria using CO_2 are either methanogens, producing methane or sulphur reducing bacteria. These CO_2 assimilating bacteria are isolated from soil, hot springs, ocean, soil and marine sediments (Hicks et al. 2017). On looking over the status of exploration of CO_2 fixing microbes, photosynthetic microbes are well studied as compared to non –photosynthetic microbes.

13.5.1 Mechanism of Microbial Carbon Fixation

CO_2 fixation in microorganisms is facilitated by biosynthetic pathways involving catalyzation by a cascade of enzymes. There are six known pathways for carbon fixation, Calvin-Benson-Bassham (CBB) cycle, 3Hydroxyproppionate bicycle, the 3-hydroxypropionate/4-hydroxybutyrate cycle, the reductive tricarboxylic acid cycle, reductive ecarb-CoA pathway, ecarboxylate/4-hydroxybutyrate cycle (Jajesniak et al. 2014). Among these, the first three pathways are aerobic and rest are anaerobic. The requirement of ATP for these CO_2 fixation pathways varies based on the pathway and higher CO_2 fixations obtained with lower ATP consumption pathways. An alternate pathway for Calvin-Benson-Bassham cycle was proposed which showed good kinetic rates and efficient CO_2 fixation in autotrophs and is identified as malonyl-CoA-oxaloacetate-glyoxalate pathway. It was also showed that it has very lesser ATP consumption with higher oxygen tolerating enzymes (carboxylases) and with better catalytic turnover rate than other pathways in autotrophs. Ample of metabolic engineering efforts are proposed to rewire central metabolism like replacing CBB cycle with alternative pathways, dividing Rubisco's catalysis with several enzymes can better enhance the CO_2 fixation efficiency (Salehizadeh et al. 2020). However, the major challenge is in vivo implementation of alternative pathways in host cells which includes localization, stability, activity and expression levels. Many metabolic engineering attempts are constantly made by researchers across the globe to increase carbon fixation. In such regard, prominent pathways for CO_2 fixations are constantly getting proposed using metabolic engineering, protein engineering and synthetic biology (Bar-Even 2018; Bar-Even et al. 2010; Hing et al. 2019). Genetic engineering plays a great role in enhancing photosynthesis such as improving the efficiency of Ribulose-1,5-bisphosphate carboxylase oxygenase, blocking competitive pathways, reducing photorespiration, increasing the carbon flux for enhancing CO_2 fixation.

Genome sequencing and genome scale models are genetic engineering tools that are paving ways for CO_2 fixating ability and for finding alternate pathways. The metabolites such as acetyl CoA and pyruvate play key roles in the synthesis of value-added products. Therefore, with potential advancements in synthetic biology, metabolic engineering and genetic engineering tools, variety of products can be produced in higher quantities (Zhou et al. 2016; Erb and Zarzycki 2016; Kanno et al. 2017). CO_2 mediated production of bioenergy and products are discussed as follows.

13.5.2 *Biomethane*

Biogas comprises about 50% of methane and is produced through anaerobic digestion or fermentation of biodegradable substances. In biomethane production, CO_2 is biologically reduced with H_2 as an electron acceptor in anaerobic state in bioreactors. The solubility of H_2 in water is less than CO_2 and hence requires more energy to accomplish the reaction process which is considered as one of the major challenges of the process. In such case, most of the H_2 is constantly agitated to get dissolved which is an energy demanding process (Claassens et al. 2018). It was observed that methane production was greatly increased by agitation of continuous stirred tank reactor. About 10% of energy consumption for agitation made this biomethane production process less energy efficient than catalytic methane production process. Thus, the source of energy supply for the reaction process is inevitable and can be supplied in the form of ATP, NADPH derived from bioelectrocatalysis, chemicals (organic/inorganic), electricity or even light. Hydrogen produced from electrolysis of water and reducing equivalents produced from gasification and pyrolysis of organic wastes are used as energy supply for the process. (Claassens et al. 2018).

Both plants and microalgae are used for biomethane production and among which microalgae are superior in terms of higher yield and biomass production. About 30 times higher yield could be achieved with microalgae over plants and some predominant microalgal species used for biomethane production includes *Phaeodactylum tricornutum, Chlorella vulgaris, Chroococcus* sp., *Chlamydomonas reinhardtii*, among which biomethane production of about 0.587 m^3 kg^{-1} VS was achieved with *Chlamydomonas reinhardtii* (Pires et al. 2012; Prajapati et al. 2013; Mussgnug et al. 2010). One advanced technology in biomethane production is using electrodes to give electrons for microbes by CO_2 reduction in anaerobic degradation and such "electron transfer from electrode to microbe" showed profound increase in methane yield (Batlle-Vilanova et al. 2019; Cai et al. 2016). A process called biogas upgradation was followed to remove CO_2 and other impurities to attain the enriched natural gas. Biogas contains varying proportions of methane, carbon dioxide, hydrogen sulphide and hydrogen. Hence upgradation is critical and microalgae are therefore employed in such process in which microalga *Scenedesmus obliquus* was co-cultivated with fungi which results in the upgraded biogas with maximum methane concentration of 90.35% vv^{-1} (Wang et al. 2017). Apart from microalgal processes, alternate bioelectrochemical technology is shown to be effective in which water electrolysis generated hydrogen and CO_2 is biologically converted to methane (Wang et al. 2019; Xu et al. 2014). However, commercial challenges exist with upgradation process too as microbial conversion efficiency needs to be improved to reduce the overall cost. To address commercial challenges, Biopower to gas strategy was followed in which H_2 was injected in situ in biomethanation plant which was initiated by commercial biogas producers in Germany (Salehizadeh et al. 2020).

13.5.3 Hydrocarbons

The important petroleum constituents are hydrogen and carbon which can be synthesized by microorganisms in their natural form as well as engineered form. Wild or engineered strains of fungi, cyanobacteria and eubacteria can synthesis renewable hydrocarbon from CO_2. However, in most biofuel production, cost effectiveness is one of the major challenges due to lower yield from microbial strains (Fu et al. 2015; Halfmann et al. 2014). CO_2 can be directly converted into alkanes and alkenes by cyanobacteria which got wider attention in recent past. Aar-ado and *ols* are the two reported native biosynthetic pathways for synthesizing hydrocarbons. In Aar-ado pathway, fatty acyl-ACP is reduced to fatty aldehyde by fatty acyl—ACP reductase followed by conversion of fatty aldehyde to alkanes/alkenes by fatty aldehyde carboxylase. A tool like metabolic engineering is exploited to increase the hydrocarbon production in microbes. For instance, expression of AAR pathway increased the olefin of heptadecane (4.2 $\mu g\ g^{-1}$) in *Synechococcus elongates* PCC7942 (Kang and Nielsen 2017). In another study, expression of ols pathway increased the nanodecadiene and nanodecane in cyanobacteria (Xie et al. 2017). A study reported that over expression of *Sy-efe* gene from *Psuedomonas syringe* resulted in the elevated production of 0.07–0.26 mmol g^{-1} DCW h^{-1} of ethylene along with the 2% usage of CO_2 (Zavřel et al. 2016).

13.5.4 Organic Acids

Organic acids are produced from carbon-di-oxide and hydrogen and the best examples are succinate, acetate, 3-hydroxypropionate lactate and butyrate (Amulya and Mohan 2019). For instance, 4 gL^{-1} of organic acid constituting formic acid and acetic acid were produced in *Clostridium ljungdahlii* from 7 pressure bars of CO_2 and H_2 (Oswald et al. 2018). Similarly, production of succinate and lactate was enhanced by 11 and 46 times in *Synechocystis* due to the presence of K^+ in anaerobic conditions (Ueda et al. 2016). Similarly, 3-hydroxypropionic acid was enhanced by the overexpression of genes such as malonyl-CoA reductase and malonate semi-aldehyde in *Synechocystis longates*. Organic acids have various potentials such as it can act as precursors for beverages, food, polymer and pharmaceutical industries. However, CO_2 fixation is not a rate limiting step in all the cases. For instance, 3-hydroxypropionic acid production was enhanced by the expression of malonyl-coA reductase and acetyl coA carboxylase genes in *Synechocystis* species showing that CO_2 fixation was not a rate limiting step in hydroxypropionic acid production (Wang et al. 2016).

13.5.5 Lipids

Fatty acids and lipids are the most demanded material for biodiesel production. Lipids in triglycerides form are the most preferential form of biodiesel production and fatty acids and glycerol are the building blocks of all kinds of lipids. Apart from biofuel sector, lipids have wider applications such as chemicals and food industries. Microalgae act as a better feedstock for lipids and fatty acids which was explored in every aspect such as energy and nutraceuticals. Microalgae with photosynthetic capability capture CO_2 and convert it into fatty acids. Transesterifying the fatty acids with alcohol results in the formation of fatty acid methyl esters (FAME) called biodiesel which is considered as promising alternate for fossil-based petrol and diesel (Esakkimuthu et al. 2020). Fatty acid synthase (FAS) are the enzymes involved in the synthesis of fatty acids from malonyl-CoA followed by chain elongation on Acyl carrier proteins and finally released by thioesterase through hydrolysis (Angermayr et al. 2015). Both microalgae and cyanobacteria are well explored in this lipid enhancement platform through metabolic engineering tools. An interesting study was made in cyanobacteria *Synechocystis elongates* in which ester synthase gene was heterologously expressed and phosphoketolase was introduced which resulted in the enhancement of fatty acid alkyl esters (biodiesel) from CO_2 (Lee et al. 2017a). In another study, 5% vv^{-1} of CO_2 was subjected to *Chlorococcum littorale* which in turn resulted in the production of 47% of intracellular lipids (Ota et al. 2015). In *Chlorella vulgaris*, varying concentration of CO_2 was subjected and maximum lipid content of about 29.5 mgL^{-1} was obtained with crucial fatty acids such as palmitic, stearic and myristic fatty acids (Ortiz Montoya et al. 2014). Fatty acid composition and chain length of the lipids is highly important for biodiesel production. The major fatty acids such as palmitic, stearic and oleic acids are desirable fatty acids to produce biodiesel with good properties. On looking from CO_2 perspective, for 1 kg of microalgal biomass production, about 1.83 kg of CO_2 is required, whereas a study reveals that about 0.4 billion m^3 of biodiesel requires 1.3 billion tons of CO_2 for feeding European biofuel transport market. Although biotechnological CO_2 fixation mediated fuel production is widely concentrated by various industries, the cost associated with microalgal derived biodiesel production is still higher and various efforts have been constantly hitting worldwide. For instance, up to 25 US$ $gallon^{-1}$ and 40 US$ $gallon^{-1}$ was required to produce algal oil and biodiesel from microalgae through CO_2 fixation respectively (Khan et al. 2018). It is noteworthy to understand that microalgal CO_2 fixation is expensive which in turn reflects the use of microalgal biodiesel production. Hence, the most common and prominent strategy for improving the cost effectiveness is improving the lipid and biomass production of microalgae. The improvement of lipid productivity and biomass was usually accomplished by manipulating the growth conditions of microalgae especially nutrients, light, pH and temperatures (Esakkimuthu et al. 2019). Hence, significant reduction in overall production cost of about 10 times is required for making this microalgal based CO_2 fixation and biofuel production cost effective for which various strategies have been constantly developing to address the challenge (Thurmond 2020).

13.5.6 Bioplastics

Bioplastics are getting attention due to the adverse effects of larger production and use of non-degradable plastics. Around 8 million metric tons of plastic wastes per year are dumped in the ocean (Jambeck et al. 2015). Such devastating effects on the environment shifted the attraction towards synthesizing bioplastics. Heterotrophic microorganisms involve in utilizing the expensive and complex carbon sources whereas cyanobacteria could utilize CO_2 for producing polyhydroxybutyrate. There are several profound advantages with cyanobacteria than heterotrophic bacteria as they are autotrophic in nature along with cheap nutrient requirements (Troschl et al. 2017). Among bioplastics, polyhydroxyalkonates production from CO_2 is useful and is enhanced in various ways especially by coculturing of microbes (Garcia-Gonzalez and De Wever 2018). *Azotobacter vinelandii* and engineered *Synechococcus elongates* were cocultured to produce polyhydroxybutyrate and resulted in simultaneous fixation of CO_2 and NH_4 (Smith and Francis 2016). A mutant (*Synechocystis* sp) was created by UV mutagenesis that resulted in elevated polyhydroxybutyrate production [37% dry cell weight (DCW)] under limited supply of phosphorous and nitrogen (Kamravamanesh et al. 2018). In an interesting study, *S. elongates cscB* and *P. putida cscA* were mixed in a culture system in which cyanobacteria converts CO_2 to sucrose which is then exported to the medium facilitated by sucrose permease *cscB* and splitting of sucrose was facilitated by *cscA* gene of *P. putida cscA* followed by the polyhydroxyalkonates accumulation (Löwe et al. 2017). PHB production under nitrogen limitation was achieved in hydrocarbon oxidizing strain by CO_2 assimilation. The resulting polyester was suitable for employing as petrol and diesel fuel. Now, industries around the globe are concentrating on PHB production as the application is wider and needed. For instance, both PHB (115 kg) and biogas (320 m^3) was produced by a power plant in Austria from 1 ton of CO_2 (Zhang 2015). Hence combination of PHB production along with biofuel production can be efficient in terms of energy and economic perspective. The above-mentioned microbial products are concentrated and produced at laboratory scale and a few of them are listed in Table 13.3.

Table 13.3 Microbial CO_2 fixation and value-added product conversion

Bioproduct	Microorganism	Optimum CO_2 concentration	Molecular strategy	Product yield	Ref
Ethanol	*Synechocystis* sp PCC6803	5%	*pdc and adh* II genes of Zymomonas mobilis expressed in the host	5.5 g L^{-1}	Gao et al. (2012)
Isobutanol	*Synechococcuselongates* PCC7942	–	*kivd AlsS, yqh and ilvCD* genes introduction	0.45 gL^{-1}	Atsumi et al. (2009)
Fatty acids	*Synechocystis* sp PCC6803	1%	Acyl carrier protein thioesterase gene introduction	0.197 g L^{-1}	Liu et al. (2011)
Lipids	*Scenedesmus dimorphus*	15%	NE	19.6% (DCW)	Ortiz Montoya et al. (2014)
Biodiesel	*Chlorella vulgaris*	10%	NE	38% (DCW)	Jain et al. (2019)
Lipids	*Chlorococcum littorale*	5%	NE	47% (DCW)	Ota et al. (2015)
Biodiesel	*Scenedesmus obliquus*	14.1%	NE	0.009 g L^{-1}	Ji et al. (2017)
Ethylene	*Synechocystis* sp PCC6803	5%	*efe* gene overexpressed in host	718 μg L^{-1}	Xiong et al. (2015)
Lactate	*Chlorella* sp.	6%	NE	0.43 g	Lee et al. (2017b)
Bioplastic (Polyhydroxybutyrate)	*Ideonella dechloratanus*	1%	NE	5.26 g L^{-1}	Tanaka et al. (2011)
Bioplastic (Polyhydroxyalkonates)	*Synechococcus elongatus* and *Psuedomonas putida*	2%	*cscB* and *cscAB* genes expression	0.023 g L^{-1}	Löwe et al. (2017)

13.6 Environmental Impact and Future Perspective of CO_2 Mediated Energy Conversion

Fixing and converting CO_2 directly from the environment using microbe is highly advantageous as the fixing agent is renewable source that also satisfies dual demands such as CO_2 reduction and value-added product synthesis. However, with present productivity status, it has to be effectively increased in terms of strain selection and engineering the strains for maximum productivity for which the platforms of genetic engineering, metabolic engineering and synthetic biology are available for exploitation (Salehizadeh et al. 2020). Such utilization of tools and increasing the efficiency of carbon capturing in microbes would directly reflect in cost reduction of the overall process. Recently, microbial electrosynthesis is an attractive process in which biofilms made of microbes are attached to electrodes which results in the conversion of CO_2 to various products such as isopropanol, acetate, ethanol, butanol, glycerol, formic acid, methane, etc. Bacterial species like *Clostridium scatologenes* are showing good efficiency in such mode of CO_2 conversion into ethanol, acetic acid and butyric acid (Mohanakrishna et al. 2018). However, lower conversion rate is again a challenge with microbial electrosynthesis which can be enhanced by using electrocatalyst and such electrocatalyst assisted microbial electrosynthesis is in vogue to address such challenge. As the field is flourishing with such novel attempts, microbial electrosynthesis is still at its infancy level and requires potential improvisation in various areas of the technology. Among microorganisms, microalgae provide plenty of opportunities to convert CO_2 into many value-added products which are demonstrated in Fig. 13.4. In general, microbial CO_2 fixation and related energy conversion would be even appropriate and more fruitful if combined with other waste water

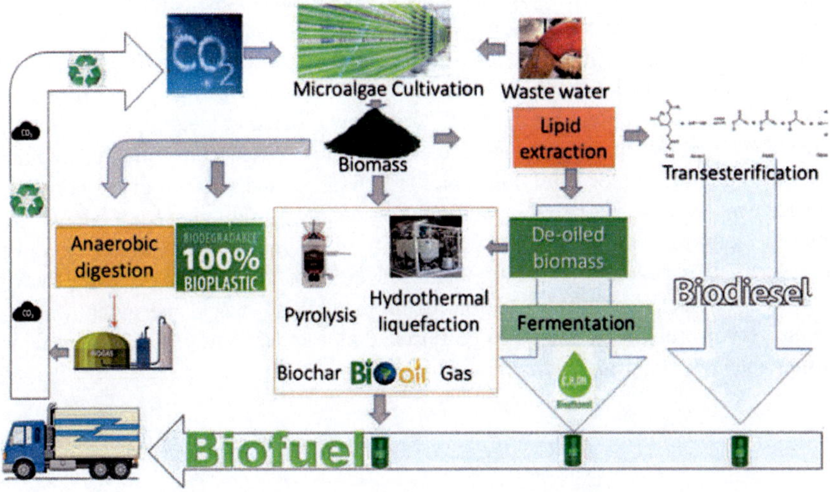

Fig. 13.4 CO_2 mediated energy conversion and recycling through microalgal application

treatment, flue gas capturing and CO_2 conversion. These integrations would make the process more efficient and multiply the advantages along with reduced cost of the overall process (Zhang 2015).

The comprehensiveness of metrics employed for environmental assessment tends to increase with multi step reactions as compared with single stoichiometric reactions. Although CO_2 is effectively converted, the overall catalyzed chemical reactions would further increase the CO_2 emission. Still, there is a lack of clear picture of the overall CO_2 utilization and emission which could be obtained by life cycle assessment with consideration of fossil depletion and global warming indices which is meagerly reported (von der Assen et al. 2013). Though a few life cycle assessment analyses supported the fact that carbon utilization powered with clean or renewable electricity could provide better environmental benefits with reduced greenhouse gas emissions. Exergy can be defined as a measure of useful power or energy available to be used. Hence exergy efficiency is an important parameter to be considered as exergy efficiency promotes sustainability whereas higher rate of exergy loss results in severe environmental impact (Ravanchi and Sahebdelfar 2020). Therefore, economic and environmental improvisations are required at a greater magnitude for both chemical and biological CO_2 mediated energy conversion.

13.7 Conclusion

To control the CO_2 emission and levels in the atmosphere, obvious choice is CO_2 fixation and mediated conversion. The economic and technological developments for the last few decades have proportionately increased the CO_2 emissions and relevant levels in the atmosphere. To capture all the emitting CO_2 and conversion is impossible with any chemical or biological method. Hence, measures for increasing the CO_2 mediated conversion strategies along with simultaneous reduction of CO_2 emissions would give a fine balance to the environment. Apart from those potential improvisation requiring areas, CO_2 based conversion provides greater advantages like producing high molecular weight PPC and low molecular weight poly carbonate ether are hot research areas which are expected to replace conventional polymer production. On the other hand, microbial CO_2 fixation and relevant energy production have made some prospecting path such as bioenergy production from microalgae such as biodiesel, biogas, etc. and opportunity of having multiple advantages like treating waste water simultaneously while converting CO_2. Precisely, irrespective of various bottlenecks and improvisation requirements, CO_2 based energy conversion seems very promising and need to be initiated as a movement globally to provide sustainable and clean environment.

References

Abomohra AE-F, Elsayed M, Esakkimuthu S, El-Sheekh M, Hanelt D (2020) Potential of fat, oil and grease (FOG) for biodiesel production: a critical review on the recent progress and future perspectives. Prog Energy Combust Sci 81:100868

Acemoglu M, Nimmerfall F, Bantle S, Stoll GH (1997) Poly (ethylene carbonate) s, part I: syntheses and structural effects on biodegradation. J Control Release 49:263–276

Aldana PAU, Ocampo F, Kobl K, Louis B, Thibault-Starzyk F, Daturi M, Bazin P, Thomas S, Roger AC (2013) Catalytic CO_2 valorization into CH_4 on Ni-based ceria-zirconia. Reaction mechanism by operando IR spectroscopy. Catal Today 215:201–207

Amouroux J, Siffert P, Massué JP, Cavadias S, Trujillo B, Hashimoto K, Rutberg P, Dresvin S, Wang X (2014) Carbon dioxide: a new material for energy storage. Prog Nat Sci Mater Int 24:295–304

Amulya K, Mohan SV (2019) Fixation of CO_2, electron donor and redox microenvironment regulate succinic acid production in Citrobacter amalonaticus. Sci Total Environ 695:133838

Angermayr SA, Rovira AG, Hellingwerf KJ (2015) Metabolic engineering of cyanobacteria for the synthesis of commodity products. Trends Biotechnol 33:352–361

Arbeláez O, Hernández E, González L-M, Bustamante F, Villa A-L (2019) Enhanced conversion in the direct synthesis of diethyl carbonate from ethanol and CO_2 by process intensification. Chem Eng Technol 42:1135–1143

Arena F, Barbera K, Italiano G, Bonura G, Spadaro L, Frusteri F (2007) Synthesis, characterization and activity pattern of Cu–ZnO/ZrO_2 catalysts in the hydrogenation of carbon dioxide to methanol. J Catal 249:185–194

Aresta M, Dibenedetto A (2003) Carbon dioxide fixation into organic compounds. In: Carbon dioxide recovery and utilization. Springer, pp 211–260

Ashok J, Falbo L, Das S, Dewangan N, Visconti CG, Kawi S (2019) Catalytic CO_2 conversion to added-value energy rich C 1 products. In: An economy based on carbon dioxide and water. Springer, pp 155–210

Atsumi S, Higashide W, Liao JC (2009) Direct photosynthetic recycling of carbon dioxide to isobutyraldehyde. Nat Biotechnol 27:1177–1180

Bar-Even A (2018) Daring metabolic designs for enhanced plant carbon fixation. Plant Sci 273:71–83

Bar-Even A, Noor E, Lewis NE, Milo R (2010) Design and analysis of synthetic carbon fixation pathways. Proc Natl Acad Sci 107:8889–8894

Basu P (2010) Biomass gasification and pyrolysis: practical design and theory. Academic press

Batlle-Vilanova P, Rovira-Alsina L, Puig S, Balaguer MD, Icaran P, Monsalvo VM, Rogalla F, Colprim J (2019) Biogas upgrading, CO_2 valorisation and economic revaluation of bioelectrochemical systems through anodic chlorine production in the framework of wastewater treatment plants. Sci Total Environ 690:352–360

Bege N, Renette T, Jansch M, Reul R, Merkel O, Petersen H, Curdy C, Müller RH, Kissel T (2011) Biodegradable poly (ethylene carbonate) nanoparticles as a promising drug delivery system with "stealth" potential. Macromol Biosci 11:897–904

Bege N, Steinmüller SO, Kalinowski M, Reul R, Klaus S, Petersen H, Curdy C, Janek J, Kissel T (2012) Drug eluting stents based on Poly (ethylene carbonate): optimization of the stent coating process. Eur J Pharm Biopharm 80:562–570

Bouarab R, Akdim O, Auroux A, Cherifi O, Mirodatos C (2004) Effect of MgO additive on catalytic properties of Co/SiO_2 in the dry reforming of methane. Appl Catal A Gen 264:161–168

Bradford MCJ, Vannice MA (1999) CO_2 reforming of CH_4. Catal Rev 41:1–42

Brunner G (2005) Supercritical fluids: technology and application to food processing. J Food Eng 67:21–33

Cai W, Liu W, Yang C, Wang L, Liang B, Thangavel S, Guo Z, Wang A (2016) Biocathodic methanogenic community in an integrated anaerobic digestion and microbial electrolysis system for enhancement of methane production from waste sludge. ACS Sustain Chem Eng 4:4913–4921

Centi G, Quadrelli EA, Perathoner S (2013) Catalysis for CO_2 conversion: a key technology for rapid introduction of renewable energy in the value chain of chemical industries. Energy Environ Sci 6:1711–1731

Çepelioğullar Ö, Pütün AE (2013) Utilization of two different types of plastic wastes from daily and industrial life. J Selcuk Univ Nat Appl Sci 2:694–706

Chae Y, An Y-J (2018) Current research trends on plastic pollution and ecological impacts on the soil ecosystem: a review. Environ Pollut 240:387–395

Cheng ZX, Zhao XG, Li JL, Zhu QM (2001) Role of support in CO_2 reforming of CH_4 over a Ni/γ-Al_2O_3 catalyst. Appl Catal A Gen 205:31–36

Choi M, Cho D (2008) Research activities on the utilization of carbon dioxide in Korea. Clean–soil Air, Water 36:426–432

Christensen PS (2014) Hydrothermal liquefaction of waste biomass: optimizing reaction parameters. Dep Chem Fac Sci Technol

Chu D, Curdy C, Riebesehl B, Beck-Broichsitter M, Kissel T (2013a) In situ forming parenteral depot systems based on poly (ethylene carbonate): effect of polymer molecular weight on model protein release. Eur J Pharm Biopharm 85:1245–1249

Chu D, Curdy C, Riebesehl B, Zhang Y, Beck-Broichsitter M, Kissel T (2013b) Enzyme-responsive surface erosion of poly (ethylene carbonate) for controlled drug release. Eur J Pharm Biopharm 85:1232–1237

Claassens NJ, Sánchez-Andrea I, Sousa DZ, Bar-Even A (2018) Towards sustainable feedstocks: a guide to electron donors for microbial carbon fixation. Curr Opin Biotechnol 50:195–205

CRI—Carbon Recycling International [WWW Document] (n.d.). https://www.carbonrecycling.is/. Accessed 24 May 21.

Dormer A, Finn DP, Ward P, Cullen J (2013) Carbon footprint analysis in plastics manufacturing. J Clean Prod 51:133–141

Dorner RW, Hardy DR, Williams FW, Willauer HD (2010) Heterogeneous catalytic CO_2 conversion to value-added hydrocarbons. Energy Environ Sci 3:884–890

Du K, Wen X, Wang Z, Liang F, Luo L, Peng X, Xu Y, Geng Y, Li Y (2019) Integrated lipid production, CO_2 fixation, and removal of SO_2 and NO from simulated flue gas by oleaginous *Chlorella pyrenoidosa*. Environ Sci Pollut Res 26:16195–16209

Carbon Engineering|Direct Air Capture of CO_2|Home [WWW Document], n.d. https://carbonengineering.com/. Accessed 24 May 21

Erb TJ, Zarzycki J (2016) Biochemical and synthetic biology approaches to improve photosynthetic CO_2-fixation. Curr Opin Chem Biol 34:72–79

Esakkimuthu S, Krishnamurthy V, Wang S, Abomohra AE-F, Shanmugam S, Ramakrishnan SG, Subrmaniam S, Swaminathan K (2019) Simultaneous induction of biomass and lipid production in *Tetradesmus obliquus* BPL16 through polysorbate supplementation. Renew Energy 140:807–815

Esakkimuthu S, Krishnamurthy V, Wang S, Hu X, Swaminathan K, Abomohra AE-F (2020) Application of p-coumaric acid for extraordinary lipid production in *Tetradesmus obliquus*: a sustainable approach towards enhanced biodiesel production. Renew Energy

EtoGas Power-to-Gas—Hitachi Zosen Inova AG [WWW Document], n.d. https://www.hz-inova.com/de/renewable-gas/etogas/. Accessed 24 May 21

Feng J, Zhuo R-X, Zhang X-Z (2012) Construction of functional aliphatic polycarbonates for biomedical applications. Prog Polym Sci 37:211–236

Fu W-J, Chi Z, Ma Z-C, Zhou H-X, Liu G-L, Lee C-F, Chi Z-M (2015) Hydrocarbons, the advanced biofuels produced by different organisms, the evidence that alkanes in petroleum can be renewable. Appl Microbiol Biotechnol 99:7481–7494

Fukuoka S, Tojo M, Hachiya H, Aminaka M, Hasegawa K (2007) Green and sustainable chemistry in practice: development and industrialization of a novel process for polycarbonate production from CO_2 without using phosgene. Polym J 39:91–114

Gao Z, Zhao H, Li Z, Tan X, Lu X (2012) Photosynthetic production of ethanol from carbon dioxide in genetically engineered cyanobacteria. Energy Environ Sci 5:9857–9865

García-Cubero R, Moreno-Fernández J, García-González M (2018) Potential of *Chlorella vulgaris* to abate flue gas. Waste Biomass Valorization 9:2015–2019

Garcia-Gonzalez L, De Wever H (2018) Acetic acid as an indirect sink of CO_2 for the synthesis of polyhydroxyalkanoates (PHA): comparison with PHA production processes directly using CO_2 as feedstock. Appl Sci 8:1416

Gnanamani MK, Jacobs G, Hamdeh HH, Shafer WD, Liu F, Hopps SD, Thomas GA, Davis BH (2016) Hydrogenation of carbon dioxide over Co–Fe bimetallic catalysts. Acs Catal 6:913–927

Gopinath KP, Nagarajan VM, Krishnan A, Malolan R (2020) A critical review on energy, environmental and economic factors on various processes used to handle and recycle plastic wastes: development of a comprehensive index. J Clean Prod 123031

Guan LT, Xiao M, Meng YZ, Li RKY (2006) Chemically foaming of biodegradable poly (propylene carbonate) derived from carbon dioxide and propylene oxide. Polym Eng Sci 46:153–159

Guan LT, Du FG, Wang GZ, Chen YK, Xiao M, Wang SJ, Meng YZ (2007) Foaming and chain extension of completely biodegradable poly (propylene carbonate) using DPT as blowing agent. J Polym Res 14:245–251

Habazaki H, Yamasaki M, Zhang B-P, Kawashima A, Kohno S, Takai T, Hashimoto K (1998) Co-methanation of carbon monoxide and carbon dioxide on supported nickel and cobalt catalysts prepared from amorphous alloys. Appl Catal A Gen 172:131–140

Halfmann C, Gu L, Gibbons W, Zhou R (2014) Genetically engineering cyanobacteria to convert CO_2, water, and light into the long-chain hydrocarbon farnesene. Appl Microbiol Biotechnol 98:9869–9877

Hicks N, Vik U, Taylor P, Ladoukakis E, Park J, Kolisis F, Jakobsen KS (2017) Using prokaryotes for carbon capture storage. Trends Biotechnol 35:22–32

Hing NYK, Liang F, Lindblad P, Morgan JA (2019) Combining isotopically non-stationary metabolic flux analysis with proteomics to unravel the regulation of the calvin-benson-bassham cycle in synechocystis sp. PCC 6803. Metab Eng 56:77–84

Home Page|Novomer [WWW Document], n.d. https://www.novomer.com/. Accessed 24 May 21

Huang C-H, Tan C-S (2014) A review: CO_2 utilization. Aerosol Air Qual Res 14:480–499

Huang X, Huang J, Wu J, Yu X, Gao Q, Luo Y, Hu H (2015) Fabrication and properties of polybutadiene rubber-interpenetrating cross-linking poly (propylene carbonate) network as gel polymer electrolytes for lithium-ion battery. RSC Adv 5:52978–52984

IEA I (2011) World energy outlook 2011. Int Energy Agency 666

Jain D, Ghonse SS, Trivedi T, Fernandes GL, Menezes LD, Damare SR, Mamatha SS, Kumar S, Gupta V (2019) CO_2 fixation and production of biodiesel by *Chlorella vulgaris* NIOCCV under mixotrophic cultivation. Bioresour Technol 273:672–676

Jajesniak P, Ali H, Wong TS (2014) Carbon dioxide capture and utilization using biological systems: opportunities and challenges. J Bioprocess Biotech 4:2

Jambeck JR, Geyer R, Wilcox C, Siegler TR, Perryman M, Andrady A, Narayan R, Law KL (2015) Plastic waste inputs from land into the ocean. Science (80–):347, 768–771

Jamradloedluk J, Lertsatitthanakorn C (2014) Characterization and utilization of char derived from fast pyrolysis of plastic wastes. Procedia Eng 69:1437–1442

Ji M-K, Yun H-S, Hwang J-H, Salama E-S, Jeon B-H, Choi J (2017) Effect of flue gas CO_2 on the growth, carbohydrate and fatty acid composition of a green microalga *Scenedesmus obliquus* for biofuel production. Environ Technol 38:2085–2092

Kamravamanesh D, Kovacs T, Pflügl S, Druzhinina I, Kroll P, Lackner M, Herwig C (2018) Increased poly-β-hydroxybutyrate production from carbon dioxide in randomly mutated cells of cyanobacterial strain Synechocystis sp. PCC 6714: mutant generation and characterization. Bioresour Technol 266:34–44

Kang M-K, Nielsen J (2017) Biobased production of alkanes and alkenes through metabolic engineering of microorganisms. J Ind Microbiol Biotechnol 44:613–622

Kanno M, Carroll AL, Atsumi S (2017) Global metabolic rewiring for improved CO_2 fixation and chemical production in cyanobacteria. Nat Commun 8:1–11

Kato K, Nomura S, Uematsu H (2003) Waste plastics recycling process using coke ovens. J Mater Cycles Waste Manag 5:98–101

Khan MI, Shin JH, Kim JD (2018) The promising future of microalgae: current status, challenges, and optimization of a sustainable and renewable industry for biofuels, feed, and other products. Microb Cell Fact 17:1–21

Klaus S, Lehenmeier MW, Herdtweck E, Deglmann P, Ott AK, Rieger B (2011) Mechanistic insights into heterogeneous zinc dicarboxylates and theoretical considerations for CO_2–epoxide copolymerization. J Am Chem Soc 133:13151–13161

Konieczynska MD, Lin X, Zhang H, Grinstaff MW (2015) Synthesis of aliphatic poly (ether 1, 2-glycerol carbonate) s via copolymerization of CO_2 with glycidyl ethers using a cobalt salen catalyst and study of a thermally stable solid polymer electrolyte. ACS Macro Lett 4:533–537

Langanke J, Wolf A, Peters M (2015) Polymers from CO_2—an industrial perspective. In: Carbon dioxide utilisation. Elsevier, pp 59–71

Lee BY, Cyriac A (2011) Making the gradient. Nat Chem 3:505–507

Lee Y, Kim D, Seo J, Han H, Khan SB (2013) Preparation and characterization of poly (propylene carbonate)/exfoliated graphite nanocomposite films with improved thermal stability, mechanical properties and barrier properties. Polym Int 62:1386–1394

Lee HJ, Choi J, Lee S-M, Um Y, Sim SJ, Kim Y, Woo HM (2017a) Photosynthetic CO_2 conversion to fatty acid ethyl esters (FAEEs) using engineered cyanobacteria. J Agric Food Chem 65:1087–1092

Lee T-M, Tseng Y-F, Cheng C-L, Chen Y-C, Lin C-S, Su H-Y, Chow T-J, Chen C-Y, Chang J-S (2017b) Characterization of a heat-tolerant Chlorella sp. GD mutant with enhanced photosynthetic CO_2 fixation efficiency and its implication as lactic acid fermentation feedstock. Biotechnol Biofuels 10:1–12

Li H, Wang J (2004) Study on CO2 reforming of methane to syngas over Al_2O_3–ZrO_2 supported Ni catalysts prepared via a direct sol–gel process. Chem Eng Sci 59:4861–4867

Liang Y, Tang J, Luo Y, Kaczmarek MB, Li X, Daroch M (2019) Thermosynechococcus as a thermophilic photosynthetic microbial cell factory for CO_2 utilisation. Bioresour Technol 278:255–265

Liu S, Wang X (2017) Polymers from carbon dioxide: polycarbonates, polyurethanes. Curr Opin Green Sustain Chem. 3:61–66

Liu X, Sheng J, Curtiss R III (2011) Fatty acid production in genetically modified cyanobacteria. Proc Natl Acad Sci 108:6899–6904

Löwe H, Hobmeier K, Moos M, Kremling A, Pflüger-Grau K (2017) Photoautotrophic production of polyhydroxyalkanoates in a synthetic mixed culture of *Synechococcus elongatus* cscB and Pseudomonas putida cscAB. Biotechnol Biofuels 10:1–11

Lwanga EH, Vega JM, Quej VK, de los Angeles Chi J, Del Cid LS, Chi C, Segura GE, Gertsen H, Salánki T, van der Ploeg M (2017) Field evidence for transfer of plastic debris along a terrestrial food chain. Sci Rep 7:1–7

Ma S, Li D, Yu Y, Li D, Yadav RS, Feng Y (2019) Application of a microalga, *Scenedesmus obliquus* PF3, for the biological removal of nitric oxide (NO) and carbon dioxide. Environ Pollut 252:344–351

Miandad R, Barakat MA, Aburiazaiza AS, Rehan M, Ismail IMI, Nizami AS (2017) Effect of plastic waste types on pyrolysis liquid oil. Int Biodeterior Biodegradation 119:239–252

Mohanakrishna G, Vanbroekhoven K, Pant D (2018) Impact of dissolved carbon dioxide concentration on the process parameters during its conversion to acetate through microbial electro synthesis. React Chem Eng 3:371–378

Morales M, Sánchez L, Revah S (2018) The impact of environmental factors on carbon dioxide fixation by microalgae. FEMS Microbiol Lett 365:fnx262

Muradov N (2014) Industrial utilization of CO_2: a win–win solution. In: Liberating energy from carbon: introduction to decarbonization. Springer, pp 325–383

Mussgnug JH, Klassen V, Schlüter A, Kruse O (2010) Microalgae as substrates for fermentative biogas production in a combined biorefinery concept. J Biotechnol 150:51–56

Nair LS, Laurencin CT (2007) Biodegradable polymers as biomaterials. Prog Polym Sci 32:762–798

Nejat P, Jomehzadeh F, Taheri MM, Gohari M, Majid MZA (2015) A global review of energy consumption, CO_2 emissions and policy in the residential sector (with an overview of the top ten CO_2 emitting countries). Renew Sustain Energy Rev 43:843–862

Nkwachukwu OI, Chima CH, Ikenna AO, Albert L (2013) Focus on potential environmental issues on plastic world towards a sustainable plastic recycling in developing countries. Int J Ind Chem 4:1–13

Nocito F, Dibenedetto A (2020) Atmospheric CO_2 mitigation technologies: carbon capture utilization and storage. Curr Opin Green Sustain Chem 21:34–43

Ortiz Montoya EY, Casazza AA, Aliakbarian B, Perego P, Converti A, de Carvalho JCM (2014) Production of *Chlorella vulgaris* as a source of essential fatty acids in a tubular photo bioreactor continuously fed with air enriched with CO_2 at different concentrations. Biotechnol Prog 30:916–922

Oswald F, Stoll IK, Zwick M, Herbig S, Sauer J, Boukis N, Neumann A (2018) Formic acid formation by *Clostridium ljungdahlii* at elevated pressures of carbon dioxide and hydrogen. Front Bioeng Biotechnol 6:6

Ota M, Takenaka M, Sato Y, Smith RL Jr, Inomata H (2015) Variation of photoautotrophic fatty acid production from a highly CO_2 tolerant alga, *Chlorococcum littorale*, with inorganic carbon over narrow ranges of pH. Biotechnol Prog 31:1053–1057

Pachauri RK, Allen MR, Barros VR, Broome J, Cramer W, Christ R, Church JA, Clarke L, Dahe Q, Dasgupta P (2014) Climate change 2014: synthesis report. Contribution of working groups I, II and III to the fifth assessment report of the Intergovernmental Panel on Climate Change (IPCC)

Pan Y, Liu C, Ge Q (2010) Effect of surface hydroxyls on selective CO_2 hydrogenation over $Ni_4/\gamma\text{-}Al_2O_3$: a density functional theory study. J Catal 272:227–234

Park SE, Chang JS, Lee KW, Joo OS, Jung KD, Jung Y (2004) Camere process for methanol synthesis from CO_2 hydrogenation. Stud Surf Sci Catal 153:67–72

Parry M, Parry ML, Canziani O, Palutikof J, Van der Linden P, Hanson C (2007) Climate change 2007-impacts, adaptation and vulnerability: working group II contribution to the fourth assessment report of the IPCC. Cambridge University Press

Pires JCM, Alvim-Ferraz MCM, Martins FG, Simões M (2012) Carbon dioxide capture from flue gases using microalgae: engineering aspects and biorefinery concept. Renew Sustain Energy Rev 16:3043–3053

Prajapati SK, Kaushik P, Malik A, Vijay VK (2013) Phycoremediation coupled production of algal biomass, harvesting and anaerobic digestion: possibilities and challenges. Biotechnol Adv 31:1408–1425

Quadrelli EA, Centi G, Duplan J, Perathoner S (2011) Carbon dioxide recycling: emerging large-scale technologies with industrial potential. Chemsuschem 4:1194–1215

Rafiee A, Khalilpour KR, Milani D (2019) CO_2 conversion and utilization pathways. In: Polygeneration with polystorage for chemical and energy hubs. Elsevier, pp 213–245

Ravanchi MT, Sahebdelfar S (2020) Catalytic conversions of CO_2 to help mitigate climate change: recent process developments. Process Saf Environ Prot

Renewables Everywhere—Sunfire [WWW Document], n.d. https://www.sunfire.de/en/. Accessed 24 May 21

Robert A, Kummert M (2012) Designing net-zero energy buildings for the future climate, not for the past. Build Environ 55:150–158

Rönsch S, Schneider J, Matthischke S, Schlüter M, Götz M, Lefebvre J, Prabhakaran P, Bajohr S (2016) Review on methanation—from fundamentals to current projects. Fuel 166:276–296

Saha GR, Das T, Handique P, Kalita D, Saikia BK (2018) Copyrolysis of low-grade Indian coal and waste plastics: future prospects of waste plastic as a source of fuel. Energy Fuels 32:2421–2431

Sahebdelfar S, Ravanchi MT (2015) Carbon dioxide utilization for methane production: a thermodynamic analysis. J Pet Sci Eng 134:14–22

Salehizadeh H, Yan N, Farnood R (2020) Recent advances in microbial CO_2 fixation and conversion to value-added products. Chem Eng J 390:124584

Samiee L, Gandzha S (2019) Power to methanol technologies via CO_2 recovery: CO_2 hydrogenation and electrocatalytic routes. Rev Chem Eng 1

Schweitzer C (2017) Small scale methanol plants: a chance for re-industrialisation. In: The international methanol conference, Copenhagen, Demark, pp 8–10

Seo J, Jeon G, Jang EuS, Bahadar Khan S, Han H (2011) Preparation and properties of poly (propylene carbonate) and nanosized ZnO composite films for packaging applications. J Appl Polym Sci 122:1101–1108

Sharifi A (2020) Co-benefits and synergies between urban climate change mitigation and adaptation measures: a literature review. Sci Total Environ 141642

Smith MJ, Francis MB (2016) A designed *A. vinelandii–S. elongatus* coculture for chemical photoproduction from air, water, phosphate, and trace metals. ACS Synth Biol 5:955–961

Solution Center CovestrolCovestro AG [WWW Document], n.d. https://solutions.covestro.com/en?utm_source=google&utm_medium=cpc&utm_campaign=Brand&utm_term=Covestro&gclid=Cj0KCQjwna2FBhDPARIsACAEc_W_GYeCoivQwKnh1RwpH6riw2nc3esi7KLJz8gc1RTqYKKRecIRNCkaAkfoEALw_wcB. Accessed 24 May 21

Stefansson B (2015) Power and CO_2 emissions to methanol. In: Presentation, 2015 European Methanol Policy Forum

Stroud T, Smith TJ, Le Saché E, Santos JL, Centeno MA, Arellano-Garcia H, Odriozola JA, Reina TR (2018) Chemical CO_2 recycling via dry and bi reforming of methane using Ni–Sn/Al_2O_3 and Ni–Sn/CeO_2–Al_2O_3 catalysts. Appl Catal B Environ 224:125–135

Tanaka K, Miyawaki K, Yamaguchi A, Khosravi-Darani K, Matsusaki H (2011) Cell growth and P (3HB) accumulation from CO_2 of a carbon monoxide-tolerant hydrogen-oxidizing bacterium, Ideonella sp. O-1. Appl Microbiol Biotechnol 92:1161–1169

Thompson RC, Moore CJ, Vom Saal FS, Swan SH (2009) Plastics, the environment and human health: current consensus and future trends. Philos Trans R Soc B Biol Sci 364:2153–2166

Thurmond W (2020) Global biofuels, drop-in fuels, biochems markets and forecasts. Algae 2:1–4

Till Z, Varga T, Sója J, Miskolczi N, Chován T (2018) Kinetic modeling of plastic waste pyrolysis in a laboratory scale two-stage reactor. In: computer aided chemical engineering. Elsevier, pp 349–354

Troschl C, Meixner K, Drosg B (2017) Cyanobacterial PHA production—review of recent advances and a summary of three years' working experience running a pilot plant. Bioengineering 4:26

Trott G, Saini PK, Williams CK (2016) Catalysts for CO_2/epoxide ring-opening copolymerization. Philos Trans R Soc A Math Phys Eng Sci 374:20150085

Ueda S, Kawamura Y, Iijima H, Nakajima M, Shirai T, Okamoto M, Kondo A, Hirai MY, Osanai T (2016) Anionic metabolite biosynthesis enhanced by potassium under dark, anaerobic conditions in cyanobacteria. Sci Rep 6:1–9

van Heek J, Arning K, Ziefle M (2017) Reduce, reuse, recycle: acceptance of CO_2-utilization for plastic products. Energy Policy 105:53–66

Visconti CG, Lietti L, Tronconi E, Forzatti P, Zennaro R, Finocchio E (2009) Fischer-Tropsch synthesis on a Co/Al_2O_3 catalyst with CO_2 containing syngas. Appl Catal A Gen 355:61–68

von der Assen N, Jung J, Bardow A (2013) Life-cycle assessment of carbon dioxide capture and utilization: avoiding the pitfalls. Energy Environ Sci 6:2721–2734

Wang Y, Sun T, Gao X, Shi M, Wu L, Chen L, Zhang W (2016) Biosynthesis of platform chemical 3-hydroxypropionic acid (3-HP) directly from CO_2 in cyanobacterium Synechocystis sp. PCC 6803. Metab Eng 34:60–70

Wang X, Bao K, Cao W, Zhao Y, Hu CW (2017) Screening of microalgae for integral biogas slurry nutrient removal and biogas upgrading by different microalgae cultivation technology. Sci Rep 7:1–12

Wang L, He Z, Guo Z, Sangeetha T, Yang C, Gao L, Wang A, Liu W (2019) Microbial community development on different cathode metals in a bioelectrolysis enhanced methane production system. J Power Sources 444:227306

Winandy JE, Stark NM, Clemons CM (2004) Considerations in recycling of wood-plastic composites. In: 5th global wood and natural fibre composites symposium, 27–28 April 2004. Kassel, Germany, [9] Page

Wittich K, Krämer M, Bottke N, Schunk SA (2020) Catalytic dry reforming of methane: insights from model systems. ChemCatChem

Xie M, Wang W, Zhang W, Chen L, Lu X (2017) Versatility of hydrocarbon production in cyanobacteria. Appl Microbiol Biotechnol 101:905–919

Xiong W, Morgan JA, Ungerer J, Wang B, Maness P-C, Yu J (2015) The plasticity of cyanobacterial metabolism supports direct CO_2 conversion to ethylene. Nat Plants 1:1–7

Xu H, Wang K, Holmes DE (2014) Bioelectrochemical removal of carbon dioxide (CO_2): an innovative method for biogas upgrading. Bioresour Technol 173:392–398

Xu Y, Lin L, Xiao M, Wang S, Smith AT, Sun L, Meng Y (2018) Synthesis and properties of CO_2-based plastics: environmentally-friendly, energy-saving and biomedical polymeric materials. Prog Polym Sci 80:163–182

Yang C, Li R, Cui C, Liu S, Qiu Q, Ding Y, Wu Y, Zhang B (2016) Catalytic hydroprocessing of microalgae-derived biofuels: a review. Green Chem 18:3684–3699

Yu X, Xiao M, Wang S, Zhao Q, Meng Y (2010) Fabrication and characterization of PEO/PPC polymer electrolyte for lithium-ion battery. J Appl Polym Sci 115:2718–2722

Zangeneh FT, Sahebdelfar S, Ravanchi MT (2011) Conversion of carbon dioxide to valuable petrochemicals: an approach to clean development mechanism. J Nat Gas Chem 20:219–231

Zavřel T, Knoop H, Steuer R, Jones PR, Červený J, Trtílek M (2016) A quantitative evaluation of ethylene production in the recombinant cyanobacterium Synechocystis sp. PCC 6803 harboring the ethylene-forming enzyme by membrane inlet mass spectrometry. Bioresour Technol 202:142–151

Zhai L, Li G, Xu Y, Xiao M, Wang S, Meng Y (2015) Poly (propylene carbonate)/aluminum flake composite films with enhanced gas barrier properties. J Appl Polym Sci 132

Zhang X (2015) Microalgae removal of CO_2 from flue gas. IEA Clean Coal Centre, UK

Zhong X, Lu Z, Valtchev P, Wei H, Zreiqat H, Dehghani F (2012) Surface modification of poly (propylene carbonate) by aminolysis and layer-by-layer assembly for enhanced cytocompatibility. Colloids Surfaces B Biointerfaces 93:75–84

Zhou D, Zhou R, Chen C, Yee W-A, Kong J, Ding G, Lu X (2013) Non-volatile polymer electrolyte based on poly (propylene carbonate), ionic liquid, and lithium perchlorate for electrochromic devices. J Phys Chem B 117:7783–7789

Zhou J, Zhu T, Cai Z, Li Y (2016) From cyanochemicals to cyanofactories: a review and perspective. Microb Cell Fact 15:1–9

Zhu Q (2019) Developments on CO_2-utilization technologies. Clean Energy 3:85–100

Chapter 14
Plastic Recycling for Energy Production

Shah Faisal, Muhammad Naveed, Sayyed Kifayatullah, and Mian Muhammad

Abstract Plastics are a type of synthetic organic polymer that is widely used in modern society and has a significant environmental impact because of its slow degradation. In addition, toxic substances are released into the soil when plastic bags perish under sunlight. Plastic waste can be converted into hydrogen, diesel, crude oil, and Sulfur using conventional methods. Plastic recycling to energy will have a significant positive impact on the global economy and waste management. It is possible to convert plastic into fuels using modern technology and systematic approaches, which will help in plastic waste management and its recycling to fuels. The most common methods for converting plastic waste into fuel are hydrothermal processing and pyrolysis. In addition, the microbial and enzymatic biodegradation of synthetic plastics got increasing attention in recent years, offering the possibility of developing biological treatment technology for plastic waste. Many enzymes that can degrade plastic and convert it into biofuel have been isolated and studied. However, there are numerous limitations in the degradation of plastic by microbes and its conversion into fuels. This chapter provides an overview of global plastic use, conventional plastic recycling, and the harmful effects of plastic recycling. In addition, the microbial conversion of plastic to energy has been briefly described. Microbial pathways and enzymes involved in the transformation of plastics into fuels have been investigated. Finally, the economic feasibility of bioconversion of plastic into energy has been discussed.

S. Faisal
Department of Environmental Engineering, School of Architecture and Civil Engineering, Chengdu University, Chengdu 610106, People's Republic of China

M. Naveed
Institute of Microbiology, School of Life Sciences, Lanzhou University, Tianshui Road No. 222, Lanzhou 730000, People's Republic of China

S. Kifayatullah
Depatment of Environmental Science, International Islamic University, Islamabad, Pakistan

M. Muhammad (✉)
Department of Chemistry, University of Malakand, Chakdara Dir Lower 18800, Khyber Pukhtunkhwa, Pakistan
e-mail: mianchem@uom.edu.pk

A. E.-F. Abomohra et al. (eds.), *Waste-to-Energy*,
https://doi.org/10.1007/978-3-030-91570-4_14

Keywords Plastic · Degradation · Microorganisms · Enzymes · Biofuel

14.1 Introduction

The term "plastic" is derived from the Greek word "plastikos," which means "fit for molding," and refers to their malleability during manufacturing and processing. Plastics are high molecular weight polymeric compounds that can be shaped into the desired shape when hot and retain their shape when cold or hardened (Seymour 1989; Ugoeze et al. 2021). Plastic has a wide range of applications due to its unique properties such as lightweight, low cost, durability, robustness, strength, corrosion resistance, thermal and electrical insulation, versatile fabrication and design capabilities, and ability to be easily molded into a variety of products (Mohanraj et al. 2017). Because of these properties, they can be cast, pressed, or extruded into a wide range of shapes, including films, bottles, tubes, fibers, boxes, and plates (Carraher 2017). They can be used for making water bottles, food packaging, clothing, medical instruments, electronics, and materials of construction. Natural gas, coal, and petroleum are the primary precursors used to produce plastic materials (Gautam et al. 2007).

Because of their versatile properties and low production costs, plastics experienced market growth of 23% between 2008 and 2015. These packaging materials are made from biodegradable plastic, such as epoxy resins, polystyrenes, expanded polystyrenes, polyolefin, polyvinyl chloride, and polycarbonates hazardous substances incorporated into them to aid in the aesthetics of the product (Ugoeze et al. 2021). Plastics have advantages over other materials (like metallic products, ceramics materials, wood, etc.), such as decreasing cost, incline durability, and have lightweight (Andrady and Neal 2009). Consequently, their overuse and improper disposal lead to numerous environmental problems. Almost 4% of the world's petrochemical production is used as feedstock for plastics materials, with the remaining 3 to 4% is used in their manufacturing to provide energy (Hopewell et al. 2009). Plastic has a multitude of benefits for society but has also posed many hazards to environments (Mwanza and Mbohwa 2017). Plastics contain a variety of toxic additives, including di-(2-Ethylhexyl) phthalate (DEHP), Bisphenol A (BPA), polyhalogenated materials, and heavy metals, all of which pose a potential health risk to humans (Halden 2010). Most of these additives are quickly immobilized in the environment, resulting in harmful effects on human health, such as endocrine system disruption (North and Halden 2013; Ali et al. 2014). Because they can be quickly and affordably replaced by other materials, plastics have become an indispensable material for humanity. Quality of life, employment, and economic growth have all been improved as a result of plastic products. Due to its mechanical stability, most plastics have a long lifespan (Koushal et al. 2014). Without plastics, modern society would be unable to function. When it comes to medical devices ranging from human artificial body organs to blood bags, plastic's specific properties determine its use.

14.2 Types of Plastic

Plastics are divided into seven major categories. These are polyethylene terephthalate (PET), high-density polyethylene (HDPE), polyvinyl chloride (PVC), low-density polyethylene (LDPE), polypropylene (PP), and polystyrene (PS), each denoted by a number from 1 to 6. In addition, there is one more type of resin to indicate plastic other than the previous six, marked by code seven and labeled "other" (Chow et al. 2016). When two or more types of plastic resin are mixed, it can harm the environment, such as when PVC and PET are combined, producing hydrochloric gas (Wahab et al. 2006; Ruj et al. 2015).

14.2.1 Polyethylene Terephthalate (PET)

Many types of plastic bags can be found in supermarkets, but PET is one of them. PET is most commonly used in food and beverage packaging because of its strong ability to prevent oxygen from entering and spoiling the product contained within it. Carbon dioxide from carbonated beverages is also prevented by PET (Thomas et al. 2015).

14.2.2 High-Density Polyethylene (HDPE)

High-density polyethylene has long, virtually unbranched polymer chains, making it extremely dense and, as a result, more robust and thicker than PET. HDPE is commonly used in grocery bags, opaque milk containers, juice, shampoo, and medicine bottles. HDPE is a more durable material than PET (Hidayat et al. 2019). Studies reported that the leaching of HPPE derivatives disturbed the estrogen that can disrupt the human hormonal system when exposed to ultraviolet light, it is considered a safer option for food and beverage use (Muthu et al. 2012).

14.2.3 Polyvinyl Chloride (PVC)

PVC is also known as vinyl, and it was once the world's second most widely used plastic resin, after polyethylene. The everyday use of PVC is in Toys, blood bags, blister wrap, detergent bottles, cling wrap, medical tubing, and loose-leaf binders (Onen Cinar et al. 2020). PVC manufacturing and disposal had previously been identified as a source of severe health risks and environmental pollution. In terms of toxicity, PVC is regarded as the most dangerous plastic. Its use may leach a wide range of toxic chemicals, including Bisphenol A (BPA), phthalates, lead, dioxins, mercury, and cadmium (Proshad et al. 2018).

14.2.4 Low-Density Polyethylene (LDPE)

Polyethylene is the most widely used plastic globally. It has any plastic polymer's basic chemical structure, which makes it very simple and cost-efficient processing. LDPE polymers have highly branching chains and have lengthy side chains; consequently, having low density and crystallinity than polyethylene, which is commonly flexible and thinner (Sen and Raut 2015). Typical uses of LDPE are in bags (grocery, garbage bags, dry cleaning, and bags for frozen foods), coatings for paper, cartons, and packs for milk, and squeezable bottles (honey, mustard), containers for the storage of food, beverage cups, container lids and coverings of cables (Jnr et al. 2018).

14.2.5 Polypropylene (PP)

Polypropylene (PP) is a stiffer, more heat-resistant material commonly used for hot food containers (Chaukura et al. 2016). It has a robust quality that falls somewhere between Low-density polyethylene and Low-density polyethylene. It is usually used in diapers, sanitary, thermal vests, and parts of cars. It is considered a safe plastic for use in the coverings of beverages and food. Despite having these properties, PP is not entirely recyclable and may cause hormone disruption and asthma when ingested as microplastic and nanoplastic (Suwanmanee et al. 2013).

14.2.6 Polystyrene (PS)

Polystyrene (PS) is a type of Styrofoam widely used in food containers, egg cartons, disposable cups and bowls, packaging, and bike helmets (Turner 2020). Styrene is a known toxin of the nervous system. PS may leach styrene when exposed to hot and oily foods. It may also impact the immunity, genes, respiratory system, and liver; in addition to all of these dangers, it is widely used worldwide (Mangalara and Varughese 2016).

14.2.7 Other

A plastic that can be layered or mixed with another plastic, such as bioplastics, is referred to as others. Because of its association with Bisphenol A (BPA), the most common type of plastic in this category is polycarbonate (PC) (Kawasaki et al. 2009). There are other names for PC, such as Lexan and Makrolon. Baby bottles and sippy cups, ketchup containers, water gallons, metal cane liners, and dental sealants are commonly made of PC. The toxic nature of PC has led some countries to ban its use in the production of baby bottles and infant formula packaging (Farcas et al. 2019).

14.3 Global Potential of Plastic Production

The production of plastic increases tremendously in the 1940s, becoming one of the booming industries in the industrial revolution (Gourmelon 2015). Global plastic production increased steadily between the 1970s and 2012 because most paper, metal, and glass packaging were increasingly replaced with plastic packaging, particularly for food products (Gourmelon 2015). In 2009, 30% of the annual sales of plastic packaging materials were accounted (Shin and Selke 2014). Similarly, the American automobile industry has benefited from this change to reduce vehicle weight, according to federal mileage standards in the United States. The traditional American vehicle was about 10% plastics by weight, which equals 336 pounds of plastic per vehicle (Swift et al. 2015).

Plastics are now used in bumpers and door panels, and engine components to replace metals. The global plastics industry now generates nearly $600 billion in revenue each year (Koo 2019). Transportation, food items covering, telecommunications, consumer goods, and health care are all businesses that use plastics (Jalal 2017). Per capita, the consumption of plastic in North America and Western Europe exceeded 100 kg (Rippy 1948). At the moment, Asian countries consume 20 kg of plastic per person. This figure, however, is expected to increase at a dramatic rate (Jalal 2017). In 2013, the Asian region generated 45.65% of all plastics produced globally, with China producing roughly one-quarter of all plastic made globally (Radenahmad et al. 2018). As a result of the country's growing population and manufacturing sector, plastic production in India has recently increased (Mutha et al. 2006). Today, the former Soviet Union and European states are responsible for 22.9% of global plastic output, and Germany is the leading producer in Europe (Kapinga and Chung 2020). Similarly, 19.4% of the global plastics are used in the United States (Gourmelon 2015), 7.3% in the Middle East and Africa, and 4.8% in Central and South America, minor shares of the global plastic production (Jalal 2017). From 22 to 43% of plastics are dumped into landfills worldwide, drain resources, take up precious space, and damage communities (Assadourian 2008).

Plastic recovered from the waste stream for recycling or energy production has the efficiency in solving these problems. However, a large portion of the plastic collected for recycling is transported to countries with lower environmental standards, creating a precarious balance between environmental protection, clean material cycles, and resource utilization. Furthermore, energy recovery from plastics is inefficient, requires air pollution controls, and produces hazardous ash (Gourmelon 2015). The actual recovery rates varied by country. In Europe, 26% of plastic wastes were recycled in 2012, having a mass of 6.6 million tons, and for energy production, 36% was burned. In Europe, 38% of plastic waste was a dump in the landfills (Jalal 2017). However, a 26% decrease in plastic wastes in landfills has occurred since 2006. Most of the European countries dump their plastic wastes into landfills (Gourmelon 2015). In nine countries of Europe, plastics' waste has been banned for landfills, including Luxembourg, Austria, Germany, Sweden, Belgium, Denmark, Netherlands, Norway, and Switzerland. While these countries have higher recycling rates than non-landfill-ban countries, most plastic is disposed of through energy recovery incineration (Wu et al. 2014). Only 9% of the wastes of plastic in the United States (2.8 million tons) was recycled in 2012. The remaining 32 million tonnes, or nearly 13% of the nation's municipal solid waste stream, were discarded (Gourmelon 2015). Although some of it is transferred to Canada and Mexico, the United States primarily depends on Hong Kong and China for its plastic waste disposition. The rest of the globe has a significantly lower rate of recycling plastic wastes. The 57% of the plastic from Africa, 40% of Asia, and 32% in Latin America is not properly collected and is burned and dumped in the open space as per the data of the UN environmental program (Assadourian 2008).

The USA is the highest exporter of plastic wastes, followed by Japan, Germany, and the UK (Gourmelon 2015). Europe is the leading recycler of plastic trash globally, and most of the plastic wastes were collected from countries of the west, with established collection systems ending up in China, which is the importer of 56% of plastic wastes to the world (Velis 2014). According to the International Solid Waste Association, indirect data shows that family-run, low-tech businesses still reprocess most plastic with few environmental safeguards (Hahladakis et al. 2018). Low-quality plastics are also discarded or recycled for energy recovery in areas with lost control of air pollution is raising global concerns.

Over the last six decades, the production of plastic products and their waste has risen steeply due to urbanization. It increases plastic consumption in high- and low-income countries and increases production of "use and discard" public behavior (Rodrigues et al. 2019). Because of their widespread use in different sectors, massive amounts of waste plastic are produced in various locations, severely harming the environment, allowing visual pollution with imminent dangers, and defacing cities (Parker 2018, 2019; Ferronato and Torretta 2019). According to the United Nations, plastic waste accounts for 60–80% of marine debris and is one of the world's most pervasive pollution problems affecting our oceans and waterways (Kumar et al. 2017).

Between 2008 and 2015, the plastics market grew by 23% due to their versatility and low production costs, in the year 2020, about 91% of all plastic produced has not been recycled, according to the Environmental Protection Agency (EPA). It's estimated that 12,000 Mt of plastic waste will end up in landfills or the natural environment by 2050 if current plastic production and waste management trends continue (Geyer et al. 2016). In 2015, the plastics industry emitted 1.78 Gt CO_2-eq. In 2050, these emissions will reach 6.5 GT. CO_2-eq (Zheng and Suh 2019). Our natural environment is put at risk as a result of the increasing amount of virgin plastic production and waste.

14.4 Conventional Methods for the Treatment of Plastics Waste

Treatment of plastic wastes and disposal methods that have been used in the past could lead to more harmful effects in the future. It takes much energy to burn plastic fractions, which results in several toxic gaseous products that have serious consequences for humans and the environment (Ali et al. 2009; Pramila and Ramesh 2011). The interaction of plastics with groundwater and moisture-rich substances present in the dump produces toxic sludge, which is harmful to the environment (Teuten et al. 2009). Because of this, plastic waste must be handled and disposed of properly. Recycling plastic waste is one of the most effective alternative treatments, but it has its limitations because it cannot be recycled once it has reached the decline stage of its life cycle (see below). Another approach is to create and use biodegradable polymers. Plastic waste treatment options are divided into three broad categories in Fig. 14.1.

14.4.1 Overview of Plastic Waste Recycling

The United States Environmental Protection Agency (USEPA) defines municipal solid waste (MSW) as any garbage or refuse sludge from a wastewater treatment

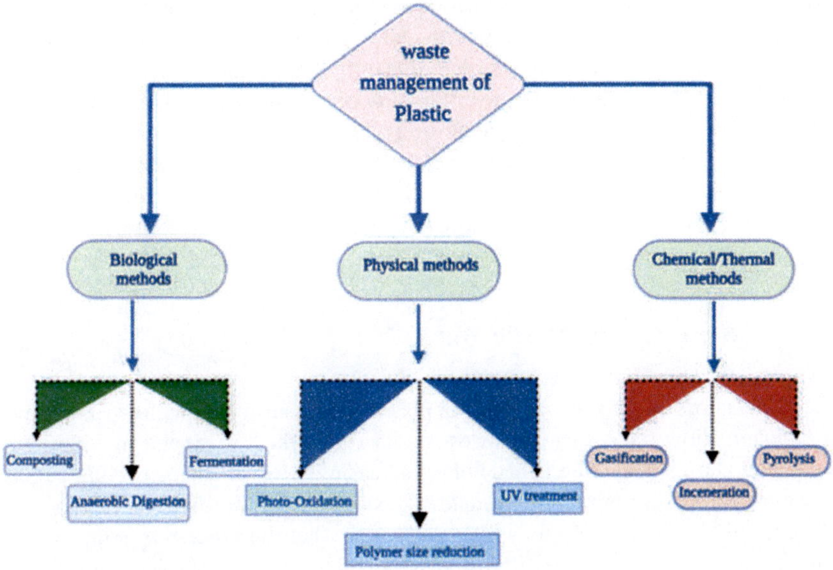

Fig. 14.1 Conventional methods for treating plastic waste

plant, a water supply treatment plant, air pollution control facility, and other discarded material solid and liquid. Semisolid or contained gaseous material produced as a result of industrial, commercial, mining, and agricultural operations, as well as community activities (Jambeck et al. 2015). MSW plastic components increased from 390,000 tons in the 1960s to 31.9 million tons in 2012. Perhaps more importantly, the percentage of plastic waste in MSW increased from less than 1% in 1960 to 12.7% in 2012. Maybe the increase in the discarded amount has made plastics an obvious target for environmentalists (Lowy 2004; Austen 2008; Kumar 2009), and plastic recycling is a widely desired alternative to landfill disposal (Merrington 2017). Plastic recycling encompasses many processes that collect and recycle plastic waste rather than simply disposing of it. Plastic recycling, when appropriately managed, can reduce reliance on landfills, conserve resources, and protect the natural environment from plastic pollution and greenhouse gas emissions (Al-Salem et al. 2009; Hopewell et al. 2009; Ignatyev et al. 2014).

Plastic recycling is defined as the process of recovering scrap or waste plastics and recycling them into valuable products that are often significantly different from their original state (Garcia and Robertson 2017). Alternative plastic recycling strategies have been further subdivided by classifying them based on the final output of any given recovery procedure. Plastic recycling can be classified as primary, secondary, tertiary, or quaternary. Closed-loop recycling entails returning recovered materials to their original state. The recovered plastic is used to make items with the same performance characteristics as those made from virgin polymers. PET recovered from post-consumer bottles makes new bottles, exemplifying primary recycling (La Mantia 2002). The recovered plastic is used in products that have lower performance requirements than the original application. Secondary recycling frequently necessitates refinement to meet the criteria of the new product. The production of flooring tiles from mixed polyolefins is a simple illustration of secondary recycling. Waste plastic is used as a feedstock in the production of chemicals and fuels. The glycolysis of PET to produce diols and dimethyl terephthalate, which can make virgin PET, exemplifies tertiary recycling. The incineration of plastic waste generates energy that can be used for other purposes. TDF (tire-derived fuels) is an excellent demonstration of quaternary recycling in action (Subramanian 2000).

14.4.2 Recycling of Plastic Wastes

The main factors limiting the recycling of plastic wastes are the difficulty in its separation and infections transmission. However, with a landfills shortage and growing environmental alarms, recycling is becoming increasingly important. Recycling begins with the identification of recyclable materials, which is followed by materials sorting. Manually the sorting of plastic wastes is possible, but the process is complex and time taking. Plastic wastes are sorted here based on the type of material, color, shape, and others. Nowadays, there is much interest in automated sorting techniques. The use of near-infrared (NIR) technology allows for rapid identification. This technique works well with transparent plastics. X-ray fluorescence was used to separate PVC from PET because of its organic nature. It's also possible to practice techniques such as density separation and froth flotation with surfactants to sort plastic wastes (Ruj et al. 2015). Figure 14.1 shows the four most common plastic recycling ways after the plastic waste has been collected, sorted, and cleaned. These include primary recycling, secondary recycling, tertiary recycling, and quaternary recycling.

Industrial recycling of plastic wastes was done as a part of the process for standard production within manufacturing companies. The plastic wastes are commonly recycled with new material in extrusion to enhance final production where material and contamination constraints allow. As a result, the focus and public perception of plastic waste have shifted late in the evolution of the plastic industry. Growing concern about the financial and environmental costs of landfilling, as well as the sheer volume of plastic waste entering that waste stream, have both contributed to this trend (Al-Salem et al. 2009).

Material can be considered reclaimable if it has both the technology for waste treatment and the market for the products. Despite increased research efforts to find technological solutions, it's still not economically feasible to separate the wide variety of plastics in the waste stream. The ideal solution would be to recycle single plastic waste that does not require technical effort. Household waste may contain a variety of plastic wastes, including packaging, carrier bags, bottles, and plastic lids, and food and household container waste. Besides being mixed, these plastics contain food waste, residue, labels, glue, and other contaminants. It is possible to use plastics in conjunction with other products, for example, aluminum or other metal. Professional recyclers, usually in large quantities, collect various types of plastic from businesses. As well as these, there's also LLDPE, which is less familiar material. The second largest is polypropylene, accounting for 15% of the total. Other plastics make up 7% of the waste stream. Consideration must also be made of different temperatures required to melt and process them.

One type of plastic may melt, while another may remain solid, depending on the type of plastic used. In this case, as well, the characteristics of the final product can be influenced. While PET melts at 245 °C, PE melts at 135 °C. When PE reaches 245 °C, it is likely to degrade. As opposed to PE, which would need to be melted to be processed at 150 °C, PET would be perfect for processing. Therefore, it is not recommended to recycle a mixture of these two materials (Rigamonti et al. 2014).

14.4.3 Recycling Techniques

Some of the most common processes include physical waste homogenization (such as shredding or shredding of waste) and reprocessing of melts (such as refining). For reprocessing machinery, the feedstock must have a compatible size, and size reduction may be required. Before flaking, large materials need to be shredded to a depth of 25–50 mm. As a result of detergent washing, residues and contamination will be removed from the product. If the raw material is pure, subsequent sorting processes can be more efficient. Material is cleaned of labels, glue, dirt, and other contaminants before being dried.

In many cases, sorting will be required. It's possible to do this manually by entering identification codes. But this is a labor-intensive process, and the final purity depends on the accuracy of the human hand. This method can be cost-effective if human resources are cheap. Nevertheless, mechanized sorting is preferred in the developed world. The different properties of plastics are used in several sorting technologies to separate them (Rozenstein et al. 2017). Polyolefins are polymers with linear C–C double bonds their waste makes up 70% of household waste, and it can be separated using density differences.

14.4.3.1 Primary Recycling

It recycled the plastic wastes into new products with similar properties as the original material. For instance, PET from used bottles can be recycled and used to make new bottles that are similar in shape and look. Mixed with virgin material, recycled scrap or waste plastic can be used to create new products (Al-Sabagh et al. 2016). A closed-loop process is a well-established method. Single-type plastics that are relatively clean or uncontaminated are subjected to this method of degradation. As a result of closed-loop recycling, wastes can be easily incorporated back into the production cycle.

14.4.3.2 Secondary Recycling

It's a process of recovering plastic waste mechanically. The collection of material, its separation and sorting, and its washing to remove organic or other contaminants, and grinding are the major steps for mechanical recycling. The mechanical characteristics are difficult to maintain after mechanical recycling. The reduced molecular weight caused by recycling and contamination with other polymers can reduce mechanical properties. Furthermore, recycled plastics are susceptible to thermomechanical degradation (Ragaert et al. 2017). Highly contaminated waste is difficult to recycle mechanically. When mixed polyolefins are used in the production of flooring tiles, secondary recycling is evident. Other methods such as screw extrusion, injection molding, and blow molding are included in secondary recycling (Singh et al. 2017).

14.4.3.3 Tertiary Recycling or Chemical Recycling

Tertiary recycling, also known as chemical recycling, is converting plastic materials into smaller molecules, usually liquids or gases, that are then used as feedstock to produce chemicals and fuels. Chemical recycling recovers the petrochemical components in plastics through a chemical process. Pyrolysis is one example of this approach, in which plastics are heated to high temperatures in the presence of a catalyst (Rahimi and Garca 2017). A high temperature and high pressure are then applied to convert the polymer into long and small-chain polymers which is prone to easy degradation. Oils, gases, and hydrochar are the three byproducts of Pyrolysis. As a furnace oil, the oil produced can be used in various applications, while the syngas produced can replace natural gas or coal, which can be used for a variety of purposes (Goodship 2007). The feedstock and the technology are used to determine the final output of Pyrolysis (Qureshi et al. 2020). The process of Pyrolysis is carried out in an oxygen-free environment. Gasification refers to the process when it is carried out in a controlled environment. Aragaw and Mekonnen (2021) recently reported pyrolyzing PP to face masks and PVC gloves to produce fuel energy. Pyrolysis converted more than 75% of the waste into bio-crude oil during the one-hour thermal treatment in a closed air system at 400 °C. Moreover, hemolysis and cracking are also best options for the recycling of plastics.

To depolymerize polymers, unlike Pyrolysis, chemolysis uses chemicals instead of heat and flames. Polyethylene and polypropylene can't be used, and neither can unsort or mixed plastics. Depolymerizing polyamides with simple chemicals are not feasible. Therefore, a catalyst is used to shorten the chain of polymers and also decline the energy and time required to depolymerize complex polymer. Fluid catalytic cracking and vapor phase catalytic cracking are two types of catalytic cracking. Catalyst comes into direct contact with molten polymers in fluid catalytic cracking (FCC). The vapor produced by cracking collides with the catalyst in vapor phase catalytic cracking.

14.4.3.4 Incineration of Quaternary Recycling

Through incineration, the volume of wastes is significantly decline by this approach, and the amount and recovery of energy from waste plastic is high. It can be applied in a condition when wastes are highly contaminated, and conventional recycling methods fail. Tire-derived fuel, or TDF recycling, is an example of quaternary recycling (Kumar 2021). For example, activated carbon can be used to reduce harmful gases, as neutralization by acid or addition of ammonia to the chamber of combustion. To effectively decompose toxic waste, the waste must be reduced to less than 1% of its original volume (Ignatyev et al. 2014).

14.5 Plastic Recycling for the Production of Value-Added Products

14.5.1 *Melt Processing of Thermoplastics*

The three simple steps used to make plastic components are: injection molding, blow molding, and film blowing (Al-Salem et al. 2009). Combining polymer with other additives is done by extrusion (called compounding). Plastic sheets and window profiles can also be made with it. More sophisticated and complex products can be produced faster and cheaper with injection molding. A few examples are mobile phone cases, automotive parts, handles of doors, and TV parts. Film blowing is a technique used to create thin films for wrapping or carrier bags. Plastic pellets the size of peas are commonly used as the feedstock for all of these operations. Materials that are recycled must usually go through a conversion mechanism before being used as feedstock (Brems et al. 2012).

14.5.2 *Heat Generation and Distribution*

As the product is being manufactured, it allows to take on a shape after that. Heating and flowing the polymer will be followed by cooling before solidification, it is high temperature required process. Due to molecular damage, chemical degradation can occur due to chain cleavage and crosslinking. During standard manufacturing processes and subsequent processes like granulation, polymers are vulnerable to shear forces. Reprocessing plastics from post-consumer products is troubled by heating and shear forces (Al-Salem 2019).

As a result, recyclable alternatives must be found to process mixed plastics to achieve acceptable qualities. The environment can also damage the polymer. In addition, as it ages, its visual properties may change due to exposure to light, heat, and corrosion. Both virgin and recycled materials have different mechanical properties. As a result, plastic waste will degrade to some extent after its initial use. Heat and shear pressure history, initial stability, and polymer type all play a role in the degree of degradation (Pinto et al. 1999). Plastic degradation can be prevented by using stabilization additives. Other additives, such as fillers can be added to the mixture to improve the properties of recycling. The degradation amount of these materials will be determined by the conditions of processing they were subjected to and the levels of stabilization of materials. It will be possible to re-stabilize them for a second application. Data of the type and amount of stabilizer used initially is a prerequisite for this to work efficiently. These materials must be kept at optimal levels to ensure its protection. If necessary, additional tests to know the stabilization (both processing and long-term) may be required (Anuar Sharuddin et al. 2016).

14.5.3 Reprocessing Thermoplastic Recycles

Mechanical recycling is the most common type of recycling. Initially, recovered plastics are crushed to a size appropriate for reprocessing. The plastics industry has long practiced this in manufacturing facilities. In-house recycling, also known as primary recycling, is beneficial from an economic standpoint because it allows manufacturers to use up their waste and increase production (Wölfel et al. 2020). Recycling used material from other sources necessitates more work for a reuse. Garbage of plastic can come in various shapes and sizes, including bales, moldings, and large plastic lumps. Secondary recycling refers to the recycling of materials in this category. A further complication is that the material's unknown origin, and thus the properties of the resultant recycle produced may differ significantly from the properties of the virgin material, even if the original grade is known (Oliveira et al. 2017).

For the high-quality products, high-quality plastic materials are needed. Since their properties can change, recycled materials have a difficult time competing in markets. Properties, for example, could be different depending on the situation in which they are used. In the past, how many times was it reprocessed? To what extent has it been harmed by thermal degradation? Any polymer will do, no matter the composition, supposing it has been contaminated with contaminants such as oil or dirt. What was it used for in the past, and why? Whether or not it has been damaged due to its long-life span (prolonged light exposure, water exposure, steam, high temperatures, etc.). These answers are usually known in a closed-loop cycle, which is why primary recycling is so typical. In the automotive, packaging, and electronic equipment industries, the creation of closed-loop infrastructures and recycling materials standardization schemes have been critical factors in recycling success. Many industries have strict regulations to improve recycling. Corporations are required to participate in these programs (Guo et al. 2020).

14.5.4 Plastic Conversion to Fuel

Many measures have been implemented to manage plastic waste in the environment. Until now, there have been four important approaches for the disposition of the wastes of plastic globally: landfill, pyrolysis, incineration, and regenerative granulation. These methods pose numerous environmental risks to the soil, water, and air (Kong et al. 2017). Solar-driven conversion, enzyme-driven conversion, and low-temperature-driven catalytic conversion are the dominant available strategies for converting plastics into fuels under mild conditions (Jiao et al. 2021). Plastic waste can be used as a feedstock for thermochemical conversion (via pyrolysis, hydrogenation, and gasification) into energy, lowering costs, landfill constraints, and carbon footprints (Nanda and Berruti 2021).

Many studies in recent years have demonstrated the effective biodegradation of plastic wastes under mild conditions, where the plastics can be modified or consumed by microorganisms and enzymes. For example, bacterial consortia isolated from plastic garbage processing areas were able to degrade LDPE strips with an 81% weight loss percentage after 120 days at 45 °C; and it was also reported that isotactic PP could be converted into hydrocarbons by bacterial consortia isolated from soils in plastic wastes after 175 days. Furthermore, under moderate conditions, other plastics such as PET, PVC, and PS could be biodegraded into carbon-based fuels by various enzymes (Cacciari et al. 1993; Skariyachan et al. 2016; Giacomucci et al. 2019). Depolymerizing enzymes, both extracellular and intracellular, play an active role in the biological degradation of polymers. Microbial exoenzymes degrade complex polymers, resulting in short chains or smaller molecules such as oligomers, dimers, and monomers. These molecules are water-soluble and can pass through the semi-permeable outer bacterial membranes as carbon and energy sources. Depolymerization is the initial process of breaking down polymers; mineralization is the degradation process when the end products are inorganic species (e.g., CO_2, H_2O, or CH_4) (Gu 2003).

Recently, new environmentally friendly waste plastic recycling methods have been introduced for the public's interest, one of them being waste plastics as a supplemental fuel with coal in the steelmaking industry (Nakanishi et al. 2000). Microbial cell factories have been designed to work with chemical processes to deconstruct PET (i.e., hybrid biochemical approach). PET-derived monomers can be biotransformed into high-value platform chemicals and biomaterials, such as bioplastic PET substitutes. It allows for the development of a circular material economy for PET. (Sohn et al. 2020). Under anoxic conditions, anaerobic microorganisms' consortia are responsible for polymer degradation. Under methanogenic (anaerobic) conditions, plastic is converted into microbial biomass, carbon dioxide, methane, and water (Barlaz et al. 1989). Additionally, chemical catalytic upgrading could be used to convert PHA into valuable fully deoxygenated hydrocarbon jet (C8–C16) or diesel (C8–C21) fuels (Linger et al. 2014).

14.5.5 Problems Associated with Plastic Recycling Using Conventional Methods

Due to its tenacity, polyester is a significant burden on the environment and is resistant to biodegradation. Discarded plastic wastes in the same state for a long period, posing serious threat to the ecosystem, plants, wildlife, and humans. Plastics play an essential role in human life due to their numerous and diverse applications. Plastic waste has become a serious issue because of its enhanced production rate and a lack of proper management, efficient treatment, and methods for disposal. India has a nearly 16% annual plastic production rate, China has a 10% yearly production rate, and the United Kingdom has a 2.5% annual production rate (http://www.bpf.co.uk/article/the-pla

stics-industry-in-india-an-overview-446). Some guidelines have been developed to deal with the challenges posed by untreated plastic wastes, which society should adhere to eliminate this waste management problem indefinitely. Every country has its own set of rules and regulations based on necessity and specific criteria that citizens must follow. One of the most pressing environmental issues we face today is the abundance of solid plastic waste. The rising consumption of goods packaged in plastic, combined with the superior economics of plastic production, has made this material an unavoidable part of our daily lives. A plastic, according to the American Society for Testing and Materials (ASTM), is a material that contains an organic substance with a high molecular weight as an essential ingredient, is solid in its finished state, and can be shaped by flow at some point during its manufacture or processing into finished articles (Merrington 2017).

14.5.6 Adverse Environmental Effects of Plastics Recycling by Conventional Methods

Today, over 300 million tons of plastic are produced each year, with 8 million tons of those ending up in the oceans. If we continue to pollute the seas in this manner, the mass of plastic in the oceans is expected to exceed the mass of fish by 2050 (Fakirov 2021). Because of non-degradability, plastic waste pollution poses a significant threat to the ecological environment. The amount of plastic waste in the seas is expected to rise further, owing primarily to increased plastic consumption (about 9% per year) and the inadequacy of its recycling, and waste management practices (Kumartasli and Avinc 2020). According to other statistics, approximately 6300 metric tons (Mt) of plastic waste was generated in 2015, with around 9% recycled, 12% incinerated, and 79% accumulating in landfills or the natural environment. If current production and waste management trends continue, by 2050, approximately 12,000 Mt of plastic waste will be in landfills or the natural environment (Geyer et al. 2017; Geyer 2020).

Traditional plastic recycling processes have several negative environmental consequences. Plastics, for example, account for approximately 10% of household trash and are primarily disposed of in landfills, even though landfilling is the most common method of waste management in many countries. As a result of the types and quantities of harmful substances present and their potential for leaching at waste sites, the consequences of landfills are unfavorable (Gouin et al. 2011). Furthermore, while incineration is a viable alternative to landfilling plastic waste, there is growing concern about the possibility of harmful chemicals being released into the atmosphere during the process. Plastic waste gases, for example, emit halogenated additives and polyvinyl chloride, whereas burning plastics emit furans, dioxins, and polychlorinated biphenyls (PCBs) (Charles 2006). The air pollution caused by toxic gases released into the environment is a disadvantage of burning plastics. This is because plastics cause irreversible damage to the combustion heater of operating systems during plastic incineration. They can evaporate directly into the air and pollute it, while some may form flammable mixtures while others may oxidize in a solid state for an extended time, depending on their variations. It is common for plastics to be burned after they have formed charks, and the amount of coking produced depends on the incineration conditions. Combustion of plastic and plastic composite products produces hazardous gas emissions. The combustion of PVC releases a variety of chemicals that have adverse health effects. As a result of the burning of plastics, ash, soot, and other powders are released into the air, soil, and possibly aquatic environments. They may leach into the soil, pollute the groundwater, or be absorbed by plants growing on the soil, resulting in their entry into the food chain. In some cases, these plastic combustion products can cause pH shifts in the water, disrupting the aquatic ecosystem. In some cases, recycling operations may be ineffective and cause more harm than good (Balakrishnan and Sreekala 2016). As a result, the overall proportion of plastics eligible for recycling may be small, and plastic recycling may not contribute to solving our environmental problems significantly. Making plastic trash suitable for recycling is also time-consuming (Balakrishnan and Sreekala 2016).

Being a major source of toxic and hazardous pollutants plastic poses serious threat to the environment. Because it is made of harmful chemicals and is non-biodegradable, plastic pollutes the environment and contributes to air and water pollution. This also interacts with the food chain, affecting both the environment and humans and animals. Plastic waste cannot be disposed of safely, and waste causes significant environmental damage during its manufacture, use, and disposal. Current consumption patterns are insecure because they cause pollution and other negative consequences, resulting in a global waste management issue (Thompson et al. 2009a, b). When waste plastic is discarded improperly, it pollutes the environment in which it is dumped. Invertebrates are more likely to absorb heavy metals into their bodies due to plastic consumption (Hodson et al. 2017). It has been reported that micro-and macroplastics are entering terrestrial food webs and the effects of plastic pollution on terrestrial geochemistry, the biophysical environment, and ecotoxicology (Lwanga et al. 2017). As a result, proper plastic waste management is a critical issue addressed globally through policy and waste management.

The environmental impact of recycling is determined by various factors, including the amount of energy used to collect the plastic waste and the type of material and application replaced by the recycled plastic (Shen and Worrell 2014). The market for recycled plastic is still small but growing. The applications of recycled materials differ, influencing the overall environmental benefits and economics of recycling. The increased generation of polymeric waste materials (plastics and rubbers) worldwide has led to the development of effective methods to reuse these waste materials and reduce the adverse effects caused by simple disposal into the environment. Traditional processes of removing polymer waste, such as combustion and landfilling, have several drawbacks, including the formation of dust, fumes, toxic gases in the air, and pollution of underground water resources (Fazli and Rodrigue 2020). Despite significant global progress in plastic waste management, treatment, and recycling over the last three decades, most plastic waste may still end up in landfills or is openly burned, emitting carbon monoxide (CO) and carbon dioxide (CO_2). They may end up in engineered landfills at best. The traditional method of waste management is landfill, but landfill space is becoming scarce in some countries. Apart from the effects of collection and transportation, a well-managed landfill site causes little immediate environmental harm; however, there are long-term risks of contamination of soils and water by some additives and breakdown byproducts in plastics, which can turn into persistent organic pollutants (Tahir et al. 2019). When plastic waste is burned, landfill space is saved, but hazardous substances could be released into the environment. There is a high probability that dioxins, other polychlorinated biphenyls, and furans will be released into the environment when mixed plastic waste is disposed of (Teuten et al. 2009). Plastic incineration is less common as a waste management strategy than landfilling and mechanical recycling because of the perceived pollution risk.

14.5.7 Limitations of Different Recycling Methods

The process of reusing waste generated regularly by industries and households is known as recycling. It has both positive and negative impacts. Plastic recycling is vital for environmental protection. Recycling is essential for lowering pollution and making the environment safe (Salhofer et al. 2021). For example, large numbers of trees are cut to make paper, and forests are being destroyed for the sake of human comfort, resulting in increased pollution and disease. Paper reusing would help to reduce the need for plant and tree cutting. The most common thermoplastic polymer recovery methods are recycling and incineration. Incineration causes problems such as releasing toxic gases and metals like lead and cadmium. Recycling has several advantages, including reducing environmental issues and saving both material and energy Fig. 14.2 (Mohammadzadeh 2009; Francis 2016, Favis and Le Corroller 2017).

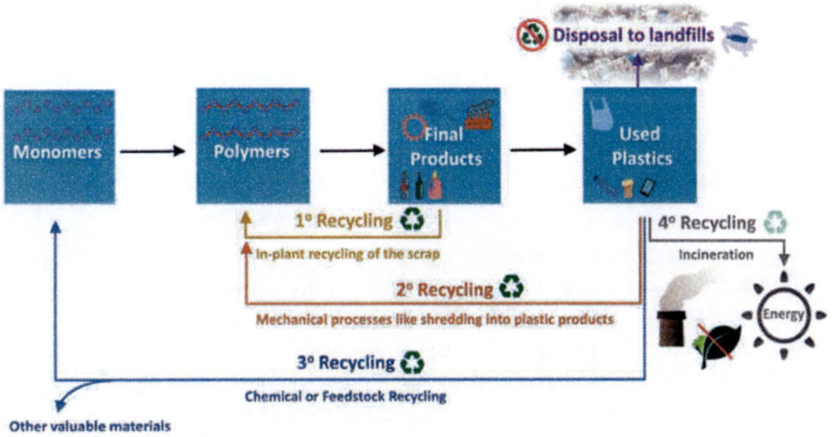

Fig. 14.2 Different plastic recycling methods (Okan et al. 2019)

Table 14.1 Advantages and disadvantages of different recycling methods

Technique	Advantages	Disadvantages	Reference
Primary	Simple and cheap to adopt	primary recycling is where PET recovered from post-consumer bottles is used in the production of new bottles, landfills all over the world are composed of non-biodegradable Variety of polymers and materials	Nkwachukwu et al. (2013), Singh et al. (2017)
Secondary, Mechanical	Low cost, great efficiency, and well known	Heterogeneity of the solid waste	Hopewell et al. (2009), Grigore (2017)
Chemical recycling	Monomers or raw materials produced which made up the original polymer	Increasing use of petroleum-based products and no biodegradability, large amounts of organic solvent	Shukla et al. (2009), Sezgin and Yalcin-Enis (2020)
Energy recovery	Polymers are used to generate a significant amount of energy	HCl and traces of highly toxic dioxins and furans, un acceptable in ecology	(Al-Salem et al. (2009), Grigore (2017)

The disadvantage of primary recycling is a limit to the number of cycles for each material. Furthermore, the drawbacks of this approach in secondary recycling are the heterogeneity of the solid waste and the decline of product properties in each cycle due to the low molecular weight of the recycled resin (Grigore 2017). Chemical recycling, also known as feedstock recycling, has the advantage of recovering the petrochemical components of polymers, which can then be used to remanufacture plastic or generate other synthetic chemicals (Hopewell et al. 2009). Each process has its own set of advantages and disadvantages Table 14.1.

14.6 Methods Used for Conversion of Waste Plastic to Energy

Plastic polymer waste is commonly recovered by recycling and incineration. Several environmental issues must be addressed when using waste-to-energy processes, including the release of toxic gases and heavy metals (quaternary recycling), which is widely adopted in Singapore and Japan (Michael 2013; Saleh & Danmaliki 2020). Many commercial companies have started initiatives to implement tertiary recycling in response to a pressing need for enhancing the recycling rates. Refers to processes that dissolve the polymer's long-chain repeating hydrocarbon backbone into mixtures of monomers and oligomers with shorter chains. De novo plastic synthesis feedstock or other applications such as fuels can be produced from the output (Olah et al. 2009). It is the first step in closing the loop on the plastic life cycle. Current recycling operations account for a tenth of total recycling operations (Hundertmark et al. 2018a, b). There are two notable examples: Eastman's investment into PET methanolysis and BP's PET depolymerization plant in the United States (Hundertmark et al. 2018a, b).

Recently, biologically mediated approaches have attracted some attention from the scientific community. In the past, researchers have looked into whether microbes can speed up the degradation of intrinsically degradable plastics (e.g., polylactic acid, polycaprolactone) or act as bioplastics factories (Keshavarz and Roy 2010; Karamanlioglu et al. 2017). Despite these advances, synthetic plastics are still a massive problem in the environment. This is due to the discovery of a microbe capable of degrading PET in 2016. Since then, the number of publications in the field of microbial plastic degradation has increased exponentially (Yoshida et al. 2016). If the hydrocarbon backbone is ruptured by gasification, pyrolysis, or depolymerization into oligomers and monomers by a targeted biochemical approach, such as hydrolysis, or methanolysis, then tertiary plastic recycling is defined as a bulk conversion. Each of these processes is described in detail in the following sections.

14.6.1 Thermal Degradation

This approach is also known as thermal cracking, pyrolysis breaks down polymer chains into smaller, intermediate products in the presence of heat and oxygen-free conditions. Inert gases can also be used to speed the process because it can treat co-mingled mixtures of different plastic types while minimizing the negative impact on the waste from various organic, inorganic, or biological residues, this process has many advantages (Al-Salem et al. 2017). On the contrary, high-temperature combustion produces toxic gaseous emissions as well as climate-altering emissions (Brems et al. 2011). The degradation efficiency of other chemical recycling technologies is dependent on the purity of the input. Depending on the plastic component, waste plastics are fed into a pressurized primary reaction chamber that has been purged of oxygen and heated to around 500 °C for varying times. There is a large amount of liquid and gaseous hydrocarbon fuel (> 90 wt%), with a small amount of solid ash residue (10 wt%). As a result of this process, it is possible to vary the liquid/gas ratios of the desired end products by modulating certain operational variables appropriately (Bridgwater 2012).

Plastic wastes such as PET, PP, PE, and PS have had their pyrolysis reaction conditions optimized for the most common plastic wastes, with PS yielding the most liquid fuel and PET producing the least (Adrados et al. 2012). However, these reactions took place in batch or semi-batch reactors rather than under continuous processing conditions, resulting in a > 90% conversion of PE and PP into light oil (Czajczyska et al. 2017). Onwudili and colleagues showed in 2009 (Onwudili and colleagues 2009) that 97% liquid oil could obtain from PS pyrolysis at 425 °C, confirmed by Liu and colleagues (Abnisa and Daud 2014). According to reports, PET pyrolysis produced only 39.9% liquid and 53% gaseous fuel. PET's lower volatile matter content results in lower liquid yields, directly related to the lower liquid yield (Abnisa & Daud 2014). Recently, researchers have also looked at plastic mixtures gathered from real-world waste streams in their research. Because they can be made from various plastics, liquid oils are more resistant to contaminants in the input stream than other recycling methods. As a result of multiple studies, PP/PE/PS waste mixtures yielded nearly 50% liquid oil. While pyrolysis of single plastics produces less liquid oil, these results are sufficient to show that other materials, such as those found in common single-use plastics such as packaging materials, can be tolerated (Demirbas 2004). When waste plastic packaging was pyrolyzed at 500 °C in a non-stirred semi-batch reactor, it yielded 40.9% liquids and 25.6% gases fuels (Adrados et al. 2012).

14.6.2 Chemical Degradation

It is an alternative to the non-specific decomposition of plastic waste. As a result, the economic value of these processes is higher (Francis 2016). By improving reaction efficiency or lowering reaction temperatures with catalyst addition, overall costs are reduced. As a result of the high demand for infrastructure investment, research and development time, and specialized personnel, industrial implementation has been hampered. However, a plastic circular economy requires the conversion of plastics into component monomers that can be used to remanufacture virgin plastics. Depolymerization reactions are compatible with condensation polymerized polymers such as PET, allowing them to return to their monomers. Lacking a targetable side chain, polyolefins synthesized via addition polymerization, such as PP and LDPE, can't be broken down in the same way as their monomers; pyrolysis is a better option (Al-Salem et al. 2017). As a result, DMT and EG are produced by PET methanolysis at elevated temperatures and pressures (180–280 °C and 2–4 MPa, respectively) (Scheirs 1998). Major manufacturers such as DuPont have used it for decades to recover clean PET scrap from the manufacturing cycle (Sinha et al. 2010). DMT purified via distillation is the desired end product, and up to 90% conversion rates are possible. As a result of the difficulty of injecting waste material into a pressurized reactor, this process is not suitable for long-term operation. In addition, the produced EG contains 11–20% dissolvable (MHET), which hinders subsequent EG purification and raises the cost of the process (Achilias and Karayannidis 2004).

There are many companies that use PET glycolysis in the commercial world. Bis(hydroxyethyl) terephthalate (BHET) can be produced by treating PET with ethylene glycol for an extended period. A milder version of the process, partial glycolysis at 190–240 °C and 0.01–0.6 MPa, will produce oligomers of varying lengths (Richard et al. 2011). A neutral or alkaline environment can hydrolyze PET flakes. TPA is extracted by acid neutralization of the reaction mixture after PET is treated for 3–5 h with an aqueous hydroxide solution of 4–20 wt% (Scheirs 1998). Alicia Lee 2020, reports that alkaline hydrolysis is highly dependent on the use of a catalyst; without a catalyst, complete degradation can only occur at 200 °C, whereas it can be reduced to 90 °C with the help of a phase transfer catalyst, such as trioctyl methyl ammonium bromide, with 90% reaction after five hours with only 0.01 mol catalyst/mol PET (Alicia Lee 2020). A high-quality TPA (monomer) is created, and the reagents are inexpensive and not needed in large quantities. It is important to note that this method can also be used to treat PET that has been contaminated with other materials (Kumagai et al. 2018). TPA and EG are extracted from the corrosive mixture using acid hydrolysis, which is less flexible and expensive because concentrated mineral acids (sulfuric, nitric acid) are used in the depolymerization process. However, even a slight reduction in sulfuric acid concentration slows down the reaction by more than 50%. As low as 70 °C is possible, the process takes more than 70 h (Yoshioka et al. 1998); increasing the reaction temperature allows for a more dilute acid to be substituted and product recovery via dialysis to be achieved Table. 14.2 (Paszun & Spychaj 1997).

14.6.3 Microbial or Biological Degradation of Waste Plastics

For a clean, and eco-friendly way to degrade plastics, biodegradation is a viable option (Leja and Lewandowicz 2010). Over time and prolonged exposure to plastic pollutants in the ecosystem, naturally occurring microorganisms in the soil, compost, or marine environments have evolved capabilities to degrade plastic, and metabolize it as an energy source (Emadian et al. 2017). According to characterization studies, degradative enzymes can act on the polymer backbone in a targeted manner under environmentally neutral conditions with little energy input (Ronkvist et al. 2009). Microorganisms found in plastic recycling could provide a more robust process implementation. Microbial degradation has led to new investments in scaling this technology by companies like Carbios and BioCellection.

Standard plastic disposal methods, such as landfilling, incineration, and recycling, are ineffective for the effective management of plastic wastes, and thus there is growing interest in the use of efficient microorganisms for biodegradation of recalcitrant synthetic polymers (Seneviratne et al. 2006). Larger polymers are more difficult to degrade than smaller ones. For example, the degradation of polyethylene (PE), used as the sole source of carbon and energy in soil microorganisms, revealed that small fragments were consumed faster than larger ones (Kawai et al. 2004). However, this cannot be used to indicate biodegradation that requires a decrease in molecular weight; other factors may influence plastic biodegradation (Sivan 2011).

In general, the biodegradation process consists of the following major stages: Adherence of the microorganism to the plastic surface, growth, and colonization of the microorganism, including biofilm formation by metabolizing the polymer as a carbon source, cleavage of the polymer through enzymatic hydrolysis into shorter-chain molecules, and ultimate degradation into low molecular weight compounds (Bhardwaj et al. 2013). Because this is primarily mediated by enzymatic action, its

Table 14.2 Comparison of the various tertiary plastic waste recycling methods and conditions, and an analysis of feasibility of implementation

Methods of recycling	Input plastic types	Condition of the reaction	Output	Advantages	Disadvantages	References
Pyrolysis	PP, PE, PET, PS	500 °C	Liquid + Gas Hydrocarbon Fuels	Liquid fuels can be directly used Does not need Pre-treatment of inputs Under pilot tests Works with mixtures of waste	High energy input needed Not a closed loop on plastics	Keshavarz and Roy (2010), Leja and Lewandowicz (2010)
Methanolysis	PET	180–280 °C 2–4 MPa	DMT, EG	Conversion up to 90% Already integrated into polymer Production lines	Hurdled in injection to reactor EG contains residual MHET Declining demand for DMT	Brems et al. (2011)
Glycolysis	PET	190–240 °C 0.1–0.6 MPa	BHET	Already integrated into polymer Production lines	End products need purification BHET declining demand	Bridgwater (2012)
Acid hydrolysis	PET	70–90 °C	TPA, EG	Short reaction time and lower operating temperature	Requires sorting High concentration of acid needed for high conversion Need corrosion-resistant reactors Extraction of products, e.g., EG is challenging	Abnisa and Daud (2014), Yoshida et al. (2016), Al-Salem et al. (2017)

(continued)

Table 14.2 (continued)

Methods of recycling	Input plastic types	Condition of the reaction	Output	Advantages	Disadvantages	References
Alkaline hydrolysis	PET, another minor Contaminants	210–250 °C 1.4–2.0 MPa 90 °C with catalyst	TPA, EG	90% conversion with catalyst Can treat PET contaminated with other materials, e.g., PVC	Requires sorting of waste Polymer crystallinity can be rate limiting Catalyst waste management	Adrados et al. (2012), Yoshida et al. (2016), Al-Salem et al. (2017)
Neutral Hydrolysis	PET	200–300 °C 1–4 MPa	TPA, EG	Eco-friendly; no inorganic salts production	Requires sorting Cannot handle impurities in PET Not suitable for polyolefin	Sulaiman et al. (2012), Al-Salem et al. (2017)
Biological organisms	PE, PP, PET	30–70 °C, ambient pressure Aerobic reactor	TPA from PET, oligomer- ers, CO2, methane	Eco-friendly; low generation of greenhouse gases	Long degradation time Management of biomass Expensive start-up investment Variability in live microbial feed-stocks	Polk et al. (1999), Kumagai et al. (2018)
Biological enzymes	PE, PP, PET	40–70 °C, ambient pressure pH 8.0	TPA, EG	Decline the emission of GHGs Genetic and protein engineering for the optimization of performance Genetic engineering Can be combined with pre-treatment, e.g., UV, heat	Scalability and overall industrial feasibility have not yet been demonstrated	Paszun and Spychaj (1997), Yoshioka et al. (1998)

effectiveness is highly dependent on the polymeric chain's accessibility. Because of their physical and chemical properties, plastics are resistant to degradation. Surface hydrophobicity limits microorganism adhesion and colonization; higher-order structures, such as crystalline regions, and higher melting temperatures prevent amorphous polymer regions, which are more susceptible to enzymatic attack (Arkatkar et al. 2009).

A recent study found that 76% of the total plastic production is disposed of as waste. 90% of the waste is recycled, 14% of it is incinerated, and the remaining 79% is disposed of in landfills or released into the environment (Garcia and Robertson 2017; Geyer et al. 2017). When microorganisms feed on polymers such as ethylene and polyurethane, they degrade them (Glass and Swift 1989). Degradation of common plastics such as HDPE, LDPE, and PP begins with photo degradation (primarily due to UV-B radiation) and progresses through thermo oxidation and hydrolysis to a lesser extent natural condition (e.g., the marine environment). Plastic fragmentation (smaller pieces) and a decrease in polymer molecular weight are two effects of degradation processes (MW). Microbes can then metabolize them. The degradation of plastic polymers can take more than 50 years (Andrady 2011; Webb et al. 2013). Environment and polymer properties have an impact on plastic degradation (Ali et al. 2021a, b). PET and polyethylene were traditionally considered non-biodegradable. On the other hand, microbes may be able to degrade, transform, and metabolize them (Alshehrei 2017). In addition to mangrove rhizosphere soil, polythene buried in soil, marine water, and plastic and soil at dumping sites, plastic-degrading microbes have been isolated. Plastic is naturally degraded by bacteria, fungi, and algae (Rutkowska et al. 2002).

As a result of plastic's high molecular weight, which makes it insoluble to microbes, microbes can quickly degrade small subunits in the form of monomeric units or oligomeric units. The substrate is accumulated in the cell membranes of microorganisms and then degraded by cellular enzymes (Shah et al. 2008). For plastic waste to biodegrade, it must first be broken down into low molecular weight compounds such as oligomers, dimers, and monomers using enzymes that bind to the polymer and catalyze its hydrolytic cleavage. In the end, these low-MW compounds are converted to CO_2 and H_2O through mineralization. Many microbes have been identified as polymer degraders, including bacteria, fungi, and algae (Ali et al. 2021a, b). The following steps are involved in polymer biodegradation: Affixation of the microorganism to the surface of the polymer as a carbon source, the polymer encourages the growth of microorganisms. The ultimate decomposition of the polymer. If a polymer is hydrophilic, microorganisms can attach to it. Using the polymer as a carbon source, the microorganism can grow once attached to the surface. First, extracellular enzymes cause cleavage, resulting in low molecular weight fragments such as dimers, oligomers, or monomers in the organism. They are then converted into carbon and energy by microbes. Assimilation of the organism's internal environment by small oligomers is possible (Premraj and Doble 2005).

It has been determined that several microbial and fungal species are capable of degrading polymers such as PP, PE, and PET. Although they did not cause as much weight loss in PE films as *Aspergillus Niger* and *Penicillium pinophilum* from the soil,

Kocuria palustris M16, *Bacillus pumilus* M27, and *Bacillus subtilis* H1584 from the Arabian Sea did (Polk et al. 1999). PE is intrinsically resistant to degradation due to its hydrophobicity, highly stable C–C and C–H covalent bonds, and lack of targetable functional groups (Harshvardhan and Jha 2013). Similar to PP, which is composed of methylene subunits, PP is highly hydrophobic and resistant to enzymatic attack, with a soil consortium that contained *Bacillus flexus*, for example, only being able to induce 10.7% weight loss, 51.8% loss in tensile strength, and 28% growth in crystallinity of the film over 12 months (Arkatkar et al. 2009). Just like with chemicals, polyolefins may require additional abiotic treatments to expose reactive end groups and promote hydrolysis, including UV or thermal oxidation. LDPE films that had been UV-treated for 30 days at 50 °C were degraded by *Brevibacillus borstelensis* strain 707. (Hadad et al. 2005).

For plastic management, bacteria biodegradation is a promising technology. Bacillus and Streptomyces spp., among others, have shown a high degradation potential against a variety of plastic polymers (Yoshida et al. 2016; Li et al. 2020). Over 90 genera of bacteria and fungi, as well as actinomycetes, can degrade plastic. As a rule, microorganisms take a long time to degrade plastics, while certain microorganisms are incapable of degrading certain plastics (Mahdiyah and Mukti 2013; Jyoti and Gupta 2014). Microorganisms, such as bacteria, fungi, and actinomycetes, degrade both natural and synthetic plastics. Plastics biodegrade aerobically in the environment, anaerobically in sediments and landfills, and partially aerobically in compost and soil, according to the Environmental Protection Agency (EPA), biodegradation in anaerobic conditions produces carbon dioxide and water, whereas aerobic biodegradation generates methane (Ishigaki et al. 2004).

Diverse bacteria capable of breaking down polyolefins (PE, PS, and PP), PVC, and PET have been isolated from the environment, including soil from a plastic-dumping site, mulch-film waste, marine water, and soil contaminated with crude oil (Ru et al. 2020). Insect larvae such as mealworms and waxworms have been found to harbor bacteria that depolymerize plastics. (Yang et al. 2014, 2015). It has been shown that Pseudomonas strain AKS2, isolated from soil, deep sea, or waste dumps, can degrade PE (Tribedi and Sil 2013; Urbanek et al. 2018). As a result, different microbes degrade various types of plastics in different ways. It was found that the maximum molecular weight for microorganisms to degrade PE was around 2000 Da (Watanabe et al. 2004). Therefore, a synergistic action of photo- or thermoxidation and microorganism biological activity was assumed to be responsible for the environmental degradation of long-chain PE (Hakkarainen and Albertsson 2004). On average, over the course of six months in shaker culture, Pseudomonas species (37.09 and 28.42%), Streptomyces species (46.16%), and Aspergillus species (20.96 and 16.84%) all degraded polythene and plastic to varying degrees (Usha et al. 2011). Over the course of a one-month study, Pseudomonas species degraded 20.54 percent of polythene and 8.16% of plastics, and Aspergillus species degraded 28.80% of polythene and 7.26% of plastics (Kathiresan 2003).

Several promising and potential sources of plastic-degrading bacteria at landfill sites, including activated sludge from wastewater treatment plants, contaminated soil and sediments, and the guts of invertebrates that feed on the waste. On the

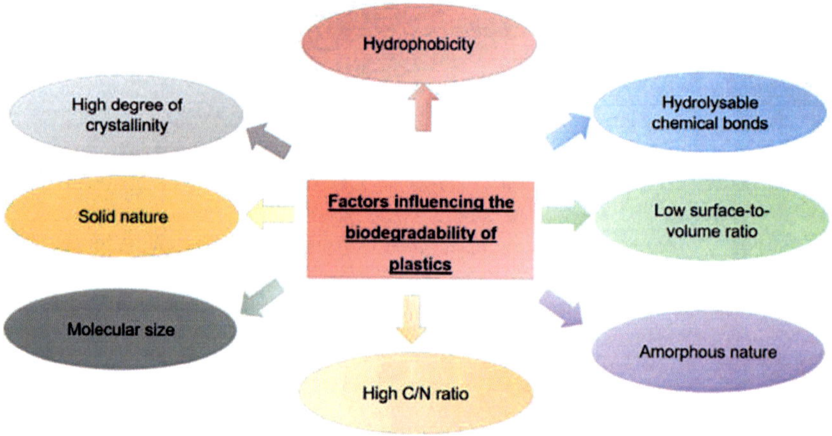

Fig. 14.3 Factors influencing the biodegradability of plastic

soil-buried polyester polyurethane surface, bacteria and fungi were found to grow, and their ability to degrade PU was then tested in the lab (Nakajima-Kambe et al. 1995). More fungi have been discovered than bacteria. Many bacteria from the genera Pseudomonas, Bacillus, and Comamonas can degrade polyester in vitro, using PU as their only carbon source (Ruiz et al. 1999; Rowe and Howard 2002). *Brevibacillus borstelensis* is a thermophillic bacterium isolated from the soil when used as the sole carbon and energy source. After incubation with *B. borstelensis* and polyethylene, the polyethylene molecular weight was reduced by 30%. (Hadad et al. 2005). Brevibacillus spp. and Bacillus spp. are two other strains that can degrade polyethylene. There is a role for proteases in the process (Sivan 2011). Because of this, they were exploring the conditions required for biodegradation processes, and the mechanisms involved are essential. The identification, interaction with plastics, and isolation of microbes that degrade plastics must be studied. Microbes can break down biodegradable synthetic plastic, but multiple factors can influence the biodegradable plastic by microbes and are presented in Fig. 14.3.

14.7 Microbial Enzymes Used in Plastic Degradation

Synthetic plastics are highly resistant to a wide range of physical, chemical, and biological factors (Thompson et al. 2009a, b). However, because of their durability, they degrade exceptionally slowly in the environment. The synthetic polymers' hydrophobicity, crystallinity, surface topography, and molecular size limit their biodegradability. In addition to those factors, many other factors and characteristics, such as hydrolyzable chemical bonds, hydrophobicity, and a low surface-to-volume ratio, made plastic resistant to biodegradability (Webb et al. 2013; Restrepo-Flórez et al. 2014). These constraints and plastic's resistance to biodegradability can be overcome through UV irradiation, oxygen, high temperatures, and the addition of chemical oxidants and peroxidation (Motta et al. 2009). Laccase, cutinase, hydrolase, esterase, protease, and urease were reported to be the most abundant enzymes from bacterial sources. *Streptococcus*, *Bacillus*, *Pseudomonas*, and *Staphylococcus* are the microbial species that have been thoroughly studied for plastic degradation (Bhardwaj et al. 2013).

Ideonella sakaiensis, a bacterium, can use polyethylene terephthalate (PET) as its primary energy and carbon source. PET is converted into its monomers terephthalic acid and ethylene glycol, by two actively participating enzymes (PETase and MHETase) in this bacterium (Yoshida et al. 2016; Palm et al. 2019). Even though a few enzymes and their activity on plastic degradation are known, the problem has received little attention. This is due to a lack of technologies for improving the efficiency and commercialization of these plastic-degrading enzymes. Plastic degradation and deterioration are caused by abiotic factors and microorganisms working together; the bulk polymer becomes fragmented, with more exposed surfaces available for a biological attack. Inducible extracellular enzymes play an essential role in the subsequent depolymerization of synthetic polymers (Sivan 2011; Bhardwaj et al. 2013).

PET, unlike PE and PP, has a large number of hydrolytic enzymes that have been identified. There are many cutinases, which have broad substrate specificity and can hydrolyze both insoluble triglycerides and esters that are soluble. Cutinase from leaf-branch compost degrades PET into TPA and EG when incubated for 24 h at 50 °C and pH 8.0. (Sulaiman et al. 2012). At 70 °C, cutinase from Humicola isolates degrades a PET film by 97% after only 96 h (Sulaiman et al. 2012). Scientists in Japan have discovered that *Ideonella sakaiensis* 201-F6 is one of the strains of bacteria that can degrade PET bottles in a bottle recycling facility in Sakai, Japan (Yoshida et al. 2016). With a 0.13 mg/cm^2 day degradation rate, *I. sakaiensis* 201-F6 preferentially degrades PET and MHET at 30 °C. PETase and MHETase from *I. sakaiensis* were identified and characterized in subsequent research, paving the way for rational redesigns to improve binding efficacy to PET, thermal stability, and hydrolytic activity in the future (Alicia Lee 2020).

The ability of two Actinomycetes sp. to degrade or modify PE films was also demonstrated, demonstrating that extracellular enzymes detected were able to lessen the polymer, albeit at a slow rate (Orhan and Büyükgüngör 2000). For six months, LDPE modified with starch was tested for biodegradability in soil microcosms. It was discovered that inoculating soil with P. chrysosporium increased degradation and biomass much more than non-inoculated soil (Orhan and Büyükgüngör 2000). As lignin-degrading fungi IZU-154, Phanerochaete chrysosporium, and Trametes Versicolor degraded polyethylene membrane, manganese peroxidase (MnP) was found to be the primary enzyme in polyethylene degradation. The treatment of polyethylene membranes with partially purified MnP in the presence of Tween 80, Mn(II), and Mn(III) chelators resulted in significant degradation (Iiyoshi et al. 1998). It was found that the fungus IZU-154, which degrades lignin, formed four end groups,

CHO, NHCHO, CH3, and CONH2, which indicated that nylon-66 was degraded by an oxidative process (Deguchi et al. 1997; Kale et al. 2015).

In lignin-degrading fungi, laccases catalyze the oxidation of a variety of polyaromatic molecules. Aside from aromatic substrates, laccase (La) is also well known for its ability to act on non-aromatic substrates (Mayer and Staples 2002). In polyethylene, laccase can oxidize the hydrocarbon backbone. Cell-free laccase reduces the average molecular weight and molecular amount of polyethylene by 20% and 15%, respectively, when incubated with polyethylene (Bhardwaj et al. 2012). Laccase produced by the actinomycete *R. ruber*, involved in polyethylene biodegradation. Laccases are most commonly found in lignin-degrading fungi, where they catalyze the oxidation of aromatic compounds. Laccase activity has been demonstrated to act on non-aromatic substrates (Mayer and Staples 2002). Plastics are being degraded by microbial enzymes capable of degrading lignin, such as laccases, manganese peroxidase, and lignin peroxidases (Krueger et al. 2015). Laccase can aid in the oxidation of polyethylene's hydrocarbon backbone. After incubation with polyethylene, cell-free laccase reduces the polyethylene's average molecular weight and molecular amount by 20% and 15%, respectively (Sivan 2011). Papain and urease are the two proteolytic enzymes that have been discovered to degrade medical polyester polyurethane. Urethane and urea linkages hydrolyzed by papain produced free amine and hydroxyl groups in the polymer degraded (Phua et al. 1987). It is a renewable plastic resource, hydrolyzed by enzymes, and the hydrolysate can be recycled as a polymer material. Lipase from Rhizopus delemer and polyurethane esterase from *Comamonas acidovorans* were investigated to degrade low molecular weight PLA, and Amycalotopsis sp. strains were found to degrade high molecular weight PLA (Masaki et al. 2005). Serine hydrolases, esterases, and lipases are the enzymes responsible for the Pseudomonas spp. Biodegradation. PHA depolymerizes are serine hydrolases that can attack the polymer's branching chains and cyclic components (Tokiwa et al. 2009).

Protein and enzyme sources such as putative polyurethanases have been isolated and characterized from several microbes, including *Pseudomonas chlororphis* and *Comomonas acidovorans*, along with the fungus Candida rugosa. Activated enzymes such as esterases, lipases, proteases, and ureases degrade polyurethane by cleaving the ester bonds. Serine hydrolase-containing fungus *Pestalotiopsis microspora* uses polyurethane as a carbon source and degrades it in a matter of days (Russell et al. 2011).

14.8 Economic Feasibility of Plastic Conversion to Fuel

Capital investment, fixed costs, production costs, and variable costs all influence producing plastic fuel. Several variables must be considered when calculating the return on investment (ROI), including plastic-type, transportation, separation of municipal waste plastic from municipal waste plastic, and labor (Westerhout et al. 1998; Al-Salem et al. 2014). Costs per tonne range from US$65 to $400. (Miskolczi et al. 2009). The cost of labor has a significant impact on the return on investment

(ROI). ROI (return on investment) is defined as profit before taxes as a percentage of total investment (Westerhout et al. 1998). The ROI of a Bubbling Fluidized Bed (BFB) reactor operating at 740 °C is 21.7%, a Rotating Cone Reactor (RCR) operating at 625 °C is 14.2%, and a Circulating Fluidized Bed (CFB) reactor operating at 840 °C is 29.5%, according to a techno-economic analysis of a mixed plastics reactor in the Netherlands A lower pyrolysis temperature is advantageous from an economic and environmental standpoint because energy recovery isn't perfect for a CFB pyrolysis plant with a 50 k ton/year capacity, the capital investment ranges from 7.84 to 11.2 million dollars (Westerhout et al. 1998). According to another study, the tipping fee was set between $35 and $45, while the reactor feeding rate was set at 200 t/day, resulting in a 15% ROI (Shelley and El-Halwagi 1999). Plastics thermochemical conversion in the United Kingdom was studied by Al-Salem et al. (2014). In comparison to low-temperature pyrolysis, hydrocracking offers a higher return on investment (LTP). ROI for LTP is 43% on an annual basis. Hydrocarking, on the other hand, has a higher net present value and pre-tax profit than LTP. Techno-economic aspects of a fluidized bed gasification unit for converting plastic into energy are being studied in Italy. According to the study, a 23.7% energy conversion efficiency yields an ROI of 8.3% at a plant cost of 4.79 k€/k Wt and an operating cost of 0.74(k€/y) per k Wt. The operational costs include all variable costs (Arena et al. 2011).

14.9 Conclusions and Recommendations

Globally a considerable amount of plastic is used because of its versatile nature and remarkable properties. Plastic is highly resistant to natural degradation due to its complex structure. Plastic wastes are accumulated continuously in the environment. It is considered one of the most severe threats to the ecosystem and the top pollutant. Various types of plastic are synthesized and used throughout the world. There are numerous methods for managing plastic wastes in the environment, but each has its limitation and hazards to the environment. Plastic wastes can be converted into value-added products and play a vital role in the circular economy. In value-added products, plastic conversion, energy is one of them that can help manage plastic wastes. Thermal degradation, chemical degradation, and biological degradation are various means for plastic to convert into energy. Each method of plastic conversion to energy has some limitations, which need to be solved and explored in all the mentioned methods. Especially microbial and biological degradation needs more attention to study novel microbes and their enzymes. In addition to that, genetic engineering and omics approaches are needed to discover and engineer a particular microbe for converting plastic to fuel. Plastic conversion to fuel is a very expansive and costly process, and it needs cheap resources and methods to recycle plastic into value-added products with special emphasis on fuel. It is concluded from the literature that plastic wastes have a high potential to be used in the circular economy and to use for the production of value-added products.

References

Abnisa F, Daud WMAW (2014) A review on co-pyrolysis of biomass: an optional technique to obtain a high-grade pyrolysis oil. Energy Convers Manage 87:71–85

Achilias D, Karayannidis G (2004) The chemical recycling of PET in the framework of sustainable development. Water Air Soil Pollut Focus 4:385–396

Adrados A, De Marco I, Caballero BM, López A, Laresgoiti MF, Torres A (2012) Pyrolysis of plastic packaging waste: a comparison of plastic residuals from material recovery facilities with simulated plastic waste. Waste Manage 32:826–832

Ali MI, Perveen Q, Ahmad B, Javed I, Razi-Ul-Hussnain R, Andleeb S, Atique N, Ghumro PB, Ahmed S, Hameed A (2009) Studies on biodegradation of cellulose blended polyvinyl chloride films. Int J Agric Biol 11:577–580

Ali M, Ahmed S, Javed I, Ali N, Atiq N, Hameed A, Robson G (2014) Biodegradation of starch blended polyvinyl chloride films by isolated *Phanerochaete chrysosporium* PV1. Int J Environ Sci Technol 11:339–348

Ali SS, Elsamahy T, Al-Tohamy R, Zhu D, Mahmoud Y, Koutra E, Metwally MA, Kornaros M, Sun J (2021a) Plastic wastes biodegradation: mechanisms, challenges and future prospects. Sci Total Environ 146590

Ali SS, Elsamahy T, Koutra E, Kornaros M, El-Sheekh M, Abdelkarim EA, Zhu D, Sun J (2021b) Degradation of conventional plastic wastes in the environment: a review on current status of knowledge and future perspectives of disposal. Sci Total Environ 771:144719

Alicia Lee MSL (2020) Tertiary recycling of plastics waste: an analysis of feedstock, chemical and biological degradation methods. J Mater Cycles Waste Manag

Al-Sabagh A, Yehia F, Eshaq G, Rabie A, ElMetwally A (2016) Greener routes for recycling of polyethylene terephthalate. Egypt J Pet 25:53–64

Al-Salem S, Lettieri P, Baeyens J (2009) Recycling and recovery routes of plastic solid waste (PSW): a review. Waste Manage 29:2625–2643

Al-Salem S, Papageorgiou L, Lettieri P (2014) Techno-economic assessment of thermo-chemical treatment (TCT) units in the Greater London area. Chem Eng J 248:253–263

Al-Salem S, Antelava A, Constantinou A, Manos G, Dutta A (2017) A review on thermal and catalytic pyrolysis of plastic solid waste (PSW). J Environ Manage 197:177–198

Al-Salem S (2019) Energy production from plastic solid waste (PSW). Plastics Energy 45–64 (Elsevier)

Alshehrei F (2017) Biodegradation of synthetic and natural plastic by microorganisms. J Appl Environ Microbiol 5:8–19

Andrady AL (2011) Microplastics in the marine environment. Mar Pollut Bull 62:1596–1605

Andrady AL, Neal MA (2009) Applications and societal benefits of plastics. Philos Trans R Soc Biol Sci 364:1977–1984

Anuar Sharuddin SD, Abnisa F, Wan Daud WMA, Aroua MK (2016) A review on pyrolysis of plastic wastes. Energy Convers Manage 115:308–326

Aragaw TA, Mekonnen BA (2021) Current plastics pollution threats due to COVID-19 and its possible mitigation techniques: a waste-to-energy conversion via Pyrolysis. Environ Syst Res 10:1–11

Arena U, Di Gregorio F, Amorese C, Mastellone ML (2011) A techno-economic comparison of fluidized bed gasification of two mixed plastic wastes. Waste Manage 31:1494–1504

Arkatkar A, Arutchelvi J, Bhaduri S, Uppara PV, Doble M (2009) Degradation of unpretreated and thermally pretreated polypropylene by soil consortia. Int Biodeterior Biodegradation 63:106–111

Assadourian E (2008) Global economic growth continues at expense of ecological systems

Austen I (2008) Canada takes steps to ban most plastic baby bottles. New York Times April 19

Balakrishnan P, Sreekala MS (2016) Recycling of plastics. Recycl Polym 115–139

Barlaz MA, Ham RK, Schaefer DM (1989) Mass-balance analysis of anaerobically decomposed refuse. J Environ Eng 115:1088–1102

Bhardwaj H, Gupta R, Tiwari A (2012) Microbial population associated with plastic degradation. Sci Rep 5:272–274

Bhardwaj H, Gupta R, Tiwari A (2013) Communities of microbial enzymes associated with biodegradation of plastics. J Polym Environ 21:575–579

Brems A, Baeyens J, Vandecasteele C, Dewil R (2011) Polymeric cracking of waste polyethylene terephthalate to chemicals and energy. J Air Waste Manag Assoc 61:721–731

Brems A, Baeyens J, Dewil R (2012) Recycling and recovery of post-consumer plastic solid waste in a European context. Therm Sci 16:669–685

Bridgwater AV (2012) Review of fast pyrolysis of biomass and product upgrading. Biomass Bioenerg 38:68–94

Cacciari I, Quatrini P, Zirletta G, Mincione E, Vinciguerra V, Lupattelli P, Giovannozzi Sermanni G (1993) Isotactic polypropylene biodegradation by a microbial community: physicochemical characterization of metabolites produced. Appl Environ Microbiol 59:3695–3700

Carraher CE Jr (2017) Introduction to polymer chemistry. CRC Press

Charles AH (2006) Handbook of plastics technologies: the complete guide to properties and performance. McGraw-Hill Education, New York

Chaukura N, Gwenzi W, Bunhu T, Ruziwa DT, Pumure I (2016) Potential uses and value-added products derived from waste polystyrene in developing countries: a review. Resour Conserv Recycl 107:157–165

Chow C-f, So W-MW, Cheung T-Y (2016) Research and development of a new waste collection bin to facilitate education in plastic recycling. Appl Environ Educ Commun 15:45–57

Czajczyńska D, Anguilano L, Ghazal H, Krzyżyńska R, Reynolds A, Spencer N, Jouhara H (2017) Potential of pyrolysis processes in the waste management sector. Therm Sci Eng Prog 3:171–197

Deguchi T, Kakezawa M, Nishida T (1997) Nylon biodegradation by lignin-degrading fungi. Appl Environ Microbiol 63:329–331

Demirbas A (2004) Pyrolysis of municipal plastic wastes for recovery of gasoline-range hydrocarbons. J Anal Appl Pyrol 72:97–102

Emadian SM, Onay TT, Demirel B (2017) Biodegradation of bioplastics in natural environments. Waste Manage 59:526–536

Fakirov S (2021) A new approach to plastic recycling via the concept of microfibrillar composites. Adv Indus Eng Polym Res

Farcas MT, Stefaniak AB, Knepp AK, Bowers L, Mandler WK, Kashon M, Jackson SR, Stueckle TA, Sisler JD, Friend SA (2019) Acrylonitrile butadiene styrene (ABS) and polycarbonate (PC) filaments three-dimensional (3-D) printer emissions-induced cell toxicity. Toxicol Lett 317:1–12

Favis BD, Le Corroller P (2017) Polymeric material and process for recycling plastic blends. Google Patents

Fazli A, Rodrigue D (2020) Waste rubber recycling: a review on the evolution and properties of thermoplastic elastomers. Materials 13:782

Ferronato N, Torretta V (2019) Waste mismanagement in developing countries: a review of global issues. Int J Environ Res Public Health 16:1060

Francis R (2016) Recycling of polymers: methods, characterization and applications. Wiley

Garcia JM, Robertson ML (2017) The future of plastics recycling. Science 358:870–872

Gautam R, Bassi A, Yanful E, Cullen E (2007) Biodegradation of automotive waste polyester polyurethane foam using *Pseudomonas chlororaphis* ATCC55729. Int Biodeterior Biodegradation 60:245–249

Geyer R, Kuczenski B, Zink T, Henderson A (2016) Common misconceptions about recycling. J Ind Ecol 20:1010–1017

Geyer R, Jambeck JR, Lavender K (2017) Law production, use, and fate of all plastics ever made. Sci Adv 3:e1700782

Geyer R (2020) Production, use, and fate of synthetic polymers. Plastic Waste Recycl 13–32 (Elsevier)

Giacomucci L, Raddadi N, Soccio M, Lotti N, Fava F (2019) Polyvinyl chloride biodegradation by *Pseudomonas citronellolis* and *Bacillus flexus*. New Biotechnol 52:35–41

Glass J, Swift G (1989) Agricultural and synthetic polymers, biodegradation and utilization, vol 433, pp 9–64
Goodship V (2007) Introduction to plastics recycling. iSmithers Rapra Publishing
Gouin T, Roche N, Lohmann R, Hodges G (2011) A thermodynamic approach for assessing the environmental exposure of chemicals absorbed to microplastic. Environ Sci Technol 45:1466–1472
Gourmelon G (2015) Global plastic production rises, recycling lags. Vital Signs 22:91–95
Grigore ME (2017) Methods of recycling, properties and applications of recycled thermoplastic polymers. Recycling 2:24
Gu J-D (2003) Microbiological deterioration and degradation of synthetic polymeric materials: recent research advances. Int Biodeterior Biodegradation 52:69–91
Guo H, Yue L, Rui G, Manas-Zloczower I (2020) Recycling poly(ethylene-vinyl acetate) with improved properties through dynamic cross-linking. Macromolecules 53:458–464
Hadad D, Geresh S, Sivan A (2005) Biodegradation of polyethylene by the thermophilic bacterium *Brevibacillus borstelensis*. J Appl Microbiol 98:1093–1100
Hahladakis JN, Velis CA, Weber R, Iacovidou E, Purnell P (2018) An overview of chemical additives present in plastics: migration, release, fate and environmental impact during their use, disposal and recycling. J Hazard Mater 344:179–199
Hakkarainen M, Albertsson A-C (2004) Environmental degradation of polyethylene. Long Term Prop Polyolefins 177–200
Halden RU (2010) Plastics and health risks. Annu Rev Public Health 31:179–194
Harshvardhan K, Jha B (2013) Biodegradation of low-density polyethylene by marine bacteria from pelagic waters, Arabian Sea, India. Mar Pollut Bull 77:100–106
Hidayat YA, Kiranamahsa S, Zamal MA (2019) A study of plastic waste management effectiveness in Indonesia industries. AIMS Energy 7:350–370
Hodson ME, Duffus-Hodson CA, Clark A, Prendergast-Miller MT, Thorpe KL (2017) Plastic bag derived-microplastics as a vector for metal exposure in terrestrial invertebrates. Environ Sci Technol 51:4714–4721
Hopewell J, Dvorak R, Kosior E (2009) Plastics recycling: challenges and opportunities. Philos Trans R Soc Biol Sci 364:2115–2126
Hundertmark T, Mayer M, McNally C, Simons TJ, Witte C (2018a) How plastics waste recycling could transform the chemical industry. McKinsey & Company, p 12
Hundertmark T, Mayer M, McNally C, Simons TJ, Witte C (2018) How plastics waste recycling could transform the chemical industry. McKinsey & Company, p 12
Ignatyev IA, Thielemans W, Vander Beke B (2014) Recycling of polymers: a review. Chemsuschem 7:1579–1593
Iiyoshi Y, Tsutsumi Y, Nishida T (1998) Polyethylene degradation by lignin-degrading fungi and manganese peroxidase. J Wood Sci 44:222–229
Ishigaki T, Sugano W, Nakanishi A, Tateda M, Ike M, Fujita M (2004) The degradability of biodegradable plastics in aerobic and anaerobic waste landfill model reactors. Chemosphere 54:225–233
Jalal N (2017) Binging on plastics containing bisphenol A. J Microbiol Biotechnol Rep 1
Jambeck JR, Geyer R, Wilcox C, Siegler TR, Perryman M, Andrady A, Narayan R, Law KL (2015) Plastic waste inputs from land into the ocean. Science 347:768–771
Jiao X, Zheng K, Hu Z, Zhu S, Sun Y, Xie Y (2021) Conversion of waste plastics into value-added carbonaceous fuels under mild conditions. Adv Mater 2005192
Jnr AK-L, Yunana D, Kamsouloum P, Webster M, Wilson DC, Cheeseman C (2018) Recycling waste plastics in developing countries: use of low-density polyethylene water sachets to form plastic bonded sand blocks. Waste Manage 80:112–118
Jyoti S, Gupta K (2014) Screening and identification of low density polyethylene (LDPE) degrading soil fungi isolated from polythene polluted sites around Gwalior city (MP). Int J Curr Microbiol App Sci 3:443–448

Kale SK, Deshmukh AG, Dudhare MS, Patil VB (2015) Microbial degradation of plastic: a review. J Biochem Technol 6:952–961

Kapinga CP, Chung SH (2020) Marine plastic pollution in South Asia. Dev Papers 20–02

Karamanlioglu M, Preziosi R, Robson GD (2017) Abiotic and biotic environmental degradation of the bioplastic polymer poly (lactic acid): a review. Polym Degrad Stab 137:122–130

Kathiresan K (2003) Polythene and plastics-degrading microbes from the mangrove soil. Rev Biol Trop 51:629–633

Kawai F, Watanabe M, Shibata M, Yokoyama S, Sudate Y, Hayashi S (2004) Comparative study on biodegradability of polyethylene wax by bacteria and fungi. Polym Degrad Stab 86:105–114

Kawasaki S, Yamada M, Kobori K, Sakamoto H, Kondo Y, Jin F, Takata T (2009) Preparation of a novel alloy composed of fluorene-based polyester and polycarbonate and their properties for the optical uses. J Appl Polym Sci 111:461–468

Keshavarz T, Roy I (2010) Polyhydroxyalkanoates: bioplastics with a green agenda. Curr Opin Microbiol 13:321–326

Kong SC, Wong TL, Yang M, Chow CF, Tse KH (2017) Emerging practices in scholarship of learning and teaching in a digital era. Springer

Koo JH (2019) Polymer nanocomposites: processing, characterization, and applications. McGraw-Hill Education

Koushal V, Sharma R, Sharma M, Sharma R, Sharma V (2014) Plastics: issues challenges and remediation. Int J Waste Resour 4:2–6

Krueger MC, Harms H, Schlosser D (2015) Prospects for microbiological solutions to environmental pollution with plastics. Appl Microbiol Biotechnol 99:8857–8874

Kumagai S, Hirahashi S, Grause G, Kameda T, Toyoda H, Yoshioka T (2018) Alkaline hydrolysis of PVC-coated PET fibers for simultaneous recycling of PET and PVC. J Mater Cycles Waste Manage 20:439–449

Kumar R (2021) Tertiary and quaternary recycling of thermoplastics by additive manufacturing approach for thermal sustainability. Mater Today Proc 37:2382–2386

Kumar S, Smith SR, Fowler G, Velis C, Kumar SJ, Arya S, Rena, Kumar R, Cheeseman C (2017) Challenges and opportunities associated with waste management in India. Royal Soc Open Sci 4:160764

Kumar A (2009) Legislators consider ban on plastic shopping bags. Washington Post, Thursday B04

Kumartasli S, Avinc O (2020) Recycling of marine litter and ocean plastics: a vital sustainable solution for increasing ecology and health problem. Sustain Text Apparel Ind 117

La Mantia F (2002) Handbook of plastics recycling. iSmithers Rapra Publishing

Leja K, Lewandowicz G (2010) Polymer biodegradation and biodegradable polymers-a review. Polish J Environ Stud 19

Li J, Kim HR, Lee HM, Yu HC, Jeon E, Lee S, Kim D-H (2020) Rapid biodegradation of polyphenylene sulfide plastic beads by Pseudomonas sp. Sci Total Environ 720:137616

Linger JG, Vardon DR, Guarnieri MT, Karp EM, Hunsinger GB, Franden MA, Johnson CW, Chupka G, Strathmann TJ, Pienkos PT (2014) Lignin valorization through integrated biological funneling and chemical catalysis. Proc Natl Acad Sci 111:12013–12018

Lowy J (2004) Plastic left holding the bag as environmental plague. Seattle Post-Intelligencer 21

Lwanga EH, Vega JM, Quej VK, de los Angeles Chi J, Del Cid LS, Chi C, Segura GE, Gertsen H, Salánki T, van der Ploeg M (2017) Field evidence for transfer of plastic debris along a terrestrial food chain. Sci Rep 7:1–7

Mahdiyah D, Mukti B (2013) Isolation of polyethylene plastic degrading-bacteria. Biosci Int 2:29–32

Mangalara SCH, Varughese S (2016) Green recycling approach to obtain nano-and microparticles from expanded polystyrene waste. ACS Sustain Chem Eng 4:6095–6100

Masaki K, Kamini NR, Ikeda H, Iefuji H (2005) Cutinase-like enzyme from the yeast Cryptococcus sp. strain S-2 hydrolyzes polylactic acid and other biodegradable plastics. Appl Environ Microbiol 71:7548–7550

Mayer AM, Staples RC (2002) Laccase: new functions for an old enzyme. Phytochemistry 60:551–565
Merrington A (2017) Recycling of plastics. In: Applied plastics engineering handbook. Elsevier, pp 167–189
Michael T (2013) Environmental and social impacts of waste to energy (WTE) conversion plants. In: Waste to energy conversion technology. Elsevier, pp 15–28
Miskolczi N, Angyal A, Bartha L, Valkai I (2009) Fuels by pyrolysis of waste plastics from agricultural and packaging sectors in a pilot scale reactor. Fuel Process Technol 90:1032–1040
Mohammadzadeh M (2009) Characterization of recycled thermoplastic polymers. University of Borås/School of Engineering.
Mohanraj C, Senthilkumar T, Chandrasekar M (2017) A review on conversion techniques of liquid fuel from waste plastic materials. Int J Energy Res 41:1534–1552
Motta O, Proto A, De Carlo F, De Caro F, Santoro E, Brunetti L, Capunzo M (2009) Utilization of chemically oxidized polystyrene as co-substrate by filamentous fungi. Int J Hyg Environ Health 212:61–66
Mutha NH, Patel M, Premnath V (2006) Plastics materials flow analysis for India. Resour Conserv Recycl 47:222–244
Muthu SS, Li Y, Hu J, Mok PY, Ding X (2012) Eco-impact of plastic and paper shopping bags. J Eng Fibers Fabr 7:155892501200700100
Mwanza BG, Mbohwa C (2017) Drivers to sustainable plastic solid waste recycling: a review. Procedia Manufact 8:649–656
Nakajima-Kambe T, Onuma F, Kimpara N, Nakahara T (1995) Isolation and characterization of a bacterium which utilizes polyester polyurethane as a sole carbon and nitrogen source. FEMS Microbiol Lett 129:39–42
Nakanishi Y, Yoshihara Y, Hiraoka M, Nishiwaki K, Soloiu V, Mitsuhara Y (2000) Application of a new fuel produced from waste plastics and heavy oil to diesel engine, pp 461–466
Nanda S, Berruti F (2021) Thermochemical conversion of plastic waste to fuels: a review. Environ Chem Lett 19:123–148
Nkwachukwu OI, Chima CH, Ikenna AO, Albert L (2013) Focus on potential environmental issues on plastic world towards a sustainable plastic recycling in developing countries. Int J Indus Chem 4:34
North EJ, Halden RU (2013) Plastics and environmental health: the road ahead. Rev Environ Health 28:1–8
Okan M, Aydin HM, Barsbay M (2019) Current approaches to waste polymer utilization and minimization: a review. J Chem Technol Biotechnol 94:8–21
Olah GA, Goeppert A, Prakash GS (2009) Chemical recycling of carbon dioxide to methanol and dimethyl ether: from greenhouse gas to renewable, environmentally carbon neutral fuels and synthetic hydrocarbons. J Org Chem 74:487–498
Oliveira TA, Oliveira RR, Barbosa R, Azevedo JB, Alves TS (2017) Effect of reprocessing cycles on the degradation of PP/PBAT-thermoplastic starch blends. Carbohyd Polym 168:52–60
Onen Cinar S, Chong ZK, Kucuker MA, Wieczorek N, Cengiz U, Kuchta K (2020) Bioplastic production from microalgae: a review. Int J Environ Res Public Health 17:3842
Onwudili JA, Insura N, Williams PT (2009) Composition of products from the pyrolysis of polyethylene and polystyrene in a closed batch reactor: effects of temperature and residence time. J Anal Appl Pyrol 86:293–303
Orhan Y, Büyükgüngör H (2000) Enhancement of biodegradability of disposable polyethylene in controlled biological soil. Int Biodeterior Biodegradation 45:49–55
Palm GJ, Reisky L, Böttcher D, Müller H, Michels EA, Walczak MC, Berndt L, Weiss MS, Bornscheuer UT, Weber G (2019) Structure of the plastic-degrading *Ideonella sakaiensis* MHETase bound to a substrate. Nat Commun 10:1–10
Parker L (2018) We depend on plastic. Now we're drowning in it. NationalGeographic com
Parker L (2019) The world's plastic pollution crisis explained. National Geographic 7

Paszun D, Spychaj T (1997) Chemical recycling of poly (ethylene terephthalate). Ind Eng Chem Res 36:1373–1383

Phua S, Castillo E, Anderson J, Hiltner A (1987) Biodegradation of a polyurethane in vitro. J Biomed Mater Res 21:231–246

Pinto F, Costa P, Gulyurtlu I, Cabrita I (1999) Pyrolysis of plastic wastes. 1. Effect of plastic waste composition on product yield. J Anal Appl Pyrol 51:39–55

Polk MB, Leboeuf LL, Shah M, Won C-Y, Hu X, Ding W (1999) Nylon 66, nylon 46, and PET phase-transfer-catalyzed alkaline depolymerization at atmospheric pressure. Polym-Plast Technol Eng 38:459–470

Pramila R, Ramesh KV (2011) Biodegradation of low density polyethylene (LDPE) by fungi isolated from municipal landfill area. J Microbiol Biotechnol Res 1:e136

Premraj R, Doble M (2005) Biodegradation of polymers. Indian J Biotechnol 4:186–193

Proshad R, Kormoker T, Islam MS, Haque MA, Rahman MM, Mithu MMR (2018) Toxic effects of plastic on human health and environment: a consequences of health risk assessment in Bangladesh. Int J Health 6:1–5

Qureshi MS, Oasmaa A, Pihkola H, Deviatkin I, Tenhunen A, Mannila J, Minkkinen H, Pohjakallio M, Laine-Ylijoki J (2020) Pyrolysis of plastic waste: opportunities and challenges. J Anal Appl Pyrol 152:104804

Radenahmad N, Rahman ISA, Morni NAH, Azad AK (2018) Acacia-polyethylene terephthalate co-gasification as renewable energy resource. Int J Renew Energy Res (IJRER) 8:1612–1620

Ragaert K, Delva L, Van Geem K (2017) Mechanical and chemical recycling of solid plastic waste. Waste Manage 69:24–58

Rahimi A, García JM (2017) Chemical recycling of waste plastics for new materials production. Nat Rev Chem 1:1–11

Restrepo-Flórez J-M, Bassi A, Thompson MR (2014) Microbial degradation and deterioration of polyethylene—a review. Int Biodeterior Biodegradation 88:83–90

Richard GM, Mario M, Javier T, Susana T (2011) Optimization of the recovery of plastics for recycling by density media separation cyclones. Resour Conserv Recycl 55:472–482

Rigamonti L, Grosso M, Møller J, Martinez Sanchez V, Magnani S, Christensen TH (2014) Environmental evaluation of plastic waste management scenarios. Resour Conserv Recycl 85:42–53

Rippy JF (1948) German investments in Argentina. J Bus Univ Chicago 21:50–54

Rodrigues M, Abrantes N, Gonçalves F, Nogueira H, Marques J, Gonçalves A (2019) Impacts of plastic products used in daily life on the environment and human health: what is known? Environ Toxicol Pharmacol 72:103239

Ronkvist ÅM, Xie W, Lu W, Gross RA (2009) Cutinase-catalyzed hydrolysis of poly (ethylene terephthalate). Macromolecules 42:5128–5138

Rowe L, Howard GT (2002) Growth of *Bacillus subtilis* on polyurethane and the purification and characterization of a polyurethanase-lipase enzyme. Int Biodeterior Biodegradation 50:33–40

Rozenstein O, Puckrin E, Adamowski J (2017) Development of a new approach based on midwave infrared spectroscopy for post-consumer black plastic waste sorting in the recycling industry. Waste Manage 68:38–44

Ru J, Huo Y, Yang Y (2020) Microbial degradation and valorization of plastic wastes. Front Microbiol 11:442

Ruiz C, Main T, Hilliard NP, Howard GT (1999) Purification and characterization of two polyurethanase enzymes from *Pseudomonas chlororaphis*. Int Biodeterior Biodegradation 43:43–47

Ruj B, Pandey V, Jash P, Srivastava V (2015) Sorting of plastic waste for effective recycling. Int J Appl Sci Eng Res 4:564–571

Russell JR, Huang J, Anand P, Kucera K, Sandoval AG, Dantzler KW, Hickman D, Jee J, Kimovec FM, Koppstein D (2011) Biodegradation of polyester polyurethane by endophytic fungi. Appl Environ Microbiol 77:6076–6084

Rutkowska M, Heimowska A, Krasowska K, Janik H (2002) Biodegradability of polyethylene starch blends in sea water. Pol J Environ Stud 11:267–272
Saleh TA, Danmaliki GI (2020) Polymer consumption, environmental concerns, possible disposal options, and recycling for water treatment. In: Sustainable infrastructure: breakthroughs in research and practice. IGI Global, pp 691–708
Salhofer S, Jandric A, Soudachanh S, Le Xuan T, Tran TD (2021) Plastic recycling practices in Vietnam and related hazards for health and the environment. Int J Environ Res Public Health 18:4203
Scheirs J (1998) Polymer recycling: science, technology and applications. Wiley, Journals, Baffins Lane, Chichester, Sussex PO 19 1 UD, UK, 1998 591
Sen SK, Raut S (2015) Microbial degradation of low density polyethylene (LDPE): a review. J Environ Chem Eng 3:462–473
Seneviratne G, Tennakoon N, Nandasena K, Weerasekara M (2006) Polyethylene biodegradation by a developed Penicillium-Bacillus biofilm
Seymour RB (1989) Polymer science before and after 1899: notable developments during the lifetime of Maurits Dekker. J Macromol Sci Chem 26:1023–1032
Sezgin H, Yalcin-Enis I (2020) Turning plastic wastes into textile products. In: Baskar C, Ramakrishna S, Baskar S, Sharma R, Chinnappan A, Sehrawat R (eds) Handbook of solid waste management: sustainability through circular economy. Springer Singapore, Singapore, pp 1–27
Shah AA, Hasan F, Hameed A, Ahmed S (2008) Biological degradation of plastics: a comprehensive review. Biotechnol Adv 26:246–265
Shelley MD, El-Halwagi MM (1999) Techno-economic feasibility and flowsheet synthesis of scrap tire/plastic waste liquefaction. J Elastomers Plast 31:232–254
Shen L, Worrell E (2014) Plastic recycling. Handbook of recycling. Elsevier, pp 179–190
Shin J, Selke SE (2014) 11-food packaging. In: Clark, S, Jung, S, Lamsal, B (eds) Food processing: principles and applications, 2nd edn, pp 249–273
Shukla SR, Harad AM, Jawale LS (2009) Chemical recycling of PET waste into hydrophobic textile dyestuffs. Polym Degrad Stab 94:604–609
Singh N, Hui D, Singh R, Ahuja I, Feo L, Fraternali F (2017) Recycling of plastic solid waste: a state of art review and future applications. Compos B Eng 115:409–422
Sinha V, Patel MR, Patel JV (2010) PET waste management by chemical recycling: a review. J Polym Environ 18:8–25
Sivan A (2011) New perspectives in plastic biodegradation. Curr Opin Biotechnol 22:422–426
Skariyachan S, Manjunatha V, Sultana S, Jois C, Bai V, Vasist KS (2016) Novel bacterial consortia isolated from plastic garbage processing areas demonstrated enhanced degradation for low density polyethylene. Environ Sci Pollut Res 23:18307–18319
Sohn YJ, Kim HT, Baritugo K-A et al (2020) Recent advances in sustainable plastic upcycling and biopolymers. Biotechnol J 15:1900489
Subramanian P (2000) Plastics recycling and waste management in the US. Resour Conserv Recycl 28:253–263
Sulaiman S, Yamato S, Kanaya E, Kim J-J, Koga Y, Takano K, Kanaya S (2012) Isolation of a novel cutinase homolog with polyethylene terephthalate-degrading activity from leaf-branch compost by using a metagenomic approach. Appl Environ Microbiol 78:1556–1562
Suwanmanee U, Varabuntoonvit V, Chaiwutthinan P, Tajan M, Mungcharoen T, Leejarkpai T (2013) Life cycle assessment of single use thermoform boxes made from polystyrene (PS), polylactic acid,(PLA), and PLA/starch: cradle to consumer gate. Int J Life Cycle Assess 18:401–417
Swift T, Moore M, Sanchez E (2015) Plastics and polymer composites in light vehicles. Econ Stat Depart/Am Chem Counc
Tahir A, Taba P, Samawi M, Werorilangi S (2019) Microplastics in water, sediment and salts from traditional salt producing ponds. Glob J Environ Sci Manag 5:431–440
Teuten EL, Saquing JM, Knappe DR, Barlaz MA, Jonsson S, Björn A, Rowland SJ, Thompson RC, Galloway TS, Yamashita R (2009) Transport and release of chemicals from plastics to the environment and to wildlife. Philos Trans R Soc Biol Sci 364:2027–2045

Thomas VA, Rugeley PB, Lau FH (2015) Digital photograph security: what plastic surgeons need to know. Plast Reconstr Surg 136:1120–1126

Thompson RC, Swan SH, Moore CJ, vom Saal FS (2009a) Our plastic age. Philos Trans R Soc Lond B Biol Sci 364:1973–1976

Thompson RC, Swan SH, Moore CJ, Vom Saal FS (2009b) Our plastic age. The Royal Society Publishing

Tokiwa Y, Calabia BP, Ugwu CU, Aiba S (2009) Biodegradability of plastics. Int J Mol Sci 10:3722–3742

Tribedi P, Sil AK (2013) Low-density polyethylene degradation by Pseudomonas sp. AKS2 biofilm. Environ Sci Pollut Res 20:4146–4153

Turner A (2020) Foamed polystyrene in the marine environment: sources, additives, transport, behavior, and impacts. Environ Sci Technol 54:10411–10420

Ugoeze K, Amogu E, Oluigbo K, Nwachukwu N (2021) Environmental and public health impacts of plastic wastes due to healthcare and food products packages: a review. J Environ Sci Public Health 5:1–31

Urbanek AK, Rymowicz W, Mirończuk AM (2018) Degradation of plastics and plastic-degrading bacteria in cold marine habitats. Appl Microbiol Biotechnol 102:7669–7678

Usha R, Sangeetha T, Palaniswamy M (2011) Screening of polyethylene degrading microorganisms from garbage soil. Libyan Agric Res Cent J Int 2:200–204

Velis C (2014) Global recycling markets-plastic waste: a story for one player–China. Int Solid Waste Assoc—Glob Waste Manag Task Force 1–66

Wahab DA, Hussain A, Scavino E, Mustafa M, Basri H (2006) Development of a prototype automated sorting system for plastic recycling. Am J Appl Sci 3:1924–1928

Watanabe M, Kawai F, Shibata M, Yokoyama S, Sudate Y, Hayashi S (2004) Analytical and computational techniques for exogenous depolymerization of xenobiotic polymers. Math Biosci 192:19–37

Webb HK, Arnott J, Crawford RJ, Ivanova EP (2013) Plastic degradation and its environmental implications with special reference to poly(ethylene terephthalate). Polymers 5:1–18

Westerhout R, Van Koningsbruggen M, Van Der Ham AG, Kuipers J, Van Swaaij WPM (1998) Techno-economic evaluation of high temperature pyrolysis processes for mixed plastic waste. Chem Eng Res Des 76:427–439

Wölfel B, Seefried A, Allen V, Kaschta J, Holmes C, Schubert DW (2020) Recycling and reprocessing of thermoplastic polyurethane materials towards nonwoven processing. Polymers 12:1917

Wu C, Nahil MA, Miskolczi N, Huang J, Williams PT (2014) Processing real-world waste plastics by pyrolysis-reforming for hydrogen and high-value carbon nanotubes. Environ Sci Technol 48:819–826

Yang J, Yang Y, Wu W-M, Zhao J, Jiang L (2014) Evidence of polyethylene biodegradation by bacterial strains from the guts of plastic-eating wax worms. Environ Sci Technol 48:13776–13784

Yang Y, Yang J, Wu W-M, Zhao J, Song Y, Gao L, Yang R, Jiang L (2015) Biodegradation and mineralization of polystyrene by plastic-eating mealworms: part 2. Role of gut microorganisms. Environ Sci Technol 49:12087–12093

Yoshida S, Hiraga K, Takehana T, Taniguchi I, Yamaji H, Maeda Y, Toyohara K, Miyamoto K, Kimura Y, Oda K (2016) A bacterium that degrades and assimilates poly (ethylene terephthalate). Science 351:1196–1199

Yoshioka T, Okayama N, Okuwaki A (1998) Kinetics of hydrolysis of PET powder in nitric acid by a modified shrinking-core model. Ind Eng Chem Res 37:336–340

Zheng J, Suh S (2019) Strategies to reduce the global carbon footprint of plastics. Nat Clim Chang 9:374–378

Chapter 15
Microbial-Mediated Lignocellulose Conversion to Biodiesel

Mahdy Elsayed, Hamed M. El-Mashad, and Mohamed Mahmoud-Aly

Abstract Biodiesel is commercially produced from edible lipids such as soybean, canola, and palm oils. Using edible lipids for biodiesel production increases the competition with food production leading to the "*food versus fuel*" debate. Therefore, alternative feedstocks are needed to produce biodiesel and other bioenergy carriers without the use of land and resources needed for food production. Single-cell oil (SCO) produced from oleaginous microorganisms (e.g., microalgae, yeast, fungi, and bacteria) has been studied as good alternatives to edible oil for biodiesel production. Recent research has been conducting on indirect lipid biosynthesis from lignocellulosic biomass (LCB) using oleaginous microbes for biodiesel production. This chapter gives a deep overview of systems employed and potential technical challenges for the production and extraction of lipids from oleaginous microorganisms cultivated on LCB and its derivatives. Different cultivation systems of oleaginous microorganisms on lignocellulosic feedstocks and their derivatives are discussed. Most economic and sustainable pretreatment methods of lignocellulosic materials for successful production of lipids are presented. Downstream processing of residues and wastes of biodiesel production from lignocellulosic feedstocks are summarized. In addition, yields of biodiesel production from lignocellulosic feedstocks are mentioned.

M. Elsayed
Department of Agricultural Engineering, Faculty of Agriculture, Cairo University, Giza 12613, Egypt
e-mail: mahdy.mahdy@agr.cu.edu.eg

H. M. El-Mashad (✉)
Department of Biological and Agricultural Engineering, University of California, Davis, CA 95616, USA
e-mail: heelmashad@ucdavis.edu

Department of Agricultural Engineering, Faculty of Agriculture, Mansoura University, Mansoura 35516, Egypt

M. Mahmoud-Aly
Plant Physiology Division, Department of Agricultural Botany, Faculty of Agriculture, Cairo University, Giza 12613, Egypt
e-mail: m_mahmoudaly@agr.cu.edu.eg

A. E.-F. Abomohra et al. (eds.), *Waste-to-Energy*,
https://doi.org/10.1007/978-3-030-91570-4_15

Keywords Lignocelluloses · Lipids · Bioenergy · Single-cell oil

15.1 Introduction

As a result of the limited supply of non-renewable fuels and its ecological problems, in particular participation in global warming, there is a broad global push to develop and implement new renewable energy careers and technologies. Biodiesel is one of the world's largest liquid biofuels (Abomohra et al. 2020a, b). In the way of establishing a sustainable source of eco-friendly renewable energy, many studies have focused their efforts on biodiesel production. Biodiesel is non-toxic fuel and environmentally friendly, i.e., emitting much less gaseous pollutants than conventional diesel and completely biodegradable (Suhud Shote 2019). It is mainly composed of monoalkyl esters of long-chain fatty acids (LCFAs) conjugated with some short-chain aliphatic alcohols, e.g., fatty acid ethyl esters (FAEEs) and fatty acid methyl esters (FAMEs).

Biodiesel's current production cost is incredibly high as compared to fossil diesel, limiting its ability to completely replace fossil diesel. So far, several recommendations have been made because of extensive studies conducted around the world. However, the viability of producing biodiesel from current feedstocks is still a major challenge, despite the fact that it offers tremendous benefits (Elsayed et al. 2020; Xu et al. 2019). To support biodiesel to substitute normal diesel in the global market, biodiesel must overcome the problems of energy security, i.e., sustainable sources, reasonable prices, and produced net energy.

Over decades, biodiesel feedstock comes from oilseeds crops, animal fats, as well as fat, oil and grease (FOG) waste (Abomohra et al. 2020a, b; Bhuiya et al. 2016). Using these traditional feedstocks which are lacking to sustainability, high production rate and reasonable prices as well as threaten food security enhances many scientists to use lignocellulosic biomass (LCB) as an advantageous alternative feedstock for microbial oil and biodiesel production subsequently (Schmidt et al. 2015; Sims et al. 2010).

Oleaginous microorganisms (e.g., microalgae, yeast, fungi, and bacteria) have been studied as good alternatives and promising approach to substitute terrestrial oilseed crops for biodiesel production. In addition, microbial-based oils are found to be reliable competitor precursor to enhance sustainability and reduce their impact on environment and bioresources. Many of these oils also have been called single cell oils (SCOs), as they show an overwhelming amount of fatty acid profile (Di Fidio et al. 2020). About 20% w/w lipids of dry weight (DW) can be accumulated under special conditions for oleaginous microorganisms (Ma et al. 2018). Different industrial applications may also be developed via the SCO's FA profile: biofuels development (Khoo et al. 2020), pharmaceuticals, and food supplements which differ depending on the microorganism and feedstock (Bharathiraja et al. 2017).

Cellulose, hemicellulose, and lignin are the main components of lignocellulosic feedstocks (Carroll and Somerville 2009). LCB, the most available biomass on the

Earth, mainly come from harvest residues of food crops, grasses, and trees. Its production rate is estimated at 181.5 billion tonnes/year (Paul and Dutta 2018). Therefore, the use of lignocellulosic materials, as sustainable low-cost substrates, may result in the development of microbial lipids becoming economically viable. This method has the dual benefit of decreasing the need for waste discharge and treatment technology while also producing lipids for various industrial applications, thus assisting in the transformation toward the circular economy (Elsayed et al. 2020).

Moreover, SCO production based on lignocellulose media presents many advantages overproduction of vegetable oils, e.g., cheapness, sustainability, rapidity, high yield productivity specially after genetic improvement (Wang et al. 2020; Aguieiras et al. 2015). However, LCB is chemically complexed material not only needs microbial fermentation but also needs upstream and downstream processes to produce biodiesel, such as, pretreatment, enzymatic hydrolysis, and lipid recovery, as concluded in Fig. 15.1. In addition, microbial genetic modifications are crucial factors in increasing lipid synthesis, decreasing lipid catabolic pathways rate, and/or improving the microbial physiology adaptability to specific environmental conditions. All of the previously mentioned processes and factors are significant challenges that must be overcome in order to make SCO manufacture economically feasible (Adegboye et al. 2021; Yellapu et al. 2018).

In this chapter, the concept of microbial-mediated lignocellulose conversion to biodiesel is deeply reviewed and discussed. Recent advances and associated challenges of biodiesel production process are covered with biological, technical, and economical prospective. Moreover, this chapter provides a high-throughput of the complete conversion of microbial-mediated lignocellulose biomass into bioenergy and bio-based products for the full use of microbial constituents toward the sustainable production of biofuels with a "zero waste" system.

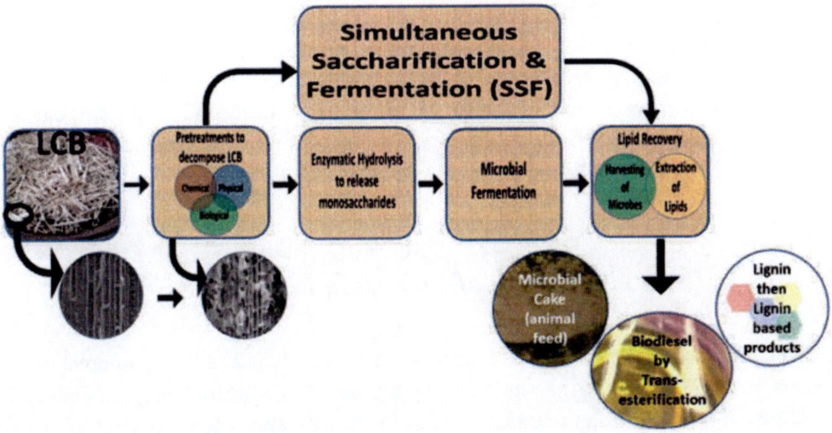

Fig. 15.1 Main processes for microbial-mediated biodiesel production (the two microscopic pictures under license: 5034700628405)

15.2 Different Biodiesel Feedstocks

15.2.1 Terrestrial Oil-Crops

The majority of biodiesel now in use comes from so-called first-generation feedstock (edible vegetable oils or oil from other food crops). This leads to the "food vs. fuel" dispute over the use of edible oil for biodiesel production, resulting in competition in water and land resources (Abomohra et al. 2020a, b). This led to the development of the second-generation feedstock such as waste cooking oil, animal fats, rice bran oil, and *Jatropha curcas* oil.

The European Parliament's all member states decided to replace crop-based biofuels by 2021 (Masri et al. 2019). However, this will not be possible without alternate production routes that can meet rising energy demand and biofuel criteria for clean energy production. Unlike existing biofuel technologies, future bioprocesses must avoid removing vulnerable ecosystems such as rainforests, which serve as a sustainable carbon sink. In this context, SCO has been acknowledged as the greatest alternatives to plant-based oil production.

15.2.2 Lipidic Wastes

Fat, oil, and grease (FOG) is a lipid-rich biowastes produced by a variety of sources, including food enterprises, hotels, restaurants, and multistory apartment buildings and housing multifamily dwellings (Husain et al. 2014). The generated FOG adheres to the pipe surfaces and results in blocking of pipes occurs on incremental deposition (He et al. 2012). The predecessors of this FOG are abundant in edible ingredients such as oils derived from all cooking processes (cooking, frying, roasting, etc.), meat in various forms (gravy, dishes, and raw wastes), baked stuffs, dairy products (cheese, milk, butter, yoghurt, paneer, etc.), beverages and ice creams (Khorsha 2011). FOG discharge clogs the system pipes, disrupts individual house plumbing, and causes property flooding. The problems of FOG discharge and eventual clogging spans from individual property to the larger sewage blockage in the city.

15.2.3 Indirect Conversion of Lignocelluloses

In an era of biotechnology with increased fuel costs, there is a renewed interest in the conversion of lignocellulosic materials into biofuels. As a biofuel and agrochemical feedstock, lignocellulosic biomass (LCB) is the most viable alternative to starch and sugar. Currently, the biorefining technology of lignocellulose is now being developed.

15.3 Oleaginous Microbial Conversion

Oleaginous microbes can transform carbon sources into lipid droplets that contain more than 20% lipid per dry biomass, which can be stored in the form of fat (Patel et al. 2020).

Generally, those microorganisms are classified to three major groups: (1) microalgae, (2) yeasts and filamentous fungi, and (3) bacteria. Single cell oil (SCO), a microbial lipid, is produced in the stationary growth phase under nitrogen limitation with simultaneous excess of a carbon source. The highest practical conversion of 100 g glucose is 20–22 g oil; i.e., 5 tonnes glucose are needed to make 1 tonne oil (Ratledge and Cohen 2008).

A wide range of industrial applications can be served by these microorganism-derived lipids because of their varying fatty acid compositions (Abomohra et al. 2020a, b). Given that crude oil is running out, plant oils for biodiesel are problematic, and the oceans are overfished; SCO are regarded acceptable oil alternatives. Many yeasts and filamentous fungi are considered to accumulate as much as 30–75% of total cell weight as lipids and have commonly been employed to produce SCO. Recently, Karamerou et al. (2021) estimated the SCO prices produced using sugars to be in the range of \$1.19 to \$1.81 per kg depending on the capacity of the production facility. These prices are lower than that produced from lignocellulosic materials (\$5.15–5.41/kg). In comparison, palm and soybean oil and costs are around \$0.5–0.68 per kg (Masri et al. 2019). Therefore, to meet the economic production conditions for SCO, new approaches are needed to increase the productivity and to reduce the production costs of SCO. For example, some yeast strains of *R. sporidium*, *R. torula*, and *Lipomyces* can accumulate lipids as much as 70% of their cells weight (Papanikolaou and Aggelis 2011). Growing under nitrogen deprivation conditions, the most efficient oleaginous yeast *Cryptococcus curvatus* can accumulate lipids up to >60% on a dry weight basis. These lipids usually consist of SOC-90% w/w triacylglycerol with about 44% of saturated fatty acids (SFA), which are similar to many crop seed oils (Tchakouteu et al. 2015).

An arachidonic acid (AA, C20:4n-6) accumulating filamentous fungus, *Mortierella alliacea* strain YN-15, mainly in the form of triglyceride in its mycelia, which produced about 46.1 g of dry cell weight, 19.5 g of total fatty acid, and 77.1 g/l AA after seven days of incubation in a 50-L bioreactor (Aki et al. 2001). From these results, oleaginous yeasts or mushrooms could be used as biodiesel substitutes. However, these oleaginous microbes produce less lipids when they grow on LCB not simple sugars, e.g., *Cryptococcus* sp. (KCTC 27583) only accumulates 34 not 60% lipids when grows on pretreated banana peel medium (Han et al. 2019). Table 15.1 presents a collection of recently studied oleaginous microorganisms grown on different lignocellulosic substrates and their lipid content %. As can be seen from the table, more existing examples show producing less lipid contents on different lignocellulosic materials compared to lipid contents of previously mentioned microbes grown on simple sugars. Also, it can be inferred that microalga *Auxenochlorella protothecoides* is a promising oleaginous microbe produces 66 and 63% accumulated

Table 15.1 Recently studied oleaginous microorganisms grown on different lignocellulosic substrates and their lipid content (%, w/w) dry weight

Oleaginous microorganisms	Lignocellulosic substrates	Lipid accumulation (%, *w/w*)	References
Microalgae			
Auxenochlorella protothecoides	Organosolv pretreated birch biomass hydrolysates	66	Patel et al. (2018a)
Auxenochlorella protothecoides	Organosolv pretreated spruce biomass hydrolysates	63	Patel et al. (2018a)
Yeasts and filamentous fungi			
Cryptococcus sp. (KCTC 27583)	Pretreated banana peel	34	Han et al. (2019)
Rhodosporidium kratochvilovae HIMPA1	*Cassia fistula* L. fruit pulp	53.18	Patel et al. (2015)
Trichosporon fermentans CICC 1368	Pretreated waste sweet potato vines under simultaneous saccharification and fermentation (SSF)	36	Shen et al. (2018)
Rhodosporidium toruloides	Brewers' spent grain	56	Patel et al. (2018b)
Fusarium oxysporum	Sweet sorghum stalks (12% w/w solid load)	22	Matsakas et al. (2017)
Aspergillus sp.	Corncob waste liquor	22.1	Venkata Subhash and Venkata Mohan (2011)
Naganishia albida	Crude glycerol derived from the biodiesel industry and onion waste hydrolysate	34	Sathiyamoorthi et al. (2020)
Rhodotorula mucilaginosa Y-MG1 strain	Acid-pretreated wheat bran	38.7	Ayadi et al. (2019)
Lipomyces starkeyi	Enzymatic hydrolysate of alkali pretreated *Arundo donax* biomass	43	Ayadi et al. (2019)
Trichosporon oleaginosus	Enzymatic hydrolysate of alkali pretreated corn cobs	31.27	Šantek et al. (2018)
Bacteria			
Rhodococcus opacus DSM 1069	Organosolv pretreated pine	26.99	Wells et al. (2015)

(continued)

Table 15.1 (continued)

Oleaginous microorganisms	Lignocellulosic substrates	Lipid accumulation (%, *w/w*)	References
Rhodococcus opacus PD630 (Xsp8 engineered strain)	Unbleached kraft hardwood pulp hydrolysate	45.8	Kurosawa et al. (2013)

lipids when grown on organosolv pretreated birch and organosolv pretreated spruce biomass hydrolysates, respectively (Patel et al. 2018a).That can be achieved by its mixotrophic behavior when using organic carbon sources from LCB and its photosynthesis ability. Also, *Rhodosporidiumkratochvilovae*HIMPA1 and *Rhodosporidium toruloides* are oleaginous yeasts produce 53 and 56% total lipids when grown on *Cassia fistula* L. fruit pulp and Brewers' spent grains, respectively (Patel et al. 2015, 2018b).

Although there are many studies on SCOs, their production is still not economically feasible so far mainly due to high capital and operation costs. Therefore, recycling of waste streams and the identification of new microorganisms with high-value products or/and higher productivities are possibilities. Zhang et al. (2019a) isolated a bacterium that they referred to as the lignolytic mycobacterium *Mycobacterium smegmatis LZ-K2* from rotten wood. The isolated bacterium showed high lipid production and high efficiency of lignin degradation. The corn straw medium-stimulated LZ-K2 lipid output was 0.083 g/l, with an alkali pretreatment of 0.9 and without the use of chemical pretreatment was 0.2 g/l^1. More than 80% of the fats (C14–C24, specifically palmitic acid) were found in the untreated corn stalks.

LCB is the most available and renewable source for SCO production. It has been a focus of many researchers in recent years. Breakdown of LCB via acid and/or enzymatic hydrolysis to generates long-chain polysaccharides, which are then converted into their corresponding monosaccharides. More studies have focused on screening suitable strains for lipid generation on LCB hydrolysates (Zhang et al. 2019a, b; Matsakas et al. 2017; Ayadi et al. 2019). To date, the reported oleaginous microorganisms that can use low-cost sources for SCO production. Five oleaginous yeast strains, *Cryptococcus curvatus, Rhodotorula glutinis, Rhodosporidium toruloides, Lipomyces starkeyi,* and *Yarrowia lipolytica*, were used for SCO production on wheat straw hydrolysates, and *Cryptococcus curvatus* showed the highest lipid yield (Jin et al. 2015). However, there are still some technical issues that must be overcome when the cultivation of oleaginous strains on LCB such as low lipid content, low lipid productivity, and low lipid titer are mainly accrued. Therefore, using proper technologies for pretreatment, hydrolysis, and detoxification could make the industrial production of SCO from LCB possible in the future.

15.4 Structure of Lignocellulosic Biomass (LCB)

Agro-industrial residues, sometimes called LCB. Physical and chemical structural aspects can often be divided into two categories. Crystallinity of the cellulose, degree of polymerization, pore volume, and accessible surface area are all physical structural characteristics. Its chemical structure includes lignin content, hemicellulose, and acetyl groups (Hendriks and Zeeman 2009). In plants, lignin is the most difficult part to hydrolysis due to its insoluble nature. In addition, cellulose is resistant to hydrolysis due to its crystallinity, hemicellulose, and lignin provide a strong barrier against cellulose hydrolysis (Elsayed et al. 2020). Due to its structural features, cellulose in the original biomass has a limited enzymatic digestibility (less than 20% yield) (Ahmad et al. 2018). LCB typically includes between 55 and 75% polysaccharides by dry weight. As like starch, cellulose is a glucose-polymer that can be used to make a variety of products. However, in contrast to starch, the structure of cellulose favors a strongly packed, highly crystalline structure, which is water insoluble and depolymerization-resistant. The hemicellulose is a heterogeneous-polymer, substituted, and branched polymer of xylose, arabinose, galactose, mannose, and glucose with lesser amounts of acetyl groups (Ahmad et al. 2018). Together, cellulose and hemicellulose form the structural backbone to plant cellular walls and are potential carbohydrate sources (Jin et al. 2015).

15.4.1 Impact of Structural Features on Fiber Hydrolysis

The hydrolysis of LCB is affected by several factors. The limiting variables were generally split into two groups: those related to the structural properties of lignocelluloses and those related to the mechanism and interactions of cellulolytic enzymes. More information on structural features is presented below.

15.4.1.1 Lignin Content

Lignin has a large impact on the rate and extent of lignocellulose hydrolysis. More studies have indicated that increasing lignin removal improves cellulose digestibility (Mosier et al. 2005). The major inhibitory role of the lignin is attributed to the non-specific uptake of the enzyme to lignin, and inaccessibility of the enzyme to cellulose due to the 3D steric barrier of lignin composition. In addition, delignification (lignin removal) causes disruption of lignin structure, and subsequently leads to biomass swelling with increasing the internal surface area and pore volume of fiber (Hendriks and Zeeman 2009).

15.4.1.2 Hemicellulose Content

Hemicellulose is a physical hindrance that encapsulates the cellulose microfibers and can protect the cellulose from degradation by enzymatic attack (Peng et al. 2019). In addition, removing hemicellulose from the intracrystalline structure, which increases the surface area and pore volume of microfibers, making cellulose more accessible to cellulase enzymes (Rivers and Emert 1987).

15.4.1.3 Crystallinity

There are varying orientations of cellulose molecules in different parts of the structure, which results in different levels of crystallization. Thus, cellulose fiber can be divided into two regions of structures: amorphous (low crystallinity) and crystalline (high crystallinity). Due to crystalline cellulose is less susceptible to degradation by enzymatic attack than amorphous cellulose part, crystallinity influences the efficacy of enzyme engagement with cellulose (Hendriks and Zeeman 2009). Many studies have reported a negative relationship between crystallinity index and biodigestibility (Rivers and Emert 1987). Recently, researchers hypothesized that the effect of reduced crystallinity index on hydrolysis rate could be due to increased surface area or decreased particle size through applying combined pretreatment tends to decrease the particle size and crystallinity of biofibers while simultaneously increasing the surface area (Ai et al. 2019).

15.4.1.4 Accessible Surface Area

The accessible surface area of LCB is a significant impact on LCB's digestibility. The accessible surface area and fiber digestibility have a positive correlation (Hendriks and Zeeman 2009). Surface area is not included as a dependent factor that impacts fiber digestibility because it may be associated with cellulose crystallinity and/or removal of lignin. Additionally, the accessible surface area is closely linked to the pore volume (Barakat et al. 2014). Fiber digestibility is correlated with a variety of structural features, as shown in Table 15.2.

Table 15.2 Summary of correlation between fiber structural features and digestibility

Structural features		Correlation between structural feature and digestibility
Physical	Surface area	Positive
	Crystallinity	Negative
	Pore volume	Positive
Chemical	Lignin	Negative
	Hemicellulose	Negative
	Acetyl group	Negative

Among all of these structural characteristics, lignin content, acetyl content, and crystallinity are crucial parameters impacting fiber digestibility since they are distinctive aspects of the three primary lignocellulose components (Hendriks and Zeeman 2009; Monlau et al. 2013).

15.4.2 The Objectives of Pretreatment

The purpose of any pretreatment technique is to modify or remove compositional and structural obstacles to accelerate hydrolysis rate to increase yields of simple sugars from cellulose and/or hemicelluloses as represented in Fig. 15.2.

As a result of various pretreatment processes, plant biomass undergoes physical and/or chemical modifications, which break the lignin seal and cause disruption of the cellulose crystal structure. It is estimated that pretreatment is the most expensive processing step in cellulosic ethanol, costing up to $0.3/gallon of ethanol produced in converting cellulosic biomass to fermentable sugars (Mosier et al. 2005). Scientists are making great potential to increase efficiency and reduce costs through research and development. The selection of an appropriate pretreatment process remains one of the most difficulties in establishing economically viable biofuel manufacturing technologies. Several LCB pretreatment technologies have been developed over the years. The optimal pretreatment technology is determined by biomass characteristics, conversion modes, economic analysis, and environmental impact (Patinvoh et al. 2017; Patowary and Baruah 2018). Several criteria define effective pretreatment, which can be summarized in the following items:

- Maximize the enzymatic convertibility.
- Minimize loss of sugars.
- Minimize the use of energy, chemicals, and capital equipment.
- Cost-effective in operation.
- Be environment friendly.
- Be scalable to industrial size.

15.5 Pretreatment Technologies

Pretreatment techniques are often divided into three categories: physical, chemical, and biological. These techniques are described in more detail below.

15.5.1 Physical Pretreatment

Physical pretreatment techniques for LCB include comminution (size reduction), steam explosion (autohydrolysis), extrusion, and irradiation (ultrasound and

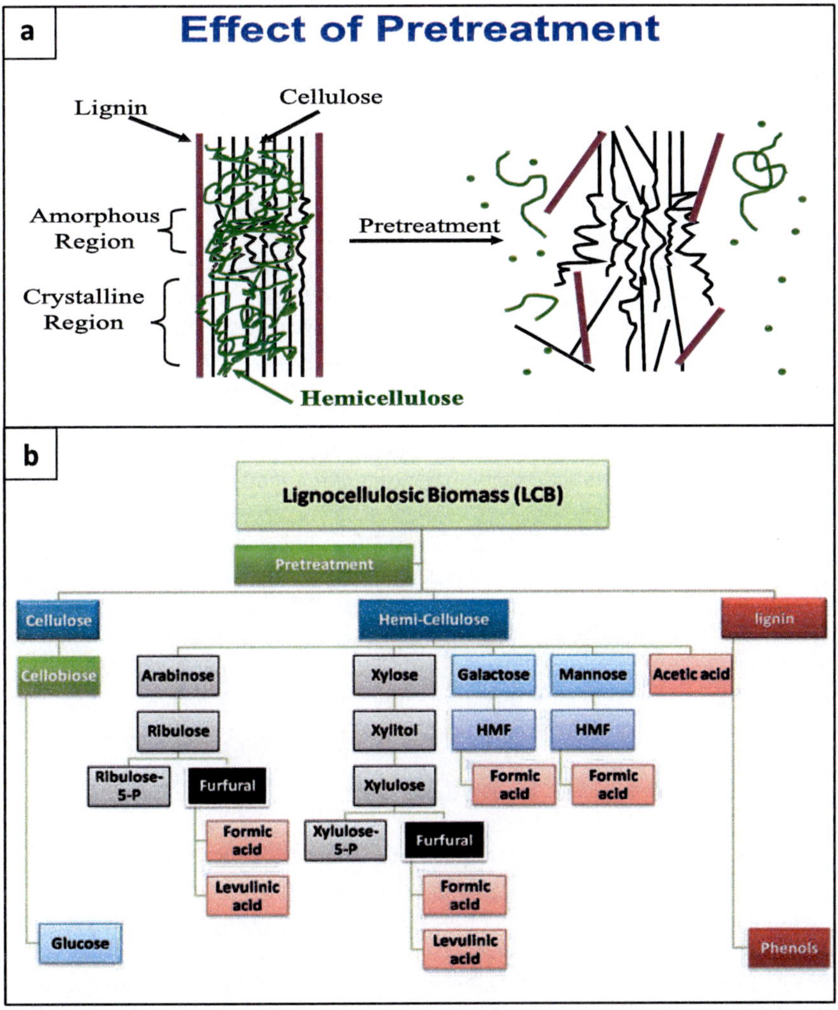

Fig. 15.2 **a** Pretreatment effect on LCB (Mosier et al. 2005) after the required copyright permission (No. 5082580398075). **b** Derived simple compounds produced by enzymatic hydrolysis of complexed compounds (cellulose, hemicellulose, and lignin) which disengaged from pretreated LCB. HMF: hydroxy-methyl furfural, Ribulose-5-P: Ribulose-5-Phosphate

microwave). The physical treatment includes fine grinding or grinding of biomass, which reducing the particle size and the degree of crystallization. As the crystallinity and particle size are reduced, this results in improved hydrolysis and mass transfer properties (Patowary and Baruah 2018).

15.5.1.1 Mechanical Comminution

Mechanical comminution of LCB through a combination of cutting, crushing, and grinding can be applied to reduce particle size and cellulose crystallinity. The size of the materials is usually 10–30 mm after cutting and 0.2–2 mm after grinding (Hakkila 1989).

Comminution of agricultural residuals is used to reduce particle size and is commonly applied in manufacturing processes of compost, biogas, and ethanol. Comminution or size reduction was applied prior to other pretreatment techniques to enhance biomass treatment, handling, and storage. Size reduction techniques can alter the inherent ultrastructure of LCB, which increases the surface area of particle, reduces the degree of cellulose crystallinity, and decreases the degree of cellulose polymerization for improved solubility, digestibility, and biodegradability (Hendriks and Zeeman 2009). Particle size influences the rheological properties and flow behavior of biomass, which has a direct impact on mixing and pumping operations (Dibble et al. 2011). In addition, comminution is frequently used to reduce particle size in order to improve the rheological characteristics for subsequent steps in any process. The energy consumption of LBC is linked to the machine category and its conditions, beginning and end particle size and biomass properties (i.e., bulk density, composition amount, and moisture content). Electricity or fuel consumption in comminution is a significant input for the processing of LCB. Therefore, reducing the cost of energy is a crucial element to reduce the total production costs. The theory is based on the premise that the energy required to bring about a change dL in a particle of a typical size dimension L is a power function of L:

$$\mathrm{d}E/\mathrm{d}L = KL^n$$

where dE is the differential energy required, dL is the change in a typical dimension; L is the magnitude of a typical length dimension; K and n are constants.

The specific comminution energy refers to the amount of energy required to grind one cubic meter or one ton of biomass into the specified particle size. The comminution energy increases linearly with LBC density. Hence, the machine power requirement and fuel consumption are generally higher for hard LBC than for soft LBC. According to Hakkila (1989), the specific cutting energy for hardwoods is on average 2.6 kWh/m^3 and for softwoods 1.8 kWh/m^3.

15.5.1.2 Steam Explosion (Autohydrolysis)

Steam explosion, also known as autohydrolysis, is a process that combines high-pressure with high-temperature (240 °C and 33.5 bar). The biomass particles are placed in a vessel and exposed to high-temperature, high-pressure steam for 5–30 min, which hydrolyzes the glycosidic linkages in the LBC (Chen et al. 2011). Following that, the steam is released, and the biomass is rapidly cooled, causing the water in the biomass to explode and opening up the structure of the lignocelluloses in the cell wall. Therefore, it makes the biomass available to the microorganisms in the subsequent bioprocess (Bauer et al. 2009). The advantages of steam explosion include:

- Increasing the biogas and ethanol yield from lignocellulose rich materials (Taherzadeh and Karimi 2008).
- Increasing the speed of the digestibility rate which enables smaller reactors and lower investments (Biswas et al. 2014).
- Reducing the risks of outing layers in the bioreactor with low density substrates like straw or feathers (Shafiei et al. 2013).
- The material will be easier to transport through pipes and the bioreactor stirring will be improved (He et al. 2015).
- The substrate homogeneity is increased (He et al. 2015).

15.5.1.3 Extrusion

Extrusion is a pretreatment process that mechanically crushes the substrate through screw extruder. This enhances the substrate's available surface area. As the biomass progresses, the pressure and temperature rise to a maximum of 2 bar and 160–180 °C, respectively. A steam explosion-like pressure and temperature drop occur when substrate exits an extruder. In addition to being able to deliver strong shear pressures, quickly transmit heat, and quickly and effective mixing (Kratky and Jirout 2011). Karunanithy and Muthukumarappan (2011) examined the impact of biomass moisture level and extruder factors such as screw compression ratio, speed, and barrel temperature on a variety of prairie grasses and switchgrass under moisture levels ranging from 15 to 45% (wet basis). Using screw speeds of 50 rpm, barrel temperatures of 150 °C, and moisture contents of 15% (wet basis), the highest sugar recovery yields were achieved with switchgrass and prairie cord grass, while screw speeds of 50 rpm, barrel temperatures of 50 °C, and moisture contents of 25% yields. They concluded that a compression ratio of 3:1 led yielded the highest sugar recovery.

15.5.1.4 Irradiation and Ultrasonic

Irradiation is a physical pretreatment method that includes microwave, ultrasound, gamma-ray, and electron beam. Energy is generated by an electromagnetic field and delivered directly to the material to provide rapid heating throughout the entire material with reduced thermal gradients as well as, capacity of heating decreases the time and consumed energy of processing (Mishra and Sharma 2016). A material's ability to be heated with microwave energy is determined by its microwave field and dielectric response (Patowary and Baruah 2018). Microwave technology might provide a powerful alternative to standard heating. It can heat a huge amount more quickly and decrease the processing time, which might result in significant energy savings (Hendriks and Zeeman 2009).

Ultrasonic pretreatment can alter the structure of the cell wall, increase the surface area, and reduce the degree of polymerization, which increases the biodegradability of LCB. Ultrasound pretreatment causes monolithic cavitation, which has physical and chemical consequences in liquid solutions (Shanthi et al. 2019).

However, limited studies have been done on the application of gamma-ray and electron beam radiation for biofuel generation from different types of LCB. In industrial applications, irradiation technologies are typically costly and demanding (challenges in handling high volumes, process scaling up, and inconsistent operation) (Patinvoh et al. 2017).

15.5.2 Chemical Pretreatment

Treatments with chemical reagents are aimed specifically to remove the predominant influence of base-pretreated lignin, while acid pretreatment removes the major effect of hemicellulose, both of which increase the pore size and surface area of LCB. In addition, chemical pretreatment is well-known have the potential to be an affordable cost (Elsayed et al. 2018).

15.5.2.1 Acidic Pretreatment

The process of treating LCB with acid may include adding diluted acids of 0.2–2.5 wt% to the solid fiber, and the continuous mixing of these solutions at 120–210 °C (Yang and Wyman 2009). The first condition, wherein a temperature of $T > 160$ °C is maintained for long periods of time, is used for low solid loading. The second condition, wherein a temperature of $T < 160$ °C is sustained for periods of time, is utilized for high solid loading (Jørgensen et al. 2007).

Mostly, H_2SO_4 and HCl are employed as concentrated strong acids in strong acid hydrolysis although no enzymatic hydrolysis is performed afterward. The key benefit of this technique is the high solubility of carbohydrate (hemicellulose and cellulose) in acid and high glucose production rate without the requirement for further enzymatic hydrolysis. However, the cost of recovering the acids used in this process and purchasing corrosion-resistant equipment is prohibitive (Hendriks and Zeeman 2009). Hydroxyl methyl furfural is another fermentation inhibitor that is generated in large amounts, limiting the efficacy of this approach. However, the approach is suitable for biofiber with low lignin concentration (El-Mashad 2015). The most crucial criterion when treating with acid is to maintain the correct pH.

15.5.2.2 Alkaline Pretreatment

In comparison with other pretreatment methods, alkaline pretreatment techniques employ lower temperatures and pressures. Calcium, potassium, sodium, and ammonium hydroxide are commonly chosen as the alkaline reagents and are used to soak LBC before mixing it with an appropriate temperature and certain amount of time (Elsayed et al. 2018). Alkaline pretreatments are among the most often applied for LCB processing. The main effect of alkaline treatment of biofibers improves surface area by particle swelling, lignin removal, a decrease in crystallinity, and polymerization, and the separation of structural connections between lignin and holocellulose (Peng et al. 2019). In comparison to the other pretreatment processes, the alkaline process necessitates fewer extreme conditions. The efficiency of this procedure has been demonstrated by the outcomes of alkali pretreatment of corn stover, bagasse, wheat, and rice straw (Elsayed et al. 2018).

Because of low cost, compatibility with oxidants, simplicity of recovery, and ease of use, lime ($Ca(OH)_2$) is commonly chosen as the alkaline agent in industrial applications. A novel pretreatment by lime, water, and an oxidizing agent (air or O2) is mixed with the biomass at temperatures ranging from 40 °C for a period ranging from hours to weeks as shown in Fig. 15.3 (Fu and Holtzapple 2011). During this process, the straw biomass pile and solution should be circulated through the pile by drawing alkaline solution from the bottom and pumping it to the top, and air can be blown through a scrubber to remove carbon dioxide and it discharged to the bottom of the pile reactor to enhance lignin removal by alkaline oxidation. The temperature of the straw pile can be controlled by regulating the temperature of the circulating solution using a heat exchanger.

Compares various alkalis, the results reveals that lime is an effective pretreatment agent because it is cheap (0.06/kg $), safe, and can be recovered by carbonating wash water with CO_2. Unfortunately, lime appears to be less effective than other alkalis since it is a weak base with a limited solubility. However, in correct pretreatment conditions, digestion of moderate-lignin biomass (e.g., switchgrass, bagasse,

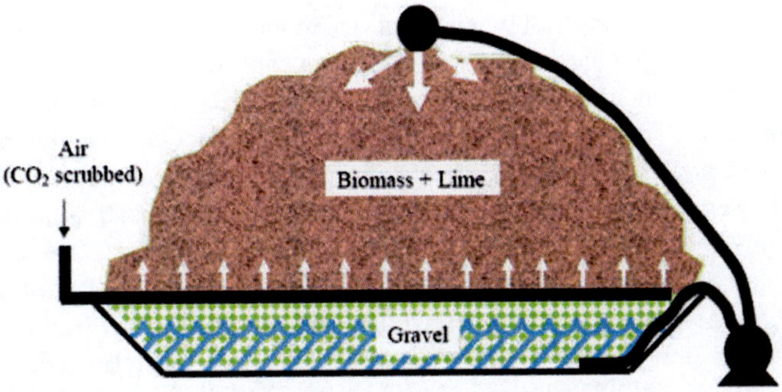

Fig. 15.3 Pile reactor for oxidative lime pretreatment of lignocellulosic biomass as proposed Fu and Holtzapple (2011)

and wheat straw) is considerably increased by removing 30–43% of lignin and all acetyl groups. For high-lignin biomass, lime alone does not remove enough lignin to appreciably increase digestibility; an oxidant should be added (Fu and Holtzapple 2011).

15.5.3 Biological Pretreatment

Biological pretreatment, as it is commonly understood, entails the use of microorganisms (mostly fungi) to degrade lignin and hemicellulose while leaving cellulose intact (Sharma et al. 2019; Wan and Li 2012). Biological pretreatment is mainly aimed at minimizing the loss of carbohydrates and maximizing lignin removal for enhanced digestibility. Pretreatment by microbial consortiums and enzymatic pretreatment has been the major topic in biological pretreatment for improvement biofuel generation. For this reason, several fungi were utilized, e.g., brown, white, and soft-red fungus. Brown rots attack cellulose mostly in the first stage, whereas white and soft rots affect both cellulose and lignin. For biological pretreatment of LCB, white-rot fungi are the most effective (Rouches et al. 2016). Biological treatment with several varieties of rot fungi, which is a safe and environmentally acceptable technology, is rapidly being promoted as a way to remove lignin from an LCB without consuming a lot of energy (Wan and Li 2012).

Biological pretreatment often uses significantly less energy than physical and chemical pretreatment procedures, and it does not involve the use of chemicals heat energy. These are the key advantages of biological pretreatment. The utilization of these methods in commercial applications has been limited due to the lengthy processing time (Elsayed et al. 2019).

15.5.4 Innovative Pretreatment Technology

This technique combines two of mechanical size reduction and chemical and/or other physiochemical pretreatments strongly indicates that digestibility is significantly increased, with a concurrent decrease in energy required while maintaining holocellulose biomass. These would enhance overall process economics, demonstrating that integrated pretreatments are critical for reducing energy demand in dry lignocellulosic biorefineries (Barakat et al. 2014). Combining physical treatment with dry chemical, physicochemical, and/or biological processing (hydrolysis and saccharification) in a continuous flow-through process can reduce energy requirements by 2–5 times, particle size by 2–5 times while increasing lignocellulosic conversion, and water utilization by 5–10 times while reducing waste generation (Barakat et al. 2014). Barakat et al. (2014) illustrated the possibilities of the proposed "dry" biorefinery concept, which might possibly apply to the transformation of a wide range of LBC for more efficient and low carbon footprint.

15.6 Cultivation of Microalgae on Lignocellulosic Material for Biodiesel Production

Cultivation of microalgae has gained wide acceptance for integrated wastewater remediation coupled with bioenergy production (biodiesel) and nutrient recovery. However, the major hinders application at scaling up are poor separability and harvesting of microalgal cells in suspended culture. This method, known as "Biofilm technology", enables the attachment of algal cells to a carrier surface while also separating the effluent from it. Mechanical scraping makes it simple to harvest algal biomass (which contains 80–90% moisture) (Wang et al. 2018). Algal biofilm systems might treat diverse wastewaters more efficiently and affordably than suspended systems due to their high penetration efficiency and large mass transfer rate.

15.6.1 Merits of Microalgae Biofilm Systems

Rice husk, rice straw, pine sawdust, and sugarcane bagasse, among other lignocellulosic biomass, are inexpensive, renewable, and widely spread. Previously, those biomaterials were shown to be effective biocarriers for algal biofilm production under poly-culture conditions (7.32–10.92 g/m/day) (Zhang et al. 2020). They also possessed a rough natural surface that could boost algal biomass productivity (Zhang et al. 2017). They also possessed a rough natural surface that could boost algal biomass productivity (Zhang et al. 2017). In addition, by using the LCB as biocarrier, the material can be processed by washing or leaching through the medium to improve its energy conversion characteristics, which is a win–win strategy for developing algal biofilms and LBC utilization.

15.6.2 Factors Affecting the Biosystems of Algal Biocarriers

Depending on the process/system and the part of the cells utilized, several methods of algal biofilm culture can be employed for algal biomass generation. The lipid part of the biomass can be extracted and converted into biodiesel through a process like that used for any other plant oil. Subsequently, the lipid-free algal biomass residues can be used for further energy production by other refining processes. Therefore, a proper biofilm system conversion implemented should be affordable and applicable with low energy consumption which are summarized in this section.

15.6.2.1 Microalgae Strains

Selection of the suitable microalgae strains for bioconversion and treatment of wastewater depends mainly on the source of wastewater, temperature, light source, and exposure time. However, the types of microalgae strains can influence that additional process steps such as cultivation, harvesting, and oil extraction, which increases the final cost of biodiesel production (Zhang et al. 2019b). The microalgae *Chlorella vulgaris*, *Hydrodictyon reticulatum*, and *Diplosphaera* sp. have used in the biosystem proposed by Zhang et al. (2020).

15.6.2.2 System Design

In technological aspects, the major types of algal culture systems are open ponds and closed photo-bioreactors with dry biomass content about 0.5–5 g l^{-1}, respectively (Zhang et al. 2020). Various lab and pilot scale microalgae biofilm systems have been designed for the wastewater treatment, e.g., algal turf scrubber, vertical biofilm reactors, rotating/revolving biofilm reactors (Zhang et al. 2019a, b). Carriers promoting algae growth in a biofilm play a vital function in the system (Gross et al. 2016). Zhang et al. (2019a, b) developed a biofilm system with various types of biocarriers through using a flat-plate algal biofilm photo-bioreactor (FPBR) that is capable of simultaneously treatment of wastewater with algal biomass production.

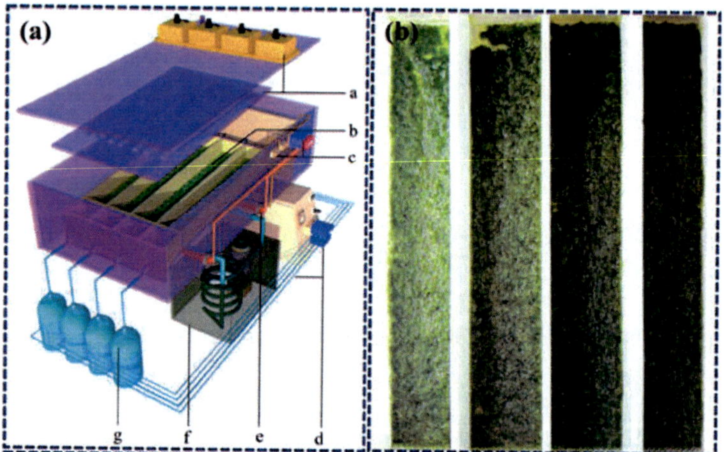

a, lighting system; b, algal biofilm with bio-carriers; c, temperature control system; d, medium circulation system; e, heating system; f, cooling system; g, medium container.

Fig. 15.4 **a** Schematic design of a flat-plate algal biofilm photo-bioreactor. **b** *C. vulgaris biofilm grown on pine sawdust as a biocarrier* (0.420–0.595 mm) for 2 days. *Source* From Zhang et al. (2019b) after the required copyright permission (No. 5085830067848)

Growing algae (*C. vulgaris*) in FPBR Fig. 15.4 can treat the wastewater and create biomass in the presence of biofilm.

15.6.2.3 Carrier Types and Conditions

Carrier is considered as the most important factor at lab and pilot scale microalgae biofilm systems because it greatly affects wastewater treatment and growing algae. The carrier used for attachment growth is a vital element. Currently, fiberglass, metal mesh/plate, cellulose acetate membrane, filter paper, cardboard, polystyrene foam, cotton rope/duct, glass, and polyvinyl chloride are among the carriers used in the described systems. The ability to increase algal biomass yield has been considered as one of the finest carriers for biofilm cultivation (Zhang et al. 2020). However, many operators employed in the current investigations are costly to get (e.g., metal mesh/plate fiberglass), inadequate mechanical strength (e.g., filter paper, carton, and polystyrene foam), or unsustainable (e.g., rope/draw cotton). The system can be run affordably with increased nutrient removal rates and significant biomass yields if the carrier is long-lasting, inexpensive, easy to obtain, and non-toxic to algal cells and especially can enhance cell attachment (Zhang et al. 2017, 2019b). These findings are particularly important for increasing the utility value of LCB, creating new carriers, lowering biofilm cultivation costs, and improving nutrient removal.

15.7 From Pretreated LCB to Lipid at Molecular Level

Different ways of pretreating LCB are previously mentioned and discussed. One of the main purposes of this process is disrupting the lignocellulose complex in the cell walls to increase the sugar yield of enzymatic hydrolysis (Jin et al. 2015), as concluded in Fig. 15.5. This will facilitate consuming simple sugars derived from these complexed compounds (cellulose, hemicellulose, and lignin) by the oleaginous microbes to produce their lipids in vivo in the way to finally in vitro produce biodiesel.

15.7.1 Lipid Production in Microalgae Compared to Other Oleaginous Microbes

Glycerol is the end product of most microalgae fatty acid synthesis. Fatty acids are often found in membrane lipids at the stereospecific numbering system (sn) sn-1 and sn-2 locations of the backbone of its glycerol, together with one of two head groups at the sn-3 position (Boudière et al. 2014). When fatty acids are esterified in all three places along the glycerol backbone, a triacylglyceride (TAG) is generated.

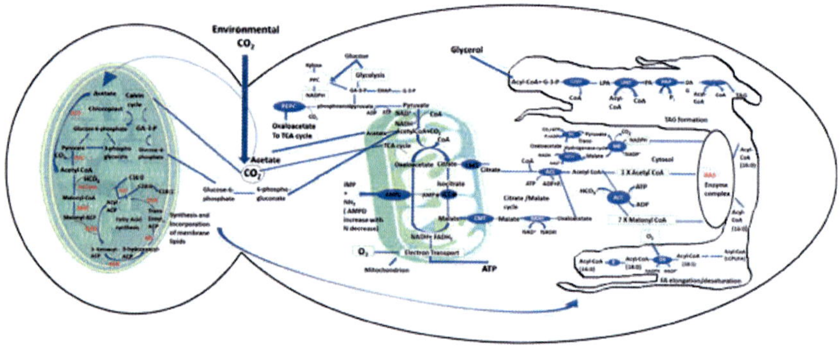

Fig. 15.5 Major cellular pathways involved in the conversion of glucose and xylose to lipids in green microalgae (lift side) with photosynthesis and yeasts (right side). ACC = ACCase: acetyl-CoA carboxylase; ACL, ATP:citrate lyase; ACP, acyl carrier protein; ACS, acyl-CoA synthase; AMPD, AMP deaminase; CMT, citrate–malate translocase; CoA, coenzyme A; DAG, diacylglyceride; DGAT, diacylglycerol acyltransferase; DHAP, dihydroxyacetone phosphate; DS, desaturase; E, elongase; ENR, enoyl-ACP reductase; FA, fatty acid; FAS, fatty acid synthetase; FAT, fatty acyl-ACP thioesterase; GA-3-P, glyceraldehyde 3-phosphate; G-3-P, glycerol 3-phosphate; GPAT, glycerol 3-phosphate acyltransferase; HD, 3-hydroxy acyl-ACP dehydratase; ICDH, isocitrate dehydrogenase; IMP, inosine monophosphate; KAR, 3-keto acyl-ACP reductase; KAS, 3-keto acyl-ACP synthase; LCPUFA, long-chain polyunsaturated fatty acid; LPA, lysophosphatidate; LPAT, lysophosphatidate transferase; MAT, malonyl-CoA: ACP transacylase; MDH, malate dehydrogenase; ME, malic enzyme; PA, phosphatidate; PAP, phosphatidate phosphatase; PC, pyruvate carboxylase; PPC, pentose phosphate cycle; TAG, triacylglyceride; TCA, tricarboxylic acid. The information is adapted from Jin et al. (2015), Kong et al. (2018), Li-Beisson et al. (2019), Li et al. (2013), and Mühlroth et al. (2013).

TAG is the main storage lipid in many algae, some of which are capable of accumulating 20–60% TAG per cell dry weight (Hu et al. 2008). Algae use de novo fatty acid synthesis in chloroplasts as a primary mechanism for glycerolipid production and membrane construction. Animals and fungi, on the other hand, mostly make their fatty acids in the cytosol (Ohlrogge and Browse 1995). Fatty acid synthetic routes primarily create 16- and 18-carbon fatty acids, which are then employed as precursors in the synthesis of membrane glycerolipids and storage TAG. TAG production in algae is hypothesized to take place via a straight glycerol pathway, where fatty acids generated in the chloroplast are successively transported from coenzyme A (CoA) to the *sn*-1, -2, and -3 positions of glycerol 3-phosphate (G-3-P). This process is usually referred to as the Kennedy pathway, which was reviewed in details by Li-Beisson et al. (2019), Li et al. (2013).

15.7.1.1 Fatty Acid Biosynthesis

Figure 15.5 shows that the first enzyme devoted to fatty acid synthesis, acetyl-CoA carboxylase, catalyzes the ATP-dependent production of malonyl-CoA from acetyl-CoA and bicarbonate in chloroplast (Cronan and Thomas 2009). Genes producing

ACCase are less variable in gene dosage and transcriptional expression during TAG accumulation in the oleaginous microalga *Nannochloropsis oceanica* IMET1 than those implicated in the Kennedy pathway (Li et al. 2013), demonstrating that the transcriptional regulation of ACCase to the division of fatty acid precursors into distinct lipid synthesis pathways is less important.

The primary carbon source for fatty acid synthesis is malonyl-CoA, which is generated by the enzyme ACCase. ACP, a cofactor protein that covalently binds all fatty acyl intermediates, receives the malonyl group and transfers it to the acyl carrier protein (AP) for further processing. As a thioester, the fatty acid's expanding acyl chain is linked to the ACP. The malonyl-thioester conducts a series of polymerization events involving acyl-ACP or acetyl-CoA acceptors (Ohlrogge and Browse 1995). Transfer of malonyl moiety to ACP and subsequent extension of the acyl chain with malonyl-ACP is catalyzed by fatty acid synthase (FAS) (Guschina and Harwood 2006). FAS can be divided into two categories: There are two types of FAS systems: type I, which has enormous enzyme complexes that carry all of the proteins involved in fatty acid synthesis on one or two long polypeptide chains, and type II, which contains distinct peptides produced by different genes (Chan and Vogel 2010).

Three more processes are required for successive elongation of the fatty acid chain: a reductive reaction catalyzed by 3-ketoacyl ACP reductase, accompanied by a dehydration step by hydroxyacyl ACP dehydratase, and then an enoyl-ACP reductase reductive step. Each cycle adds two carbons to the precursor fatty acid. FAS's end products are frequently saturated (palmitic-ACP = C16:0-ACP and stearic-ACP = C18:0-ACP) (Ohlrogge and Browse 1995). The 3-ketoacyl ACP synthase (KAS) enzymes may control the rate of FAS (Li et al. 2013). ACP is ended either by an acyltransferase in the chloroplast removing the acyl group and moving freshly generated fatty acids directly from ACP to G-3-P, or by an acyl-ACP thioesterase hydrolyzing acyl-ACP and releasing free fatty acids, which ends fatty acid elongation (Ohlrogge and Browse 1995). First, chloroplast acyltransferases catalyze a reaction known as the "prokaryotic route" to create glycerolipids containing C16 at the sn-2 position of the glycolipid backbone (Browse and Somerville 1991). GPAT competes with acyl-ACP thioesterase for acyl-ACP in chloroplasts.

The thioesterase catalyzes a process that hydrolyzes acyl-ACPs to create free fatty acids in the chloroplast inner envelope. Long-chain fatty acyl-CoA synthetases reintegrate free fatty acids into acyl-CoAs for use in cytosolic glycerolipid synthesis after they are exported to the outer envelope membrane (Chapman and Ohlrogge 2012). The route of fatty acid export from chloroplasts is still a mystery; however, evidence suggests that fatty acid transport occurs by protein-mediated transfer of nascent fatty acids released on the chloroplast inner envelope to long-chain acyl-CoA synthetases outside the chloroplast (Koo et al. 2004). It's also possible that in the chloroplast membrane there is a direct membrane contact point where acyl groups can be added to phosphatidylcholine (PC) before it is delivered to the endoplasmic reticulum (Karki et al. 2019). Some algal species, including *Chlamydomonas reinhardtii*, PC is not present in their lipidomes; therefore, more research is needed to determine how fatty acids are exported by algae. Cytosolic malonyl-CoA can be used by fatty acid elongase complexes to elongate existing long-chain acyl-CoA, resulting in very long-chain fatty acids with 20 carbons (Guschina and Harwood 2006). De novo generated saturated fatty acids can be desaturated by either plastidic or ER desaturases.

15.7.1.2 The Kennedy Pathway

There are three consecutive acylation stages of G-3-P required to generate TAG in eukaryotic organisms via the Kennedy pathway (Fig. 15.5). Also, the Kennedy pathway uses PA and DAG as intermediates to synthesize membrane polar lipids that influence microalgal development and reproduction. It is therefore critical to comprehend the Kennedy route in order to create genetic engineering tactics and procedures for enhancing microalgae growth and TAG output. Few Kennedy pathway enzymes, however, have been discovered in microalgae. This section relies heavily on information gleaned from higher plants and other species and will serve as a foundation for future research.

As the first Kennedy pathway enzyme, GPAT facilitates the transfer of a fatty acid from an acyl-CoA pool (or an acyl-ACP pool in plastids) to the G-3-P sn-1 or sn-2 site, resulting in lysophosphatidic acid (lyso-PA). Under stressful conditions like strong light or nitrogen deprivation, *C. reinhardtii* accumulates TAG along with an increase in the expression of the GPAT gene (Jia et al. 2009). A crucial precursor for the synthesis of TAG and membrane phospholipids is formed via esterification of lyso-PA at the sn-2 position of the glycerol backbone, which is mediated by LPAT (Chungjatupornchai et al. 2019; Coleman and Lee 2004). DAG, a precursor for the production of TAG and phospholipids, is produced by phospholipid phosphatase (PAP), which dephosphorylates PA (Coleman and Lee 2004) as well as galactolipids (Benning and Ohta 2005). The Kennedy pathway's final step of TAG synthesis is catalyzed by diacylglycerol acyltransferase (DGAT), which acylates an acyl-CoA with a sn-1, 2-DAG (Li et al. 2013). DGAT may be critical in regulating the flow of fatty acids into the fatty acid storage protein TAG. Type 1 DGAT (DGAT1) and type 2 DGAT are two examples of DGATs (DGAT2) (Liu et al. 2016; Lung and Weselake 2006).

15.7.2 Biotechnological Implications and Prospective

In the field of biofuel generation, oleaginous microalgae have been identified as a promising biomass feedstock (Hu et al. 2008). However, the availability of algae-derived biofuels is restricted. Current methods for producing algal biomass and lipid feedstocks, which use naturally occurring strains, are plagued by low cellular lipid content and poor biomass productivity (Mahmoud-Aly et al. 2018). Metabolic engineering strategies have been introduced to either increase TAG productivity or improve biomass yield in order to improve algal-biofuel production efficiency. It has also been suggested that lipase-based enzymatic hydrolysis and transesterification techniques be used to convert algae oils directly to the biodiesel components FAMEs and fatty acids via transesterification for efficient fuel synthesis.

Lenka et al. (2016) and Poliner et al. (2018) reviewed numerous studies of lipid manipulation in microalgae to increase TAG content. Some of those studies focused on, (1) fatty acid synthesis manipulation, e.g., (Chen et al. 2017), (2) overexpression of the genes encoding the Kennedy pathway enzymes, e.g., (Liu et al. 2016), (3) blocking the competing pathway thus seems to be an effective strategy for overproduction of TAG in microalgae, e.g., (Li et al. 2010), (4) overexpression of the genes encoding kinases and transcription factors, e.g., (Li-Beisson et al. 2019), and (5) over-expression of TAG storage and LB formation genes, e.g., (Xu et al. 2020).

Chemical recovery of microalgal lipids, followed by processing and transformation of the lipids into biodiesel, i.e., FAMEs are prohibitively expensive. However, Martinez-Silveira et al. (2019) reported a new biodiesel manufacturing protocol based on quick in situ transesterification from oleaginous yeasts. Considering that the biodiesel produced here is economically competitive with that produced from fossil fuels, and this procedure could be quite promising. Enzymatic hydrolysis and transesterification of plant oils using lipase have been harnessed and commercialized for efficient fuel production (Nielsen et al. 2008; Vasudevan and Briggs 2008; Vyas et al. 2010). Algal crude oils, on the other hand, differ significantly from plant oils in that they often include substantially larger levels of membrane phosphoglycerolipids and glycoglycerolipids, making the conversion of algal crude oils to biodiesels with commercially available lipases challenging (Wang et al. 2014). This is because commercial lipases, in particular glycoglycerolipid lipases, hydrolyze weak membrane lipids poorly. But in a water-based reaction system, PDAT from *C. reinhardtii* can easily hydrolyze neutral lipids (TAG, DAG, and MAG) as well as phospholipids (with strong lipase activity toward PG, PI, PS, and PA) and galactolipids (with high lipase activity toward MGDG) to create free fatty acids. This enzyme can convert different lipids (TAG, DAG, MAG, phospholipids, and MGDG) into FAMEs in a hydrophobic solvent solutions, suggesting that this enzyme can be used as a biocatalyst for the production of biodiesel from algal oils (Yoon et al. 2012).

15.8 Conclusions and Future Perspectives

Biodiesel production from microbial-mediated lignocellulose is a good alternative to the biodiesel produced from food-grade lipids. More research is needed to optimize and reduce the production and extraction of lipids from different oleaginous microbes. The costs of biodiesel production from oleaginous microorganisms need to be reduced to values comparable to the biodiesel produced from other lipids. There is dual-species cultivation of microalgae and yeast for wastewater treatment/remediation have been presented and discussed with simultaneous production of biofuels. In addition, sequential biofuel production has been employed to enhance the efficiency of wastewater treatment and biofuel industry. A modern approach to full energy recovery from dual-species biomass is biomass conversion into biofuels at lower waste production rate. Finally, microbial genetic modifications, co-cultivation strategy, and introducing new oleaginous microbial species are the way to enhance lipid productivity as well as producing a combination of high- and low-value lipids is the way to enhance the whole process profitability.

References

Abomohra AE-F, Elsayed M, Esakkimuthu S, El-Sheekh M, Hanelt D (2020a) Potential of fat, oil and grease (FOG) for biodiesel production: a critical review on the recent progress and future perspectives. Prog Energy Combust Sci 81:100868

Abomohra AE, Wang Q, Huang J, Saad-Allah K (2020b) A sustainable approach for bioconversion of food- and lignocellulosic-wastes into liquid biofuel using a new Metschnikowia pulcherrima isolate. Int J Energy Res. https://doi.org/10.1002/er.6028

Adegboye MF, Ojuederie OB, Talia PM, Babalola OO (2021) Bioprospecting of microbial strains for biofuel production: metabolic engineering, applications, and challenges. Biotechnol Biofuels 14:1–21. https://doi.org/10.1186/s13068-020-01853-2

Aguieiras ECG, Cavalcanti-Oliveira ED, Freire DMG (2015) Current status and new developments of biodiesel production using fungal lipases. Fuel 159:52–67. https://doi.org/10.1016/j.fuel.2015.06.064

Ahmad S, Pathak VV, Kothari R, Singh RP (2018) Prospects for pretreatment methods of lignocellulosic waste biomass for biogas enhancement: opportunities and challenges. Biofuels 9:575–594

Ai P, Zhang X, Ran Y, Meng L, Elsayed M, Fan Q, Abomohra AEF (2019) Biomass briquetting reduces the energy loss during long-term ensiling and enhances anaerobic digestion: a case study on rice straw. Bioresour Technol 292:121912. https://doi.org/10.1016/j.biortech.2019.121912

Aki T, Nagahata Y, Ishihara K, Tanaka Y, Morinaga T, Higashiyama K, Akimoto K, Fujikawa S, Kawamoto S, Shigeta S (2001) Production of arachidonic acid by filamentous fungus, Mortierella alliacea strain YN-15. J Am Oil Chem Soc 78:599–604

Ayadi I, Belghith H, Gargouri A, Guerfali M (2019) Utilization of wheat bran acid hydrolysate by Rhodotorula mucilaginosa Y-MG1 for microbial lipid production as feedstock for biodiesel synthesis. Biomed Res Int. https://doi.org/10.1155/2019/3213521

Barakat A, Mayer-Laigle C, Solhy A, Arancon RAD, De Vries H, Luque R (2014) Mechanical pretreatments of lignocellulosic biomass: towards facile and environmentally sound technologies for biofuels production. Rsc Adv 4:48109–48127

Bauer A, Bösch P, Friedl A, Amon T (2009) Analysis of methane potentials of steam-exploded wheat straw and estimation of energy yields of combined ethanol and methane production. J Biotechnol 142:50–55

Benning C, Ohta H (2005) Three enzyme systems for galactoglycerolipid biosynthesis are coordinately regulated in plants. J Biol Chem 280:2397–2400. https://doi.org/10.1074/jbc.R400032200

Bharathiraja B, Sridharan S, Sowmya V, Yuvaraj D, Praveenkumar R (2017) Microbial oil—a plausible alternate resource for food and fuel application. Bioresour Technol 233:423–432. https://doi.org/10.1016/j.biortech.2017.03.006

Bhuiya MMK, Rasul MG, Khan MMK, Ashwath N, Azad AK (2016) Prospects of 2nd generation biodiesel as a sustainable fuel—part: 1 selection of feedstocks, oil extraction techniques and conversion technologies. Renew Sustain Energy Rev 55:1109–1128. https://doi.org/10.1016/j.rser.2015.04.163

Biswas R, Uellendahl H, Ahring BK (2014) Wet explosion pretreatment of sugarcane bagasse for enhanced enzymatic hydrolysis. Biomass Bioenergy 61:104–113

Boudière L, Michaud M, Petroutsos D, Rébeillé F, Falconet D, Bastien O, Roy S, Finazzi G, Rolland N, Jouhet J, Block MA, Maréchal E (2014) Glycerolipids in photosynthesis: composition, synthesis and trafficking. Biochim Biophys Acta Bioenerg 1837:470–480. https://doi.org/10.1016/J.BBABIO.2013.09.007

Browse J, Somerville C (1991) Glycerolipid synthesis: biochemistry and regulation. Annu Rev Plant Physiol Plant Mol Biol 42:467–506. https://doi.org/10.1146/annurev.pp.42.060191.002343

Carroll A, Somerville C (2009) Cellulosic biofuels. Annu Rev Plant Biol. https://doi.org/10.1146/annurev.arplant.043008.092125

Chan DI, Vogel HJ (2010) Current understanding of fatty acid biosynthesis and the acyl carrier protein. Biochem J 430:1–19. https://doi.org/10.1042/BJ20100462

Chapman KD, Ohlrogge JB (2012) Compartmentation of triacylglycerol accumulation in plants. J Biol Chem 287:2288–2294. https://doi.org/10.1074/jbc.R111.290072

Chen H, Yu B, Jin S (2011) Production of levulinic acid from steam exploded rice straw via solid superacid, S_2O_{82}-/ZrO_2–SiO_2–Sm_2O_3. Bioresour Technol 102:3568–3570

Chen BY, Zhao BC, Li MF, Liu QY, Sun RC (2017) Fractionation of rapeseed straw by hydrothermal/dilute acid pretreatment combined with alkali post-treatment for improving its enzymatic hydrolysis. Bioresour Technol 225:127–133. https://doi.org/10.1016/j.biortech.2016.11.062

Chungjatupornchai W, Areerat K, Fa-Aroonsawat S (2019) Increased triacylglycerol production in oleaginous microalga Neochloris oleoabundans by overexpression of plastidial lysophosphatidic acid acyltransferase. Microb Cell Fact 18:1–12. https://doi.org/10.1186/s12934-019-1104-2

Coleman RA, Lee DP (2004) Enzymes of triacylglycerol synthesis and their regulation. Prog Lipid Res 43:134–176. https://doi.org/10.1016/S0163-7827(03)00051-1

Cronan JE, Thomas J (2009) Chapter 17 bacterial fatty acid synthesis and its relationships with polyketide synthetic pathways. Methods Enzymol. http://doi.org/10.1016/S0076-6879(09)04617-5

Di Fidio N, Dragoni F, Antonetti C, De Bari I, Raspolli Galletti AM, Ragaglini G (2020) From paper mill waste to single cell oil: enzymatic hydrolysis to sugars and their fermentation into microbial oil by the yeast Lipomyces starkeyi. Bioresour Technol 315. https://doi.org/10.1016/j.biortech.2020.123790

Dibble CJ, Shatova TA, Jorgenson JL, Stickel JJ (2011) Particle morphology characterization and manipulation in biomass slurries and the effect on rheological properties and enzymatic conversion. Biotechnol Prog 27:1751–1759

El-Mashad HM (2015) Biomethane and ethanol production potential of Spirulina platensis algae and enzymatically saccharified switchgrass. Biochem Eng J 93:119–127

Elsayed M, Abomohra AEF, Ai P, Wang D, El-Mashad HM, Zhang Y (2018) Biorefining of rice straw by sequential fermentation and anaerobic digestion for bioethanol and/or biomethane production: comparison of structural properties and energy output. Bioresour Technol 268:183–189. https://doi.org/10.1016/j.biortech.2018.07.130

Elsayed M, Abomohra AEF, Ai P, Jin K, Fan Q, Zhang Y (2019) Acetogenesis and methanogenesis liquid digestates for pretreatment of rice straw: a holistic approach for efficient biomethane production and nutrient recycling. Energy Convers Manag 195:447–456. https://doi.org/10.1016/j.enconman.2019.05.011

Elsayed M, Ran Y, Ai P, Azab M, Mansour A, Jin K, Zhang Y, Abomohra AE-F (2020) Innovative integrated approach of biofuel production from agricultural wastes by anaerobic digestion and black soldier fly larvae. J Clean Prod 263:121495. https://doi.org/10.1016/j.jclepro.2020.121495

Fu Z, Holtzapple MT (2011) Anaerobic thermophilic fermentation for carboxylic acid production from in-storage air-lime-treated sugarcane bagasse. Appl Microbiol Biotechnol 90:1669–1679

Gross M, Zhao X, Mascarenhas V, Wen Z (2016) Effects of the surface physico-chemical properties and the surface textures on the initial colonization and the attached growth in algal biofilm. Biotechnol Biofuels 9:1–14

Guschina IA, Harwood JL (2006) Lipids and lipid metabolism in eukaryotic algae. Prog Lipid Res 45:160–186. https://doi.org/10.1016/j.plipres.2006.01.001

Hakkila P. (1989) Comminution of Residual Forest Biomass. In: Utilization of Residual Forest Biomass. Springer Series in Wood Science. Springer, Berlin, Heidelberg. https://doi.org/10.1007/978-3-642-74072-5_5

Han S, Kim G-Y, Han J-I (2019) Biodiesel production from oleaginous yeast, Cryptococcus sp. by using banana peel as carbon source. Energy Rep 5:1077–1081. https://doi.org/10.1016/j.egyr.2019.07.012

He X, Osborne J, de los Reyes FL (2012) Physico-chemical characterization of grease interceptors with and without biological product addition. Water Environ Res 84:195–201

He L, Huang H, Zhang Z, Lei Z (2015) A review of hydrothermal pretreatment of lignocellulosic biomass for enhanced biogas production. Curr Org Chem 19:437–446

Hendriks A, Zeeman G (2009) Pretreatments to enhance the digestibility of lignocellulosic biomass. Bioresour Technol 100:10–18

Hu Q, Sommerfeld M, Jarvis E, Ghirardi M, Posewitz M, Seibert M, Darzins A (2008) Microalgal triacylglycerols as feedstocks for biofuel production: perspectives and advances. Plant J 54:621–639. https://doi.org/10.1111/j.1365-313X.2008.03492.x

Husain IAF, Alkhatib F, Jammi MS (2014) Problems, control, and treatment of fat, oil, and grease (FOG): a review. http://doi.org/10.5650/jos.ess13182

Jia Y, Xue L, Liu H, Li J (2009) Characterization of the glyceraldehyde-3-phosphate dehydrogenase (GAPDH) gene from the halotolerant alga Dunaliella salina and inhibition of its expression by RNAi. Curr Microbiol 58:426–431. https://doi.org/10.1007/s00284-008-9333-3

Jin M, Slininger PJ, Dien BS, Waghmode S, Moser BR, Orjuela A, da Costa Sousa L, Balan V (2015) Microbial lipid-based lignocellulosic biorefinery: feasibility and challenges. Trends Biotechnol 33:43–54

Jørgensen H, Kristensen JB, Felby C (2007) Enzymatic conversion of lignocellulose into fermentable sugars: challenges and opportunities. https://doi.org/10.1002/bbb.4

Karamerou EE, Parsons S, McManus MC, Chuck CJ (2021) Using techno-economic modelling to determine the minimum cost possible for a microbial palm oil substitute. Biotechnol Biofuels 14:57

Karki N, Johnson BS, Bates PD (2019) Metabolically distinct pools of phosphatidylcholine are involved in trafficking of fatty acids out of and into the chloroplast for membrane production. Plant Cell 31:2768. https://doi.org/10.1105/TPC.19.00121

Karunanithy C, Muthukumarappan K (2011) Influence of extruder and feedstock variables on torque requirement during pretreatment of different types of biomass—a response surface analysis. Biosyst Eng 109:37–51

Khoo KS, Chew KW, Yew GY, Leong WH, Chai YH, Show PL, Chen W-H (2020) Recent advances in downstream processing of microalgae lipid recovery for biofuel production. Bioresour Technol 304. https://doi.org/10.1016/j.biortech.2020.122996

Khorsha G (2011) The production and fate of fats, oils, and grease (FOG) from small dairy-based food service establishments

Kong F, Romero IT, Warakanont J, Li-Beisson Y (2018) Lipid catabolism in microalgae. New Phytol 218:1340–1348. https://doi.org/10.1111/nph.15047

Koo AJK, Ohlrogge JB, Pollard M (2004) On the export of fatty acids from the chloroplast. J Biol Chem 279:16101–16110. https://doi.org/10.1074/jbc.M311305200

Kratky L, Jirout T (2011) Biomass size reduction machines for enhancing biogas production. Chem Eng Technol 34:391–399

Kurosawa K, Wewetzer SJ, Sinskey AJ (2013) Engineering xylose metabolism in triacylglycerol-producing Rhodococcus opacus for lignocellulosic fuel production. Biotechnol Biofuels 6. https://doi.org/10.1186/1754-6834-6-134

Lenka SK, Carbonaro N, Park R, Miller SM, Thorpe I, Li Y (2016) Current advances in molecular, biochemical, and computational modeling analysis of microalgal triacylglycerol biosynthesis. Biotechnol Adv 34:1046–1063. https://doi.org/10.1016/j.biotechadv.2016.06.004

Li Y, Han D, Hu G, Dauvillee D, Sommerfeld M, Ball S, Hu Q (2010) Chlamydomonas starchless mutant defective in ADP-glucose pyrophosphorylase hyper-accumulates triacylglycerol. Metab Eng 12:387–391

Li Y, Han D, Yoon K, Zhu S, Sommerfeld M, Hu Q (2013) Molecular and cellular mechanisms for lipid synthesis and accumulation in microalgae: biotechnological implications. In: Handbook of microalgal culture: applied phycology and biotechnology, 2nd edn., pp 545–565. http://doi.org/10.1002/9781118567166.ch28

Li-Beisson Y, Thelen JJ, Fedosejevs E, Harwood JL (2019) The lipid biochemistry of eukaryotic algae. Prog Lipid Res 74:31–68. https://doi.org/10.1016/j.plipres.2019.01.003

Liu J, Han D, Yoon K, Hu Q, Li Y (2016) Characterization of type 2 diacylglycerol acyltransferases in Chlamydomonas reinhardtii reveals their distinct substrate specificities and functions in triacylglycerol biosynthesis. Plant J 86:3–19. https://doi.org/10.1111/tpj.13143

Lung S-C, Weselake RJ (2006) Diacylglycerol acyltransferase: a key mediator of plant triacylglycerol synthesis. Lipids 41:1073–1088. https://doi.org/10.1007/s11745-006-5057-y

Ma Y, Gao Z, Wang Q, Liu Y (2018) Biodiesels from microbial oils: opportunity and challenges. Bioresour Technol 263:631–641. https://doi.org/10.1016/j.biortech.2018.05.028

Mahmoud-Aly M, Li Y, Shanab SMM, Amin AY, Ahmed AHH (2018) Physiological characterization of a Chlamydomonas reinhardtii vacuolar transporter chaperon1 mutant under phosphorus deprivation condition

Martinez-Silveira A, Villarreal R, Garmendia G, Rufo C, Vero S (2019) Process conditions for a rapid in situ transesterification for biodiesel production from oleaginous yeasts. Electron J Biotechnol 38:1–9. https://doi.org/10.1016/j.ejbt.2018.11.006

Masri MA, Garbe D, Mehlmer N, Brück TB (2019) A sustainable, high-performance process for the economic production of waste-free microbial oils that can replace plant-based equivalents. Energy Environ Sci 12:2717–2732

Matsakas L, Giannakou M, Vörös D (2017) Effect of synthetic and natural media on lipid production from Fusarium oxysporum. Electron J Biotechnol 30:95–102. https://doi.org/10.1016/j.ejbt.2017.10.003

Mishra RR, Sharma AK (2016) Microwave–material interaction phenomena: heating mechanisms, challenges and opportunities in material processing. Compos Part A Appl Sci Manuf 81:78–97. https://doi.org/10.1016/J.COMPOSITESA.2015.10.035

Monlau F, Barakat A, Trably E, Dumas C, Steyer J-P, Carrère H (2013) Lignocellulosic materials into biohydrogen and biomethane: impact of structural features and pretreatment. Crit Rev Environ Sci Technol 43:260–322

Mosier N, Wyman C, Dale B, Elander R, Lee YY, Holtzapple M, Ladisch M (2005) Features of promising technologies for pretreatment of lignocellulosic biomass. Bioresour Technol 96:673–686. https://doi.org/10.1016/j.biortech.2004.06.025

Mühlroth A, Li K, Røkke G, Winge P, Olsen Y, Hohmann-Marriott MF, Vadstein O, Bones AM (2013) Pathways of lipid metabolism in marine algae, co-expression network, bottlenecks and candidate genes for enhanced production of EPA and DHA in species of chromista. Mar Drugs 11:4662–4697. https://doi.org/10.3390/md11114662

Nielsen PM, Brask J, Fjerbaek L (2008) Enzymatic biodiesel production: technical and economical considerations. Eur J Lipid Sci Technol 110:692–700. https://doi.org/10.1002/ejlt.200800064

Ohlrogge J, Browse J (1995) Lipid biosynthesis. Plant Cell 7:957–970. https://doi.org/10.1105/tpc.7.7.957

Papanikolaou S, Aggelis G (2011) Lipids of oleaginous yeasts. Part I: biochemistry of single cell oil production. Eur J Lipid Sci Technol 113:1031–1051

Patel A, Sindhu DK, Arora N, Singh RP, Pruthi V, Pruthi PA (2015) Biodiesel production from non-edible lignocellulosic biomass of Cassia fistula L. fruit pulp using oleaginous yeast Rhodosporidium kratochvilovae HIMPA1. Bioresour Technol 197:91–98. https://doi.org/10.1016/j.biortech.2015.08.039

Patel A, Matsakas L, Rova U, Christakopoulos P (2018a) Heterotrophic cultivation of Auxenochlorella protothecoides using forest biomass as a feedstock for sustainable biodiesel production. Biotechnol Biofuels 11. https://doi.org/10.1186/s13068-018-1173-1

Patel A, Mikes F, Bühler S, Matsakas L (2018b) Valorization of brewers' spent grain for the production of lipids by oleaginous yeast. Molecules 23. https://doi.org/10.3390/molecules23123052

Patel A, Karageorgou D, Rova E, Katapodis P, Rova U, Christakopoulos P, Matsakas L (2020) An overview of potential oleaginous microorganisms and their role in biodiesel and omega-3 fatty acid-based industries. Microorganisms 8. https://doi.org/10.3390/microorganisms8030434

Patinvoh RJ, Osadolor OA, Chandolias K, Horváth IS, Taherzadeh MJ (2017) Innovative pretreatment strategies for biogas production. Bioresour Technol 224:13–24

Patowary D, Baruah DC (2018) Effect of combined chemical and thermal pretreatments on biogas production from lignocellulosic biomasses. Ind Crops Prod 124:735–746. https://doi.org/10.1016/j.indcrop.2018.08.055

Paul S, Dutta A (2018) Challenges and opportunities of lignocellulosic biomass for anaerobic digestion. Resour Conserv Recycl 130:164–174. https://doi.org/10.1016/j.resconrec.2017.12.005

Peng J, Abomohra AEF, Elsayed M, Zhang X, Fan Q, Ai P (2019) Compositional changes of rice straw fibers after pretreatment with diluted acetic acid: towards enhanced biomethane production. J Clean Prod 230:775–782. https://doi.org/10.1016/j.jclepro.2019.05.155

Poliner E, Farré EM, Benning C (2018) Advanced genetic tools enable synthetic biology in the oleaginous microalgae Nannochloropsis sp. Plant Cell Rep 37:1383–1399. https://doi.org/10.1007/s00299-018-2270-0

Ratledge C, Cohen Z (2008) Microbial and algal oils: do they have a future for biodiesel or as commodity oils? 20:155–160. http://doi.org/10.1002/lite.200800044

Rivers DB, Emert GH (1987) Lignocellulose pretreatment: a comparison of wet and dry ball attrition. Biotechnol Lett 9:365–368

Rouches E, Herpoël-Gimbert I, Steyer JP, Carrere H (2016) Improvement of anaerobic degradation by white-rot fungi pretreatment of lignocellulosic biomass: a review. Renew Sustain Energy Rev 59:179–198. https://doi.org/10.1016/j.rser.2015.12.317

Šantek MI, Lisičar J, Mušak L, Špoljarić IV, Beluhan S, Šantek B (2018) Lipid production by yeast Trichosporon oleaginosus on the enzymatic hydrolysate of alkaline pretreated corn cobs for biodiesel production. Energy Fuels 32:12501–12513. https://doi.org/10.1021/acs.energyfuels.8b02231

Sathiyamoorthi E, Dikshit PK, Kumar P, Kim BS (2020) Co-fermentation of agricultural and industrial waste by Naganishia albida for microbial lipid production in fed-batch fermentation. J Chem Technol Biotechnol 95:813–821. https://doi.org/10.1002/jctb.6271

Schmidt JH, Weidema BP, Brandão M (2015) A framework for modelling indirect land use changes in life cycle assessment. J Clean Prod 99:230–238. https://doi.org/10.1016/j.jclepro.2015.03.013

Shafiei M, Kabir MM, Zilouei H, Horváth IS, Karimi K (2013) Techno-economical study of biogas production improved by steam explosion pretreatment. Bioresour Technol 148:53–60

Shanthi M, Rajesh Banu J, Sivashanmugam P (2019) Synergistic effect of combined pretreatment in solubilizing fruits and vegetable residue for biogas production: hydrolysis, energy assessment. Fuel 250:194–202. https://doi.org/10.1016/J.FUEL.2019.04.001

Sharma HK, Xu C, Qin W (2019) Biological pretreatment of lignocellulosic biomass for biofuels and bioproducts: an overview. Waste Biomass Valorization 10:235–251. https://doi.org/10.1007/s12649-017-0059-y

Shen Q, Chen Y, Lin H, Wang Q, Zhao Y (2018) Agro-industrial waste recycling by Trichosporon fermentans: conversion of waste sweetpotato vines alone into lipid. Environ Sci Pollut Res 25:8793–8799. https://doi.org/10.1007/s11356-018-1231-z

Sims REH, Mabee W, Saddler JN, Taylor M (2010) An overview of second generation biofuel technologies. Bioresour Technol 101:1570–1580. https://doi.org/10.1016/j.biortech.2009.11.046

Suhud Shote A (2019) Biofuel: an environmental friendly fuel. In: Banu JR (ed) Anaerobic digestion. IntechOpen, pp 1–14. http://doi.org/10.5772/intechopen.82856

Taherzadeh MJ, Karimi K (2008) Pretreatment of lignocellulosic wastes to improve ethanol and biogas production: a review. Int J Mol Sci 9:1621–1651

Tchakouteu SS, Chatzifragkou A, Kalantzi O, Koutinas AA, Aggelis G, Papanikolaou S (2015) Oleaginous yeast Cryptococcus curvatus exhibits interplay between biosynthesis of intracellular sugars and lipids. Eur J Lipid Sci Technol 117:657–672

Vasudevan PT, Briggs M (2008) Biodiesel production—current state of the art and challenges. J Ind Microbiol Biotechnol 35:421–430. https://doi.org/10.1007/s10295-008-0312-2

Venkata Subhash G, Venkata Mohan S (2011) Biodiesel production from isolated oleaginous fungi Aspergillus sp. using corncob waste liquor as a substrate. Bioresour Technol 102:9286–9290. https://doi.org/10.1016/j.biortech.2011.06.084

Vyas AP, Verma JL, Subrahmanyam N (2010) A review on FAME production processes. Fuel 89:1–9. https://doi.org/10.1016/j.fuel.2009.08.014

Wan C, Li Y (2012) Fungal pretreatment of lignocellulosic biomass. Biotechnol Adv. https://doi.org/10.1016/j.biotechadv.2012.03.003

Wang Y, Liu J, Gerken H, Zhang C, Hu Q, Li Y (2014) Highly-efficient enzymatic conversion of crude algal oils into biodiesel. Bioresour Technol 172:143–149. https://doi.org/10.1016/j.biortech.2014.09.003

Wang J-H, Zhuang L-L, Xu X-Q, Deantes-Espinosa VM, Wang X-X, Hu H-Y (2018) Microalgal attachment and attached systems for biomass production and wastewater treatment. Renew Sustain Energy Rev 92:331–342

Wang J, Ledesma-Amaro R, Wei Y, Ji B, Ji X-J (2020) Metabolic engineering for increased lipid accumulation in Yarrowia lipolytica—a review. Bioresour Technol 313. https://doi.org/10.1016/j.biortech.2020.123707

Wells T, Wei Z, Ragauskas A (2015) Bioconversion of lignocellulosic pretreatment effluent via oleaginous Rhodococcus opacus DSM 1069. Biomass Bioenerg 72:200–205. https://doi.org/10.1016/j.biombioe.2014.11.004

Xu S, Elsayed M, Ismail GA, Li C, Wang S, Abomohra AE-F (2019) Evaluation of bioethanol and biodiesel production from Scenedesmus obliquus grown in biodiesel waste glycerol: a sequential integrated route for enhanced energy recovery. Energy Convers Manag 197:111907. https://doi.org/10.1016/j.enconman.2019.111907

Xu C, Fan J, Shanklin J (2020) Metabolic and functional connections between cytoplasmic and chloroplast triacylglycerol storage. Prog Lipid Res 101069

Yang B, Wyman CE (2009) Dilute acid and autohydrolysis pretreatment. Methods Mol Biol 581:103–114. https://doi.org/10.1007/978-1-60761-214-8_8

Yellapu SK, Kaur R, Kumar LR, Tiwari B, Zhang X, Tyagi RD (2018) Recent developments of downstream processing for microbial lipids and conversion to biodiesel. Bioresour Technol 256:515–528. https://doi.org/10.1016/j.biortech.2018.01.129

Zhang Q, Liu C, Li Y, Yu Z, Chen Z, Ye T, Wang X, Hu Z, Liu S, Xiao B (2017) Cultivation of algal biofilm using different lignocellulosic materials as carriers. Biotechnol Biofuels 10:1–16

Zhang K, Xu R, Abomohra AE-F, Xie S, Yu Z, Guo Q, Liu P, Peng L, Li X (2019a) A sustainable approach for efficient conversion of lignin into biodiesel accompanied by biological pretreatment of corn straw. Energy Convers Manag 199:111928

Zhang Q, Yu Z, Jin S, Zhu L, Liu C, Zheng H, Zhou T, Liu Y, Ruan R (2019b) Lignocellulosic residue as bio-carrier for algal biofilm growth: effects of carrier physicochemical proprieties and toxicity on algal biomass production and composition. Bioresour Technol 293:122091

Zhang Q, Wang L, Yu Z, Zhou T, Gu Z, Huang Q, Xiao B, Zhou W, Ruan R, Liu Y (2020) Pine sawdust as algal biofilm biocarrier for wastewater treatment and algae-based byproducts production. J Clean Prod 256:120449

Chapter 16
Insect-Mediated Waste Conversion

Abdelaziz M. Shaboon, Xuewei Qi, and Mohamed A. A. Omar

Abstract Insects are the most prolific animals in the world due to their broad adaptability to a great variety of food feedstocks. Many insect species exhibit high conversion rates for organic wastes such as food waste, animal byproducts, and agricultural waste. Fly larvae are considered as a promising source of high-value substances due to their richness in proteins and lipids, where their biomass can be utilized being protein supplements and bioenergy substrates the black soldier fly, house fly, and yellow mealworm has been extensively studied with corresponding to high suitability for biodegradation involving organic waste. The maggots of flesh fly and blowfly are found to develop well in meat production waste. Moreover, large-scale industrial larvae production from organic waste comprises plenty of technological obstacles. Likewise, current international legislation about scaling-up of insect rearing. In this chapter, we summarized the methods, advantages, and limitations of using insects during waste conversion regarding global legislation. The information could strengthen the capacity of waste industrial transformation to larval proteinaceous and lipidic biomass.

A. M. Shaboon (✉)
Department of Economic Entomology and Pesticides, Faculty of Agriculture, Cairo University, Giza 12613, Egypt
e-mail: Abdelaziz.mansour@agr.cu.edu.eg

Department of Integrated Biological Science, College of Natural Science, Pusan National University, Busan 46241, South Korea

X. Qi
Hubei Key Laboratory of Insect Resource Application and Sustainable Pest Control, College of Plant Science and Technology, Huazhong Agricultural University, Wuhan, China
e-mail: qixuewei77@163.com

M. A. A. Omar
Department of Plant Protection, Faculty of Agriculture (Saba Basha), Alexandria University, Alexandria 21531, Egypt
e-mail: dr.abdelwanees@alexu.edu.eg

Ministry of Agriculture Key Lab of Molecular Biology of Crop Pathogens and Insects/Institute of Insect Science, Zhejiang University, 866 Yuhangtang Road, Hangzhou 310058, China

A. E.-F. Abomohra et al. (eds.), *Waste-to-Energy*,
https://doi.org/10.1007/978-3-030-91570-4_16

Keywords Waste bioconversion · Edible insects · Food and feed · Industry · Protein · Biofuel

16.1 Introduction

Due to the wide range of insect species that grow and consumed large amounts of several food sources and different types of organic matter, offering great help to facilities bioconversion of various food wastes whether they came from agriculture or during food processes. Biological characteristics of resource insects are very important to be reported, as well as introducing some key insect species such as the black soldier fly and its role in bioconversion of organic matter and raw material to proteins and other high-value components during its living and feeding and it can be used as a substitute for protein resources. Furthermore, the house fly, the flower chafer beetle, yellow mealworm, flesh fly, and blowfly have the ability to feed on various organic waste and their immature stages can be used as a good substitute for cheap protein sources. Subsequently, discussing how insects can utilize and digest different organic wastes, including food waste, industrial waste, agricultural waste, livestock manure, and so on. Thus, all insects obtained from such waste materials could be used as a significant source of protein as well as biofuel production. Much research attention is given to study the possible methods for upscaling insect-based waste bioconversion to help in solving the problem of protein shortage and keep up with the growing populations, as well as, offering novel paths to sustainable and cheap food at present and future. However, the knowledge gaps of insect cultivation and commercialization consider the big challenge facing this new field. In the same context, how regulatory affairs and authorization are among the most important factor affecting the success and acceptance of insects as a rich source of protein, fat, and other important nutrients around the world. This Chapter is focused in detail on all the above aspects and concluded that insect-mediated waste conversion shows noteworthy promise for solving many problems related to increasing the global population, and their nutritional and energy requirements.

16.2 Biological Characteristics of Resource Insects

16.2.1 Black Soldier Fly

The black soldier fly (BSF), *Hermetia illucens*, belongs to order: Diptera, family: Stratiomyidae. It was originally found in America and spread globally in temperate and tropical regions now (Sheppard et al. 1994). The life cycle duration of BSF is about 35–45 days including, four stages: egg, larva, pupa, and adult (Grossule and Lavagnolo 2020). Eggs are glossy, creamy oval with egg stage is about 4 days (Samayoa and Hwang 2018). The larval stage is approximately 15–26 days and has six instars on account of molting (Wang et al. 2020). Larvae are carrion feeders, mainly feeding on organic waste such as livestock and poultry manure and household waste (Zheng et al. 2012). The main feeding period is from the 2nd to the 5th instar. The 6th instar larvae, also known as prepupae, cease to feed and find a dry place to pupate (Diener et al. 2011). The pupal stage is about 11 days, and the pupal epidermis is dark brown and hard. Adults are usually found in green bushes and are about 15 mm long, mainly black in appearance with a slightly reddish abdomen, a white translucent spot on each side of the anterior abdomen, and white tibial segments. Because of the short life span of the adult stage, it is sustained during this period by the consumption of its body fat stores (Booth and Sheppard 1984). One female can lay eggs from 120 to 1000 (Tomberlin and Sheppard 2002) after looking for suitable crevices to lay eggs, rather than laying directly in food (Booth and Sheppard 1984). Thus, they decrease the probability of egg mass being contaminated and facilitate the collection of BSF eggs.

The mass rearing of BSF has been promoted and applied in treatment of animal manure and household garbage (Li et al. 2011). Because of its high nutritional value, the black soldier fly larvae (BSFL) can be used as a substitute for protein raw materials of livestock, poultry breeding, and aquaculture (Loponte et al. 2017; Ipema et al. 2020). The lack of raw materials is an important factor causing developmental limitation of animal facilities that possess a prominent shortage of protein resources; therefore, insects have been targeted to be the top priority as a substitute for protein resources (Klonick 2017). BSFL contains a considerable amount of protein, trace elements, and various amino acids such as methionine, lysine, and arginine. These proteins can be effectively replacing the conventional feed materials for aquaculture. Previous studies showed that the crude protein content of BSFL was more than 40%. While the crude fat content was more than 30% of the total content. The content of trace minerals such as calcium, iron, and zinc for animal growth was high, as well as the high content of essential amino acids (Wang et al. 2020).

The BSFL which lives in the environment of organic matter for a long time, can kill a variety of pathogenic bacteria, and resist the invasion of various harmful factors in the environment (Liu et al. 2008; Fan et al. 2020). They can produce a lot of antibacterial peptides, lauric acid, protease, detoxification enzymes (P450), hydrolase, and other antibacterial substances. The antimicrobial peptides obtained from BSFL had high bactericidal activity, antiviral and anti-tumor activity. Pupa shells of BSF can also be used as raw materials for the production of chitin and chitosan (Bhavsar et al. 2021). Therefore, many antimicrobial and health care medical products can be developed and manufactured using BSF. BSFL comprises several fat types which are novel raw materials for biodiesel production. Oil components have been successfully extracted from BSFL, and the biodiesel produced from BSF has more advantages than biodiesel produced from rapeseed (Hadj Saadoun et al. 2020).

16.2.2 House Fly

The house fly (*Musca domestica* L.) belongs to order: Diptera and has a wide regional distribution in the world. It is a fly with long-term and large-scale symbiosis with domestic animals and human beings. The growth period of the housefly is short, ranging from 16 to 20 days for each generation. According to that, it is divided into four-stages: egg, larva, pupa, and adult stages (Kökdener and Kiper 2021). The egg stage is 1–2 days with 1.00 mm length. The larval stage is 4–6 days, and the pupal stage is about 5 days. Pupae are generally brown and its length is about 3.0 0–7.00 mm. Flies body length is about 4.00–9.00 mm, dark red eyes, and a pale-yellow abdomen. They start laying eggs 5–7 days after emergence. Their habitats are in manure, litter, and organic matter, and feed on liquid matter, including milk, sweet water, rotting fruit, and protein-rich liquid manure (Issa 2019). Flies generally lay eggs in suitable substrates, such as animal or human manure, and various food wastes.

The house fly larvae (HFL) are rich in nutrients like proteins, fatty acids, amino acids, chitin, multivitamins, and minerals. They can be used as a good alternative source of protein (Cheng et al. 2021). The crude protein content of the HFL dry powder was 53%, 45.1% of which were containing eight essential amino acids needed by the human body. The total amount of essential amino acids was higher than the reference value of 40%, according to the Food and Agriculture Organization (FAO), and the World Health Organization (WHO). The ratio of total essential amino acids to nonessential amino acids (E/N) was 0.83, which was quite higher than the reference value (0.60) proposed by FAO-WHO (Cheng et al. 2021). Most fatty acid of larvae contains a 16–18 carbon chain, which is like the chemical structure of petroleum, and the main fatty acids were palmitic acid, oleic acid, palmitoleic acid, linoleic acid, and myristic acid (Niu et al. 2016).

16.2.3 Flower Chafer Beetle

The flower chafer beetle *Protaetia brevitarsis* L. belongs to order Coleoptera: family Cetoniinae. It has one generation per year. The larvae are saprophagous and often feed on decaying plants and animal dung. The mature larvae generally have a body length of about 30 mm with a small, brownish head, and a thick, yellowish-white carcass, and three pairs of short pectorals on the chest. They often overwinter where the soil is rich in organic matter. The length of the adult is generally about 20 mm, and the adult is elliptical with bronze on its back. There are irregular white scattered patterns on the body surface of the adult which mainly feed on fruits and vegetables (Li et al. 2010).

The protein content in dry matter of the *P. brevitarsis* larvae reached 58%, and the crude fat content was 16% (Noh et al. 2018; Ghosh et al. 2017), including seven kinds of human essential amino acids and essential amino acids comprised 39.43% of the total amino acids. The proportion of essential to nonessential amino acids was 65.11%. Methionine was the first restricting amino acid in larval protein, indicating that the larva is such a valuable protein source (Cheng et al. 2014).

16.2.4 Yellow Mealworm

The yellow mealworm (YMW), *Tenebrio molitor* L. that belongs to Tenebrionidae, Coleoptera is thoroughly distributed worldwide (Liu et al. 2010). The life cycle of YMW is divided into four stages: egg, larva, pupa, and adult. The egg length is 1–1.5 mm, round, and milky white. The shell wall of the egg is thin and easy to be damaged. The adult stages secrete mucus to cover the outer wall of the eggshell. The incubation period was 12–18 days (Liu 2006). The length of newly-hatching larvae is 2.5–3.5 mm. The larvae body color is milky white when are hatched and then generally changes to yellowish-brown with growth and development. The period of the larvae is generally 70–120 days which are divided into 15 instars. The length of mature larvae is about 29–32 mm. The larvae feed on various biomass materials. Nutrition and water are the keys to the growth of YMW larvae. When the feed and water supply are insufficient, the larvae kill each other (Huang 2013). The pupa length is 1.5–2.1 cm, and the pupal duration is 7–12 days. The length of adults is 13–20 mm, and they are white after emergence, with soft wings. After 5 days, their wings are hard and black, indicating that they are mature and ready for mating and laying eggs. The adult duration is ranged from 40 to 120 days, while most of them are 60–90 days (Qi et al. 2019). The adults can mate and lay eggs several times in their life. Female adults can lay up to 40 eggs in a day, and a pair of adults can lay about 800 eggs (Ma et al. 2001).

The total content of crude protein of YMW larvae contains an average of 52.4% (ranges from 47.0 to 60.2%). The crude fat content of larvae has an average of 30.8% and ranges from 19.1 to 36.7% based on the processing method. In addition, it also consists of multiple vitamins and inorganic elements, which have high nutritional value (Hong et al. 2020).

16.2.5 Fresh Fly and Blowfly

The blowfly (BF), *Chrysomya megacephala,* belonging to Diptera, Calliphoridae, is an important forensic insect and resource insect. The BF originated from Southeast Asia and is widely distributed in all parts of the world with human activities, except Antarctica (Wang et al. 2015a, b). They pass through four stages: egg, larva, pupa, and adult, each with a different morphology. The BF eggs were banana-shaped, white, about 1.30 ± 0.02 mm in length and 0.32 ± 0.03 mm in width. The development time was 8–16 h. Under suitable conditions, the larval stage lasts for 5–8 days, during which the body color gradually changes from transparent and milky white to milky yellow. The mature larva is about 5–13 mm in length. Larvae like to drill holes, and they are afraid of strong light. They can feed on various organic wastes such as rotten manure, animal carcasses, food waste, and so on (Rabêlo et al. 2011). The pupa belongs to the peri-pupa with a period of 5–6 days. The BF body length adults

Table 16.1 Key performance of resource insect

Species	Larva period (days)	Protein content (%)	Fat content (%)	Number of eggs per female	References
Musca domestica	4–6	~57	~15	500	Niu et al. (2017)
Protaetia brevitarsis	One generation/year	58	~16	–	Noh et al. (2018)
Chrysomya megacephala	5–8	50–65	15–20	500–800	Wang et al. (2018a, b)
Hermetia illucens	15–26	41–43 (prepupae)	35–38 (prepupae)	700–900	Wang et al. (2020)
Tenebrio molitor	70–120	47–60.2	33	400–800	Hong et al. (2020)

is about 8.00–11.0 mm. They like to lay eggs in animal carcasses, garbage sites, and other places, but also visit flowers (Qiu et al. 2015; Wang et al. 2016a, b).

BF larvae (BFL) contain high protein, oil, and amino acid, which have better nutritional value. The crude protein contents of BFL, pupae, and adults were 63.7, 76.8, and 70.5%, respectively. Crude fat contents were 16.4, 12.8, and 7.23%, respectively. Total amino acids were 56.4, 55.6, and 59.1%, respectively. The essential amino acids and nonessential amino acids ratios were 0.94, 1.13, and 0.80%, respectively (Zhao et al. 2006). Furthermore, insects have different habits due to living in various environments, which lead to a series of biological characteristics for insect species. Many biological characteristics of insects can be used for the management or optimization of organic waste Table 16.1.

16.3 Insect Utilization of Organic Waste

16.3.1 Food Waste

With the improvement of living standards, the annual production of food waste is gradually increasing worldwide, which is up to 1.3 billion tons (FAO 2018). Food waste is characterized by high water content, high salt content, high organic matter, and oil. It is highly susceptible to decay and odor and breeds pathogenic microorganisms, which seriously threaten the environment and human health. Anaerobic digestion, composting, landfilling, feed production, and organic fertilizers are the major methods to treat food waste (Li et al. 2017a, b, c; Liu et al. 2020). However, these methods are suffering from low utilization efficiency, so it is particularly important to develop resource-oriented, safe, and thorough treatment methods with high economic value. Different methods have been successfully employed to sustainable utilization of food waste and ease the harmful effects on the environment and economy. About the selection, this investigation offers one of the most novel food waste utilization approaches, where food waste is converted to insect larval biomass Fig. 16.2.

It is a good choice to transform the mixture of food waste and other organic waste by insects. For example, sawdust and food waste could be transformed by BSFL, which reduced the composting time to only 9 days. Compared with initial mixture materials, BSFL decreased organic matter, total Kjeldahl nitrogen, and volatile fatty acids from 97.41 to 85.96%, 23.01 to 17.77 g/kg, and 3.25 to 1.69 g/kg, respectively. Meanwhile, BSFL increased the total amount of phosphorous and potassium to 3.8, 5 folds, respectively (Liu et al. 2021). Total greenhouse gas emissions from food waste treated by larvae composting are lower than those from non-treated food waste (Ermolaev et al. 2019). In the transformation process of 80% Food waste and 20% rice straw powder, material flow analysis performed on a dry weight basis indicated that 6% of the diet was turned into BSF prepupae, and 51% was stored in ripen compost, and 43% was released to the air. The contents of crude protein and fat of larvae fed on food waste were 45.1% and 48.1% (Yin et al. 2021).

The HFL plays a vital role in the disposal of food waste due to its scavenging nature and rapid reproduction rate. 42.95% of food waste diet was consumed then; the residues may be used as bio-fertilizers. 53.08 g dried maggots were produced from 1000 g culture mediums, and they had a high content of protein and oil (57.06% and 15.07%, respectively) (Niu et al. 2016). Conversion processes of food waste by BSFL and HFL were compared in Fig. 16.1.

Food waste can be transformed by BFL, and the whole transformation process can be completed within 5–6 days. About 120 kg of fresh maggot and 300 kg of organic fertilizer (10% water content) can be obtained by adding the appropriate amount of auxiliary materials for every 1ton of food waste. The results indicate that the crude protein content of BF meal reared with food waste was over 50%, and the crude protein content of BF meal after degreasing was over 70%, which was significantly

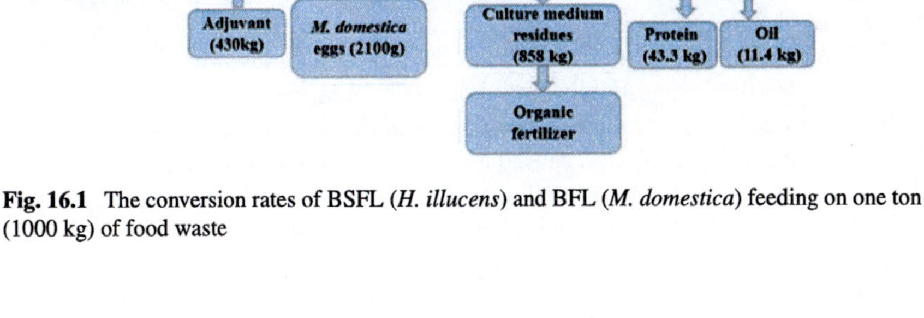

Fig. 16.1 The conversion rates of BSFL (*H. illucens*) and BFL (*M. domestica*) feeding on one ton (1000 kg) of food waste

Fig. 16.2 Insect-based bioconversion for various types of waste and its valuable outputs

higher than that one in the fish meal (about 65%). The effect of organic fertilizer reached or exceeded the national technical index of organic fertilizer (Hu 2012).

16.3.2 Livestock Manure

In the development process of animal facilities, the traditional animal facilities mode has been changed and the modernization construction has been gradually strengthened. Large amounts of manure are produced by concentrated animal facilities worldwide that must be managed, utilized, and eliminated properly (Zhu et al. 2014). Applying insect transformation to animal facilities can effectively deal with breeding pollution and promote the resource utilization of livestock manure.

Although manures contain a large amount of cellulose, hemicellulose, and lignin which obstacle the transformation process, BSFL are efficiently converted pig, chicken, and horse manures (Xiao et al. 2018). Meanwhile, the addition of chicken dung for co-transformation improves the digestion of other manure (Rehman et al. 2017a, b). The conversion of houseflies and BSF into livestock and poultry manure has been reported. Pig, chicken, and cow dung could be transformed efficiently by houseflies and BSFL (Hussein et al. 2017). 12.5% of dry pig manure, 13% of carbon, 25% of nitrogen, 14% of energy, 8.5% of phosphorus, and 9% of potassium were stored in BSFL mass (Parodi et al. 2021), and larvae could survive in horse and donkey dung well. BSFL degraded the alcohol, carboxylic and aliphatic components in pig manure and chicken manure, while it causes an increase in carboxylic materials in cow manure. After the digestion by BSF, the proteinous ingredients in all types of manure are decreased; however, the molecular weight, aromatic polycondensation, as well as polar functional groups of organic matter are increased. The simple structure of organic content (5.99–29.50%) was degraded and transformed to humic-like materials after BSFL digestion (Wang et al. 2021). The resource utilization of BSF in livestock and poultry manure treatment is summarized in the following aspects. Firstly, decreasing the pollution caused by manure, and reducing the animal manure accumulation, to avoid odor pollution to the air. BSFL feeding on manure could decrease the accumulation of cattle and chicken manure 48.7% within 19 days with bioconversion rate of 10.8%, food conversion ratio of 4.5%, the conversion activity of digestion 22.3%, cellulose 72.9%, hemicellulose 68.5%, lignin 32.8%, and nutritional contents protein, 71.2% and fat, 67.8% (Rehman et al. 2020). Secondly, the breeding of BSF in animal manure can effectively inhibit the harmful microbial activities in animal manure and reduce its damage to the environment (Awasthi et al. 2020).

BFL can transform swine manure with high efficiency by small scale trials (Wang et al. 2018b). Under laboratory conditions, BFL-transformed swine manure had an average loss of 37–52% in fresh weight and loss of 49–61% in dry matter compared with raw manure, and 30–120 kg of fresh larvae were produced from per ton of swine manure. BFL-transformed swine manure had significantly lower water content and turned to granulation. Fertility parameters including total nitrogen (2.0%), total phosphorus (4.7%), and total potassium (1.3%). Bacterial diversity analysis and quantitative PCR showed that the microbial community is notably changed, and the content of *salmonella* sp. declined sharply in the swine manure after BFL conversion. Moreover, the toxicity of Cd was decreased after transformation during BFL composting of manure and highly decreased the CH_4 and N_2O emission rates in comparison with natural accumulation. The composition and content of the Volatile organic compounds (VOCs) changed significantly. Harmful benzene series phenol, 3-methyl-phenol, and 1, 2, 4, 5-tetramethyl-benzene decreased significantly (Wang et al. 2018b).

Pig manure could be transformed by HFL, the treatment of 0.5% larva inoculum led to a high yield of larval late stage, 11.6% larval weight/manure weight. HFL fed on pig manure contained 5.32% protein, 1.34% methionine, 4.15% lysine. The total content of nitrogen, phosphorus, and potassium in pig manure residue from HFL treatment was 10.72% (dry weight), far higher than the biofertilizer in previous study reported (Zhu et al. 2015). In another paper, the bioconversion cycle produced HFL yield of 95–120 kg/m^3 pig manure. This process offered an alternative ingredient for animal feed with rich content of protein 56.9% and total fat of 23.8% as dry weight. In addition, bioconversion process helps in reducing the odor emission that is likely caused by 3-methylindole and *E. coli*, as well as, reductions in the total weight 67.2%, moisture content 80%, and Kjeldahl total nitrogen 76% (Wang et al. 2013). Compared with traditional pig manure degradation without larval aid, the stronger biodegradation led to a higher level of aromaticity and humification under bioconversion of HFL (Wang et al. 2016a, b).

Antibiotic degradation and waste reduction were verified during the 6-day composting process by HFL (Zhang et al. 2014). Livestock based compost provides a balanced ingredient for HFL growth, and the consumed manure appeared the reductions in the proportion of the total nitrogen and phosphorus 24.9% and 6.2%, respectively. Larva yield was approximately 2% of the total weight of the compost. Larva diet contains 60% protein with a well-balanced profile of amino acids, and 20% fat that contains monounsaturated fatty acids and saturated fatty acids 57% and 39%, respectively (Hussein et al. 2017).

16.3.3 Industrial Waste

Industrial organic waste defines as solid, liquid, and gaseous waste that contains organic matter disposed in artificial production. They commonly exist in many industrial production fields related to the national economy and people's livelihood, such as chemical industry, pharmaceutical chemical industry, fine chemical industry, mechanical processing, resource exploitation, and so on. A large amount of organic wastes were discharged every year. Industrial organic waste is usually characterized by complex composition, poor biodegradability, toxicity, and other characteristics which lead to that they being treated with poor efficiency, high cost, and easy to cause secondary pollution (Tsai and Chou 2004).

The integration of waste treatment and valuable product development, instead of disposing wastes in street's open plots or rivers, is a good choice to reach a win–win situation. The insect transformation technology is a high potential method for waste management Table 16.2. BSFL fed with soybean residues had a bioconversion rate 6.9%, dry mass reduction 49.0%, crude protein content 52.8%, and fat content 26.1% (Somroo et al. 2019). BSFL have good growth rates when it grew on fruit waste (0.52 g/d) and palm decanter cake (0.23 g/d) (Leong et al. 2016). Larvae grown on okara and brewer's grains showed the highest indexes for waste elimination and efficient conversion rate of the ingested feed (Bava et al. 2019). In addition, sawdust can increase the wet and dried larval yields feed on brewer's spent grain and cause increasing in nitrogen and phosphorus in frass compost by 21 and 15%, respectively (Beesigamukama et al. 2021). Winery by-products can be used as rearing materials for BSFL mass production, which induce high larval performance and nutritional composition, such as lowest saturated and the highest polyunsaturated fatty acid composition (Meneguz et al. 2018). BSFL can effectively convert materials affected with up to 20% biochar to a high-quality substrate for animal feeding and rich nutritional fertilizer can be used for organic farming (Beesigamukama et al. 2020).

At present, BSF is mainly used in the treatment of industrial organic waste, but some other insects are also used. The flesh fly showed high yield when used in the conversion of poultry manure and fish waste 22.47% and 35.34%, respectively which is significantly higher than the yield of houseflies feeding on regular diet ingredients (Marchaim et al. 2003). Gibberellin fermentation residue and rice straw could be transformed by HFL with a high conversion rate and the HF meal contained 56.4% protein, 21.6% fat (Yang et al. 2015b). Recently it was shown that YMW larvae are able to chew and eat expanded polystyrene foam (Yang et al. 2015a, b, c) and vulcanized rubber residues (Aboelkheir et al. 2019), which provides a theoretical basis for the degradation of polymer waste in the future.

Table 16.2 Conversion rate of several insects feeding on different types of wastes

Insect	Type	Waste type	Conversion rate (%)	References
Chrysomya megacephala	Industrial waste	Poultry waste	22.47 (wet/wet)	Marchaim et al. (2003)
	Food waste	Food waste	12 (wet/wet)	Hu (2012)
Protaetia brevitarsis	Agricultural waste	Fermented corn stalks	5.4 (wet/wet)	Yang et al. (2015a)
Musca domestica	Industrial waste	Gibberellin fermentation residue and rice straw	10.9 (dry/wet)	Yang et al. (2015b)
Tenebrio molitor	Agricultural waste	Wheat bran and straw	8.13 (dry/dry)	Li et al. (2015a, b)
Musca domestica	Food waste	70% food waste and 30% adjuvants	5.3 (dry/wet)	Niu et al. (2016)
	Livestock manure	Cattle manure	2 (wet/wet)	Hussein et al. (2017)
Chrysomya megacephala	Livestock manure	Pig manure	1.2 (wet/wet)	Wang et al. (2018a, b)
Hermetia illucens	Industrial waste	Soybean residues	6.9 (dry/dry)	Somroo et al. (2019)
Musca domestica	Agricultural waste	50% corn straw and 50% wheat bran	16 (dry/dry)	Qi et al. (2019)
Hermetia illucens	Food waste	80% food waste and 20% rice straw	6 (dry/dry)	Yin et al. (2021)
	Livestock manure	Pig manure	12.5 (dry/dry)	Parodi et al. (2021)

16.3.4 Agricultural Waste

The agricultural production process requires an input of resources to seek greater output. The output material contains two parts: one part is what people expect to get, while the other part is something that has no direct value for people that is collectively called agricultural waste, such as crop straw. In fact, agricultural waste also belongs to the category of resources (Martínez-Blanco et al. 2010) and must be properly treated. The optimal treatment method is to promote the conversion of these wastes into useful energy. If not handled properly, the soil will be seriously polluted and the surrounding atmospheric environment will also be greatly damaged. So the big problem we are facing now is how to properly deal with all kinds of agricultural wastes produced in the process of agricultural production, and how to use advanced science and technology to convert these wastes into energy that is useful for industrial production and people's life is the priority issue (Kashif et al. 2020).

More than 8000 million tons of crop straws were produced after crop production (FAO 2019). Unfortunately, its high lignocellulose content and poor palatability are bottlenecks to being potentially recycled. The cellulosic materials degradation using rumen microbes is hindered by the lignin encrustation in crop straw that makes crop straw an inappropriate substrate for animal feed (Van-Kuijk et al. 2015). Consequently, most farmers resort to burning corn straws in the field after each harvest which could result in severe air pollution that could potentially affect human health (Santiago-De la Rosa et al. 2017; Wang et al. 2018a, b). A few farmers retain these as a means of promoting soil nutrients and improving crop yields (Wang et al. 2019). Unfortunately, this practice generally stimulates greenhouse gas emissions and consequently increases the incidence of plant diseases (Qi et al. 2015; Romasanta et al. 2017). In addition to the techniques mentioned above, different physical and chemical methods have also been exploited to degrade lignin-rich materials (Zhao et al. 2018). However, these techniques are either expensive or environmentally unfriendly (Monlau et al. 2013). As such, sustainable alternatives need to be explored to make full use of crop straw.

As the largest group of animals, insects have a wide adaptability to varied food materials (Chapman 2013). Insects have been shown to play important roles in the bioconversion of agricultural byproducts. To make full use of all kinds of agricultural waste, apple, banana, and spent grain was used for rearing BSFL. Growth rate remarkably differed among different food. Rearing larvae on spent grain grew twice faster than larvae-fed apples (Scala et al. 2020). Corncob was treated by two-step treatment containing restaurant wastewater fermentation and BSFL feeding, and this was a high biogas production with degradation of cellulose, hemicellulose, and lignin reached about 27.3%, 45.1%, and 29.3%, respectively (Li et al. 2018). In a recent study, the effective conversion of rice straw via BSFL could be improved by alkaline peroxide pretreatment.

Rice straw, corn straw, and wheat straw can be transformed into feed resources and organic biofertilizer by HF Table 16.2 (Qi et al. 2019). *P. brevitarsis* larvae have a strong ability to transform straw, and the organic waste after conversion is dry and has no peculiar smell. The most important advantage is that the excrement is granular, which eliminates the subsequent molding steps. After 25 days of fermentation, the conversion rate of straw was as high as 63.82% ± 30.90%, and the utilization rate was as high as 17.51% ± 8.5%. 1.00 kg larva could be produced from fermented corn stalks of 18.75 kg (Yang et al. 2015a, b, c). *P. brevitarsis* larvae were found to be able to digest both sawdust and maize straw with 24.37% and 14.46% of digestion rate, respectively (Li et al. 2019). They were also capable of feeding on the remnants of the mushroom substrate of *Auricularia auricula* and *Lentinula edodes* and forming nutrient-enriched biofertilizers with low plant toxicity and high humic acid content (Wei et al. 2020). Crop residues of rice straw, rice bran, and corn straw supported the YMW life activity and growth with 90% consumption or more of waste materials and degraded the lignocellulosic content through 32 days, and residue could be as high-quality bio adsorbent for the environmental remediation and soil conditioning (Yang et al. 2019). The bioconversion rate of YMW fed on wheat straw was 8.13%, and 78.43% has resulted in frass (Li et al. 2015a, b). The proportion of eight essential amino acids in larvae was 41.30% of the amino acids profile. The mushroom, *Lentinus edodes* may be partially used as an alternative substrate of the traditional feed for YMW (<40%), and mushroom waste could be recycled as feed substrate for resource insects (Li et al. 2020). Larvae could grow successfully on diets composed of chicken waste, and the diet did not affect survival, growth, and development (Silva et al. 2021).

Overall, all kinds of organic waste could be transformed by insects with a high conversion rate (Fig. 16.2), but the conversion rate varies with the insect species Table 16.2. To make the diet more suitable for growth, adjuvants, such as rice hull and wheat bran, were usually used to adjust the aeration, water content, and nutrient ratio of feed (Qi et al. 2019; Guo et al. 2021). Meanwhile, the mixture of different organic wastes was usually for increasing the conversion rate (Rehman et al. 2017b). All insects obtained from waste management have high nutritional value, containing 40–60% protein and 15–35% fat, so they can be used as a potential protein feed and biodiesel source.

16.4 Upscaling Insect-Based Waste Bioconversion

16.4.1 Advantages and Limitations

The growth of the global population, as well as food and feed shortage, are increasingly attracting attention to alternative sources, such as insects, that have nutritional value and environmental sustainability. Insects have the advantage of rapid maturity, high ability to reproduce huge offspring individuals, growing in narrow rearing spots, and consuming a wide range of feed ingredients, including different wastes (Eriksson and Picard 2021). In the last decade, the scientific publications focusing on edible insects have dramatically increased (Van Huis 2020). This field is highly expanding by raising interest and investments, leading to the incorporation of further insect farming and processing companies in several regions of the world. According to the current estimation, there are about 300 startups worldwide (BugBurger 2021), and millions of dollars have been invested by some companies to automate the production process. Usually, insects are sampled from the wild habitats, but recently they have mass reared through production facilities. Insect production offers many significant benefits to the agri-food process and good opportunities to execute a circular economy as well. Moreover, BSF has a high potential as an alternative for feeding livestock and aquaculture (Wang and Shelomi 2017) and active performance in manure management and biofuel production (2017a, b; Elsayed et al. 2020).

In addition, many environmental benefits are reported to be linked with insect rearing Fig. 16.3 (Dobermann et al. 2017; Guiné et al. 2021). For instance, YMW production requires water fewer than livestock animals (Miglietta et al. 2015). Insects are able to fulfill their water requirements from their diet ingredients. Also, some edible species, YMW for example, have more drought resistance than cow animals (Van Huis 2013). The greenhouse gas (GHG) emissions through insect production

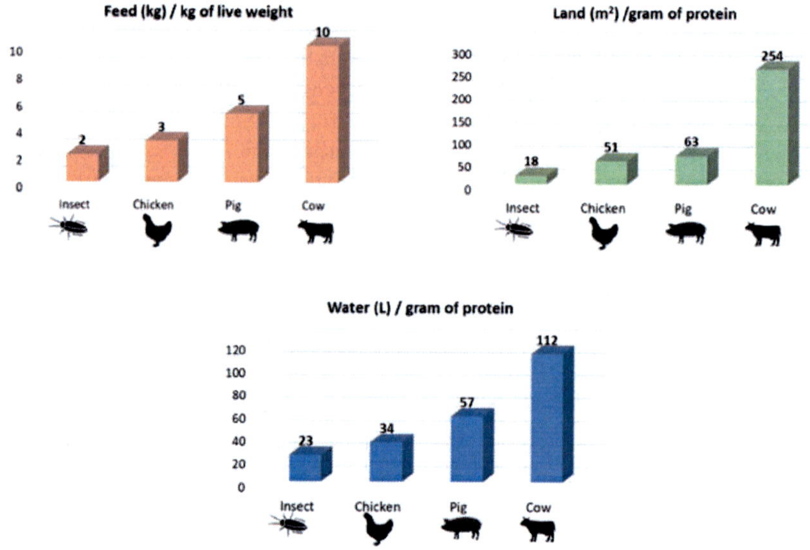

Fig. 16.3 Some measures of environmental impact of animal production, according to the species. *Source* From Guiné et al. (2021) after the required copyright permission via E-mail contact with the publishers

are much lower than that of traditional livestock (Oonincx et al. 2010). For example, farmed pigs produce GHG per kg of weight 10–100 times more than mealworms (FAO 2013). Furthermore, as the life cycle assessment (LCA) is shown in Fig. 16.3, the land-use and GHG emissions of farmed mealworm (Oonincx and de Boer 2012) and the superworm, are lower than poultry, cattle, and pigs per protein weight (kg) (Van Huis and Oonincx 2017). This aspect serves the climate objective of the UN Sustainable Development Goal No. 13 (Dicke 2018). Therefore, upscaling insect farming will be necessary to compete with the conventional sources of food and feed.

However, several challenges still hinder the upscaling of insect production and commercialization, like the general lack of regulatory frameworks due to several knowledge gaps. There is a need to develop suitable regulations (e.g., standards, codes, laws, etc.) and harmonize them through the national borders to help this emerging sector take a place at the national and global levels. Moreover, acceptance of insect consumption, especially in Western countries, is an issue that will need intensive efforts from the producers to encourage people to consume insect-based meals and insect food products.

16.4.2 Insect Breeding and Genetic Manipulation

Breeding insects as alternative diet sources are still in the primary stage. Recent reports have estimated the nutritional values and appropriateness for animal feed, mass rearing challenges, genetic improvements, and management processes (Makkar et al. 2014; Jensen et al. 2017). Several insects are globally farmed for feed and food purposes, and the highest potential three species include the YMW, the house cricket, and BSF that have been grabbing the most attention. Other remarkable species include the house fly, the blowfly, the oak moth, and the palm weevil (Hall et al. 2018; Chinarak et al. 2020; Li et al. 2020). It is found that BSFL yielded higher biomass within a shorter development duration was associated with supplying more raw rice straw (Manurung et al. 2016). However, many insects mainly those that have closed populations, face the longstanding problem; inbreeding drooping, a mating consequence between close relatives that is aggravated in captive groups caused by small founders (Rhode et al. 2020). Therefore, insect development and performance are important for straw successfully bioconversion (Gao et al. 2019b).

Genetic selection tools Fig. 16.4 are focusing on identifying the genetic loci or genes that are related to an individual's phenotypes, then selectively crossing them with the suitable methods to produce the next generations. In the beginning, it was basically happening through the genetic inference depending on ancestors and breeding programs, Mendel peas, for example, in which the specific genes associated with the traits were unknown. Although this method works well, it consumes so long

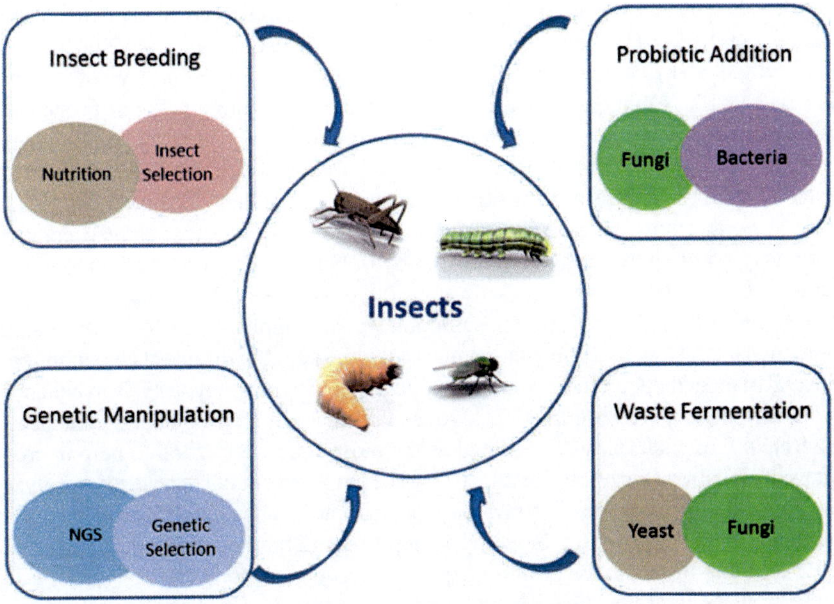

Fig. 16.4 The diagram of improving factors for edible insects used in waste bioconversion

time for optimization, especially when the phenotypes can only be estimated after a long time, mainly in agricultural livestock. Besides, the potential occurrence of other genes that could be not fully desirable may lead to potential optimization for one phenotype substitute of another inside the animal.

Next-generation sequencing (NGS) tools have recently been employed for several collected insects to characterize the nuclear genomes (Ferguson et al. 2020), such as honeybee (Honeybee Genome Sequencing 2006), Silkworm (The Silkworm Genome Consortium 2008), RNA sequencing (Liu et al. 2015; Oppert et al. 2020), and associated microbiomes (Zhan et al. 2020). Furthermore, the number of genomic loci has been increasingly characterized in individuals, including genetic markers, coding genes (Solignac et al. 2003; Gao et al. 2019a), and proteins (Park et al. 2015; Giannetto et al. 2017). For instance, in the YMW, many researchers focused on identifying the immunity-linked genes and their regulations (Johnston et al. 2014; Jo et al. 2017; Keshavarz et al. 2020). We expect that many novel investigations will be implemented by the next-generation techniques, especially with the CRISPR/Cas 9 technology (Ran et al. 2013). This so recent tool is already in the action of several insects mentioned above and has achieved novel findings.

16.4.3 Waste Fermentation

Crop straw and wheat bran are considered as diet ingredients in farming of several insects (Clariza-Samayoa et al. 2016; Zhao et al. 2017) despite, their high lignocellulosic contents (Liu et al. 2017; Li et al. 2017a, b, c). The powder of rice straw can be supplied to HFL diet to decrease the rearing cost (Yang et al. 2015b). The lignocellulose of rice straw in BSF diets was transformed into lipids within larvae (Li et al. 2015a, b) that are used for biofuel production. Corn straw was fermented then used for the farming of YMW (Wang et al. 2018a, b) and *P. brevitaris* (Yang et al. 2015a). The residues of gibberellin fermentation and rice straw were converted to larval biomass, and the other residues were used as an organic fertilizer (Yang et al. 2015b). Furthermore, several saprophagous insects have the ability to actively convert various types of straw (Yang et al. 2015a, b, c). For instance, BSF had a reduction ratio of corn and rice straw about 49% and 32%, respectively (Manurung et al. 2016; Bulak et al. 2018). Straw direct consumption would complicate the food conversion ratio of insects, so straw fermentation makes it more suitable for insect consumption. Several management methods were employed, like enzymatic hydrolysis to enhance crop straw quality (Karnaouri et al. 2016) and decrease its environmental influences. Microbial fermentation was presented as the most effective method to help in crop straw degradation (Shrivastava et al. 2011). The employment of bacteria to hydrolyze the lignocellulosic contents in straw to sugars that can be consumed directly becomes particularly essential for the bioconversion process (Binod et al. 2010; Wang et al. 2015a, b). For instance, the bacteria *Corynebacterium variabile* assisted the HFL to yield biomass and biofertilizer by converting the fermented gibberellin residue and crop straw powder (Yang et al. 2015b). The BSF efficiency in degrading rice straw was increasingly improved after the addition of a commercial microbial product (Rid-X) to the bioconversion process (Zheng et al. 2012). Straw pretreatment led to an increase in its digestibility and palatability (Ware et al. 2005).

In addition, it is reported that various fungal strains are able to produce a significant considerable amount of cellulases that actively degrade cellulosic materials (Fig. 16.4) (Kasprzycka et al. 2018). For instance, the fungi; *Trichoderma viride* degrade cellulosic content to glucose and enhance the crop straw digestion through the fungal degradation (Ghorbani et al. 2015). Also, yeast can ingest simple sugar to produce some secondary metabolites, including amino acids, B vitamins, and organic acids (Mao et al. 2013) that are useful for other microorganisms and insects. The combining of the fungi *T. viride* and yeast helps in increasing the digestibility of crop straw and improving the bioconversion rate of HFL (Qi et al. 2019). Therefore, the pretreatment of crop straws by microorganisms inoculation is essential to maximize straw utilization and exploit its nutritional benefits through facilitating the bioconversion of lignocellulosic materials into soluble sugars that are easily absorbed by insects.

16.4.4 Probiotic Addition

Insect associated microorganisms influence their host biology in numerous ways (Gould et al. 2018; Akami et al. 2019; Miller et al. 2019). Intestinal bacterial fauna plays a critical role in insect reproduction and development (Fig. 16.4) (Owings et al. 2014). Based on their location, as intra or extracellular symbionts, they established several types of associations including, mutualistic, parasitic, and in some cases pathogenic (Sazama et al. 2019). Additionally, it is well known that insect endosymbionts can contribute to their host fitness by attacking their pathogens (Paniagua Voirol et al. 2018; Cammack et al. 2021).

In transformation system, while converting the organic wastes to BSFL biomass the large variance is a big challenge (for example, feed conversion ratio, bioconversion rate, and waste reduction). As well as, larvae characteristics, like macronutrient structure, amino acids, and fatty acid profiles. These challenges are likely because of the large variance in the nutrient properties of various organic wastes. The co-digestion of BSFL with microorganisms is considered a promising tool to increase the system stability of BSFL in different organic wastes Table 16.3. For example, researchers have explored that bacterial inoculation to poultry manure is beneficial to the larval development of BSF and manure reduction as well. For instance, larval growth and development of BSF were evaluated when fed on chicken manure supplemented by four strains of *Bacillus subtilis* isolated from both BSFL guts and feed (Yu et al. 2011). The bacterial inoculation increased larval growth and shortened the development time. Furthermore, a researcher studied the impact of bacteria associated with the conversion of chicken manure to larval content. Nine species of bacteria were isolated from BSF eggs and larval gut and then inoculated to chicken manure along with BSFL. The findings highlighted that BSFL gained weight with

Table 16.3 The microorganisms use in waste fermentation and probiotics for insect farming

Microorganism	Insect	Feed Ingredients	Source
Bacillus subtilis	*Hermetia illucens*	Chicken manure	Yu et al. (2011)
The commercial product of microbes (Rid-X)		Rice straw	Zheng et al. (2012)
Effective Microorganisms (EM)	*Protaetia brevitaris*	Mixture of corn straw and cow manure	Yang et al. (2015a)
Corynebacterium variabile	*Musca domestica*	Mixture of gibberellin residue and rice straw	Yang et al. (2015b)
Bacillus subtilis	*Hermetia illucens*	Chicken manure	Xiao et al. (2018)
Trichoderma viride and *Saccharomyces cerevisiae*	*Musca domestica*	Crop straws	Qi et al. (2019)
Bacillus subtilis, and *Lactobacillus buchneri*	*Hermetia illucens*	Soybean curd residues	Somroo et al. (2019)
Bacillus subtilis, Kocuria marina, Proteus mirabilis, and *Lysinibacillus boronitolerans*		Chicken manure	Mazza et al. (2020)
Arthrobacter AK19, Bifidobacterium breve, and *Rhodococcus rhodochrous*		Standard plant based diet	Kooienga et al. (2020)
Rhodococcus rhodochrous		Gainesville diet	Franks et al. (2021)

higher protein content and weight gain as well as increased the manure reduction rates when it farmed on manure inoculated by *B. subtilis, Kocuria marina, Proteus mirabilis,* and *Lysinibacillus boronitolerans*, individually or in combination (Mazza et al. 2020).

Moreover, changing substrates may cause changes in the microbial structure relying on diet source and the chemical composition of substrate (Myers et al. 2008; Van Huis 2020). However, an innovative method could be transformed to lignocellulose-rich waste, like dairy and chicken manures, by employing lignocellulosic degradation bacteria (Rehman et al. 2019). This study assumes that the fiber structural and chemical justifications can be beneficial for the gut associated bacteria; as it enhanced larval growth and waste reduction. The co-digestion of BSF and microorganisms has been tested with *B. subtilis* on chicken manure and *L. buchneri* on soybean curd residues (Somroo et al. 2019). The *B. subtilis* supplementation to poultry manure increases BSFL bioconversion rate 12.7%, weight gain 15.9%, and waste reduction 13.4% as compared to the control one (Xiao et al. 2018).

16.5 Regulatory Affairs and Authorization

Novel methods of using edible insects in food and feed have gained high attention from the public and global media. Furthermore, the European Consumer Organisation (BEUC) studied consumer acceptance across 11 EU countries and found that 10.3% of consumers would be going to replace meat with insects in their diets (76.8% would not, 12.9% are unsure) (BEUC 2020). Insects are complex organisms and are a challenge for the characterization of insect-derived food products. Therefore, it is essential to understand their microbiology, and also consider that the whole insect parts could be consumed. Additionally, several food allergies are proteins related, so we need to evaluate whether the insect consumption could induce any allergic reactions. These could be likely due to individual's sensitivity to insect proteins, residual allergens from insect diet, like gluten, or cross-reactivity with other types of allergens. However, insect consumption is partially or fully approved in many regions of the world, including; North America, the European region, Kenya, Botswana, South Africa, several Asian countries, e.g., China, Australia, and New Zealand Fig. 16.5. This approval passes through many processes by the authoritative agencies as described in detail in the following section.

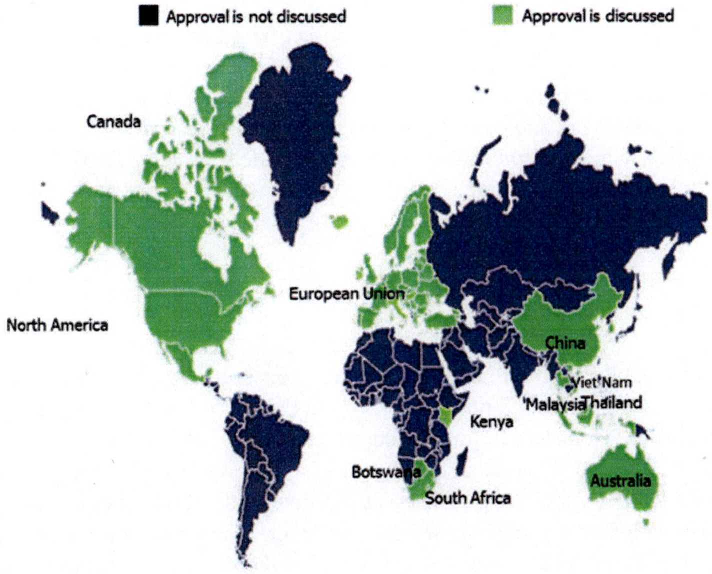

Fig. 16.5 The global approvals of insect consumption under the current legislations. This map is designed according to the data published by Food and Agriculture Organization of The United Nations, Rome 2021 (FAO 2021)

16.5.1 The Global Situation

The international standards regarding edible insects do not currently exceed the Codex level. The Asian FAO/WHO Codex Coordinating Committee has authorized a proposal to create a Codex standard for the edible house crickets and their products (FAO 2010). However, they have decided to stop working on this subject in 2014 due to the lack of information that supports the new work on edible insects, especially those regarding farming, food consumption, and commercialization of these products at both the international and local levels, as well as inappropriate data of food safety evaluation (CX/ASIA 14/19) (FAO 2014). In the following, we will discuss the regional movements towards decision making on the side of edible insect commercialization.

16.5.1.1 North America and Canada

The commercial uses of insects are fallen under the observation of Food and Drug Administration (FDA), USA. For instance, insect and insect-based products must be approved by FDA and follow its regulations to be legally used for food consumption or as color additives (GovInfo). In Canada, silkworms, YMW, and house crickets can be listed under the Non-novel Ingredients for Feed and Food (Government of Canada 2012) as reported by Health Canada, insect-based pet diet is also available in the Canadian market. The use of whole dried BSFL was authorized as a feed product for chicken broilers in 2016 (Globe Newswire 2019). However, before the product is presented for approval, the producers of insect-based animal food are required to present the assessment data of pre-market safety, growth materials, and all industrialization processes included for every insect to the Canadian Food Inspection Agency (CFIA) (Einstein-Curtis 2019).

16.5.1.2 The European Union

Insect products could soon hit Europe's supermarkets and might be a part of their daily food according to the scientific report of the European Food Safety Authority (EFSA) this year. The scientific report mentioned that the thermally dried YMW was explored to be safe for human consumption, either consumed entirely or as a powder (EFSA NDA Panel 2021). This was the first report released by EFSA as a complete evaluation of the insect-based food product. It was very important to see EFSA's scientific opinion for several reasons, including the policymakers who need to decide whether to permit these products or not before they reach the EU market (EFSA 2021). According to European Union regulations (EU Regulation 2015) insect-based products whether containing whole insect, its parts, or extracts used for human consumption are following the novel food regulation since January 2018. If genetic modifications were applied to insects as food and feed, the products would be fallen under the European regulations of genetically modified organism (GMO) food and feed (EU Regulation 2003). Additionally, in the European Union, cultured insects follow the regulations of farmed livestock animals and health biosafety standards on transmissible diseases (EU Regulation 2009, 2016). However, there are some exceptions, like protein products from the seven insect species (*Acheta domesticus, T. molitor, H. illucens, M. domestica, Gryllodes sigillatus, Alphitobius diaperinus,* and *Gryllus assimilis*) that are currently used as part of the diet formulation for pet food and aquaculture feed in the European countries (EU Regulation 2017; Belluco et al. 2017). The application includes flours made from YMW, house crickets, and migratory locusts (EU Regulation 2015; EFSA NDA Panel 2016; IPIFF 2019). The International Platform for Insects as Food and Feed (IPIFF) predicts increasing the number of Europeans consuming insect-based products and will skyrocket within this decade, reaching about 390 million in total by 2030 (IPIFF 2020).

16.5.1.3 Africa

Generally, in Africa, it seems to be a shortage of regulatory affairs for authorizing insects for consumption. However, some exceptions are found, such as mopane caterpillars which are listed as edible insects in the food law of Botswana (Grabowski et al. 2020). Additionally, some insects, including BSFL and termites, are mainly farmed for animal feed in South Africa (Niassy et al. 2018). Furthermore, the Kenya Bureau of Standards (KEBS) has recently developed some standards that will lead to the essential production of insects and their derivatives within the nation. For instance, guidelines for insect farmers to apply production security by determining the minimum limits of necessary tools and environmental conditions are presented by KS 2921:2020 (KEBS 2020).

16.5.1.4 Asia

A longstanding tradition of insect farming and consumption is carried out in many Asian countries including Malaysia, Cambodia, South Korea, Thailand, Laos, and Viet Nam (Durst and Hanboonsong 2015; Reverberi 2020). In South Korea, the Government has developed some legal standards to financially support the insect farmers with the objective of providing financial benefits to individuals as well as the national economy (FAO-Republic of Korea). In 2014, Silkworm pupae were involved in the list of allowed food in China by the Ministry of Health (Lähteenmäki-Uutela et al. 2017). Additionally, the Thai National Bureau of Agricultural Commodity and Food Standards (ACFS) released guidelines for cricket farmers in 2017 that includes information about farming crickets in a safe and effective manner as well as processing facilities falling under the correct standards (ACFS 2017).

16.5.1.5 Australia and New Zealand

The Food Standards Australia New Zealand (FSANZ) released a report on the edible insects; *A. domestica, T. molitor,* and *Z. morio*. The Advisory Committee on Novel Foods (ACNF) has considered these insect species as a non-conventional food, but also not novel food (Food Standards 2020). This indicates that they need to be compatible with the regular code of food standards, and it presents some degree of freedom for the requirements of pre-market authorization (Food Standards 2011).

16.6 Conclusions and Perspective Work

Improper disposal of animal manure, food waste, crop straw, and other organic wastes could cause ecological pollution, consequently affect human health. Organic waste conversion by insects can realize mass reduction and harvest insect biomass and biofertilizer. Different insect species have different biological characteristics, which result in different transformation periods, feed preferences, and waste-conversion rates. In this chapter, the biological characteristics and applications of several resource insects were described and compared, which provide a high potential options for transforming organic waste. Meanwhile, to upgrade the industrialization, factors of enhancing insect bioconversion for different wastes were discussed under the current regulatory framework. Further studies need to be implemented on the feasibility of employing insect biomass as a source for energy production, as well as, the risk assessment of using insects as food and feed ingredients are required to help this sector emerge safely to the public, which will be beneficial for decreasing the gap between the global population and their nutritional and energy needs.

References

Aboelkheir MG, Visconte LY, Oliveira GE, Toledo Filho RD, Souza FG Jr (2019) The biodegradative effect of *Tenebrio molitor* Linnaeus larvae on vulcanized SBR and tire crumb. Sci Total Environ 649:1075–1082

ACFS (2017) Good agricultural practices for cricket farming. https://www.acfs.go.th/standard/download/GUIDANCE_GAP_CRICKET_FARM.pdf (in Thai). An unofficial English translation of the document was made by Bugsolutely, a cricket pasta producer, and it can be found at https://www.bugsolutely.com/good-agricultural-practices-gap-cricketfarming/. Accessed on 12 May 2021

Akami M, Ren X, Qi X, Mansour A, Gao B, Cao S, Cao S, Niu C (2019). Symbiotic bacteria motivate the foraging decision and promote fecundity and survival of *Bactrocera dorsalis* (Diptera: Tephritidae). BMC Microbiol 1–13

Awasthi MK, Liu T, Awasthi SK, Duan Y, Pandey A, Zhang Z (2020) Manure pretreatments with black soldier fly *Hermetia illucens* L. (Diptera: Stratiomyidae): a study to reduce pathogen content. Sci Total Environ 737:139842

Bava L, Jucker C, Gislon G, Lupi D, Savoldelli S, Zucali M, Colombini S (2019) Rearing of *Hermetia Illucens* on different organic by-products: influence on growth, waste reduction, and environmental impact. Animals (Basel) 9(6):289

Beesigamukama D, Mochoge B, Korir NK, Fiaboe KKM, Nakimbugwe D, Khamis FM, Dubois T, Subramanian S, Wangu MM, Ekesi S, Tanga CM (2020) Biochar and gypsum amendment of agro-industrial waste for enhanced black soldier fly larval biomass and quality frass fertilizer. PLoS One 15(8):e0238154

Beesigamukama D, Mochoge B, Korir NK, Fiaboe KKM, Nakimbugwe D, Khamis FM, Subramanian S, Wangu MM, Dubois T, Ekesi S, Tanga CM (2021) Low-cost technology for recycling agro-industrial waste into nutrient-rich organic fertilizer using black soldier fly. Waste Manage 119183–119194. https://doi.org/10.1016/j.wasman.2020.09.043

Belluco S, Halloran A, Ricci A (2017) New protein sources and food legislation: the case of edible insects and EU law. Food Secur 9:803–814

BEUC (2020) One bite at a time: consumers and the transition to sustainable food. Analysis of a survey of European consumers on attitudes towards sustainable food. https://www.beuc.eu/publications/beuc-x-2020-042_consumers_and_the_transition_to_sustainable_food.pdf. Accessed on 12 May 2021

Bhavsar PS, Dalla Fontana G, Zoccola M (2021) Sustainable superheated water hydrolysis of black soldier fly exuviae for chitin extraction and use of the obtained chitosan in the textile field. ACS Omega 6(13):8884–8893

Binod P, Sindhu R, Singhania R, Vikram S, Devi L, Nagalakshmi S, Kurien N, Sukumaran R, Pandey A (2010) Bioethanol production from rice straw: an overview. Bioresour Technol 101:4767–4774

Booth DC, Sheppard C (1984) Oviposition of the black soldier fly, *Hermetia illucens* (Diptera: Stratiomyidae): eggs, masses, timing, and site characteristics. Environ Entomol 13:421–423

BugBurger (2021) The eating insects startups: here is the list of entopreneurs around the world! https://www.bugburger.se/foretag/the-eating-insects-startups-here-is-the-list-of-entopreneurs-around-the-world/. Accessed 19 May 2021

Bulak P, Polakowski C, Nowak K, Wasko A, Wiacek D, Bieganowski A (2018) *Hermetia illucens* as a new and promising species for use in entomoremediation. Sci Total Environ 633:912–919

Cammack JA, Miranda CD, Jordan HR, Tomberlin JK (2021) Upcycling of manure with insects: current and future prospects. 1–15

Chapman RF (2013) The insects: structure and function. Cambridge University Press. New York

Cheng Y, Liu Y, Xu X, Zhang J (2014) Analysis and evaluation of resource components of *Protaetia brevitarsis* (Lewis) larvae. J Shandong Agr Univ 45(2):166–170

Cheng Z, Yu L, Li H, Xu X, Yang Z (2021) Use of housefly (*Musca domestica* L.) larvae to bioconversion food waste for animal nutrition and organic fertilizer. Environ Sci Pollut Res Int 28(35):48921–48928

Chinarak K, Chaijan M, Panpipat W (2020) Farm-raised sago palm weevil (*Rhynchophorus ferrugineus*) larvae: potential and challenges for promising source of nutrients. J Food Compos Anal 92:103542

Clariza-Samayoa A, Chen WT, Hwang SY (2016) Survival and development of *Hermetia illucens* (Diptera: Stratiomyidae): a biodegradation agent of organic waste. J Econ Entomol 109(6):2580–2585

Dicke M (2018) Insects as feed and the sustainable development goals. J Insects Food Feed 4:147–156

Diener S, Solano NMS, Gutiérrez FR, Zurbrügg C, Tockner K (2011) Biological treatment of municipal organic waste using black soldier fly larvae. Waste Biom Valor 2:357–363

Dobermann D, Swift JA, Field LM (2017) Opportunities and hurdles of edible insects for food and feed. Nutri Bull 42:293–308

Durst PB, Hanboonsong Y (2015) Small-scale production of edible insects for enhanced food security and rural livelihoods: experience from Thailand and Lao People's Democratic Republic. J Insects Food Feed 1:25–31

EC summary of applications and notifications (2015). https://ec.europa.eu/food/safety/novel_food/authorisations/summary-applications-and-notifications_en. Accessed on 12 May 2021

EFSA NDA Panel (EFSA Panel on Dietetic Products, Nutrition and Allergies) (2016) Guidance on the preparation and presentation of an application for authorisation of a novel food in the context of regulation (EU) 2015/2283. EFSA J 14:4594

EFSA NDA Panel (EFSA Panel on Nutrition, Novel Foods and Food Allergens), Turck D, Castenmiller J, De Henauw S, Hirsch-Ernst KI, Kearney J, Maciuk A, Mangelsdorf I, McArdle HJ, Naska A, Pelaez C, Pentieva K, Siani A, Thies F, Tsabouri S, Vinceti M, Cubadda F, Frenzel T, Heinonen M, Marchelli R, Neuhauser-Berthold M, Poulsen M, Prieto Maradona M, Schlatter JR, van Loveren H, Ververis E, Knutsen HK (2021) Scientific opinion on the safety of dried yellow mealworm (*Tenebrio molitor* larva) as a novel food pursuant to regulation (EU) 2015/2283. EFSA J 19(1):6343, 29 pp

Einstein-Curtis A (2019) CFIA: increasing interest in insect feed ingredients prompt guidance check. Feed Navigator, 21 June 2019. Accessed on 16 May 2021. https://www.feednavigator.com/Article/2019/06/21/CFIA-Interest-in-insect-feed-ingredients-prompts-guidance-check

Elsayed M, Ran Y, Ai P, Azab M, Mansour A, Jin K, Zhang Y, Abomohra AE-F (2020) Innovative integrated approach of biofuel production from agricultural wastes by anaerobic digestion and black soldier fly larvae. J Clean Prod 263:121495

Eriksson T, Picard CJ (2021) Genetic and genomic selection in insects as food and feed abstract. J Insects Food Feed 1–22

Ermolaev E, Lalander C, Vinnerås B (2019) Greenhouse gas emissions from small-scale fly larvae composting with *Hermetia illucens*. Waste Manage 96:65–74

EU Regulation (2003) No 1829/2003 of the European Parliament and of the Council of 22 September 2003 on genetically modified food and feed

EU Regulation (2009) No 1069/2009 of the European Parliament and of the Council of 21 October 2009 laying down health rules as regards animal by-products and derived products not intended for human consumption and Repealing Regulation (EC) No 1774/2002 (animal by-products regulation). EU, Brussels, Belgium

EU Regulation (2015) No 2015/2283 of the European Parliament and of the Council of 25 November 2015 on novel foods, Amending Regulation (EU) No 1169/2011 of the European Parliament and of the Council and Repealing Regulation (EC) No 258/97 of the European Parliament and of the Council and Commission Regulation (EC) No 1852/2001. EU, Brussels, Belgium

EU Regulation (2016) No. 2016/429 of the European Parliament and of the Council of 9 March 2016 on transmissible animal diseases and amending and repealing certain acts in the area of animal health ("Animal Health Law"). EU, Brussels, Belgium

EU Regulation (2017) No 2017/893 of 24 May 2017 amending annexes I and IV to Regulation (EC) No 999/2001 of the European Parliament and of the Council and annexes X, XIV and XV to Commission Regulation (EU) No 142/2011 as regards the provisions on processed animal protein. EU, Brussels, Belgium

Fan M, Liu N, Wu X, Zhang J, Cai M (2020) Tolerance and removal of four polycyclic aromatic hydrocarbon compounds (PAHs) by black soldier fly (Diptera: *Stratiomyidae*). Environ Entomol 49(3):667–672. https://doi.org/10.1093/ee/nvaa043. PMID: 32333018

FAO (2010) Report of the seventeenth session of the FAO/WHO Coordinating Committee for Asia Bali, Indonesia (2010). http://www.fao.org/fao-who-codexalimentarius/committees/codex-regions/ccasia/meetings/en/. Accessed 10 May 2021

FAO (2013) Edible insects. Future prospects for food and feed security. FAO Forestry Paper 171. Rome, FAO, p 201. Also available at http://www.fao.org/3/i3253e/i3253e.pdf

FAO (2014) Report of the nineteenth session of the FAO/WHO Coordinating Committee for Asia. Tokyo, Japan, 3–7 Nov 2014. http://www.fao.org/fao-who-codexalimentarius/committees/codex-regions/ccasia/meetings/en/. Accessed 10 May 2021

FAO (2018) Food loss and waste reduction. Food Agric Organ

FAO (2019) AMIS agricultural management information system statistics. http://statistics.amis-outlook.org/data/index.html

FAO (2021) Looking at edible insects from a food safety perspective, challenges and opportunities for the sector. Rome

FAO-Republic of Korea. Act on fosterage and support of the insect industry. http://extwprlegs1.fao.org/docs/pdf/kor113220.pdf. Accessed 19 May 2021

Ferguson KB, Kursch-Metz T, Verhulst EC, Pannebakker BA (2020) Hybrid genome assembly and evidence-based annotation of the egg parasitoid and biological control agent *Trichogramma brassicae*. G3: Genes Genomes Genet 10:3533–3540

Food Standards (2011) https://www.foodstandards.gov.au/code/Pages/default.aspx. Accessed 12 May 2021

Food Standards (2020) Record of views formed in response to inquiries. https://www.foodstandards.gov.au/industry/novel/novelrecs/Documents/Record%20of%20views%20updated%20July%202020.pdf. Accessed 12 May 2021

Franks K, Kooienga E, Sanders M, Pendarvis K, Yang F, Tomberlin JK, Jordan HR (2021) The effect of *Rhodococcus rhodochrous* supplementation on black soldier fly (Diptera: Stratiomyidae) development, nutrition, and waste conversion. J Insects Food Feed 1–12

Gao Z, Deng W, Zhu F (2019a) Reference gene selection for quantitative gene expression analysis in black soldier fly (*Hermetia illucens*). PLoS One 14:e0221420

Gao Z, Wang W, Lu X, Zhu F, Liu W, Wang X, Lei C (2019b) Bioconversion performance and life table of black soldier fly (*Hermetia illucens*) on fermented maize straw. J Clean Prod 974–980

Ghorbani F, Karimi M, Biria D, Kariminia HR, Jeihanipour A (2015) Enhancement of fungal delignification of rice straw by *Trichoderma viride* sp. to improve its saccharification. Biochem Eng J 101:77–84

Ghosh S, Lee SM, Jung C, Meyer-Rochow VB (2017) Nutritional composition of five commercial edible insects in South Korea. J Asia-Pac Entomol 20(2):686–694

Giannetto A, Oliva S, Mazza L, Mondello G, Savastano D, Mauceri A, Fasulo S (2017) Molecular characterization and expression analysis of heat shock protein 70 and 90 from *Hermetia illucens* reared in a food waste bioconversion pilot plant. Gene 627:15–25

Globe Newswire (2019) $7.95 Billion edible insects market: global forecast to 2030. Research and markets, 1 April 2019 [Online]. Globe Newswire. Accessed on 16 May 2021. https://www.globenewswire.com/news-release/2019/04/01/1790970/0/en/7-95-Billion-Edible-Insects-Market-Global-Forecast-to-2030.html

Gould AL, Zhang V, Lamberti L, Jones EW, Obadia B, Korasidisc N, Gavryushkin A, Carlsone JM, Beerenwinkel N, Ludington WB (2018) Microbiome interactions shape host fitness. Proc Natl Acad Sci 115:E11951–E11960

Government of Canada (2012) List of non-novel determinations for food and food ingredients. https://www.canada.ca/en/health-canada/services/food-nutrition/genetically-modified-foods-other-novel-foods/requesting-novelty-determination/list-non-novel-determinations.html. Accessed on 16 May 2021

GovInfo. https://www.govinfo.gov/content/pkg/USCODE-2010-title21/pdf/USCODE-2010-title21-chap9-subchapIV-sec348.pdf. Accessed on 12 May 2021

Grabowski NT, Tchibozo S, Abdulmawjood A, Acheuk F, M'Saad Guerfali M, Sayed WAA Plötz M (2020) Edible insects in Africa in terms of food, wildlife resource, and pest management legislation. Foods 9:502

Grossule V, Lavagnolo MC (2020) The treatment of leachate using Black Soldier Fly (BSF) larvae: adaptability and resource recovery testing. J Environ Manage 253:109707

Guiné R, Correia P, Coelho C, Costa C (2021) The role of edible insects to mitigate challenges for sustainability. Open Agr 6(1):24–36

Guo H, Jiang C, Zhang Z, Lu W, Wang H (2021) Material flow analysis and life cycle assessment of food waste bioconversion by black soldier fly larvae (*Hermetia illucens* L.). Sci Total Environ 750:141656

Hadj Saadoun J, Montevecchi G, Zanasi L, Bortolini S, Macavei LI, Masino F, Maistrello L, Antonelli A (2020) Lipid profile and growth of black soldier flies (*Hermetia illucens*, Stratiomyidae) reared on by-products from different food chains. J Sci Food Agric 100(9):3648–3657

Hall HN, Masey O'Neill HV, Scholey D, Burton E, Dickinson M, Fitches EC (2018) Amino acid digestibility of larval meal (*Musca domestica*) for broiler chickens. Poult Sci 97:1290–1297

Hong J, Han T, Kim YY (2020) Mealworm (*Tenebrio molitor* Larvae) as an alternative protein source for monogastric animal: a review. Animals (Basel) 10(11):2068

Hu X (2012) Bio-translating food waste by *Chrysomya megacephala* larvae. Doctor's thesis of Sun Yat-sen University

Huang W (2013) Study on *sorghum* as feed for yellow mealworm. Shandong Agricultural University

Hussein M, Pillai VV, Goddard JM, Park HG, Kothapalli KS, Ross DA, Ketterings QM, Brenna JT, Milstein MB, Marquis H, Johnson PA, Nyrop JP, Selvaraj V (2017) Sustainable production of housefly (*Musca domestica*) larvae as a protein-rich feed ingredient by utilizing cattle manure. PLoS One 12(2):e0171708

Ipema AF, Bokkers EAM, Gerrits WJJ, Kemp B, Bolhuis JE (2020) Long-term access to live black soldier fly larvae (*Hermetia illucens*) stimulates activity and reduces fearfulness of broilers, without affecting health. Sci Rep 10(1):17428

IPIFF (International Platform for Insects as Food and Feed) (2019) Regulation (EU) 2015/2283 on novel foods. Briefing paper on the provisions relevant to the commercialization of insect-based products intended for human consumption in the EU. Brussels, IPIFF. Also available at https://ipiff.org/wp-content/uploads/2019/08/ipiff_briefing_update_03.pdf

IPIFF (International Platform for Insects as Food and Feed) (2020) Edible insects on the European market. Factsheet. Brussels, IPIFF. Also available at https://ipiff.org/wp-content/uploads/2020/06/10-06-2020-IPIFF-edible-insects-market-factsheet.pdf. Accessed 19 May 2021

Issa R (2019) *Musca domestica* acts as transport vector hosts. Bull Nat Res Cent 43:1–5

Jensen K, Kristensen TN, Heckmann L-H, Sørensen JG (2017) Breeding and maintaining high-quality insects. In: Van Huis A, Tomberlin JK (eds) Insects as food and feed: from production to consumption. Wageningen Academic Publishers, Wageningen, The Netherlands, pp 174–198

Jo YH, Kim YJ, Park KB, Seong JH, Kim SG, Park S, Noh MY, Lee YS, Han YS (2017) TmCactin plays an important role in gram-negative and -positive bacterial infection by regulating expression of 7 AMP genes in *Tenebrio molitor*. Sci Rep 7:46459

Johnston PR, Makarova O, Rolff J (2014) Inducible defenses stay up late: temporal patterns of immune gene expression in *Tenebrio molitor*. G3: Genes Genomes Genet 4:947–955

Karnaouri A, Matsakas L, Topakas E, Rova U, Christakopoulos P (2016) Development of thermophilic tailor-made enzyme mixtures for the bioconversion of agricultural and forest residues. Front Microbiol 7:177

Kashif M, Awan MB, Nawaz S, Amjad M, Talib B, Farooq M, Nizami AS, Rehan M (2020) Untapped renewable energy potential of crop residues in Pakistan: challenges and future directions. J Environ Manage 256:109924

Kasprzycka A, Lalak-Kańczugowska J, Tys J (2018) *Flammulina velutipes* treatment of non-sterile tall wheat grass for enhancing bio-degradability and methane production. Bioresour Technol 263:660–664

KEBS (2020) Production and handling of insects for food and feed-code of practice (KS 2921:2020). http://onlinecatalogue.kebs.org/webquery.dll. Consulted the 12th of May 2021

Keshavarz M, Jo YH, Edosa TT, Bae YM, Han YS (2020) TmPGRP-SA regulates antimicrobial response to bacteria and fungi in the fat body and gut of *Tenebrio molitor*. Int J Mol Sci 21:2113

Kökdener M, Kiper F (2021) Effects of larval population density and food type on the life cycle of *Musca domestica* (Diptera: *Muscidae*). Environ Entomol 50(2):324–329

Klonick A (2017) Bug ideas: assessing the market potential and regulation of insects. Master's Thesis, Duke University, Durham, NC, USA

Kooienga EM, Baugher C, Currin M, Tomberlin JK, Jordan HR (2020) Effects of bacterial supplementation on black soldier fly growth and development at benchtop and industrial scale. Front Microbiol 11:587979

Lähteenmäki-Uutela A, Grmelová N, Hénault-Ethier L, Deschamps M-H, Vandenberg GW, Zhao A, Zhang Y, Yang B, Nemane V (2017) Insects as food and feed: laws of the European Union, United States, Canada, Mexico, Australia, and China. Eur Food Feed Law Rev 12:22–36

Leong SY, Kutty SR, Malakahmad A, Tan CK (2016) Feasibility study of biodiesel production using lipids of *Hermetia illucens* larva fed with organic waste. Waste Manag 47:84–90

Li T, Ma D, Qiang S, Wang X, Wei Y (2010) A study on hosts and the occurrence regularity of *Postosia.* Xinjiang Agri Sci 47(2):320–324

Li Q, Zheng L, Cai H, Garza E, Yu Z, Zhou S (2011) From organic waste to biodiesel: black soldier fly, *Hermetia illucens*, makes it feasible. Fuel (Lond) 90:1545–1548

Li W, Li Q, Zheng L, Wang Y, Zhang J, Yu Z, Zhang Y (2015a) Potential biodiesel and biogas production from corncob by anaerobic fermentation and black soldier fly. Bioresour Technol 194:276–282. https://doi.org/10.1016/j.biortech.2015.06.112

Li W, Li MS, Zheng L, Liu YS, Zhang YL, Yu ZN, Ma Z, Li Q (2015b) Simultaneous utilization of glucose and xylose for lipid accumulation in black soldier fly. Biotechnol Biofuels 8:117

Li L, Peng X, Wang X, Wu D (2017a) Anaerobic digestion of food waste: a review focusing on process stability. Bioresour Technol 248:20–28

Li W, Zhang Z, Lin L, Terenius O (2017b) *Antheraea pernyi* (Lepidoptera: Saturniidae) and its importance in sericulture, food consumption, and traditional Chinese medicine. J Econ Entomol 110:1404–1411

Li ZH, Qi XW, Liu Y, Chen ZZ, Niu CY (2017c) Effect of corn straw fermentation by Bacillus subtilis on rearing *Musca domestica* (Diptera: Muscidae). Biotic Resour 39(6):448–454 (in Chinese)

Li W, Li Q, Wang Y, Zheng L, Zhang Y, Yu Z, Chen H, Zhang J (2018) Efficient bioconversion of organic wastes to value-added chemicals by soaking, black soldier fly (*Hermetia illucens* L.) and anaerobic fermentation. J Environ Manage 227:267–276

Li Y, Fu T, Geng L, Shi Y, Chu H, Liu F, Liu C, Song F, Zhang J, Shu C (2019) *Protaetia brevitarsis* larvae can efficiently convert herbaceous and ligneous plant residues to humic acids. Waste Manag 83:79–82

Li TH, Che PF, Zhang CR, Zhang B, Ali A, Zang LS (2020) Recycling of spent mushroom substrate: utilization as feed material for the larvae of the yellow mealworm *Tenebrio molitor* (Coleoptera: Tenebrionidae). PLoS One 15(8):e0237259

Liu Y (2006) Production and comprehensive application technology of yellow mealworm *Tenebrio molitor*. China Agriculture Press, Beijing

Liu Y, Wang F, Cui J, Zhang L (2010) Recent status and advances on study and utilization of *Tenebrio molitor*. J Environ Entomol 32(1):106–114

Liu S, Shi X-X, Jiang Y-D, Zhu Z-J, Qian P, Zhang M-J, Yu H, Zhu Q-Z, Gong Z-J, Zhu Z-R (2015) De novo analysis of the *Tenebrio molitor* (Coleoptera: Tenebrionidae) transcriptome and identification of putative glutathione S-transferase genes. Appl Entomol Zool 50:63–71

Liu Y, Qi XW, Li ZH, Niu CY (2017) Effects of three fermented crop straws on mass rearing of *Musca domestica* (Diptera: Muscidae). J Huazhong Agric Univ 36(1):55–60 (in Chinese)

Liu T, Awasthi MK, Awasthi SK, Duan YM, Zhang ZQ (2020) Effects of black soldier fly larvae (Diptera: Stratiomyidae) on food waste and sewage sludge composting. J Environ Manag 256:109967

Liu T, Awasthi SK, Qin S, Liu H, Awasthi MK, Zhou Y (2021) Conversion food waste and sawdust into compost employing black soldier fly larvae (Diptera: Stratiomyidae) under the optimized condition. Chemosphere 272:129931

Loponte R, Nizza S, Bovera F, De Riu N, Fliegerova K, Lombardi P, Vassalotti G, Mastellone V, Nizza A, Moniello G (2017) Growth performance, blood profiles and carcass traits of Barbary partridge (*Alectoris barbara*) fed two different insect larvae meals (*Tenebrio molitor* and *Hermetia illucens*). Res Vet Sci 115:183–188

Ma J, Lv C, Zhang X, Sun T (2001) Reproductive biology of *Tenebrio molitor*. J Shandong Inst Educ 5:81–83

Makkar HP, Tran G, Heuzé V, Ankers P (2014) State-of-the-art on use of insects as animal feed. Anim Feed Sci Tech 197:1–33

Manurung R, Supriatna A, Esyanthi RR, Putra RE (2016) Bioconversion of rice straw waste by black soldier fly larvae (*Hermetia illucens* L.): optimal feed rate for biomass production. J Entomol Zool Stud 4–6

Mao HL, Wang JK, Liu JX, Yoon I (2013) Effects of *Saccharomyces cerevisiae* fermentation product on in vitro fermentation and microbial communities of low-quality forages and mixed diets. J Anim Sci 91:3291–3298

Marchaim U, Gelman A, Braverman Y (2003) Reducing waste contamination from animal-processing plants by anaerobic thermophilic fermentation and by flesh fly digestion. Appl Biochem Biotechnol 109(1–3):107–115

Martínez-Blanco J, Colón J, Gabarrell X, Font X, Sánchez A, Artola A, Rieradevall J (2010) The use of life cycle assessment for the comparison of biowaste composting at home and full scale. Waste Manag 30(6):983–994

Mazza L, Xiao X, Ur Rehman K, Cai M, Zhang D, Fasulo S, Tomberlin JK, Zheng L, Soomro AA, Yu Z, Zhang J (2020) Management of chicken manure using black soldier fly (Diptera: Stratiomyidae) larvae assisted by companion bacteria. Waste Manag 102:312–318. https://doi.org/10.1016/j.wasman.2019.10.055

Meneguz M, Schiavone A, Gai F, Dama A, Lussiana C, Renna M, Gasco L (2018) Effect of rearing substrate on growth performance, waste reduction efficiency and chemical composition of black soldier fly (*Hermetia illucens*) larvae. J Sci Food Agric 98(15):5776–5784

Miglietta P, De Leo F, Ruberti M, Massari S (2015) Mealworms for food: a water footprint perspective. Water 7:6190–6203

Miller DL, Parish AJ, Newton IL (2019) Transitions and transmission: behavior and physiology as drivers of honey bee-associated microbial communities. Curr Opin Microbiol 50:1–7

Monlau F, Barakat A, Trably E, Dumas C, Steyer JP, Carrère H (2013) Lignocellulosic materials into biohydrogen and biomethane: impact of structural features and pretreatment. Crit Rev Environ Sci Technol 43:260–322

Myers HM, Tomberlin JK, Lambert BD, Kattes D (2008) Development of black soldier fly (Diptera: Stratiomyidae) larvae fed dairy manure. Environ Entomol 37:11–15

Niassy S, Ekesi S, Hendriks SK, Haller-Barker A (2018) Legislation for the use of insects as food and feed in the South African context. In: Halloran A, Flore R, Vantomme P, Roos N (eds) Edible insects in sustainable food systems. Springer, Cham, pp 457–470

Niu Y, Zheng D, Yao BH, Cai ZZ, Zhao ZM, Wu SQ (2016) A novel bioconversion for value-added products from food waste using *Musca domestica*. Waste Manag 61:455–460

Niu Y, Zheng D, Yao B, Cai Z, Zhao Z, Wu S, Cong P, Yang D (2017) A novel bioconversion for value-added products from food waste using *Musca domestica*. Waste Manage Mar;61:455–460

Noh JH, Jeong JS, Park SJ, Yun EY, Hwang JS, Kim JY, Jung KJ, Park HJ, Son HY, Moon KS (2018) Toxicological safety evaluation of freeze-dried *Protaetia brevitarsis* larva powder. Toxicol Rep 2(5):695–703

Oonincx DGAB, De Boer IJM (2012) Environmental impact of the production of mealworms as a protein source for humans—a life cycle assessment. PLoS One 7:e51145

Oonincx DGAB, van Itterbeeck J, Heetkamp MJW, van Den Brand H, van Loon JJA, van Huis A (2010) An exploration on greenhouse gas and ammonia production by insect species suitable for animal or human consumption. PLoS One 5:e14445

Oppert B, Perkin LC, Lorenzen M, Dossey AT (2020) Transcriptome analysis of life stages of the house cricket, *Acheta domesticus*, to improve insect crop production. Sci Rep 10:3471

Owings C, Spiegelman C, Tarone A, Tomberlin J (2014) Developmental variation among *Cochliomyia macellaria* Fabricius (Diptera: Calliphoridae) populations from three ecoregions of Texas, USA. Int J Leg Med 128:709–717

Paniagua Voirol LR, Frago E, Kaltenpoth M, Hilker M, Fatouros NE (2018) Bacterial symbionts in lepidoptera: their diversity, transmission, and impact on the host. Front Microbiol 9:556

Park SI, Kim JW, Yoe SM (2015) Purification and characterization of a novel antibacterial peptide from black soldier fly (*Hermetia illucens*) larvae. Develop Compar Immunol 52:98–106

Parodi A, Gerrits WJJ, Van Loon JJA, De Boer IJM, Aarnink AJA, Van Zanten HHE (2021) Black soldier fly reared on pig manure: bioconversion efficiencies, nutrients in the residual material, greenhouse gas and ammonia emissions. Waste Manage 126:674–683

Qi Y, Zhen W, Li H (2015) Allelopathy of decomposed maize straw products on three soil-born diseases of wheat and the analysis by GC-MS. J Integr Agr 14:88–97

Qi X, Li Z, Akami M, Mansour A, Niu C (2019) Fermented crop straws by *Trichoderma viride* and *Saccharomyces cerevisiae* enhanced the bioconversion rate of *Musca domestica* (Diptera: Muscidae). Environ Sci Pollut Res Int 26(28):29388–29396

Rabêlo KC, Thyssen PJ, Salgado RL, Araújo MS, Vasconcelos SD (2011) Bionomics of two forensically important blowfly species *Chrysomya megacephala* and *Chrysomya putoria* (Diptera: Calliphoridae) reared on four types of diet. Forensic Sci Int 210:257–262

Ran FA, Hsu PD, Wright J, Agarwala V, Scott DA, Zhang F (2013) Genome engineering using the CRISPR-Cas9 system. Nat Prot 8:2281–2308

Rehman Ku, Cai M, Xiao X, Zheng L, Wang H, Soomro AA, Zhou Y, Li W, Yu Z, Zhang J (2017a) Cellulose decomposition and larval biomass production from the co-digestion of dairy manure and chicken manure by mini-livestock (*Hermetia illucens* L.). J Environ Manag 196:458–465

Rehman KU, Rehman A, Cai M, Zheng L, Xiao X, Somroo AA, Wang H, Li W, Yu Z, Zhang J (2017b) Conversion of mixtures of dairy manure and soybean curd residue by black soldier fly larvae (*Hermetia illucens* L.). J Clean Prod 154:366–373

Rehman KU, Rehman RU, Somroo AA, Cai M, Zheng L, Xiao X, Rehman AU, Rehman A, Tomberlin JK, Yu Z, Zhang J (2019) Enhanced bioconversion of dairy and chicken manure by the interaction of exogenous bacteria and black soldier fly larvae. J Environ Manag 237:75–83

Rehman KU, Rehman RU, Somroo AA, Cai MM, Zheng LY, Xiao XP, Rehman AU, Rehman A, Tomberlin JK, Yu Z, Zhang J (2020) Enhanced bioconversion of dairy and chicken manure by the interaction of exogenous bacteria and black soldier fly larvae. J Environ Manag 237:75–83

Reverberi M (2020) Edible insects: cricket farming and processing as an emerging market. J Insects Food Feed 6:211–220

Rhode C, Badenhorst R, Hull KL, Greenwood MP, Bester-van der Merwe AE, Andere AA, Picard CJ, Richards C (2020) Genetic and phenotypic consequences of early domestication in black soldier flies (*Hermetia illucens*). Animal Genet 51:752–762

Romasanta RR, Sander BO, Gaihre YK, Alberto MC, Gummert M, Quilty J et al (2017) How does burning of rice straw affect CH_4 and N_2O emissions? A comparative experiment of different on-field straw management practices. Agric Ecosyst Environ 239:143–153

Samayoa AC, Hwang S-Y (2018) Degradation capacity and diapause effects on oviposition of hermetia illucens (Diptera: Stratiomyidae). J Econ Entomol 111:1682–1690

Santiago-De la Rosa N, Mugica-Álvarez V, Cereceda-Balic F, Guerrero F, Yáñez K, Lapuerta M (2017) Emission factors from different burning stages of agriculture wastes in Mexico. Environ Sci Pollut Res 24:24297–24310

Sazama EJ, Ouellette SP, Wesner JS (2019) Bacterial endosymbionts are common among, but not necessarily within, insect species. Environ Entomol 48:127–133

Scala A, Cammack JA, Salvia R, Scieuzo C, Franco A, Bufo SA, Tomberlin JK, Falabella P (2020) Rearing substrate impacts growth and macronutrient composition of *Hermetia illucens* (L.) (Diptera: *Stratiomyidae*) larvae produced at an industrial scale. Sci Rep 10(1):19448

Sheppard DC, Newton GL, Thompson SA, Savage SA (1994) Value-added manure management-system using the black soldier fly. Bioresour Technol 50:275–279

Shrivastava B, Thakur S, Khasa YP, Gupte A, Puniya AK, Kuhad RC (2011) White-rot fungal conversion of wheat straw to energy rich cattle feed. Biodegradation 22(4):823–831

Silva LB, de Souza RG, da Silva SR, Feitosa ADC, Lopes EC, Lima SBP, Dourado LRB, Pavan BE (2021) Development of *Tenebrio molitor* (Coleoptera: Tenebrionidae) on poultry litter-based diets: effect on chemical composition of larvae. J Insect Sci 21(1):7

Solignac M, Vautrin D, Loiseau A, Mougel F, Baudry E, Estoup A, Garnery L, Haberl M, Cornuet J-M (2003) Five hundred and fifty microsatellite markers for the study of the honeybee (*Apis mellifera* L.) genome. Molec Ecol Notes 3:307–311

Somroo AA, Ur Rehman K, Zheng L, Cai M, Xiao X, Hu S, Mathys A, Gold M, Yu Z, Zhang J (2019) Influence of *Lactobacillus buchneri* on soybean curd residue co-conversion by black soldier fly larvae (*Hermetia illucens*) for food and feedstock production. Waste Manag 86:114–122

The Honeybee Genome Sequencing Consortium (2006) Insights into social insects from the genome of the honeybee *Apis mellifera.* Nature 443:931
The International Silkworm Genome Consortium (2008) The genome of a lepidopteran model insect, the silkworm *Bombyx mori.* Insect Biochem Mol Biol 38:1036–1045
Tomberlin JK, Sheppard DC (2002) Factors influencing mating and oviposition of black soldier flies (Diptera: Stratiomyidae) in a colony. J Entomol Sci 37:345–352
Tsai WT, Chou YH (2004) A review of environmental and economic regulations for promoting industrial waste recycling in Taiwan. Waste Manag 24(10):1061–1069
Van Huis A (2013) Potential of insects as food and feed in assuring food security. Ann Rev Entomol 58:563–583
Van Huis A (2020) Insects as food and feed, a new emerging agricultural sector: a review. J Insects Food Feed 6:27–44
Van Huis A, Oonincx DGAB (2017) The environmental sustainability of insects as food and feed. A review. Agron Sustain Dev 37
Van-Kuijk SJA, Sonnenberg ASM, Baars JJP, Hendriks WH, Cone JW (2015) Fungal treated lignocellulosic biomass as ruminant feed ingredient: a review. Biotechnol Adv 33:191–202
Wang YS, Shelomi M (2017) Review of black soldier fly (*Hermetia illucens*) as animal feed and human food. Foods 6(10):91
Wang H, Zhang Z, Czapar GF, Winkler MK, Zheng J (2013) A full-scale house fly (Diptera: Muscidae) larvae bioconversion system for value-added swine manure reduction. Waste Manag Res 31(2):223–231
Wang C, Xiong X, Li T, Zhang X, Li N (2015a) Fermented crop straw feed *Tenebrio molitor*. Feed Res 11:66–69 (in Chinese)
Wang XY, Xiong M, Lei CL, Zhu F (2015b) The developmental transcriptome of the synanthropic fly *Chrysomya megacephala* and insights into olfactory proteins. BMC Genomics 16:20
Wang H, Wang S, Li H, Wang B, Zhou Q, Zhang X, Li J, Zhang Z (2016a) Decomposition and humification of dissolved organic matter in swine manure during housefly larvae composting. Waste Manag Res 34(5):465–473
Wang X, Li K, Zhu F, Lei C (2016b) Progress in the integrated application of *Chrysomya megacephala* (Diptera: Calliphoridae). Acta Entomol Sinica 59(10):1143–1150
Wang S, Luo L, Ding Y, Zhang D (2018a) Effects of agricultural straws on the development and the content of trehalose of *Tenebrio molitor*. J Environ Entomol 40(1):52–57 (in Chinese)
Wang X, Wang W, Gao Q, Wang X, Lei C, Zhu F (2018b) *Chrysomya megacephala* larvae feeding favourably influences manure microbiome, heavy metal stability and greenhouse gas emissions. Microb Biotechnol 11(3):498–509
Wang L, Yuan X, Liu C, Li Z, Chen F, Li S et al (2019) Soil C and N dynamics and hydrological processes in a maize-wheat rotation field subjected to different tillage and straw management practices. Agric Ecosyst Environ 285:106616
Wang SY, Wu L, Li B, Zhang D (2020) Reproductive potential and nutritional composition of *Hermetia illucens* (Diptera: Stratiomyidae) prepupae reared on different organic wastes. J Econ Entomol 113:527–537
Wang Q, Ren XN, Sun Y, Zhao JC, Awasthi MK, Liu T (2021) Improvement of the composition and humification of different animal manures by black soldier fly bioconversion. J Clean Prod 278:123397
Ware RA, Torrentera N, Zinn RA (2005) Influence of maceration and fibrolytic enzymes on the feeding value of rice straw. J Anim Vet Adv 4:6
Wei P, Li Y, Lai D, Geng L, Liu C, Zhang J, Shu C, Liu R (2020) *Protaetia brevitarsis* larvae can feed on and convert spent mushroom substrate from Auricularia auricula and *Lentinula edodes* cultivation. Waste Manag 1(114):234–239
Xiao X, Mazza L, Yu Y, Cai M, Zheng L, Tomberlin JK, Yu J, van Huis A, Yu Z, Fasulo S, Zhang J (2018) Efficient co-conversion process of chicken manure into protein feed and organic fertilizer by *Hermetia illucens* L. (Diptera: Stratiomyidae) larvae and functional bacteria. J Environ Manag 217:668–676

Yang C, Liu Y, Xu X, Zhao L (2015a) The study on the effect of *Protaetia brevitaris* Lewis larvae transformation the corn straw. J Environ Entomol 37(1):122–127 (in Chinese)

Yang S, Xie J, Hu N, Liu Y, Zhang J, Ye X et al (2015b) Bioconversion of gibberellin fermentation residue into feed supplement and organic fertilizer employing housefly (*Musca domestica* L.) assisted by corynebacterium variabile. PLoS One 10(5):e0110809

Yang Y, Yang J, Wu WM, Zhao J, Song Y, Gao L, Yang R, Jiang L (2015c) Biodegradation and mineralization of polystyrene by plastic-eating mealworms: part 1. Chemical and physical characterization and isotopic tests. Environ Sci Technol 49(20):12080–12086

Yang SS, Chen YD, Zhang Y, Zhou HM, Ji XY, He L, Xing DF, Ren NQ, Ho SH, Wu WM (2019) A novel clean production approach to utilize crop waste residues as co-diet for mealworm (*Tenebrio molitor*) biomass production with biochar as byproduct for heavy metal removal. Environ Pollut 252(Pt B):1142–1153

Yin J, Gong X, Sun L, Han M, Yang Y, Xu X, Wang X (2021) Study on the transformation of nutrients in kitchen waste by black soldier fly. J Agric Sci Technol 23(6):154–159. https://doi.org/10.13304/j.nykjdb.2020.0263

Yu G, Cheng P, Chen Y, Li Y, Yang Z, Chen Y, Tomberlin JK (2011) Inoculating poultry manure with companion bacteria influences growth and development of black soldier fly (Diptera: Stratiomyidae) larvae. Environ Entomol 40:30–35

Zhan S, Fang G, Cai M, Kou Z, Xu J, Cao Y, Bai L, Zhang Y, Jiang Y, Luo X, Xu J, Xu X, Zheng L, Yu Z, Yang H, Zhang Z, Wang S, Tomberlin JK, Zhang J, Huang Y (2020) Genomic landscape and genetic manipulation of the black soldier fly *Hermetia illucens*, a natural waste recycler. Cell Res 30:50–60

Zhang Z, Shen J, Wang H, Liu M, Wu L, Ping F, He Q, Li H, Zheng C, Xu X (2014) Attenuation of veterinary antibiotics in full-scale vermicomposting of swine manure via the housefly larvae (*Musca domestica*). Sci Rep 30(4):6844

Zhao F, Wang J, Tian J, Lei C (2006) Determination of nutritional components in *Chrysomya megacephala*. Chin Bull Entomol 43(5):688–690

Zhao Y, Wang WQ, Zhu F, Wang X, Wang XY, Lei CL (2017) The gut microbiota in larvae of the housefly *Musca domestica* and their horizontal transfer through feeding. AMB Express 7:147

Zhao SG, Li GD, Zheng N, Wang JQ, Yu ZT (2018) Steam explosion enhances digestibility and fermentation of corn stover by facilitating ruminal microbial colonization. Bioresour Technol 253:244–251

Zheng LY, Hou Y, Li W, Yang S, Li Q, Yu ZN (2012) Biodiesel production from rice straw and restaurant waste employing black soldier fly assisted by microbes. Energy 47:225–229

Zhu JC, Zhang ZQ, Fan ZM, Li HR (2014) Biogas potential, cropland load and total amount control of animal manure in China. J Agrometeoro 33:435–445

Zhu FX, Yao YL, Wang SJ, Du RG, Wang WP, Chen XY, Hong CL, Qi B, Xue ZY, Yang HQ (2015) Housefly maggot-treated composting as sustainable option for pig manure management. Waste Manag 35:62–67

Chapter 17
Phycoremediation: Role of Microalgae in Waste Management and Energy Production

Paulo Alexandre S. da Silva, José V. C. Vargas, André B. Mariano, and Ihana A. Severo

Abstract The indiscriminate disposal of industrial waste has aggravated environmental pollution problems and is an imminent concern for industrial facilities. These issues can be circumvented by conventional effluent treatment techniques, however, there are several bottlenecks associated that make them unattractive. Given this scenario, phycoremediation can be an alternative and are under scrutiny by researchers, developers, and industrialists. Microalgae-mediated processes are a promising approach for the direct removal of polluting compounds from residues, such as organic matter, phosphorus, and nitrogen. Microalgae are also responsible for producing biomass and obtaining various valuable bioproducts. However, some factors have limited the choice of the ideal microalgae cultivation system, which can complicate biomass recovery. Thus, the adoption of cell immobilization techniques can be an economically and environmentally favorable option for removing waste from effluents. In this sense, the objective of this chapter is to present the role of microalgae in waste management and energy production as phycoremediation agents. The main highlights of the chapter include an overview of microalgae and culture conditions, phycoremediation, microalgae in wastewater treatment systems, biological immobilization systems, and the cultivation of immobilized microalgae. Finally, the potential bioenergy products from microalgae and some recommendations are introduced and discussed.

Keywords Microalgae · Wastewater treatment · Photobioreactors · Cell immobilization · Biofuels · Bioeconomy

---

P. A. S. da Silva · J. V. C. Vargas · A. B. Mariano · I. A. Severo (✉)
Technology Sector, Postgraduate Program in Materials Science and Engineering (PIPE), and Sustainable Energy Research and Development Center (NPDEAS), Federal University of Paraná (UFPR), P.O. box 19011, Curitiba, PR 81531-980, Brazil

P. A. S. da Silva
e-mail: paulo.alexandre@ufpr.br

J. V. C. Vargas
e-mail: viriato@ufpr.br

A. B. Mariano
e-mail: andrebmariano@ufpr.br

A. E.-F. Abomohra et al. (eds.), *Waste-to-Energy*,
https://doi.org/10.1007/978-3-030-91570-4_17

17.1 Microalgae and Culture Conditions: Overview

Microalgae are chlorophyllic, eukaryotic, or prokaryotic microorganisms, responsible for approximately half of all carbon dioxide (CO_2) fixed in the Earth. In the presence of sunlight, these microorganisms are capable of converting CO_2 (inorganic carbon present in the atmosphere) and water into oxygen and biomass through photosynthesis. As primary producers, microalgae are the basis of the ocean food chain as there are more than 500 thousand species, which play a very important role in ecology, and increasingly for biotechnology (Borowitzka et al. 2016; Sajjadi et al. 2018). These distinct microorganisms can be easily found in different environments, such as rivers, lakes, and oceans. Besides, due to their fast growth rate, they can thrive in saline and non-potable water, without the need for a nutrient-rich medium, and require a relatively small area for their growth (Gouveia and Oliveira 2009). These advantages allow great commercial applicability of these microorganisms, thus attracting the interest of researchers and industries, in areas such as pharmacy, chemical engineering, biology, environment, and energy (Daneshvar et al. 2021).

Usually, microalgae are cultivated in a photoautotrophic regime, but they can thrive in photoheterotrophic, heterotrophic, and mixotrophic media, depending on the luminosity and carbon source to which they are submitted Fig. 17.1. In photoautotrophic conditions, microalgae require inorganic carbon and light as energy sources. However, if the provided carbon is organic, in the presence of light, it thus is called photo-heterotrophic. In heterotrophic conditions there is no light source, all the energy required for the growth is provided by an exogenous carbon source of organic

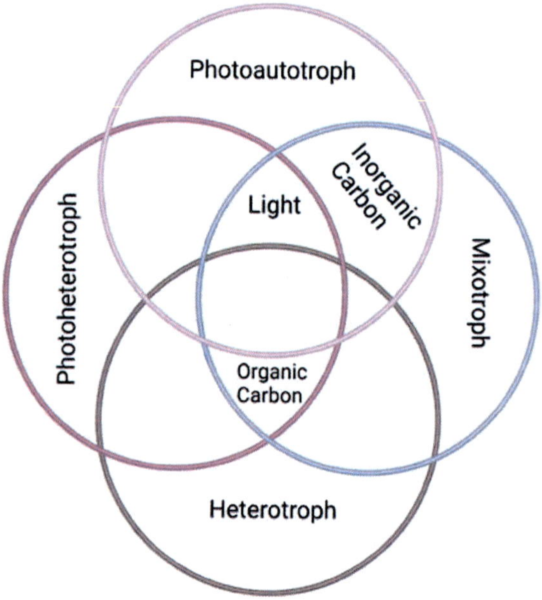

Fig. 17.1 Correlations between the energy and carbon sources demand with the type of microalgae cultivation

origin. At last, if there is a source of light and the medium is supplemented with organic and inorganic carbon, it is called mixotrophic (Verma et al. 2020).

Due to the multiple metabolic diversities of the microalgae, the elucidation of the preferential routes of carbon and energy assimilation must be well identified to favor their adequate growth and achieve the desired results. Associated with these aspects, several operational factors and process parameters influence the successful performance of microalgal cultivations and, therefore, the biomass final composition (Richmond 2004). These factors have been broadly studied over the years and will be presented and discussed below.

17.1.1 Light

Microalgae cultivation presents a unique demand in industrial biotechnology, which is the adequate supply of light to the cells. This characteristic substantially modifies the configuration of the bioreactors for microalgal production (Sivakaminathan et al. 2018). Light is the main source of energy for photo-dependent cultures. It is converted by the photosynthetic system of microalgae into chemical energy. In nature, the primary source of light energy is solar radiation, but in lab conditions, it can be replaced by fluorescent lamps or Light Emitting Diodes (LEDs) (Prajapati et al. 2013). Light is a critical factor for cultivation and can be characterized by three parameters: intensity, wavelength, and photoperiods.

The luminous intensity is defined by the amount of light received on a surface in a given time and has a direct relation to the photosynthetic rate, thus resulting in a curve with three phases: photo-limitation, photo-saturation, and photo-inhibition. The photo-limitation (up to 300 μmol m^{-2} s^{-1}) is a phase of low microalgal development due to the simple fact that there is not enough light to carry out photosynthesis and cellular growth. It may happen when the culture has an already high cell density, therefore preventing light from penetrating within the medium, or when there is not enough light. Photo saturation (up to 300–1600 μmol m^{-2} s^{-1}) is the ideal stage for cultivation, as there is an increase in the cell growth rate. In photo-inhibition (above 1600 μmol m^{-2} s^{-1}), the photosynthetic system of microalgae is damaged, resulting in the cell growth low and, therefore, culture death (Straka and Rittmann 2018; Daneshvar et al. 2021).

The radiation spectrum of sunlight is formed by countless wavelengths, nevertheless, microalgae are generally only able to use the 400–700 nm range of visible light, considered to be the photosynthetically active radiation (PAR). The ideal wavelength is defined by the pigments present in the photosystems considering that each species of microalgae can have different types of them. The main ones are Chlorophyll *a* (present in all microalgae) and Chlorophyll *b* (665 and 650 nm respectively), Xanthophyll (500–550 nm), and Phycoerythrin (490–650 nm) (Borowitzka et al. 2016; Nwoba et al. 2019). Finally, the photoperiod plays a key role in the growth and distribution of microalgae on large scale, although requirements vary widely between species, cell density, and cultivation conditions. Photoperiod is defined by the cycle of light exposure of the culture through time. In photoautotrophic conditions, cell growth is related to light: dark cycles (Che et al. 2019). According to some studies, fluctuations in photoperiods and light intensity influence the biomass biochemical composition. As an example, light cycles of 12: 12 h (light: dark) have caused levels of neutral lipids above 70% (George et al. 2014). This means that the maximum exposure to light does not produce maximum results in all microalgae species, however, the photoperiod regime must be optimized depending on the cultivation system.

17.1.2 Temperature

Temperature is another critical factor associated with microalgae cultivation. It can determine the bioavailability of certain nutrients, as in the solubility of inorganic nitrogen sources, having a direct impact on the growth rate of microorganisms. Besides, this parameter can also produce biochemical changes in cell structure and the formation and quality of microalgae lipids (Yuan et al. 2020). In general, the optimum temperature for cultivation occurs in the range of 25–35 °C, although some strains can withstand intervals of high temperatures, in the range of 60 °C, or snow temperatures. Therefore, this factor must be optimized for each process and species for it governs how fast reactions may occur and determines the metabolism rate. Importantly, the vast majority of outdoor culture systems assume temperature variation depending on local climatic conditions. Thus, the use of external heat exchanger systems must be considered for the control of this parameter (Schmidell et al. 2001; Barten et al. 2020; Daneshvar et al. 2021).

17.1.3 pH

The pH is a highly important aspect that directly impacts the metabolic activity of microalgae. The solubility of each compound in the medium is subject to pH variations, thus impacting the bioavailability of carbon, phosphate, and nitrogen, and gas transfer. In conditions in which $pH > 9$, growth inhibition and the volatilization of ammonia may occur (Lu et al. 2020; Rossi et al. 2020). Furthermore, variations in the pH beyond the optimal range ($6 < pH < 9$) can change the physiological and morphological characteristics of microalgae and alter the permeability and selectivity of the cell membrane, thus allowing for the passage of ions to change the internal biochemical composition of cells (Galès et al. 2020).

17.1.4 Aeration and Agitation

Aeration is a necessary operation in microalgae cultivation because it ensures uniform mixing of the reaction vessels, favoring the exposure of cells to light, heat transfer, and gas exchange. When this factor is properly adjusted, it minimizes the formation of cellular aggregates, which can affect the bioreactor's overall performance (Zhao et al. 2011; Daneshvar et al. 2021). In addition to the main role of providing CO_2, sometimes aeration is responsible for the promotion of agitation, in cases where impellers and propellers are not allowed or unviable. Usually, at the bottom of the culture tank, spargers are inserted, equipment with tiny orifices, responsible for the air dispersion in the system. The formed bubbles carry mechanical energy, disturbing the culture medium through their path to the surface and being an ideal solution in cultures sensitive to significant shear forces (Schmidell et al. 2001; Doran 2013).

17.1.5 Nutrients

Microalgae biomass is predominantly composed of carbon, nitrogen, and phosphorus, in proportions of approximately 50, 8, and 1%, respectively. Thus, the availability of these nutrients is essential to sustain robust cellular performance in bioreactors. Although there are many chemical elements present in aquatic environments, their distribution is not homogeneous. Thus, while some compounds are in high concentrations (macronutrients), others have low availability (micronutrients). There is a correlation between nutrients and their importance in cell metabolism Table 17.1. Consequently, nutritional deficiency can cause several problems to the culture, such as the appearance of morphologically defective cells containing low concentrations of starches, lipids, and chlorophyll, or even preventing its growth depending on which nutrient is in short supply (Grobbelaar 2007; Borowitzka et al. 2016).

17.2 Phycoremediation

Microalgae have been used in a plethora of ways, especially in environmental biotechnology applications. Bioremediation is a branch of biotechnology, which applies microorganisms to environmental treatment, but when microalgae are used, it receives the nomination of phycoremediation (Emparan et al. 2019). The goal of phycoremediation is to remove or treat contaminants and to degrade or transform toxic components into non-reactive chemical species. Concerning microalgae, this procedure is carried out through its application for the consumption of nutrients (e.g., nitrogen and phosphorus) in wastes or wastewater for subsequent biomass

Table 17.1 Relationship between nutrients and metabolic functions

Element	Functions
Macronutrients	
C	Structural component of macromolecules
N	Structural component of enzymes and proteins
P	Component of nucleic acids, phospholipids, and osmotic regulation
S	Structural component of enzymes and proteins
K	Osmotic regulation
Mg	Structural component of chlorophylls
Micronutrients	
Fe	Cofactor for some enzymatic reactions and components of cytochrome
Mn	Electron transport and maintenance of chloroplasts
Zn	Ribosome structure, nucleic acid replication, and polymerization
Cu	Electron transport
Co	Photosynthesis pathways
Mo	Nitrogen reduction (nitrate and nitrite reduction to ammonium)

Adapted from: Grobbelaar (2007)

production. The process is performed mainly in three ways: (i) biosorption, (ii) biocoagulation, and, (iii) biodegradation.

Biosorption consists of the aggregation of compounds on the cell surface, allowing the removal of the contaminants with biomass recovery. During cell growth, some microorganisms, including microalgae, are capable of secreting extracellular substances (nucleic acids, proteins, polysaccharides, and phospholipids), which can precipitate. The contaminants are adsorbed in the polymeric chain and dragged out of the solution during decantation, which results in the biocoagulation process. Biodegradation is based on the cleavage of the contaminant chemical bonds, which are metabolized during cell growth, metabolizing the initial compound into CO_2 (Anjaneyulu et al. 2005; Emparan et al. 2019; Sarkar and Dey 2021).

By way of example, algae are generally used in textile dye degradation and discoloration processes. Porous diatoms, such as *Odontella aurita*, with a structure and size of 1–500 μm have been reported as efficient biosorption agents in the treatment of dye-rich effluents. On the other hand, the green alga of the genus *Spirogyra* has been applied for the biocoagulation of the Reactive Yellow 22 and Acid Red 274 dye, with removal rates exceeding 90%. The cyanobacterial species *Oscillatoria* and the microalgae *Chlorella* are used in biodegradation processes of azo coloring agents, such as Tectilon Yellow 2 G, producing aniline as an intermediate compound, which allowed the removal of color by almost 70% (Sarkar and Dey 2021). The metabolic pathways involved in the removal of recalcitrant pollutants, thus minimizing their bioavailability and toxicity, depending especially on the production of exopolysaccharides that can measure the absorption of toxic compounds on the cell surface and/or their complexation in less bioavailable forms. The compound adhered to the membrane or cell wall of the microalgae can remain in it or internalized and chelated by molecules belonging to the phytochelatin classes (Dell' Anno et al. 2021).

17.2.1 The Role of Microalgae in the Effluent Treatment

The increase in the generation of urban wastewater is one of the biggest environmental challenges today. Effluents must be well treated to achieve a minimum level of safety before being discharged into water bodies. Conventional effluent treatment is conducted in treatment stations that are divided into three stages: primary, secondary, and tertiary treatment implementing physical, biological, or chemical procedures (Abdel-Raouf et al. 2012).

The primary treatment is responsible for the removal of suspended solids through physicochemical processes, such as flocculation and decantation, that remove up to 40% of all organic materials in effluents. The secondary treatment uses mainly biological processes to remove all organic compounds and as much inorganic matter as possible, rich in nitrogen and phosphorous. When it is not possible to remove all the impurities in the effluent, it requires a tertiary treatment through chemical or biological processes. This progressing step targets specific components that are still in the water and involves denitrification, where nitrate is reduced to nitrite in the first process and, thus, nitrite will be reduced to nitrogen gas which will release into the atmosphere (von Sperling 2007a).

The wastewaters requiring treatment are constituted mostly of carbonaceous organic compounds, determined by the biochemical demand of oxygen (BOD), and inorganic, such as phosphates, ammonium, bicarbonate, sodium, among others. When organic matter is the main pollutant it's preferable to use biological treatments, because they are simpler, more efficient, and cheaper when compared to physicochemical methods (Mohsenpour et al. 2021).

The biological treatment of wastewater can involve: (i) aerobic bioremediation (oxidation ponds, aeration lagoons, aerobic bioreactors, activated sludge, biological filters, rotating biological counters, biological nutrient removal) and anaerobic bioremediation (bioreactors and anaerobic ponds); (ii) phytoremediation of wastewater; and (iii) mycoremediation of wastewater. These procedures use microorganisms such as bacteria, fungi, yeasts, and algae (microalgae and cyanobacteria). However, biological processes that do not involve algae usually have some failures as removal inefficiency, costly, and energy-intensive (Samer 2015). For this reason, the microalgae-based wastes treatment is particularly attractive due to the competitive advantages of these microorganisms in converting organic material, such as nitrogen and phosphorus responsible for eutrophication, into biomass for the production of valuable bioproducts of market interest. Commercial use of microalgae for effluents bioremediation is not new. It started in the 1950s and has been improving over the decades with research in several plants around the world for the secondary and tertiary treatment of waste as a substrate. The main sources include agro-industrial wastes, livestock,

Table 17.2 Removal rates of pollutants present in different wastewaters and growth of microalgae species in treatment processes

Specie	Wastewater sources	Treatment process	Microalgal growth	Nutrient removal	Reference
Tetradesmus obliquus	Swine manure, domestic sewage, and biodigested cattle manure	Airlift photobioreactors	1.2, 0.75, and 0.85 g L^{-1}	NH_4^+: 99.3% P: 99.2%	(Miyawaki et al. 2021)
Arthrospira platensis	Dairy wastewater	Activated sludge	4.1 g L^{-1}	COD: 98,4% PO_4^{3-}: 98.8% NO_3^-: 99.6% NH_4^+: ~100%	(Hena et al. 2018)
Hydrodictyon reticulatum	Secondary treatment effluent	Raceway pond	1700 mg L^{-1}	TN: 75.5% TP: 86%	(Min et al. 2021)
Chlamydomonas Mexicana	Piggery wastewater	Activated sludge	0.56 g L-1	TN: 62% TP: 28% IC: 29% Ca: 33%	(Abou-Shanab et al. 2013)
Chlorella vulgaris	Tertiary treatment effluent	Sodium alginate beads (lab-scale)	0.18 d^{-1}	NO_3^-: ~100% NO_2^-: ~100% PO_4^{3-}: 90% TN: 90%	(Filippino et al. 2015)
	Poultry slaughterhouse wastewater	Sodium alginate beads (lab-scale)	-	NH_4^+: 92.4, 72.3, and 65.5% PO_4^{3-}: 94.7, 76 and 67% NO_3^-: 96.4, and 55.5%	(Hameed 2007)
	Municipal wastewater	Membrane photobioreactor	0.65 g L^{-1}	TN: 40.1% TP: 95.8%	(Gao et al. 2014)
Chlorella sp. and *Scenedesmus* sp.	Anaerobically digested piggery effluent	Raceway pond	800 and 1600 mg L^{-1}	NH_4^+: 1.97 g m^{-2} d^{-1} COD: 5.83 g m^{-2} d^{-1}	(Moheimani et al. 2018)

(continued)

Table 17.2 (continued)

Specie	Wastewater sources	Treatment process	Microalgal growth	Nutrient removal	Reference
Chlorella sp. and *Phormidium* sp.	Gray water	Algal biofilm reactor	3.64 g m^{-2} d^{-1}	NH_4^+: 93% P: 88% COD: 80%	(Choudhary et al. 2017)
Halochlorella rubescens	Secondary treatment effluent	Twin-layer photobioreactor	54 g m^{-2} (6.3 g m^{-2} d^{-1})	NH_4^+: 85% PO_4^{3-}: 78.9% NO_3^-: 83%	(Shi et al. 2014)
Tetraselmis suecica	Fish farm wastewater	Tubular photobioreactor	0.75 and 1.0 g L^{-1}	TN: 95.7% TP: 99.7%	(Michels et al. 2014)
Scenedesmus sp.	Secondary treatment effluent	Tubular photobioreactor	2560, 1585, and 1661 mg L^{-1}	TN: 42–95% TP: 16–100%	(Arbib et al. 2013)

* (−): not presented; *COD* chemical oxygen demand; *TN* total nitrogen; *TP* total phosphorus; *IC* inorganic carbon; NH_4^+ ammonia; NO_3^- nitrate; NO_2^- nitrite; *Ca* calcium

domestic sewage, and industrial (Lima et al. 2007; Lutzu et al. 2021). Table 17.2 summarizes the microalgae cultivated in different waste sources.

17.2.2 Facultative Ponds

Stabilization ponds are the most commonly used effluent treatment processes, which are based on systems that naturally treat wastewater. There are many kinds of stabilization ponds, nevertheless the most complex and used is the optional ponds. Optional ponds consist of a set of aerobic and anaerobic treatments employing bacteria and microalgae usage (Abdel-Raouf et al. 2012). The treatment process is quite simple, in which the effluent that must be treated enters on one side of the lagoon and exits on the other side, being treated throughout the course, considering the time spent in the lagoon (an average of 20 days). For better efficiency, it is possible to use multiple stagger ponds. The pond can be split into 3 parts, being anaerobic, aerobic, and facultative zones (von Sperling 2007b).

At the bottom of the pond, organic solids (particulate BOD) make the sludge (anaerobic area). At this part, organic matter is degraded by anaerobic bacteria into CO_2, methane, hydrogen sulfate, among others. Dissolved organic compounds (soluble BOD) stay on top of the pond and make the aerobic area, where microalgae are responsible for organic matter oxidation. Both phases complement each other, having an equilibrium on the production and consumption of CO_2 and O_2. During the day, microalgae perform photosynthesis. At night, bacteria, which can develop both in the presence and absence of O_2, are responsible to stabilize organic matter. These bacteria on anoxic conditions use nitrate as the final acceptor of electrons, making the optional area, and thus naming the main stabilization pond. Microalgae are vital in this treatment method, and the concentration can reach up to 200 mg/L. Among the most found algae are the green algae of the genus *Chlamydomonas* e *Chlorella*, and cyanobacteria of the genus *Oscillatoria* and *Anabaena* (von Sperling 2007b; Abdel-Raouf et al. 2012).

Some noticeable advantages of optional ponds include the simplicity and reliability of the process, dispensing machinery or workforce, in addition to low capital and operating cost. In contrast, the downside is associated with the influence of the environment on the process, such as controlling sunlight, temperature, and climate oscillation, and extensive occupied areas.

17.2.3 Activated Sludge Systems

The active sludge system is the most used conventional effluent treatment process, especially for household effluent. The goal of the system is to withdraw organic matter and nitrogenous components through biochemical aerobic reactions. The treatment process is more complex, which requires two tanks and activated flocculant sludge enough for inoculation (Mohsenpour et al. 2021). The activated sludge flakes are formed from the decantation and oxidation of effluent, through intense aeration, forming a gelatinous matrix that possesses bacteria, protozoa, microalgae among other organisms. The flakes are added to the first tank, named the aeration tank. The function of the sludge on the aeration tank is to promote the generation of newly activated sludge flakes, and to input, oxygen to induce flocculation. The effluent that contains the active sludge is sent to the second tank, called a decantation tank, which sends the flakes to the bottom of the tank, remaining cleared liquid on the surface. The cleared liquid can be sent to the receiver water body, although if there are still contaminants it must undergo further treatment, for example, anaerobic treatment to remove phosphorus or a tertiary treatment to remove more specific components. For the remaining sludge, part of it is recycled to the aeration tank, to work as a new inoculum to the next process. The whole cycle takes an average of 10–18 days (Andreoli et al. 2007).

The principal advantage of the process is the removal efficiency, which approaches 70% in some settings, with a well-defined method and many alternatives to complement and optimize. Among the main disadvantages of the process, is the need for an auxiliary anaerobic process, to remove the remaining components, energy expenditure, and aeration of the system, besides the formation of toxic residue (sludge) (von Sperling 2015).

The final destination of the sludge is a major inconvenience and, generally, many systems incinerate this material. However, for environmental reasons, this procedure causes the generation of secondary pollutants, such as the emission of harmful gases. Therewith, alternatives for the destination were made, such as using the sludge as a substrate to cultivate a mixture of microalgae and bacteria. The effluent treatment that contains the sludge is processed in a single stage, causing the mixture of microalgae/bacteria to minimize the complexity and necessary energy to treat the residue. The nitrogen is easily assimilated by microalgae, turning into nitrogen oxide, while phosphorus and carbonaceous are consumed in the growth of microalgae and bacteria. The process generates fewer greenhouse gases because it exists a balance between the aerobic and anaerobic processes (Nguyen et al. 2020; Mohsenpour et al. 2021).

17.2.4 Microalgae Cultivation Systems

Many research efforts have been directed toward the design and operation of microalgae cultivation systems for wastewater treatment to achieve high efficiency and cost–benefit. There are many types of these systems, which differ between open, closed, and hybrid systems. All of these approaches can be associated with effluent treatment.

Open systems are the most used commercially, which include raceway ponds and circular tanks. They have an extensive surface area, reaching up to 5000 m^2, and low depth, between 0.2 and 0.4 m, which facilitates light penetration. They are low cost and easy to implement, operate and scale. Among the disadvantages, are highlighted the influence of the environment, exposure to contaminants, and high evaporation rate. All of these bottlenecks influence the low productivity of the system and high harvest costs (Acién et al. 2017). Thus, the design of the microalgae cultivation system for waste treatment and management is a determining factor for successful large-scale applications aimed at high nutrient removal and biomass productivity (Li et al. 2019). Closed systems, which normally include tubular photobioreactors, flat plates, and bubble columns, are designed to achieve this goal. They play a crucial role in the continuous and stable performance of the process in the long term. Tubular photobioreactors, for example, have the most promising configuration on an industrial scale (Assunção and Malcata 2020). They are equipment built by long transparent tubes of plastic or glass, with an internal diameter of approximately 0.1 m arranged in loops, and aligned horizontally or vertically. Under this arrangement, tubular photobioreactors are designed and optimized to improve light capture, hydrodynamic properties, and productivity. Besides, it must have a proper structure so O_2 does not accumulate in the tubes and, therefore, many models are equipped with a retention tank (degassing zone) that must disperse the dissolved O_2 and homogenize the medium. Perceptible advantages include high biomass production, system control, and linear scaling. However, the disadvantages are purely economic, which can be circumvented in several ways (Satyanarayana et al. 2011; Abdel-Raouf et al. 2012; Acién et al. 2017; Vo et al. 2018).

Figure 17.2 exemplifies a compact tubular photobioreactor on an industrial scale. The system is installed at Sustainable Energy Research and Development Center (NPDEAS), located in Curitiba-PR, Brazil, where the authors of this work acting. The plant has 5 photobioreactors (2 m × 5 m × 8 m) with 12 m^3 of capacity, 3.5 km of transparent tubes, and an area of 10 m^2 each, in addition to pilot-scale photobioreactors. They are operating under the input of several types of wastes, such as CO_2 emissions, solid and liquid effluents. The focus of the research group is the renewable energy generation from microalgae (e.g., biofuels) as well as the parallel production of other commercial value products (Vargas et al. 2014; Corrêa et al. 2017; Telles et al. 2018; Miyawaki et al. 2021). The engineering conception of the photobioreactors was patented under registration US2012088296A1 and WO2012050608A1.

17.3 Biological Immobilization Systems

The waste treatment by microalgae is something that still has some bottlenecks, such as its recovery from the treated effluent. New technologies have been considered, such as cellular immobilization. This technique consists of trapping or fixing cells in a matrix through physical–chemical interactions. The use of immobilized biomass to treat effluents is being increasingly studied, as shown in Table 17.3. The immobilization tends to improve the intrinsic characteristics of microorganisms, such as the removal of BOD retention, solids, in addition to eliminating the sedimentation

Fig. 17.2 Large-scale compact tubular photobioreactors for microalgae cultivation at NPDEAS (Curitiba, Paraná, Brazil)

Table 17.3 Wastes treatment by immobilized microalgae

Species	Wastewater source	Matrix	Reference
Scenedesmus intermedius	Pig manure	Sodium alginate	(Jiménez-Pérez et al. 2004)
Synechococcus elongatus	Artificial wastewater	Chitosan	(Aguilar-May and Sánchez-Saavedra 2009)
Chlorella vulgaris	Poultry slaughterhouse wastewater	Sodium alginate	(Hameed 2007)
Scenedesmus abundans; Chlorella vulgaris	Secondary effluent	Sodium alginate	(Kube et al. 2020)
Desmodesmus sp	Domestic wastewater	Sodium alginate	(Wang et al. 2020)
Lobosphaera sp.	Secondary effluent	Chitosan	(Vasilieva et al. 2021)
Pseudomonas putida	Artificial wastewater	Polyvinyl alcohol	(Al-Zuhair and El-Naas 2011)
Chlorella pyrenoidosa	Secondary effluent from Activated sludge	Polyvinyl alcohol	(Huang and Wang 2003)

step. Therefore, the immobilized cell system can increase efficiency and decrease biological process costs (Giese 2015).

The immobilization matrix commonly called support, must have some basic characteristics of the application, being the main ones: not toxic to cells, resistant to chemical and biological degradation, high capacity of cell retention, presenting malleability to mechanical loads (tension shear, compression, and internal gas pressure), and allow the diffusion of substrates and products.

There are three main categories as to the origin of the material that can be used as a support, which can be defined as (i) natural polymeric materials, using alginate, cellulose, and collagen; (ii) synthetic polymeric materials, using polyacrylamide, polystyrene, and polyurethane; and (iii) inorganic materials, using alumina, zirconia, and silica (Schmidell et al. 2001). In addition, the main techniques of cell immobilization can be classified into two main groups, such as immobilization with support material (adsorption, entrapment, and encapsulation) and self-immobilization (self-aggregation of cells and self-flocculating), as exemplified in Fig. 17.3. In general, immobilization confers some advantages, the main ones being: the possibility of recycling or disposal, resistance to contamination, and greater mechanical, chemical, and thermal resistance. Microalgae cell immobilization approaches improve the cost–benefit of phycoremediation processes (Wittmann and Liao 2016).

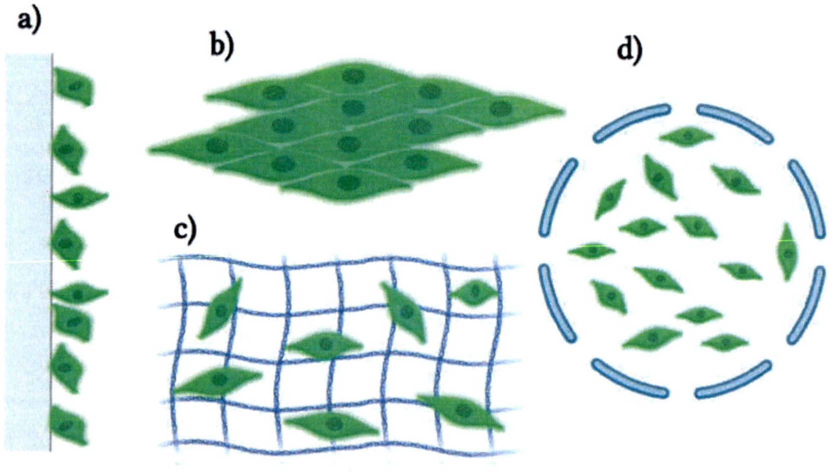

Fig. 17.3 Microalgae cell immobilization systems. *Note* **a** microalgae bonded in support by adsorption; **b** biofilm formation by self-immobilization; **c** microalgae entrapped within the matrix; **d** microalgae encapsulated in the matrix

17.3.1 Attachment or Adsorption

The adhesion of microorganisms to surfaces is something common in nature, however, the adsorption method tends to force an interaction between the microorganism and the desired surface. The bonds between cells and the support are modulated by physical–chemical processes, such as pH and ionic strength, mainly by covalent bonds that result in hydrophobic or hydrophilic interactions. Among immobilization methods, adsorption appears to be relatively the cheapest, depending mainly on the cost of the support. Despite the economic advantage, the process has some disadvantages, such as reduced viability and damage to the cell surface, and loss of biomass according to the use of the support after immobilization. The efficiency of the immobilization process can be improved by the formation of pores on the surface of the support, or by washing the support with electrolytic or polycationic solutions, the purpose of which is to promote electrostatic interactions between the cell surface and the support (Schmidell et al. 2001; Moreno-Garrido 2008; Giese 2015).

17.3.2 Self-Immobilization

Many cells have can self-flocculate or form biofilms naturally or stress-induced. Some microalgae species, such as Scenedesmus obliquus, have this ability to bioflocculate, to form cellular aggregates associated with rapid sedimentation, and to form biofilms, forming an abstract surface through connections between cells. Normally, these mechanisms are activated due to (i) excretion of organic macromolecules, such as glucose, mannose, galactose, and fructose; (ii) growth of adjacent daughter cells; (iii) environmental stresses, such as pH, temperature, and nutrient shortages (Wittmann and Liao 2016; Mantzorou and Ververidis 2019).

The self-immobilization mechanism has several advantages compared to traditional immobilization techniques, including the fact that it does not require the use of supports, which makes the technique the most promising and important in the bioprocess industry. Besides, simplicity, low energy consumption and cost with support material, recovery of biomass without contamination by the support, better microbiological propagation due to the absence of the physical barrier, and better adaptation to the production of metabolites primary products such as ethanol (Wittmann and Liao 2016).

17.3.3 Entrapment Within a Matrix

The entrapment methodology is one of the most studied techniques for cell immobilization and is divided into two main categories, entrapment in gel matrices and entrapment in porous particles. In this method, the cells get contained in a rigid complex or semi-rigid. For gel immobilization, polysaccharides can be used (alginates, chitosan, and agar) or polymeric matrices (collagen and polyvinyl alcohol). Usually, microorganisms' growth is partially inhibited or limited to support porosity, the formation of a concentration gradient between products and subtracts in and out of the matrix, which forces mass transport through diffusion. Notwithstanding, immobilization in matrical pores is also an option when is needed support with high compressive strength and abrasion or demands a large-scale process. Nevertheless, aggregate cells in a matrix can present low productivity problems due to the inevitable cell detachment from the support and overuse (Mallick 2002; Wittmann and Liao 2016).

17.3.4 Cells Contained Behind a Barrier

This technique, normally denominated encapsulation, is based on the entrapment of cells by a membrane in a capsule shape, with porous characteristics or semi-permanent properties. The membrane separates the cells from the aqueous external, allowing small molecules to enter (gas and nutrients) into the aqueous or semi-solid, inside the capsule. Cell growth is limited to capsule size, which is made of alginate or Poly-L-lysine (PLL). Initially, the technique was used to develop medicines in pharmaceutical companies. Nevertheless, the ease of reproduction of the technique allowed the application in other areas, such as fermentative processes (Guisan 2006).

The process is considered relatively cheap depending on the material used on capsule formation. Besides cost, the technique shows other advantages, such as convenience to withdraw the capsules from the cultivation, reuse possibility of the spheres or capsules with the microorganism. Besides, they are transparent and are non-toxic (Moreno-Garrido 2008). On the other hand, the capsules do not have mechanical resistance and can break, but this deficiency can be overcome with a second polymeric layer on the capsule surface, normally a treatment with chitosan is enough (Giese 2015).

17.4 Cultivation of Immobilized Microalgae

Once the cell immobilization process occurs, the microalgae culture can be used in unconventional bioreactors. Most immobilized microalgae photobioreactors are made up of continuously operated columns, such as fluidized beds and membrane bioreactors. Normally, the cultivation of immobilized microalgae has the objective of producing biomass and extracting energy products and, in parallel, the effluent treatment (Schmidell et al. 2001; Vasilieva et al. 2016). Yet, the growth rate of immobilized microalgae is slightly lower when compared to free-grown microalgae. Studies by Zhou et al. (2012) have shown that the duration of the lag phase (cell adaptation phase in a new environment, with a growth rate $\mu \approx 0$) was longer in the cultivation of *S. obliquus* immobilized in carrageenan. However, based on cell growth, this difference disappeared and the maximum cultivation density reached similar values.

The cultivation of microorganisms can be operated in different configurations, including commonly in batch, fed-batch, continuous, and plug-flow (specific method for fixed-bed bioreactors, in continuous regime). In batch-operated cultivations, the substrate is added along with the inoculum of microorganisms, and no material is inserted in the process, except O_2 or CO_2. While in the batch-fed regime, the substrate is added at regular time intervals. Both processes are finalized after the consumption of all the substrate and filling of the bioreactor, for final product removal. In a bioreactor operated in continuous mode, there are continuous inlets and outlets of the same proportion, containing fresh medium and medium with formed products, respectively (Schmidell et al. 2001). Each mode of operation has its advantages and disadvantages, as presented in Table 17.4.

Table 17.4 Advantages and disadvantages of different modes of cell culture operation

Operational method/system	Advantages	Disadvantages
1. Continuous mode (Open system)	(a) It is the best option for large-scale production of low-value products; (b) High productivity; (c) Constant product quality; (d) Easier automation and use of advanced controls;	(a) Contamination risk (b) The lineage can mutate after a long time of production, resulting in an inefficient strain (c) The operation downstream can be complicated in continuous mode (d) Non-flexible process
2. Batch (Closed system)	(a) Adaptation to different processes is easy and low cost; (b) Low risk of infection (through constant sterilizations) and mutation (short process time);	(a) High operating cost (b) Considerable downtime for sterilization, start of production, and cleaning (c) Delicate processes for filling and emptying the reactor
3. Fed-batch (Open system)	(a) The same advantages of the Batch regime (b) Decreased inhibition effects by-product and substrate	(a) Higher costs compared to the simple batch regime (b) Large volume to be processed in downstream
Plug flow		
1. Continuous mode (Open system)	(a) High substrate conversion (b) Fixed bed operation (for immobilized cells or enzymes) or film (c) High conversion of substrates in the gas phase	(a) Requires immobilized cells (b) The difference between the retention times of the liquid and gas phases requires multiple passes in the reactor

Adapted from: Villadsen et al. (2011)

17.4.1 Fluidized Bed Photobioreactors

One of the most studied photobioreactors worldwide applied to immobilized microalgae is the fluidized bed photobioreactor, which is ideal for cells immobilized by entrapment or encapsulation techniques. The photobioreactor can have several variations, but generally, it consists of a vertical column of circular section, where immobilized microalgae are contained. The culture medium or effluent is then pumped into the bioreactor and the treated product or effluent is removed from the opposite side. In some configurations, it is possible to perform the recirculation of the medium to obtain better results (Schmidell et al. 2001; Moreno-Garrido 2008).

Usually, fluidized bed bioreactors are designed on small scales, but it is used mainly in the ethanol production by yeasts. This model of bioreactor has almost complete homogeneity, which makes it a reliable and replicable system (Schmidell et al. 2001). According to Cheirsilp et al. (2017), with batch cultivation for the effluent treatment and lipid production, using *Nannochloropsis* sp., it was possible to remove more than 90% of nitrogenous and phosphate residues.

17.4.2 Biofilm Photobioreactors

The biofilm photobioreactor consists of an immobilized microalgae film, in which effluent must pass by capsular, along with the film or by section, as a membrane. Microalgae are immobilized by adsorption techniques on materials such as membranes, filters, cotton, or foams. Normally used on a laboratory scale, this bioreactor is not homogeneous and produces a concentration gradient along with the film. This continuous mode of operation, common in fixed bed bioreactors, is called Plug-flow (Villadsen et al. 2011; Assunção and Malcata 2020).

The microalgae that make up the biofilm have high efficiency in effluent treatment. However, they need some surface area on the biofilm to perform photosynthesis, with better light use, when compared to free algae. The easiness of biomass harvest and high productivity are the best advantages of this photobioreactor (Mantzorou and Ververidis 2019).

17.5 Bioenergy from Microalgae

Waste is a source of a large number of nutrients, rich mainly in nitrogen, phosphorus, and organic matter. The use of microalgae-based processes has allowed the removal of about 70% of nutrients from these effluents, which are converted into several products, especially energetic. Many microalgae used for phycoremediation, such as *Chlorella* sp., *Botryococcus braunii*, and *Scenedesmus* sp., have a high concentration of lipids and carbohydrates, among which the main fuels that can be produced are

Table 17.5 Biofuels produced by microalga biomass and process

Biofuel	Process	Microalgae	References
Biodiesel	Chemical transesterification by acid/alkali catalyzes	*Chlamydomonas mexicana*	(Abou-Shanab et al. 2013)
		Scenedesmus obliquus	(Han et al. 2016)
		Nannochloropsis sp.	(Jiang et al. 2011)
Bioethanol	Fermentation by *Saccharomyces cerevisiae*	*Chlorella vulgaris*	(Abdulla et al. 2020)
			(Ho et al. 2013a)
		Scenedesmus obliquus	(Ho et al. 2013b)
Biohydrogen	Photolysis, fermentation	*Chlamydomonas reinhardtii*	(Hwang et al. 2014)
		Scenedesmus sp.	(Kumar et al. 2016)
Biogas	Anaerobic digestion of biomass	*Chlamydomonas subcaudata*	(Hernández et al. 2016)
		Chlorella kessleri	(Caporgno et al. 2015)
Biobutanol	Fermentation by *Clostridium acetobutylicum*	*Chlorella vulgaris*	(Gao et al. 2016)
		Scenedesmus sp.	(Castro et al. 2015)

biodiesel, bioethanol, biohydrogen, biogas, and biobutanol (Deprá et al. 2018; Kumar et al. 2021; Severo et al. 2021). Some research examples can be in Table 17.5.

Additionally, there are some emerging sources of microalgae energy products that are under investigation, such as bioelectricity from microalgae-based microbial fuel cells, direct combustion of biomass to generate heat, and the biogeneration of volatile organic compounds with energy potential for use as gaseous biofuels in thermal systems (Deprá et al. 2018; Severo et al. 2018, 2020).

17.5.1 Biodiesel

Biodiesel is the biofuel that has gained the most visibility as an alternative to the use of fossil diesel. It may be produced from several raw materials, such as microorganisms (yeast and algae), animal fat, and vegetable oils. However, the use of microalgae as a biodiesel source offers some advantages, such as relatively high productivity, efficient lipid biosynthesis, and not compete with terrestrial oil crops. Besides, biodiesel is characterized by produce of low CO_2 levels, SO_2, and unburned hydrocarbons than petroleum-based diesel fuel, and is biodegradable, renewable, and non-toxic. Another critical factor is the amount and lipid composition because they determine the oxidative stability of biodiesel together with the performance properties in the combustion engines (Koberg et al. 2011; Gupta and Bux 2019).

Additionally, the microalgae biomass has considerable lipid content, ranging from 10–50% of their dry weight. Lipids, also known as fatty acids, are converted into biodiesel (monoalkyl esters of fatty acids) via transesterification (Kumar et al. 2021). For instance, *Botryococcus braunii* has the capability to synthesize and store high content of hydrocarbons, which can surpass 70% of its dry biomass. But the bottleneck related to this microalga is its low growth rate, resulting in low biomass and lipid productivities. The values found in the literature are between 0.1 and 0.3 g L^{-1} day^{-1} when compared to other strains of microalgae. According to Francisco et al. (2010), the lipid content of different microalgae species ranged from 6.3 to 27%, in which *Chlorella vulgaris* was the microalgae with better performance in terms of lipid productivity, cell growth rate, and fatty acids profile with quality for biodiesel production. These comparisons represent that lipid content and productivity are not directly related.

If a microalga produced a high amount of biomass and average oil content of 30% by dry weight, this would represent almost 100 m^3 per hectare of biodiesel generated (Deprá et al. 2018). This relationship, however, is not feasible. This is because microalgae biodiesel production costs are excessively high (ranging from USD 0.42 to 22.60 per liter). To meet current energy demands, it would have to be manufactured in high volumes and this limitation depends almost entirely on the microalgal oil yield and the cultivation system to operate on a large scale (Severo et al. 2019).

17.5.2 Bioethanol

Microalgae can be used as a carbohydrates source (starch, glucose, and cellulose), however in general its total concentration does not reach 50% of the biomass weight. Nevertheless, it does not have lignin in its cell component, eliminating pre-treatment to ease the access to carbohydrates (Jambo et al. 2016). The microalgal biomass undergoes physical, chemical, and biological processes, to transform carbohydrates into fermentable sugars. After that, fermentable sugars serve as a substrate to alcoholic fermentation, usually promoted by yeasts, such as *Saccharomyces cerevisiae*. The productivity of bioethanol depends on the carbohydrate concentration present in microalgae cells and the alcohol conversion rate of the fermenter microorganism. As they do not have a high carbohydrate content in their composition, their biomass is not competitive enough for commercialization. Therefore, an alternative to this inconvenience would be the genetic modification of the strain to increase productivity (Jambo et al. 2016; Kumar et al. 2021).

17.5.3 Biomethane and Biohydrogen

Biomethane can be produced from the anaerobic digestion of microalgae biomass. Generally, this procedure takes place in two stages. First, any carbohydrate or fermentable sugar is converted into alcohol via the fermentation process, promoted by anaerobic bacteria. In the second part of the process, methanogenic bacteria use the product from the previous step to produce biomethane. One of the main sources of biomass for the biomethane production is microalgae sludge, which comes from effluent treatment processes. Using the knowledge of biorefineries, it is possible to use the residual biomass of the microalgae, after extracting lipids for the production of biodiesel, obtaining the maximum use of energy (Gupta and Bux 2019). More than 60% of biomethane can be produced from microalgal biomass because of the low energy requirements during its processing (Deprá et al. 2018).

Another gaseous biofuel produced by microalgae is biohydrogen. Unlike other biofuels, no disruption to the biomass process is necessary. The microalgae under certain stress conditions can produce the gas without the use of fermentative processes. Hwang et al. (2014) observed that the photoheterotrophic culture using *Chlamydomonas reinhardtii*, containing acetate or butyrate, both substances rich in wastewater, was capable of producing large concentrations of biohydrogen during cell growth. The microalgae *Chlorella* sp. and *S. obliquus* are also capable of producing the compound by the same mechanism. The production process occurs by direct biophotolysis, in which microalgae use sunlight as an energy source to break down water molecules into protons (H^+), electrons, and oxygen (Gupta and Bux 2019; Kumar et al. 2021).

17.5.4 Biobutanol

Biobutanol is four-carbon alcohol produced by the fermentation of biomass. As a production source, microalgae-based carbohydrates and residual glycerol from biodiesel production can be used. It is commonly obtained from the fermentation of *Chlorella vulgaris* biomass by the bacteria *Clostridium acetobutylicum.* Among the main advantages of biobutanol, it can stand out for its high density and low volatility, when compared to bioethanol. Also, biobutanol has several applications in the industry, such as chemical solvents. The production of biobutanol from microalgae is considered commercially viable, and its potential as a candidate for biofuel may attract the attention of researchers in the coming years (Gupta and Bux 2019; Kumar et al. 2021).

17.6 Conclusions and Recommendations

Although phycoremediation has enormous potential to solve environmental problems, up to now, few technologies have been raised from years of research to deliberately use microalgae for waste management. Many investigations have been conducted to achieve the best potential for industrial application in the forthcoming, such as microalgae immobilization, and their combination with other processes.

There are two noticeable bottlenecks for treating waste with microalgae. Firstly, it is a purely economic criterion, that is, using a conventional method based on chemical products, which is cheaper and simple, practically precludes the biological method application which, in turn, is more expensive. In this case, taking into account the volume of effluent to be treated, a slight increase in operating costs may not be selling an ecological technology for traditional effluent treatment stations. Second, the criterion to be considered is in terms of the removal efficiency of the polluting compounds and their quantity. This represents a gap of days when using microalgae, which often depends on oscillatory environmental parameters, compared to the hours that a conventional and stable method takes.

Today, one of the best options is the adoption of the process integration approach in the biorefinery context, in which the microalgae immobilization can be a complementary technology to conventional chemical routes, in addition to the potential generation of energy inputs, which could be a financial return point. It is worth mentioning that there is pressure on polluting companies, not only in legal terms but also in presenting in their portfolio the use of environmentally correct techniques, which would benefit their image and public acceptance.

The expectation is that microalgae immobilization strategy can be competitive in the market when the following aspects are made feasible: the concomitant waste treatment, the improvement of photobioreactors and operational parameters for specific applications of immobilized cells, the upgrading of the polymeric support materials, and the metabolic and genetic modifications of the strains for better adherence to the matrix and obtaining metabolites for conversion into energy products. These are the main recommendations for action, which in the future will make immobilized microalgae technology useful for industrial bioremediation.

References

Abdel-Raouf N, Al-Homaidan AA, Ibraheem IBM (2012) Microalgae and wastewater treatment. Saudi J Biol Sci 19:257–275. https://doi.org/10.1016/j.sjbs.2012.04.005

Abdulla R, King TK, Jambo SA, Faik AA (2020) Microalgae *Chlorella* as a sustainable feedstock for bioethanol production. In: Green Engineering for Campus Sustainability. Springer, Singapore. https://doi.org/10.1007/978-981-13-7260-5_7

Abou-Shanab RAI, Ji MK, Kim HC et al (2013) Microalgal species growing on piggery wastewater as a valuable candidate for nutrient removal and biodiesel production. J Environ Manage 115:257–264. https://doi.org/10.1016/j.jenvman.2012.11.022

Acién FG, Molina E, Reis A et al (2017) Photobioreactors for the production of microalgae. In Woodhead Publishing Series in Energy, Microalgae-Based Biofuels and Bioproducts. https://doi.org/10.1016/B978-0-08-101023-5.00001-7

Aguilar-May B, Sánchez-Saavedra MDP (2009) Growth and removal of nitrogen and phosphorus by free-living and chitosan-immobilized cells of the marine cyanobacterium *Synechococcus elongatus*. J Appl Phycol 21:353–360. https://doi.org/10.1007/s10811-008-9376-7

Al-Zuhair S, El-Naas M (2011) Immobilization of *Pseudomonas putida* in PVA gel particles for the biodegradation of phenol at high concentrations. Biochem Eng J 56:46–50. https://doi.org/10.1016/j.bej.2011.05.005

Andreoli CV, von Sperling M, Fernandes F (2007) Sludge treatment and disposal. IWA Publishing. https://doi.org/10.2166/9781780402130

Anjaneyulu Y, Sreedhara Chary N, Samuel Suman Raj D (2005) Decolourization of industrial effluents—available methods and emerging technologies—a review. Rev Environ Sci Biotechnol 4:245–273. https://doi.org/10.1007/s11157-005-1246-z

Arbib Z, Ruiz J, Álvarez-Díaz P et al (2013) Photobiotreatment: influence of nitrogen and phosphorus ratio in wastewater on growth kinetics of *Scenedesmus Obliquus*. Int J Phytorem 15:774–788. https://doi.org/10.1080/15226514.2012.735291

Assunção J, Malcata FX (2020) Enclosed "non-conventional" photobioreactors for microalga production: a review. Algal Res 52. https://doi.org/10.1016/j.algal.2020.102107

Barten RJP, Wijffels RH, Barbosa MJ (2020) Bioprospecting and characterization of temperature tolerant microalgae from Bonaire. Algal Res 50. https://doi.org/10.1016/j.algal.2020.102008

Borowitzka MA, Beardall J, Raven JA (2016) The physiology of microalgae. In: Developments in applied phycology. Springer International Publishing. https://doi.org/10.1007/978-3-319-24945-2

Caporgno MP, Taleb A, Olkiewicz M et al (2015) Microalgae cultivation in urban wastewater: nutrient removal and biomass production for biodiesel and methane. Algal Res 10:232–239. https://doi.org/10.1016/j.algal.2015.05.011

Castro YA, Ellis JT, Miller CD, Sims RC (2015) Optimization of wastewater microalgae saccharification using dilute acid hydrolysis for acetone, butanol, and ethanol fermentation. Appl Energy 140:14–19. https://doi.org/10.1016/j.apenergy.2014.11.045

Che CA, Kim SH, Hong HJ et al (2019) Optimization of light intensity and photoperiod for Isochrysis galbana culture to improve the biomass and lipid production using 14-L photobioreactors with mixed light emitting diodes (LEDs) wavelength under two-phase culture system. Biores Technol 285:121323. https://doi.org/10.1016/j.biortech.2019.121323

Cheirsilp B, Thawechai T, Prasertsan P (2017) Immobilized oleaginous microalgae for production of lipid and phytoremediation of secondary effluent from palm oil mill in fluidized bed photobioreactor. Biores Technol 241:787–794. https://doi.org/10.1016/j.biortech.2017.06.016

Choudhary P, Prajapati SK, Kumar P et al (2017) Development and performance evaluation of an algal biofilm reactor for treatment of multiple wastewaters and characterization of biomass for diverse applications. Biores Technol 224:276–284. https://doi.org/10.1016/j.biortech.2016.10.078

Corrêa DO, Santos B, Dias FG et al (2017) Enhanced biohydrogen production from microalgae by diesel engine hazardous emissions fixation. Int J Hydrogen Energy 42:21463–21475. https://doi.org/10.1016/j.ijhydene.2017.05.176

Daneshvar E, Sik Ok Y, Tavakoli S et al (2021) Insights into upstream processing of microalgae: a review. Bioresour Technol 329. https://doi.org/10.1016/j.biortech.2021.124870

Dell' Anno F, Rastelli E, Sansone C, Brunet C, Ianora A, Dell' Anno A (2021) Bacteria, fungi and microalgae for the bioremediation of marine sediments contaminated by petroleum hydrocarbons in the omics era. Microorganisms 9, 1695. https://doi.org/10.3390/microorganisms9081695

Deprá MC, dos Santos AM, Severo IA et al (2018) Microalgal biorefineries for bioenergy production: can we move from concept to industrial reality? Bioenergy Res 11:727–747. https://doi.org/10.1007/s12155-018-9934-z

Doran PM (2013) Bioprocess engineering principles, 2nd ed. Academic Press. https://doi.org/10.1016/C2009-0-22348-8

Emparan Q, Harun R, Danquah MK (2019) Role of phycoremediation for nutrient removal from wastewaters: a review. Appl Ecol Environ Res 17, 889–915. https://doi.org/10.15666/aeer/1701_889915

Filippino KC, Mulholland MR, Bott CB (2015) Phycoremediation strategies for rapid tertiary nutrient removal in a waste stream. Algal Res 11:125–133. https://doi.org/10.1016/j.algal.2015.06.011

Francisco ÉC, Neves DB, Jacob-Lopes E, Franco TT (2010) Microalgae as feedstock for biodiesel production: carbon dioxide sequestration, lipid production and biofuel quality. J Chem Technol Biotechnol 85:395–403. https://doi.org/10.1002/jctb.2338

Galès A, Triplet S, Geoffroy T et al (2020) Control of the pH for marine microalgae polycultures: a key point for CO_2 fixation improvement in intensive cultures. J CO_2 Utilization 38, 187–193. https://doi.org/10.1016/j.jcou.2020.01.019

Gao F, Yang ZH, Li C et al (2014) Concentrated microalgae cultivation in treated sewage by membrane photobioreactor operated in batch flow mode. Biores Technol 167:441–446. https://doi.org/10.1016/j.biortech.2014.06.042

Gao K, Orr V, Rehmann L (2016) Butanol fermentation from microalgae-derived carbohydrates after ionic liquid extraction. Biores Technol 206:77–85. https://doi.org/10.1016/j.biortech.2016.01.036

George B, Pancha I, Desai C et al (2014) Effects of different media composition, light intensity and photoperiod on morphology and physiology of freshwater microalgae *Ankistrodesmus falcatus*—a potential strain for bio-fuel production. Biores Technol 171:367–374. https://doi.org/10.1016/j.biortech.2014.08.086

Giese EC (2015) Potencial biotecnológico do uso de micro-organismos imobilizados em gel de alginato de cálcio. CETEM/MCTI

Gouveia L, Oliveira AC (2009) Microalgae as a raw material for biofuels production. J Ind Microbiol Biotechnol 36:269–274. https://doi.org/10.1007/s10295-008-0495-6

Grobbelaar JU (2007) Algal nutrition—mineral nutrition. In: Handbook of microalgal culture 95–115. Blackwell Publishing Ltd. https://doi.org/10.1002/9780470995280.ch6

Guisan JM (2006) Immobilization of enzymes and cells, 2nd ed. Humana Press. https://doi.org/10.1007/978-1-62703-550-7

Gupta SK, Bux F (2019) Application of microalgae in wastewater treatment, 1st edn. Springer International Publishing. https://doi.org/10.1007/978-3-030-13913-1

Hameed MSA (2007) Effect of algal density in bead, bead size and bead concentrations on wastewater nutrient removal. Afr J Biotech 6:1185–1191

Han SF, Jin W, Tu R et al (2016) Optimization of aeration for biodiesel production by *Scenedesmus obliquus* grown in municipal wastewater. Bioprocess Biosyst Eng 39:1073–1079. https://doi.org/10.1007/s00449-016-1585-x

Hena S, Znad H, Heong KT, Judd S (2018) Dairy farm wastewater treatment and lipid accumulation by *Arthrospira platensis*. Water Res 128:267–277. https://doi.org/10.1016/j.watres.2017.10.057

Hernández D, Riaño B, Coca M et al (2016) Microalgae cultivation in high rate algal ponds using slaughterhouse wastewater for biofuel applications. Chem Eng J 285:449–458. https://doi.org/10.1016/j.cej.2015.09.072

Ho SH, Huang SW, Chen CY et al (2013a) Characterization and optimization of carbohydrate production from an indigenous microalga *Chlorella vulgaris* FSP-E. Biores Technol 135:157–165. https://doi.org/10.1016/j.biortech.2012.10.100

Ho SH, Huang SW, Chen CY et al (2013b) Bioethanol production using carbohydrate-rich microalgae biomass as feedstock. Biores Technol 135:191–198. https://doi.org/10.1016/j.biortech.2012.10.015

Huang G, Wang Y (2003) Nitrate and phosphate removal by co-immobilized *Chlorella pyrenoidosa* and activated sludge at different pH value. Water Qual Res J Can 38:541–551. https://doi.org/10.2166/wqrj.2003.035

Hwang JH, Kabra AN, Kim JR, Jeon BH (2014) Photoheterotrophic microalgal hydrogen production using acetate- and butyrate-rich wastewater effluent. Energy 78:887–894. https://doi.org/10.1016/j.energy.2014.10.086

Jambo SA, Abdulla R, Mohd Azhar SH et al (2016) A review on third generation bioethanol feedstock. Renew Sustain Energy Rev 65:756–769. https://doi.org/10.1016/j.rser.2016.07.064

Jiang L, Luo S, Fan X et al (2011) Biomass and lipid production of marine microalgae using municipal wastewater and high concentration of CO_2. Appl Energy 88:3336–3341. https://doi.org/10.1016/j.apenergy.2011.03.043

Jiménez-Pérez MV, Sánchez-Castillo P, Romera O et al (2004) Growth and nutrient removal in free and immobilized planktonic green algae isolated from pig manure. Enzyme Microb Technol 34:392–398. https://doi.org/10.1016/j.enzmictec.2003.07.010

Koberg M, Cohen M, Ben-Amotz A, Gedanken A (2011) Bio-diesel production directly from the microalgae biomass of *Nannochloropsis* by microwave and ultrasound radiation. Biores Technol 102:4265–4269. https://doi.org/10.1016/j.biortech.2010.12.004

Kube M, Mohseni A, Fan L, Roddick F (2020) Energy and nutrient recovery by treating wastewater with fluidised-beds of immobilised algae. J Water Process Eng 38:101585. https://doi.org/10.1016/j.jwpe.2020.101585

Kumar G, Sivagurunathan P, Thi NBD et al (2016) Evaluation of different pretreatments on organic matter solubilization and hydrogen fermentation of mixed microalgae consortia. Int J Hydrogen Energy 41:21628–21640. https://doi.org/10.1016/j.ijhydene.2016.05.195

Kumar R, Mehariya S, Verma P et al (2021) Microalgae-based biorefineries for sustainable resource recovery from wastewater. J Water Process Eng 40:101747. https://doi.org/10.1016/j.jwpe.2020.101747

Li K, Liu Q, Fang F et al (2019) Microalgae-based wastewater treatment for nutrients recovery: a review. Biores Technol 291:121934. https://doi.org/10.1016/j.biortech.2019.121934

Lima U de A, Aquarone E, Borzani W, Schmidell W (2007) Biotecnologia industrial: processos fermentativos e enzimáticos, 2nd edn. E. Blücher, São Paulo

Lu W, Asraful Alam M, Liu S et al (2020) Critical processes and variables in microalgae biomass production coupled with bioremediation of nutrients and CO_2 from livestock farms: a review. Sci Total Environ 716:135247. https://doi.org/10.1016/j.scitotenv.2019.135247

Lutzu GA, Ciurli A, Chiellini C et al (2021) Latest developments in wastewater treatment and biopolymer production by microalgae. J Environ Chem Eng 9:104926. https://doi.org/10.1016/j.jece.2020.104926

Mallick N (2002) Biotechnological potential of immobilized algae for wastewater N, P and metal removal: a review. Biometals 15:377–390. https://doi.org/10.1023/A:1020238520948

Mantzorou A, Ververidis F (2019) Microalgal biofilms: a further step over current microalgal cultivation techniques. Sci Total Environ 651:3187–3201. https://doi.org/10.1016/j.scitotenv.2018.09.355

Michels MHA, Vaskoska M, Vermuë MH, Wijffels RH (2014) Growth of *Tetraselmis suecica* in a tubular photobioreactor on wastewater from a fish farm. Water Res 65:290–296. https://doi.org/10.1016/j.watres.2014.07.017

Min KJ, Oh DY, Park KY (2021) Pilot-scale cultivation of water-net in secondary effluent using an open pond raceway for nutrient removal and bioethanol production. Chemosphere 277:130129. https://doi.org/10.1016/j.chemosphere.2021.130129

Miyawaki B, Mariano AB, Vargas JVC et al (2021) Microalgae derived biomass and bioenergy production enhancement through biogas purification and wastewater treatment. Renewable Energy 163:1153–1165. https://doi.org/10.1016/j.renene.2020.09.045

Moheimani NR, Vadiveloo A, Ayre JM, Pluske JR (2018) Nutritional profile and in vitro digestibility of microalgae grown in anaerobically digested piggery effluent. Algal Res 35:362–369. https://doi.org/10.1016/j.algal.2018.09.007

Mohsenpour SF, Hennige S, Willoughby N et al (2021) Integrating micro-algae into wastewater treatment: a review. Sci Total Environ 752:142168. https://doi.org/10.1016/j.scitotenv.2020.142168

Moreno-Garrido I (2008) Microalgae immobilization: current techniques and uses. Biores Technol 99:3949–3964. https://doi.org/10.1016/j.biortech.2007.05.040

Nguyen TTD, Nguyen TT, An Binh Q et al (2020) Co-culture of microalgae-activated sludge for wastewater treatment and biomass production: exploring their role under different inoculation ratios. Biores Technol 314:123754. https://doi.org/10.1016/j.biortech.2020.123754

Nwoba EG, Parlevliet DA, Laird DW et al (2019) Light management technologies for increasing algal photobioreactor efficiency. Algal Res 39:101433. https://doi.org/10.1016/j.algal.2019.101433

Prajapati SK, Kaushik P, Malik A, Vijay VK (2013) Phycoremediation coupled production of algal biomass, harvesting and anaerobic digestion: possibilities and challenges. Biotechnol Adv 31:1408–1425. https://doi.org/10.1016/j.biotechadv.2013.06.005

Richmond A (2004) Handbook of microalgal culture: biotechnology and applied phycology. Blackwell Publishing Ltd, Oxford, UK. https://doi.org/10.1002/9780470995280

Rossi S, Casagli F, Mantovani M et al (2020) Selection of photosynthesis and respiration models to assess the effect of environmental conditions on mixed microalgae consortia grown on wastewater. Biores Technol 305:122995. https://doi.org/10.1016/j.biortech.2020.122995

Sajjadi B, Chen WY, Raman AAA, Ibrahim S (2018) Microalgae lipid and biomass for biofuel production: a comprehensive review on lipid enhancement strategies and their effects on fatty acid composition. Renew Sustain Energy Rev 97:200–232. https://doi.org/10.1016/j.rser.2018.07.050

Samer M (2015) Biological and chemical wastewater treatment processes, wastewater treatment engineering. IntechOpen. https://doi.org/10.5772/61250

Sarkar P, Dey A (2021) Phycoremediation—an emerging technique for dye abatement: an overview. Process Saf Environ Prot 147:214–225. https://doi.org/10.1016/j.psep.2020.09.031

Satyanarayana KG, Mariano AB, Vargas JVC (2011) A review on microalgae, a versatile source for sustainable energy and materials. Int J Energy Res 21. 10.1002/er

Schmidell W, Lima U de A, Aquarone E, Borzani W (2001) Biotecnologia industrial: engenharia bioquímica, 2nd edn. E. Blücher, São Paulo

Severo IA, Deprá MC, Dias RR et al (2020) Bio-combustion of petroleum coke: the process integration with photobioreactors. Part II—sustainability metrics and bioeconomy. Chem Eng Sci **213**, 115412. https://doi.org/10.1016/j.ces.2019.115412

Severo IA, Deprá MC, Barin JS et al (2018) Bio-combustion of petroleum coke: the process integration with photobioreactors. Chem Eng Sci 177:422–430. https://doi.org/10.1016/j.ces.2017.12.001

Severo IA, Dias RR, Sartori RB, Maroneze MM, Zepka LQ, Jacob-Lopes E (2021) Chapter 4: technological bottlenecks in establishing microalgal biorefineries. In: Shekh A, Schenk P, Sarada R (eds) Microalgal biotechnology: recent advances, market potential, and sustainability. Royal Society of Chemistry, pp 118–134

Shi J, Podola B, Melkonian M (2014) Application of a prototype-scale twin-layer photobioreactor for effective N and P removal from different process stages of municipal wastewater by immobilized microalgae. Biores Technol 154:260–266. https://doi.org/10.1016/j.biortech.2013.11.100

Sivakaminathan S, Hankamer B, Wolf J, Yarnold J (2018) High-throughput optimisation of light-driven microalgae biotechnologies. Sci Rep 8:1–13. https://doi.org/10.1038/s41598-018-29954-x

Straka L, Rittmann BE (2018) Light-dependent kinetic model for microalgae experiencing photoacclimation, photodamage, and photodamage repair. Algal Res 31:232–238. https://doi.org/10.1016/j.algal.2018.02.022

Telles EC, Yang S, Vargas JVC et al (2018) A genset and mini-photobioreactor association for CO_2 capturing, enhanced microalgae growth and multigeneration. Renewable Energy 125:985–994. https://doi.org/10.1016/j.renene.2018.03.027

Vargas JVC, Mariano AB, Corrêa DO, Ordonez JC (2014) The microalgae derived hydrogen process in compact photobioreactors. Int J Hydrogen Energy 39:9588–9598. https://doi.org/10.1016/j.ijhydene.2014.04.093

Vasilieva SG, Lobakova ES, Lukyanov AA, Solovchenko AE (2016) Immobilized microalgae in biotechnology. Mosc Univ Biol Sci Bull 71:170–176. https://doi.org/10.3103/S0096392516030135

Vasilieva S, Lobakova E, Grigoriev T et al (2021) Bio-inspired materials for nutrient biocapture from wastewater: microalgal cells immobilized on chitosan-based carriers. J Water Process Eng 40:101774. https://doi.org/10.1016/j.jwpe.2020.101774

Verma R, Kusuma Kumari KVL, Srivastava A, Kumar A (2020) Photoautotrophic, mixotrophic, and heterotrophicculture media optimization for enhanced microalgae production. J Environ Chem Eng 8:104149. https://doi.org/10.1016/j.jece.2020.104149

Villadsen J, Nielsen J, Lidén G (2011) Bioreaction engineering principles, 3rd Ed. Springer US. https://doi.org/10.1007/978-1-4419-9688-6

Vo HNP, Ngo HH, Guo W et al (2018) A critical review on designs and applications of microalgae-based photobioreactors for pollutants treatment. Sci Total Environ 651:1549–1568. https://doi.org/10.1016/j.scitotenv.2018.09.282

von Sperling M (2007a) Wastewater characteristics, treatment and disposal. IWA Publishing UFMG

von Sperling M (2007b) Waste stabilisation ponds. IWA Publishing. https://doi.org/10.2166/9781780402109

von Sperling M (2015) Activated sludge and aerobic biofilm reactors. IWA Publishing. https://doi.org/10.2166/9781780402123

Wang R, Li F, Ruan W et al (2020) Removal and degradation pathway analysis of 17β-estradiol from raw domestic wastewater using immobilised functional microalgae under repeated loading. Biochem Eng J 161:107700. https://doi.org/10.1016/j.bej.2020.107700

Wittmann C, Liao JC (2016) Industrial biotechnology: products and processes. Wiley-VCH & Co

Yuan S, Hu J, Liu Z et al (2020) Modeling microalgae growth in continuous culture: parameters analysis and temperature dependence. Energy 195:116961. https://doi.org/10.1016/j.energy.2020.116961

Zhao B, Zhang Y, Xiong K et al (2011) Effect of cultivation mode on microalgal growth and CO_2 fixation. Chem Eng Res Des 89:1758–1762. https://doi.org/10.1016/j.cherd.2011.02.018

Zhou M, He H, Jin T, Wang H (2012) Power generation enhancement in novel microbial carbon capture cells with immobilized *Chlorella vulgaris*. J Power Sources 214:216–219. https://doi.org/10.1016/j.jpowsour.2012.04.043

Chapter 18
Waste to Energy Plant in Spain: A Case Study Using Technoeconomic Analysis

Rubí Medina-Mijangos, Sergi Contelles-Rodríguez, Hilda Guerrero-García-Rojas, and Luis Seguí-Amórtegui

Abstract The population growth and the new consumption models contribute significantly to a greater generation of waste, which is generally incorrectly managed because a large percentage of the waste generated is sent to landfills. Waste to energy (WtE) plants play a fundamental role in managing and treating municipal waste because they reduce the amount of waste sent to landfills and reduce dependence on imported fossil fuels; however, these facilities can also cause negative impacts. This case study evaluates the technical–economic feasibility of an incineration plant by using a social cost–benefit analysis, which considers economic, social, and environmental impacts taking into account the 3 pillars of sustainability and allowing policymakers to have a complete view of the impacts generated by the facility. The WtE facility is in Barcelona (Spain). It produces energy from municipal solid waste (MSW) with a total capacity of more than 350,000 tons of waste treated per year, which means the generation of more than 180,000 MWh of electricity and 110,000 tons of steam per year. The positive and negative impacts generated by this facility are identified, discussed, and monetarily valued to carry out this economic analysis. Some of the impacts considered are the sale of energy, the decrease in waste disposal in landfills, the reduction of greenhouse gas (GHG) emissions, and the generation of dioxin emissions. The results show that the facility is profitable from a private point of view (BP = 15.97) and an economic, environmental, and social perspective (BT = 37.48). Finally, the same impacts can be considered by

R. Medina-Mijangos (✉) · S. Contelles-Rodríguez
Department of Civil and Environmental Engineering, Universitat Politècnica de Catalunya, Carrer de Jordi Girona, 1, 08034 Barcelona, Spain
e-mail: rubi.alejandra.medina@upc.edu

H. Guerrero-García-Rojas
Department of Economics, Universidad Michoacana de San Nicolás de Hidalgo (UMSNH), Av. Fco J. Múgica, 58040 Morelia, Michoacan, Mexico
e-mail: hilda.guerrero@umich.mx

L. Seguí-Amórtegui
Faculty of Business and Communication, Universidad Internacional de la Rioja, Av. de la Paz, 137, 26006 Logroño, Spain
e-mail: luisalberto.segui@unir.net

A. E.-F. Abomohra et al. (eds.), *Waste-to-Energy*,
https://doi.org/10.1007/978-3-030-91570-4_18

researchers in future economic analyzes of other WtE projects or waste management systems.

Keywords Technical–economic analysis · MSW · Social impacts · Environmental impacts · Waste to energy · Case study

18.1 Introduction

The energy recovery from Municipal Solid Waste (MSW) in incineration plants represents an opportunity to reduce the amount of waste that is sent to landfills and, therefore, to be able to meet the objectives set by the European Commission on waste, which indicates that the share of Municipal Solid Waste (MSW) deposited in landfills will be limited to 10% by 2035 (Medina-Mijangos and Seguí-Amórtegui 2021). Furthermore, it is an alternative to reduce dependence on energy generated from fossil fuels, which are generally imported (Jamasb and Nepal 2010). Waste to energy (WtE) has several positive effects since this process avoids methane (CH_4) emissions from landfills and carbon dioxide (CO_2) emissions from fossil fuels (Scarlat et al. 2019). Moreover, Lim et al. (2014) identify some benefits related to WtE plants: (1) improvement of energy security defined as the uninterrupted availability of energy at an affordable price; (2) reduction of GHG emissions because the WtE facilities allow reducing the use of landfills and fossil fuels; (3) creation of employment and (4) extension of landfill life expectancy.

However, WtE facilities can also cause negative impacts due to the emissions of pollutants such as particulates (PM_{10}), nitrogen oxides (NO_x), acid gases (SO_2, HF, HC1), carbon dioxide (CO_2), volatile organic compounds (VOCs), dioxins (PDDC/PFs), heavy metal (Cr, Pb, Cu, Ni, Zn, Cd, Hg and As). These pollutants can have undesirable effects on public health, agriculture, buildings, ecosystems, and promote climatic change (European Commission 2000). In addition, these facilities can affect the price of houses located near due to the disamenities generated and the "Not In My Back Yard" syndrome.

The European Parliament establishes a waste hierarchy for legislation and policy on the prevention and management of waste (European Parliament 2008), where prevention, reuse, and recycling are prioritized over other types of recovery (including energy recovery) and deposit in landfills. However, other options that deviate from the hierarchy may be considered, as long as it is justified by technical viability, economic viability and environmental protection.

18.1.1 Waste to Energy Facilities in Spain

In the case of Spain, in 2019, 22,438 thousand tons of MSW were generated, of which 54% was sent to landfills, 11.29% was converted into energy, and the remaining percentage was recycled or treated biologically to obtain compost. In 2018, 122 landfills were registered, where 11,917,233 tons were deposited (MITERD 2018). Additionally, Spain has ten incineration plants that treat an average of 2,527,000 tons/year, as shown in Table 18.1. In Spain, the incineration plants are responsible for the management and thermal treatment of non-hazardous municipal waste. Specifically, the residual waste from the non-selective collection of the gray container. This fraction is previously sent to mechanical–biological treatment (MBT) facilities. The waste is subjected to different physical and biological processes to recover materials (organic matter, plastic, cardboard, steel, among others). Waste that can no longer be materially recovered (rejected waste) are sent to incineration plants for energy recovery, obtaining electricity and steam.

On the one hand, Spain shows a significant dependence on energy generated from fossil fuels. In 2018, 44% of primary energy consumption came from oil and petroleum products and 20.75% from natural gas (INE 2020). On the other hand, Spain presents a critical problem with foreign energy dependence since 73.3% of primary energy was acquired outside the country. Specifically, a 78% dependence

Table 18.1 Waste to energy facilities located in Spain

Facility	ID	Opening year	Location	No of furnaces	Nominal capacity (ton/year)
Energy recovery facility of Meruelo	TIRCANTABRIA	2006	Cantabria	1	120,500
Energy recovery facility of Sant Adrià de Besòs	TERSA	1975	Barcelona	3	360,000
Mataró incinerator	TRM	1994	Barcelona	2	160,000
Girona incinerator	TRARGISA	1984	Girona	2	35,000
Tarragona incinerator	SIRUSA	1991	Tarragona	2	140,000
Cerceda thermoelectric plant	SOGAMA	2000	La Coruña	2	360,000
Energy recovery facility of Mallorca	TIRME	1997	Balearic Islands	4	730,000
Las Lomas energy recovery plant	MADRID	1993	Madrid	3	328,500
Energy recovery facility of Melilla	REMESA	1994	Melilla	1	47,000
Zabalgarbi facility	ZABALGARBI	2005	Vizkaia	1	246,000
Total				21	2,527,000

Source Adapted from MITERD (2018)

on imports of solid fossil fuels and 99% of oil and petroleum products is shown (European Commission 2020).

18.2 Methodology and Data

The data were obtained from public documents (such as annual accounts, environmental and technical studies, production data, among others), available on the company's website. Additionally, studies published in indexed journals about the analyzed facility and studies about environmental and social impacts of other MSW treatment facilities are used.

The methodology presented in Medina-Mijangos et al. (2012b) was used to carry out the technical–economic analysis. It considers the private and external impacts (revenues and costs) generated by the MSW management projects. Additionally, several of the impacts described in this paper are contemplated in this case study.

This methodology is based on cost–benefit analysis and considers that the systems or projects evaluated must comply with sustainability principles and its three pillars; therefore, the methodology used considers the project's economic, social, and environmental dimensions, as shown in Fig. 18.1. In addition, it is implied that the best option is the one that meets the needs of society, and it is environmentally and economically viable, socially and environmentally bearable, as well as economically and socially equitable (Mensah 2019).

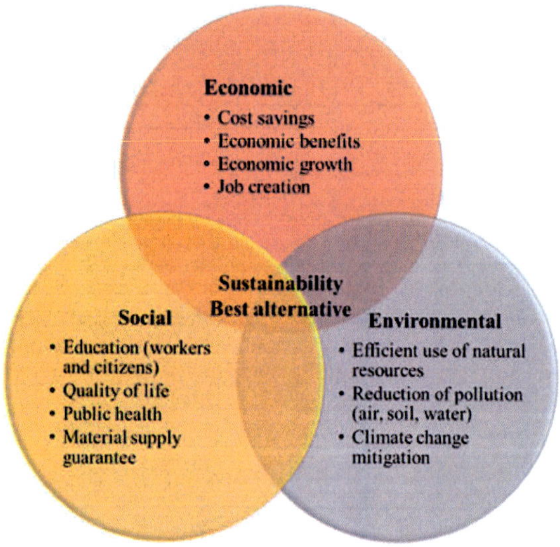

Fig. 18.1 Evaluation of MSW management systems considering pillars of sustainability. *Source* Medina-Mijangos et al. (2021a). Creative Commons Attribution 4.0 International License

18.2.1 Objective Definition

The aim of this study is to determine the Private Benefit (BP) and the Total Benefit (BT) using Eqs. 18.1 and 18.2, based on the determination of both private and external revenues and costs generated by the Energy Recovery Facility (in this case study, it is identified as ERF). Therefore, the results are expressed in € per ton, where it is necessary to divide the annual results by the total waste treated.

$$B_P = \sum_{n=0}^{N} [(\mathrm{AV}_n * \mathrm{SP}) - (\mathrm{IC}_n + \mathrm{OMC}_n + \mathrm{FC}_n + T_n)] \tag{18.1}$$

$$B_T = \sum_{n=0}^{N} [(\mathrm{AV}_n * \mathrm{SP}) - (\mathrm{IC}_n + \mathrm{OMC}_n + \mathrm{FC}_n + T_n) + (\mathrm{PE} - \mathrm{NE}) - \mathrm{OC}] \tag{18.2}$$

where AV: Annual volume sold; FC: Financial Costs; IC: Investment Costs; N: Total project duration; n: Project year index ($n = 0, \ldots, N$); NE: Negative Externalities; OC: Opportunity Cost; OMC: Operational and Maintenance Costs; PE: Positive Externalities; SP: Price of Sale; T: Taxes.

In this way, it can be concluded whether the facility is profitable from a private point of view (if BP is greater than 0) and profitable from an economic, environmental, and social perspective (if BT is greater than 0).

18.2.2 Description of the Scope of the Study

In this case study, an Energy Recovery Facility (ERF) located in Sant Adrià de Besòs, Barcelona (Spain) is analyzed. This facility is managed by TERSA (by its Catalan name *Tractament i Selecció de Residus S.A.*) and was inaugurated in 1975, being the oldest incinerator in Spain. This facility performs the process of minimizing the volume of waste through combustion, taking advantage of the energy generated by this process to produce steam and electricity (Medina-Mijangos and Seguí-Amórtegui 2021). The ERF manages the waste generated in the Barcelona Metropolitan Area (AMB), an area made up of 36 municipalities such as Barcelona City, Badalona, Sant Adrià de Besòs, among others, with approximately 3.24 million inhabitants (AMB 2021). Additionally, the ERF and the mechanical–biological treatment plant known as Ecoparc 3 (managed by the *Ecoparc del Mediterrani S.A.*) are part of the Integral Waste Recovery Plant.

The costs and revenues generated by this ERF are evaluated, considering only one year, 2019. Considering only the processes carried out by the ERF, without considering the impacts generated during the previous processes as in the case of the collection, transport or treatment carried out in the mechanical–biological treatment

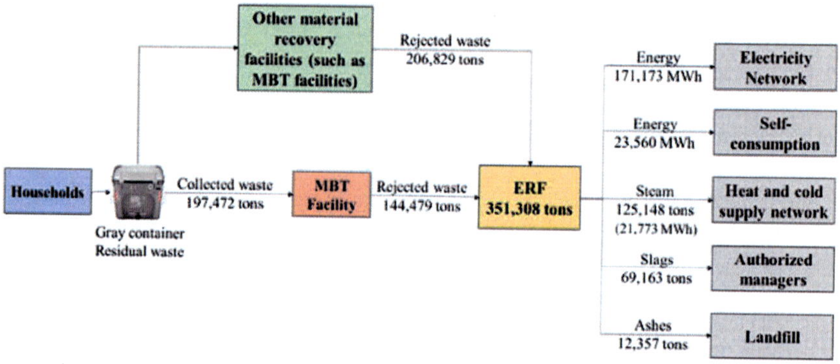

Fig. 18.2 Waste management processes of the ERF in 2019. *Source* Authors elaboration. *MBT* Mechanical–Biological Treatment; *ERF* Energy Recovery Facility

plant. Previously, this facility was economically analyzed in Medina-Mijangos and Seguí-Amórtegui (2021), but considering the year 2017. This case study also includes other impacts that had not been analyzed.

This facility receives rejected waste from Ecoparc 3, as well as other waste classification and treatment facilities. Through the thermal process, energy is obtained for self-consumption and sale to the electricity grid. On the other hand, steam is also obtained that is sold for the city's hot and cold network Fig. 18.2. Finally, the slags are sold to authorized managers to produce ecological concrete, and the ashes are sent to controlled landfills.

Specifically, in 2019, 351,308 tons of rejected waste from Ecoparc 3 and other treatment facilities were treated, obtaining 197,733 MWh of energy, of which 23,560 MWh were used for self-consumption, and 171,173 was sold to the electric power grid. Additionally, 23,560 tons of steam generated was sold to Districlima (the company in charge of managing the urban heat and cold distribution network of Barcelona city). Finally, 69,163 tons of slag were sold to authorized managers and 12,357 tons of ash were sent to landfills.

18.2.3 Waste to Energy Technology

The ERF incorporates heat recovery and power generation. In addition, this facility has implemented advanced process controls and exhaust gas cleaning measures to ensure that the levels of contaminants are under the legal limits established by regulatory bodies as the European Commission and the Spanish legislation.

The technology used for the thermal treatment of waste in the ERF is described below.

Waste reception: The rejected waste from Ecoparc 3 is transported through an underground conveyor that directly discharges the waste into the pit. The rest of the facilities transport waste using trucks, which are weighed before accessing the facilities. Once in the pit, the waste treatment furnaces are fed through an overhead crane.

Energy generation. The rejected waste is burned in the furnace. Combustion is carried out in a controlled manner in three combustion grates with a nominal capacity of 15 tons/h per furnace. The gases produced are conducted through a boiler, where water is heated to steam. The steam produced moves two turbines, responsible for producing electricity. The equipment present in energy generation is described in detail below.

- Furnace. A feed hopper introduces the waste into one of the three furnaces from the top. Inside the furnaces, three groups of fixed and movable grates lower the waste at a controlled speed to burn it. Air is injected to maintain the fire. Above the combustion grates, a natural gas burner automatically ignites if the temperature of the gases drops below 850 °C. Next, to neutralize the nitrogen oxides that appear as a result of combustion, urea is injected.
- Tubular boiler. The water in the boiler is heated to steam with the hot gases from the furnace. The gases emitted by the furnace at 850 °C are conductive around a circuit filled with water. As it passes the circuit, the heat is transmitted to the water, heated to 400 °C generating superheated steam.
- Turbines. The steam from the water boiler is conducted through the turbine. As it passes, it spins the rotor blades. This movement is transmitted through the shaft to an alternator, which in turn rotates magnets along with electrical cables. This movement of the magnets generates a variable magnetic field around the cables, and with it, an electric current.
- Condensers. Water vapor passes through a tank filled with cold seawater through a circuit. It transmits heat to seawater and cools down to a liquid state.

Flue gas cleaning system. After combustion, the incineration gases are scrubbed to avoid emitting pollutants into the atmosphere. The treatment process allows the removal of solid particles, acid gases, dioxins, heavy metals, and fine particles to reach levels well below the legal limits. The ERF has a continuous measurement system to guarantee the quality of the treatment, which continuously controls the levels of these and other substances. The equipment present in flue gas treatment is described below.

- Electrofilter. The combustion gases in the furnace horizontally pass through a chamber with vertical electrodes, which electrically charge the solid particles in suspension. Next to the electrodes, there are flat metal plates that attract the particles and retain them. Periodically, the metal plates are shaken, and with the resulting vibration. Consequently, the particles fall from the plates into a hopper at the bottom of the chamber.
- Atomizer and gas absorber. The gases emitted from the electrofilter are sprayed with hydrated lime. Hydrochloric and hydrofluoric acids react with lime, resulting in a mixture of water and salts.
- Selective Catalytic Reduction system. The ERF has a Selective Catalytic Reduction (SCR) system based on ammonia injection as a reducing agent for combustion

gases such as nitrous oxides (NO_x). The catalyst requires a working temperature between 220 and 340 °C to be effective. The SCR is located at the outlet of the flue gas cleaning system to treat acid gases and their particles present in the combustion gases.
- Activated carbon injection. At the outlet of the absorber, solid activated carbon is injected into the gas flow. Carbon absorbs dioxins and heavy metals.
- Bag filter. It removes fine particles, including combustion particles, micronized lime and micronized activated carbon. The air passes through bag filters, which retain fine particles.
- Stack. It expels the gases produced by incineration under the conditions required by law, that is, without exceeding the required concentration thresholds. Then, it releases the purified gases. This technology has a continuous measurement system that allows always knowing the levels of pollutants.

Slag separation. The solid materials (residues) that come out of the furnace are collected, cooled, and separated to be recycled (metals) or used as a basis for roads and other civil works. The ashes are disposed of at landfills.

- Slag extractor. It collects residues that fall from the furnace or reaches the end of the grates without being burned (such as metals). Then, it transports them to the slag and ash separator. The slag extractor is a conduit with water, where the burned residues fall. It carries water to extinguish objects that are still incandescent, and because of this way, the finest materials dissolve in it and do not disperse during transport. Finally, there is a conveyor that ejects the largest objects from the extractor.
- Slag and ash separator. It separates residues into metallic and non-metallic. The residues collected by the slag and ash extractor are dropped into a pit. A worker loads them with a crane onto a conveyor. With the movement, the conveyor separates the largest metal objects. An electromagnet then separates the rest of the metal objects.

18.2.4 Stakeholders Involved

Waste treatment facilities generally involve different stakeholders with different (sometimes opposite) points of view and interests. The stakeholders can positively or negatively support the facilities, depending on the negative or positive impact generated by the installation. The technical and economic analysis was performed from the viewpoint of the ERF, a public company owned by the Barcelona City Council. The ERF stakeholders are listed below.

- Shareholders/investors
- Workers
- European/national/local government
- Health authorities (i.e., Agència de Salut Pública de Catalunya)
- Environmental authorities (i.e., Miteco, Agència de Residus de Catalunya)

- Non-governmental organizations (NGOs)
- Community Groups (*Aire Net*)
- Spanish electrical network
- Other treatment facilities (i.e., Ecoparc 3)
- Authorized slag managers
- District cooling and heating company (Districlima)
- Power/energy consumers
- Population living near the facility
- Barcelona citizens.

A sustainable waste management system can only be achieved by involving all stakeholders. According to Contreras et al. (2008), the role of stakeholders has transformed over time from being merely receivers of impacts to playing an essential function in the design, implementation and promotion of MSW management systems.

18.2.5 Analysis of Private Revenues and Costs

Internal or private impacts refer to the revenues and costs associated with the investment, operation and maintenance of waste treatment facilities (Jamasb and Nepal 2010). These are costs incurred by the investor or the project developer (public or private entity) and, therefore, are restricted to the spatial boundary of a waste treatment facility (Aleluia and Ferrão 2017). Waste to energy facilities require highly complex and advanced technologies, implying significant investments and high operating and maintenance costs.

The ERF private costs and revenues were calculated directly from the information provided in the annual accounts. Table 18.2 presents the private costs related to operational and maintenance costs (OMC), including labor costs, equipment maintenance and repair costs, provision costs, depreciation of fixed assets and other costs. Total private costs are 144.09 €/ton, considering that 351,308 tons of waste were treated in 2019.

In Table 18.3, private revenues are presented, related to the sale of energy and steam generated by the ERF and the sale of other materials such as slag. Also, revenues are obtained due to the gate fees, which correspond to the amount paid by local authorities for each ton of waste received for treatment in a specific facility. Also, other revenues are taken into account.

The amount of energy, steam, slag, and water sold (AV) is multiplied by the sale price (SP), which corresponds to the market price for these goods. On the other hand, the revenues due to gate fees are obtained by multiplying the total amount of waste treated at the facility by the rate set per ton. As a result, total private revenues are equivalent to 156.60 €/ton.

The sale of energy is the revenue that shows the most significant variability as shown in Fig. 18.3, since the Spanish electricity market regulates the price, showing

Table 18.2 Summary of the ERF private costs in 2019

Concept		Annual Costs (€/year)	Cost per ton (€/ton)
Labor cost			
	Salaries and Wages	4,867,153	13.85
	Social security	1,401,801	3.99
	Other labor costs	473,957	1.35
Equipment repair and maintenance costs		2,326,893	6.62
Provision costs			
	Raw materials and inputs	1,507,621	4.29
	Provision of services (subcontracting)	30,046,827	85.53
Depreciation of fixed assets		2,036,141	5.80
Other costs		7,960,717	22.66
Total		50,621,110	144.09

Source Authors elaboration based on Faura-Casas Auditors Consultors (2019)

Table 18.3 Summary of the ERF private revenues in 2019

Concept	Description	Quantity	Unit	Unitary Price €/tons	Annual revenues (€/year)	Revenues per ton (€/ton)
Sales	Energy	171,173	MWh	50.54	8,650,424	24.62
	Steam	125,148	Tons	7.60	951,271	2.71
	Water	15,300	m^3	1.02	15,606	0.04
	Ashes and slags	81,520	Tons	0.15	12,228	0.03
Gate fees	MSW treatment fee	351,308	Tons	29.00	10,186,748	29.00
Other revenues	Provision of services				30,059,537	85.56
	Other Revenues				5,137,887	14.64
	Total				55,013,701	156.60

Source Authors elaboration based on Faura-Casas Auditors Consultors (2019)

its lowest level in 2016 where the price was 41.02 €/MWh. The highest level was in 2018 where the price was 60.28 €/MWh.

From the results obtained, the Private Benefit (BP) is calculated through Eq. 18.1. The ERF is economically analyzed, considering a specific year (2019); therefore, N is equal to 1. Investment costs (IC) are equal to 0 because they are included in the depreciation values of the fixed asset.

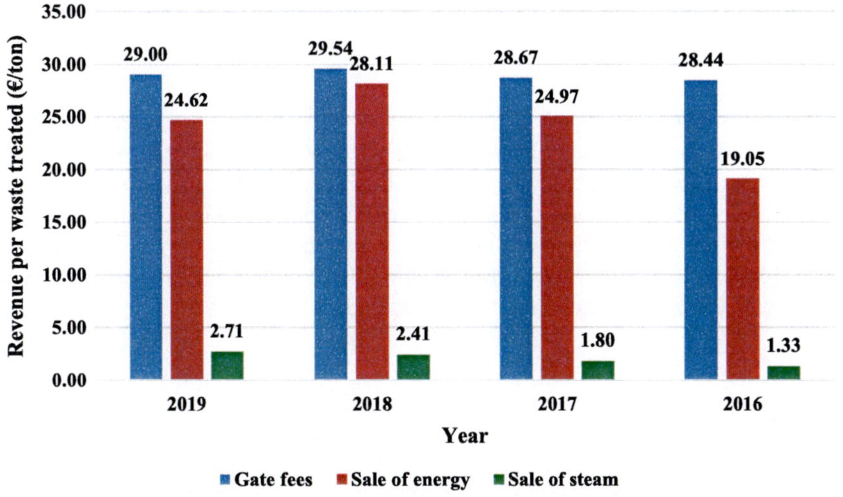

Fig. 18.3 Comparison of private revenues related to the sale of energy and steam and the gate fees. *Source* Authors elaboration

In the case of *FC*, both financial costs and revenues are considered. Therefore, according to the company's annual accounts, it has revenues due to the investment in financial instruments and costs due to third-party debts, having total financial revenues of 2,077,236 €/year, that is, 5.91 €/ton. In the case of *T*, the 25% corporate tax is considered minus the bonuses received for the provision of local public services (BOE 2014), obtaining a tax value of 862,307 €/year, that is, 2.45 €/ton.

Finally, a BP of 15.97 €/ton is obtained (Eq. 18.3), which means that the facility is profitable from a private perspective. This result is slightly higher than that presented in Medina-Mijangos and Seguí-Amórtegui (2021), where a BP of 9.86 €/ton was obtained.

$$B_P = \sum_{n=1}^{N}[(156.60) - (0 + 144.09 - 5.91 + 2.45)] = 15.97\,€/\text{ton} \tag{18.3}$$

Considering only Private Benefits can bias against alternatives such as recycling and even incineration, which may be more expensive than landfills from a purely private (financial) perspective, but preferable from an economic, social and environmental point of view (Nahman 2011). Therefore, it is advisable to evaluate projects and facilities considering their private and external impacts.

18.3 Overview of Environmental and Social Impacts of the ERF

External revenues and costs or externalities refer to those impacts caused directly or indirectly by the operation of a treatment plant but whose effects are assumed by a party other than its operator or owner (Aleluia and Ferrão 2017). These revenues and costs are essentially related to social and environmental impacts.

This section describes and discusses the main external impacts generated by the ERF. Impacts associated with waste, environment, public health, quality of life, education, and economic development are included.

18.3.1 Use of Waste

This impact group is associated with the benefits obtained from the use of waste. For example, the reduction of the quantity of waste sent to landfills and, consequently, achieving the objectives set by the European Commission of limiting the deposit of waste in landfills to 10% (European Commission 2015) and reducing the environmental and social impacts generated by the landfills. Furthermore, the generation of renewable energy that allows increasing the participation of these sources in the Spanish electricity mix and reducing the use of fossil fuels ensures an uninterrupted supply by having a continuous generation of waste, reducing environmental impacts due to energy production from fossil sources.

18.3.2 Reduce Waste Sent to Landfill

The ERF is shown as a facility capable of managing a large amount of waste generated by the Metropolitan Area of Barcelona (AMB). Due to its capacity, each year, more than 350,000 tons of municipal waste are incinerated. Consequently, an added value of the ERF is to provide the AMB with waste treatment capacity since, without its presence, this waste would end up in the landfill (Medina-Mijangos et al. 2021b).

Landfills can cause various impacts due to the risk of air, water and soil contamination through the emission of leachate, landfill gases and other pollutants such as methane (CH_4), carbon dioxide (CO_2), nitrogen oxides (NO_x), sulfur dioxide (SO_2), volatile organic compounds (VOCs), and particulates (PM_{10}) that have the potential to cause environmental damage (Nahman 2011). Furthermore, these facilities are related to social impacts such as the depreciation of the adjacent property (due to odors, dust, windblown trash, vermin, noise, traffic/congestion, visual intrusion), and the opportunity costs of the land where the landfill is located (Hirshfeld et al. 1992). According to Jamasb and Nepal (2010), the cost of landfilling waste is likely to increase due to land scarcity, further thus making energy recovery from waste even more cost-effective. In addition to health damage due to the risk of fires and explosions and the emissions of contaminants.

The economical amount saved per canon paid per ton of waste sent to landfills is considered to quantify the benefit obtained by this facility. It is important to note that this value must be subtracted from the canon paid per ton of waste sent to incineration, which has already been included in the private costs. The tax rate of 47.10 € per ton of municipal waste destined for controlled deposit is set, and a tax of 23.60 € per ton of municipal waste incinerated (BOE 2017a). Consequently, a saving of 23.50 €/ton of waste is considered. In 2019, 351,308 tons of waste had been treated, preventing 338,951 tons of waste from being sent to landfills; finally, only 12,357 tons of ash were sent to controlled landfills.

18.3.3 Willingness to Pay for Renewable Energy

Renewable electricity, also called green electricity, is generated from renewable energy sources (solar, hydro, biomass, wind, geothermal) (Guo et al. 2014). Green electricity has significant environmental benefits and can reduce greenhouse gas emissions, while meeting energy needs and decreasing dependence on fossil fuels (Midilli et al. 2006).

Several studies show that there is a Willingness To Pay (WTP) a premium for renewable energy. For example, in Soliño et al. (2009), the WTP for biomass energy in Spain was calculated using the contingent valuation method. The results show that the WTP vary from 3.79 to 5.71 €/household/month depending on whether it is a single bounded or a double-bounded dichotomous format and the periodicity of the payment (annual or bimonthly). The authors highlighted that society would experience a positive change in welfare if a renewable energy program were implemented.

Hanemann et al. (2011) conducted a study using the contingent valuation method, showing that Spanish households strongly favor applying green electricity programs that make electricity more expensive to reduce carbon dioxide emissions. The average willingness to pay per month and household is 29.91 € over the current electricity bill. The results also show that people living in the Mediterranean area are more likely to pay for green electricity programs and are willing to pay higher electricity prices to prevent climate change effects. Gracia et al. (2012) identify the willingness to pay (WTP) for renewable energy in Spain through the choice experiment method. The findings suggest that most consumers are not willing to pay a premium for increases in the share of renewable energies in the electricity mix. In the case of energy from biomass, a discount of 1.51 €/month would be necessary.

Because the results of individual studies are often inconclusive or even contradictory, with considerable variations in the magnitude, sign, and importance of their WTP estimates, Soon and Ahmad (2015) made a summary estimate of the WTP from numerous studies using a meta-analytic approach, where a WTP of 7.16 USD was obtained.

The summary WTP obtained (7.16 USD_{2013}) was adjusted to the reference year (2019) and currency (EUR_{2019}), applying the annual inflation rate (CPI) and the exchange rate between USD and EUR (World Bank Group 2021; OECD 2021). A WTP of 7.02 €/month was obtained over the current electricity bill by renewable energy use. In Spain, approximately 235.88 kWh was consumed per month and household in 2019 (INE 2021), resulting in a WTP per kWh de 0.02976 €/kWh. In 2019, the ERF sold 171,173 MWh of energy electricity, but it is considered that only 50% of the energy produced by the ERF is renewable, that is, 85,586.5 MWh, giving a benefit of 2,547,054.24 €/year.

18.3.4 Dependence of Other Companies

The Districlima Company in charge of managing the urban heat and cold distribution network of the Barcelona city depends on the supply of steam generated by the ERF for heating, air conditioning and sanitary hot water of more than 100 buildings connected to the network, made up of hotels, offices, homes, schools, shopping centers, among others. In 2019, Districlima served 117 buildings connected through a 20.2 km network distributed throughout the Barcelona city, supplying 113 MW of cold power and 79 MW of heat power.

The investments made in total exceed €64.7 million in a network that has more than 68 km of pipes, which run, for the most part, through the subsoil of the city—providing the company with direct economic benefits of approximately 2,615,000 € per year. In addition to other environmental, economic and social advantages such as reduction of CO_2 emissions, mitigation of the "heat island" effect (managing to lower the ambient temperature between 1 °C and 2 °C, thanks to the replacement of hundreds of air conditioning units), the continuous guarantee of supply, savings in the user's energy bill, esthetic effects, among others (Districlima 2020).

Districlima depends mainly on the ERF since its activity depends entirely on the supply of the steam generated. Therefore, the closure of the ERF would affect Districlima and the citizens and consumers of Barcelona.

As there is no financial information for the 2019 year, to calculate the revenue generated per ton of steam sold, the average of the last three years available is taken. Considering that in 2019, the ERF sold 125,148 tons of steam to Districlima, a net profit of €2,978,423 is obtained Table 18.4.

According to Vlachokostas et al. (2020), WtE facilities can be economically viable when they are located close to domestic or industrial consumers to benefit from energy and steam production, as is the case of the ERF.

Table 18.4 Summary about the activity of Districlima

Concept	Information year									
	2019	2018	2017	2016	2015	2014	2013	2012	2011	2010
Steam sold (tons)	125,148	111,674	95,509	78,012	75,822	75,102	78,611	68,042	68,263	66,382
Operating income (thousands of €)	–	15,044	13,527	11,276	10,286	10,086	9,186	8,651	8,361	6,985
Annual net profit (thousands of €)	2,978[a]	2,615	2,423	1,764	1,239	1,118	786	629	730	1,306
Nº of consumers	117	109	104	95	89	84	81	78	67	59
Km of network	20.2	19.5	18.6	16.8	15.6	15	15	14.4	13.4	13.1

[a] calculated from the net profit per ton of steam sold in the last three years

Source Adapted from Districlima (2020)

Table 18.5 Main pollutants emitted by the waste to energy facilities

Impact	Pollutants
Air emissions	Particulates (PM_{10}) NO_x SO_2 CO_2 CO VOCs HCl, HF (acid gases) PCDD/Fs Heavy metals N_2O
Water emissions	Dioxins/dibenzofurans (PCDD/Fs) Heavy metals Salts
Soil emissions	Heavy metals (Cr, Ni, Pb, Zn, Cu, Cd, As and Hg)
Avoided emissions	CO_2, SO_2, NO_x (emitted by electric power generation plants)

Source Adapted from Medina-Mijangos et al. (2021b)

18.3.5 Environmental

This impact group is associated with the negative effects on the environment caused by the waste facilities due to pollutants emitted into the air, water, and soil. Furthermore, the emission of contaminants avoided due to the production of steam and electricity are included. Table 18.5 shows the main pollutants emitted by the waste to energy facilities.

18.3.6 Climate Change

According to the Intergovernmental Panel on Climate Change (IPCC), climate change and its most visible manifestation, global warming, are fundamentally anthropic and is essentially caused by greenhouse gas (GHG) emissions by using fossil fuels. Therefore, CO_2 emissions eq. are an essential element when analyzing external impacts from the ERF.

First, direct CO_2 emissions generated by the energy recovery process and by the consumption of fossil fuels (natural gas and diesel) are considered to determine CO_2 eq. emissions. It is calculated that 34.37% of the total direct emissions are of biogenic origin, from organic matter, and the remaining 65.63% are anthropogenic origin, from other materials present in municipal waste. Second, indirect emissions related to the consumption of electrical energy from the electrical network are considered.

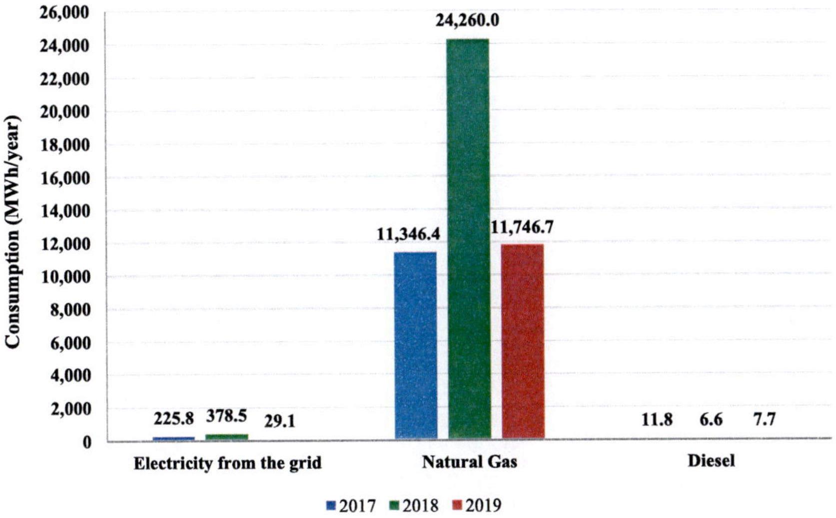

Fig. 18.4 Comparison of consumption of the ERF. *Source* Adapted from TERSA (2020a)

Electrical energy is used mainly for the operation of the plant. This energy usually comes from self-consumption, less in shutdown periods due to maintenance or breakdown in which electricity from the grid is used. Natural gas is used as an auxiliary fuel for combustion and as a fuel for emergency engines. Diesel is consumed in trucks, as well as in generator sets and fire pumps.

Figure 18.4 shows the electricity, diesel, and natural gas consumption made between 2017 and 2019. In 2018, the increase in electricity and natural gas consumption had been caused by various plant shutdowns/starts. Consequently, no maintenance shutdowns have been made at the ERF in 2019, which has led to the normalization of consumption.

Table 18.6 shows the emissions generated in 2019 by the ERF, where a distinction is made between CO_2 eq. of biogenic and anthropogenic origin. Biogenic CO_2 (CO_2 emissions associated with the natural degradation of organic matter) was excluded because biogenic carbon is a short-term emission derived from the biosphere, completing a typical biological carbon cycle (Edwards et al. 2018; Medina-Mijangos and Seguí-Amórtegui 2021). In this case, the emissions generated by the ERF were 208,931.49 tons of CO_2 eq. or 0.595 tons of CO_2 eq./ton treated.

Table 18.7 shows the emissions of CO_2 eq. avoided by the generation of steam and energy from waste. The energy generated by the ERF was sold to the electricity grid and used in the ERF (self-consumption). The steam generated was sold to Districlima for the urban network of cooling and heating. It was used for air conditioning, central heating and hot sanitary water (Medina-Mijangos and Seguí-Amórtegui 2021). The CO_2 eq. emission factor was considered, assuming that if the energy is generated from waste, it would have to come from the electricity grid, meaning an emission factor of 0.241 kg CO_2/MWh (Generalitat de Catalunya 2020). In this case, the emissions

Table 18.6 Emissions of CO_2 eq. generated by the ERF in 2019

Concept	Consumption (MWh)	Emission factor (kg CO_2/MWh)	Emissions of CO_2 eq. (tons)	Waste treated (tons)	Emission of CO_2 eq. per ton of waste (ton/ton treated)
Direct CO_2 emissions (anthropogenic origin)	–	1	206,781.00	351,308	0.589
Direct CO_2 emissions (biogenic origin)	–	1	109,397.00		0.311
Natural gas consumption	11,746.7	0.180	2,141.42		0.006
Diesel consumption	7.7	0.270	2.05		0.000
Indirect emissions related to electricity consumption	29.1	0.241	7.01		0.000
Total (with CO_2 emissions of biogenic origin)			**318,328.49**		**0.906**
Total (without CO_2 emissions of biogenic origin)			**208,931.49**		**0.595**

Source Adapted from TERSA (2020a), Generalitat de Catalunya (2020)

Table 18.7 Avoided Emissions of CO_2 eq. by the ERF in 2019

Concept	Energy production (MWh)	Emission Factor (kg CO_2/kWh)	Emissions of CO_2 eq. (tons)	Waste treated (tons)	Emission of CO_2 eq. per ton of waste (ton /ton treated)
Electric energy for self-consumption	23,560	0.241	5,677.96	351,308	0.016
Electric energy sold to the grid	171,921	0.241	4,1432.961		0.118
Steam sold to Districlima	21,773	0.241	5,247.293		0.015
Total	**217,254**		**52,358.21**		**0.149**

Source Adapted from TERSA (2020a), Generalitat de Catalunya (2020) avoided by the ERF were 52,358.21 tons of CO_2 eq. or 0.149 tons of CO_2 eq./ton treated.

The tax set by Catalan legislation on emissions from various industrial activities is considered to calculate the cost due to CO_2 emissions. According to its industrial activity, the ERF is classified as a municipal waste incineration facility with a capacity greater than 3 tons per hour. The CO_2 eq. emission price has been set at an average value of about 10 €/ton of CO_2 eq., which should increase to a value of about 30 €/ton CO_2 eq. in 2025 (BOE 2017b).

The objective of these taxes is that the damage caused by greenhouse gas emissions falls on those who generate them and therefore reduce the emissions through new technologies and innovation. Therefore, emitters have an incentive to reduce emissions as long as it is cheaper than paying the price per ton of CO_2 emitted. 30 €/ton is considered a minimum estimate of the damage currently caused by carbon emissions. Pricing emissions above 30 €/ton do not guarantee that polluters pay for the total damage they cause or that prices are high enough to decarbonize economies (OECD 2018). However, a price below 30 €/ton means that polluters do not directly face the cost of emissions and possible damage to society. The incentives for a profitable reduction are too weak. According to OECD (2018), it is considered that carbon prices should amount to at least USD 40–80 (35–70 €) per ton of CO_2 by 2020, and USD 50–100 (44–88 €) per ton of CO_2 by 2030.

18.3.6.1 Air Emissions

Regarding atmospheric emissions, several strategic projects have been carried out to reduce emissions, setting limits much lower than those established in the current regulations at a European level. The ERF has different filter systems and smoke and gas catalysis to avoid the local deterioration of air quality. Initially, in 2004, the ERF installed NO_x and HCl emission reduction systems. Later in 2018, upgraded the NO_x emission reduction system with a catalytic filter, which reduces NO_x emissions to 50 mg/Nm^3, representing an investment of €14.5 million.

Although the projects represent high-investment costs, they produce benefits for the ERF due to avoiding damage to both the environment and the public health. Table 18.8 shows the results of the 2019 checks, where the mean values are lower than the legal limits.

18.3.6.2 Emissions to Water

The ERF performs two different releases. On the one hand, the sanitary water and rainwater (without treatment) are released directly into the municipal sewers; on the other, the cooling water, which is taken from the sea and, after passing through the thermal process, is returned to the sea with the only variation being a slight increase in temperature. The ERF carries out three-monthly checks on the two emission points.

Table 18.8 Atmospheric emissions of the ERF in 2019

Contaminants	Mean values			Legal Limits
	2017	2018	2019	
Particulates (mg/Nm^3)	3.23	3.02	3.17	10
CO (mg/Nm^3)	19.84	29.32	26.47	50
HCl (mg/Nm^3)	5.15	4.20	5.10	10
SO_2 (mg/Nm^3)	12.82	10.20	10.58	50
HF (mg/Nm^3)	0.07	0.08	0.098	1
NO_x (mg/Nm^3)	125.16	100.48	109.39	200
TOC (mg/Nm^3)	1.74	1.90	1.17	10
Hg (μg/Nm^3)	1.15	0.600	0.324	50
Various (Sb + Cr + Co + Cu + Mn + Ni + V + As + Pb) (mg/Nm^3)	0.0345	0.0465	0.0240	0.5
Cd + Tl (mg/Nm^3)	0.0020	0.0042	0.00300	0.05
PCDD/PCDFs (nmg/Nm^3)	0.0288	0.0171	0.0174	0.1

Source Adapted on TERSA (2020a)

Table 18.9 Emissions to water by the ERF in 2019

Concept	Mean Values	Legal Limits
pH	7.63	Between 6 and 10
Chemical oxygen demand (mg/l)	279.17	1500
Chlorides (mg/l)	208.33	2500
Soluble Salts (mg/l)	1516.67	6000
Suspended matter (mg/l)	36.90	750
Inhibitory Matter (equitox/m^3)	22.20	25
Total phosphorus (mg/l)	5.22	50
Nitrogen (mg/l)	45.62	90

Source Adapted from TERSA (2020a)

Table 18.9 shows the results of the 2019 checks, where the mean values are below the legal limits; therefore, only the costs associated with the periodic checks, which have already been included in the operating costs, are considered.

18.3.7 Public Health

This group of impacts includes damage to the health of the ERF workers or the population living near the facility due to pollutant emissions. Also, physical accidents to workers caused by activities carried out in the ERF are considered.

18.3.7.1 Chemical Risks

In the MSW incineration process, fumes are produced because of combustion. These fumes are mixtures of oxides, heavy metals, carbon particles, dioxins and furans (PCDD/Fs), among other elements that generate danger to human health. The results of García-Pérez et al. (2013) show that there is an excess risk for all cancers combined and for lung cancer, in particular, there are marked increases in the risk of tumors of pleura and gallbladder (in men) and tumors of the stomach (in women) for people around incinerators.

Specifically, PCDD/Fs constitute a group of persistent organic chemical compounds. PCDD/Fs can enter the body via ingestion, skin absorption, and inhalation pathways. The possible health effects of dioxin emission are detailed below.

- Short-term exposure to high levels of PCDD/Fs may cause skin lesions known as chloracne, which is persistent (World Health Organization 2010).
- Longer-term exposure may cause a range of toxicity, including immunotoxicity, developmental and neurodevelopmental effects, and effects on thyroid and steroid hormones and reproductive function; the most sensitive life stage is considered to be the neonate or fetus (World Health Organization 2010).
- PCDD/Fs are environmental pollutants that have raised considerable concern, especially due to the potential carcinogenic effects (Domingo et al. 2017).

In Domingo et al. (2017), air and soil samples were collected in locations near the ERF to determine the levels of PCDD/Fs and the possible risks to human health. It was determined that the main route of human exposure to PCDD/Fs in the study area is air inhalation. The hazardous quotient (HQ) is used to evaluate the non-carcinogenic effects of exposure to a specific contaminant. HQ values below unity are considered safe. The HQ for the area was 0.01, indicating that there are no significant non-cancer risks due to human exposure to PCDD/Fs in the vicinity of the ERF (Domingo et al. 2017).

On the other hand, Agència de Salut Pública de Barcelona (2018) has carried out a study that explores the risk of mortality due to causes associated with the exposure of PCDD/Fs in the area of Barcelona city for the period from 1991 to 2015. They have included diseases with an origin related to dioxin exposure such as malignant neoplasia of the liver, malignant neoplasm of the trachea, bronchi and lungs, neoplasia of connective tissue and other soft tissues, Non-Hodgkin's lymphoma, leukemia and diseases of the circulatory system. This study aims to analyze if the proximity to the incinerator could lead to increased exposure to PCDD/Fs in the air. This environmental exposure could lead to an increased risk of suffering from certain cancers and diseases of the circulatory system that would be reflected in higher mortality from these causes.

Next, the study results are shown according to the Standardized Mortality ratio map—SMR Fig. 18.5 and the Probability map of exceeding the Barcelona city's mean mortality—PEM Fig. 18.6 depending on the proximity to the plant. SMR is the ratio of the observed number of deaths (or incidents) to the number of deaths (or

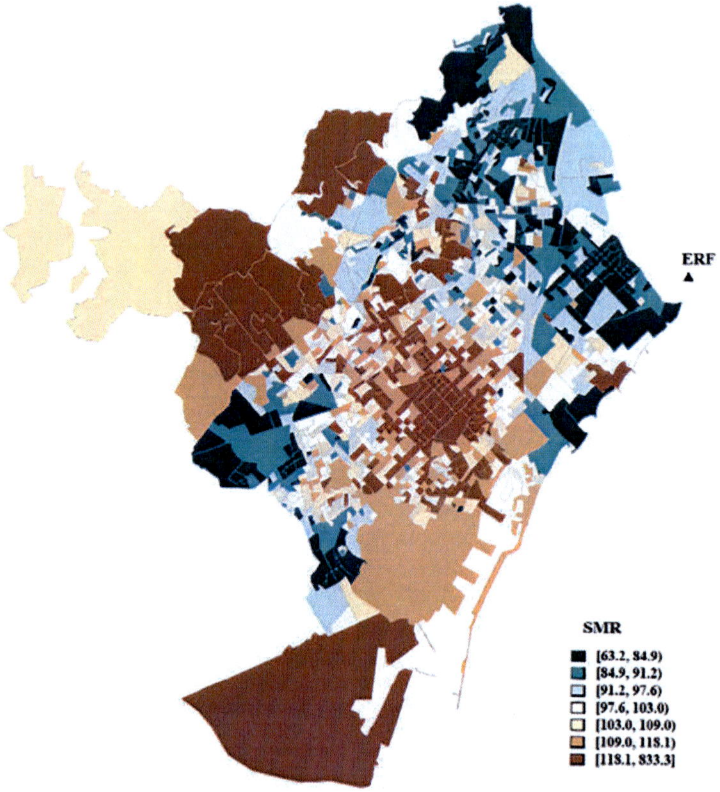

Fig. 18.5 Map of the areas of Barcelona city according to the Standardized Mortality Ratio (SMR) for all mortality causes. *Source* Agència de Salut Pública de Barcelona (2018). Creative Commons Attribution 4.0 International License

incidents) that would be expected in a reference population or area (Kelsey and Gold 2017). SMR for the entire city is 100, and values above 100 indicate higher mortality than in Barcelona city.

These descriptive maps show that the areas closest to the incineration plant do not have a higher mortality ratio than the Barcelona average. In both cases, in the vicinity of the ERF, the best results are observed compared to other areas of the city of Barcelona. The areas with the worst results are the areas colored in brown in the Fig. 18.5 and the areas colored in red in the Fig. 18.6, which do not coincide with the areas close to the ERF. Therefore, it can be concluded that no groupings of areas have been detected in the vicinity of the incineration plant with a mortality rate above the city average. Furthermore, no significant association has been found between proximity and mortality to the incineration plant (Agència de Salut Pública de Barcelona 2018).

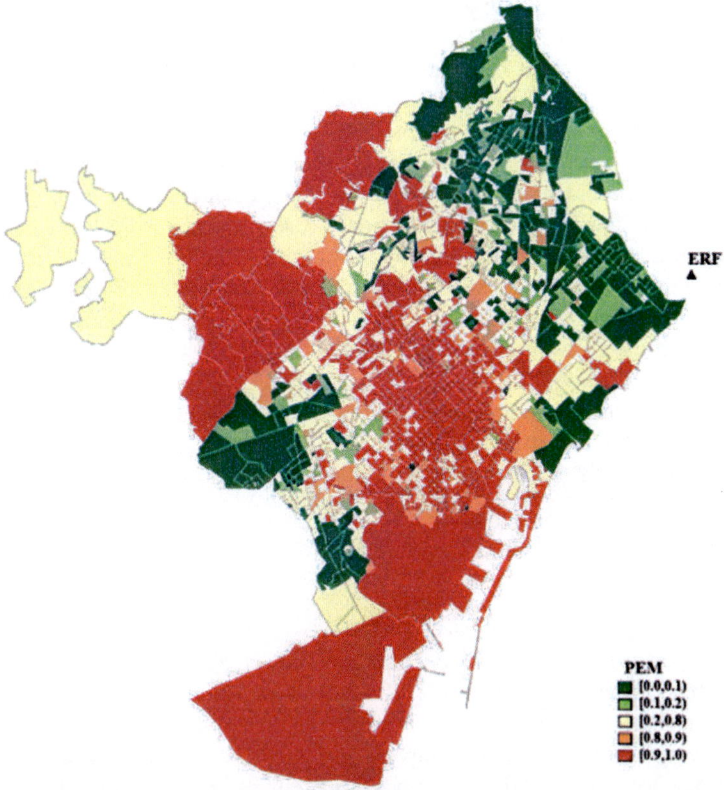

Fig. 18.6 Map of the areas of Barcelona city according to the probability of exceeding Barcelona mean mortality (PEM) for all causes of mortality. *Source* Agència de Salut Pública de Barcelona (2018). Creative Commons Attribution 4.0 International License

The study considers a period of time in the past in which the environmental levels of dioxins were higher than those of today. In recent years, a series of catalysts and particulate filters have been installed and renewed to prevent the emission of substantial amounts of toxic fumes. These improvements and innovations are reflected in the investment costs (included in the private impacts) and avoid the costs generated by damage to health. In this case, it can be concluded that the cost due to impacts on public health caused by chemical pollutants is equal to 0 €.

18.3.7.2 Physical Risks

Damages to the health of the ERF workers are mainly associated with injuries caused by minor accidents and include dislocations and sprains, fractures and superficial injuries (TERSA 2020b; Medina-Mijangos and Seguí-Amórtegui 2021).

According to Gil Fisa and Pujol Senovilla (2009), these work accidents would cause various costs such as (1) Cost of time lost during the accident; (2) Costs for material damage; (3) Costs due to production losses; (4) General and medical expenses; (5) Time spent investigating the accident by other company personnel. Only the expenses incurred by mutual or public entities are considered to avoid double-counting since the salary payment, the social security fee, has already been considered in the company's annual accounts as part of the labor costs (Medina-Mijangos and Seguí-Amórtegui 2021). In this case, only the medical care costs and the worker salary paid by the public administration during sick leave are considered (generally, in Spain, 25% of the worker salary is paid by the company and 75% by Social Security). According to Medina-Mijangos and Seguí-Amórtegui (2021), in 2017, there were six accidents in the ERF, and the cost for physical risks was 13,660.50 €. However, in 2019, no accidents were recorded in the treatment facility, which means that there is no cost related to physical damage, more than the costs incurred for risk prevention, which in 2019 were 12,837.96 € compared to 2018, which were 1,236.79 €, these costs are included in the annual accounts.

18.3.8 Quality Life

Generally, treatment facilities generate various disamenities such as dust, odors, visual intrusion (smokestack) and noise. In the case of incinerators, they can generate changes in environmental quality associated with the emissions of pollutants.

In order to assess the economic impacts due to the disamenities generated, several authors have carried out studies to analyze the effects on the quality of life of the households that live in the vicinity of incinerators and their negative effect on house prices. For example, in the case of Sun et al. (2017), a study was carried out in Shenzhen city, China using the hedonic price method, where it is concluded that for each additional kilometer that the property moves away from the WTE plants, the value of the properties can increase by 1.30%. On the other hand, Rivas Casado et al. (2017) point out that the impact of incinerators on local UK house prices ranges between approximately 0.4% and 1.3%.

Many projects were significantly delayed or even abandoned, mainly due to opposition from the local community and the "Not In My Back Yard" (NIMBY) syndrome, which is often exacerbated when facilities are located near dense urban areas (Vlachokostas et al. 2020). In the case of the ERF, the *Aire Net* platform (by its name in Catalan) was created, made up of numerous entities and associations from the municipalities of Barcelona, Sant Adrià de Besòs and Badalona, to inform citizens about environmental pollution that cause industries and service infrastructures.

In this case study, the figure of 8 € per ton of waste treated was used to monetary value the disamenities generated by incinerators; this is slightly lower than the impacts caused by the landfill disamenities, that is, 10 €/ton (European Commission 2000).

18.3.9 Education

This impact group refers to the change of behavior of citizens and workers through training and awareness programs to obtain benefits related to improving the processes of the treatment facilities.

Waste incorrectly classified by households (prior to waste collection processes) increases the risk of spontaneous fires, higher operating costs, production errors, possible damage to equipment, as well as an increase in workplace accidents (Ibrahim 2020). Therefore, in this study, it is considered that the incorrect classification of waste by citizens does not affect the ERF. Despite this, the ERF makes annual investments in developing environmental education programs aimed at citizens to benefit other treatment facilities.

On the other hand, it is considered that the training programs for workers allow improving the skills of the workforce with which it is possible to achieve greater effectiveness and efficiency of the manufacturing process and the quality of the goods produced. According to Mital et al. (1999), the economic benefits of worker training include significant productivity improvements through reduction of waste, reduction of production time, improvements in quality, greater flexibility to respond to needs, and an advantage competitive for employers and countries as a whole. However, these training programs require investments, which are reflected in the annual accounts. In 2019, 101,925.24 € was invested for the training of workers, compared to the 52,732.1 € registered in 2018.

In this case study, the increase in energy efficiency is evaluated Table 18.10 due to greater investment in the training of workers. In 2018, there was an energy efficiency of 526 kWh/ton treated, compared to 554 kWh/ton in 2019. Therefore, two different scenarios are analyzed to compare the benefits obtained. Firstly, considering the revenues obtained if the efficiency had remained the same as in 2018 (i.e., 526 kWh/ton). The second scenario considers the revenues obtained in 2019 due to the increase in electrical efficiency (i.e., 554 kWh/ton). In both cases, the sale price of energy corresponds to 50.54 €/MWh, and it is considered that 23,560 MWh of

Table 18.10 Information about the benefits due to increased energy efficiency

Increase in energy efficiency		
Concept	Scenario 1	Scenario 2
Energy efficiency (kWh/ton)	526	554
Waste treated (ton)	351,308	351,308
Total production (MWh)	184,959	194,740
Energy sold to grid (MWh)	161,399	171,180
Price electricity (€/MWh)	50.54	
Total revenue (€)	8,157,129.56	8,651,437.20
Benefit (€)	494,307.64	

Source Adapted from TERSA (2020b)

Table 18.11 Energy efficiency of TERSA, TIRME, SIRUSA and MADRID energy recovery facilities in 2019

Facility	Year	Waste treated (ton)	Energy produced (MWh)	Energy efficiency (MWh/ton)
TERSA	2019	351,308	194,740	0.554
TIRME	2018 [a]	573,788	326,804	0.570
SIRUSA	2019	129,815	49,649	0.382
MADRID	2019	331,955	228,263	0.687

[a]Information about 2019 activity is not available

the total energy production was used for self-consumption. The remaining was sold to the electricity grid.

Comparing the energy efficiency of the ERF with other recovery plants located in Spain Table 18.11, we can see that there are facilities with better results than the ERF analyzed, so it is necessary for this facility to improve its processes to achieve greater energy efficiency and therefore better economic and environmental results.

18.3.10 Economic Development of the Area

It is important to note that two vastly different ecosystems coexist in the vicinity of the ERF. On the one hand, the ERF is located in a highly industrialized area that provides urban services to the Catalan capital, such as waste treatment, electricity production, heat production, and wastewater treatment. On the other hand, the industrial area (where the ERF is located) is surrounded by an urban area with good quality public transportation services, a new university campus, shopping malls, along other services.

Despite the benefits obtained from the ERF related to the management of MSW, it avoids urbanization and the growth of the tourism, financial and real estate sector, having a "conflict of interest" between the land of industrial use and the land of urban use where industrial investments are losing ground to urban development and its associated investments.

18.4 Monetary Valuation of Externalities

In Table 18.12, the results obtained from the different external impacts are presented, where the results are expressed in annual costs and revenues (€/year) and costs and revenues per ton of waste treated (€/ton). As 351,308 tons have been treated, a total external cost of 13.95 €/ton and total external revenue of 41.30 €/ton.

Table 18.12 Economic analysis of external impacts related to the ERF in 2019

Impact group	Impact Identification		Impact Quantification	Impact Valuation (€/year)		Impact Valuation (€/ton)	
	Costs	Revenues		Costs	Revenues	Revenues	Revenues
Use of waste		Reduce waste sent to landfill	338,951 tons of waste		7,965,348.50		22.67
		Quality of energy (renewable energy)	85,586.5 MWh of energy		2,547,054.24		7.25
		Districlima dependence	125,148 tons of steam		2,978,423.00		8.48
Environment	Emissions to air (CO_2)		208,931.49 tons CO_2 eq	2,089,314.90		5.95	
		Avoided emissions to air (CO_2)	52,358.21 tons CO_2 eq		523,582.10		1.49
Public Health	Physical injuries		0 people affected	0		0	
	Chemical risk (Cancer by emission of PCDD/Fs)		0 people affected	0		0	
Education		Technique of workers (increase in energy efficiency)	% productivity (change 526 to 554 kWh/ton)	494,307.64			1.41
Quality Life	Disamenities		Price of households	2,810,464.00		8	

(continued)

Table 18.12 (continued)

Impact group	Impact Identification		Impact Quantification	Impact Valuation (€/year)		Impact Valuation (€/ton)	
	Costs	Revenues		Costs	Revenues	Revenues	Revenues
Total external impacts				4,899,778.90	14,508,715.48	13.95	41.30

Source Authors elaboration

As can be seen, this facility generates several positive impacts. The most representative positive impact (revenue) is related to reducing waste that is sent to landfills. In contrast, the negative impact (cost) with the most significant effect is related to the disamenities generated by the ERF. The results show no costs related to health damage due to chemical risks; however, it is essential to closely monitor dioxin emissions to detect abnormal situations and continue investing in innovative projects and advanced technology.

Once the impacts described above have been monetarily valued, it is possible to add the costs and revenues to obtain the Total Benefit through Eq. 18.2. In the case of opportunity cost, it is considered as the value of the waived alternative share. Under the concept of sustainable development and its three pillars, the best alternative is the one that provides not only the best economic performance but also the best environmental and social performance (Medina-Mijangos and Seguí-Amórtegui 2021).

As it is not considered that there is a better alternative for the treatment and use of rejected waste, because the alternative treatment would be the disposal in landfills, facilities that entail various negative environmental and social impacts; it is determined that the opportunity cost is that provided by a financial instrument when the company's capital and reserves are invested in them (68,336,034 €). The interest on financial instruments in 2019 was 3% (Banco de España 2021); therefore, the opportunity cost is 2050,081 €, the equivalent of 5.84 €/ton. Finally, the Total Benefit of 37.48 €/ton is obtained, as shown in Eq. 18.4.

$$B_T = \sum_{n=1}^{N}[(156.60) - (0 + 144.09 - 5.91 + 2.45) + (41.30 - 13.95) - 5.84] = 37.48\ €/\text{ton} \quad (18.4)$$

Therefore, it can be concluded that the ERF is profitable from a private point of view (BP = 15.97) and an economic, environmental, and social point of view (BT = 37.48).

18.5 Sensitivity Analysis

This section analyzes the robustness of the management system by considering and evaluating different scenarios and variables such as CO_2 emissions, the impacts of dioxins on public health and the opportunity cost of the land where the ERF is located.

Table 18.13 Effect of biogenic emissions and the increase in the CO_2 emission tax on the Total Benefit of the ERF

Concept	Emission of CO_2 per ton treated (ton CO_2/ton)	Cost per ton treated with a tax of 10 € (€/ton)	Cost per ton treated with a tax of 30 € (€/ton)	Total Benefit with a tax of 10 € (€/ton)	Total Benefit with a tax of 30 € (€/ton)
Without emissions of Biogenic origin	0.595	5.95	17.85	37.34	28.42
With emissions of biogenic origin	0.906	9.06	27.18	34.23	19.09

Source Authors elaboration

18.5.1 CO_2 Emissions

An important factor is related to emissions of biogenic origin (basically due to the organic matter contained in the waste), with the entry into operation of the previous selection of waste, through Ecoparc 3, and the consequent decrease in organic matter reaching the ERF, there is generally a tendency in recent years for the percentage values of biogenic CO_2 to decrease. Some studies consider biogenic emissions as a critical sensitivity factor, noting that whether or not biogenic carbon is included as an externality can make a significant difference in the total cost of the project (Edwards et al., 2018).

In this case study, if emissions of biogenic origin are considered, the total emissions of CO_2 eq. It would be 0.906 ton CO_2 eq./ton of waste instead of 0.595 ton CO_2 eq./ton of waste. This value becomes more important if we consider the payment imposed in Catalonia per ton of CO_2 eq., in 2025, it will be 30 € instead of 10 €.

When considering biogenic emissions, the Total Benefit decreases, reaching its lowest level when the tax reaches 30 €/ton CO_2 eq. as shown in Table 18.13. Including biogenic emissions can incentivize the ERF and other waste management companies to reduce total CO_2 emissions through innovative projects and advanced technology.

18.5.2 Public Health

Another sensitivity factor is related to the possible damage to health from the emissions of pollutants, specifically from the emission of PCDD/Fs. Carcinogenic risks are expressed in terms of the probability of developing cancer due to exposure throughout life (estimated at 70 years); the carcinogenic risk of < 10 − 6 is considered significant (Domingo et al. 2017). The carcinogenic risks due to exposure to PCDD/Fs for residents in the vicinity of the ERF were 2.3 × 10 − 6 in 2017, exceeding the threshold of 10–6, which is why it is considered a significant risk. The results indicate that residents living in the vicinity of the ERF are 3–4 times more likely to develop cancer throughout their lives (due to exposure to PCDD/Fs) than residents of cities such as Girona, Mataró and Tarragona Fig. 18.7, where there are also other incineration plants operating (Domingo et al. 2017).

These results have caused great concern among the population. However, the authors note that the most critical limitation of the current study is the small number of air and soil samples. Consequently, the results should be viewed with caution.

Although the previous economic results did not reflect costs related to the impacts on public health, because the study presented by the Agència de Salut Pública de Barcelona (2018) was taken as a reference; if the results of Domingo et al. (2017) are considered, the results of the economic analysis could vary.

For the calculation of the total costs of cancer in Spain, the costs presented by Badia and Tort (2015) were taken as reference, where (a) direct costs composed of hospital costs, costs of the consumption of antineoplastic drugs and the primary care costs; (b) indirect costs made up of premature mortality costs and disability costs (both temporary and permanent), and (c) informal care costs. Thus, obtaining a total cost of €12,216 million through the human capital method that supposes that when a worker leaves the labor market, his productivity is lost until he returns to work and a total cost of €7168 million according to the friction method that supposes that when a worker leaves the labor market, his productivity is lost until he is replaced.

A most recent study realized by Wyman (2020) considers (a) direct medical costs composed of treatment cost, follow-up cost, pharmacy cost paid by the patient, palliative care; (b) direct non-medical costs consisting of transportation, accommodation and subsistence paid by the patient, equipment and works, formal and informal care; transport to radiation therapy subsidized by the State, (c) indirect costs composed of loss of income after cancer and demand for productivity due to premature death. It is estimated that cancer costs for Spanish society around €19.3 billion for the total people diagnosed each year, equivalent to 1.6% of Spanish GDP.

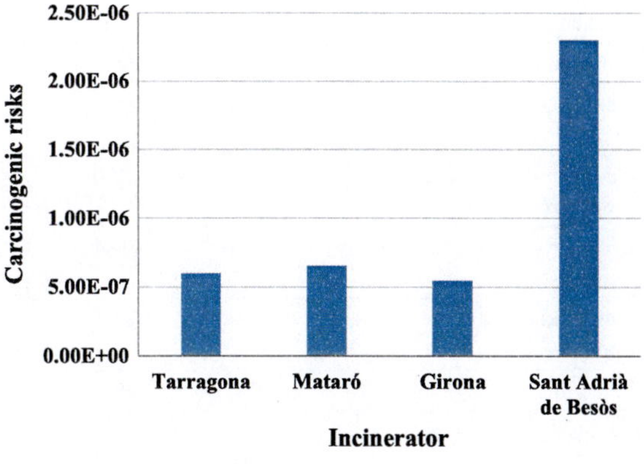

Fig. 18.7 Carcinogenic risks due to PCDD/Fs exposure for residents living near Catalan incinerators. *Source* Adapted from Domingo et al. (2017)

The incidence (new cancer cases) was considered to calculate the total number of patients in Spain in 2019, 275,617 people (Observatorio AECC 2021); this would mean an approximate cost of 70,024.71 €/patient. These results coincide with a study carried out in France where it is established that the total cost of cancer in France is 10 billion €/year for treatment and 15 billion €/year, including lost productivity. Therefore, the cancer incidence is considered to be 240,000 new cases per year in France, which implies a cost per case of approximately 42,000 €/cancer per treatment and 63,000 €/cancer, including lost productivity (Rabl et al. 2010).

In Table 18.14, the incidence of cancer in different geographical areas is presented among the populations where incinerators are located (Tarragona, Girona and Barcelona). First, we can see that the incidence rate (per 100,000 inhabitants) is below the national average. Based on the Barcelona incidence, the incidence by type of cancer in Sant Adrià de Besòs has been calculated considering only diseases with an origin related to dioxin exposure such as malignant neoplasia of the liver, malignant neoplasia of the trachea, bronchi and lungs, connective tissue and other soft tissue neoplasia, Non-Hodgkin's lymphoma, leukemia. If we assume that all the incidents of these five types of cancers (58 patients) are due to the presence of the incinerator, the total cost for cancer in the area would be 4,061,433.18 €/year, that is, 11.56 €/ton. In this case, the Total Benefit obtained by the system is 25.92 €/ton treated, showing that the system continues to be economically profitable because the condition of BT > 0 is met.

On the other hand, if it is considered that in the city of Sant Adrià de Besòs there are 3–4 times more likely to develop cancer than residents of cities such as Girona, Mataró and Tarragona for the five types of cancer considered, we would have an incidence of 174 people considering the probability of 3 times more than in the other communities. This represents a public health expenditure for cancer

Table 18.14 Cancer incidence in geographical areas of Spain where incinerators are located

Geographic area	Tarragona	Girona	Barcelona	Spain	Sant Adrià de Besòs
Incidence	4,644	4,291	32,164	275,617	211
Population	804,664	771,044	5,664,579	47,105,358	37,097
Incidence rate [a]	577	557	568	586	568
Neoplasms of the liver	115	106	785	6,768	5
Neoplasm of trachea, bronchus and lung	476	438	3,230	27,945	21
Connective and soft tissue neoplasm	455	416	3,184	27,197	21
Non-Hodgkin's Lymphoma	134	124	934	7,947	6
Leukemia	100	93	698	5,941	5

[a] incidence per 100,000 inhabitants

Source Adapted from Observatorio AECC (2021)

of 12,184,299.54 €, that is, 34.68 €/ton of waste treated. This would mean that the Total Benefit of the system, considering the data presented by Domingo et al. (2017), would be 2.80 €/ton treated, showing that the system continues to be economically profitable because the condition of BT > 0 is met.

The economic results obtained are preliminary and with a global vision because this analysis has been carried out with secondary data from public statistics. There are no specific data on the areas where the incinerators are located.

18.5.3 Opportunity Cost of Land

The ERF is located in industrial land, but if it is considered for other uses the land where the ERF is located, such as urban development in the area, the Total Benefit obtained from the system would considerably change.

It is necessary to consider the available land where the ERF is located as urban land instead of industrial land to calculate the cost associated with this impact. It is estimated that in 2019, the average price of urban land in the municipality of Sant Adrià del Besòs is equivalent to 2735 €/m^2 (Idealista 2021). Finally, according to the AMB, the price of industrial land is 730 €/m^2 (AMB 2019). Therefore, given the alternative in land use, a cost of 2005 €/m^2 is established. The total area of the ERF is 10,044 m^2, obtaining an opportunity cost of 20,138,220 €, that is, 57.32 €/ton treated.

In this case, the Total Benefit obtained by the system is −14.00 €/ton treated, showing that the system becomes economically unprofitable because the condition of BT > 0 is not met. However, this industrial zone is essential for the proper functioning of the AMB; therefore, the change from industrial to urban land is not viable since the ERF limits this change and other facilities.

18.6 Conclusions and Recommendations

Waste to energy facilities emerges as an alternative to landfilling of rejected waste (waste that can no longer be materially recovered), reducing the environmental and social impacts that the landfills generate. Although the European Parliament establishes a waste hierarchy for legislation and policy on the prevention and management of waste, prevention, reuse, and recycling are prioritized over other types of recovery (including energy recovery) and deposit in landfills. It is considered that rejected waste can only be managed by incineration or landfilling. In the case of Spain, incineration is prioritized over landfilling, complying with the European waste hierarchy principle. Consequently, the ERF fulfills a fundamental function for the AMB because it allows the energy recovery of more than 360,000 tons of rejected waste, which would otherwise end up in landfills.

Besides, there is a strong dependence on other companies such as Districlima, which has made significant investments in the heating and cooling network, and whose activity is based on the supply of steam generated by the ERF.

In the present case study, the infrastructure is profitable from a private and external point of view. We can even observe that externalities make this infrastructure more profitable and reliable since even in pessimistic scenarios, the infrastructure continues to generate economic benefits, as shown by the sensitivity analysis, except in the case of the assessment of the opportunity cost of land, where the result becomes negative (BT < 0). Although the ERF limits investment, it weighs down the local attractiveness, preventing urbanization and the growth of the tourism, financial and real estate sectors. The ERF is located in a highly industrialized area that provides urban services to the Catalan capital, such as waste treatment, electricity production, heat production, and wastewater treatment. This industrial zone is essential for the proper functioning of the AMB; therefore, the change from industrial to urban land is not viable since the ERF limits this change and other facilities.

As mentioned previously, the results of Domingo et al. (2017) should be taken with care due to the small number of samples taken, also because there may be other sources of pollutant emissions in the area (other industrial facilities or even traffic). Moreover, other studies have shown that no groupings of areas in the vicinity of the incineration plant with mortality above the city average have been detected. Despite this, it is essential to make investments in strategic projects that allow the reduction of pollutant emissions through new technologies and innovation, which has already been done for several years, such as the implementation of a catalytic NO_x reduction system, which allow reducing NO_x emissions to 50 mg/Nm^3, representing an investment of €14.5 million. Furthermore, periodic measurements of contaminants are also crucial to ensure that the legal maximums are met. Additionally, to detect abnormal situations and that there are no risks to public health.

It is essential to include externalities in the technical–economic analysis of waste treatment facilities because sometimes, if an analysis is carried out from a purely financial perspective, infrastructures such as landfills may seem less expensive than incinerators; however, by including externalities, the results are reversed, demonstrating that incineration plants are profitable from an environmental and social perspective. For this reason, researchers and policymakers should be interested in the economic values of externalities to allow the internalization of external costs related to incineration through instruments such as regulations, taxes, subsidies, compensations, and negotiable emission permits to avoid direct damages to society. Spanish legislation by including gate fees to landfills and incinerators aims to incorporate externalities into private costs. Additionally, a CO_2 emission tax has been set in Spain, which corresponds to a tax of 10 €/ton of CO_2 and will reach a value of 30 €/ton of CO_2 by 2025. Despite this, this value may prove to be insufficient to motivate the decarbonization of economies. According to the OECD (2018), the carbon prices should amount to at least 35–70 €/ton of CO_2 by 2020 and 44–88 €/ton of CO_2 by 2030. Therefore, policymakers should set taxes and fees that ensure the minimum cost of the damage that carbon emissions currently cause.

Finally, it is essential to invest in new technologies and innovation systems to improve the process controls to ensure that pollutant levels are below the legal limits established by regulatory bodies such as the European Commission and Spanish legislation. Consequently, reduce the concern and rejection of the population that is in the vicinity of these facilities.

References

Agència de Salut Pública de Barcelona (2018) Anàlisi del risc de mortalitat associat a la proximitat de la incineradora de Sant Adrià del Besòs. http://hdl.handle.net/11703/116087. Accessed 15 Feb 2020

Aleluia J, Ferrão P (2017) Assessing the costs of municipal solid waste treatment technologies in developing Asian countries. Waste Manag 69:592–608. https://doi.org/10.1016/j.wasman.2017.08.047

AMB (2019) Informe del 2n semestre de 2019. In: Mercat immobiliari Ind. a l'àrea Metrop. Barcelona. https://www.amb.cat/web/desenvolupament-socioeconomic/actualitat/publicacions/detall/-/publicacio/mercat-immobiliari-industrial-2019--2n-semestre-/9201523/11708. Accessed 15 Apr 2021

AMB (2021) Municipios metropolitanos. In: Àrea Metrop. Barcelona. https://www.amb.cat/s/es/web/area-metropolitana/municipis-metropolitans.html. Accessed 31 Jan 2021

Badia X, Tort M (2015) La carga del cáncer en España

Banco de España (2021) Tabla tipos de interés legal—cliente Bancario, Banco de España. In: Eurosistema. https://clientebancario.bde.es/pcb/es/menu-horizontal/productosservici/relacionados/tiposinteres/guia-textual/tiposinteresrefe/Tabla_tipos_de_interes_legal.html. Accessed 8 Mar 2021

BOE (2014) Ley 27/2014, de 27 de noviembre, del Impuesto sobre Sociedades. https://www.boe.es/buscar/doc.php?id=BOE-A-2014-12328. Accessed 16 Apr 2021

BOE (2017a) Ley 5/2017, de 28 de marzo, de medidas fiscales, administrativas, financieras y del sector público y de creación y regulación de los impuestos sobre grandes establecimientos comerciales, sobre estancias en establecimientos turísticos... https://www.boe.es/buscar/doc.php?id=BOE-A-2017-7353. Accessed 15 Apr 2021

BOE (2017b) Ley 16/2017, de 1 de agosto, del cambio climático. https://www.boe.es/buscar/pdf/2017/BOE-A-2017-11001-consolidado.pdf. Accessed 15 Apr 2021

Contreras F, Hanaki K, Aramaki T, Connors S (2008) Application of analytical hierarchy process to analyze stakeholders preferences for municipal solid waste management plans, Boston, USA. Resour Conserv Recycl 52:979–991. https://doi.org/10.1016/j.resconrec.2008.03.003

Districlima (2020) Un DHC consolidado. http://www.districlima.com/districlima/uploads/descargas/presentaciones/y/otros/documentos/2020_01. Presentación estándar Districlima (CAST).pdf. Accessed 3 Mar 2021

Domingo JL, Rovira J, Nadal M, Schuhmacher M (2017) High cancer risks by exposure to PCDD/Fs in the neighborhood of an integrated waste management facility. Sci Total Environ 607–608:63–68. https://doi.org/10.1016/j.scitotenv.2017.06.272

Edwards J, Burn S, Crossin E, Othman M (2018) Life cycle costing of municipal food waste management systems: the effect of environmental externalities and transfer costs using local government case studies. Resour Conserv Recycl 138:118–129. https://doi.org/10.1016/j.resconrec.2018.06.018

European Commission (2000) A study on the economic valuation of environmental externalities from landfill disposal and incineration of waste. Final Main Report

European Commission (2015) Proposal for a directive of the european parliament and of the council amending directive 2008/98/EC on waste

European Commission (2020) EU energy in figures 2020. In: Energy Stat. Pocketb. https://ec.europa.eu/energy/en/data/energy-statistical-pocketbook. Accessed 10 Mar 2021

European Parliament (2008) Directive 2008/98/EC of the European parliament and the council of 19 November 2008 on waste and repealing centain directives. In: Official Journal of the European Union. http://ec.europa.eu/environment/waste/framework/. Accessed 15 Apr 2021

Faura-Casas Auditors Consultors (2019) Comptes anuals 2019. https://www.tersa.cat/media/2857/comptes-anuals-2019.pdf. Accessed 14 Mar 2021

García-Pérez J, Fernández-Navarro P, Castelló A et al (2013) Cancer mortality in towns in the vicinity of incinerators and installations for the recovery or disposal of hazardous waste. Environ Int 51:31–44. https://doi.org/10.1016/j.envint.2012.10.003

Generalitat de Catalunya (2020) Calculadora de emisiones de una actividad 2019. In: Cambio climático. https://canviclimatic.gencat.cat/es/actua/calculadora_demissions/. Accessed 23 Apr 2021

Gil Fisa A, Pujol Senovilla L (2009) Metodología para la evaluación económica de los accidentes de trabajo. In: tituto Nac. Segur. e Hig. en el Trab. http://www.oect.es/InshtWeb/Contenidos/Documentacion/TextosOnline/Rev_INSHT/2000/7/artFondoTextCompl.pdf. Accessed 15 Apr 2020

Gracia A, Barreiro-Hurlé J, Pérez y Pérez L (2012) Can renewable energy be financed with higher electricity prices? Evidence from a Spanish region. Energy Policy 50:784–794. https://doi.org/10.1016/j.enpol.2012.08.028

Guo X, Liu H, Mao X et al (2014) Willingness to pay for renewable electricity: a contingent valuation study in Beijing, China. Energy Policy 68:340–347. https://doi.org/10.1016/j.enpol.2013.11.032

Hanemann M, Labandeira X, Loureiro ML (2011) Climate change, energy and social preferences on policies: exploratory evidence for Spain. Clim Res 48:343–348. https://doi.org/10.3354/cr00994

Hirshfeld S, Vesilind PA, Pas EI (1992) Assessing the true cost of landfills. Waste Manag Res 10:471–484. https://doi.org/10.1177/0734242X9201000602

Ibrahim MA (2020) Modeling of risk for improper sorting of waste at recycling centers. Waste Manag 102:550–560. https://doi.org/10.1016/j.wasman.2019.11.017

Idealista (2021) Histórico de precios de venta en Sant Adriá de Besós. In: Histórico. https://www.idealista.com/sala-de-prensa/informes-precio-vivienda/venta/cataluna/barcelona-provincia/sant-adria-de-besos/historico/. Accessed 26 Apr 2021

INE (2020) España en Cifras 2020. In: Inst. Nac. Estadística. https://www.ine.es/ss/Satellite?L=es_ES&c=INEPublicacion_C&cid=1259924856416&p=1254735110672&pagename=ProductosYServicios%2FPYSLayout&param1=PYSDetalleGratuitas. Accessed 4 Apr 2021

INE (2021) Datos. In: Cantidad Total y cantidades medias Consum. Aliment. bebidas, tabaco, Combust. y otras fuentes energía(25168). https://www.ine.es/jaxiT3/Datos.htm?t=25168. Accessed 23 Apr 2021

Jamasb T, Nepal R (2010) Issues and options in waste management: a social cost-benefit analysis of waste-to-energy in the UK. Resour Conserv Recycl 54:1341–1352. https://doi.org/10.1016/j.resconrec.2010.05.004

Kelsey JL, Gold EB (2017) Observational epidemiology. In: International encyclopedia of public health, pp 295–307. Elsevier

Lim SY, Lim KM, Yoo SH (2014) External benefits of waste-to-energy in Korea: a choice experiment study. Renew Sustain Energy Rev 34:588–595. https://doi.org/10.1016/j.rser.2014.03.045

Medina-Mijangos R, Seguí-Amórtegui L (2021) Technical-economic analysis of a municipal solid waste energy recovery facility in Spain: a case study. Waste Manag 119:254–266. https://doi.org/10.1016/j.wasman.2020.09.035

Medina-Mijangos R, De Andrés A, Guerrero-Garcia-Rojas H, Seguí-Amórtegui L (2021b) A methodology for the technical-economic analysis of municipal solid waste systems based on social cost-benefit analysis with a valuation of externalities. Environ Sci Pollut Res 28:18807–18825. https://doi.org/10.1007/s11356-020-09606-2

Medina-Mijangos R, Ajour El Zein S, Guerrero-García-Rojas H, Seguí-Amórtegui L (2021a) The economic assessment of the environmental and social impacts generated by a light packaging

and bulky waste sorting and treatment facility in Spain: a circular economy example. Environ Sci Eur 33. https://doi.org/10.1186/s12302-021-00519-6
Mensah J (2019) Sustainable development: meaning, history, principles, pillars, and implications for human action: literature review. Cogent Soc Sci 5. https://doi.org/10.1080/23311886.2019.1653531
Midilli A, Dincer I, Ay M (2006) Green energy strategies for sustainable development. Energy Policy 34:3623–3633. https://doi.org/10.1016/j.enpol.2005.08.003
Mital A, Pennathur A, Huston RL et al (1999) The need for worker training in advanced manufacturing technology (AMT) environments: a white paper. Int J Ind Ergon 24:173–184. https://doi.org/10.1016/S0169-8141(98)00024-9
MITERD (2018) Memoria anual de generación y gestión de residuos 2018. In: Minist. para la Transic. ecológica y el reto demográfico. https://www.miteco.gob.es/es/calidad-y-evaluacion-ambiental/publicaciones/memoriaresiduosmunicipales2018_tcm30-521965.pdf. Accessed 20 Apr 2021
Nahman A (2011) Pricing landfill externalities: emissions and disamenity costs in Cape Town, South Africa. Waste Manag 31:2046–2056. https://doi.org/10.1016/j.wasman.2011.05.015
Observatorio AECC (2021) Incidencias. https://app.powerbi.com/view?r=eyJrIjoiNDVmNjdlOTQtOTE5Zi00YmJmLWJhMjgtNTgxMGJkM2FjYTE3IiwidCI6ImJjYTNjYTJlLTYyNGMtNDNhYS05MTgxLWY2N2YxYzI3OTAyOSIsImMiOjh9. Accessed 26 Apr 2021
OECD (2018) Effective carbon rates 2018: pricing carbon emissions through taxes and emissions trading
OECD (2021) Data. In: Convers. rates - Exch. rates. https://data.oecd.org/conversion/exchange-rates.htm. Accessed 23 Apr 2021
Rabl A, Spadaro JV, Bachmann TM (2010) Estimating environmental health costs: monetary valuation of trace pollutants. Environnement, Risques & Santé 9:136–150. https://doi.org/10.1684/ers.2010.0338
Rivas Casado M, Serafini J, Glen J, Angus A (2017) Monetising the impacts of waste incinerators sited on brownfield land using the hedonic pricing method. Waste Manag 61:608–616. https://doi.org/10.1016/j.wasman.2016.10.036
Scarlat N, Fahl F, Dallemand JF (2019) Status and opportunities for energy recovery from municipal solid waste in Europe. Waste Biomass Valorization 10:2425–2444. https://doi.org/10.1007/s12649-018-0297-7
Soliño M, Vázquez MX, Prada A (2009) Social demand for electricity from forest biomass in Spain: does payment periodicity affect the willingness to pay? Energy Policy 37:531–540. https://doi.org/10.1016/j.enpol.2008.10.002
Soon JJ, Ahmad SA (2015) Willingly or grudgingly? A meta-analysis on the willingness-to-pay for renewable energy use. Renew Sustain Energy Rev 44:877–887. https://doi.org/10.1016/j.rser.2015.01.041
Sun C, Meng X, Peng S (2017) Effects of waste-to-energy plants on china's urbanization: evidence from a hedonic price analysis in shenzhen. Sustain 9:1–18. https://doi.org/10.3390/su9030475
TERSA (2020a) Declaració Ambiental 2019. In: Declar. Ambient. https://www.tersa.cat/media/2941/declaració-ambiental-pve.pdf. Accessed 16 Mar 2021
TERSA (2020b) Memoria de sostenibilidad 2019. https://www.tersa.cat/media/2847/tersa_2019_memoria_online_es.pdf. Accessed 15 Mar 2021
Vlachokostas C, Achillas C, Michailidou AV et al (2020) Externalities of energy sources: the operation of a municipal solid waste-to-energy incineration facility in the greater Thessaloniki area, Greece. Waste Manag 113:351–358. https://doi.org/10.1016/j.wasman.2020.06.015
World Health Organization (2010) Exposure to dioxins and dioxins-like substances: a major public health concern.

World Bank Group (2021) Data. In: Inflation, Consum. prices (annual %). https://data.worldbank.org/indicator/FP.CPI.TOTL.ZG?end=2019&locations=US&start=2013. Accessed 23 Apr 2021
Wyman O (2020) El impacto económico y social del cáncer en España. In: Asoc. Española contra el Cáncer. https://www.aecc.es/sites/default/files/content-file/Informe-Los-costes-cancer.pdf. Accessed 1 Apr 2021

Chapter 19
Case Study in Arid and Semi-arid Regions

Yassir Makkawi, Fatemeh Hassan Pour, and Omar Moussa

Abstract This chapter aims to discuss the potential and future prospective of waste-to-energy in arid and semi-arid regions. The main focus will be on the conversion of waste biomass and organic matter to biofuel and biochar through thermochemical and biological processes. The chapter starts by identifying the various types of waste biomass sources that are particularly pertaining to these regions. This will be followed by providing the most recent data on the quantities of waste biomass in some identified worldwide arid and semi-arid regions. Systematic assessment of the biomass and organic matter characteristics (physical, chemical, and thermal) will be presented to evaluate their potential for biofuels and biochar production. Apart from the great potential of biofuels as renewable energy sources, this chapter demonstrates the environmental benefit of biochar in countering land degradation, improving soil fertility, besides highlighting the potential of the by-product water in stretching the limited water resources to support the growth of plants and animal life in arid and semi-arid regions. The chapter concludes by stating the main challenges and recommendations for sustainable bioenergy technologies in arid and semi-arid regions.

Keywords Waste-to-energy · Arid and semi-arid regions · Biochar · Biofuel · Thermochemical conversion · Biological conversion

19.1 Introduction

Organic waste management is one of the main challenges to authorities in urban and rural areas. Biomass waste is considered a major feedstock for the sustainable production of energy fuels and bio-products through bioenergy systems. Waste-to-energy (WtE), which includes various forms of bioenergy systems, offers an ideal solution and is nowadays fast developing worldwide. Unfortunately, arid and semi-arid regions are still behind in the application of modern WtE, mainly hindered by the limited alternative sources of organic matter and the lack of understanding of

Y. Makkawi (✉) · F. H. Pour · O. Moussa
Bioenergy and Solar Conversion Research Group (BSCRG), College of Engineering, American University of Sharjah, P.O. Box 26666, Sharjah, United Arab Emirates
e-mail: ymakkawi@aus.edu

A. E.-F. Abomohra et al. (eds.), *Waste-to-Energy*,
https://doi.org/10.1007/978-3-030-91570-4_19

the potential of WtE in these regions. The first and second generations of bioenergy systems started in the developing countries depending on organic sources coming from agricultural waste, forestry residues, and energy crops. In most arid and semi-arid regions, green waste is not available in sufficient quantities due to the low rainfall and the associated problems of poor soil, desertification, and lack of suitable environment for plant and animal life. Figure 19.1 shows the global map of arid and semi-arid regions, identified based on the amount of rainfall. The regions receiving less than 25 cm of rain precipitation per year are classified as arid, while that of precipitation within the range of 25–50 cm rain precipitation per year is classified as a semi-arid region (Karmaoui 2019). The development of WtE technology in such regions, such as the MENA region and part of the west coast of the USA, necessarily requires the identification of suitable quantities of locally available feedstock.

The Middle East, which represents more than 75% of the arid and semi-arid land in the world, has traditionally relied heavily on fossil fuels to satisfy its energy needs. The majority of the waste is disposed of by landfilling, which is causing serious environmental concerns due to the air pollution and contamination of the groundwater by seeping leachate. The diminishing resources of fossil fuels and the negative impact of waste disposal by landfilling encouraged the shift towards an alternative solution using locally available organic waste.

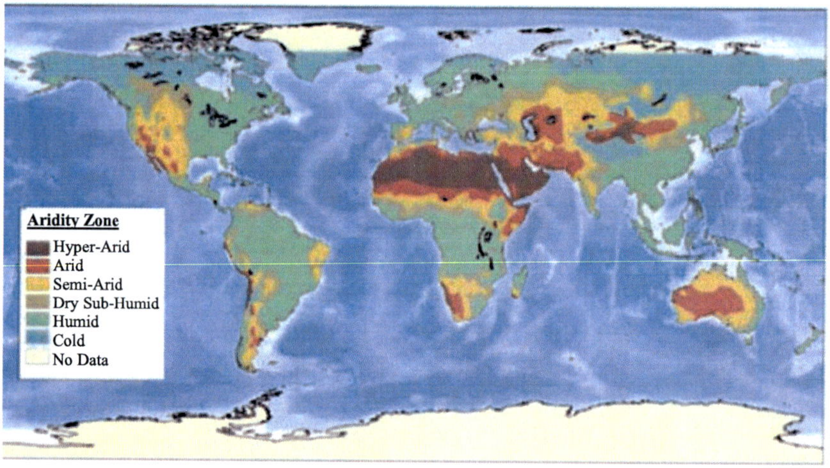

Fig. 19.1 Map of the world distribution of arid and semi-arid regions (Berhane 2015)

19.2 Waste Feedstock

19.2.1 Food Loss

Food loss in the literature is defined in multiple ways, for ease of understanding, it can be broadly categorized under two types: (i) pre-consumption, where the waste is produced in the farmland, during manufacturing/processing, storage, and supply chain (ii) post-production, where the waste mainly comes from food leftovers in restaurants and households. Food waste is found in abundant quantiles in almost every large city around the world. It is particularly pertaining to arid and semi-arid regions as it constitutes a unique source of organic waste in the absence of or limited plantation waste in such regions. One of the primary reasons behind the generation of post-production food waste is the lack of proper planning in food management, as various sectors (e.g. manufacturers, hospitality, lodging, travel and tourism, recreation, and wholesalers) try to satisfy the consumers' requirements of high quality and quantity supplies (Thi et al. 2015). Post-production in the household and hospitability sectors, especially in highly populated cities, buffets in hotels and restaurants are major contributors to food waste, because in many cases, the extra food cannot legally be reused nor donated without following strict health regulations (Pirani and Arafat 2016).

According to the United Nations Environment Programme (UNEP), 931 million tonnes of food sold to households, retailers, restaurants, and other food services were wasted in 2019. The Food and Agriculture Organization (FAO), has estimated that around 936 billion USD worth of food is lost annually, in addition to billions of dollars spent on transportation and proper disposal (Jouhara et al. 2018). The FAO has also stated that the global food loss is nearly one-third of the total food produced, amounting to 1.3 billion tonnes per year. Considering that 690 million people went hungry in 2019, such values are of high significance in order to have an idea about the generated food waste, Table 19.1 presents data of food waste generated in some countries falling under the category of arid or semi-arid regions. Currently, only a small fraction of food waste is converted to energy, as the majority is disposed of by landfilling. As a demonstration, Fig. 19.2 shows the percentage of the implemented management/disposal techniques for food waste from various sources in the USA [Adapted from EPA for the year 2018 in the USA (United States Environmental Protection Agency 2020)].

19.2.2 Sewage Sludge

Sewage sludge is a biosolid by-product in wastewater treatment plants (WWTPs). It is produced in abundant quantities in most of the major cities around the world. Figure 19.3 shows an overview of the various steps employed in a WWTP. Prior to the final discharge, the sludge usually goes through multiple stages of treatment in

Table 19.1 Household food waste in some arid and semi-arid countries (Forbes et al. 2021)

Country	tonnes/year	kg/capita/year
Saudi Arabia	3,594,080	105
Australia	2,563,110	102
UAE	923,675	95
Egypt	9,136,941	91
Iran	5,884,842	71
Namibia	229,344	92
Chad	1,637,656	103
Tunisia	1,064,407	91
Somalia	1,585,898	103
Ethiopia	10,327,236	92
Oman	470,322	95
Iraq	4,734,434	120
Jordan	939,897	93
Niger	2,393,877	103
Mali	2,018,765	103
Chile	1,401,043	74
Morocco	3,319,524	91
Libya	513,146	76
Qatar	267,739	95

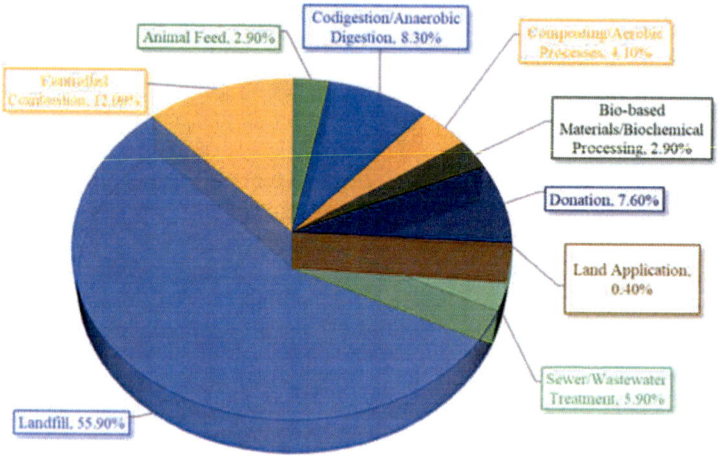

Fig. 19.2 Demonstration of food waste disposal methods in the USA. Adapted from the EPA report for the year 2018 (United States Environmental Protection Agency 2020)

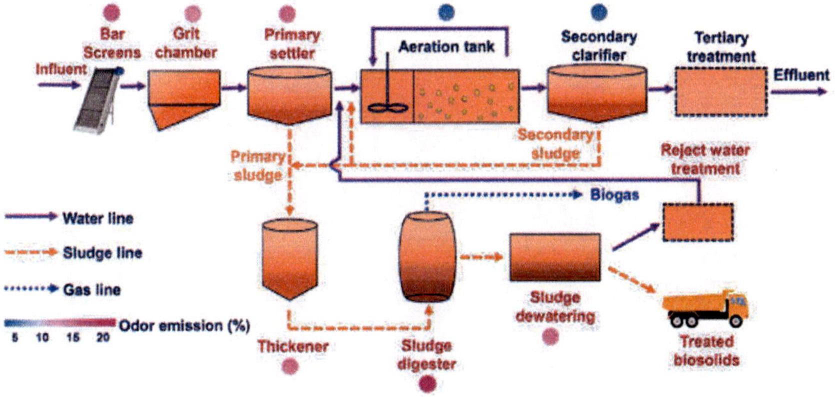

Fig. 19.3 Typical WWTP process sequence (Fan et al. 2020)

order to stabilize it and increase the solid content. The sludge may be collected after thickening/dewatering, or it may be further treated via anaerobic digestion and then dried. Therefore, the characteristics of the sewage sludge in any WWTP may vary depending on which treatment stage they are collected (Kim and Parker 2008).

In arid and semi-arid regions, sewage sludge produced in wastewater treatment plants (WWTP) is one of the main sources of organic waste. Because its' quantity depends on the amount of wastewater treated, it is natural that larger cities with a higher population and numbers of WWTPs will produce a larger amount of sludge than others. For example, the total dried sludge production in the United Arab Emirates, which falls under the category of arid region, is approximately 167,000 tonnes/year with nearly 50% of this produced in the main cities of Abu Dhabi and Dubai (FCSA UAE 2019). In North Africa, Egypt produces 2.1 million tonnes/year of dried sewage sludge, with the capital Cairo contributing more than 400,000 tonnes/year (Abdel Wahaab et al. 2020). Comparatively, large cities produce higher quantities of sewage sludge because they are expected to have a higher population. For example, the highly populated Shandong province in China, which is classified as cold semi-arid, produces 3.86 million tonnes/year. Which is almost double the entire sewage sludge production in Egypt. At least three other provinces in China produce a similar amount (Wei et al. 2020). Table 19.2 shows the recent statistics on the amount of sewage sludge produced in some major arid and semi-arid countries.

The amount of sludge production is expected to grow with an increasing world population; therefore, it is crucial that such waste is properly disposed of or recycled. In the UAE, nearly 80% of the produced sludge in 2019 was dumped into landfills, although the emirates of Dubai and Sharjah recycle ~40% and 56%, respectively, of the sludge as a solid biofuel or for composting (FCSA UAE 2019).

Table 19.2 Quantiles of sewage sludge in some arid and semi-arid countries

Country	Sewage sludge (tonnes/year)	Source
Australia	225,000	Pickin et al. (2020)
UAE	167,000	FCSA UAE (2019)
Egypt	2,100,000	Abdel Wahaab et al. (2020)
Morocco	106,000	Ghacha et al. (2020)
USA (California)	665,000	Kester (2017)
Chile (Santiago)	300,000	(United Nations Framework Convention on Climate Change, n.d.)

19.2.3 Halophytes

The long-term technical and economic sustainability of waste-to-energy technology essentially requires the availability of a good quantity and quality of organic waste. In arid regions, especially areas that extend along coastal deserts, the availability of waste plant biomass and organic residues is limited. Recently, halophytic crops have been reportedly studied for their adaptation to marginal areas or salt-affected land. *Salicornia bigelovii*, abbreviated by *S. bigelovii*, is one of the most interesting halophytic species experimentally tested in several arid regions for multiple potential uses. The *S. bigelovii* seeds have a high oil content (~30%), making it a promising source of vegetable oil. There are also several studies on the utilization of *Salicornia* for both edible and non-edible purposes. The leaf tips of the plant can be consumed as human food either fresh or as a pickled vegetable. The seedcake can be used as animal feed due to its high protein content (~45%). In addition, the green biomass stage of the plant can also be consumed as livestock forage. Figure 19.4 shows photos of the *Salicornia* plant in the plantation field and after cultivation and drying. Field studies have shown that *S. bigelovii* irrigated with high saline water (24 ppt) can achieve high biomass yields reaching approximately 3.5 kg m^{-2}, whereas the seed yield is approximately 80 g m^{-2}. Therefore, it is considered an attractive feedstock, especially after extraction of the oil for the seed, whereas the remaining of the plant can serve as a biomass feed in waste-to-energy applications. Further details about this type of halophytes can be found in Makkawi et al. (2021).

19.2.4 Date Palm Waste

Date palm, scientifically known by the name *Phoenix dactylifera*, is widely cultivated in arid and semi-arid countries, especially in the Middle East and North Africa (MENA) regions and the low desert region of California in the USA. The date palm tree produces date fruit, which is considered to be one of the most important fruits in aid and semi arid countries within the MENA region (see Table 19.3). It is highly tolerant to heat and can grow in sandy soil and under various irrigation-water salinity.

Fig. 19.4 *Salicornia bigelovii* **a** plant after drying in open air **b** and **c** plant grown in filed (Makkawi et al. 2021)

Currently, there are around 100–120 million date palm trees worldwide. In addition to the trunk, each tree has around 6–10 fruit bunches, 12−15 stems, and 120−240 leaves (Barreveld 1993; Nasser et al. 2016), collectively producing around 35 kg of waste during seasonal pruning and harvesting. Figure 19.5 shows a date palm tree demonstrating its anatomic parts that constitute the major waste.

In the MENA region, which shares around 70–90% of the worldwide date palm trees, the waste is burnt or left at landfill sites, hence causing a challenging disposal problem and environmental pollution (El-Juhany 2010). There are some efforts on the commercial utilization of the date palm waste as a source of wood or as additives in some industries such as cement and paper mills (Hegazy and Aref 2010; Nasser 2014), however, these efforts remain at the experimental level or have not yet been realized for large scale implementation. Similar to general woody biomass, the date palm waste is rich in lignin, cellulose, and hemicellulose, which makes it a very attractive feedstock in waste-to-energy applications, as discussed here (Table 19.3).

Fig. 19.5 Date palm tree and some of its anatomic parts. The arrows point to leaves, fruit bunch, and steams after cutting and drying, which constitute the major parts of the waste (Makkawi et al. 2019)

Table 19.3 Quantiles of date fruit produced in 2019 in some countries within the arid and semi-arid region (The Science Agriculture 2019)

Country	Date palm trees (tonnes)
Saudi Arabia (Gulf)	1,539,756
UAE (Gulf)	323,478
Egypt (N Africa)	1,603,762
Iran (Asia)	1,307,908
Tunisia (N Africa)	288,700
Sudan (Africa)	438,700
Oman (Gulf)	372,572
Iraq (Middle east)	639,315
Pakistan (Asia)	483,071
Algeria (N Africa)	1,136,025

19.2.5 Municipal Solid Waste (MSW)

Municipal solid waste (MSW) is a natural side product of urbanization, economic development, and population growth (Kaza et al. 2018). According to the Environmental Protection Agency (EPA), MSW encompasses significant portions of waste derived from organic matter, such as paper/paperboard, food, plastic, yard trimming, wood, textiles, glass, rubber/leather, in addition to some inorganic matters such as metals and glass. Same as with sewage sludge and food waste, MSW is available in large quantities in arid and semi-arid regions, and it increases with the increase in population and prosperity. The estimated total global MSW produced in 2018 was 292.4 million tonnes (*National Overview: Facts and Figures on Materials, Wastes and Recycling. Facts and Figures about Materials, Waste and Recycling. US EPA* n.d.) and this is expected to rise to 3.40 billion tonnes by 2050 (Zhang et al. 2021b). The most common current practices of MSW management are through landfilling, recycling, composting and combustion for energy recovery (*National Overview: Facts and Figures on Materials, Wastes and Recycling. Facts and Figures about Materials, Waste and Recycling. US EPA* n.d.). Table 19.4 presents some statistical data on each of these methods for the year 2018. Each of the MSW disposal methods

Table 19.4 Global data on MSW management and disposal

Disposal method	Amount (million tonnes)	Component
Recycle	69.1	Paper/Paperboard: 66.54% Metal: 12.62% Rubber/Leather/Textiles: 6.05% Wood: 4.49% Plastic: 4.47% Glass: 4.43% Other: 1.4%
Compost	25	Yard trimming: 22.3 million tonnes Food waste: 2.6 million tonnes
Combustion	34.6	Food: 21.85% Plastics: 16.27% Paper/Paperboard: 12.16% Textile: 9.32% Metals: 8.54% Wood: 8.22% Yard trimming: 7.44% Rubber/Leather: 7.24% Glass: 4.75% Misc. inorganic waste: 2.32% Other: 1.91%
Landfill	146.1	Food: 24.14% Plastics: 18.46% Paper/Paperboard: 11.78% Metals: 9.53% Wood: 8.32% Textile: 7.73% Yard trimming: 7.21% Glass: 5.17% Rubber/Leather: 3.42% Misc. inorganic waste: 2.24% Other: 2.01%

comes with various environmental and health hazards. For example, plastics being persistent for long periods are a global concern and may be digested by the organism, and hence creating health hazards within the food chain (Chen et al. 2020). In addition, it was estimated that in 2016, the fugitive emissions from MSW treatment and disposal generated 1.6 billion tonnes of CO_2 equivalent greenhouse gases (5% of global emissions) (The World Bank 2018). Nitrogen pollution in the MSW leachate is another risk associated with MSW causing disease and nutrient imbalances in the surrounding water bodies. The combustion of MSW can imply a risk for human health as it can release a significant amount of air pollutants (Chen et al. 2020). Interestingly, MSW can adversely affect economic development by diminishing tourism (Kaza et al. 2018).

A critical challenge in MSW management is the waste collection step which has varying rates according to the income level. Upper-middle and high-income countries provide universal waste collection (Kaza et al. 2018). For example, Europe and Central Asia, and North America manage to collect 90% of their waste. However, low-income countries have 48% of their waste collected within the cities and 26% outside of urban areas. Sub-Saharan Africa collects only about 44% of waste streams. Another issue is related to the composition of the MSW: High-income countries tend to generate more dry waste (plastic, paper, cardboard, metal, and glass) that has a higher potential for recycling compared to food and green waste (32%). In low-income countries, recyclable waste accounts for only 16% of the waste (Kaza et al. 2018).

Just like food waste, the available data on MSW generation suffer from inconsistencies in definitions, data collection methodologies, and availability, and their reliability is significantly influenced by undefined words or phrases; incomplete or inconsistent definitions; lack of dates, methodologies, or original sources; inconsistent or omitted units; and estimates based on assumptions (Kaza et al. 2018). Table 19.5 presents data on MSW generated per capita in some countries falling within the arid and semi-arid regions. It is worth noting that, the UAE, which has a limited population of around 7 million, comes on top of the list for food waste production per capita, followed by Australia and Saudi Arabia. Countries of low Gross National Income (GNI), such as Sudan and Ethiopia appear to be at the bottom of the list. This suggests that the potential of food waste as a feedstock in WtE is relatively low in developing counties, particularly that of low GNI.

19.2.6 *Other Synthetic and Industrial Waste*

19.2.6.1 Waste Tires and Plastics

The production of tires is an industry that has been consistently growing for the past few decades. It is reported that the number of waste tires produced globally is projected to reach 500 million units by 2030 (Buss et al. 2019). China is the largest producer of waste tires (40.2% by mass), followed by the US (13.7% by mass). In

Table 19.5 MSW in countries falling within the arid and semi-arid regions (Kaza et al. 2018)

Country	Waste generation rates (kg/capita/day)
Paraguay	0.76
United Arab Emirates	1.6
Iran	0.6
Tunisia	0.66
Australia	1.54
Saudi Arabia	1.4
Egypt	0.67
Pakistan	0.43
Mali	0.34
Mauritania	0.36
Morocco	0.55
Sudan	0.2
Niger	0.57
Ethiopia	0.18
India	0.57

arid and semi-arid regions, waste tires present an excellent opportunity to increase sources of organic waste for energy conversion.

The introduction of waste management directives that recommend the use of post-treatment waste tires as a fuel has shown great potential in reducing the number of waste tires going to landfills (Grammelis et al. 2021). Standard car tires consist mostly of vulcanized rubber, which is supplemented with steel and fiber for strength/support (Landi et al. 2016). Since different types of tires may consist of different types and amounts of rubber, along with the aforementioned additives, it can be very difficult to establish and implement a wide-scale solution for tire recycling. The practice of landfilling the waste tires is completely unsustainable and burning them carries a great risk of pollutant release. One way of reusing them is through the conversion to polymer rubber composites (Sienkiewicz et al. 2017). Tires may also be used to produce biochar, bio-oil, and gas through pyrolysis. The bio-oil and gas can be used for energy generation as they have reportedly high heating values close to conventional fuels (up to 40 MJ/Nm^3), whereas the char may be used as activated carbon or as a soil conditioner (Grammelis et al. 2021).

19.2.6.2 Waste Plastics

Plastic is another type of waste that exists in large quantities in major cities around the world. It may consist of various objects made of polystyrene and polyethylene polymers, such as beverage bottles, grocery bags, wrapping films, plastic cups, and labeling materials. It is currently causing serious environmental concern especially when it reaches the oceans in countries with extended coastal areas. Organic waste that is causing serious concern in recent years as it correlates to the amount of plastics produced is also concerning as the statistics show that it has reached 368 million metric tonnes in 2019 and is predicted to reach 12 billion metric tons by 2050 (Tiseo, 2020). For example, plastic waste in the UAE accounts for 20% of the total produced waste (Belpoliti et al. 2018). As a result, the research on plastic waste management and recycling is now receiving increasing attention. Most obviously, it is possible to reuse some discarded plastic waste several times (e.g., from packaging) before moving on to more complicated methods such as mechanical recycling, chemical recycling, or energy generation. In mechanical recycling, the use is limited to plastics made from one polymer, and even then, it is not possible to make products of the same quality as those produced from first-use plastics. In chemical recycling, the polymers in the waste plastics may be broken down to be used in the production of other plastics (Ncube et al. 2021). Lastly, in energy generation, such as thermochemical conversion through catalytic pyrolysis and gasification can be incredibly useful to produce biofuels (Al-Salem et al. 2017). Pyrolysis is generally more beneficial as it avoids the possibility of release of dioxins observed in gasification (Chen et al. 2014).

19.3 Waste Characteristics

19.3.1 Physical Characteristics

In waste-to-energy applications, the physical characteristics of the feedstock dictate the extent of pretreatment required and may strongly affect the quality of the end product. For example, studies have shown that organic waste of small particle size undergoing a thermal conversion process is likely to produce a solid residue (biochar) of a smaller size and lower density (Zaman et al. 2017). In arid and semi-arid regions, especially deserts and sandy soil, such small size biochar is highly desirable for enhancing the soil properties through biological processes and increasing the water retention due to increasing the particle nutrients content and pore space (de Jesus Duarte et al. 2019).

The particle size of the feedstock also plays an important role in waste-to-energy through thermal conversion, especially in terms of heat transfer. In fluidized bed reactors, which are the most commonly used, it is recommended to use small particle sizes to ensure a rapid heat transfer rate because small particles have lower resistance to thermal degradation (Akhtar and Saidina Amin 2012). Biomass pyrolysis in a bubbling fluidized bed with small particles will lead to an increased yield of biofuel, whereas a larger particle size will increase the amount of biochar produced (Islam et al. 1999). While in downdraft reactors, auger reactors, and most of the reactors used in the chemical and biological waste conversion, the process is relatively less sensitive to the particle size and can handle a highly heterogeneous mixture of a wide range of particle sizes and shapes. The sewage sludge and horse manure are most likely to be reasonably uniform in shape and size, hence, they can be used directly in most waste-to-energy technologies without the requirement of size reduction. Other organic matters, such as food waste, MSW, car tires, plastic waste, date palm waste, and *Salicornia* plants must be subjected to size reduction to satisfy the requirement in most energy conversion reactors and guarantee a good level of uniformity before processing.

Studies have shown that the solid by-product from a feedstock undergoing pyrolysis would produce lower density biochar. Askeland et al. (2019) reported that light biochar added to the soil would cause ramification by raising the soil aeration and porosity and improving microbial respiration. Lanh et al. (2019) also reported a direct relationship between the water retention capacity of the biochar and the improvement in the water retention capacity of the soil. Nonetheless, as stated by Allaire et al. (2015) lower bulk density can reduce the bagging and transportation cost. Water retention capacity and surface area are other further factors related to the biochar that can determine its application in soil amendment. It is worth noting that, as shown in Fig. 19.6, the sewage sludge is the highest in density, most likely due to its high ash content (minerals), as will be demonstrated later.

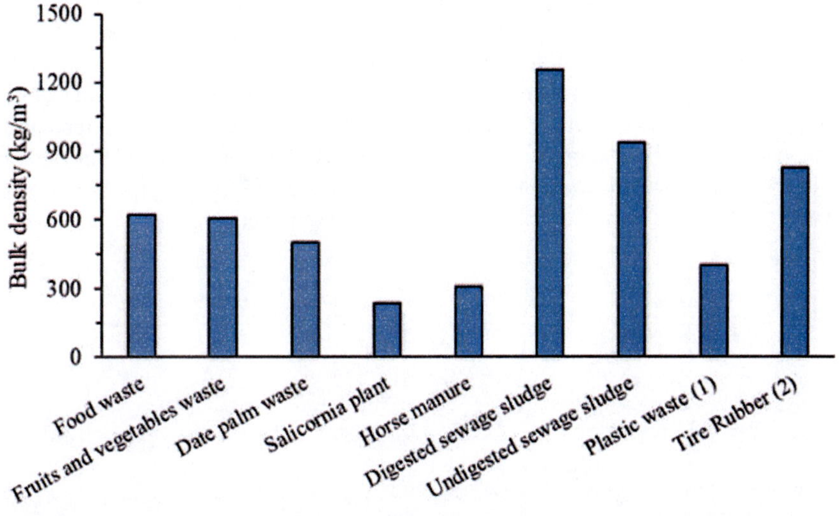

Fig. 19.6 Average bulk density of organic waste of relevance to arid and semi-arid regions. (1) Al-Salem et al. (2020), (2) Bekhiti et al. (2014)

19.3.2 Chemical Characteristics

The chemical characteristics of a feedstock in waste-to-energy processes are commonly assessed through proximate and ultimate analysis. The former analysis provides the feedstock composition in terms of moisture, volatile, fixed carbon, and ash, while the latter analysis provides a composition in terms of carbon, hydrogen, nitrogen, oxygen, and sulfur. The amount of lipids, cellulose, hemicellulose, starch, lignin, and protein are also commonly reported as part of the chemical characteristics as these represent the major constituents of the biomass volatile.

Figure 19.7 shows the proximate analysis of some organic waste of high potential for waste-to-energy in arid and semi-arid regions. This includes sewage sludge, date palm, and *Salicornia* from the UAE, plastic waste from Kuwait, and tire rubber waste from Palestine. Clearly, sewage sludge has the highest ash content, mainly due to the high content of heavy metals and minerals. The ash may have a negative or positive impact depending on the energy conversion method. It may negatively impact the energy production in a boiler via the formation of undesirable slag (Uddin et al. 2018) or reduce the bio-oil yield in pyrolysis via the catalytic cracking of the heavy hydrocarbons in the condensable vapors (Khanmohammadi et al. 2015). On the other hand, the ash may positively impact gasification by reducing the undesirable tar through the catalytic cracking of heavy hydrocarbons. In soil enhancement, using a biosolid residue of high ash content may add nutrients such as potassium (K) and phosphorus (P), which are vital for the growth and reproduction of plants. However, in the long run, the ash may increase the metal concentration beyond the permissible limits in soil. This is especially critical when it comes to elements such as mercury, zinc, and copper. For example, the concentrations of the aforementioned metals in sewage sludge biochar were found to be higher by factors of 2, 9, and 13, respectively, than the maximum soil pollutant limits (Tomasi Morgano et al. 2018).

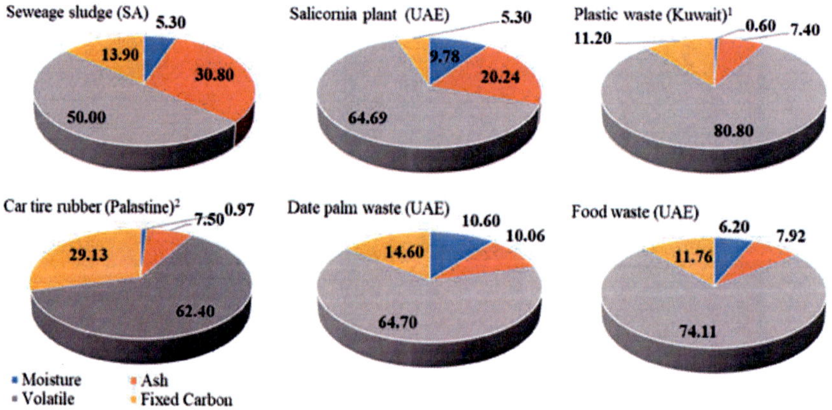

Fig. 19.7 Proximate analysis of organic waste of potential use for waste-to-energy technology. (1) Al-Salem et al. (2020), (2) Abdallah et al. (2020)

Plastic and food waste have the highest amounts of volatiles. This is one of the important characteristics for the production of biofuel through waste-to-energy. In food waste, these volatiles are mainly of lipids, cellulose, hemicellulose, starch, lignin, and protein origin. Compared to other organic or carbonaceous solid waste commonly used in bioenergy applications (e.g., woody biomass, agricultural and general municipal solid waste), food waste has a higher fraction of lipids (saturated fat) and protein. In addition, studies have shown that food waste or its biochar is very rich in nutrients that may greatly benefit plant growth if applied as a soil conditioner (O'Connor et al. 2021). However, food waste as received from sources may contain an excessively high fraction of water (80% by mass on average), which requires energy-intensive pretreatment to reduce it to below 10%. Another problem associated with food waste is that it is prone to biological degradation if left in open air for a long time, thus making its storage challenging. In general, the high moisture content in any feedstock undergoing thermochemical conversion will adversely affect the energy content in the final project and increase the energy input during processing (Uddin et al. 2018). On the other hand, some energy conversion methods, such as hydrothermal, can process a feedstock of 75–90% moisture content (Lucian and Fiori 2017).

The *Salicornia* and date palm discussed here are classified as succulent plants, i.e., high ability to retain water, hence, can survive in harsh environments with minimum water as in the case of most arid and semi-arid regions. Therefore, they are similar in proximate analysis to some extent, except that the *Salicornia* is higher in ash because it was irrigated by seawater of full salinity, while the date palm was irrigated by groundwater. The minerals coming with the high salinity water usually end up in the plant in the form of ash.

Figure 19.8 shows the ultimate analysis of some organic waste that has potential for waste-to-energy in arid and semi-arid regions. Noticeably, the carbon and hydrogen content in the car tire rubber and the food waste are the highest, with the formed exceptionally high in carbon. This indicates that these two waste materials are high in calorific value, hence better energy recovery potential through thermochemical treatment (Opatokun et al. 2016). Oxygen is the second-highest component in most organic matter. The effect of oxygen content is debatable, as it is reported to increase the liquid yields in the pyrolysis process (Grycová et al. 2016) but at the same time, it may negatively impact the bio-oil quality by increasing the oxygenated compounds, hence lowering the bio-oil stability (Bridgwater, 2003). The sulfur mass content is usually very low (<1%), so these feedstocks have a limited corrosion effect in the storage and processing units. Some studies have also reported the chlorine content as a part of the ultimate analysis because it is expected to have relatively high food waste compared to woody biomass. A study by Caton et al. (2010) reported the mass content of chlorine in food waste to be in the range of 0.9–1.4%, which is slightly higher than sulfur. Car tire rubber has a relatively lower chlorine content in the range of 0.05%. In general, chlorine is usually ignored in the ultimate although it may have a dangerous impact on the environment if released into the open atmosphere in large quantities (Guo et al. 2021).

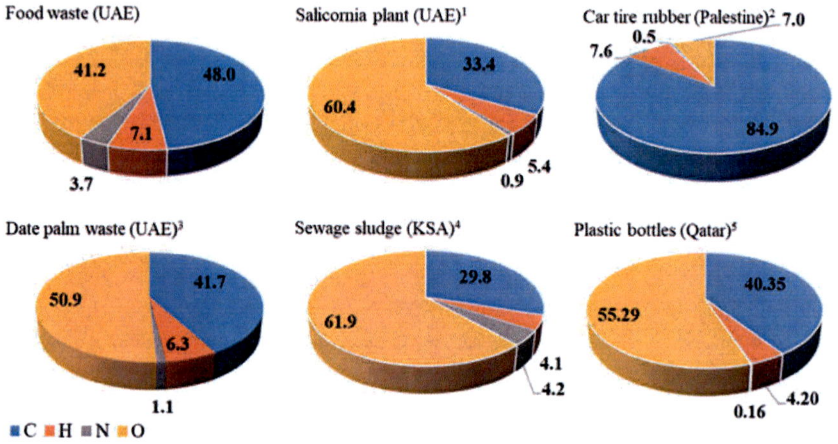

Fig. 19.8 Ultimate analysis of some waste biomass materials in arid and semi-arid regions. (1) Makkawi et al. (2021), (2) Abdallah et al. (2020), (3) Makkawi et al. (2019), (4) Abdel-Aziz et al. (2017), (5) Baawain et al. (2017)

The car tire rubber and plastic waste both are the lowest in terms of the H/C molar ratio. Several reported studies have suggested a strong relationship between the H/C ratio and the biosolid stability, such that the lower the H/C the stronger the stability of the organic carbon (Makkawi et al. 2021). This makes the car tires and plastic bottles and their by-product biosolid the best for long-term carbon sequestration and enhancing the life and growth of plants through soil amendment.

19.3.3 Thermal Characteristics

The potential use of a solid biofuel as a feedstock depends on its thermal characteristics. It is commonly reported in terms of the feedstock higher heating value (HHV) and thermal degradation assessed by Thermogravimetric Analysis (TGA).

The range of HHV varies considerably, mainly depending on the amount of carbon and other combustible material in the feedstock. Figure 19.9 shows the HHV of samples of feedstock available in arid and semi-arid regions. The fossil-based waste, e.g., car tire rubber and plastic bottles, are naturally high in HHV, while the sewage sludge and *Salicornia* plant have the lowest HHV values (12.5 MJ/kg and 13.3 MJ/kg, respectively). This is a consequence of the significantly high ash content found in both feedstocks, as shown earlier in Fig. 19.7. The ash content also negatively affects the yield of bio-oil in pyrolysis by catalytically cracking the pyrolysis gas through secondary reactions (Arazo et al. 2017). On the positive side, the ash may also act as a catalyst for enhancing hydrogen production through the water–gas shift reaction, which in turn, greatly increases the HHV of the non-condensable gases (Gómez et al. 2018). Although primary sludge is known for its incredibly high moisture content

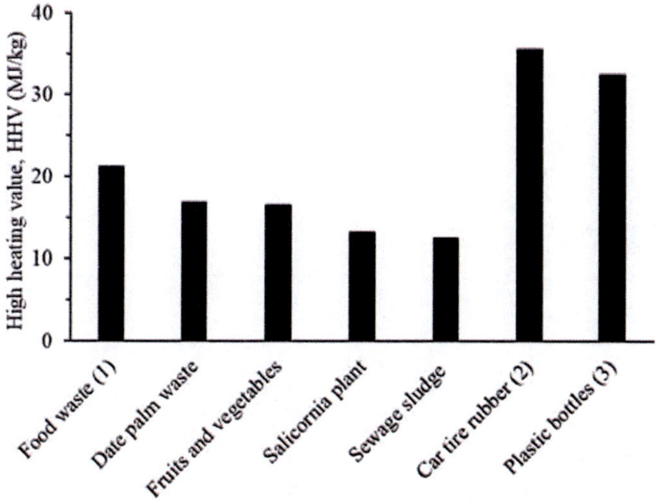

Fig. 19.9 High heating value of some organic waste relevant to arid and semi-arid regions. (1) Abdelaal et al. (2019), (2) Abdallah et al. (2020), (3) Baawain et al. (2017)

(Mortula et al. 2016), the sludge must be dewatered and then dried to be reduced to less than 10 wt% moisture in order to produce biofuels efficiently (Zaker et al. 2019). Lastly, it is observed that the high ash content comes at the expense of decreasing the volatile fraction in the sludge, which is another reason for the observed low HHV in the sewage sludge (Fonts et al. 2009). The food waste, which comes third in the hierarchy of HHV, is consistently reported in the literature within the range of 17.0–25.0 MJ/kg (Corder, 1976). The high HHV in food waste is usually associated with high fractions of meat and dairy products in the sample (e.g., HHV of meat > 25.2 MJ/kg (Jo et al. 2017)). Having biomass with a high heating value is an advantage for waste-to-energy application as it is most likely to result in the improvement of energy output.

Figure 19.10 shows the Thermogravimetric Analysis (TGA) profiles demonstrating the thermal degradation behavior of sewage sludge, date palm, and *Salicornia* plant as reported in the literature. The TGA data presented here are produced by the gradual increase of temperature in an inert environment, thus, to some extent, mimicking the thermal degradation behavior in waste-to-energy through pyrolysis. The initial stage of mass loss is attributed to the release of moisture from the sample, the second stage of mass loss, which is shown to occur roughly within the range of 250–500 °C is attributed to the major loss of volatiles in the samples. In the last stage, the mass loss occurs at a slower rate signifying the cease of devolatilization process. In waste-to-energy application, such information is important to identify the recommended range of operating temperature for thermochemical conversion of organic waste. For example, the date palm appears to complete the release of major volatiles at the temperature ~450 °C, while the sewage sludge requires a higher temperature of around 500 °C for the same degree of degradation. The initiation of the plastic

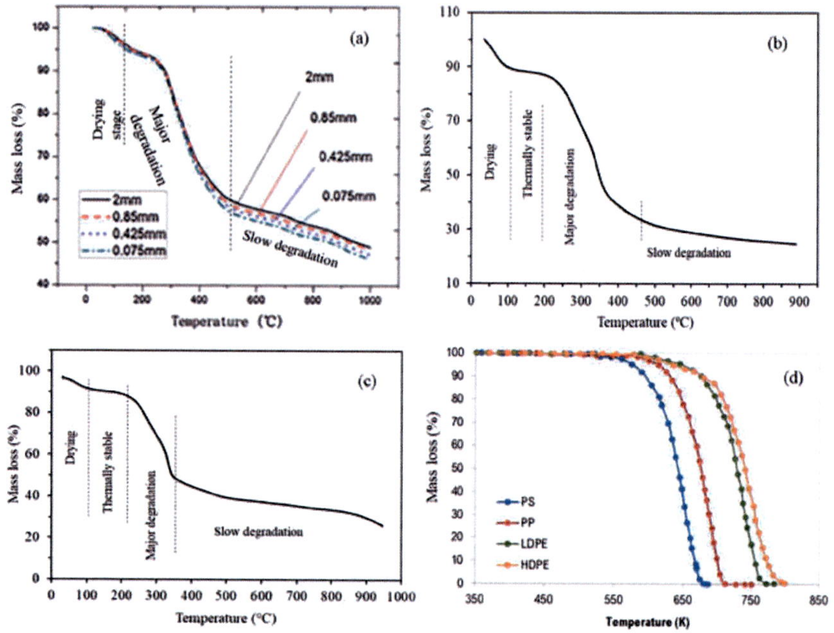

Fig. 19.10 Thermogravimetric profiles of **a** sewage sludge at different particle sizes (Dai et al. 2015) **b** and **c** date palm waste and *Salicornia* plant from the UAE (Makkawi et al. 2019) and (Makkawi et al. 2021), respectively **d** various types of plastic waste polymers (Dubdub and Al-Yaari 2020)

waste thermal degradation occurs at a relatively much higher temperature and the major devolatilization may extend up to 800 °C.

19.4 Waste-to-Energy Technologies

19.4.1 Thermochemical Conversion

19.4.1.1 Incineration

Incineration is the most prevalent waste-to-energy technology targeting 255 million tonnes of waste per year worldwide (Escamilla-García et al. 2020). It is also considered the main competitor to landfilling regarding environmental burden and economic criteria (Vlachokostas et al. 2020). In incineration plants, the waste is burnt in the presence of oxygen at a temperature of at least 850 °C for 2 s generating, ash, flue gas, and heat. Small particles carried over in the flue gas need to be separated in an air pollution control system making up to 3–4% of the waste input in the MSW incineration process. These solids are referred to as fly ash, which mainly contains mineral material and is below 1 mm in size (Huber et al. 2020). On the other hand, bottom ash, comprising 20–25% of the waste input, contains pieces of minerals, glass, and metals (Huber et al. 2020). For every ton of MSW incinerated, 15–40 kg of hazardous waste in the form of emissions of carbon oxides, sulfur oxides, particulates, heavy metals, and other pollutants are generated which necessitates strict secondary treatment solutions (Abbasi 2018). Flue gas emits a significant amount of dust, gaseous, and vaporous organic substances, NO_x, hydrogen chloride (HCl), hydrogen fluoride (HF), sulfur dioxide (SO_2), heavy metals, and dioxins/furans. Proper flue gas control techniques include mechanical and electrostatic filters for particle removal (e.g., fabric filters and electrostatic precipitators), scrubbers for acid gases neutralization through injection of bases (e.g., calcium-based systems, bicarbonate system), and dioxins/furans removal through injection of carbon-based adsorbers, reduction systems for NO_x abatement (ammonia-based selective catalytic and non-catalytic reduction systems) (Stabile et al. 2020). Incineration can help in reducing waste volume by almost 90% while avoiding water contamination, soil mitigation, and methane release. It can help with energy recovery for the ongoing energy demand. The amount of energy generated depends upon the type of waste and its calorific value to justify the amount of energy utilized (Escamilla-García et al. 2020).

19.4.1.2 Pyrolysis

Pyrolysis is a thermal process where waste/biomass is decomposed into bio-oil, biochar, and pyrolytic gas (none-condensable gases) in an inert environment and at a high temperature. Depending on the feedstock's residence type and heating rate, pyrolysis can be categorized as slow and fast pyrolysis. Slow pyrolysis is characterized by low operating temperature (300–700 °C), slow heating rate (10–200 °C/s), and high vapor residence time (0.5–10 s), and it mainly favors the production of biochar (Hu and Gholizadeh 2019). Biochar is the residual solid product of pyrolysis and has significant application pollution remediation, soil fertility improvement, and carbon sequestration (Tomczyk et al. 2020). Fast pyrolysis takes place with a high heating rate (103–104 °C/s), and short residence time for the vapors (less than 0.5 s) that allows for the production of up to 75 wt% of bio-oil (Hu and Gholizadeh 2019; Wang et al. 2017). The bio-oil produced has the potential to be used in boilers and furnaces, diesel engines, and turbines. It can also be utilized in the production of chemicals like resin precursors, additives in fertilizing and pharmaceutical industries, flavoring agents (e.g., glycolaldehyde) in food industries, acetic acid, hydroxy acetaldehyde, levoglucosan, levoglucosenone, and maltol. Nonetheless, if properly treated and upgraded via hydrotreating or catalytic vapor cracking, it can also be used as a transport fuel (Lakshman et al. 2021).

19.4.1.3 Gasification

Gasification is another thermochemical conversion technique in which biomass is converted to syngas at an elevated temperature (>700 °C). The syngas mainly consists of hydrogen, carbon monoxide, carbon dioxide, and methane. It has significant applications in gas turbine, proton fuel cells, petrochemical industry hydrogenation reaction in refinery applications, Fischer–Tropsch process, and polyolefin industry (Janajreh et al. 2021). Gasification also results in a particulate solid product, which mainly consists of minerals and metals with a small fraction of carbon. If high in carbon, it may be referred to as char and if high in metal and mineral content, it is usually classified as ash (Klinghoffer et al. 2011). Compared to other hydrogen production techniques such as fermentation, photosynthesis, water–gas shift reactions, and electrolytic hydrogen production, gasification has proven to be more efficient, with higher hydrogen yield and the feasibility of scale-up (Cao et al. 2020).

19.4.2 Biological and Chemical Conversion

19.4.2.1 Anaerobic Digestion

Anaerobic digestion (AD) is a well-established waste-to-energy technology. The process disintegrates and stabilizes the organic matter to energy-rich biogas through some naturally occurring microorganisms under oxygen-depleted conditions (Deepanraj et al. 2021; Munir et al. 2021). The main factors affecting the biogas production rate and quality are reactor design, feedstock characteristics, type of microorganism used, pretreatment process, carbon to nitrogen (C:N) ratio, organic loading rate, hydraulic retention time, and operating conditions (such as pH and temperature) (Deepanraj et al. 2021; Khan 2020; Zamri et al. 2021). Biogas, being the main product of the AD process, is 55–70% methane and 30–40% carbon dioxide and can generate heat and energy upon combustion (Zhang et al. 2021a). AD has been proven to be efficient when treating liquid waste and high Chemical Oxygen Demand (COD) content such as distillery and food industry wastewaters. However, when it comes to MSW and biodegradable solid waste such as leaf litter, weeds, vegetable, and fruit peels, food waste, there is an additional cost involved to ensure smooth feeding, digestion, and movement of the final digested product out of the reactor (Abbasi 2018). This difficulty in mass transport dictates a lot of pre-processing, pretreatment, and also post-digestion processing leading to higher costs for the whole operation. In addition, variations caused in the nature of the waste due to the time and location can also have adverse effects on the C/N ratio and hence the final biogas production rate (Abbasi 2018). On the other hand, organic waste such as food waste which is normally shown to have a moisture content higher than 50%, is not suitable for management methods that require sustainable self-combustion (Khan 2020). Besides the main gas product, the AD produces a digestate that can be added to soil to help with plant growth and disease resistance as a replacement of chemical fertilizers, hence, the process is particularly attractive for WtE in arid and semi-arid regions (Zamri et al. 2021).

19.4.2.2 Fermentation

Fermentation is a biochemical process for the production of bioethanol from organic matter. It mainly involves a pretreatment step to increase the digestibility of the waste, followed by hydrolysis, fermentation, and product separation/distillation (Balat 2011). Pretreatment of the organic matter is conventionally carried out by acid, alkali, thermal and enzymatic processes (Pham et al. 2015) and it is this step is responsible for the increased cost of the entire process (Matsakas et al. 2014). Typical difficulties related to bioconversion of waste to bioethanol are related to the resistance of the biomass to break down along with the variety in the types of sugar released that would cause a rise in the need to find or genetically engineer organisms to efficiently ferment these sugars (Balat 2011). Generally, bioethanol is produced from energy-rich crops and lignocellulosic biomass, however, feedstock with high carbon content such as food waste has also proven to be a good candidate for ethanol fermentation (Kim et al. 2011). Bioethanol can be used as a gasoline additive, gasoline subsistent, octane enhancer in the form of ETBE (ethyl tertiary butyl ether), and to reduce the emissions of exhaust gases in bioethanol–diesel blends (Balat 2011).

19.5 Case Studies

19.5.1 Pyrolysis of Sewage Sludge, Salicornia, and Date Palm

This section is focused on a waste-to-energy case study utilizing sewage sludge, *Salicornia*, and date palm waste. These feedstocks are available in abundant quantities in the UAE. In general, biomass can be converted to biofuel through pyrolysis to produce three forms of energy: bio-oil, fuel gas, and biochar. While the first two products are highly valuable as a source of clean and renewable energy, the biochar can be used as a solid biofuel in addition to a soil conditioner, thus extending the benefit of waste-to-energy technology to countering desertification that may exist in arid and semi-arid regions.

Figure 19.11 shows typical data on the pyrolysis yield from their waste feedstocks collected in the UAE. The conversion of the Sewage sludge and *Salicornia* was carried out in an auger reactor at the temperature of 550 °C, while that of the date palm was carried out in a bubbling fluidized bed reactor at 520 °C. Clearly, the bio-oil yield varies considerably due to a number of factors, including the variation in the reactor type, pyrolysis temperature, and feedstock characteristics. The latter is believed to have the dominant effect. The sewage sludge produces the highest bio-oil yield but with a relatively large fraction of the aqueous phase. This bio-oil was

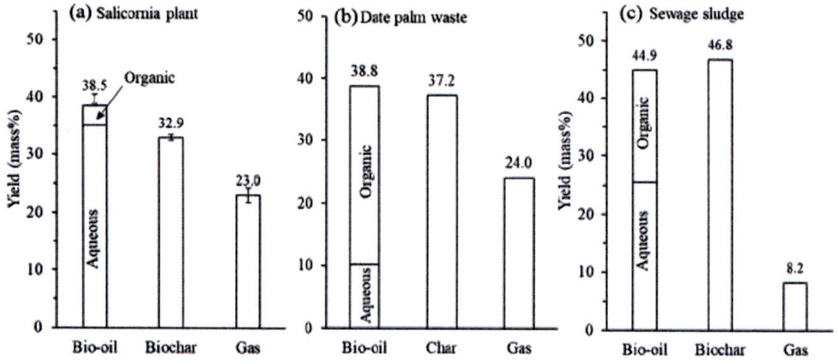

Fig. 19.11 Examples of the pyrolysis yield from various feedstocks of relevance to arid and semi-arid regions **a** *Salicornia* plant data (Makkawi et al. 2021) **b** date palm waste data (Makkawi et al. 2019) **c** sewage sludge (Moussa et al. 2021)

found to possess a quite low water content (~7%) and tends to be basic in nature (pH 9.27). However, it is also found to be quite viscous, which could be problematic during application. Additionally, the gas produced (non-condensable fraction of the pyrolysis gas) was found to contain a high amount of hydrogen (up to 23 wt% at the pyrolysis temperature of 600 °C). This significantly increased the HHV of the gas to nearly 43.1 MJ/kg. The increased hydrogen product is likely to result from the water–gas shift reaction inside the reactor driven by the high amount of sewage sludge ash. On the other hand, the bio-oil from *Salicornia* consisted of a large fraction of the aqueous phase. While this is undesirable, a recent study indicated that the aqueous fraction from the *Salicornia* pyrolysis possesses good qualities (>70% water and pH ~7), hence may have the potential to be utilized to stretch the water resources in arid regions where the freshwater sources are limited (Makkawi et al. 2021). Finally, Table 19.6 shows the HHVs of bio-oil produced from some waste feedstocks found in arid and semi-arid regions. It is interesting to note that the HHV of the biofuel derived from the sewage sludge is higher than that derived from the plant biomass.

Table 19.6 Comparison of the HHVs of bio-oil from different feedstocks

Feedstock	Source	Bio-oil HHV (MJ/kg)	References
Sewage sludge	UAE	32.84 (at 550 °C)	Moussa et al. (2021)
Sewage sludge	Germany	32.08–33.52 (at 400–500 °C)	Tomasi Morgano et al. (2018)
Raw pine wood	Georgia, USA	24.57–25.29 (at 400–600 °C)	Ben et al. (2019)
Salicornia plant	UAE, Dubai	29.27 (at 550 °C)	Makkawi et al. (2021)
Date palm waste	UAE, Sharjah	27.23 (at 520 °C)	Makkawi et al. (2019)
Date palm waste	UAE, Abu Dhabi	29.06 (at 500 °C)	Hai et al. (2021)

19.5.2 Anaerobic Digestion of Food Waste

Methane production from organic waste through anaerobic digestion has been reported in several studies (Ren et al. 2018). This is one of the leading waste-to-energy technologies under the classification of biological conversion. As noted earlier, food waste is an important organic waste available in abundant quantities in arid and semi-arid regions. (Beschkov 2017). As stated earlier, food waste mainly consists of lipids, carbohydrates, starch, lignin, and protein. Generally, the protein and carbohydrate portion of the organic matter goes through a faster transformation to biogas (Beschkov 2017; Morales-Polo et al. 2018). On the other hand, the lipids have a lower degradability rate but result in higher and more quality of the biogas (Morales-Polo et al. 2018). Therefore, the anaerobic digestion of food waste requires careful attention to multiple factors to achieve the maximum benefits and make the process more economically viable. These factors include, but are not limited to, the reactor type, loading rate, hydraulic retention time temperature, and pH (Pham et al. 2015). Ren et al. (2018) presented a full review of the ongoing research on anaerobic digestion of food waste, along with the required pretreatment, co-digestion, inhibition, and mitigation techniques related to this technology. The study suggested that a two-stage anaerobic digestion system combining hydrogen or ethanol production along with methane fermentation is a promising technology with high potential to enhance energy recovery and overall process efficiency.

A number of studies have investigated methods to improve the quantity and quality of the methane gas produced from anaerobic digestion. For example, Tayyab et al. (2019) attempted to improve the biogas quality by mixing different ratios of food waste (catering section in Pakistan) with pretreated *Parthenium hysterophorus* weed. Maximum biogas of 559 ml/L.day was produced at a 60:40 ratio of food waste to *Parthenium* weed. Zhang et al. (2019) utilized encapsulated metal additives obtained from high-speed mixing of nickel ions solution with hydrophobic silica nanoparticles to prevent metal precipitation in food waste anaerobic digestion processes. This study is particularly important because of the critical effects of metals on the digestion process. Elements like nickel, iron, and cobalt act as donors/acceptors and help in the synthesis of enzymes, and affect enzymatic methanogenic pathways. These additives improved the bioavailability of nickel and hence resulted in a considerable increase in the methane yield. In another experimental anaerobic digestion study using mixed organic waste, Panigrahi et al. (2020) investigated the effect of thermally pretreated yard waste (grass, dried leaves, and wood chips) and co-digestion with food waste (from students hostels in India). The yard waste, which is expected to be available in most of the populated cities in arid and semi-arid regions, consisted of grass, dried leaves, and wood chips. The results showed that methane production was highest (431 mL/g) at a 1.5 ratio of food waste to microorganisms. Table 19.7 shows literature data on the methane yield from anaerobic digestion of food waste. It is interesting to note that despite the variations in the food waste sources and the inoculums used, the data show that the methane yield remains almost consistent.

Table 19.7 Examples of methane yield from anaerobic digestion of food waste (Pour and Makkawi 2021)

Food waste source	Inoculum	Methane yield (L/g VS)	References
Mixture of sources	Not reported	0.40	Banks et al. (2011)
Restaurant waste	Mesophilic digested sludge	0.49	Forster-Carneiro et al. (2008)
Synthetic food	Seed sludge	0.52	Park et al. (2008)
Restaurant waste	Anaerobic seed sludge	0.45	Zhang and Jahng (2012)
Synthetic waste	Seed sludge	0.46	Nagao et al. (2012)
Mixture of sources	Anaerobic sludge	0.44	Zhang et al. (2007)

VS volatile solid

19.5.3 Transesterification and Fermentation of Food Waste

The food was an attractive feedstock for WtE application due to its high content of fatty acids (lipids). There have been reports on experiments on the conversion of lipid-derived from waste collected from restaurants to biodiesel (Carmona-Cabello et al. 2018, 2019). However, the produced biodiesel was found to be of poor quality and did not fulfill the criteria set by European Standard EN 14214 in terms of fatty acid methyl esters (FAME) yield, oxidation stability, and glyceride content. For this reason, it was recommended to use additives or to blend biodiesel with fossil fuel diesel. Further optimization of the process only improved the process in terms of the reaction time and energy saving but did not provide any benefit to the oxidation stability, FAME yield, and glyceride content. To improve the lipid extraction from the food waste (collected from hostel kitchen in India), another study included a drying step before transesterification (Barik et al. 2018). The product biodiesel physical properties were found to be within the acceptable range set by ASTM D6751 standards, but the calorific value (31.38 MJ/kg) was lower than the EN14214 standard (35 MJ/kg).

An alternative approach on food waste esterification, which has shown reasonable success, involved the use of potassium hydroxide (KOH) and lipase as catalysts and methanol (Karmee et al. 2015). The results showed 100% biodiesel yield with a lipid to methanol ratio of 1:10 M, a temperature of 60 °C, and 2 h of processing. Table 19.8 presents the properties of the biodiesel produced from food waste reported in some of the studies discussed above. It is clear that the food waste biodiesel can be very similar to petroleum diesel in terms of density and viscosity, only 10% less in HHV, but relatively high in ignition point.

There are several studies on the fermentation of food waste to produce ethanol. Most of the reported studies focused on the effect of operating conditions on the ethanol produced. The parameters commonly investigated in fermentation include the type of microorganism (e.g., yeast and bacteria), pH, temperature, inoculum size, moisture content, and incubation time (Anwar Saeed et al. 2018). Other studies have been focused on process optimization and cost reduction. For example, attempts have

Table 19.8 Characteristics of biodiesel produced from food waste

Food waste source	Kinematic viscosity[a] (mm^2/s)	Density[b] (kg/m^3)	Heating value (MJ/kg)	Flash point (°C)	Acid value (mg KOH/g)	References
Restaurant	4.8 ± 0.1	887 ± 1	39.6 ± 0.1[d]	164 ± 3	–	Carmona-Cabello et al. (2018)
	4.1 ± 0.1	870 ± 1	39.5 ± 0.1[d]	166 ± 1	0.16 ± 0.01	Carmona-Cabello et al. (2019)
Hostel kitchen	2.2	872	31.4[c]	164	0.60	Barik et al. (2018)
Food waste ferment	4.0	870	39.4[d]	–	–	Bhatia et al. (2019)
Restaurant-India	4.4	875–882	39.2–41.5[d]	180	0.13 ± 0.05	Priyadarshi and Paul (2018)

[a]at 40 °C
[b]at 15 °C
[c]Lower Heating Value (LHV)
[d]Higher Heating Value (HHV)

been made to reduce the initial cost of the process by using thermophilic fungus *Myceliophthora thermophila* produced from food waste (Matsakas and Christakopoulos 2015). This was utilized in enzymatic saccharification of the food waste prior to fermentation. As a result, the mass concentration of the ethanol produced was found to significantly increase by around 222%, hence, emphasizing the effectiveness of the enzymes on the fermentation process. Another theoretical study on the optimization of food waste fermentation predicted the feasibility of effective process optimization through changing the saccharification and fermentation pH, reaction and fermentation of enzyme, temperature, enzyme concentration, and fermentation time (Kim et al. 2008). Table 19.9 summarizes some of the important literature data related to the production of ethanol from food waste. It is clear that the ethanol concentration and yield are highly sensitive to the variations in the food waste composition and the fermentation conditions (e.g., processing time and type of fermentation organism).

19.6 Challenges and Recommendations

In this chapter, a number of potentially suitable and, to some extent, sustainable sources of organic waste have been identified and recommended as feedstocks for waste-to-energy. However, there remains a number of challenges associated with each feedstock. It was clearly noted in the previous sections that every type of feedstock discussed here is different in terms of characteristics (physical, chemical, and thermal). The feedstocks have been shown to differ greatly in composition, structure, and complexity. Wastes such as plastic, for example, can be composed of multiple polymers of different shapes and sizes that may sometimes be easy to separate but is

Table 19.9 Ethanol production from food waste—literature review summary

Food waste source	Microorganism	Time (h)	Yield (g EtOH /g)	EtOH concentration (g /l)	References
Supermarket—Japan	*Zymomonas mobilis*	48	0.5	99.8	Ma et al. (2016)
Dining hall—China	*S. cerevisiae*	48	0.5	–	Yan et al. (2011)
Cafeteria—South Korea	*S. cerevisiae*	48	0.4	–	Hong and Yoon (2011)
Cafeteria—South Korea	*S. cerevisiae*	14	–	57.5	Kim et al. (2008)
Cafeteria and household—Turkey	*S. cerevisiae*	58.8	0.4	32.2	Uncu and Cekmecelioglu (2011)
Dining room hall FW—China	*S. cerevisiae*	67.6	0.2	33.1	Wang et al. (2008)

not always the case. Moreover, different lignocellulosic waste biomass, such as the date palm or *Salicornia* plant, contain different ratios of cellulose, hemicellulose, and lignin (Irmak 2019). The lack of homogeneity in the feedstock makes the implementation of a singular system of waste biomass utilization very difficult. The problem may in particular be critical with food waste and sewage sludge. For example, sewage sludge and food waste are both high in moisture when received from the source, which makes it challenging to apply in waste-to-energy without drying, especially in thermochemical conversion. On the other hand, logistically, the use of plant waste, such as *Salicornia* and date palm requires an adequate collection method, which adds to the overall operational cost. Many improvements may be made to thermochemical conversion methods to enhance the production of biofuel and biochar. The feedstock may undergo multiple resource extraction methods to obtain various products. Sewage sludge samples may undergo lipid extraction, carbohydrate extraction, and then pyrolysis in order to obtain biodiesel, bioethanol, and bio-oil, respectively (Supaporn et al. 2019).

The use of food waste as a feedstock for fermentation would significantly reduce the raw material cost in the bioethanol production industry (Yukesh Kannah et al. 2020). However, there are two main challenges associated with food waste fermentation. One is related to the composition heterogeneity, which is a general problem with all types of food waste recycling methods, as noted earlier. The other challenge is related to the need for further processing of the liquid effluent (wastewater) and the solid residue remaining after the fermentation process (Anwar Saeed et al. 2018; Poe et al. 2020).

Finally, it is worth noting that, in arid and semi-arid regions, it is likely that one type of feedstock may not be sufficient to satisfy the requirement of a particular waste-to-energy technology. In such situations, it is highly recommended to consider

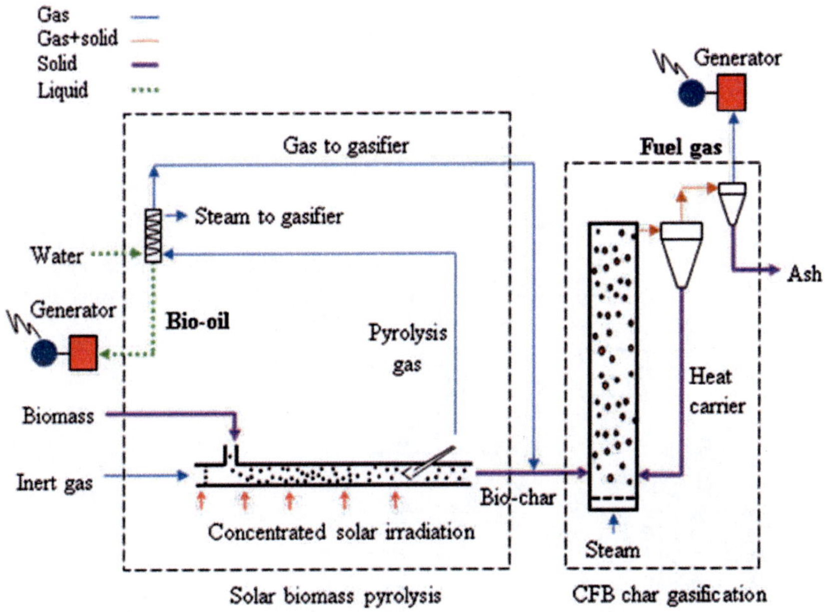

Fig. 19.12 Waste-to-energy generation through integrated biomass pyrolysis and gasification (Bashir et al. 2017)

co-feeding, such as food waste with sewage sludge, or multi-feeding by creating a tailored refused derived fuel (RDF). For example, the UAE is potentially well suited to create an RDF system consisting of date palm waste, *Salicornia* plant, food waste, and sewage sludge for use as a solid fuel in WtE through thermochemical conversion. In such a case, multiple products can be utilized, such that the biofuel can benefit the renewable energy sector, the biochar can be utilized in improving the desert soil, while the water from the feedstock drying may be utilized in irrigation or as a source of water, after minimum pretreatment. Another potentially suitable approach to maximize the products from WtE is through the integration of different processing methods. Figures 19.12 and 19.13 show an integrated transesterification and anaerobic digestion, which could convert food waste to several forms of energy, in addition to a solid by-product (biochar) of multiple benefits.

19.7 Conclusions

The increased concern over waste management and the urgent need to diversify energy sources led to the development of waste-to-energy (WtE) technologies all over the world. This has also been pushed by the strict regulation of carbon emissions associated with the use of fossil fuels. However, the potential of this technology in arid

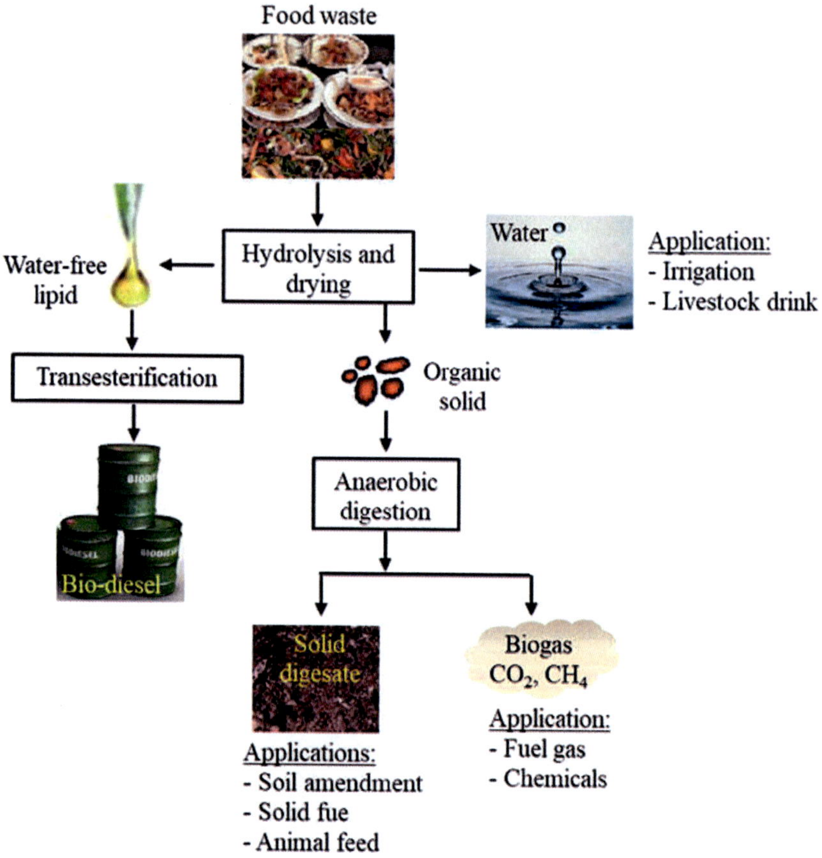

Fig. 19.13 Waste-to-energy generation through integrated transesterification and anaerobic digestion of food waste (Pour and Makkawi 2021)

and semi-arid regions has been hampered by various reasons, including the limited resources of organic matter from natural sources. The WtE technology represents an ideal solution and offers an excellent opportunity for arid and semi-arid regions to catch up with the rest of the world and contribute to the global effort on sustainable clean energies.

The most abundant waste organic matters that may have potential as a feedstock for WtE in arid and semi-arid regions include, but are not limited to, general municipal solid waste (MSW), sewage sludge, industrial organic waste, and plantation that can survive harsh environments such as *Salicornia.* While these feedstocks have shown encouraging results when converted to biofuels and biochar, some may also be useful in extending the water resources by utilizing the effluent water produced during drying and processing.

The successful implementation of WtE in arid and semi-arid regions is highly recommended based on the co-feeding of multiple organic wastes to avoid the shortage or seasonable availability of a single source. It will also depend on the adaptation and integration of energy conversion methods, such as gasification and pyrolysis or transesterification and anaerobic digestion. For example, in gulf regions, which have an extended coast and large deserts, the waste generated from the plantation of halophytes has proven to be promising for the production of multiple products through various WtE technologies. Sewage sludge is another highly potential feedstock in arid and semi-arid regions. Despite the fact that it is relatively low in heating value, recent experiments conducted in the UAE, have shown that the bio-oil produced from this feedstock is of high quality, besides the benefit of producing a gas stream rich in hydrogen. On the other hand, the bio-chars produced from organic waste through pyrolysis are also expected to play a major role in enhancing the soil in arid and semi-arid regions.

References

Abbasi SA (2018) The myth and the reality of energy recovery from municipal solid waste. Energy, Sustain Soc 8(1). https://doi.org/10.1186/s13705-018-0175-y

Abdallah R, Juaidi A, Assad M, Salameh T, Manzano-Agugliaro F (2020) Energy recovery from waste tires using pyrolysis: Palestine as case of study. Energies 13(7):1–13. https://doi.org/10.3390/en13071817

Abdel-Aziz MH, Bassyouni M, Soliman MF, Gutub SA, Magram SF (2017) Removal of heavy metals from wastewater using thermally treated sewage sludge adsorbent without chemical activation. J Mater Environ Sci 8(5):1737–1747

Abdel Wahaab R, Mahmoud M, van Lier JB (2020) Toward achieving sustainable management of municipal wastewater sludge in Egypt: the current status and future prospective. Renew Sustain Energy Rev 127(May):109880. https://doi.org/10.1016/j.rser.2020.109880

Abdelaal AH, McKay G, Mackey HR (2019) Food waste from a university campus in the Middle East: drivers, composition, and resource recovery potential. Waste Manage 98(Oct 2020):14–20. https://doi.org/10.1016/j.wasman.2019.08.007

Akhtar J, Saidina Amin N (2012) A review on operating parameters for optimum liquid oil yield in biomass pyrolysis. Renew Sustain Energy Rev 16(7):5101–5109. https://doi.org/10.1016/j.rser.2012.05.033

Al-Salem SM, Antelava A, Constantinou A, Manos G, Dutta A (2017) A review on thermal and catalytic pyrolysis of plastic solid waste (PSW). J Environ Manage 197(1408):177–198. https://doi.org/10.1016/j.jenvman.2017.03.084

Al-Salem SM, Yang Y, Wang J, Leeke GA (2020) Pyro-oil and wax recovery from reclaimed plastic waste in a continuous Auger pyrolysis reactor. Energies 13(8):1–10. https://doi.org/10.3390/en13082040

Allaire S, Lange S, Auclair I, Quinche M, Greffard L (2015) Report: analyses of biochar properties. In: Centre de Recherche sur les Matériaux Renouvelables (Issue Sept). https://doi.org/10.13140/RG.2.1.2789.4241

Anwar Saeed M, Ma H, Yue S, Wang Q, Tu M (2018) Concise review on ethanol production from food waste: development and sustainability. Environ Sci Pollut Res 25(29):28851–28863. https://doi.org/10.1007/s11356-018-2972-4

Arazo RO, Genuino DAD, de Luna MDG, Capareda SC (2017) Bio-oil production from dry sewage sludge by fast pyrolysis in an electrically-heated fluidized bed reactor. Sustain Environ Res 27(1):7–14. https://doi.org/10.1016/j.serj.2016.11.010

Askeland M, Clarke B, Paz-ferreiro J (2019) Comparative characterization of biochars produced at three selected pyrolysis temperatures from common woody and herbaceous waste streams. 1–20. https://doi.org/10.7717/peerj.6784

Baawain M, Al-Mamun A, Omidvarborna H, Al-Amri W (2017) Ultimate composition analysis of municipal solid waste in Muscat. J Clean Prod 148:355–362. https://doi.org/10.1016/j.jclepro.2017.02.013

Balat M (2011) Production of bioethanol from lignocellulosic materials via the biochemical pathway: a review. Energy Convers Manage 52(2):858–875. https://doi.org/10.1016/j.enconman.2010.08.013

Banks CJ, Chesshire M, Heaven S, Arnold R (2011) Anaerobic digestion of source-segregated domestic food waste: performance assessment by mass and energy balance. Biores Technol 102(2):612–620. https://doi.org/10.1016/j.biortech.2010.08.005

Barik S, Paul KK, Priyadarshi D (2018) Utilization of kitchen food waste for biodiesel production. IOP Conf Ser: Earth Environ Sci 167(1). https://doi.org/10.1088/1755-1315/167/1/012036

Barreveld WH (1993) Date palm products. FAO Agri Serv Bull 101

Bashir M, Yu X, Hassan M, Makkawi Y (2017) Modeling and performance analysis of biomass fast pyrolysis in a solar-thermal reactor. ACS Sustain Chem Eng 5(5):3795–3807. https://doi.org/10.1021/acssuschemeng.6b02806

Bekhiti M, Trouzine H, Asroun A (2014) Properties of waste tire rubber powder. Eng Technol Appl Sci Res 4(4):669–672. https://doi.org/10.48084/etasr.439

Belpoliti V, Abbas HS, Ali MAA, Al Khulaifi AM (2018) Design from waste: review of the UAE waste management sector and experimentation on recycled materials to support the development of a sustainable and energy-efficient building market. In: 2018 Advances in science and engineering technology international conferences, ASET 2018, pp 1–6.https://doi.org/10.1109/ICASET.2018.8376765

Ben H, Wu F, Wu Z, Han G, Jiang W, Ragauskas AJ (2019) A comprehensive characterization of pyrolysis oil from softwood barks. Polymers 11(9). https://doi.org/10.3390/polym11091387

Berhane D (2015) Determination of losing and gaining reaches in arid and semi-arid environments of NSW (Issue Nov). The University of Sydney New South Wales

Beschkov V (2017) Biogas, biodiesel and bioethanol as multifunctional renewable fuels and raw materials. In: Jacob-Lopes E, Zepka LQ (eds) Frontiers in bioenergy and biofuels. InTech. https://doi.org/10.5772/65734

Bhatia SK, Gurav R, Choi TR, Jung HR, Yang SY, Song HS, Kim YG, Yoon JJ, Yang YH (2019) Effect of synthetic and food waste-derived volatile fatty acids on lipid accumulation in Rhodococcus sp. YHY01 and the properties of produced biodiesel. Energy Convers Manage 192:385–395. https://doi.org/10.1016/j.enconman.2019.03.081

Bridgwater AV (2003) Renewable fuels and chemicals by thermal processing of biomass. Chem Eng J 91(2–3):87–102. https://doi.org/10.1016/S1385-8947(02)00142-0

Buss AH, Kovaleski JL, Pagani RN, da Silva VL, de Matos Silva J (2019) Proposal to reuse rubber waste from end-of-life tires using thermosetting resin. Sustainability (Switzerland) 11(24). https://doi.org/10.3390/su11246997

Cao L, Yu IKM, Xiong X, Tsang DCW, Zhang S, Clark JH, Hu C, Ng YH, Shang J, Ok YS (2020) Biorenewable hydrogen production through biomass gasification: a review and future prospects. Environ Res 186(April):109547. https://doi.org/10.1016/j.envres.2020.109547

Carmona-Cabello M, Leiva-Candia D, Castro-Cantarero JL, Pinzi S, Dorado MP (2018) Valorization of food waste from restaurants by transesterification of the lipid fraction. Fuel 215:492–498. https://doi.org/10.1016/j.fuel.2017.11.096

Carmona-Cabello M, Sáez-Bastante J, Pinzi S, Dorado MP (2019) Optimization of solid food waste oil biodiesel by ultrasound-assisted transesterification. Fuel 255:115817. https://doi.org/10.1016/j.fuel.2019.115817

Caton PA, Carr MA, Kim SS, Beautyman MJ (2010) Energy recovery from waste food by combustion or gasification with the potential for regenerative dehydration: a case study. Energy Convers Manage 51(6):1157–1169. https://doi.org/10.1016/j.enconman.2009.12.025

Chen D, Yin L, Wang H, He P (2014) Pyrolysis technologies for municipal solid waste: a review. Waste Manage 34(12):2466–2486. https://doi.org/10.1016/j.wasman.2014.08.004

Chen DMC, Bodirsky BL, Krueger T, Mishra A, Popp A (2020) The world's growing municipal solid waste: trends and impacts. Environ Res Lett 15(7). https://doi.org/10.1088/1748-9326/ab8659

Corder SE (1976) Properties and uses of bark as an energy source, p 21

Dai Q, Jiang X, Lv G, Ma X, Jin Y, Wang F, Chi Y, Yan J (2015) Investigation into particle size influence on PAH formation during dry sewage sludge pyrolysis: TG-FTIR analysis and batch scale research. J Anal Appl Pyrol 112:388–393. https://doi.org/10.1016/j.jaap.2014.09.015

de Jesus Duarte S, Glaser B, Cerri CEP (2019) Effect of biochar particle size on physical, hydrological and chemical properties of loamy and sandy tropical soils. Agronomy 9(4). https://doi.org/10.3390/agronomy9040165

Deepanraj B, Senthilkumar N, Ranjitha J, Jayaraj S, Ong HC (2021) Biogas from food waste through anaerobic digestion: optimization with response surface methodology. Biomass Convers Biorefinery 11(2):227–239. https://doi.org/10.1007/s13399-020-00646-9

Dubdub I, Al-Yaari M (2020) Pyrolysis of mixed plastic waste: I. kinetic study. Materials 13(21):1–15. https://doi.org/10.3390/ma13214912

El-Juhany LI (2010) Degradation of date palm trees and date production in Arab countries: causes and potential rehabilitation. Aust J Basic Appl Sci 4(8):3998–4010. https://doi.org/10.1016/j.radphyschem.2006.10.004

Escamilla-García PE, Camarillo-López RH, Carrasco-Hernández R, Fernández-Rodríguez E, Legal-Hernández JM (2020) Technical and economic analysis of energy generation from waste incineration in Mexico. Energy Strat Rev 31.https://doi.org/10.1016/j.esr.2020.100542

Fan F, Xu R, Wang D, Meng F (2020) Application of activated sludge for odor control in wastewater treatment plants: approaches, advances and outlooks. Water Res 181:115915. https://doi.org/10.1016/j.watres.2020.115915

FCSA UAE (2019) F treated wastewater 2019 المياه العادمة المعالجة.https://fcsc.gov.ae/en-us/Pages/Statistics/Statistics-by-Subject.aspx#/%3Ffolder=Agriculture

Fonts I, Azuara M, Gea G, Murillo MB (2009) Study of the pyrolysis liquids obtained from different sewage sludge. J Anal Appl Pyrol 85(1–2):184–191. https://doi.org/10.1016/j.jaap.2008.11.003

Food waste is a massive problem—here's why. FoodPrint. (n.d.). FoodPrint

Forbes H, Quested T, O'Connor C (2021) Food waste index. United Nations Environment Programme. In: Wrap

Forster-Carneiro T, Pérez M, Romero LI (2008) Influence of total solid and inoculum contents on performance of anaerobic reactors treating food waste. Biores Technol 99(15):6994–7002. https://doi.org/10.1016/j.biortech.2008.01.018

Ghacha A, Ammari M, Ben Allal L (2020) Sustainable sewage sludge management in Morocco: constraints and solutions. J Water Land Dev 46:71–83. https://doi.org/10.24425/jwld.2020.134199

Gómez N, Banks SW, Nowakowski DJ, Rosas JG, Cara J, Sánchez ME, Bridgwater AV (2018) Effect of temperature on product performance of a high ash biomass during fast pyrolysis and its bio-oil storage evaluation. Fuel Process Technol 172(Nov 2017):97–105. https://doi.org/10.1016/j.fuproc.2017.11.021

Grammelis P, Margaritis N, Dallas P, Rakopoulos D, Mavrias G (2021) A review on management of end of life tires (ELTs) and alternative uses of textile fibers. Energies 14(3):571. https://doi.org/10.3390/en14030571

Grycová B, Koutník I, Pryszcz A (2016) Pyrolysis process for the treatment of food waste. Biores Technol 218:1203–1207. https://doi.org/10.1016/j.biortech.2016.07.064

Guo Z, Guo F, Miao Z, Guo Y, Zhao X, Liu H, Wu J, Zhang Y (2021) Migration pattern of chlorine during co-pyrolysis for herb residue and coal. J Environ Chem Eng 9(4):105196. https://doi.org/10.1016/j.jece.2021.105196

Hai A, Bharath G, Rambabu K, Kannan P, Banat F, Taher H, Jayaraman R, Show PL (2021) Pyrolysis of different date palm industrial wastes into high-quality bio-oils: a comparative study. Clean Technol Environ Policy 23(1):55–64. https://doi.org/10.1007/s10098-020-01888-x

Hegazy SS, Aref IM (2010) Suitability of some fast-growing trees and date palm fronds for particleboard production. For Prod J 60(7–8):599–604. https://doi.org/10.13073/0015-7473-60.7.599

Hong YS, Yoon HH (2011) Ethanol production from food residues. Biomass Bioenerg 35(7):3271–3275. https://doi.org/10.1016/j.biombioe.2011.04.030

Hu X, Gholizadeh M (2019) Biomass pyrolysis: a review of the process development and challenges from initial researches up to the commercialisation stage. J Energy Chem 39(x):109–143. https://doi.org/10.1016/j.jechem.2019.01.024

Huber F, Blasenbauer D, Aschenbrenner P, Fellner J (2020) Complete determination of the material composition of municipal solid waste incineration bottom ash. Waste Manage 102:677–685. https://doi.org/10.1016/j.wasman.2019.11.036

Irmak S (2019) Challenges of biomass utilization for biofuels. In: Biomass for bioenergy—recent trends and future challenges, pp 1–11.https://doi.org/10.5772/intechopen.83752

Islam MN, Zailani R, Ani FN (1999) Pyrolytic oil from fluidised bed pyrolysis of oil palm shell and its characterisation. Renew Energy 17(1):73–84. https://doi.org/10.1016/s0960-1481(98)00112-8

Janajreh I, Adeyemi I, Raza SS, Ghenai C (2021) A review of recent developments and future prospects in gasification systems and their modeling. Renew Sustain Energy Rev 138(Sept 2020):110505. https://doi.org/10.1016/j.rser.2020.110505

Jo JH, Kim SS, Shim JW, Lee YE, Yoo YS (2017) Pyrolysis characteristics and kinetics of food wastes. Energies 10(8):1–14. https://doi.org/10.3390/en10081191

Jouhara H, Ahmad D, Czajczyńska D, Ghazal H, Anguilano L, Reynolds A, Rutkowski P, Krzyżyńska R, Katsou E, Simons S, Spencer N (2018) Experimental investigation on the chemical characterisation of pyrolytic products of discarded food at temperatures up to 300 °C. Therm Sci Eng Prog 5:579–588. https://doi.org/10.1016/j.tsep.2018.02.010

Karmaoui A (ed) (2019) Climate change and its impact on ecosystem services and biodiversity in arid and semi-arid zones, vol i. IGI Global. https://www.researchgate.net/publication/331235517

Karmee SK, Linardi D, Lee J, Lin CSK (2015) Conversion of lipid from food waste to biodiesel. Waste Manage 41:169–173. https://doi.org/10.1016/j.wasman.2015.03.025

Kaza S, Yao L, Bhada-Tata P, Van Woerden F (2018) What a waste. World Bank Publications

Kester G (2017) California bioresources alliance report

Khan I (2020) Waste to biogas through anaerobic digestion: hydrogen production potential in the developing world—a case of Bangladesh. Int J Hydrogen Energy 45(32):15951–15962. https://doi.org/10.1016/j.ijhydene.2020.04.038

Khanmohammadi Z, Afyuni M, Mosaddeghi MR (2015) Effect of pyrolysis temperature on chemical and physical properties of sewage sludge biochar. Waste Manage Res 33(3):275–283. https://doi.org/10.1177/0734242X14565210

Kim Y, Parker W (2008) A technical and economic evaluation of the pyrolysis of sewage sludge for the production of bio-oil. Biores Technol 99(5):1409–1416. https://doi.org/10.1016/j.biortech.2007.01.056

Kim JK, Oh BR, Shin HJ, Eom CY, Kim SW (2008) Statistical optimization of enzymatic saccharification and ethanol fermentation using food waste. Process Biochem 43(11):1308–1312. https://doi.org/10.1016/j.procbio.2008.07.007

Kim JH, Lee JC, Pak D (2011) Feasibility of producing ethanol from food waste. Waste Manage 31(9–10):2121–2125. https://doi.org/10.1016/j.wasman.2011.04.011

Klinghoffer N, Castaldi MJ, Nzihou A (2011) Beneficial use of ash and char from biomass gasification. In: 19th Annual North American waste-to-energy conference, NAWTEC19, April 2016, pp 13–17. https://doi.org/10.1115/nawtec19-5421

Lakshman V, Brassard P, Hamelin L, Raghavan V, Godbout S (2021) Pyrolysis of miscanthus: developing the mass balance of a biorefinery through experimental tests in an auger reactor. Bioresour Technol Rep 14(Jan). https://doi.org/10.1016/j.biteb.2021.100687

Landi D, Vitali S, Germani M (2016) Environmental analysis of different end of life scenarios of tires textile fibers. Procedia CIRP 48:508–513. https://doi.org/10.1016/j.procir.2016.03.141

Lucian M, Fiori L (2017) Hydrothermal carbonization of waste biomass: process design, modeling, energy efficiency and cost analysis. Energies 10(2). https://doi.org/10.3390/en10020211

Ma K, Ruan Z, Shui Z, Wang Y, Hu G, He M (2016) Open fermentative production of fuel ethanol from food waste by an acid-tolerant mutant strain of Zymomonas mobilis. Biores Technol 203:295–302. https://doi.org/10.1016/j.biortech.2015.12.054

Makkawi Y, El Sayed Y, Salih M, Nancarrow P, Banks S, Bridgwater T (2019) Fast pyrolysis of date palm (Phoenix dactylifera) waste in a bubbling fluidized bed reactor. Renew Energy 143(May):719–730. https://doi.org/10.1016/j.renene.2019.05.028

Makkawi Y, El Sayed Y, Lyra DA, Hassan Pour FH, Khan M, Badrelzaman M (2021) Assessment of the pyrolysis products from halophyte Salicornia bigelovii cultivated in a desert environment. Fuel 290(January):119518. https://doi.org/10.1016/j.fuel.2020.119518

Matsakas L, Christakopoulos P (2015) Ethanol production from enzymatically treated dried food waste using enzymes produced on-site. Sustainability (switzerland) 7(2):1446–1458. https://doi.org/10.3390/su7021446

Matsakas L, Kekos D, Loizidou M, Christakopoulos P (2014) Utilization of household food waste for the production of ethanol at high dry material content. Biotechnol Biofuels 7(1):1–9. https://doi.org/10.1186/1754-6834-7-4

Morales-Polo C, del Mar Cledera-Castro M, Yolanda Moratilla Soria B (2018) Reviewing the anaerobic digestion of food waste: from waste generation and anaerobic process to its perspectives. Appl Sci (Switzerland) 8(10). https://doi.org/10.3390/app8101804

Mortula M, Ahmad A, Shah SA (2016) Assessment of mixing potential of sewage sludge, green waste and food waste for co-composting in Sharjah, UAE. Int J Environ Sustain 5(2). https://doi.org/10.24102/ijes.v5i2.670

Moussa O, Makkawi Y, Sayed YE, Fattah K (2021) Comparative assessment of the pyrolysis of anaerobically digested and undigested sewage sludge

Munir MT, Mohaddespour A, Nasr AT, Carter S (2021) Municipal solid waste-to-energy processing for a circular economy in New Zealand. Renew Sustain Energy Rev 145(April):111080. https://doi.org/10.1016/j.rser.2021.111080

Nagao N, Tajima N, Kawai M, Niwa C, Kurosawa N, Matsuyama T, Yusoff FM, Toda T (2012) Maximum organic loading rate for the single-stage wet anaerobic digestion of food waste. Biores Technol 118:210–218. https://doi.org/10.1016/j.biortech.2012.05.045

Nasser RA (2014) An evaluation of the use of midribs from common date palm cultivars grown in Saudi Arabia for energy production. BioResources 9(3):4343–4357

Nasser RA, Salem MZM, Hiziroglu S, Al-Mefarrej HA, Mohareb AS, Alam M, Aref IM (2016) Chemical analysis of different parts of date palm (Phoenix dactylifera L.) using ultimate, proximate and thermo-gravimetric techniques for energy production. Energies 9(5). https://doi.org/10.3390/en9050374

National overview: facts and figures on materials, wastes and recycling. Facts and figures about materials, waste and recycling. US EPA. (n.d.)

Ncube LK, Ude AU, Ogunmuyiwa EN, Zulkifli R, Beas IN (2021) An overview of plastic waste generation and management in food packaging industries. Recycling 6(1):1–25. https://doi.org/10.3390/recycling6010012

O'Connor J, Hoang SA, Bradney L, Dutta S, Xiong X, Tsang DCW, Ramadass K, Vinu A, Kirkham MB, Bolan NS (2021) A review on the valorisation of food waste as a nutrient source and soil amendment. Environ Pollut 272:115985. https://doi.org/10.1016/j.envpol.2020.115985

Opatokun SA, Kan T, Al Shoaibi A, Srinivasakannan C, Strezov V (2016) Characterization of food waste and its digestate as feedstock for thermochemical processing. Energy Fuels 30(3):1589–1597. https://doi.org/10.1021/acs.energyfuels.5b02183

Panigrahi S, Sharma HB, Dubey BK (2020) Anaerobic co-digestion of food waste with pretreated yard waste: a comparative study of methane production, kinetic modeling and energy balance. J Clean Prod 243:118480. https://doi.org/10.1016/j.jclepro.2019.118480

Park Y, Hong F, Cheon J, Hidaka T, Tsuno H (2008) Comparison of thermophilic anaerobic digestion characteristics between single-phase and two-phase systems for kitchen garbage treatment. J Biosci Bioeng 105(1):48–54. https://doi.org/10.1263/jbb.105.48

Pham TPT, Kaushik R, Parshetti GK, Mahmood R, Balasubramanian R (2015) Food waste-to-energy conversion technologies: current status and future directions. Waste Manage 38(1):399–408. https://doi.org/10.1016/j.wasman.2014.12.004

Pickin J, Wardle C, O'Farrell K, Piya Nyunt SD (2020) National waste report 2018. Blue Environment Pty Ltd (Issue Nov). https://www.environment.gov.au/system/files/resources/7381c1de-31d0-429b-912c-91a6dbc83af7/files/national-waste-report-2018.pdf

Pirani SI, Arafat HA (2016) Reduction of food waste generation in the hospitality industry. J Clean Prod 132:129–145. https://doi.org/10.1016/j.jclepro.2015.07.146

Poe NE, Yu D, Jin Q, Ponder MA, Stewart AC, Ogejo JA, Wang H, Huang H (2020) Compositional variability of food wastes and its effects on acetone-butanol-ethanol fermentation. Waste Manage 107:150–158. https://doi.org/10.1016/j.wasman.2020.03.035

Pour FH, Makkawi YT (2021) A review of post-consumption food waste management and its potentials for biofuel and biochar production (Submitted to Energy reports, June 2021)

Priyadarshi D, Paul KK (2018) Single phase blend: an advanced microwave process for improved quality low-cost biodiesel production from kitchen food waste. Biochem Eng J 137:273–283. https://doi.org/10.1016/j.bej.2018.06.006

Ren Y, Yu M, Wu C, Wang Q, Gao M, Huang Q, Liu Y (2018) A comprehensive review on food waste anaerobic digestion: research updates and tendencies. Biores Technol 247:1069–1076. https://doi.org/10.1016/j.biortech.2017.09.109

Sienkiewicz M, Janik H, Borzędowska-Labuda K, Kucińska-Lipka J (2017) Environmentally friendly polymer-rubber composites obtained from waste tyres: a review. J Clean Prod 147:560–571. https://doi.org/10.1016/j.jclepro.2017.01.121

Stabile L, Scungio M, Frattolillo A, Buonanno G (2020) Effects of the flue gas treatment of incinerator plants on sub-micron particle concentrations at the stack. Waste Manage 101:9–17. https://doi.org/10.1016/j.wasman.2019.09.038

Supaporn P, Ly HV, Kim SS, Yeom SH (2019) Bio-oil production using residual sewage sludge after lipid and carbohydrate extraction. Environ Eng Res 24(2):202–210. https://doi.org/10.4491/eer.2017.178

Tayyab A, Ahmad Z, Mahmood T, Khalid A, Qadeer S, Mahmood S, Andleeb S, Anjum M (2019) Anaerobic co-digestion of catering food waste utilizing Parthenium hysterophorus as co-substrate for biogas production. Biomass Bioenerg 124:74–82. https://doi.org/10.1016/j.biombioe.2019.03.013

The Science Agriculture (2019) 10 world's biggest countries in date production of 2019. https://scienceagri.com/10-worlds-biggest-countries-in-date-production-of-2019/

The World Bank (2018) What a waste 2.0: trends in solid waste management. https://datatopics.worldbank.org/what-a-waste/trends_in_solid_waste_management.html

Thi NBD, Kumar G, Lin CY (2015) An overview of food waste management in developing countries: current status and future perspective. J Environ Manage 157:220–229. https://doi.org/10.1016/j.jenvman.2015.04.022

Tiseo I (2020) Plastic waste worldwide—statistics and facts. https://www.statista.com/topics/5401/global-plastic-waste/

Tomasi Morgano M, Leibold H, Richter F, Stapf D, Seifert H (2018) Screw pyrolysis technology for sewage sludge treatment. Waste Manage 73:487–495. https://doi.org/10.1016/j.wasman.2017.05.049

Tomczyk A, Sokołowska Z, Boguta P (2020) Biochar physicochemical properties: pyrolysis temperature and feedstock kind effects. Rev Environ Sci Biotechnol 19(1):191–215. https://doi.org/10.1007/s11157-020-09523-3

Uddin MN, Techato K, Taweekun J, Rahman MM, Rasul MG, Mahlia TMI, Ashrafur SM (2018) An overview of recent developments in biomass pyrolysis technologies. Energies 11(11):1–23. https://doi.org/10.3390/en11113115

UN: 17% of all food available at consumer levels is wasted—world. ReliefWeb (2021)

Uncu ON, Cekmecelioglu D (2011) Cost-effective approach to ethanol production and optimization by response surface methodology. Waste Manage 31(4):636–643. https://doi.org/10.1016/j.wasman.2010.12.007

United Nations Framework Convention on Climate Change (n.d.) Santiago biofactory, Chile. Web article. https://unfccc.int/climate-action/momentum-for-change/planetary-health/santiago-biofactory-chile

United States Environmental Protection Agency (2020) 2018 wasted food report (Issue Nov). EPA 530-R-20-004

Van Lanh N, Bich NH, Hung BN, Quyen NN, Preston TR (2019) Water retention capacity of biochar and its effect on growth of maize. Livestock Res Rural Dev 31

Vlachokostas C, Achillas C, Michailidou AV, Tsegas G, Moussiopoulos N (2020) Externalities of energy sources: the operation of a municipal solid waste-to-energy incineration facility in the greater Thessaloniki area, Greece. Waste Manage 113:351–358. https://doi.org/10.1016/j.wasman.2020.06.015

Wang Q, Ma H, Xu W, Gong L, Zhang W, Zou D (2008) Ethanol production from kitchen garbage using response surface methodology. Biochem Eng J 39(3):604–610. https://doi.org/10.1016/j.bej.2007.12.018

Wang S, Dai G, Yang H, Luo Z (2017) Lignocellulosic biomass pyrolysis mechanism: a state-of-the-art review. Prog Energy Combust Sci 62:33–86. https://doi.org/10.1016/j.pecs.2017.05.004

Wei L, Zhu F, Li Q, Xue C, Xia X, Yu H, Zhao Q, Jiang J, Bai S (2020) Development, current state and future trends of sludge management in China: based on exploratory data and CO_2-equivaient emissions analysis. Environ Int 144(May):106093. https://doi.org/10.1016/j.envint.2020.106093

Yan S, Wang P, Zhai Z, Yao J (2011) Fuel ethanol production from concentrated food waste hydrolysates in immobilized cell reactors by Saccharomyces cerevisiae H058. J Chem Technol Biotechnol 86(5):731–738. https://doi.org/10.1002/jctb.2581

Yukesh Kannah R, Sivashanmugham P, Kavitha S, Rajesh Banu J (2020) Valorization of food waste for bioethanol and biobutanol production. In: Food waste to valuable resources. Elsevier, pp 39–73. https://doi.org/10.1016/B978-0-12-818353-3.00003-1

Zaker A, Chen Z, Wang X, Zhang Q (2019) Microwave-assisted pyrolysis of sewage sludge: a review. Fuel Process Technol 187(Aug 2018):84–104. https://doi.org/10.1016/j.fuproc.2018.12.011

Zaman CZ, Pal K, Yehye WA, Sagadevan S, Shah ST, Adebisi GA, Marliana E, Rafique RF, Johan RB (2017) Pyrolysis: a sustainable way to generate energy from waste. In: Samer M (ed) Pyrolysis. InTech. https://doi.org/10.5772/intechopen.69036

Zamri MFMA, Hasmady S, Akhiar A, Ideris F, Shamsuddin AH, Mofijur M, Fattah IMR, Mahlia TMI (2021) A comprehensive review on anaerobic digestion of organic fraction of municipal solid waste. Renew Sustain Energy Rev 137(Nov 2020):110637. https://doi.org/10.1016/j.rser.2020.110637

Zhang L, Jahng D (2012) Long-term anaerobic digestion of food waste stabilized by trace elements. Waste Manage 32(8):1509–1515. https://doi.org/10.1016/j.wasman.2012.03.015

Zhang R, El-Mashad HM, Hartman K, Wang F, Liu G, Choate C, Gamble P (2007) Characterization of food waste as feedstock for anaerobic digestion. Biores Technol 98(4):929–935. https://doi.org/10.1016/j.biortech.2006.02.039

Zhang L, Zhang J, Loh KC (2019) Enhanced food waste anaerobic digestion: an encapsulated metal additive for shear stress-based controlled release. J Clean Prod 235:85–95.https://doi.org/10.1016/j.jclepro.2019.06.301

Zhang J, Gu D, Chen J, He Y, Dai Y, Loh KC, Tong YW (2021a) Assessment and optimization of a decentralized food-waste-to-energy system with anaerobic digestion and CHP for energy utilization. Energy Convers Manage 228(Nov 2020):113654. https://doi.org/10.1016/j.enconman.2020.113654

Zhang Y, Wang L, Chen L, Ma B, Zhang Y, Ni W, Tsang DCW (2021b) Treatment of municipal solid waste incineration fly ash: state-of-the-art technologies and future perspectives. J Hazar Mater 411:125132. https://doi.org/10.1016/j.jhazmat.2021.125132

Chapter 20
Integrated Approaches and Future Perspectives

Abdelrahman S. Zaky, Santosh Kumar, and Andrew J. Welfle

Abstract Many conversion technologies and processes for bioenergy generation from wastes have been reported and discussed in the previous chapters. These conversion technologies are being selected and applied depending on the type of wastes, chemical composition of the available waste and the desired energy vector. For example, anaerobic digestion (AD) is used for biogas production from mixed biological wastes with varied chemical compositions. Fermentation for bioethanol production is used for wastes that are rich in simple sugars and/or starch. Pyrolysis, combustion and gasification are used for crude bio-oil and/or syngas production from a wide range of wastes, especially those that contain high lignocellulosic compounds. These are environmentally friendly technologies for waste management and bioenergy production, but their economic feasibility is usually limited using a process of a single conversion route. Recent research and prospects suggested that integrating processes for bioenergy production from waste could increase the efficiency of the system in terms of economy, energy recovery and beneficial impact on the environment. This chapter discusses waste biorefinery as a recent trend towards circular bioeconomy. The chapter provides suggestions for future integrated systems for the simultaneous production of multiple energy vectors and high-value chemicals from different types of wastes. In the integrated system, the by-products from the first conversion process are used as substrates for the subsequent conversion process and so on. The importance of catalysis in offering flexibility in an integrated biorefinery system by providing novel routes and downstream environmental solutions for flue gas and exhaust gas cleaning was also covered in the chapter. The chapter concludes

A. S. Zaky (✉)
School of Biological Sciences, University of Edinburgh, King's Buildings, Edinburgh E9 3FF, UK
e-mail: A.Zaky@ed.ac.uk

S. Kumar
Diamond Light Source, Beamline B07, Harwell Campus, Didcot OX11 0QX, Oxfordshire, UK
e-mail: santosh.kumar@diamond.ac.uk

A. J. Welfle
Tyndall Centre for Climate Change Research, Department of Mechanical, Aerospace and Civil Engineering, Pariser Building, The University of Manchester, Sackville Street, Manchester M13 9PL, UK
e-mail: andrew.welfle@manchester.ac.uk

A. E.-F. Abomohra et al. (eds.), *Waste-to-Energy*,
https://doi.org/10.1007/978-3-030-91570-4_20

with a discussion of the role of models with bioenergy and biomass resource decision making, with focus on bioenergy from waste case studies. This includes assessment of the key issues that determine the economic feasibility and environmental impacts of feedstock choices and technology options. Also covering the social and political frameworks that will enable and drive transitions towards increased bioenergy from waste activities. The chapter presents policy case studies from the EU, China, USA and India, and highlights how social acceptance will be key to the success of any bioenergy from waste sector.

Keywords Integrated biorefinery · Economic feasibility · Environmental impact · Sustainable bioenergy · Policy frameworks · Social acceptance · Catalysis · Photocatalysis · Materials

20.1 Waste Biorefinery as a Recent Trend Towards Circular Bioeconomy

The world population exceeded 7.8 billion people as of April 2021 and it is projected to reach 10 billion people by 2057 (Worldometer 2021). They generate more than 2 billion tonnes of municipal solid waste annually, mainly biological wastes. The annual global waste is projected to grow to 3.4 billion tonnes by 2050 which is more than double the population growth over the same period (The World Bank 2021). Each tonne of waste is a huge loss of resources and although reducing wastes remains the most sustainable option, the generation of wastes is inevitable. Utilising biological wastes in biorefineries for bioenergy and co-products production is a sustainable approach for waste management and a good example of the circular bioeconomy.

The term circular bioeconomy is an integrated concept of bioeconomy and circular economy (Leong et al. 2021). This concept focuses on the conception of reuse, recycle and maintaining a sustainable manufacturing process to generate valuable bioproducts. Hence, circular bioeconomy can be considered as a low carbon economy since it exhibits the potential for developing a greener and more sustainable environment (Venkata Mohan et al. 2016b; Leong et al. 2021). The main benefits of circular bioeconomy include (**a**) improved resource and eco-efficiency, (**b**) lower GHGs footprints, (**c**) reduced reliance on fossil resources and (**d**) valorisation of waste materials from numerous sources such as agro-industrial, agriculture, aquaculture, fishery and many other sources (Leong et al. 2021).

Biorefining is a term that characterises the sustainable bioprocessing of biomass resources for the production of various bioproducts (e.g. lipids, proteins, carbohydrates, biomaterials and bioactive compounds), biofuels and direct bioenergy. It is a primary mechanism targeted within many circular bioeconomy strategies as it can close the loop of the basic resources, water, minerals and carbon (Dahiya et al. 2018; Leong et al. 2021). Waste biorefinery is currently receiving great interest as it attractive waste management approach that can provide many wider economic, environmental and social benefits (Dahiya et al. 2018; Leong et al. 2021).

Circular bioeconomy based on waste biorefinery processes may provide additional low carbon economy benefits by integrating CO_2 sequestration which can resolve many global issues. This strategy also represents a sustainable and economical mode for waste disposal (Venkata Mohan et al. 2016a; b). The valorisation of biological wastes into value-added bioproducts, such as biopolymers and biofuels, through bioprocessing, could potentially replace the utilisation of fossil fuels which supports the flow of ecologically friendly carbon. This approach often goes by many names included 'waste-as-a-value', 'waste-to-wealth' or 'zero-waste'—each being highly regarded as attractive, environmentally friendly and low-cost waste disposal pathways (Leong et al. 2021).

Shifting the economy from a petroleum refinery model towards a waste biorefinery model reflects a great effort on carbon management and GHGs mitigation. This can be achieved using biofuels-produced from waste-for various purposes (e.g. transportation). The production of biofuels and generation of bioenergy from waste materials through biological, biochemical and thermochemical roots is also regarded as a sustainable and low carbon alternative to existing practices and, in addition, can increase energy security (Almarashi et al. 2020; Abomohra et al. 2020). Waste biorefinery is a sustainable circular bioeconomy approach because it is based on the philosophy of recycle, reuse, remanufacture and maintain, unlike the linear economy approach which is based on the principle of take, make and dispose (Jorgensen and Pedersen 2018; Taherzadeh et al. 2019).

20.2 Integrated Waste Biofuel Production Systems

Biorefinery technologies provide high flexibility through the application of different conversion technologies allowing broad ranging categories of waste biomass to be converted to the full range of biofuels (biogas, bioethanol, biodiesel, etc.) and co-products. Many of these conversion technologies are described in Chap. 4. Traditionally, a system of single conversion technology is applied on a waste biomass to produce a single type of biofuel as the main product accompanied by one or more co-products, mainly fertilisers or animal feed. The traditional approach (single route) provides many environmental, social and economic advantages, albeit with limited energy recovery efficiencies. Recent research suggests that applying an integrated system (multiple routes) that applies consecutive conversion processes can increase efficiencies of biomass conversion to bioenergy (Kumar et al. 2018; Xu et al. 2019; Abomohra et al. 2020; Alherbawi et al. 2021). The integrated system produces more biofuels, more energy recovery, more co-products and less waste. The following are some examples of potential integrated systems for the integrated production of biofuels from waste biomass recently reported in the literature.

20.2.1 *Jet Biofuel, Ethanol and Power Co-production from Lipid-Cane Whole Crop*

Sugarcane is one of the most productive crops on the planet with gross yields ranging from 36 to 134 tonne ha^{-1} (Johnson and Richard 2005). Lipid-producing sugarcane, also known as lipid-cane, is the genetically modified sugarcane that has been engineered to accumulate lipids while producing sugars. In theory, sugarcane can accumulate up to 20% lipid by weight in the stems if all energy from the sucrose, is diverted to triacylglycerols (TAGs) (Kumar et al. 2018). Kumar et al. (2018) proposed an integrated production system for jet biofuel (JBF), bioethanol and power from lipid-cane. In this system, the whole plant of lipid-cane is utilised for the co-production of bioethanol, jet fuel and electricity. As shown in Fig. 20.1, the juice is extracted from lipid-cane and subjected to suitable treatment followed by a process for oil and sugar separation. The sugar is concentrated then subjected to fermentation by yeast for bioethanol production. The lipids are processed for jet fuel production as the primary product and diesel and naphtha as co-products. Bagasse can be used in combustion for steam and electricity generation. The system is projected to produce 43.7 L ethanol, 15.8 L JBF and about 100 kWh surplus electricity per metric tonne of lipid-cane containing 10% lipid. Based on lipid-cane containing 20% lipid, the system is projected to produce 31.5 L JBF and 156.9 kWh surplus electricity per metric tonne of lipid-cane with no ethanol production. However, the system could achieve a minimum JBF selling price of 0.58–1.43 $/kg for an annual plant capacity of 1.6 million tonnes (Kumar et al. 2018).

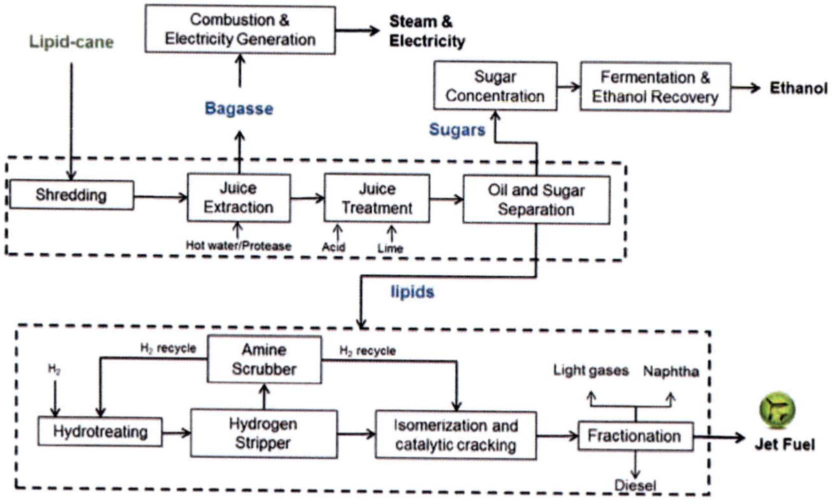

Fig. 20.1 Schematic diagram illustrating the major sections of biorefinery of lipid-cane biomass. Sugars are fermented to bioethanol, lipids are processed to produce jet fuel and bagasse is utilised to produce steam and electricity (Kumar et al. 2018)

20.2.2 Power Generation and Bio-oil Co-production from Jatropha Whole Fruit and Wastes

Alherbawi et al. (2021) investigated a novel integrated system that implements five integrated techniques including conventional hydro process, gasification, Fischer–Tropsch, reforming and power generation technologies for the production of a cost-effective jet biofuel and power as illustrated in Fig. 20.2. The system demonstrates different concepts, including valorisation of waste, incorporation of co-products and integration of heat and power. The modelled system revealed interesting results as 49% (wt) of Jatropha fruit was converted into liquid fuels, of which jet biofuel accounts for 64.5%. In addition, 7 wt% of jatropha fruit was converted into syngas. The generated electricity in this system covered 97.6% of the power requirement for the whole system. Furthermore, the system is water self-sufficient with a significant amount of additional produced water. At the economic and environmental levels, the proposed system could significantly lower the production cost of JBF to a minimum selling price of 0.445 $/kg which was below the market price for the base year of analysis (2019). The proposed system also managed to reduce the GHG emissions by 23% compared to the conventional Jet-A fuel and to reduce the land required to grow the feedstock by 50% as compared to utilising the jatropha oil alone (Alherbawi et al. 2021). This system utilises the whole fruit of jatropha which consist of 75 wt% residues (35 wt% shells and 40 wt% seedcakes) while oil is only 25 wt% (Singh et al. 2008), therefore, it could be considered an integrated system for bioenergy generation from wastes.

20.2.3 Biodiesel and Bioethanol Co-production from Waste Glycerol

In this system, the green microalgae *Scenedesmus obliquus* was cultivated on waste glycerol (WG) to enhance the accumulation of lipids and carbohydrates in the algal biomass. The lipids were firstly extracted from the microalgae biomass for biodiesel production by esterification. The carbohydrates in the lipid-free biomass were then used for bioethanol production by fermentation. The results confirmed that the use of WG as an organic carbon supplement in the growth medium of *S. obliquus* at 2.5 gL^{-1} has enhanced lipids and carbohydrates accumulation and subsequently enhanced the biodiesel and bioethanol yields. This has led to a significant impact on energy recovery, with biodiesel yield of 127.1 mgg^{-1} dw and bioethanol yield of 0.558 gg^{-1} dw. This integrated system recorded enhanced gross bioenergy output of 21.4 $GJton^{-1}$ and improved energy conversion efficiency of 62.9% (Xu et al. 2019).

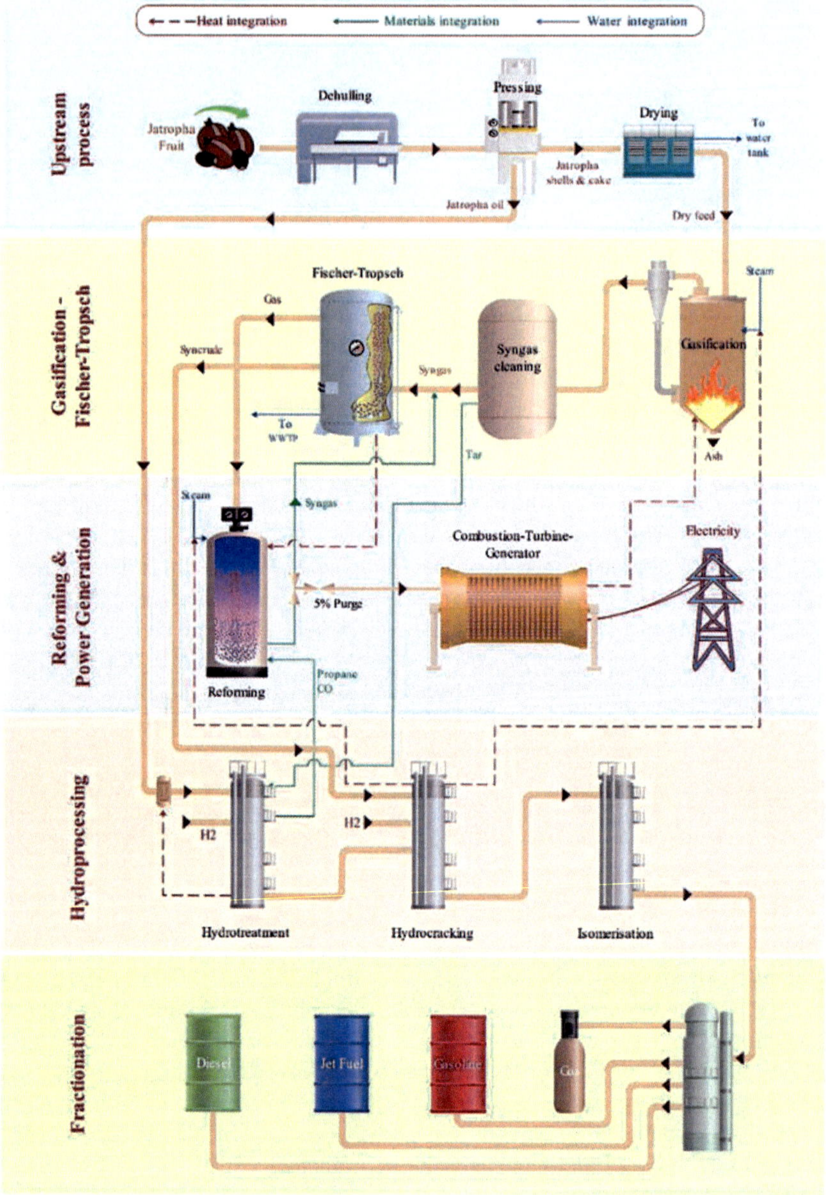

Fig. 20.2 The process flowsheet of the system of power generation and bio-oil co-production from Jatropha whole fruit and wastes (Alherbawi et al. 2021)

20.2.4 Biodiesel Followed by Biogas Production from Fat, Oil and Grease (FOG)

Fat, oil and grease (FOG) is a lipid-rich waste that includes yellow grease (used and refined cooking oil), brown grease (trap/interceptor grease) and animal fats (edible and inedible tallow, lard, white grease and poultry fat). The precursors of FOG are various edible foods such as oils obtained from cooking practices, meat in various forms, baked stuff, dairy products and beverages. Hence, FOG is generated from many sources including multi-storied apartments, food industries, restaurants, hotels and many other premises. FOG mainly consists of free fatty acids (FFAs), triacylglycerols (TAGs), esters, waxes, phospholipids, sterols and sterol esters. Yellow grease contains less than 15% FFAs while brown grease contains more than 15% FFAs (Canakci 2007; Abomohra et al. 2020). One of the most effective processes for FOG management is biodiesel production. This approach has been pursued using many conversion technologies including biological conversion (Badoei-dalfard et al. 2019), nanocatalytic technology (Sahabdheen and Arivarasu 2020), acid esterification (Sumannakarn et al. 2009), supercritical esterification (Ghoreishi and Moein 2013) and steam stripping (Usseglio et al. 2019). Biogas production from FOG was also investigated but the AD process faced major challenges, due to the high lipid content of FOG, resulting in a limited biogas yield. The accumulation of LCFAs in AD system changes the cellular morphology of the methanogens, decreases their cell permeability and limits the mass transfer within the system (Dasa et al. 2016; Amha et al. 2018). Therefore, the sequential energy recovery from FOG through biodiesel production followed by AD for biogas production could enhance the total energy yield (Abomohra et al. 2020).

Kobayashi et al. (2014) investigated an integrated production system for the co-production of biodiesel and biogas from restaurant grease trap waste (GTW). In this system, GTW was heated at 60 °C for 6 h to extract FOG from the other GTW components. FOG was then pumped away and used to produce biodiesel by trans-esterification. The GTW collected after FOG layer extraction was used as a single substrate for the AD to produce biogas. The system showed that the total energy produced from biogas and biodiesel in the integrated production system was about 40% higher than that produced from the co-digestion system which produces biogas only (Kobayashi et al. 2014). In addition, AD of the solids from the GTW significantly reduces both GHGs emissions and energy consumption (Tu and McDonnell 2016).

20.3 Coastal Integrated Marine Biorefinery (CIMB) System for the Production of Biofuels, High Value Chemicals and Co-products

Conventional biofuel production takes place based on an Inland-Freshwater-Biorefinery (IFB) system model and mainly utilises food crops that rely on freshwater and arable land. This has contributed to the ongoing discussions related to the food vs fuel debate (Zaky et al. 2021). In summary, there are constraints in the extent that we can sustainably produce food to feed ourselves in addition to growing feedstocks for the bioenergy sector. There are physical constraints on the levels of suitable land, water and resources for feedstock production. Activities that result in land use change where land is converted for use for feedstock produced can result in large net GHG emissions to the atmosphere as carbon locked within natural sinks is released. This can far outweigh any emission benefits gained through the production of the feedstock. A potential solution to this could be the Coastal Integrated Marine Biorefinery (CIMB) systems that rely solely on marine resources that can be converted into biofuels, high value chemicals (HVC) and other co-products. CIMB systems refrain from using arable land and freshwater throughout the production chain. This would increase the economic and environmental values of the products, potentially achieving negative Water Footprint (WF) and negative Carbon Footprint (CF) values (Zaky 2021).

CIMB systems are based on coastal locations with direct access to seawater and marine biomass. They rely on marine elements including seawater, marine biomass (seaweed) and marine microorganisms (marine yeast and marine microalgae). Seas and oceans cover more than 70% of the Earth's surface and contain more than 97% of the earth's water and the all minerals needed for biomass and biofuel production (Zaky et al. 2014). Marine biomass (seaweed) grows much faster than land plants and is more efficient in CO_2 fixation compared to terrestrial crops. Marine biomass represents a hugely abundant resource worldwide and can wash up on shores in millions of tons—biomass waste called 'green tide' for the green seaweed and 'golden tide' for the brown seaweed. Marine biomass can be also planted and cultivated to form sea forests of seaweed that potentially provide great advantages to the marine environment and they do not burn as the terrestrial forests do. Carbohydrate is the main chemical component in seaweed and therefore it was investigated primarily as a suitable candidate for bioethanol production (Kostas et al. 2017, 2020). Marine yeast is highly tolerant to osmotic and halotolerant, so they are suitable candidates for bioethanol production especially when seawater and salty biomass are used in the fermentation medium (Zaky et al. 2018, 2020). They are also more tolerant to the inhibitors generated during biomass hydrolysis compared to terrestrial yeast (Zaky et al. 2016; Greetham et al. 2019). Microalgae are able to fix CO_2 to produce biodiesel and HVC (Mondal et al. 2017; Ashour et al. 2019; Darwish et al. 2020). CIMB system can be also linked with other renewable energy systems such as offshore and inshore wind farms, solar energy and tidal energy (Zaky 2021).

The following are three proposed models for biofuel production using CIMB systems.

20.3.1 Bioethanol Followed by Biodiesel Production from Macroalgal Blooms Using a CIMB System

In this system, seaweed blooms are firstly processed to extract the HVC and hydrolyse the complex sugars into fermentable sugars for the next step. Seaweed hydrolysates, after HVC extraction, are subjected to fermentation process using seawater and marine yeast to produce bioethanol. The biogenic CO_2 generated during bioethanol production and the non-fermentable sugars left in the media are used as carbon substrates to produce marine microalgae. Biodiesel is produced from microalgae biomass after the extraction of HVC. The photobioreactor for microalgae production can be optimised to produce biohydrogen. Besides biofuel and HVC, the system produces several co-products such as distilled water, sea salts and salted animal feed (Fig. 20.3).

Seaweed and microalgae have been studied separately for their potential in biofuels production based on inland sites, but current research suggests freshwater

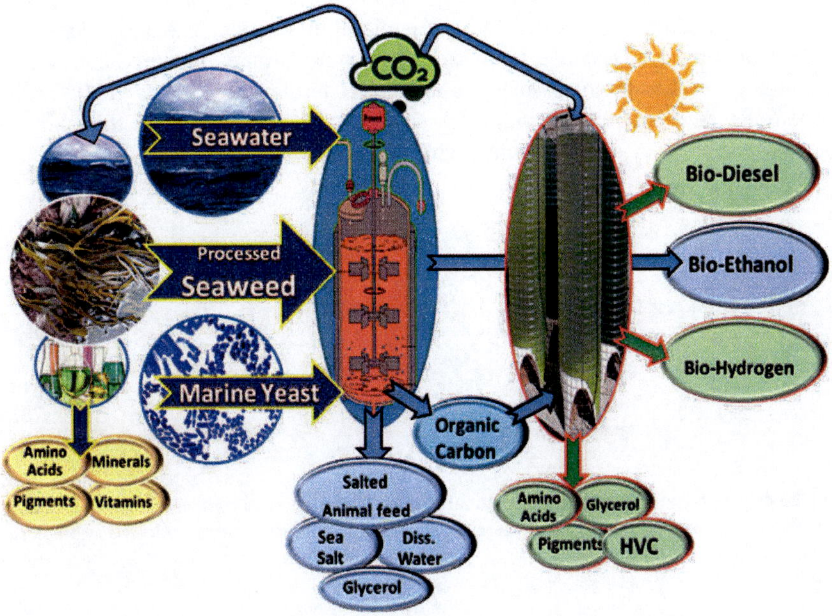

Fig. 20.3 Coastal integrated marine biorefinery (CIMB) system (Zaky 2021)

systems may be economically unviable. This system, therefore, propose the utilisation of both marine biomass (seaweed and marine microalgae) for the complementary production of biofuels (bioethanol, biodiesel and biohydrogen), HVC and co-products in a complete marine system based on coastal sites. The system has a great potential to reduce the water and carbon footprints of biofuels by involving seawater in the production and utilising the biogenic CO_2 released in the system. The system has also a great potential to enhance the economic viability and environmental impact of biofuels. Also, it can be advanced further by involving anaerobic digestion and integrating other inshore and offshore renewables into the system.

20.3.2 Bioethanol Followed by Biodiesel then Biogas Production from Macroalgal Blooms Using Advanced CIMB System

In this scenario, seaweed bloom is subjected to hydrolysis process to extract the valuable compounds and liberate sugars for the fermentation process. The fermentable sugars left in the seaweed hydrolysates are fermented to bioethanol using marine yeast in a dark bioreactor. The biogenic CO_2 resulting from the fermentation process is used as a carbon substrate to produce marine microalgae in a photobioreactor. Microalgal biomass is harvested, dewatered and processed to extract the HVC and lipids. The lipids are then processed for biodiesel production. The photobioreactor can be optimised to produce biohydrogen as well. Besides biofuels, the system produces several co-products such as animal feed, fertilisers, sea salt and HVC. The system is linked with inshore and offshore renewable energy systems to maximise the energy efficiency and economic viability of the system (Fig. 20.4a).

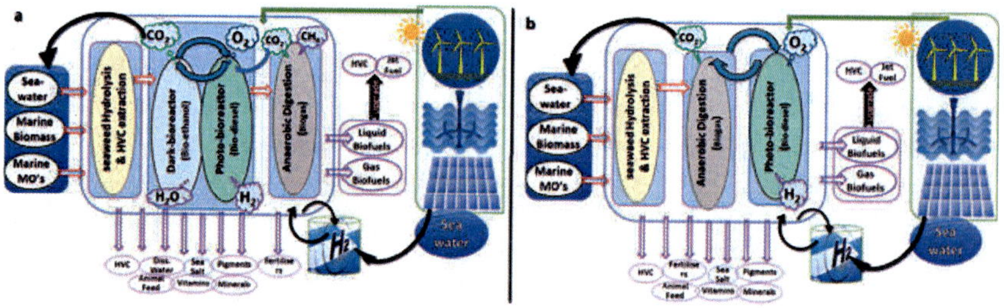

Fig. 20.4 Coastal integrated marine biorefinery (CIMB) system linked with other renewable energy systems. **a** Bioethanol followed by biodiesel then biogas production. **b** Biogas followed by biodiesel production (Zaky 2021)

20.3.3 Biogas Followed by Biodiesel Production from Macroalgal Blooms Using Advanced CIMB System

In this scenario, seaweed bloom is subjected to hydrolysis processing to extract the valuable compounds and break down the complex components of seaweed. Treated seaweed bloom is used as a feedstock for anaerobic digestion to produce biogas. Part of the produced biogas and/or the CO_2 resulting from biogas purification is used as a carbon substrate to cultivate marine microalgae. Algal biomass is processed to extract the HVC then the lipids are processed for biodiesel production. As in the previous scenario, the photobioreactor can be optimised to produce biohydrogen and the system can be linked with other renewable energy systems to maximise the energy efficiency and economic viability of the CIMB system (Fig. 20.4b).

20.4 Role of Catalysis in Bioenergy Production

Biomass feedstocks, carriers/platform molecules and conversions are inter-linked in an integrated biorefinery system. Catalysis-based technologies play an important role in creating flexibility in an integrated biorefinery system with an alternative strategy and downstream environmental solutions for the cleaning of flue and exhaust gases (Ragauskas et al. 2006; Lin and Huber 2009). It is critical for an integrated biorefinery system that the feasibility of the various strategies must be assessed using full life cycle assessments, including catalysis role on energy balance, cost and environmental impact (Huber et al. 2006).

20.4.1 Catalysis Role in Pre-treatment of Biomass Waste

Most biomass feedstocks contain significant amounts of heteroatoms including oxygen (O), nitrogen (N) and sulphur (S), unlike petroleum feedstocks, which have mostly C, H and some O in elementary distribution (Serrano-Ruiz et al. 2011). Therefore, when it comes to biomass processing, the biomass chemical and structural complexity is an important element to consider. Biomass can be broken down into small fractions to improve process ability through various pre-treatment steps which greatly reduce complexity by removal of O and adjustment of molecular weight and then facilitate efficient processing. Although several treatment methods are available for biomass processing, the chemical based catalytic pre-treatment methods outperform in terms of efficiency over the traditional pre-treatment methods, such as steam pre-treatment, torrefaction, microwave and hydrothermal pre-treatments in a full life cycle assessment of the biorefinery (Kumar and Sharma 2017). For example, acid-catalysed liquid hot water (LHW) treatment followed by enzymatic hydrolysis of cellulose for fermentable glucose is critical for ethanol and other HVC production from lignocellulosic biomass because it allows the efficient production of sugars from carbohydrates via the enzymatic route. (Suriyachai et al. 2020).

In general, the pre-treatment step is combined with further processes such as enzymatic hydrolysis to liberate the fermentable sugars for the fermentation process. Since enzymatic hydrolysis requires a rather long residence time, often takes days, these conventional pretreatment methods can be costly and a time-consuming procedure (Lindedam et al. 2014). Therefore, ionic liquids containing cations or anions are a new class of materials that have been widely used for the pre-treatment of lignocellulose in recent years (Zavrel et al. 2009). The chemistry between ionic liquids and biomass is greatly influenced by ions, temperature and time of the pre-treatment. In particular, the ionic liquids participate in hydrogen bonding with lignocellulosic molecules, which leads to disruption of lignocellulosic network. For example, Dadi et al. observed a two-fold increase in yield and rate of hydrolysis when they used 1-Butyl-3-methylimidazolium chloride for pre-treatment (Dadi et al. 2006).

Ionic liquid pre-treatment is promising over other methods due to their low vapour pressure of solvents, high thermal and chemical stability, relatively safe and simple processing conditions and wide temperature window for presence of liquid state. More importantly, these ionic liquids are easily recyclable and non-derivatizing. As other conventional pre-treatment techniques consume a lot of energy and generate a lot of waste during acid/base neutralisation or enzyme hydrolysis, biomass pre-treatment technologies based on catalysis can enable the production of fuel with better-defined specifications and increased energy densities providing great benefit to the bioeconomy as a whole. However, there are some drawbacks to using ionic liquid pre-treatment, including (a) the incompatibility of cellulase and ionic liquids, which results in cellulase unfolding and inactivation and (b) the cost. Furthermore, extremely high temperatures cause side reactions that have negative consequences such as reducing ionic liquid stability (Mäki-Arvela et al. 2010).

20.4.2 Catalysis Role in Lignocellulosic Biomass Waste Conversion

20.4.2.1 Catalytic Conversion of Lignocellulose Derived Platform Molecules

Following biomass pre-treatment, varieties of biomass-derived compounds, so-called platform molecules can be produced. For example, some of the sugar-based platform molecules are succinic acid, fumaric acid, malic acid, 3-hydroxy glucaric acid, 2,5-furan dicarboxylic acid, propionic acid, levulinic acid, aspartic acid, itaconic acid, glutamic acid, 3-hydroxybutyrolactone, arabinitol/xylitol, sorbitol and glycerol. These compounds have been classified based on their potential to generate value-added chemicals and fuels (Bozell and Petersen 2010).

Platform chemicals, also known as chemical building blocks, are complex molecules composed of a number of multifunctional structural molecules that can be easily converted into a wide range of value-added products within a biorefinery using catalysis technology as shown in Fig. 20.5a (Lin and Huber 2009). These molecules and catalysts are carefully selected based on a variety of criteria, including market availability and their potential for producing value-added fuels and chemicals in biorefineries (Lin and Huber 2009; Bozell and Petersen 2010). Liquid-phase, lower temperature cascade catalytic routes involving reactions such as hydrolysis, isomerization, dehydration, oxidation, aldol condensation, hydrogenation and dehydrogenation will direct selective biomass processing (Lin and Huber 2009; Ruppert et al. 2012).

The overall challenge with biomass conversion is selective removal of oxygen (and nitrogen in the case of biomass like algae) from biomass feedstocks using above cascade catalytic routes while producing a molecule with a high energy density and good combustion properties as shown in Fig. 20.5b (Lin and Huber 2009). In general, oxygen is removed as a combination of CO_2 and H_2O in multiple steps during the biomass conversion processes. The first step, for example, is the deconstruction of solid lignocelluloses into smaller hydrophilic products. During the second conversion step, more oxygen is removed from the hydrophilic materials and the biomass is converted into the value-added products CO_2 and H_2O. As a result, biomass conversion necessitates multi-stage chemistry. The ideal fuel will almost certainly be one that is compatible with existing infrastructure and has properties similar to gasoline, diesel and jet fuel as shown in Fig. 20.5b (Xu et al. 2014a). The ideal catalysis-based conversion process within a biorefinery would maximise biofuel production yield while remaining as cost-effective as possible for a sustainable bioeconomy.

20.4.2.2 Considerations for Catalyst Design and Selection

Catalyst design and selection for biomass conversions necessitates careful pore structure tailoring to minimise mass transport limitations. This should improve the hydrothermal stability in aqueous conditions with wide range of pH, resistance to any kind of leaching and tuneable hydrophobicity which is a key factor for reactant/product adsorption on the catalyst surface (Wilson et al. 2012). For example, biomass platform molecules obtained from fermentation are frequently present in aqueous solutions with typical concentrations of 10% or less, in addition to some of the most common polar molecules (Cheng et al. 2012). Corma Canos et al. (2007) conducted a thorough review of conventional methods for converting these platform molecules into chemicals and suggested that catalyst development should prioritise the use of tuneable and high surface area porous material as supports to improve reactant adsorption on active acid/basic sites (Corma Canos et al. 2007). Porosity of the catalyst is also important in allowing very bulky and viscous molecules to diffuse to active sites on catalyst surface (Sudarsanam et al. 2018). Various strategies for handling biomass based platform molecules will differ significantly from those used in petroleum processing, necessitating reverse chemical conversions in

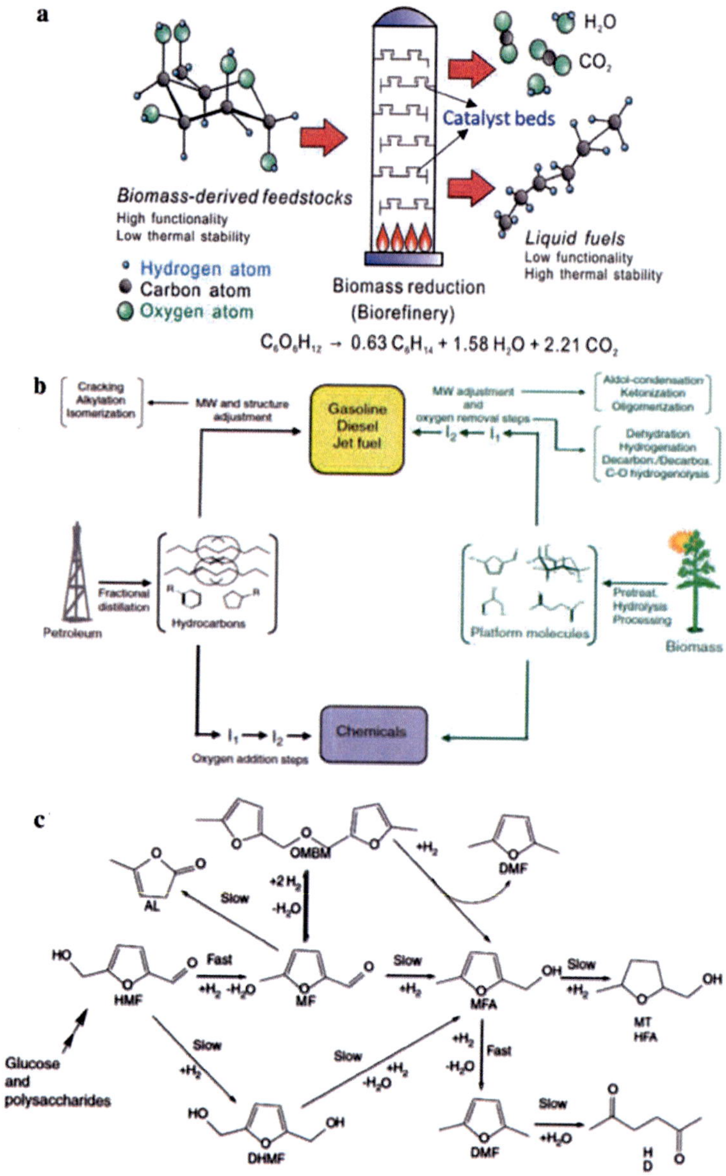

Fig. 20.5 **a** Oxygen removal from biomass platform molecules in biorefinery using catalysis technology (Lin and Huber 2009). **b** Petroleum vs biomass processing to chemicals and fuels (Xu et al. 2014a). **c** Possible reaction products obtained from cascade-type reaction dehydration/hydrogenation of Glucose/HMF (Yepez et al. 2013). Reproduced by permission of the Royal Society of Chemistry

which highly functional biomolecules are deoxygenated to their target product rather than oxygenated, as is typical process when we start with crude oil (Schlaf 2006). To facilitate the transition away from currently available short and medium chain hydrocarbon supplies, there is great demand for novel catalyst that is compatible with support that possesses hydrophilic and bulky in nature.

20.4.2.3 Multifunctional Catalysts

Multifunctional catalysts are great assets to facilitate various cascade reactions such as hydrolysis/hydrogenation and dehydration/hydrogenolysis and play significant role in catalytic processing for biomass conversion (Rinaldi and Schüth 2009; Sudarsanam et al. 2018). The design of multi-step reactions has the added benefit of allowing optimised conditions to be translated into a more scalable catalytic process with continuous flow in which various reaction parameters can be controlled to maximise biomass conversion.

Several studies have identified various pathways to key chemical intermediates from biomass feedstock, with succinic acid being one such valuable platform molecule from which a diverse range of value-added chemicals and fuels can be derived (Sudarsanam et al. 2018). Succinic acid may also enable the development of new biopolymers-based polyamides, polyesteramides and polyesters. Cu-based aluminosilicates, for example, were microwave-irradiated catalysts that were compared to counterpart Pd systems and other commercially available catalysts (Yepez et al. 2013). The reactions, as shown in Fig. 20.5c, involve multiple steps, the first of which is the dehydration of sugars to HMF and the second of which is hydrogenation/hydrogenolysis, which produces a variety of products. Succinic acid is one of the main products. Furthermore, carbon-based solid acid catalysts for succinic acid esterification with ethanol have been demonstrated to be effective (Clark et al. 2008).

A case study based on a market responsive biorefinery concept that produces high-octane gasoline (HOG) and jet fuel blend stocks from biomass using a multifunctional catalyst and methanol as an important intermediate chemical (Fig. 20.6a). In particular, Ruddy et al., linked both fundamental and applied catalyst development research with process models and techno-economic analysis to quantify the impact of catalyst advancements on process economics (Ruddy et al. 1929). They discovered that by facilitating the reincorporation of by product isobutene during dimethyl ether homologation via acid-catalysed coupling, a beta zeolite supported Cu catalyst enabled a 38% increase in HOG product yield and a 35% reduction in conversion cost (Fig. 20.6b) when compared to the benchmark beta zeolite catalyst (Fig. 20.6c, d) (Ruddy et al. 1929; Simonetti et al. 2011). This demonstrates an alternative pathway that improved the fuel properties of the resulting distillate-range product. They also show how C4 by-products can be used to make synthetic kerosene that meets five specifications for typical jet fuel at a minor cost increase over the HOG alone case. The cost of fuel synthesis for HOG and distillate products from methanol was $69

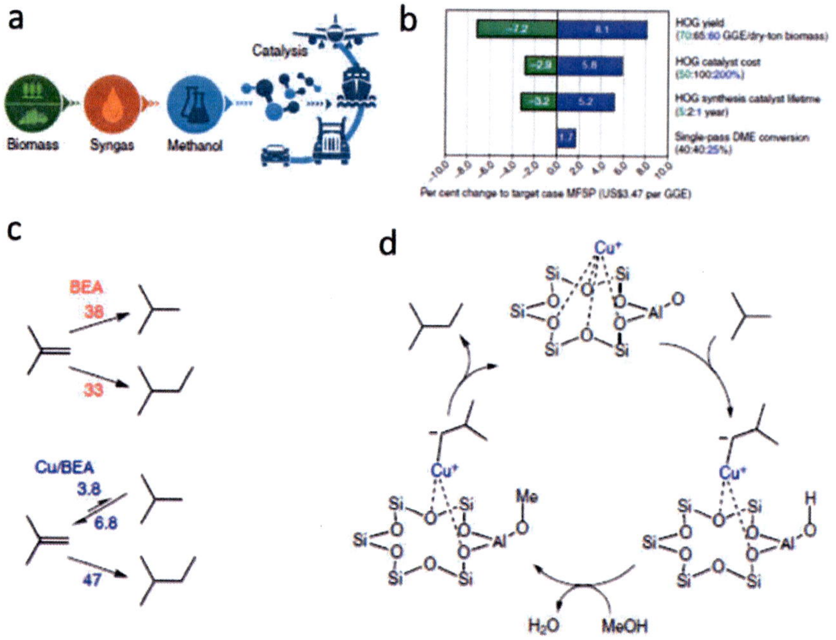

Fig. 20.6 **a** Schematic diagram for a market responsive biorefinery. **b** Analysis of critical catalyst parameters in the synthesis of HOG reactor. MFSP and DME stand for minimum fuel-selling price dimethyl ether, respectively. **c** Isobutane and isobutene conversion rates (μmol $mol^{-1}s^{-1}$) on Beta zeolite (BEA) and Beta zeolite supported. **d** Proposed reaction mechanism of direct methylation of activated isobutane via reaction of a nucleophilic copper isobutyl intermediate with an electrophilic zeolite-bound methyl. Reproduced by permission of Nature Publishing Group (Ruddy et al. 1929)

per gasoline-gallon equivalent (GGE), which is comparable to the cost of synthesis for HOG only, which was $66 per GGE.

20.4.3 *Catalysis Role in Algae Biomass Waste Conversion*

Over the last few decades, microalgal biomass contributing to the development of various high-end (food supplements, nutraceuticals, colourants and pharmaceuticals) and low-end (biogas, biodiesel and bioethanol) products because of its continuous growth and high carbon fixing efficiency (Costa et al. 2019). As a result, microalgae are regarded as a valuable and renewable resource for achieving net zero targets while reducing our current usage of fossil fuels. Thermochemical processes such as hydrothermal liquefaction (HTL), pyrolysis, gasification and have emerged as more viable among the technologies available in the market for producing bioenergy from microalgae biomass as they use biomass in full capacity (Chen et al. 2015; Raheem et al. 2015; Ong et al. 2019). However, the complex structure of algae biomass which contains carbohydrates, lipids and proteins can provide challenges that need to be overcome to increase the sustainability of algal-based bioenergy feedstocks. Essentially, algae-derived fuels require additional catalytic processing to remove the O and/or N. Because of the low sulphur concentration in algal biofuels, sulphur removal, also known as hydrodesulfurization (HDS) is frequently not an issue. Due to the thermodynamic limitations of the aliphatic C–N bond hydrogenolysis reaction, removing N from heterocyclic compounds, also known as hydrodenitrogenation (HDN) is a more difficult process than removing S (Girgis et al. 1991). As previously stated, O removal from algae occurs in several steps, including catalytic cracking, hydrodeoxygenation (HDO) and decarboxylation/decarbonylation. The overall mechanism is quite complex because three mechanisms can occur concurrently in a single biomass processing reaction.

Various microalgae based biomass have been studied using heterogeneous catalysts such as supported metal catalysts including Ni, Cu, Pd, Au, Pt and R) and various metal oxide catalysts (Duan and Savage 2011a; Tian et al. 2014; Djandja et al. 2020). However, the impact of metal catalysts on biocrude production yield is very complex as all of the metal catalysts cannot improve yield; in fact, in most instances, they significantly reduce HTL's performance. For example, Nava Bravo et al. (2019) reported that the algal biomass composition, (carbohydrates, lipids, protein and ash content) also plays a critical role in determining bio-oil production rate and yield (Nava Bravo et al. 2019). Interestingly the Ni-based catalyst can improve overall biomass conversion by catalysing bond cleavages and thereby the process of depolymerization. In another study, the catalytic efficiency of HZSM-5 type zeolites and Raney-Ni in the presence of *C. pyrenoidosa* biomass was assessed using ethanol as a solvent (Yang et al. 2011). The results show that the catalyst has no effect on bio-oil production rate and yield under the various conditions tested. The catalyst used, on the other hand, increased the production yield of other reaction products, like gasoline different hydrocarbons. A few more zeolite-based catalysts, such as Ce/H-ZSM-5 and H-ZSM-5 have also shown increasing production yields of bio-oil from 32 to 38% and 52%, respectively (Xu et al. 2014b).

According to some reports HDN mechanisms include intermediate saturation, elimination and nucleophilic substitution using model nitrogenated chemicals such as pyridine/piperidine, quinoline/tetrahydroquinoline and indole/indoline (Duan and Savage 2011b; Ho 1988). However, the underlying reaction pathway of algal biomass conversion on heterogeneous catalysts remains unknown. Furthermore, specific challenges, such as the presence of a high concentration of heteroatoms such as oxygen, nitrogen and sulphur in biofuels, potential corrosion of reactor material, uncertainty in recyclability and catalyst deactivation, impede the catalyst's long-term application in algae conversion to biofuels. As a result, developing novel catalysts through understanding is a critical step for the selective conversion of microalgae into bioenergy products in order to boost process efficiency and product selectivity in biorefinery.

20.4.4 Photocatalysis Role in Biomass Waste Conversion

Nature-inspired artificial photosynthesis or photocatalysis is a clean way of converting biomass to chemicals and fuels and is believed to be one of the best methods to overcome both the energy crisis and global climate change (Faunce et al. 2013). Photocatalytic biomass conversion to sustainable hydrogen at room temperature is potentially the most affordable route compared to energy intensive thermal processes (e.g. Gasification), which require high temperature (>750 °C) to decompose organic structure and release hydrogen (Wakerley et al. 2017). In 1981 a Pt–TiO_2 photocatalyst was first used to generate H_2 from various plastics such as Teflon, polyethylene and polyvinyl chloride from biomass such as wood, grass, flowers, rice plant, and seaweed; also food wastes such as olive oil and sweet potato, under both neutral (H_2O) and alkaline (5 M NaOH) conditions (Kawai and Sakata 1981; Sakata and Kawai 1981). In addition to H_2 various oxidative products such as (CH_3CHO, ethanol (C_2H_5OH), acetone (CH_3COCH_3) and acetaldehyde were seen. Pt–TiO_2 photocatalysts were further used in several other studies for photocatalytic reforming of paper, wood, grass, bamboo and rice (Caravaca et al. 2016; Uekert et al. 2021). In recent years CdS and carbon nitride have emerged as an alternative to proto-type TiO_2 photocatalyst due to their ability to utilise visible light and work without an expensive metal co-catalyst such as Pt and Au. For example, CdS quantum dots were used in aqueous alkaline solution for photocatalytic conversion of biomass (Kadam et al. 2014), plastics (Uekert et al. 2018) and food waste (Uekert et al. 2020) with production rates up to 9350 μmol H_2 $g^{-1}h^{-1}$. Interestingly, various oxidation products such as formats and acids, have also been observed (Kadam et al. 2014; Uekert et al. 2018, 2020).

Finally, while photocatalytic biomass conversion is now mostly employed for lab-scale H_2 production, the method has the potential to extend large-scale biohydrogen production; nonetheless, the process is limited in its practical application due to efficiency and the CO_2 produced in the process. Photocatalytic biomass conversion could theoretically be combined with CO_2 reduction to CO, CH_4 or CH_3OH, nitrogen reduction to ammonia, toxic metal reduction in water body. However, this coupled photocatalytic process faces additional challenges such as poor adsorption over many inorganic photocatalysts, low biomass solubility and low yield (Uekert et al. 2021). Furthermore, direct biomass and CO_2 reduction necessitates a higher negative reduction potential and multi-electrons sluggish kinetics to produce CO, oxygenate products such as formic acid and methanol and hydrocarbon products such as methane, ethane and ethylene) (Ran et al. 2018; Stolarczyk et al. 2018). Although preliminary techno-economic and life cycle assessments on photocatalytic biomass conversion revealed that its carbon footprint is significantly lower than or comparable to currently available methods for H_2 production and waste management, the production cost and energy balance need to be improved further before industrial application can be considered (Table 20.1). A lack of understanding of the photocatalytic process and its underlying mechanism has hampered the development of optimal materials and the production of selective products. As a result, developing

Table 20.1 Comparison of photocatalytic process to alternative waste-to-fuel technologies. Reproduced by permission of Nature Publishing Group (Uekert et al. 2021)

Conversion method	Gasification	Pyrolysis	Fermentation	Waste to chemical	Photocatalytic process
Feedstock	Biomass/Plastic	Biomass/Plastic	Biomass/Food	Biomass/Food	Biomass/Polar plastic/Food
Temperature (°C)	600–1000	300–900	25–70	37–220	10–60
Products (and side products)	H_2 (main), CO, CH_4, CO_2, ash and tar	Bio/synthetic crude oil (main), ash and char	H_2 (main), CH_4, CO_2, NH_3 and organics, solids	Levulinic acid	H_2 (CO_2, organics)
Price (£ GJ^{-1} product)	9–42	1–31	11–380	1.3–6.5 (£ kg^{-1} acidic product)	60–290 (pilot scale)
Carbon footprint (g CO_2e)	13–124 MJ^{-1} H_2	10–85 MJ^{-1} crude	28–5000 MJ^{-1} H_2	800–5300 kg^{-1} acid	35–125 MJ^{-1} H_2
Technology readiness level (1–9)	6–8	5–8	5–6	7–8	3–4
Advantages	Fast, high yields and scalable	Compatible with many wastes and also product compatible with existing fuel Technology	Bio-based and low energy use, compatible with wet waste and low capital cost	High yields and high value products and fast	Compatible with mixed and wet wastes, solar energy, low temperature; High purity H_2 and simple process
Disadvantages	High temperature, dry waste, post purification of gases and high capital cost	High temperature, different wastes yield different oil compositions and post-purification of oils	Not compatible with plastic/mixed waste, pilot scale and slow process	Well-defined waste streams and low usage of plastic waste	Lab scale, low productivity and; slow process

photocatalytic processes with a thorough understanding to avoid secondary pollution, as well as achieving high conversion efficiency and selectivity, remain major challenges in biomass conversions for solar-biorefinery towards a circular economy. Future research directions include careful reactor/cell design for operando studied liquid–solid interface for better understanding reaction mechanism and for large scale scald capability. Furthermore, the coupling of photocatalytic oxidation organic waste or biomass with other reduction reaction strategies.

20.5 The Importance of Modelling for Bioenergy and the Role of Wastes

20.5.1 The Role of Modelling in Energy Strategy Development

Each country's decision makers are faced with unique challenges when designing policy frameworks and strategies to drive progress towards decarbonisation (Welfle et al. 2020a). Energy models are key tools used by decision makers when developing energy strategies. They are particularly valuable when used to road-test plans before they are implemented, potentially highlighting the impacts of policy/ interventions to enable better informed decisions (Savvidis et al. 2019).

Models may provide quantitative insights into alternative designs, energy strategies or technologies, decreasing the unknowns and uncertainties of different options. As the suite of available energy technologies has increased and become more complex, there has been a corresponding increase in the number and approaches of energy focused models. Energy models range from macro Integrated Assessment Models (IAMs) that assess the performances of energy systems within the context of global earth and human systems; to Energy Systems Models that are designed with specific focus on energy technologies and how they may be deployed to meet energy targets and; Specialist Models that are typically highly bespoke in design, focusing of analysing specific technologies, issues or value chains.

20.5.1.1 Bioenergy Within Energy Models

Energy models focused on bioenergy typically analyse one of more of the intrinsic stages that are key to all bioenergy systems:

- **Biomass Resources**—that fuel bioenergy technologies. Models focus on themes such as the sustainable supply of biomass resources, feedstock characteristics, temporality and timings and the spatial dimensions of resource availability. Determining the types and extent that different categories of biomass resources are available and the characteristics of resources are fundamental in influencing the success of the overall bioenergy systems (Welfle et al. 2020a).

- **Conversion Technologies**—that generate bioenergy, biofuels and other bioproducts from different biomass feedstocks. Choices between conversion technology options and their performances will influence the forms and extent of bioenergy vectors that may be produced, in addition to influencing the environmental and economic performance of the system (Welfle et al. 2020a).
- **Systems Issues**—the sustainability of bioenergy systems and biomass resources used as feedstocks will be linked to natural systems, industry sectors, society, etc. In contrast to other renewable technologies, bioenergy systems may present unique challenges. For example, many bioenergy feedstocks are the waste by-products of wider processes such as agriculture and industrial wastes. There may be potential competition for these resources with wider sectors, there may be wider sustainability implications to consider such as impacts on natural systems. It is important to identify and evaluate both the potential benefits and impacts of bioenergy systems in order to determine sustainability performances (Welfle et al. 2020a).

20.5.1.2 Different Approaches of Bioenergy Models

There are countless models that have been designed to assess bioenergy themes. These may be broadly categorised as either: (i) Integrated Assessment Models (IAM) that analyse the interactions between natural and human systems, (ii) Energy System Models (ESM) that are typically technology focused and (iii) Specialist Modelling that provide assessment on specific themes, technologies or processes (Welfle et al. 2020a). Specialised bottom-up modelling applies methods to test the performance of specific case studies such as life cycle assessment (LCA) analysis, techno-economic analysis (TEA). Further models may use software and techniques such as geographic information systems (GIS) to measure the spatial contexts; or may use Process Model methods and software to analyse questions around specific bioenergy technologies or supply chains.

Figure 20.7 provides a summary of the characteristics and typical approaches of the main categories of models used to analyse energy and bioenergy systems. The axes of Fig. 20.7 reflect the scale and dimensions on which the different categories of model focus. IAMs are shown to focus on large global scale analyses, as a consequence, they have to keep narrow dimensions when assessing issues. At the other end of the scale, specialist models are shown to be highly focused on specific bioenergy issues, achieved through analysing a large number of dimensions for the target case study or project.

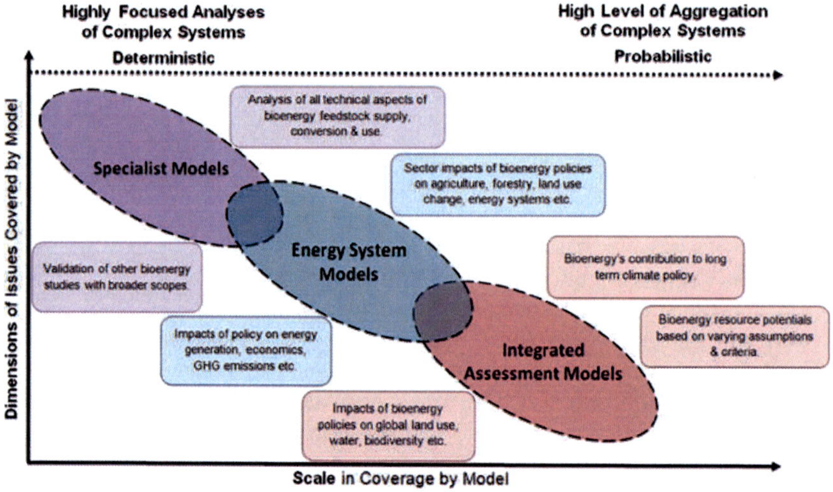

Fig. 20.7 Energy modelling categories and their approaches and capability for bioenergy analyses (Welfle et al. 2020a)

20.5.2 Role of Resource Modelling Within Bioenergy Strategy Development

Biomass resource models are typically developed with the primary objective of estimating the availability of biomass resources from different sources within a chosen geography. Biomass availability calculated through resource models is typically categorised in terms of a hierarchy of 'availability potentials'. Depending on the focus of the analyses undertaken and the choices of constraints applied, biomass resource forecasts may be extremely large or highly focused:

- **Theoretical Availability Potentials**—provide estimates of the extent that biomass resources may be grown/harvested/mobilised, restricted only by physical and/or biological barriers. Theoretical forecasts have limited use for estimating realistic biomass production levels for bioenergy, but can be used to provide comparative indicators of production potential (Welfle 2014).
- **Technical Availability Potentials**—provide estimates of the extent that biomass resources may be grown/harvested/collected restricted by technical constraints such as technology efficiencies. Technical forecasts may be useful when evaluating how resource availability may change as a consequence of changing technology dynamics such as the introduction of technological advances (Welfle 2014).
- **Economic Availability Potentials**—provide estimates of the extent that biomass resources may be grown/harvested/collected restricted by economic constraints, such as those driven by supply–demand curves. This can be a highly variable modelling method as the markets and resulting economic conditions can change

widely over time—the key variable is often the cost associated with accessing the bioenergy feedstock. Economic forecasts are typically useful for evaluating the feasibility of bioenergy supply chains and wider systems compared to alternative energy pathways (Welfle 2014).

- **Sustainable Availability Potentials**—provide estimates of the extent that biomass resources may be grown/harvested/collected restricted by environmental constraints. Sustainable forecasts are typically useful to evaluate the scales of resources that may be sustainably mobilised without risking the likelihood of continuing to source future resources (Welfle 2014).

When comparing the outputs from biomass resource models, it is vital to consider the approach of the methodology and, specifically the choices of constraints considered. Only the outputs derived from models with comparable approaches can be realistically compared on a like-for-like basis.

A further important distinction when comparing the outputs of different biomass resource models is the concepts of 'biomass potential' and 'bioenergy potential'. Biomass potential typically refers to an estimation of the amount (weight, volume, etc.) of a biomass resource over a given timeframe. Studies may reference biomass potential as the gross amount of energy contained within the biomass quantified. The term 'bioenergy potential' typically reflects the extent of that energy may be generated from a given quantity of biomass resource, taking into consideration of conversion efficiencies and losses of a chosen conversion pathway. These distinctions are not always clear but should be investigated when comparing outputs from biomass resource models and resulting studies.

20.5.3 Biomass Resource Modelling—Assessing the Potential of Waste Resources

Analysing the availability of different types of wastes and assessing their suitability as potential bioenergy feedstocks is the fundamental objective for bioenergy from waste resource models. The classification of different wastes is a key factor that will influence how waste resources may be collected, transported, managed, disposed and if/how they may be used as bioenergy feedstocks (UK Government 2018). This 'fuel classification' may also determine whether biomass is listed as a product/co-product, waste or a residue—an important differentiation in some countries. For example, Table 20.2 lists the categories of waste and residue resources in the UK that are supported as potential feedstocks for generating bioenergy or biofuels.

Estimating the potential availability of different categories of wastes for the bioenergy sector is typically calculated based on the extent that wastes are generated within a chosen geography and the characteristics of the existing waste management strategies. As a starting point, it is typically assumed that organic waste materials currently sent to landfills should be prioritised for potential use for bioenergy. The potential levels of bioenergy that may be generated from these wastes can be estimated by

Table 20.2 Waste and residue resources supported as potential bioenergy feedstocks within the UK (DfT 2018)

Waste and residue feedstocks		
Products	• Acid ester • Brown/sulphite liquor • Corn or wheat dried distillers grain (DDGS) • Corn oil • Crude tall oil • Glycerol (refined) from virgin oils • Meal from virgin oil production • Molasses	• Palm fatty acid distillate (PFAD) Palm kernel oil • Palm oil olein • Palm stearin • Slaughter products • Starch slurry regular • Sugar beet pulp • Tallow • Virgin oils
Agricultural residues	• Arboriculture residues • Bagasse • Cobs • Forestry residues	• Husks • Nut shells • Straw
Wastes and processing residues	• Brown grease • Cashew nutshell liquid • Crude glycerine • Empty palm fruit bunches • Ethanol used in the cleaning/extraction of blood plasma • Food wastes (unsuitable for animal feed) • Grape marc and wine lees • Low grade starch slurry • Waste starch slurry • Manure • Organic municipal solid waste (MSW) • Palm oil mill effluent (POME) • Poultry feather acid oil	• Rapeseed residue • Renewable component of end-of-life tyres • Roadside grass cuttings • Sewage sludge • Sewage system FOG • Soap stock acid oil contaminated with sulphur • Spent bleaching earth • Sugar beet tops, tails, chips and process water • Tall oil pitch • Tallow • Used cooking oil • Waste pressings from production of vegetable oils • Waste wood
Non-food cellulosic and lignocellulosic material	• Miscanthus	• Short rotation coppice (SRC)
Renewable fuels of non-biological origin	• Carbon dioxide	• Water
Other material	• Free fatty acids • Acid oils • Yellow grease	• Soapstocks • Used cooking oil (UCO) mixed with animal fats

applying assumed calorific energy content values and conversion efficiencies for the chosen bioenergy technologies (Welfle and Alawadhi 2021).

20.5.4 *Role of Life Cycle Assessment Modelling Within Bioenergy Strategy Development*

Life cycle assessment (LCA) is a popular and widely implemented approach for evaluating the environmental impacts of products, services, or systems. LCAs undertaken for bioenergy systems focus on assessing the activities and processes at each stage of a bioenergy value chain. Within a bioenergy GHG emission LCA, the net impact of each life cycle step is analysed by calculating the net flux of GHG emissions (predominantly CO_2, CH_4 and N_2O) within the atmosphere. Figure 20.8 provides an example of a typical bioenergy value chain. In the case of forest residue biomass, key life cycle steps may include: growth of the forest biomass, collection and harvesting of the resource, transportation to a site where the resource is converted into a fuel such as wood pellets. These pellets may then be transported to a bioenergy facility where the fuels are converted to bioenergy. There may also be post-conversion processes such as management of ash material. At each step, there will likely be energy expended that may generate GHG emissions in addition to direct capture or release of GHGs to the atmosphere. An LCA will aim to map and estimate the flows of GHG emissions over the whole life cycle in order to evaluate the overall GHG performance of the system (Welfle et al. 2020b).

To determine the comparative GHGs performance of bioenergy systems it is important to make comparisons with reference energy systems, for example comparing bioenergy GHG performance to that of fossil fuels. This reference energy system should be chosen to provide a comparison between the bioenergy system with the energy systems that it is likely to replace, for example comparing the GHG performance of energy generated from an anaerobic digestion biogas system with that of a natural gas fossil fuel system.

A further important consideration when undertaking an LCA is the scope of analysis, typically defined as the 'system boundary'. The LCA calculations will assess activities taking place within the system boundary, so it is vital that this also includes vital upstream processes relating to the collection of biomass resources, also end-of-life processes such as management of wastes. The system boundary needs to be designed so the bioenergy system analysed can be related to that of the comparable reference energy systems.

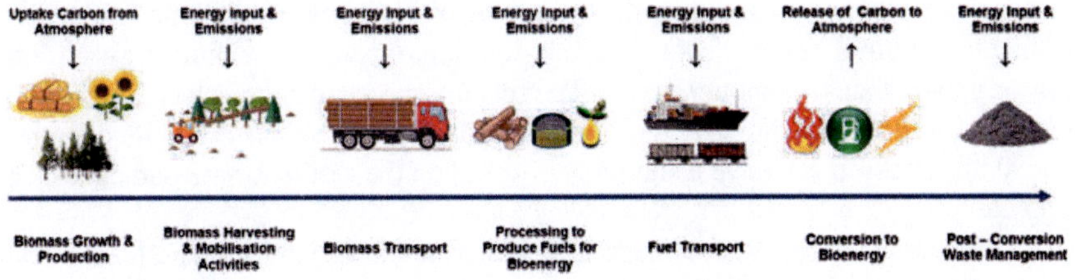

Fig. 20.8 Example life cycle steps for a bioenergy pathway (Welfle et al. 2020b)

Many LCA studies have been completed and the majority of these demonstrate that bioenergy systems can deliver energy with GHGs performances (sometimes far) exceeding that of fossil fuel systems. However, there are circumstances when bioenergy may generate more GHGs emissions than fossil fuels. Bioenergy systems calculated to have high GHGs emissions footprints are more often than not caused by an unsustainable sourcing of biomass, use of inefficient technologies or bad use of land resulting in land use change (Welfle et al. 2017).

LCA provides a very flexible tool that can be adapted to address a wide variety of different research questions. LCA can provide overall performance values to allow systems to be compared and benchmarked against targets. The analyses of each individual life cycle step can also provide a mechanism to identify the extent that specific activities contribute to the overall performance of the system. Where specific activities are found to generate large impacts, actions can be made to change the design of the system to reduce these impacts (Thornley et al. 2015).

To undertake LCA analysis it is important to be aware of the potential limitations of the approach. The accuracy of any LCA analysis will be reliant on the data and assumptions used to model the system (Aguilar 2014). For example, when modelling natural systems such as the carbon dynamics of land, there will always be certain levels of variability, this 'aleatory uncertainty' is largely irreducible to models (Hutton et al. 2010). This tends to lead to analyses with either multiple scenarios to reflect performance ranges or use of average or mean values to reflect certain parameters. In both cases, there will be uncertainty, either through the presentation of multiple values or a single value with error margins (Welfle et al. 2017).

20.5.4.1 Importance of Counterfactuals When Assessing the Performance of Bioenergy Systems

To analyse bioenergy systems, it is important to also analyse counterfactuals that describe what may otherwise have happened. For example, what would otherwise have happened to waste materials if not used as a feedstock for bioenergy technologies. For waste materials, the question is, how would the wastes have been managed? If this is through a potentially high environmental impact pathway such as being sent to landfills, using the wastes for bioenergy could result in the mitigation of large impacts associated with landfilling. Alternatively, if the waste materials may otherwise have been recycled and reused, their use as bioenergy feedstock may not result in the mitigation of counterfactual impacts and the impact of the bioenergy system may be greater than the counterfactual. The specific choice of counterfactual scenario and the scale and time horizon over which the bioenergy scenario and its counterfactual are assessed can have a significant impact on the results of assessments such as LCA.

Example bioenergy and counterfactual scenarios for agricultural and food waste resources are presented in Fig. 20.9. Various lifecycle steps are documented for both bioenergy and counterfactual scenarios. It is important to calculate the impacts associated with each stage of the bioenergy scenario, these should be compared to

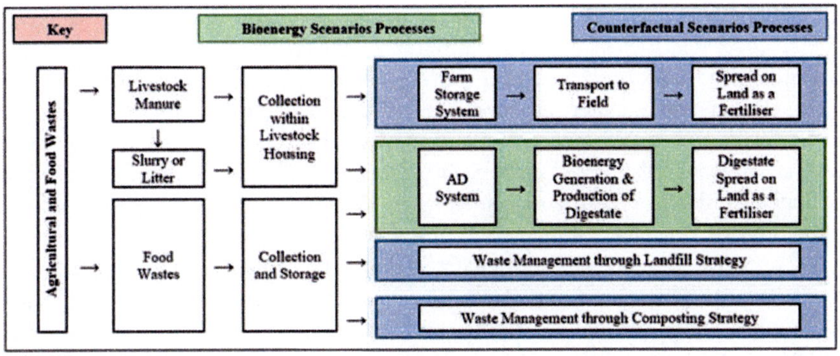

Fig. 20.9 Example life cycle steps for an agricultural residue and food waste bioenergy pathway (Welfle et al. 2016)

the corresponding impacts of the counterfactual scenarios. For example, the overall GHGs performance of generating energy from food wastes via an anaerobic digestion pathway should be compared against alternative counterfactual uses for the waste materials. The wastes could otherwise be managed by either a landfill or composting strategy. As a general rule, use of wastes for bioenergy that would have been sent to landfills will likely result in the mitigation of potentially large GHG emissions (Lou et al. 2020).

20.5.5 Role of Techno-economic Assessment Modelling Within Bioenergy Strategy Development

Techno-economic assessment (TEA) is an analysis method carried out to evaluate the economic performance of processes or systems. TEA analyses generate estimates based on assumptions of capital costs, operating costs, revenues and provide overall assessments of economic performances given financial parameters and timeframes. A primary reason for carrying out TEAs is to assess the economic feasibility of specific processes or overall systems.

TEA analyses follow a bottom-up approach that is typically undertaken in several stages. Capital costs are estimated reflecting the levels of capital investment that may be required for designing, purchasing, building and installing plant and associated auxiliary facilities. These capital costs may be divided into 'fixed capital costs' for once only expenses required to set up the project and 'working capital costs' which are estimated additional capital costs that may be required to start the operation before income is generated. Capital costs associated with bioenergy projects may include 'Direct Fixed Capital Costs' for purchasing equipment, installation, instrumentation and controls, piping and electrics, facility servicing. Also, 'Indirect Capital Costs' such as purchasing engineering skills, contractors' fees and costs to compensate for unpredictable events. When assessing the overall economic feasibility of projects, capital costs are assessed on an annualised basis, the cost is spread over an estimated period of time reflecting the project's active lifespan (Garcia-Freites 2019).

Operation and maintenance costs are further estimations built within TEAs reflecting the expenses incurred to operate the plant to obtain the final product/ service/ revenue. The costs are typically accounted for over an annual cycle and include 'fixed operational costs' such as maintenance and labour. Also 'variable operating costs' such as raw materials, utilities and transport requirements.

A key factor for any TEA of bioenergy projects is the cost of the biomass feedstocks over the lifetime of the plant. Varying feedstock costs can have potentially large implications for the long-term economic viability of projects (Garcia-Freites 2019).

Carrying out a TEA can sometimes be overlooked when projects are developed, although the potential benefits are clear. TEAs provide a quick and cost-effective method of testing designs before they are implemented. TEA's may be used to identify key areas where designs may need further development, identifying processes and activities that should be avoided and those that should be implemented to increase the likelihood of the economic viability of projects (Welfle et al. 2019). As with all projects, the economic attractiveness and viability of bioenergy projects will be determined by the levels of revenues that may be generated and when/ if the project will achieve a point where profits can be achieved.

20.6 Influence of Policy, Legislation and Social Acceptance on Bioenergy from Waste Projects

The development of energy strategies for many countries is framed by their commitments for achieving international targets such as those laid out by the Paris Agreement (UNFCCC 2015). Every country is currently being asked about the future of their energy sectors and choices have to be made about the extent that different low carbon technologies may be deployed (Welfle et al. 2020a).

These are complex decisions with many intricacies given the diversity in the forms of energy required (heat, power, fuels, etc.), the energy demands (buildings, industry, etc.) and the multiple wider activities that the energy sector drives (transportations, agriculture, etc.). The success or failure of any energy strategy will be largely reliant on the framework of policies and support legislation that has been designed to aid the deployment of alternative energy technologies and transition away from fossil fuels. A further crucial factor to success will be gaining the acceptance of the society, both to support the necessary policy framework but also to embrace alternative energy technologies. This is especially true for energy from waste projects as support will be needed to ensure that waste management strategies are aligned with the energy strategy (Welfle et al. 2014).

20.6.1 Policy for Bioenergy from Waste Projects in Different Countries

Alignment of waste policy with the energy strategy is vital to ensure success of energy from waste projects. Policy can influence the types and extent that different wastes are generated, the characteristics and composition of waste materials, the ways in which wastes must be managed and the extent that different waste materials may be used for energy end uses. The following sections summarise the main policy interventions and the implications for bioenergy projects in the European Union, China, United States and India; describing their respective waste policies and providing insight into the potential opportunities for the bioenergy from waste sector.

20.6.1.1 Waste Policy and Implications for Bioenergy in the European Union

Waste policy in the European Union (EU) is built on the foundations of a number of EU Directives. The EU's 'Waste Framework Directive' (2008/98/EC) (European Commission 2008) provides the legal framework for the treatment of waste with the aim of protecting the environment and human health through the prevention of the harmful effects of waste generation and waste management. It also introduces the concept of the 'waste hierarchy' which is key to determining how wastes should be managed, including for energy end uses. The 'Directive on Environmental Liability with Regard to the Prevention and Remedying of Environmental Damage' (2004/35/EC) (European Commission 2004) established a framework based on the principle that 'polluter pays' to prevent and remedy environmental damage to species, habitats, water, soils, etc. This policy is important in enforcing how waste materials are managed and providing protection to the environment. The 'Directive on the Incineration of Waste' (2000/76/EC) (European Commission 2000) provides a policy framework developed to prevent or reduce the negative effects resulting from the incineration and co-incineration of wastes. This is achieved through stringent operational conditions that prevent and limit the types and extent that different emissions may be released into the atmosphere. This policy provides the minimum performance requirements that must be met by energy from waste systems.

Under the requirements of the Waste Framework Directive, the key issues for potential energy from waste projects are the classifications determining which materials may be categorised as a 'waste' and how the management of wastes is prioritised in accordance with the waste hierarchy. There are specific criteria that determine when a waste material ceases to be a 'waste' and gains the status of a 'product' or a 'secondary raw material'. This includes circumstances where: the material is commonly used for specific purposes; where there is an existing market or demand for the material; use of the material is lawful and will not lead to overall adverse environmental or human health impacts (European Commission 2008). Where materials are classified as wastes the available management options are ordered in preference of environmental and waste reduction performance. Recovery of energy from waste is the appropriate waste management option for wastes that cannot be prevented, reused, or recycled with a lower GHG impact. Wastes with no other uses that may otherwise be diverted to landfills are thus resources highly suitable for energy recovery processes (Welfle et al. 2019).

20.6.1.2 Waste Policy and Implications for Bioenergy in China

In 1996 China enacted the 'Solid Waste Environmental Pollution Prevention and Control Law' (People's Republic of China 1996) which provides the legislative framework to prevent and control pollution from industrial, household, construction, agricultural and hazardous wastes. This has subsequently been periodically updated by the Ministry of Ecology and Environment and other relevant agencies culminating as the 2020 'Solid Waste Law' (People's Republic of China 2020). The 2020 law emphasises the solid waste handling principles of 'reduction', 'recycling' and 'harmlessness' while strengthening the supervision and management responsibilities of the government and its relevant departments. The overarching aim is to achieve 'zero' solid waste imports and integrate solid waste management into current environmental programs.

Energy recovered from the incineration of wastes is the leading energy from waste approach deployed in China, with large investments over the past decade resulting in a rapid rise in the quantities of waste incinerated each year. However, choices about waste management options and whether wastes should be routinely managed through incineration plants is a topic that has gained traction with policymakers in China. Better use of wastes is a focus for strategic plans for zero-waste cities, waste treatment and infrastructure and initiatives to promote learning and participating in activities such as recycling (Zhou 2020). This could result in less waste materials being available for large scale incineration plants.

There has also been a growth in bioenergy facilities focused on biofuel production. These are subject to varying levels of regulation in China. Bioethanol plants can only be built with direct government approval and only official facilities are entitled to subsidies and incentives, as a result, almost all facilities are owned and operated by state-owned enterprises. Agricultural waste materials are amongst the list of suitable feedstocks eligible for bioethanol production. In contrast, the biodiesel industry is mostly unregulated and dominated by a large number of small, private producers. Waste cooking oils (or oil) rendered from animal fats are the primary feedstock used for biodiesel production (van Dyk et al. 2016).

20.6.1.3 Waste Policy and Implications for Bioenergy in the United States

In the United States, the Environmental Protection Agency (EPA) regulates household, industrial, manufacturing and commercial solid and hazardous wastes under the 1976 'Resource Conservation and Recovery Act' (94th United States Congress 1976). This provides the framework for the effective management of solid wastes involving federal, state, regional and local entities.

All levels of government are actively involved in regulating solid wastes including legislation at the national, federal, state scale with further requirements developed by individual agencies. These cover all steps within waste management strategies including waste collection, segregation, transportation, storage, treatment, disposal labelling and internal and international movement of waste materials (Portney and Stavins 2000). There is also much legislation at each governance level focused specifically on how waste materials may be used within energy and biofuel applications.

Interest in waste materials for energy end uses increased following the 'Energy Policy Act of 1992' (102nd United States Congress 1992) that provided guidance for federal programs for the increased production of bioenergy and biofuels. The subsequent 'Energy Policy Act of 2005' (109th United States Congress 2005) included specific amendments to previous policy to allow the production of biofuels from waste materials including animal wastes, MSW, sludges and oils derived from wastewater treatment processes. This was expanded further through the introduction of the 'Energy Independence and Security Act of 2007' (110th United States Congress 2007), which was developed with the aim of moving the United States towards increased energy security and increasing the contribution of renewable technologies was a core theme. A key element of this intervention was the amendment of the definition of 'renewable fuel' to require that it be made from feedstocks that qualify as 'renewable biomass', including wastes from forestry, agricultural residues, food and garden wastes and materials generated by the food processing sector.

The priority goals of the EPA may be categorised through their 'Resource Conservation Challenge' that favours waste management strategies that: (i) prevent pollution and promote reuse and recycling; (ii) reduce priority and toxic chemicals within products and wastes; (iii) conserve energy and materials; and (iv) increase recycling for key categories of MSW. Reuse and recycling of waste materials is therefore the primary waste management strategy of choice ahead of potential use as bioenergy feedstocks.

20.6.1.4 Waste Policy and Implications for Bioenergy in India

Management of wastes in India comes under the 2006 'National Environment Policy' (NEP) (Government of India 2006) that focuses on the collection and treatment systems required to use, recycle and dispose waste materials in an environmentally safe way. The implementation of this policy is governed by various subordinate legislations such as the Ministry of Environment, Forest and Climate Change and administered at the state level by the State Pollution Control Boards.

Indian waste management rules are grounded in the principles of 'sustainable development', 'precaution' and 'polluter pays' (Tripathy 2015). The classification of materials as 'wastes' in India is the subject of focused legislation that categorises waste materials and specifies the appropriate waste management options.

In 2018 the Indian Government announced a 'National Policy on Biofuels' (Government of India 2018), developed to encourage innovation in biofuels and to better use feedstocks. This policy provides a suite of financial initiatives to support the use of feedstocks including waste materials such as MSW, plastics, industrial wastes, food wastes and sewage water. India is targeting the use of waste feedstocks within its bioenergy sector to aid waste management and to transition key sectors of the economy such as industry and agriculture towards cleaner environmental practices (Gupta et al. 2020).

20.6.2 *Gaining Public Acceptance for Energy for Waste Projects*

One of the primary barriers to the implementation of energy from waste projects is social opposition, often led by individual local communities and through the lobbying of regional and national environmental-action groups (Cheeseman and Velis 2010). A study carried out in 2011 (SITA UK 2011) found that 79% of 1000 participants interviewed about their views towards energy from waste projects, were accepting and supportive of the concepts. The research concludes that the voices of minority groups often overshadow the opinions of the majority. Regardless of the level of education surrounding such a scheme, opposition will perhaps always exist.

For individuals, the base of potential opposition will likely focus on the size and cost of the scheme, questions about the environmental credentials and always issues about location and logistics of the site—nobody wanting a plant that would devalue their homes or result in increased traffic and congestion. To gain acceptance of specific projects it is vital that the concepts, reasoning, details of the project are made transparent and assessments of both the potential positive and negative issues are clearly communicated. When communicating the plans for a project and highlighting its value it is important to explain the reasoning why the plant is needed; to explain plans around its location; and to explain the chosen technologies emphasising what they do and why they are the best options for that specific project (UNTHA 2013). Public perceptions potentially being softened when discussions about energy from waste projects are linked with energy security, waste minimisation and climate change themes. These are emotive issues that the public may not automatically associate with a specific project.

There are also a number of case studies where energy from waste projects have been embraced by communities and where public opinions are far less hostile. These provide potential roadmaps that could be replicated. A leading example can be found in Denmark and Sweden where there is a long tradition of waste incineration and where the populations are sympathetic to the technologies. This relationship has been nurtured over time, through transparency and by the public gaining first-hand experience of the value of the projects, in many cases energy from waste plants provides cheap heating for communities facilitated by waste powered district heating systems. There is also a close relationship that energy from waste operators develop with local populations, starting at a young age where school children and the public are invited to tour energy from waste plants.

20.7 Conclusions

This chapter discussed the role of biological wastes in a circular bioeconomy. Biological wastes should be regarded as a valuable resource for the production of biofuels and bio-based products through biorefinery rather than an environmental hazard or financial loss. The production technologies should be chosen wisely based on the waste type and the targeted products. The integration of different technologies to produce multiple products from the waste source was suggested in several research to be an efficient production approach for waste biorefinery, this was also covered in this chapter. The chapter also proposed a new approach (CIMB systems) for the integrated production of biofuels and bio-based products based on marine waste resources. This should reduce the water footprint and land use of the products and enhance the environmental and economic value of the products.

The chapter further discussed the role of catalysis in creating flexibility in an integrated biorefinery system by providing new routes and by downstream environmental solutions for flue gas and exhaust gas cleaning. In particular, biomass pre-treatment technologies based on catalysis can enable the production of fuel with better-defined specifications and increased energy densities compared to conventional pre-treatment techniques that usually consume a lot of energy and generate a lot of waste during acid/base neutralisation or enzyme hydrolysis. Furthermore, careful catalyst design and selection are key for effective removal of heteroatoms such as oxygen, nitrogen and sulphur in biofuels, in order to minimise mass transport limitations, improve hydrothermal stability in aqueous operation with wide range of pH, resistant to any kind of leaching and tuneable hydrophobicity to aid product/reactant adsorption. The ideal catalyst within a biorefinery would produce the highest yield of biofuels while remaining as cost-effective as possible for a sustainable bioeconomy. On other hand, preliminary techno-economic and life cycle assessments of photo reforming of biomass waste revealed that its carbon footprint is already lower than or comparable to existing methods for H_2 production, but waste-to-fuel conversion and waste management, the production cost and energy balance need to be improved further before industrial application can be considered. Developing photocatalytic processes with a thorough understanding to avoid secondary pollution, as well as achieving high conversion efficiency and selectivity, could be game changer in photo reforming of biomass waste for solar-biorefinery towards a circular economy.

The chapter also discussed the important role that modelling has in influencing the energy from waste systems. Models can be applied to explore the potential role of bioenergy technologies within wider energy systems. To identify the availability of suitable biomass resources and wastes that could be used to generate bioenergy. To undertake focused GHGs life cycle assessments to ensure that designed bioenergy projects deliver that desired GHG emission performances and techno-economic assessments to ensure that projects are technologically and financially feasible. Modelling allows the performance of technologies and energy systems to be tested and optimised before they are built. Finally, the chapter discussed the importance of policy, legislation and social acceptance on the success of any bioenergy scheme. Without a supportive policy landscape and 'buy-in' from key stakeholders, the long-term success of any bioenergy project will be less likely. Waste policy and the implication for bioenergy are presented for four key case studies, India, China, the United States and European Union. Each developing unique policy landscapes to support their growing bioenergy sectors.

Acknowledgement Santosh Kumar thanks Horizon 2020 grant 101017928 (HYSOLCHEM project) for funding. Also, Andrew Welfle thanks EPSRC, BBSRC, and UK Supergen Bioenergy Hub (EP/S000771/1) who funded time to complete this research.

References

94th United States Congress (1976) Resource Conservation and Recovery Act. United States of America

102nd United States Congress (1992) Energy Policy Act of 1992. United States of America

109th United States Congress (2005) Energy Policy Act of 2005. United States of America

110th United States Congress (2007) Energy Independence and Security Act of 2007. United States of America

Abomohra AEF, Elsayed M, Esakkimuthu S et al (2020) Potential of fat, oil and grease (FOG) for biodiesel production: a critical review on the recent progress and future perspectives. Prog Energy Combust Sci 81:100868

Aguilar F (2014) Wood energy in developed economies: resource management, economics and policy. Routledge Publishing

Alherbawi M, McKay G, Mackey HR, Al-Ansari T (2021) A novel integrated pathway for jet biofuel production from whole energy crops: a Jatropha curcas case study. Energy Convers Manag 229:113662. https://doi.org/10.1016/j.enconman.2020.113662

Almarashi JQM, El-Zohary SE, Ellabban MA, Abomohra AEF (2020) Enhancement of lipid production and energy recovery from the green microalga Chlorella vulgaris by inoculum pretreatment with low-dose cold atmospheric pressure plasma (CAPP). Energy Convers Manag 204:112314. https://doi.org/10.1016/j.enconman.2019.112314

Amha YM, Anwar MZ, Brower A et al (2018) Inhibition of anaerobic digestion processes: applications of molecular tools. Bioresour Technol 247:999–1014

Ashour M, Elshobary ME, El-Shenody R et al (2019) Evaluation of a native oleaginous marine microalga Nannochloropsis oceanica for dual use in biodiesel production and aquaculture feed. Biomass Bioenerg 120:439–447. https://doi.org/10.1016/j.biombioe.2018.12.009

Badoei-dalfard A, Malekabadi S, Karami Z, Sargazi G (2019) Magnetic cross-linked enzyme aggregates of Km12 lipase: a stable nanobiocatalyst for biodiesel synthesis from waste cooking oil. Renew Energy 141:874–882. https://doi.org/10.1016/j.renene.2019.04.061

Bozell JJ, Petersen GR (2010) Technology development for the production of biobased products from biorefinery carbohydrates—the US Department of Energy's "top 10" revisited. Green Chem 12:539–555. https://doi.org/10.1039/b922014c
Canakci M (2007) The potential of restaurant waste lipids as biodiesel feedstocks. Bioresour Technol 98:183–190. https://doi.org/10.1016/j.biortech.2005.11.022
Caravaca A, Jones W, Hardacre C, Bowker M (2016) H_2 production by the photocatalytic reforming of cellulose and raw biomass using Ni, Pd, Pt and Au on titania. In: Proceedings of the Royal Society A: Mathematical, Physical and Engineering Sciences. Royal Society of London
Cheeseman C, Velis C (2010) WTERT-UK opportunities and barriers. London
Chen WH, Lin BJ, Huang MY, Chang JS (2015) Thermochemical conversion of microalgal biomass into biofuels: a review. Bioresour Technol 184:314–327
Cheng KK, Zhao XB, Zeng J et al (2012) Downstream processing of biotechnological produced succinic acid. Appl Microbiol Biotechnol 95:841–850
Clark JH, Budarin V, Dugmore T et al (2008) Catalytic performance of carbonaceous materials in the esterification of succinic acid. Catal Commun 9:1709–1714. https://doi.org/10.1016/j.catcom.2008.01.037
Corma Canos A, Iborra S, Velty A (2007) Chemical routes for the transformation of biomass into chemicals. Chem Rev 107:2411–2502
Costa JAV, de Freitas BCB, Lisboa CR et al (2019) Microalgal biorefinery from CO_2 and the effects under the blue economy. Renew Sustain Energy Rev 99:58–65
Dadi AP, Varanasi S, Schall CA (2006) Enhancement of cellulose saccharification kinetics using an ionic liquid pretreatment step. Biotechnol Bioeng 95:904–910. https://doi.org/10.1002/bit.21047
Dahiya S, Kumar AN, Shanthi Sravan J et al (2018) Food waste biorefinery: sustainable strategy for circular bioeconomy. Bioresour Technol 248:2–12
Darwish R, Gedi MA, Akepach P et al (2020) Chlamydomonas reinhardtii is a potential food supplement with the capacity to outperform chlorella and spirulina. Appl Sci 10
Dasa KT, Westman SY, Millati R et al (2016) Inhibitory effect of long-chain fatty acids on biogas production and the protective effect of membrane bioreactor. Biomed Res Int
DfT (2018) RTFO guidance—feedstocks including wastes and residues. London
Djandja OS, Wang Z, Chen L et al (2020) Progress in hydrothermal liquefaction of algal biomass and hydrothermal upgrading of the subsequent crude bio-oil: a mini review. Energy Fuels 34:11723–11751. https://doi.org/10.1021/acs.energyfuels.0c01973
Duan P, Savage PE (2011a) Hydrothermal liquefaction of a microalga with heterogeneous catalysts. Ind Eng Chem Res 50:52–61. https://doi.org/10.1021/ie100758s
Duan P, Savage PE (2011b) Catalytic hydrothermal hydrodenitrogenation of pyridine. Appl Catal B Environ 108–109:54–60. https://doi.org/10.1016/j.apcatb.2011.08.007
European Commission (2000) Waste incineration directive. Brussels
European Commission (2004) Directive on environmental liability with regard to the prevention and remedying of environmental damage. Brussels, Belgium
European Commission (2008) Directive on waste. Brussels, Belgium
Faunce T, Styring S, Wasielewski MR et al (2013) Artificial photosynthesis as a frontier technology for energy sustainability. Energy Environ Sci 6:1074–1076
Garcia-Freites S (2019) Using agricultural residues to support sustainable development: a case study of coffee stems gasification to supply energy demands in rural areas of Colombia. University of Manchester
Ghoreishi SM, Moein P (2013) Biodiesel synthesis from waste vegetable oil via transesterification reaction in supercritical methanol. J Supercrit Fluids 76:24–31. https://doi.org/10.1016/j.supflu.2013.01.011
Girgis MJ, Gates BC, Girgis MJ (1991) Reactivities, reaction networks, and kinetics in high-pressure catalytic hydroprocessing. Ind Eng Chem Res 30:2021–2058. https://doi.org/10.1021/ie00057a001
Government of India (2006) National Environment Policy (NEP). Delhi, India
Government of India (2018) National policy on biofuels. Delhi, India

Greetham D, Zaky AS, Du C (2019) Exploring the tolerance of marine yeast to inhibitory compounds for improving bioethanol production. Sustain Energy Fuels 6:1545–1553
Gupta RP, Puri SK, Ramakumar SS V (2020) India biofuel economy roadmap—challenges and prospects. New Delhi, India
Ho TC (1988) Hydrodenitrogenation catalysis. Catal Rev 30:117–160.https://doi.org/10.1080/01614948808078617
Huber GW, Iborra S, Corma A (2006) Synthesis of transportation fuels from biomass: chemistry, catalysts, and engineering. Chem Rev 106:4044–4098.https://doi.org/10.1021/cr068360d
Hutton C, Vamvakeridou-Lyroudi L, Kapelan Z, Savic D (2010) Real-time modelling and data assimilation techniques for improving the accuracy of model predictions. Scientific report. Brussels
Johnson RM, Richard EP (2005) Sugarcane yield, sugarcane quality, and soil variability in Louisiana. Agron J 97
Jorgensen S, Pedersen LJT (2018) The circular rather than the linear economy. RESTART Sustain Bus Model Innov:103–120
Kadam SR, Mate VR, Panmand RP et al (2014) A green process for efficient lignin (biomass) degradation and hydrogen production via water splitting using nanostructured C, N, S-doped ZnO under solar light. RSC Adv 4:60626–60635. https://doi.org/10.1039/c4ra10760h
Kawai T, Sakata T (1981) Photocatalytic hydrogen production from water by the decomposition of poly-vinylchloride, protein, algae, dead insects, and excrement. Chem Lett 10:81–84. https://doi.org/10.1246/cl.1981.81
Kobayashi T, Kuramochi H, Maeda K et al (2014) Dual-fuel production from restaurant grease trap waste: bio-fuel oil extraction and anaerobic methane production from the post-extracted residue. Bioresour Technol 169:134–142. https://doi.org/10.1016/j.biortech.2014.06.071
Kostas ET, White DA, Cook DJ (2017) Development of a bio-refinery process for the production of speciality chemical, biofuel and bioactive compounds from Laminaria digitata. Algal Res 28:211–219. https://doi.org/10.1016/j.algal.2017.10.022
Kostas ET, White DA, Cook DJ (2020) Bioethanol production from UK seaweeds: investigating variable pre-treatment and enzyme hydrolysis parameters. BioEnergy Res 13:271–285
Kumar AK, Sharma S (2017) Recent updates on different methods of pretreatment of lignocellulosic feedstocks: a review. Bioresour. Bioprocess. 4:7
Kumar D, Long SP, Singh V (2018) Biorefinery for combined production of jet fuel and ethanol from lipid-producing sugarcane: a techno-economic evaluation. GCB Bioenergy 10:92–107. https://doi.org/10.1111/gcbb.12478
Leong HY, Chang CK, Khoo KS et al (2021) Waste biorefinery towards a sustainable circular bioeconomy: a solution to global issues. Biotechnol Biofuels 14
Lin YC, Huber GW (2009) The critical role of heterogeneous catalysis in lignocellulosic biomass conversion. Energy Environ Sci 2:68–80. https://doi.org/10.1039/b814955k
Lindedam J, Bruun S, Jørgensen H et al (2014) Evaluation of high throughput screening methods in picking up differences between cultivars of lignocellulosic biomass for ethanol production. Biomass Bioenerg 66:261–267. https://doi.org/10.1016/j.biombioe.2014.03.006
Lou ECWECW, Lee A, Welfle AJAJ, Abdullahi ALAL (2020) Testing the Nexus between C&D waste management strategies & GHG emission performances: the case of UK student accommodation refurbishment projects. J Build Eng. https://doi.org/10.1016/j.jobe.2020.101812
Mäki-Arvela P, Anugwom I, Virtanen P et al (2010) Dissolution of lignocellulosic materials and its constituents using ionic liquids-a review. Ind Crops Prod 32:175–201
Mondal M, Goswami S, Ghosh A et al (2017) Production of biodiesel from microalgae through biological carbon capture: a review. 3 Biotech 7
Nava Bravo I, Velásquez-Orta SB, Cuevas-García R et al (2019) Bio-crude oil production using catalytic hydrothermal liquefaction (HTL) from native microalgae harvested by ozone-flotation. Fuel 241:255–263. https://doi.org/10.1016/j.fuel.2018.12.071
Ong HC, Chen WH, Farooq A et al (2019) Catalytic thermochemical conversion of biomass for biofuel production: a comprehensive review. Renew Sustain Energy Rev 113:109266

People's Republic of China (1996) Law of the People's Republic of China on the prevention and control of solid wastes pollution to the environment. Beijing, China

People's Republic of China (2020) Law of the People's Republic of China on the prevention and control of solid wastes pollution to the environment. Beijing, China

Portney PR, Stavins RN (2000) Public policies for environmental protection, 2nd ed. Boston Public Library, Washington D.C

Ragauskas AJ, Williams CK, Davison BH, et al (2006) The path forward for biofuels and biomaterials. Science 80(311):484–489

Raheem A, Wan Azlina WAKG, Taufiq Yap YH et al (2015) Thermochemical conversion of microalgal biomass for biofuel production. Renew Sustain Energy Rev 49:990–999

Ran J, Jaroniec M, Qiao SZ (2018) Cocatalysts in Semiconductor-based photocatalytic CO_2 reduction: achievements, challenges, and opportunities. Adv Mater 30:1704649

Rinaldi R, Schüth F (2009) Design of solid catalysts for the conversion of biomass. Energy Environ Sci 2:610–626. https://doi.org/10.1039/B902668A

Ruddy DA, Hensley JE, Nash CP et al (1929) Methanol to high-octane gasoline within a market-responsive biorefinery concept enabled by catalysis. Nat Catal 2019 27(2):632–640. https://doi.org/10.1038/s41929-019-0319-2

Ruppert AM, Weinberg K, Palkovits R (2012) Hydrogenolysis goes bio: from carbohydrates and sugar alcohols to platform chemicals. Angew Chemie—Int Ed 51:2564–2601

Sahabdheen AB, Arivarasu A (2020) Synthesis and characterization of reusable heteropoly acid nanoparticles for one step biodiesel production from high acid value waste cooking oil—performance and emission studies. Mater Today Proc 22:383–392

Sakata T, Kawai T (1981) Photodecomposition of water by using organic compounds—hydrogen evolution by powdered semiconductor photocatalysts. Yuki Gosei Kagaku Kyokaishi/journal Synth Org Chem 39:589–602. https://doi.org/10.5059/yukigoseikyokaishi.39.589

Savvidis G, Siala K, Weissbart C et al (2019) The gap between energy policy challenges and model capabilities. Energy Policy 125:503–520. https://doi.org/10.1016/J.ENPOL.2018.10.033

Schlaf M (2006) Selective deoxygenation of sugar polyols to α,ω-diols and other oxygen content reduced materials—a new challenge to homogeneous ionic hydrogenation and hydrogenolysis catalysis. J Chem Soc Dalt Trans:4645–4653.https://doi.org/10.1039/b608007c

Serrano-Ruiz JC, Luque R, Sepúlveda-Escribano A (2011) Transformations of biomass-derived platform molecules: from high added-value chemicals to fuels via aqueous-phase processing. Chem Soc Rev 40:5266–5281. https://doi.org/10.1039/c1cs15131b

Simonetti DA, Ahn JH, Iglesia E (2011) Mechanistic details of acid-catalyzed reactions and their role in the selective synthesis of triptane and isobutane from dimethyl ether. J Catal 277:173–195. https://doi.org/10.1016/j.jcat.2010.11.004

Singh RN, Vyas DK, Srivastava NSL, Narra M (2008) SPRERI experience on holistic approach to utilize all parts of Jatropha curcas fruit for energy. Renew Energy 33:1868–1873. https://doi.org/10.1016/j.renene.2007.10.007

SITA UK (2011) Public attitudes to community buy in for waste and resource infrastructure. Maidenhead

Stolarczyk JK, Bhattacharyya S, Polavarapu L, Feldmann J (2018) Challenges and prospects in solar water splitting and CO_2 reduction with inorganic and hybrid nanostructures. ACS Catal 8:3602–3635.https://doi.org/10.1021/acscatal.8b00791

Sudarsanam P, Zhong R, Van Den Bosch S et al (2018) Functionalised heterogeneous catalysts for sustainable biomass valorisation. Chem Soc Rev 47:8349–8402

Sumannakarn K, Lotero E, Ngaosuwan K, Goodwin JG (2009) Simultaneous free fatty acid esterification and triglyceride transesterification using a solid acid catalyst with in situ removal of water and unreacted methanol. Ind Eng Chem Res 48:2810–2818

Suriyachai N, Weerasai K, Upajak S et al (2020) Efficiency of catalytic liquid hot water pretreatment for conversion of corn stover to bioethanol. ACS Omega 5:29872–29881. https://doi.org/10.1021/acsomega.0c04054

Taherzadeh M, Bolton K, Wong J, Pandey A (2019) Sustainable resource recovery and zero waste approaches, 1st edn. Elsevier, London

The World Bank (2021) Trends in solid waste management. Washington D.C

Thornley P, Gilbert P, Shackley S, Hammond J (2015) Maximizing the greenhouse gas reductions from biomass: the role of life cycle assessment. Biomass Bioenergy 81:35–43. https://doi.org/10.1016/j.biombioe.2015.05.002

Tian C, Li B, Liu Z et al (2014) Hydrothermal liquefaction for algal biorefinery: a critical review. Renew Sustain Energy Rev 38:933–950

Tripathy A (2015) Waste management in India: an overview. New Delhi, India

Tu Q, McDonnell BE (2016) Monte Carlo analysis of life cycle energy consumption and greenhouse gas (GHG) emission for biodiesel production from trap grease. J Clean Prod 112:2674–2683. https://doi.org/10.1016/j.jclepro.2015.10.028

Uekert T, Kuehnel MF, Wakerley DW, Reisner E (2018) Plastic waste as a feedstock for solar-driven H_2 generation. Energy Environ Sci 11:2853–2857. https://doi.org/10.1039/c8ee01408f

Uekert T, Dorchies F, Pichler CM, Reisner E (2020) Photoreforming of food waste into value-added products over visible-light-absorbing catalysts. Green Chem 22:3262–3271. https://doi.org/10.1039/d0gc01240h

Uekert T, Pichler CM, Schubert T, Reisner E (2021) Solar-driven reforming of solid waste for a sustainable future. Nat Sustain 4:383–391

UK Government (2018) Waste classification: guidance on the classification and assessment of waste. UK

UNFCCC (2015) Paris agreement. Bonn

UNTHA (2013) Understanding opposition to new energy from waste plants. York, UK

Usseglio M, Salvadori V, Siri G (2019) Advanced modeling of vegetable oils steam stripping with structured packing columns. Comput Chem Eng 121:654–669. https://doi.org/10.1016/j.compchemeng.2018.12.007

van Dyk S, Li L, Baroos Leal D et al (2016) The potential of biofuels in China. Vancouver, Canada

Venkata Mohan S, Modestra JA, Amulya K et al (2016a) A circular bioeconomy with biobased products from CO_2 sequestration. Trends Biotechnol 34:506–519

Venkata Mohan S, Nikhil GN, Chiranjeevi P et al (2016b) Waste biorefinery models towards sustainable circular bioeconomy: critical review and future perspectives. Bioresour Technol 215:2–12

Wakerley DW, Kuehnel MF, Orchard KL et al (2017) Solar-driven reforming of lignocellulose to H_2 with a CdS/CdO_x photocatalyst. Nat Energy 2:17021. https://doi.org/10.1038/nenergy.2017.21

Welfle AJ (2014) Biomass resource analyses and future bioenergy scenarios. University of Manchester

Welfle AJ, Alawadhi A (2021) Bioenergy opportunities, barriers and challenges in the Arabian Peninsula—resource modelling, surveys and interviews. Biomass Bioenergy TBC. TBC

Welfle AJ, Gilbert P, Thornley P (2014) Increasing biomass resource availability through supply chain analysis. Biomass Bioenerg 70:249–266. https://doi.org/10.1016/j.biombioe.2014.08.001

Welfle AJ, Gilbert P, Thornley P (2016) Greenhouse gas performance of UK biomass resources for heat bioenergy pathways. Manchester

Welfle AJ, Gilbert P, Thornley P, Stephenson A (2017) Generating low-carbon heat from biomass: life cycle assessment of bioenergy scenarios. J Clean Prod 149:448–460. https://doi.org/10.1016/j.jclepro.2017.02.035

Welfle AJ, Wagland S, Longhurst P et al (2019) Establishing viable pathways for increasing biofuel production from UK wastes and residues to meet transport energy targets. Manchester

Welfle A, Thornley P, Röder M (2020a) A review of the role of bioenergy modelling in renewable energy research and policy development. Biomass Bioenerg 136:105542. https://doi.org/10.1016/j.biombioe.2020.105542

Welfle AJ, Röder M, Cooper S, McManus M (2020b) Accounting whole life cycle bioenergy emissions within the UNFCCC emission accounting framework. Manchester

Wilson K, Lee AF, Dacquin J-P (2012) Heterogeneous catalysts for converting renewable feedstocks to fuels and chemicals, pp 263–304. https://doi.org/10.1007/978-1-4614-0344-9_7

Worldometer (2021) World population clock: 7.86 billion people

Xu C, Arancon RAD, Labidi J, Luque R (2014a) Lignin depolymerisation strategies: towards valuable chemicals and fuels. Chem Soc Rev 43:7485–7500

Xu Y, Zheng X, Yu H, Hu X (2014b) Hydrothermal liquefaction of Chlorella pyrenoidosa for bio-oil production over Ce/HZSM-5. Bioresour Technol 156:1–5. https://doi.org/10.1016/j.biortech.2014.01.010

Xu S, Elsayed M, Ismail GA et al (2019) Evaluation of bioethanol and biodiesel production from Scenedesmus obliquus grown in biodiesel waste glycerol: a sequential integrated route for enhanced energy recovery. Energy Convers Manag 197:111907. https://doi.org/10.1016/j.enconman.2019.111907

Yang C, Jia L, Chen C et al (2011) Bio-oil from hydro-liquefaction of Dunaliella salina over Ni/REHY catalyst. Bioresour Technol 102:4580–4584. https://doi.org/10.1016/j.biortech.2010.12.111

Yepez A, Pineda A, Garcia A et al (2013) Chemical transformations of glucose to value added products using Cu-based catalytic systems. Phys Chem Chem Phys 15:12165–12172. https://doi.org/10.1039/c3cp50707f

Zaky AS (2021) Introducing a marine biorefinery system for the integrated production of biofuels high-value-chemicals and co-products: a path forward to a sustainable future. Processes 9(10):1841. https://doi.org/10.3390/pr9101841

Zaky AS, Tucker GA, Daw ZY, Du C (2014) Marine yeast isolation and industrial application. FEMS Yeast Res 14:813–825

Zaky AS, Greetham D, Louis EJ et al (2016) A new isolation and evaluation method for marine-derived yeast spp. with potential applications in industrial biotechnology. Biotechnol Bioeng 26:1891–1907

Zaky AS, Greetham D, Tucker GA, Du C (2018) The establishment of a marine focused biorefinery for bioethanol production using seawater and a novel marine yeast strain. Sci Rep 8

Zaky AS, French CE, Tucker GA, Du C (2020) Improving the productivity of bioethanol production using marine yeast and seawater-based media. Biomass Bioenerg 139:105615. https://doi.org/10.1016/j.biombioe.2020.105615

Zaky AS, Carter CE, Meng F, French CE (2021) A preliminary life cycle analysis of bioethanol production using seawater in a coastal biorefinery setting. Processes 9

Zavrel M, Bross D, Funke M et al (2009) High-throughput screening for ionic liquids dissolving (ligno-)cellulose. Bioresour Technol 100:2580–2587. https://doi.org/10.1016/j.biortech.2008.11.052

Zhou Y (2020) Two birds, one stone: China's potential for zero waste and nationwide ethanol. Beijing, China